Macmillan's Chemical and Physical Data

Arthur M. James & Mary P. Lord

M

© The Macmillan Press Ltd

All rights reserved. No reproduction, copy or transmission of this publication may be made without written permission.

No paragraph of this publication may be reproduced, copied, or transmitted save with written permission or in accordance with the provisions of the Copyright, Designs and Patents Act 1988, or under the terms of any licence permitting limited copying issued by the Copyright Licensing Agency, 90 Tottenham Court Road, London W1P 9HE.

Any person who does any unauthorized act in relation to this publication may be liable to criminal prosecution and civil claims for damages.

First published 1992 by
THE MACMILLAN PRESS LTD
London and Basingstoke

Associated companies in Auckland, Delhi, Dublin, Gaborone, Hamburg, Harare, Hong Kong, Johannesburg, Kuala Lumpur, Lagos, Manzini, Melbourne, Mexico City, Nairobi, New York, Singapore, Tokyo.

British Library Cataloguing in Publication Data
James, Arthur M. (Arthur Montague)
  Chemical and physical data.
  1. Chemical compounds
  I. Title II. Lord, Mary P.
  540

ISBN 0-333-51167-0

While every care has been taken in compiling the information contained in this publication, the publishers and authors accept no responsibility for any errors or omissions.

Typeset in Malaysia and printed in Great Britain

Contents

Preface	vi
Foreword	vii
Commonly Used Abbreviations, Acronyms and Greek Alphabet	ix
List of Sections	x
List of Tables contained in each Section	xii
Index	551

Preface

For many years there has been a clear need for a concise and manageable reference data book. This is it! This book is suitable for purchase by scientists and students in various disciplines as diverse as Astronomy and Biological Chemistry. Furthermore lay persons and librarians seeking a concise source of reliable data will find this book useful. The choice of the material is ours alone, no doubt we shall be criticized for omissions or overtreatment; readers are invited to contact the publishers with their comments.

To minimize costs this work, in marked contrast to most other data source books, is mainly restricted to data, on the assumption that readers are familiar with the underlying theory or have access to appropriate textbooks. Tabulation, almost exclusively in SI units, is by the most appropriate method, e.g. alphabetical, numerical, proton number. Section I includes lists of basic and derived SI units together with conversion tables for other systems. We have sourced the material from a wide range of original papers, critical review articles, recognised standard reference works and manufacturers' specifications. Reference to further information and/or additional data in the tables enables the reader to delve more deeply.

Outside North America, the extensive compilations of evaluated and referenced specialised data are available only in very large libraries; moreover, such compilations are unwieldy for regular use. Recommended examples of such collections are: the data series of BS, CODATA, DIPPR, IUPAC, NBS, NIST, TRC, Pure and Applied Chemistry, Journal of Physical Chemical Reference Data.

The authors much appreciate the help and encouragement of Rosemary Foster; they are also indebted to the editors and referees for their helpful comments and to various colleagues for useful discussions.

Arthur M. James
Mary P. Lord
October 1992

Foreword

What is Macmillan's Chemical and Physical Data?

The objective of this book is to bring together fundamental data in the physical sciences to serve the ongoing needs of a wide range of investigators. They are students, teachers, practicing scientists, information specialists, managers, and laymen. They need to access data for particles, atoms, elements, compounds, alloys, common materials and naturally occurring substances.

The volume is not patterned after existing handbooks. It answers the questions typically encountered in classrooms, on the job, and in the lab — all in a concise, easy-to-read format. The information it contains is culled from the great data compilations of major libraries. Since these are not necessarily near to hand and often inaccessible, it will serve the vast majority of scientists who are currently provided with less than the desired set of numeric data.

Contents

This book is a composite collection of data in many disciplines. The physics section is particularly intriguing in that its coverage ranges from nuclide information in mass, lifetime and decay schemes all the way to acoustic information for the world's leading concert halls. Key astronomical and geophysical data one frequently needs on stars, planets and Earth have been included.

The chemistry section reports physical property data for inorganic and organic compounds as well as gases of industrial interest. A particularly useful section here offers flash point values, solubility in multiple solvents, references to Beilstein volumes and Aldrich spectral data. A section on chemical safety gives toxicity and exposure limits for common chemicals.

There is thorough treatment of conversion units and fundamental constants. Of note is a section on the ITS 90 temperature scale. SI units and IUPAC nomenclature are used throughout.

Tables covering the temperature and composition dependence of properties allow development of equations to express property dependence on the independent variable. Used with care, this feature even permits moderate extrapolation.

A strong attribute is the liberal use of citations to the primary and secondary literature as an insight to greater detail and broader coverage. Referencing is current in most cases, but older, more traditional sources are also given where warranted.

What can the Handbook do for you?

Many outstanding books in common use for classroom and reference have woefully weak and obsolete tables of data. This book enables you to locate the appropriate tables to back up concepts for study or experiments in progress. As such it will serve as an excellent supplement for the classic texts.

At a recent Engineering Foundation Conference convened to explore the concept of a National Engineering Information Service, attendees stressed the growing gap between available information literature and the end user's awareness of what it is and where to find it. By employing this handbook as a sort of guide or directory to initial values, with a pointer to further resources, the gap can be bridged.

What can't the Handbook do for you?

Looking up a value in a databook is a lot like looking up a word in a dictionary. The definition may be there in black and white, but the meaning can remain elusive. Since values don't come with directions for use, creativity and sensibility become part of the equation — and those are values the user provides.

Beware of over-reliance on data in any handbook of this nature. The field of numeric data in science is still an active one, in a state of constant change leading to newer and more reliable data with time. Since this handbook contains a broad sweep of generally useful data, it has not been critically evaluated in the same sense as CODATA (Committee on Data for Science and Technology), DIPPR (Design Institute for Physical Property Data), TRC (Thermodynamics Research Center), etc. Rather, it has been selected from results achieved by many different data-gathering agencies, each of which has its own evaluation process. In any given case, values proffered may not represent a result accumulated and critically evaluated from the sum total of knowledge on the subject.

Since some tables come from current refereed journals and authoritative compendia, the user can accept that data with no hesitation. Other data must be accepted on the basis that they give the user what engineers call 'a number they can run with', allowing calculations to be performed with reasonable accuracy.

What can you do for the Handbook?

User acceptance and need for new as well as revised tables will keep this volume current and useful to expanding audiences. The cost has been set low so it will not be an impediment to use. Whether you are a student or a Nobelist, your feedback means that future editions will better reflect the way you work and the questions you most often ask.

As for this first edition, editors Dr Mary Lord, physicist, and Professor Arthur James, chemist, are to be commended for producing this well-conceived and valuable addition to our libraries. They have proved that husband and wife scientists can collaborate on a work embracing no less than their entire professional careers.

Theodore B. Selover, Jr.
Technical Director
Design Institute for
Physical Property Data
AIChE

17 August 1992

Commonly Used Abbreviations

For the definition of symbols see:
Table I.1 for base SI units;
Table I.2 for derived SI units with special names;
Table I.3 and I.4 for some physical quantities;
Table 1.7 for Fundamental Units.

Other symbols

A	Debye-Hückel constant
A_r	relative atomic mass
$a_+, a_-, a_\pm$	activity of cation, anion, mean ionic activity
K	equilibrium constant
K_a, K_b	acid and base ionization constants
K_w	ionization constant of water
M_r	relative molecular mass
$m_+, m_-, m_\pm$	molality of cation, anion, mean ionic molality
t_+, t_-	transport number of cation, anion
u, u_+, u_-	mobility of ion, cation, anion
$Y_i, Y_+, Y_-, Y_\pm$	activity coefficient of ion, cation, anion, mean ionic activity coefficient
Δ	increase in value of thermodynamic function, e.g. ΔG
δ	chemical shift/ppm
π	ratio of circumference to diameter of circle

Superscripts

$\ominus$	indicating standard value of a property
$\ddagger$	indicating value of a property in the transition state

Subscripts

A, B	referring to a species A, B
i	referring to typical ionic species i
P, V, T, S	indicating constant pressure, volume, temperature, entropy
+, −	referring to positive, negative ion

Other abbreviations

c	circa (approximately)
g, l, s (cr)	referring to gaseous, liquid, solid (crystalline) state
aq	aqueous
bp	boiling point
mp	melting point
fp	flash point
vp	vapour pressure

pH	$-\lg a_{H+}$
pK	$-\lg K$

Mathematical symbols

$\sum_i$	sum of i terms
$\ln x$	natural logarithm of x, i.e. $\log_e x$
$\lg x$	common logarithm of x (to base 10), i.e. $\log_{10} x$
$e^x, \exp(x)$	exponential of x

Acronyms for data organisations

BS	British Standard
CODATA	Committee on Data for Science and Technology
DIPPR	Design Institute for Physical Properties Research
IUPAC	International Union of Pure and Applied Chemistry
NBS	National Bureau of Standards (to 1989)
NIST	National Institute of Standards in Technology
TRC	Thermodynamics Research Center, Texas, A&M.

Greek alphabet

Α	α	alpha
Β	β	beta
Γ	γ	gamma
Δ	δ	delta
Ε	ε	epsilon
Ζ	ζ	zeta
Η	η	eta
Θ	θ	theta
Ι	ι	iota
Κ	κ	kappa
Λ	λ	lambda
Μ	μ	mu
Ν	ν	nu
Ξ	ξ	xi
Ο	ο	omicron
Π	π	pi
Ρ	ρ	rho
Σ	σ	sigma
Τ	τ	tau
Υ	υ	upsilon
Φ	φ	phi
Χ	χ	chi
Ψ	ψ	psi
Ω	ω	omega

List of Sections

Section I	**Units, Conversion Factors and Fundamental Constants**		1
	Base and derived SI units, conversion factors for other systems of units and fundamental constants		
Section II	**Properties of Materials**		11
	Density, pressure, humidity, elastic moduli, tensile strength, viscosity and surface tension		
Section III	**Properties of the Elements**		33
	Periodic table, standard atomic weights, electronic configuration, atomic, ionic and van der Waals radii, electron and proton affinities, Earth's elemental abundance and atmospheric composition		
Section IV	**General Properties of Molecules**		51
	Bond lengths and angles, dipole moments, electron affinities, character tables, crystal system characteristics		
Section V	**Electricity and Magnetism**		67
	Resistivities, relative permittivites, remanence; magnetic flux density, Curie temperature; relative permeabilities, loss factor, super-conductivity		
Section VI	**Electrochemistry**		87
	Molar and limiting ionic conductivities, activity and osmotic coefficients, transport numbers, pH, buffer solutions, standard reduction and redox potentials; dissociation constants, solubility products, half-wave potentials		
Section VII	**Optics**		125
	Light sources, CIE specifications for standard observer and for illuminants, hue discrimination, colour deficiency; geometrical optical properties and chromatic aberration of eye; refractive indices, double refraction, thin films, specific rotation, Verdet constants, electro-optic substances; scintillators; lasers		
Section VIII	**Spectroscopy**		143
	Electromagnetic spectrum, calibration standards; molecular parameters of gases, rotational constants and moments of inertia of di- and tri-atomic molecules, fundamental vibrational frequencies, nuclear spin moments, chemical shifts, mass spectroscopy		
Section IX	**Atomic and Nuclear Physics**		169
	Free electrons and ions in gases, work function, electrons in atoms, X-rays, absorption of particles and dosimetry, radioactive elements, nuclear fission, nuclear fusion, sub-atomic particles		
Section X	**Acoustics**		271
	Speed and attenuation of sound, vibration of rods; architectural, musical and physiological acoustics		

Section XI	**Physical Properties of Inorganic and Organic Compounds**	283
	Inorganic, organic, organo-metallic and biochemical compounds, polymers, plastics, fluorocarbons; critical and van der Waals constants; solubilities, vapour pressures, critical micelle concentrations, densities of aqueous solutions	
Section XII	**Heat and Temperature**	429
	Expansion coefficients, thermal conductivities, specific heat capacities, emissivities, thermoelectricity, 1990 International Temperature Scale	
Section XIII	**Thermodynamic Properties of Elements and Compounds**	447
	Thermodynamic data (including CODATA key values) for elements and inorganic, organic and biochemical compounds; enthalpies of fusion, vaporization and formation; bond dissociation enthalpies, molar heat capacities, cryoscopic and ebullioscopic constants, Henry's law constants, calorific value of fuels, critical solution temperature, eutectic systems	
Section XIV	**Astronomy and Geophysics**	503
	Solar system, Earth's atmosphere, oceans, geomagnetic field, cosmic ray flux; time	
Section XV	**Chemical Kinetics**	511
	First and second order reactions, Arrhenius parameters, entropy and volume of activation, Hammett reaction and substituent constants	
Section XVI	**Health and Safety**	519
	Definition of exposure and toxicity limits; their values for a range of chemicals and materials	
Section XVII	**Ancillary Topics**	533
	Screw threads, percentage composition of alloys, resistor colour code, integrals, series and error functions	
Index		551

List of Tables in each Section

Section I Units, Conversion Factors and Fundamental Constants

I.1	Base SI Units	3
I.2	Derived SI Units with special names	3
I.3	Symbols and SI Units for some physical quantities	4
I.4	SI Conversion Tables	5
	Concentration Units	5
	Electricity and Magnetism	5
	Mechanics	5
	Radiation, Spectrophotometric and other optical quantities	6
	Space and Time	6
	Thermodynamics	6
I.5	Prefixes used with SI units	6
I.6	Conversion factors for some other units	7
	A. Length	7
	B. Area	7
	C. Volume	7
	D. Speed	7
	E. Mass	8
	F. Force	8
	G. Pressure	9
	H. Work and Energy	9
	I. Concentration Units	9
I.7	Fundamental Constants	10

Section II Properties of Materials

II.1	Variation of buoyancy correction with mass for weighings in air.	13
II.2	Variation of density of ambient air of 50% relative humidity, with temperature, for various air pressures.	13
II.3	Changes in density for some values given in Table II.2 for other relative humidities.	14
II.4	Temperature variation of quotient of container volume and apparent mass of water, or mercury, filling it.	14
II.5	Changes in the quotient given in Table II.4 for water at other air pressures.	14
II.6	Values of reference pressures associated with phase transitions of pure substances at a temperature of 298 K, unless otherwise stated.	14
II.7	1981 International Practical Pressure Scale: calculated values of pressure for various linear and volume compressions for NaCl at 298 K.	15
II.8	Variation with pressure of melting point for Ag, Au and Cu.	15
II.9	Variation with temperature in mass of H_2O in 1 m^3 of saturated air at standard pressure as determined by chemical hygrometer.	16
II.10	Relative humidity values, determined by forced ventilated wet- and dry-bulb hygrometer, corresponding to various wet- and dry-bulb temperatures.	16
II.11	Variation with temperature of relative humidity for air above various saturated salt solutions sited in a uniform constant-temperature enclosure.	16
II.12	Variation with temperature of density of pure air-free H_2O at standard pressure.	17
II.13	Variation with temperature of density of heavy H_2O at standard pressure.	17
II.14	Variation with temperature of density of Hg at standard pressure.	18
II.15	Typical density for some everyday substances at about STP.	19
II.16	Values of Young's, shear and bulk elastic moduli, the Poisson ratio and tensile strength for some metals at temperature 293 K.	20

II.17	Values of Young's shear and bulk elastic moduli, the Poisson ratio and tensile strength for some alloys at temperature 293 K.	21
II.18	Values of Young's, shear and bulk elastic moduli, the Poisson ratio and tensile strength for glass and quartz at temperature 293 K.	21
II.19	Values of Young's modulus of elasticity and tensile strength for some plastics at temperature 293 K.	22
II.20	Values of Young's modulus of elasticity, along the grain and in radial and tangential directions, and tensile strength for some woods.	22
II.21	Values of temperature coefficient of Young's modulus and shear moduli of elasticity of some substances at temperature 288 K.	22
II.22	Values of bulk modulus of elasticity of some non-metallic elements.	22
II.23	Values of bulk modulus of elasticity of some liquids at stated temperature and pressure.	23
II.24	Values of tensile strength of some miscellaneous substances.	23
II.25	Values of coefficient of viscosity for various gases at about standard pressure.	24–25
II.26	Values of coefficient of viscosity for some gases at various high pressures at temperature 300 K, unless otherwise stated.	25
II.27	Values of coefficient of viscosity for water at about standard pressure at stated temperatures.	25
II.28	Values of coefficient of viscosity for various glycerol/H_2O solutions at about standard pressure.	25
II.29	Values of coefficient of viscosity for various sucrose/H_2O solutions at about standard pressure.	26
II.30	Values of coefficient of viscosity for some liquid elements, molten salts, glasses and minerals at about standard pressure at stated temperatures.	26
II.31	Values of coefficient of viscosity for various organic liquids at about standard pressure at stated temperatures.	27–28
II.32	Variation of the coefficient of viscosity of water with pressure and temperature.	29
II.33	Variation of the coefficient of viscosity of various liquids with pressure at stated temperatures.	29
II.34	Values of STP molecular mean speed, mean free path and mean time between collisions, for various gases.	30
II.35	Variation with temperature of surface tension of water against air at about standard pressure.	30
II.36	Variation with percentage solute of surface tension of various aqueous solutions at stated temperatures at about standard pressure.	30–31
II.37	Surface tension γ at about standard pressure, for some liquid elements and molten salts at stated temperatures, T, together with some values of A and B in the formula $\gamma = A - BT$, and temperature range of formula applicability.	31
II.38	Surface tension γ at about standard pressure, for some liquids at temperature 293 K, unless otherwise stated, together with some values of A and B in the formula $\gamma = A - BT$ and temperature range of formula applicability.	32
II.39	Interfacial tension γ at about standard pressure, for some liquids at temperature 293 K.	32

Section III Properties of the Elements

III.1	32-column Form of the Periodic Table (IUPAC 1989)	35
III.2	Standard Atomic Weights of the Elements (1989)	36
	Alphabetical order	36
	Atomic Number order	37
III.3	Electronic Configurations of the Elements (ground states)	38
III.4	Physical Properties of the Elements	40
III.5	Crystal Structure, Atomic and Ionic Radii of the Elements	42
III.6	van der Waals radii of the Elements	44
III.7	Ionization Energies of the Elements	44
III.8	Atomic Electron Affinities at 0 K	46
III.9	Proton Affinities of Atoms	46
III.10	Electronegativities of the Elements	47
III.11	Abundance of the Elements in the Earth's Crust and in Sea Water	48

xiv List of Tables

III.12	Average Composition of the Atmosphere (exclusive of water Vapour)	48
III.13	Elemental Composition of the Adult Human Body	49

Section IV General Properties of Molecules

IV.1	Hybrid Orbitals	53
IV.2	Average bond lengths of diatomic molecules	53
IV.3	Bond Lengths of Elements	54
IV.4	Bond Lengths and Angles in Polyatomic Molecules	54
	Inorganic compounds	54
	Organic compounds	55
IV.5	Bond Lengths and Angles for Some Representative Polymers	56
IV.6	Shapes of Molecules and Ions	57
IV.7	Electric Dipole Moments of Organic Compounds	58
IV.8	Electric Dipole Moments of Inorganic Compounds	59
IV.9	Electric Dipole Moments, Polarizabilities and Polarizability Volumes of Some Inorganic and Organic Molecules	59
IV.10	Electron Affinities of Some Diatomic Molecules and Radicals	60
IV.11	Lennard-Jones (12–6) Potential Parameters	60
IV.12	Madelung Constants	60
IV.13	Character Tables	61
IV.14	Bravais Lattices	64
IV.15	Characteristics of Seven Crystal Systems	65

Section V Electricity and Magnetism

V.1	Resistivity values for some liquid metallic elements at stated temperatures.	69
V.2	Resistivity values for single crystals of some non-cubic metallic elements along a- and c- axes at standard temperature, unless otherwise stated.	69
V.3	Variation with temperature of resistivity of some polycrystalline solid metallic elements.	70–71
V.4	Variation with temperature of resistivity of some alloys.	71
V.5	Temperature coefficient of resistivity for resistance-wire alloys	71
V.6	Resistances per metre length of various standard wire gauge resistance wires for materials of Table V.5, together with copper	72
V.7	Approximate values of electric current required to maintain temperature rises, $\triangle T$, of 500 K and 1000 K in straight horizontal Ni/Cr resistance wires, of various standard wire gauges, free to radiate in air.	72
V.8	Resistivity values for semiconducting elements at standard temperature.	72
V.9	Resistivity values one minute after measurement-voltage application for some ceramics at standard temperature and 50–60% relative humidity.	73
V.10	Resistivity values one minute after measurement-voltage application for some plastics at standard temperature and 50–60% relative humidity.	73
V.11	Resistivity values one minute after measurement-voltage application for some other insulators at standard temperature and 50–60% relative humidity.	73
V.12	Relative permittivity of some gases and vapours for frequencies less than infrared at stated temperatures and, except where otherwise stated, standard pressure.	74
V.13	Relative permittivity and loss tangent, at stated temperatures and frequencies, for some liquids of controlled purity used for electrical purposes.	74
V.14	Relative permittivity and its temperature rate of change for various inorganic liquids at stated temperatures and standard pressure, unless otherwise stated.	75
V.15	Values of relative permittivity and its temperature rate of change for various organic liquids at stated temperatures and standard pressure.	76–78

V.16	Values of relative permittivity and loss tangent for some inorganic solids at stated temperatures and frequencies.	77–78
V.17	Values of relative permittivity and loss tangent for some organic solids at stated temperatures and frequencies.	78–79
V.18	Representative variation of magnetic flux density with magnetic field strength for various low-coercivity magnetic substances, together with initial and maximum relative permeability, saturation polarization, coercivity, remanence, Curie temperature, resistivity and specific total power loss.	80–81
V.19	Representative values for some core materials of initial relative permeability, frequency range, loss factor at maximum frequency, Curie temperature and resistivity, together with some other information.	82
V.20	Representative values of remanence, maximum of product of magnetic field strength and magnetic flux density (with field strength given in brackets), induction and magnetization coercive forces, resistivity, Curie temperature and maximum operating temperature, for some permanent-magnet materials.	83
V.21	Representative values of relative permeability and resistivity for some feebly magnetic steels and cast irons for various conditions.	84
V.22	Mass susceptibility per kilogram at temperature 293 K for some paramagnetic and diamagnetic materials.	84–85
V.23	Values of critical temperature and critical magnetic field(s) at temperature 0 K for some superconducting elements.	86
V.24	Values of critical temperature and critical magnetic field at temperature of 4.2 K for some superconducting compounds and alloys.	86

Section VI   Electrochemistry

VI.1	Conductivity of Potassium Chloride Solutions	89
VI.2	Molar Conductivity of Electrolytes in Aqueous Solution at 25°C	89
VI.3	Limiting Molar Ionic Conductivities in Aqueous Solution at 25°C	90
VI.4	Limiting Molar Ionic Conductivities in Aqueous Solution at Various Temperatures	91
VI.5	Limiting Molar Ionic Conductivities at 25°C in a Range of Solvents	92
VI.6	Electrical Properties of Fused salts	92
	1. Equivalent Conductivities of Single Salt Melts	92
	2. Cation Transport Numbers for Fused Salts	93
VI.7	Debye-Hückel-Onsager coefficients for 1:1 electrolytes at 25°C	93
VI.8	Ionic Strength as a Function of Concentration for Various Valence Type Electrolytes	93
VI.9	Relationship between Molality, Mean Ionic Molality, Activity and Mean Ionic Activity Coefficient for Various Valence Type Electrolytes	93
VI.10	Mean Stoichiometric Ionic Activity Coefficients at 25°C for Electrolytes in Aqueous Solution	94
VI.11	Mean Ionic Activity Coefficients for Hydrochloric Acid in water	95
VI.12	Mean Ionic Activity Coefficients for Hydrochloric Acid at 25°C in Dioxane-Water Mixtures	95
VI.13	Activity Coefficients Calculated using the Debye-Hückel Equation	96
VI.14	Transport Numbers of Cations in Aqueous Solution at 25°C	97
VI.15	Transport Numbers of Cations of 2:1 Electrolytes in Aqueous Solution at 25°C, showing the Effect of Auto-complex Formation	97
VI.16	Cation Transport Numbers of Hydrochloric Acid at 25°C in Dioxane-Water Mixtures	97
VI.17	Cation Transport Numbers in Aqueous Solution at Various Temperatures	98
VI.18	Diffusion Coefficients at Aqueous Solutions of Selected Electrolytes at 25°C	98
VI.19	Osmotic and Activity Coefficients of Sodium, Potassium and Calcium Chloride and Sulphuric Acid Solutions at 25°C	99
VI.20	Osmotic and Activity Coefficients of Sucrose Solutions at 25°C	99
VI.21	pH Values of Operational Reference Standard Solutions	100
VI.22	pH Values of Useful Buffer Solutions	100

VI.23	Biological Buffer Solutions	102
VI.24	pH Values of Some Common Buffer Solutions	102
VI.25	Approximate pH Values of Some Common Solutions, Fluids and Foodstuffs	105
VI.26	Acid-Base Indicators	107
VI.27	Standard (reduction) Potentials at 298 K in Aqueous Solution	107
VI.28	Formal Electrode Potentials at 298 K	109
VI.29	Standard Redox Potentials for Some Biological Half Reactions at 298 K and pH = 7.0	109
VI.30	Standard (reduction) Potentials in Non-aqueous Solution at 298 K (except liquid ammonia 265 K)	109
VI.31	Redox Indicators for Volumetric Use	110
VI.32	Biological Redox Indicators	110
VI.33	Dissociation Constants of Inorganic Acids in Aqueous Solution at 298 K	110
VI.34	Dissociation Constants of Organic Acids in Aqueous Solution at 298 K	111
VI.35	Dissociation Constants of Inorganic Bases in Aqueous Solution at 298 K	113
VI.36	Dissociation Constants of Organic Bases in Aqueous Solution at 298 K	113
VI.37	Ionization Constants for Water and Deuterium Oxide	114
VI.38	Autoprotolysis Constants for Some Solvents at 298 K	115
VI.39	Solubility Products in Water at 298 K	115
VI.40	Stepwise Formation (Stability) Constants	116
VI.41	Formation Constants for EDTA Complexes at 293 K and an Ionic Strength of 0.1 mol dm^{-3}	117
VI.42	Hammett Acidity Function	117
VI.43	Half-wave Potentials of Selected Inorganic Species	118
	1. Half-wave Potentials of Metals	118
	2. Half-wave Potentials of Some Other Inorganic Species	119
	3. Half-wave Potentials of Some Cations in Non-aqueous Solvents	120
VI.44	Half-wave Potentials of Selected Organic Compounds	122
VI.45	Isoelectric points (or range) for Several Proteins	123

Section VII   Optics

VII.1	Luminance and correlated colour temperature for some light sources.	127
VII.2	1931 Commission International de l'Eclairage (CIE) specification of colorimetric standard observer, for 0.5° to 4° field subtenses, in terms of variation of spectral tristimulus values with wavelength.	128
VII.3	1951 CIE scotopic relative luminous efficiency function.	129
VII.4	CIE standard illuminants for colorimetry.	129
VII.5	Relative spectral power distribution of the CIE standard illuminants A, C, and D.	130
VII.6	Hue discrimination for the normal observer.	131
VII.7	Types of colour deficiency characteristics and inherited percentage incidence in the population.	131
VII.8	Some substances which may produce significant red/green colour perception disturbance.	132
VII.9	Some substances which may produce significant blue/yellow colour perception disturbance.	132
VII.10	Variation of visual depth of focus with pupil diameter of eye.	132
VII.11	Range and mean of main geometrical optical properties for 1000 eyes in vivo.	132
VII.12	General specification of the Gullstrand-Emsley schematic eye.	132
VII.13	Chromatic aberration of the Gullstrand-Emsley schematic eye.	133
VII.14	International Commission for Optics recommended wavelengths for refractometry.	133
VII.15	Some commonly used wavelengths in refractometry.	133
VII.16	Variation of refractive index of standard air with wavelength.	134
VII.17	Refractive indices, relative to a vacuum, of various gases for wavelength 589.3 nm and for radio wavelengths at standard pressure and temperature, unless otherwise stated.	134

VII.18	Variation with wavelength of refractive index and rate of change of index with temperature for the refractometer-calibrating liquids distilled water and toluene 'Ultra' grade.	134
VII.19	Refractive indices for various liquids for wavelength 589.3 nm at or near temperature 293 K.	135
VII.20	Variation with wavelength of refractive index of some Chance Pilkington optical glasses at temperature 293 K and standard pressure.	135
VII.21	Variation with wavelength of light transmittance of 25 mm thickness of some Chance Pilkington optical glasses at temperature 293 K and standard pressure.	136
VII.22	Variation of reflectance for normally incident light in air with refractive index of a nonabsorbing reflecting substance.	136
VII.23	Refractive and absorption indices and reflectance for normally incident light in air for various metals and semiconductors at stated wavelengths.	136
VII.24	Variation with wavelength of refractive index of synthetic fused silica, calcium fluoride and lithium fluoride at temperature 293 K.	137
VII.25	Variation with wavelength of refractive index of commonly used optical plastics.	137
VII.26	Refractive indices at or near temperature 293 K of infrared optical materials at stated wavelengths, together with transmission range.	138
VII.27	Refractive indices of some optical cements at wavelength 589.3 nm and temperature 293 K.	138
VII.28	Variation with wavelength of ordinary and extraordinary refractive indices of calcite at temperature 291 K (293 K) for wavelength range 200–706 nm (801–3324 nm).	139
VII.29	Variation with wavelength of ordinary and extraordinary refractive indices of crystal quartz at temperature 293 K.	139
VII.30	Ordinary and extraordinary refractive indices at or near temperature 293 K of various birefringent optical materials at stated wavelengths, together with transmission range.	139
VII.31	Variation with angle of incidence of light of the percentage reflectance, at interface between air and glass (refractive index 1.52) for electric vector respectively parallel and perpendicular to the plane of incidence.	140
VII.32	Refractive indices for common thin film materials and their percentage reflectances in air for light of wavelength 550 nm and for thickness to form a quarter-wave film on glass of refractive index 1.52.	140
VII.33	Percentage reflectance in air for light of wavelength 550 nm of quarter-wave multilayer stacks of zinc sulphide and magnesium fluoride alternately for different numbers of layers on glass of refractive index 1.52.	140
VII.34	Variation with wavelength of specific rotation for quartz at temperature 293 K.	140
VII.35	Values of Verdet constants for various gases at wavelength 589.3 nm and STP.	140
VII.36	Variation with wavelength of Verdet constant for water at 293 K.	140
VII.37	Verdet constants for various liquids at stated temperature for wavelength 589.3 nm.	141
VII.38	Verdet constants for various solids at stated temperatures and wavelengths.	141
VII.39	Refractive indices, half-wave voltage and transmission range for some electro-optic substances.	141
VII.40	Values of wavelengths of maximum emission and cut-off, decay constant and light conversion efficiency, relative to that of NaI(Tl) for some solid scintillators.	142
VII.41	Operational mode, vacuum wavelength, average and peak powers, rate of change of output power relative to input power, and pulse duration for some representative lasers.	142

Section VIII Spectroscopy

VIII.1	Electromagnetic Spectrum	145
VIII.2	Types of Optical Spectra	146
VIII.3	The Effective Cut-off Wavelengths of some Common Solvents in the UV and Visible Regions	146
VIII.4	Cut-off Wavelengths of Materials for use in the UV and Visible Regions.	147
VIII.5	Wavelength Calibration Standards in UV and Visible Regions	147
	1. Rare Earth Filters	147
	2. Discharge Tube Emission Lines	147
VIII.6	Photometric Calibration Standards for use in the Ultraviolet and Visible Regions	148

VIII.7	Spectroscopic Solvents for use in the Infrared Region	150
VIII.8	Transmission Regions of Various Optical Materials for use as Cell Windows in the Infrared Region	152
VIII.9	Wavelength Calibration Standards in Infrared region	152
VIII.10	Wavelength Calibration Standards for Spectroscopes	153
VIII.11	Spectral Series of Hydrogen	153
VIII.12	Wavelengths of the Fraunhofer lines (Sun's Spectrum)	153
VIII.13	Absorption Characteristics of Some Common Organic Chromophores	154
VIII.14	Molecular Parameters of Gases	155
VIII.15	Spectroscopic Data on the Properties of Diatomic Molecules in their Ground State	156
VIII.16	Spectroscopic Data on the Properties of Triatomic Molecules	156
VIII.17	Rotational Constants and Moments of Inertia of Symmetric Top Molecules in their Ground States	157
VIII.18	Rotational Constants and Moments of Inertia of Asymmetric Rotor Molecules	157
VIII.19	Fundamental Vibrational Frequencies of Polyatomic Molecules	158
VIII.20	Potential Energy Barriers to Internal Rotation about Single Bonds	158
VIII.21	Nuclear Spins, Moments, Resonant Frequencies and Receptivities of Naturally Occurring Magnetically Active Isotopes	159
VIII.22	Chemical Shifts of Some Common NMR Solvents	160
VIII.23	Ranges of ^{1}H Chemical Shifts of some Common Functional Groups	161
VIII.24	Ranges of ^{13}C Chemical Shifts in Functional Groups	162
VIII.25	Ranges of ^{15}N Chemical Shifts in some Common Functional Groups	163
VIII.26	Some Chemical Shifts of ^{17}O, ^{19}F, ^{27}Al, ^{31}P and ^{35}Cl for Substances Indicated	164
VIII.27	Mass-composition of Most useful Fragments to Mass 99	164
VIII.28	Common Losses from Molecular Ions	166
VIII.29	Recommended Fluorescent Quantum Yield References in Various Emission Ranges	167

Section IX Atomic and Nuclear Physics

IX.1	Ionic mobilities of various atomic and molecular ions in some gases at standard pressure and, unless otherwise indicated, standard temperature.	171
IX.2	Electronic drift speeds in some gases for various values of applied field strength per unit pressure.	172
IX.3	Variation with pressure of the ionic recombination coefficient for air at temperature 291 K.	172
IX.4	Variation with pressure and temperature of the ionic recombination coefficient for oxygen.	172
IX.5	Ionic recombination coefficient for some gases at stated temperatures and pressures.	173
IX.6	Variation of ionic recombination coefficient for air, at temperature 288 K and standard pressure, with number of ions per cubic metre.	173
IX.7	Variation of the ratio of mean agitation energy of a free electron in a gas to that of a gas molecule with applied field strength per unit pressure, for various gases.	173
IX.8	Variation with electron energy of ratio of total electron-atom (or -molecule) collision cross-section area to area of first Bohr orbit of hydrogen atom, for various gases.	173
IX.9	Work function of various solid elements obtained from contact potential difference and/or photoelectric and/or thermionic measurements.	174
IX.10	Auger-electron kinetic energies resulting from the most intense transitions of some common elements.	175
IX.11	Kinetic energy of photoelectrons ejected from atomic energy levels 1s to $4d_{5/2}$ due to excitation with AlK$\alpha_{1,2}$ radiation, energy 1486.6 eV, for elements of atomic numbers 1–92.	176–177
IX.12	Kinetic energy of photoelectrons ejected from atomic energy levels $4f_{5/2}$ to $6d_{5/2}$ due to excitation with AlK$\alpha_{1,2}$ radiation energy 1486.6 EV, for elements of atomic numbers 49–92.	178
IX.13	Kinetic energy of photoelectrons ejected from atomic energy levels 1s to $4d_{5/2}$ due to excitation with MgK$\alpha_{1,2}$ radiation energy 1253.6 eV, for elements of atomic numbers 1–92.	179–180

IX.14	Kinetic energy of photoelectrons ejected from atomic energy levels $4f_{5/2}$ to $6d_{5/2}$ due to excitation with $MgK\alpha_{1,2}$ radiation energy 1253.6 eV, for elements of atomic numbers 49–92.	181
IX.15	Energy values of X-ray absorption edges and of characteristic X-radiation types for K series, for elements of atomic numbers 4–103.	182–183
IX.16	Energy values of X-ray absorption edges and of characteristic X-radiation types for L series, for elements of atomic numbers 4–103.	184–185
IX.17	Energy values of X-ray absorption edges and of characteristic X-radiation types for M series, for elements of atomic numbers 29–103.	186–187
IX.18	Approximate intensities of some characteristic X-radiations relative to that of the most intense of its series.	187
IX.19	Variation with photon energy of values of mass absorption coefficient for various elements and concrete for a collimated incident photon beam which emerges undeflected without photon energy loss.	188
IX.20	Inelastic mean free path of electrons, usually of 1 keV energy, in various substances, and n values.	189
IX.21	Range in aluminium of electrons for various electron energies.	189
IX.22	Range in various substances of protons for various proton energies	190
IX.23	Range in various substances of alpha particles for various particle energies.	190
IX.24	Range in various substances of pions for various pion energies.	190
IX.25	Range in various substances of muons for various muon energies.	191
IX.26	Range in aluminium of some heavy ions for various ion energies.	191
IX.27	Range in carbon of some heavy ions for various ion energies.	191
IX.28	Range in germanium of some heavy ions for various ion energies.	192
IX.29	Range in gold of some heavy ions for various ion energies.	192
IX.30	Range in silver of some heavy ions for various ion energies.	192
IX.31	Range in titanium of some heavy ions for various ion energies.	193
IX.32	Range in uranium of some heavy ions for various ion energies.	193
IX.33	Mass stopping power in various substances for protons of various energies.	193
IX.34	Mass stopping power in various substances for alpha particles of various energies.	194
IX.35	Mass stopping power in various substances for pions of various energies.	194
IX.36	Mass stopping power in various substances for muons of various energies.	194
IX.37	Air kerma-rate constant for gamma radiation of energy greater than 50 keV for some common radionuclides.	195
IX.38	International Commission on Radiation Protection risk and risk-weighting factors of fatal cancer and serious hereditary defect-inducing radiation in adult humans for a dose equivalent of 1 sievert.	195
IX.39	Values of atomic number, mass number, half-life and/or natural abundance, decay scheme, mass excess, cross section, parity and spin for various nuclides.	196–262
IX.40	Average values of the components of the total energy released by slow-neutron induced fission in some fission materials.	263
IX.41	Half-life of principal activity, principal resonance energy, average activation cross section and resonance integral for some neutron foil resonance detectors.	263
IX.42	Half-life of principal activity, mean response energy in fission spectrum, average activation cross section and resonance integral for some neutron foil threshold detectors.	263
IX.43	Average values, for neutron thermal spectrum, of cross section for elastic scattering and for radioactive capture by cladding and structural elements in nuclear reactors.	264
IX.44	Values, for neutron slowing down region, of cross section for elastic scattering by, and of capture resonance integral for, cladding and structural elements in nuclear reactors.	264
IX.45	Average values, for neutron fission spectrum, of cross section for elastic scattering, for absorption, and for inelastic scattering by cladding and structural elements in nuclear reactors.	264
IX.46	Average values, for neutron thermal spectrum, of cross section for elastic scattering and for absorption for coolant components in nuclear reactors.	264

IX.47	Values, for neutron slowing down region, of cross section for elastic scattering by, and of absorption resonance integral for, coolant components in nuclear reactors.	264
IX.48	Average values, for neutron fission spectrum, of cross section for elastic and inelastic scattering and for radiative capture by coolant components in nuclear reactors.	265
IX.49	Average values, for neutron thermal spectrum, of cross section for elastic scattering and for absorption for nuclide or element poisons and absorbers in nuclear reactors.	265
IX.50	Values, for neutron slowing down region, of cross section for elastic scattering by, and of absorption resonance integral for, nuclide or element poisons and absorbers in nuclear reactors.	265
IX.51	Average values, for neutron thermal spectrum, of number of neutrons produced per fission, of cross section for elastic scattering, for radiative capture and for fission by fissile and fertile nuclides in nuclear reactors.	265
IX.52	Values, for neutron slowing down region, of cross section of elastic scattering by, and of capture and fission resonance integrals for, fissile and fertile nuclides in nuclear reactors, together with average number of neutrons per fission.	266
IX.53	Average values, for neutron fission spectrum, of number of neutrons produced per fission, of cross section for elastic and inelastic scattering, for radiative capture and for fission by fissile and fertile nuclides in nuclear reactors.	266
IX.54	Average values, for core and blanket, of number of neutrons produced per fission and of neutron fission cross section for nuclides in a typical fast nuclear reactor fuelled by plutonium-uranium oxide and with liquid metal coolant.	266
IX.55	Thermal diffusion length and average diffusion coefficient over thermal neutron spectrum for neutrons in some moderators and shielding materials at temperature 293 K, together with mean thermal-neutron lifetimes.	266
IX.56	Some fusion reactions of possible significance for electrical energy generation.	267
IX.57	Half-life of principal activity, threshold response energy and activation cross section for 14 MeV fusion neutrons for some neutron foil detectors.	267
IX.58	The more important radionuclides produced in each chemical element likely to occur in the structural materials of a fusion-reactor vacuum vessel and breeding blanket.	267
IX.59	Values of average reaction cross section over energy spectrum of fusion neutrons emitted from a deuterium-tritium plasma at 20KeV ion temperature, for various target elements and reaction types.	268
IX.60	Charge, mass, mean life and spin for various leptons together with their principal decay modes.	268
IX.61	Baryon number, charge, mass and Z-component of isospin for the various quarks together with name and value of other quantum numbers.	268
IX.62	Baryon number, mass, mean life, quark content and spin of non-strange hadrons together with their principal decay modes.	269
IX.63	Baryon number, strangeness, mass, mean life, quark content and spin of strange hadrons, together with their principal decay modes.	269
IX.64	Baryon number, charm, strangeness, mass, mean life, quark content and spin of charmed hadrons together with their principal decay modes.	269
IX.65	Mass, spin and width of beautyonium states together with their $b\bar{b}$ configuration and principal modes of decay.	270
IX.66	Mass, spin and width of charmonium states together with their $c\bar{c}$ configuration and principal modes of decay.	270
IX.67	Charge, mass and width of gauge bosons of spin 1 together with the force transmitted and principal decay modes.	270

Section X Acoustics

X.1	Speed of sound, at about standard pressure, in some gases at stated temperature.	273
X.2	Variation of the attenuation of sound in air with relative humidity at temperature 293 K and standard pressure, for various frequencies.	273
X.3	Quotient of sound absorption coefficient and square of sound frequency for some gases at temperature 293 K and standard pressure.	274

X.4	Speed of sound in distilled and sea water at various temperatures and standard pressure.	274
X.5	Quotient of sound absorption coefficient and square of sound frequency for distilled water at various temperatures and standard pressure.	274
X.6	Variation of the attenuation of sound with temperature and frequency at standard pressure in sea water of 3.5% salinity.	274
X.7	Speed of sound and its rate of change with temperature for various liquids at stated temperature and standard pressure.	275
X.8	Quotient of sound absorption coefficient and square of sound frequency for various liquids at stated temperature and standard pressure.	275
X.9	Typical values of the speed of longitudinal bulk sound waves, irrotational rod sound waves, shear sound waves and Rayleigh sound waves in various solids, at temperature of about 293 K (unless otherwise stated).	276–277
X.10	Typical values of absorption coefficient for longitudinal sound waves in various solids at a temperature of about 293 K.	277
X.11	Characteristics of transverse vibrations of a rod free at each end.	277
X.12	Characteristics of transverse vibrations of a rod free at one end and clamped at the other.	278
X.13	Preferred octave-band, ½-octave band and ⅓-octave band centre frequencies for acoustic measurements.	278
X.14	Typical octave-band analysis of sound in various types of environment.	278
X.15	Variation of minimum audible field with sound frequency.	278
X.16	Classification of hearing impairment by average permanent threshold shift for best ear for sound frequencies 500 Hz, 1000 Hz and 2000 Hz.	278
X.17	Recommended maximum noise levels at various frequencies.	279
X.18	Variation of loudness level of pure tones with sound pressure level and frequency.	279
X.19	Approximate loudness sensation equivalents for various sound pressure levels.	279
X.20	Noise loudness in some common types of environment	279
X.21	Typical variation of reverberation absorption coefficients with sound frequency for various materials.	280
X.22	Variation of typical reverberation times of a room for various ⅓-octave bandwidths.	280
X.23	Variation of sound reduction index with sound frequency for some types of partition.	281
X.24	Variation of sound reduction index with sound frequency for some types of window.	281
X.25	Variation of typical ambient noise level, at which speech interference occurs for a normal male voice, with distance between speaker and hearer.	281
X.26	Characteristics of some concert halls.	282
X.27	Consonant musical frequency intervals.	282
X.28	Frequency ratios of musical intervals in the diatonic major scale.	282
X.29	Frequencies of notes in equal temperament.	282

Section XI Physical Properties of Inorganic and Organic Compounds

XI.1	Physical Constants of Selected Inorganic Compounds	285
XI.2	Physical Constants of Selected Organic Compounds	307
XI.3	Physical Constants of Selected Organo-metallic Compounds	360
XI.4	Physical Constants of Amino Acids and Derivatives	366
XI.5	Physical Constants of Carbohydrates and Related Compounds	376
	1. Trioses, tetroses and related alcohols	376
	2. Pentoses and related alcohols	376
	3. Hexoses and related alcohols	380
	4. Di- and Tri-saccharides	384
	5. Amino Sugars	386
	6. Uronic Acids	388
XI.6	Properties of Selected Steroids (bile acids, corticosteroids, sex hormones, sterols)	392

xxii List of Tables

XI.7	Properties of Constituents of Nucleic acids and Related Compounds (nucleosides, nucleotides, purines, pyrimidines)	395
XI.8	Physical Constants of Polymers and Plastics	398
XI.9	Physical Constants of Fluorocarbon Refrigerants	400
XI.10	Physical Constants of Some Common Refrigerants	402
XI.11	Properties of Gases Available in Cylinders (bottled gases)	402
XI.12	Critical Constants	408
	1. Elements and Inorganic Compounds	408
	2. Organic Compounds	409
XI.13	van der Waals Constants and Boyle Temperatures for Some Gases	410
XI.14	Second Virial Coefficients of Selected Gases	411
XI.15	Solubility of Gases in Water	412
XI.16	Solubility of Some Gases in Sea Water as a Function of Temperature and Salinity	413
XI.17	Vapour Pressure of Water at Different Temperatures	414
	1. Vapour Pressure of Ice; -90 to $0°$ C	414
	2. Vapour Pressure of Water; -15 to $100°$ C	414
	3. Vapour Pressure of Water above $100°$ C	414
XI.18	Vapour Pressure of Mercury from -38 to $400°$ C	415
XI.19	Boiling Point of Water at Different Pressures	415
XI.20	Ionization Potentials of Selected Molecules	416
XI.21	Hardness of Materials	417
	1. Mohs Hardness Scale	417
	2. Hardness of Materials (on Mohs Original Scale)	417
XI.22	Critical Concentrations for Micelle Formation	418
	1. Ionic Surfactants	418
	2. Non-ionic Surfactants	419
XI.23	Physical Constants of Proteins in Water at 293 K	419
XI.24	Solubility of Some Compounds in Water	419
	1. Inorganic Compounds	419
	2. Organic Compounds	419
XI.25	Density of Aqueous Ethanol Solutions at $20°$ C	420
XI.26	Density of Some Aqueous Solutions at $20°$ C	420
	1. Acids	420
	2. Bases	420
	3. Alcohols, Dioxane and Other Hydroxylic Compounds	420
	4. Some Electrolyte Solutions	421
	5. Some Sugar Solutions	421
XI.27	Vapour Pressures of Selected Elements and Inorganic Compounds	422
XI.28	Vapour Pressures of Selected Organic Compounds	424
XI.29	Data on the Properties of Commercial Laboratory Reagents	427
XI.30	Preparation of Volumetric Solutions	428

Section XII Heat and Temperature

XII.1	Volume and pressure coefficients of cubic expansion for some gases over temperature interval 273–373 K (unless otherwise stated) and various initial pressures.	431
XII.2	Coefficient of cubic expansion for various liquids at temperature 293 K.	431
XII.3	Variation with temperature of the coefficient of linear expansion of some solid elements.	432
XII.4	Variation with temperature of the coefficient of linear expansion of some solid alloys and compounds.	432
XII.5	Guide values for the coefficient of linear expansion of some miscellaneous solid substances.	432
XII.6	Variation of thermal conductivity with temperature for various gases at standard pressure.	433
XII.7	Thermal conductivity of some vapours and liquids at equilibrium saturation pressure.	433

XII.8	Thermal conductivity for various liquids at stated temperatures	433
XII.9	Thermal conductivity of solid elements at various temperatures	434–435
XII.10	Thermal conductivity along various axes of some non-cubic metal crystals at standard temperature, unless otherwise stated	435
XII.11	Typical values of thermal conductivity of some alloys at various temperatures.	436–437
XII.12	Thermal conductivity of some refractory materials at various temperatures.	438
XII.13	Typical values of thermal conductivity of some everyday materials at standard temperature, unless otherwise stated.	438
XII.14	Ratio of specific heat capacity at constant pressure to that at constant volume for various gases at stated temperatures and, unless otherwise stated, standard pressure.	439
XII.15	Variation with temperature of the specific heat capacity at constant pressure of water.	439
XII.16	Specific heat capacity at constant pressure of some metallic elements at various temperatures.	440
XII.17	Typical values of the specific heat capacity at constant pressure of some alloys at various temperatures.	441
XII.18	Typical values of the specific heat capacity at constant pressure of some miscellaneous substances at various temperatures.	442
XII.19	Addition correction to observed surface temperature as recorded by spectral pyrometer (effective wavelength $0.65\mu m$) to give true surface temperature as a function of spectral emissivity	443
XII.20	Typical values of the normal emissivity of a smooth polished surface at several wavelengths for various substances at stated temperature.	443
XII.21	Variation with temperature of the thermal electromotive force for various elements relative to platinum, one junction being at a constant temperature of 273 K.	444
XII.22	Variation with temperature of the thermal electromotive force for some alloys relative to platinum, one junction being at a constant temperature of 273 K.	445
XII.23	Variation with temperature of the absolute negative thermoelectric power of platinum.	445
XII.24	Fixed points of the International Temperature Scale of 1990	445
XII.25	Definition of the International Practical Temperature Scale of 1990.	446

Section XIII Thermodynamic Properties of Elements and Compounds

XIII.1	Thermodynamic Data for the Elements	449
XIII.2	Thermodynamic Data for Selected Inorganic Compounds	451
XIII.3	CODATA Recommended Values for Thermodynamic Functions of various Chemical Species at 101 325 Pa	466
XIII.4	Thermodynamic Data for Selected Organic Compounds at 298.15 K	470
XIII.5	Thermodynamic Data for Selected Biological Molecules at 298.15 K and 1 atmosphere pressure	474
XIII.6	Enthalpies of Combustion for Selected Organic Compounds	477
XIII.7	Enthalpies of Fusion and Vaporization at the Transition Temperature of Some Inorganic Compounds	478
XIII.8	Enthalpies of Fusion and Vaporization at the Transition Temperature for Some Organic Compounds	479
XIII.9	Enthalpies of Formation of Gaseous Atoms from Elements in their Standard States at 298.15 K	481
XIII.10	Enthalpies of Formation of Free Radicals at 298.15 K	482
XIII.11	Enthalpy of Solution of Selected Electrolytes at 298.15 K	482
XIII.12	Enthalpy of Neutralization at 298.15 K	483
XIII.13	Integral Enthalpies of Solution at 298.15 K	483
XIII.14	Mean Bond Dissociation Enthalpies (single bond enthalpies)	484
XIII.15	Bond Dissociation Enthalpies of Diatomic Molecules at 298.15 K	484
XIII.16	Bond Dissociation Enthalpies in Some Polyatomic Molecules at 298.15 K	485
XIII.17	Enthalpy of Hydration of Some Ions at 298.15 K	486
XIII.18	Free Energy Function $[(G^{\ominus}(T) - H^{\ominus}(0))]/T$ for Selected Gaseous Carbon Compounds based on $H^{\ominus}(0)$	486

XIII.19	Molar Thermodynamic Functions of a Harmonic Oscillator	487
XIII.20	Lattice Enthalpies of Some Metal Halides	488
XIII.21	Thermodynamic Data for Some Ion Association Reactions at 298.15 K	488
XIII.22	Molar Heat Capacity of Elements	489
XIII.23	Molar Heat Capacity of Some Solid Inorganic Compounds	490
XIII.24	Molar Heat Capacity of Some Gaseous Organic Compounds	491
XIII.25	Temperature Dependence of Heat Capacities	491
XIII.26	Debye Characteristic Temperatures	493
XIII.27	Standard Gibbs (free) Energy Function of Hydrolysis of Some Phosphate and other Biologically Important Compounds at pH 7 and 298.15 K	493
XIII.28	Thermodynamic Constants for Phase Changes of Water and Deuterium Oxide	493
XIII.29	Molal Cryoscopic and Ebullioscopic Constants for Some Common Solvents	494
XIII.30	Henry's Law Constants for Gases in Solution	495
	1. Henry's Law Constant for Inorganic gases at 25° C	495
	2. Henry's Law Constant for Nitrogen in Water at Different Temperatures	495
	3. Henry's Law Constant for Some Organic Compounds in Aqueous Solution at 25° C	495
XIII.31	Joule-Thompson Coefficients for Some Gases at 273 K and 1 Atmosphere pressure	496
XIII.32	Standard Molar Gibbs Energies of transfer of selected ions from water into aqueous alcohols at 298.15 K	496
XIII.33	Calorific Values of Solid, Liquid and Gaseous Fuels	497
XIII.34	Properties of Some Common Binary Azeotopes at 1 atmosphere Pressure	498
XIII.35	Critical Solution Temperatures for Some Partially Miscible Binary Liquid Mixtures	499
XIII.36	Systems with Simple Eutectic Points	500
XIII.37	Cryohydrates	501
XIII.38	Equilibrium Constants for Gaseous Reactions at Various Temperatures	501

Section XIV Astronomy and Geophysics

XIV.1	Approximate values of galactic distances from Earth.	505
XIV.2	Distance and visual magnitude of the brightest stars from Earth.	505
XIV.3	Diameter, semimajor axis, eccentricity, inclination and rotation period of solar system asteroids of diameter exceeding 200 km	505
XIV.4	Values of diameter, density, semimajor axis, eccentricity, inclination and sidereal period of known satellites of solar system planets.	506
XIV.5	Values of density, equatorial diameter, distance from Sun, surface gravity, ellipticity, eccentricity, inclination and sidereal and rotational periods for the Sun and its planets.	507
XIV.6	Values of some miscellaneous properties of the Sun and Moon.	507
XIV.7	Values of some miscellaneous properties of Earth.	507
XIV.8	Variation of model Earth's density, gravity, pressure and elastic characteristics with depth beneath surface.	507
XIV.9	Possible values of duration and time of start from present of various geological periods for Earth.	508
XIV.10	Variation with height above Earth of atmospheric pressure density and temperature.	508
XIV.11	Variation with height above Earth of upper atmospheric pressure, mean molecular weight and temperature.	508
XIV.12	Values of particle flux in vertical direction for various cosmic-ray particles at sea level, for kinetic energies above various threshold values, at geomagnetic latitudes greater than about 40°.	509
XIV.13	Average values of declination and components of the geomagnetic field parallel to the Earth's surface and vertically downwards, at various observatories in 1975.	509
XIV.14	Values of the electrical conductivity of sea water at various temperatures and salinities for standard atmospheric pressure.	510

XIV.15	Density of sea water at various temperatures and salinities for standard atmospheric pressure.	510
XIV.16	Density of sea water of salinity 35g kg^{-1} at various pressures for standard temperature.	510
XIV.17	Relationships between some time quantities.	510

Section XV Chemical Kinetics

XV.1	Kinetic Data for Some First and Second Order Reactions	513
	1. First Order Reactions	513
	2. Second Order Reactions	513
XV.2	Arrhenius Parameters for Rate Coefficients	514
	1. Thermal Unimolecular Gas Phase Reactions	514
	2. Bimolecular Reactions	514
XV.3	Rate Coefficients for Some Second Order Solution Reactions at 298 K	515
XV.4	Arrhenius Pre-exponential Factors, Entropies and Volumes of Activation for a Selection of Reactions	516
	1. Unimolecular Reactions	516
	2. Bimolecular Reactions	516
	3. Ionic Reactions in Water	516
XV.5	Activation Energies for Some Catalysed Reactions	517
XV.6	Hammett Substituent Constants, σ	517
XV.7	A Selection of Hammett Reaction Constants, ρ, from Correlations of Reaction Rate Constants	518

Section XVI Health and Safety

XVI.1	Toxicity and Approved Occupational Exposure Limits for some Organic Compounds	521
XVI.2	Approved Occupational Exposure Limits for some Inorganic Compounds	527
XVI.3	Toxicity Data for Some Inorganic Compounds	530
XVI.4	Toxicity and Approved Occupational Exposure Limits for some Pesticides and other Materials in Common Use	531

Section XVII Ancillary Topics

XVII.1	International Organization for Standardization specification for screw threads in inches and millimetres.	535
XVII.2	Whitworth and British Association specifications for standard series screw threads.	535
XVII.3	Typical percentage composition of various Alloys.	536–537
XVII.4	Colour Code for wire-wound and other Resistors.	537
XVII.5	Some commonly occurring indefinite integrals.	538–543
XVII.6	Some commonly occurring definite integrals.	544
XVII.7	Some useful formulae for approximate integration.	545
XVII.8	Some useful series.	546–548
XVII.9	The error function $\mathrm{Erf} x = 2\pi^{-\frac{1}{2}} \int_0^x e^{-x^2} dx$	549–550

Section I
Units, Conversion Factors and Fundamental Constants

		Page
I.1	**Base SI units**	3
I.2	**Derived SI units with special names**	3
I.3	**Symbols and SI units for some physical quantities**	4
I.4	**SI conversion tables**	5
	Concentration units	5
	Electricity and magnetism	5
	Mechanics	5
	Radiation, spectrophotometric and other optical quantities	6
	Space and time	6
	Thermodynamics	6
I.5	**Prefixes used with SI units**	6
I.6	**Conversion factors for some other units**	7
	A Length	7
	B Area	7
	C Volume	7
	D Speed	7
	E Mass	8
	F Force	8
	G Pressure	9
	H Work and energy	9
	I Concentration units	9
I.7	**Fundamental constants**	10

Section I
Units, Conversion Factors and Fundamental Constants

Reference for Tables I.1–I.7:
Horvath, *Conversion Tables of Units for Science and Engineering*, Macmillan, 1986
IUPAC, *Quantities, Units and Symbols in Physical Chemistry*, Blackwell, 1988

Table I.1 Base SI units.

Physical quantity	Symbol	Unit name	Unit symbol
amount of substance	n	mole	mol
current	I	ampere	A
length	l	metre	m
luminous intensity	I_v	candela	cd
mass	m	kilogram	kg
thermodynamic temperature	T	kelvin	K
time	t	second	s

Table I.2 Derived SI units with special names.

Quantity	Symbol	Unit	Unit name	Unit symbol
† plane angle			radian	rad
† solid angle			steradian	sr
absorbed dose	D	$m^2\,s^{-2}$	gray	Gy
activity	A	s^{-1}	becquerel	Bq
capacitance	C	$A^2\,s^4\,kg^{-1}\,m^{-2}$	farad	F
Celsius temperature	t, θ	K	degree Celsius	°C
charge	Q	$A\,s$	coulomb	C
conductance	G	$A^2\,s^3\,kg^{-1}\,m^{-2}$	siemens	S
dose equivalent	H	$J\,kg^{-1}$	sievert	Sv
electromotive force	E	$kg\,m^2\,s^{-3}\,A^{-1}$	volt	V
energy	E	$m^2\,kg\,s^{-2}$	joule	J
force	F	$kg\,m\,s^{-2}$	newton	N
frequency	ν, f	s^{-1}	hertz	Hz
illuminance	E_v, E	$cd\,sr\,m^{-2}$	lux	lx
inductance, self	L	$kg\,m^2\,s^{-2}\,A^{-2}$	henry	H
inductance, mutual	M	$kg\,m^2\,s^{-2}\,A^{-2}$	henry	H
kerma	K	$m^2\,s^{-2}$	gray	Gy
luminous flux	Φ_v, Φ	$cd\,sr$	lumen	lm
magnetic flux	Ψ	$kg\,m^2\,s^{-2}\,A^{-1}$	weber	Wb
magnetic induction	B	$kg\,s^{-2}\,A^{-1}$	tesla	T
potential	V, ϕ	$kg\,m^2\,s^{-3}\,A^{-1}$	volt	V
potential difference	U, V	$kg\,m^2\,s^{-3}\,A^{-1}$	volt	V
power	P	$kg\,m^2\,s^{-3}$	watt	W
pressure	P	$kg\,m^{-1}\,s^{-2}$	pascal	Pa
quantity of heat	Q	$m^2\,kg\,s^{-2}$	joule	J
radiant flux	Φ_e, Φ	$m^2\,kg\,s^{-3}$	watt	W
resistance	R, r	$kg\,m^2\,s^{-3}\,A^{-2}$	ohm	Ω
work	W	$m^2\,kg\,s^{-2}$	joule	J

† Plane angle and solid angle are usually considered to be dimensionless derived quantities; in some situations however the steradian must be treated as a base unit.

Table I.3 Symbols and SI Units for some physical quantities.

Name of quantity	Symbol	SI unit
absorptance	α	1
acceleration	a	m s^{-2}
angular frequency	ω	Hz = s^{-1}
angular momentum	L	J s
angular velocity	ω	rad s^{-1}
area	A, S	m^2
atomic number, proton number	Z	1
Bohr magneton	μ_B	A m^2
bulk modulus	K	N m^{-2}
charge, volume density of	ρ	C m^{-3}
compressibility	κ, k	N^{-1} m^2
conductivity	K	S m^{-1}
cross section	σ	m^2
Curie temperature	θ_C, T_C	K
decay constant	λ	s^{-1}
degeneracy (multiplicity) of an energy level	g	1
density	ρ	kg m^{-3}
diffusion coefficient	D	m^2 s^{-1}
Dirac constant	$\hbar$	J s
efficiency	η	1
electric current density	J	A m^{-2}
electric displacement	D	C m^{-2}
electric field strength	E	V m^{-1}
electric flux	ψ	C
electric polarization	P	C m^{-2}
electric susceptibility	χ_e	1
emissivity	ε	1
enthalpy	H	J
entropy	S	J K^{-1}
fluence	Φ	m^{-2}
Gibbs function	G	J
half life	$T_{1/2}, t_{1/2}$	s
Hamiltonian function	H	J
heat capacity: at constant pressure	C_p	J K^{-1}
heat capacity: at constant volume	C_V	J K^{-1}
Helmholtz function	A, F	J
impedance, electric	Z	Ω
internal energy	U	J
irradiance	E_e, E	J m^{-2}
Joule–Kelvin coefficient	μ	N^{-1} m^2 K
kinematic viscosity	υ	m^2 s^{-1}
kinetic energy	T, E_k, K	J
Lagrangian function	L	J
linear absorption coefficient	α	m^{-1}
linear attenuation (extinction) coefficient	μ	m^{-1}
linear energy transfer	L_Δ	m^{-1} eV m^{-1}
linear expansivity	α	K^{-1}
luminance	L_v, L	cd m^{-2}
luminous emittance	M_v, M	lm m^{-2}
magnetic field strength	H	A m^{-1}
magnetic moment	m	A m^2
magnetic quantum number	M, m_l	1
magnetic susceptibility	χ, χ_m	1
magnetization	M	A m^{-1}

Table I.3 (Continued)

Name of quantity	Symbol	SI unit
magnetomotive force	F_m	A
mass number, nucleon number	A	1
mean free path	λ, l	m
mean life	τ	s
moment of inertia	I	kg m^2
momentum	p	N s
Néel temperature	θ_N, T_N	K
neutron number	N	1
nuclear magneton	μ_N	A m^2
nuclear radius	r	m
nuclear spin quantum number	I, J	1
osmotic pressure	Π	N m^{-2}
packing fraction	f	1
period	T	s
permeability	μ	H m^{-1}
permittivity	ε	F m^{-1}
Planck function	Y	J K^{-1}
polarizability	α, γ	C m^2 V^{-1}
potential energy	E_p, V, Φ	J
principal quantum number	n, n_i	1
radiance	L_e, L	W m^{-2} sr^{-1}
radiant excitance	M_e, M	W m^{-2}
radiant intensity	I_e, I	W sr^{-1}
reactance	X	Ω
reflection factor	ρ	1
refractive index	n	1
relative atomic mass	A_r	1
relative density	d	1
relative permeability	μ_r	1
relative permittivity (dielectric constant)	ε_r	1
relaxation time	τ	s
resistivity	ρ	Ω m
Reynolds number	(Re)	1
shear modulus	G	N m^{-2}
specific heat capacity: at constant pressure	c_p	J kg^{-1} K^{-1}
specific heat capacity: at constant volume	c_V	J kg^{-1} K^{-1}
specific optical rotatory power	α_m	rad m^2 kg^{-1}
specific volume	v	m^3 kg^{-1}
speed	u	m s^{-1}
spin quantum number	S, s	1
surface tension	γ, σ	N m^{-1}
susceptance	B	S
thermal conductivity	λ	W m^{-1} K^{-1}
thermal diffusivity	α	m^2 s^{-1}
torque	T	N m
transmittance	τ	1
viscosity, coefficient of	η	kg m^{-1} s^{-1}
volume	V, v	m^3
wavelength	λ	m
wavenumber	σ	m^{-1}
weight	G	N
work function	Φ	J
Young's modulus	E	N m^{-2}

Table I.4 SI conversion tables. Factor relating a physical quantity in SI units to the value in cgs units (or other commonly used units). Value in SI units = value in cgs units × conversion factor.

Physical quantity (symbol)	SI unit	Definition of SI unit	cgs unit	Conversion factor	Dimensions
CONCENTRATION UNITS					
concentration (c)	mol m^{-3}		mol l^{-1} (M)	1.0×10^3	l^{-3}
	mol dm^{-3}		mol l^{-1} (M)	1.0	l^{-3}
ionic strength (I)	mol m^{-3}		mol l^{-1} (M)	1.0×10^3	l^{-3}
	mol dm^{-3}		mol l^{-1} (M)	1.0	l^{-3}
molality (m)	mol kg^{-1}		mol kg^{-1}	1.0	m^{-1}
mole fraction (x_i)			dimensionless qty	1.0	1
ELECTRICITY AND MAGNETISM					
capacitance (C)	F	A^2 s^4 kg^{-1} m^{-2} = A s V^{-1} = C V^{-1}	esu^2 erg^{-1}	1.112×10^{-12}	$\varepsilon\, l$
conductance (G)	Ω^{-1}, S	kg^{-1} m^{-2} s^3 A^2 = A V^{-1}	ohm^{-1}, mho	1.0	$\varepsilon\, l\, t^{-1}$
conductivity (κ)	Ω^{-1} m^{-1}, S m^{-1}	A V^{-1} m^{-1}	ohm^{-1} cm^{-1}	1.0×10^2	$\varepsilon\, t^{-1}$
electric charge, quantity of electricity (Q)	C	A s	esu	3.335×10^{-10}	$\varepsilon^{1/2} m^{1/2} l^{3/2} t^{-1}$
			emu	10.0	$\mu^{-1/2} m^{1/2} l^{1/2}$
electric current (I)	A		esu sec^{-1}	3.335×10^{-10}	$\varepsilon^{1/2} m^{1/2} l^{3/2} t^{-2}$
			emu sec^{-1}	10.0	$\mu^{-1/2} m^{1/2} l^{1/2} t^{-1}$
electric dipole moment (p)	C m	A s m	esu cm	3.335×10^{-12}	$\varepsilon^{1/2} m^{1/2} l^{5/2} t^{-1}$
electric displacement (D)	C m^{-2}		$1/4\pi$ esu cm^{-2}	2.654×10^{-7}	$\varepsilon^{1/2} m^{1/2} l^{-1/2} t^{-1}$
electric field strength (E)	V m^{-1}		dyne esu^{-1}	2.998×10^4	$\varepsilon^{-1/2} m^{1/2} l^{-1/2} t^{-1}$
electric mobility of ion i (u_i)	m^2 V^{-1} s^{-1}		cm^2 V^{-1} sec^{-1}	1.0×10^{-4}	$\varepsilon^{1/2} m^{-1/2} l^{3/2}$
electric polarization (P)	C m^{-2}		esu cm^{-2}	3.335×10^{-6}	$\varepsilon^{1/2} m^{1/2} l^{-1/2} t^{-1}$
electric potential (V)	V	kg m^2 s^{-3} A^{-1} = J A^{-1} s^{-1}	erg esu^{-1}	2.998×10^2	$\varepsilon^{-1/2} m^{1/2} l^{1/2} t^{-1}$
			erg emu^{-1}	1.0×10^{-8}	$\mu^{1/2} m^{1/2} l^{3/2} t^{-2}$
inductance (L)	H	kg m^2 s^{-2} A^{-2} = V A^{-1} s	emu of inductance	1.0×10^{-9}	$\mu\, l$
magnetic field strength (H)	A m^{-1}		oersted (Oe)	79.578	$\mu^{-1/2} m^{1/2} l^{-1/2} t^{-1}$
magnetic flux (Φ)	Wb	kg m^2 s^{-2} A^{-1} = V s	maxwell (Mx)	1.0×10^{-8}	$\varepsilon^{-1/2} m^{1/2} l^{1/2}$
magnetic flux density (B)	T	Wb m^{-2} = kg s^{-2} A^{-1} = V s m^{-2}	gauss (G)	1.0×10^{-4}	$\varepsilon^{-1/2} m^{1/2} l^{-3/2}$
magnetic susceptibility (χ_m)			dimensionless qty	4π	1
molar conductivity (Λ)	Ω^{-1} m^2 mol^{-1}		ohm^{-1} cm^2 mol^{-1}	1.0×10^{-4}	$\varepsilon\, l^3\, t^{-1}$
permeability (μ)	H m^{-1}	kg m s^{-2} A^{-2}	dimensionless qty	$4\pi \times 10^{-7} = 1.257 \times 10^{-6}$	μ
permittivity (ε)	F m^{-1}	C V^{-1} m^{-1}	dimensionless qty	8.854×10^{-12}	ε
Poynting vector (S)	W m^{-2}		erg cm^{-2} sec^{-1}	1.0×10^{-3}	$\varepsilon^{-1/2} m^{1/2} t^{-3/2}$
relative permeability (μ_r)			dimensionless qty	1.0	1
relative permittivity (ε_r)			dimensionless qty	1.0	1
resistance (R)	Ω	kg m^2 s^{-3} A^{-2} = V A^{-1}	ohm	1.0	$\varepsilon^{-1} l^{-1} t$
resistivity (ρ)	Ω m	kg m^3 s^{-3} A^{-2} = V A^{-1} m	ohm cm	1.0×10^{-2}	$\varepsilon^{-1} t$
surface charge density (σ)	C m^{-2}		esu cm^{-2}	3.335×10^{-6}	$\varepsilon^{1/2} m^{1/2} l^{-1/2} t^{-1}$
velocity of ion i (v_i)	m s^{-1}		cm sec^{-1}	1.0×10^{-2}	$l\, t^{-1}$
MECHANICS					
Density (ρ)	kg m^{-3}		g ml^{-1}, g cm^{-3}	1.0×10^3	$m\, l^{-3}$
Dynamic viscosity (η)	N s m^{-2}	kg m^{-1} s^{-1} = J m^{-3} s	dyne sec cm^{-2}	0.1	$m\, l^{-1}\, t^{-1}$
			poise (P)	0.1	$m\, l^{-1}\, t^{-1}$
energy (E)	J	kg m^2 s^{-2}	erg	1.0×10^{-7}	$m\, l^2\, t^{-2}$
			cal	4.184	$m\, l^2\, t^{-2}$
force (F)	N	kg m s^{-2} = J m^{-1}	dyne	1.0×10^{-5}	$m\, l\, t^{-2}$
kinematic viscosity (v)	m^2 s^{-1}		stokes (St)	1.0×10^{-4}	$l^2\, t^{-1}$
mass (m)	kg		g	1.0×10^{-3}	m
power (P)	W	kg m^2 s^{-3} = J s^{-1}	erg sec^{-1}	1.0×10^{-7}	$m\, l^2\, t^{-3}$
pressure (p, P)	Pa	N m^{-2} = kg m^{-1} s^{-2} = J m^{-3}	dyne cm^{-2}	0.1	$m\, l^{-1}\, t^{-2}$
specific volume (V)	m^3 kg^{-1}		cm^3 g^{-1}	1.0×10^{-3}	$l^3\, m^{-1}$
surface tension (γ)	N m^{-1}	kg s^{-2} = J m^{-2}	dyne cm^{-1}	1.0×10^{-3}	$m\, t^{-2}$
work (W)	J	kg m^2 s^{-2}	erg	1.0×10^{-7}	$m\, l^2\, t^{-2}$
			cal	4.184	$m\, l^2\, t^{-2}$

Continued

Table I.4 (Continued)

Physical quantity (symbol)	SI unit	Definition of SI unit	cgs unit	Conversion factor	Dimensions
RADIATION, SPECTROPHOTOMETRIC AND OTHER OPTICAL QUANTITIES					
angle of rotation (α)	rad		degree	$\pi/180 =$ 1.745 × 10^{-2}	
decadic absorbance, extinction (A)				dimensionless qty	1.0
illumination (E)	lx	lm m^{-2} = cd sr m^{-2}	lm cm^{-2}	1.0 × 10^4	
luminance (L)	cd m^{-2}		cd cm^{-2} (stilb)	1.0 × 10^4	
luminous flux (Φ)	lm	cd sr	lm	1.0	
molar extinction coefficient (ε)	m^2 mol^{-1}		l mol^{-1} cm^{-1}	0.1	l^2
molar optical rotatory power (α_n)	rad m^2 mol^{-1}		deg dm^{-1} l mol^{-1}	1.745 × 10^{-2}	l^2
molar refraction (R_m)	m^3 mol^{-1}		cm^3 mol^{-1}	1.0 × 10^{-6}	l^3
refraction (R)	m^3		cm^3	1.0 × 10^{-6}	l^3
refractive index (n)			dimensionless qty	1.0	
specific optical rotatory power (α_m)	rad m^2 kg^{-1}		deg dm^{-1} cm^3 g^{-1}	1.745 × 10^{-4}	$l^2 m^{-1}$
SPACE AND TIME					
acceleration (a)	m s^{-2}		cm sec^{-2}	1.0 × 10^{-2}	$l t^{-2}$
angular frequency (ω)	Hz	s^{-1}	sec^{-1}	1.0	t^{-1}
area (A, S)	m^2		cm^2	1.0 × 10^{-4}	l^2
frequency (ν, f)	Hz	s^{-1}	sec^{-1}	1.0	t^{-1}
length (l)	m		cm	1.0 × 10^{-2}	l
			angstrom (Å)	1.0 × 10^{-10}	l
molar volume (V)	m^3 mol^{-1}		cm^3 mol^{-1}	1.0 × 10^{-6}	l^3
period (T)	s		sec	1.0	t
relaxation time (τ)	s		sec	1.0	t
rotational frequency	s^{-1}		sec^{-1}	1.0	t^{-1}
speed (v)	m s^{-1}		cm sec^{-1}	1.0 × 10^{-2}	$l t^{-1}$
volume (V)	m^3		cm^3 (cc)	1.0 × 10^{-6}	l^3
wavelength (λ)	m		cm	1.0 × 10^{-2}	l
wavenumber ($\bar{\nu}$)	m^{-1}		cm^{-1}	1.0 × 10^2	l^{-1}
THERMODYNAMICS					
enthalpy† (H)	J	kg m^2 s^{-2}	cal	4.184	$m l^2 t^{-2}$
entropy† (S)	J K^{-1}	kg m^2 s^{-2} K^{-1}	cal deg^{-1}	4.184	$m l^2 t^{-2} \theta^{-1}$
Gibbs function† (G)	J	kg m^2 s^{-2}	cal	4.184	$m l^2 t^{-2}$
heat capacity (C_p, C_V)	J K^{-1}	kg m^2 s^{-2} K^{-1}	cal deg^{-1}	4.184	$m l^2 t^{-2} \theta^{-1}$
Helmholtz function† (A)	J	kg m^2 s^{-2}	cal	4.184	$m l^2 t^{-2}$
internal energy† (U)	J	kg m^2 s^{-2}	cal	4.184	$m l^2 t^{-2}$
Joule–Thomson coefficient (μ)	N^{-1} m^2 K	kg^{-1} m s^2 K = J^{-1} m^3 K	deg atm^{-1}	1/101 325	$m^{-1} l t^2 \theta$
molar entropy (S_m)	J K^{-1} mol^{-1}	kg m^2 s^{-2} mol^{-1} K^{-1}	cal mol^{-1} deg^{-1}	4.184	$m l^2 t^{-2} \theta^{-1}$
quantity of heat (Q)	J	kg m^2 s^{-2}	cal	4.184	$m l^2 t^{-2}$
specific heat capacity (c_p, c_V)	J kg^{-1} K^{-1}	m^2 s^{-2} K^{-1}	cal g^{-1} deg^{-1}	4.184 × 10^3	$l^2 t^{-2} \theta^{-1}$
thermal conductivity (λ)	W m^{-1} K^{-1}	J s^{-1} m^{-1} K^{-1}	cal sec^{-1} cm^{-1} deg^{-1}	4.184 × 10^2	$m l t^{-3} \theta^{-1}$

†The corresponding specific quantity is denoted by corresponding lower case letter; units: add kg^{-1}; conversion factor: 4.184 × 10^3. The corresponding molar quantity is denoted by subscript m (e.g., S_m, G_m); units: add mol^{-1} (see S_m); conversion factor: 4.184.

Table I.5 Prefixes used with SI units.

Factor	Prefix	Symbol	Factor	Prefix	Symbol
10^{-1}	deci-	d	10	deca-	da
10^{-2}	centi-	c	10^2	hecto-	h
10^{-3}	milli-	m	10^3	kilo-	k
10^{-6}	micro-	μ	10^6	mega-	M
10^{-9}	nano-	n	10^9	giga-	G
10^{-12}	pico-	p	10^{12}	tera-	T
10^{-15}	femto-	f	10^{15}	peta-	P
10^{-18}	atto-	a	10^{18}	exa-	E

Table I.6 Conversion factors for some other units.

A

Length	m	cm	in	ft	yd
1 metre	1	100	39.3701	3.280 84	1.093 61
1 centimetre	0.01	1	0.393 701	0.032 8084	0.010 9361
1 inch	0.0254	2.54	1	0.083 3333	0.027 7778
1 foot	0.3048	30.48	12	1	0.333 333
1 yard	0.9144	91.44	36	3	1

	km	mile	n. mile		
1 kilometre	1	0.621 371	0.539 957		
1 mile	1.609 34	1	0.868 976		
1 nautical mile	1.852 00	1.150 78	1		

1 light year = $9.460\ 70 \times 10^{15}$ metres = $5.878\ 48 \times 10^{12}$ miles
1 astronomical unit = 1.496×10^{11} metres
1 parsec = 3.0857×10^{16} metres = 3.2616 light years
1 fathom = 6 feet

B

Area	m²	acre	ha
1 square metre	1	$2.471\ 05 \times 10^{-4}$	10^{-4}
1 acre	4046.856	1	0.404 686
1 hectare	10^4	2.471 05	1
1 barn	10^{-28}	—	—

C

Volume			
1 minim	1/9600 pint		
1 fluid drachm	60 minims		
1 fluid ounce	8 fluid drachms		
1 gill	5 fluid ounces		
1 pint	4 gills	0.568 26 dm³	
1 quart	2 pints		
1 gallon	4 quarts	4.546 09 dm³	
1 gallon	0.160 544 cubic feet		
1 cubic foot	6.228 82 gallons	0.028 3168 m³	

1 litre is 1000.028 cm³; its use is discouraged
1 gallon (US) is equivalent to 0.83268 gallon (UK)

D

Speed	m s⁻¹	km h⁻¹	mile h⁻¹	ft s⁻¹
1 metre per second	1	3.6	2.236 94	3.280 84
1 kilometre per hour	0.277 778	1	0.621 371	0.911 346
1 mile per hour	0.447 04	1.609 344	1	1.466 67
1 foot per second	0.3048	1.097 28	0.681 817	1

1 knot = 1 nautical mile per hour = 0.514 444 metre per second

Continued

Table I.6 (Continued)

E

Mass	kg	g	lb	long ton
1 kilogram	1	1000	2.204 62	$9.842\ 07 \times 10^{-4}$
1 gram	10^{-3}	1	$2.204\ 62 \times 10^{-3}$	$9.842\ 07 \times 10^{-7}$
1 pound	0.453 592	453.592	1	$4.464\ 29 \times 10^{-4}$
1 long ton	1016.047	$1.016\ 047 \times 10^{6}$	2240	1

Avoirdupois units of mass		
1 grain	1/7000 pound	0.064 799 g
1 dram	1/256 pound	
1 ounce	16 drams	
1 pound	16 ounces	0.453 592 kg
1 stone	14 pounds	
1 quarter	2 stones	
1 hundredweight	4 quarters	50.802 kg
1 ton	20 hundredweights	1016.047 kg
1 gram	0.0353 ounces	
1 kilogram	2.204 62 pounds	

The avoirdupois hundredweight and ton are sometimes called the *long hundredweight* and *long ton* to distinguish them from the US measures, the *short hundredweight* (100 pounds) and *short ton* (2000 pounds).

Apothecaries' units of mass	
1 grain	1/7000 pound (avoirdupois)
1 scruple	20 grains
1 drachm	3 scruples
1 ounce (apoth)	8 drachms
1 pound (apoth)	12 ounces (apoth)

The grain (1/7000 pound avoirdupois) has the same value in the avoirdupois, troy and apothecaries' systems. Apothecaries' units should no longer be used in the UK.

Troy units of mass	
1 grain	1/7000 pound (avoirdupois)
1 carat	4 grains
1 pennyweight	6 carats
1 ounce (tr)	20 pennyweights
1 pound (tr)	12 ounces (tr)
1 hundredweight (tr)	100 pounds (tr)
1 ton (tr)	20 hundredweights (tr)

Troy units should no longer be used in the UK.

F

Force	N	kg	dyne	poundal	lb
1 newton	1	0.101 972	10^{5}	7.233 00	0.224 809
1 kilogram force	9.806 65	1	$9.806\ 65 \times 10^{5}$	70.9316	2.204 62
1 dyne	10^{-5}	$1.019\ 72 \times 10^{-6}$	1	$7.233\ 00 \times 10^{-5}$	$2.248\ 09 \times 10^{-6}$
1 poundal	0.138 255	$1.409\ 81 \times 10^{-2}$	$1.382\ 55 \times 10^{4}$	1	0.031 081
1 pound force	4.448 22	0.453 592	$4.448\ 23 \times 10^{5}$	32.174	1

Continued

Table I.6 (Continued)

G

Pressure	Pa (N m^{-2})	kg cm^{-2}	lb in^{-2}	atm
1 pascal	1	1.01972×10^{-5}	1.45038×10^{-4}	9.86923×10^{-6}
1 kilogram per square centimetre	980.655×10^2	1	14.2234	0.967 841
1 pound per square inch	6.89476×10^3	0.070 3068	1	0.068 046
1 atmosphere	1.01325×10^5	1.033 23	14.6959	1

1 pascal = 1 newton per square metre = 10 dynes per square centimetre
1 bar = 10^5 pascal = 0.986 923 atmosphere
1 torr = 133.322 pascal = 1/760 atmosphere
1 atmosphere = 760 mmHg = 29.92 in Hg = 33.90 ft water (all at 0 °C)

H

Work and Energy	J	†cal$_{IT}$	kW h	btu$_{IT}$
1 joule	1	0.238 846	2.77778×10^{-7}	9.47813×10^{-4}
†1 calorie (IT)	4.1868	1	1.16300×10^{-6}	3.96831×10^{-3}
1 kilowatt hour	3.6×10^6	8.59845×10^5	1	3412.14
1 British thermal unit (IT)	1055.06	251.997	2.93071×10^{-4}	1

1 joule = 1 newton metre = 1 watt second = 10^7 erg = 0.737 561 ft lb
1 electronvolt = 1.60210×10^{-19} joule
†cal$_{IT}$ denotes international calorie whose SI conversion factor is 4.1868 rather than 4.184 for the thermochemical calorie of Table I.4.

I

Concentration Units Subscripts: 1, 2 solvent and solute respectively; w_2: mass per cent (g solute per 100 g solution); x_2: mole fraction; m_2: molality (mol of solute per 1000 g solvent); c_2: molar concentration (mol dm^{-3}); G_2: g of solute per 1000 cm^3 solution; ρ: density of solution (g cm^{-3}); M: molar mass.

Concentration required	Concentration given				
	w_2	x_2	m_2	c_2	G_2
w_2	1	$\dfrac{100 x_2 M_2}{x_2 M_2 + M_1(1-x_2)}$	$\dfrac{100 m_2 M_2}{1000 + m_2 M_2}$	$\dfrac{c_2 M_2}{10\rho}$	$\dfrac{G_2}{10\rho}$
x_2	$\dfrac{w_2/M_2}{w_2/M_2 + (100-w_2)/M_2}$	1	$\dfrac{m_2 M_1}{1000 + m_2 M_1}$	$\dfrac{c_2 M_1}{1000\rho + c_2(M_1 - M_2)}$	$\dfrac{G_2 M_1}{G_2(M_1 - M_2) + 1000\rho M_2}$
m_2	$\dfrac{1000 w_2}{M_2(100 - w_2)}$	$\dfrac{1000 x_2}{M_1(1 - x_2)}$	1	$\dfrac{1000 c_2}{1000\rho - c_2 M_2}$	$\dfrac{1000 G_2}{M_2(1000\rho - G_2)}$
c_2	$\dfrac{10 w_2 \rho}{M_2}$	$\dfrac{100 \rho x_2}{x_2 M_2 + M_1(1 - x_2)}$	$\dfrac{1000 \rho m_2}{1000 + m_2 M_2}$	1	$\dfrac{G_2}{M_2}$
G_2	$10 w_2 \rho$	$\dfrac{1000 \rho x_2 M_2}{x_2 M_2 + M_1(1 - x_2)}$	$\dfrac{1000 \rho m_2 M_2}{1000 + m_2 M_2}$	$c_2 M_2$	1

Table I.7 Fundamental Constants.

Constant	Symbol	Value
acceleration due to gravity (standard value)	g_n	9.806 65 m s^{-2} exactly
atomic mass constant (unified atomic mass unit)	m_u	1.660 540 2(10) × 10^{-27} kg
Avogadro constant	N_A or L	6.022 136 7(36) × 10^{23} mol^{-1}
Bohr magneton	μ_B	9.274 015 4(31) × 10^{-24} J T^{-1}
Bohr radius	a_0	5.291 772 49(24) × 10^{-11} m
Boltzmann constant	k	1.380 658(12) × 10^{-23} J K^{-1}
charge on electron	e	−1.602 177 33(49) × 10^{-19} C
charge on proton	$-e$	1.602 177 33(49) × 10^{-19} C
Compton wavelength of electron	λ_C	2.426 308 9 × 10^{-12} m
electronic radius	r	2.817 939 × 10^{-15} m
electron magnetic moment	μ_e	9.284 770 1(31) × 10^{-24} J T^{-1}
Faraday constant	F	9.648 530 9(29) × 10^4 C mol^{-1}
fine structure constant	α	7.297 353 08(33) × 10^{-3}
gravitational constant	G	6.672 59(85) × 10^{-11} m^3 kg^{-1} s^{-2}
Hartree energy	E_h	4.359 748 2(26) × 10^{-18} J
Landé g factor for free electron	g_e	2.002 319 304 386(20)
Loschmidt constant	n_0	2.687 19 × 10^{25} m^{-3}
molar gas constant	R	8.314 510(70) J K^{-1} mol^{-1}
molar volume of ideal gas at standard temperature		
for standard pressure		0.022 414 m^3 mol^{-1}
for pressure of 1 bar		0.022 711 72(2) m^3 mol^{-1}
nuclear magneton	μ_N	5.050 786 6(17) × 10^{-27} J T^{-1}
permeability of vacuum	μ_0	4π × 10^{-7} H m^{-1} exactly
permittivity of vacuum	ε_0	8.854 187 816 ... × 10^{-12} F m^{-1}
Planck constant	h	6.626 075 5(40) × 10^{-34} J s
	$\hbar = h/2\pi$	1.054 572 66(63) × 10^{-34} J s
proton magnetic moment	μ_p	1.410 607 61(47) × 10^{-26} J T^{-1}
proton magnetic moment in H$_2$O	μ'_p	1.410 571 38(47) × 10^{-26} J T^{-1}
	μ'_p/μ_B	1.520 993 129(17) × 10^{-3}
proton magnetogyric ratio	γ_p	2.675 221 28(81) × 10^8 s^{-1} T^{-1}
proton resonance frequency per field in H$_2$O	$\gamma'_p/2\pi$	4.257 637 5(13) × 10^7 Hz T^{-1}
radiation constants, first	c_1	3.741 774 9(22) × 10^{-16} W m^2
second	c_2	1.438 769(12) × 10^{-2} m K
rest mass of electron	m_e	9.109 389 7(54) × 10^{-31} kg
rest mass of neutron	m_n	1.674 928 6(10) × 10^{-27} kg
rest mass of proton	m_p	1.672 623 1(10) × 10^{-27} kg
Rydberg constant	R_∞	1.097 373 153 4(13) × 10^7 m^{-1}
speed of light in vacuum	c	299 792 458 m s^{-1} exactly
standard atmosphere	atm	101 325 Pa exactly
Stefan–Boltzmann constant	σ	5.670 51(19) × 10^{-8} W m^{-2} K^{-4}
molar gas constant	R	8.314 510(70) JK^{-1} mol^{-1}
zero of Celsius scale		273.15 K exactly

Section II
Properties of Materials

		Page
II.1	Variation of buoyancy correction with mass for weighings in air	13
II.2	Variation of density of ambient air of 50% relative humidity, with temperature, for various air pressures	13
II.3	Changes in density for some values given in Table II.2 for other relative humidities	14
II.4	Temperature variation of quotient of container volume and apparent mass of water, or mercury, filling it	14
II.5	Changes in the quotient given in Table II.4 for water at other air pressures	14
II.6	Values of reference pressures associated with phase transitions of pure substances at a temperature of 298 K, unless otherwise stated	14
II.7	1981 International Practical Pressure Scale: calculated values of pressure for various linear and volume compressions for NaCl at 298 K	15
II.8	Variation with pressure of melting point for Ag, Au and Cu	15
II.9	Variation with temperature in mass of H_2O in 1 m^3 of saturated air at standard pressure as determined by chemical hygrometer	16
II.10	Relative humidity values, determined by forced ventilated wet- and dry-bulb hygrometer, corresponding to various wet- and dry-bulb temperatures	16
II.11	Variation with temperature of relative humidity for air above various saturated salt solutions sited in a uniform constant-temperature enclosure	16
II.12	Variation with temperature of density of pure air-free H_2O at standard pressure	17
II.13	Variation with temperature of density of heavy H_2O at standard pressure	17
II.14	Variation with temperature of density of Hg at standard pressure	18
II.15	Typical density for some everyday substances at about STP	19
II.16	Values of Young's, shear and bulk elastic moduli, the Poisson ratio and tensile strength for some metals at temperature 293 K	20
II.17	Values of Young's, shear and bulk elastic moduli, the Poisson ratio and tensile strength for some alloys at temperature 293 K	21
II.18	Values of Young's, shear and bulk elastic moduli, the Poisson ratio and tensile strength for glass and quartz at temperature 293 K	21
II.19	Values of Young's modulus of elasticity and tensile strength for some plastics at temperature 293 K	22
II.20	Values of Young's modulus of elasticity, along the grain and in radial and tangential directions, and tensile strength for some woods	22
II.21	Values of temperature coefficient of Young's modulus and shear moduli of elasticity of some substances at temperature 288 K	22
II.22	Values of bulk modulus of elasticity of some non-metallic elements	22
II.23	Values of bulk modulus of elasticity of some liquids at stated temperature and pressure	23
II.24	Values of tensile strength of some miscellaneous substances	23
II.25	Values of coefficient of viscosity for various gases at about standard pressure	24
II.26	Values of coefficient of viscosity for some gases at various high pressures at temperature 300 K, unless otherwise stated	25
II.27	Values of coefficient of viscosity for water at about standard pressure at stated temperatures	25
II.28	Values of coefficient of viscosity for various glycerol/H_2O solutions at about standard pressure	25
II.29	Values of coefficient of viscosity for various sucrose/H_2O solutions at about standard pressure	26
II.30	Values of coefficient of viscosity for some liquid elements, molten salts, glasses and minerals at about standard pressure at stated temperatures	26
II.31	Values of coefficient of viscosity for various organic liquids at about standard pressure at stated temperatures	27
II.32	Variation of the coefficient of viscosity of water with pressure and temperature	29
II.33	Variation of the coefficient of viscosity of various liquids with pressure at stated temperatures	29
II.34	Values of STP molecular mean speed, mean free path and mean time between collisions, for various gases	30
II.35	Variation with temperature of surface tension of water against air at about standard pressure	30
II.36	Variation with percentage solute of surface tension of various aqueous solutions at stated temperatures at about standard pressure	30
II.37	Surface tension, γ, at about standard pressure, for some liquid elements and molten salts at stated temperatures, T, together with some values of A and B in the formula $\gamma = A - BT$, and temperature range of formula applicability	31
II.38	Surface tension, γ, at about standard pressure, for some liquids at temperature 293 K, unless otherwise stated, together with some values of A and B in the formula $\gamma = A - BT$, and temperature range of formula applicability	32
II.39	Interfacial tension, at about standard pressure, for some liquids at temperature 293 K	32

Section II
Properties of Materials

Table II.1 Variation of buoyancy correction with mass for weighings in air.
Assumptions: density of weights used, 8000 kg m^{-3}; density of air, 1.2 kg m^{-3}.

density of mass/kg m^{-3}	500	1000	1500	2000	2500	3000	4000
correction/g kg^{-1}	2.25	1.05	0.65	0.45	0.33	0.25	0.15
density of mass/kg m^{-3}	6000	8000	10 000	12 500	15 000	20 000	22 000
correction/g kg^{-1}	0.05	0	−0.03	−0.054	−0.07	−0.09	−0.096

Reference: Kaye and Laby Tables of Physical and Chemical Constants, 15 ed., Longmans, 1986

Table II.2 Variation of density of ambient air of 50% relative humidity, with temperature, T, for various air pressures, p.
Assumption: carbon dioxide content of air, 0.04% by volume.

T/K	279	281	283	285	287	289	291	293	295	297	299	301	303
p/kPa						density/kg m^{-3}							
80	0.997	0.989	0.982	0.974	0.967	0.960	0.953	0.946	0.939	0.931	0.924	0.917	0.910
81	1.009	1.002	0.994	0.987	0.979	0.972	0.965	0.958	0.950	0.943	0.936	0.929	0.922
82	1.022	1.014	1.006	0.999	0.991	0.984	0.977	0.969	0.962	0.955	0.948	0.941	0.933
83	1.034	1.026	1.019	1.011	1.004	0.996	0.989	0.981	0.974	0.967	0.959	0.952	0.945
84	1.046	1.039	1.031	1.023	1.016	1.008	1.001	0.993	0.986	0.978	0.971	0.964	0.956
85	1.059	1.051	1.043	1.036	1.028	1.020	1.013	1.005	0.998	0.990	0.983	0.975	0.968
86	1.071	1.064	1.056	1.048	1.040	1.032	1.025	1.017	1.009	1.002	0.994	0.987	0.979
87	1.084	1.076	1.068	1.060	1.052	1.044	1.037	1.029	1.021	1.014	1.006	0.998	0.991
88	1.096	1.088	1.080	1.072	1.064	1.056	1.049	1.041	1.033	1.025	1.018	1.010	1.002
89	1.109	1.101	1.093	1.084	1.076	1.068	1.061	1.053	1.045	1.037	1.029	1.022	1.014
90	1.121	1.113	1.105	1.097	1.089	1.081	1.073	1.065	1.057	1.049	1.041	1.033	1.025
91	1.134	1.126	1.117	1.109	1.101	1.093	1.085	1.076	1.068	1.061	1.053	1.045	1.037
92	1.146	1.138	1.129	1.121	1.113	1.105	1.096	1.088	1.080	1.072	1.064	1.056	1.048
93	1.159	1.150	1.142	1.133	1.125	1.117	1.108	1.100	1.092	1.084	1.076	1.068	1.060
94	1.171	1.163	1.154	1.146	1.137	1.129	1.120	1.112	1.104	1.096	1.088	1.079	1.071
95	1.184	1.175	1.166	1.158	1.149	1.141	1.132	1.124	1.116	1.107	1.099	1.091	1.083
96	1.196	1.188	1.179	1.170	1.161	1.153	1.144	1.136	1.128	1.119	1.111	1.103	1.094
97	1.209	1.200	1.191	1.182	1.174	1.165	1.156	1.143	1.139	1.131	1.122	1.114	1.106
98	1.221	1.212	1.203	1.195	1.186	1.177	1.168	1.160	1.151	1.143	1.134	1.126	1.117
99	1.234	1.225	1.216	1.207	1.198	1.189	1.180	1.172	1.163	1.154	1.146	1.137	1.129
100	1.246	1.237	1.228	1.219	1.210	1.201	1.192	1.184	1.175	1.166	1.157	1.149	1.140
101	1.259	1.250	1.240	1.231	1.222	1.213	1.204	1.195	1.187	1.178	1.169	1.160	1.152
102	1.271	1.262	1.253	1.243	1.234	1.225	1.216	1.207	1.198	1.190	1.181	1.172	1.163
103	1.284	1.274	1.265	1.256	1.246	1.237	1.228	1.219	1.210	1.201	1.192	1.184	1.175
104	1.296	1.287	1.277	1.268	1.259	1.249	1.240	1.231	1.222	1.213	1.204	1.195	1.186
105	1.309	1.299	1.290	1.280	1.271	1.261	1.252	1.243	1.234	1.225	1.216	1.207	1.198
106	1.321	1.312	1.302	1.292	1.283	1.273	1.264	1.255	1.246	1.236	1.227	1.218	1.209

Reference: Giacomo, *Metrologia*, 18, 33, 1982

Table II.3 Changes in density for some values given in Table II.2 for other relative humidities, U.

T/K	283	293	303
$U/\%$	density change/kg m^{-3}		
20	0.002	0.003	0.006
25	0.001	0.002	0.005
30	0.001	0.002	0.004
35	0.001	0.001	0.003
40	0	0.001	0.002
45	0	0	0.001

T/K	283	293	303
$U/\%$	density change/kg m^{-3}		
55	0	−0.001	−0.001
60	−0.001	−0.001	−0.002
65	−0.001	−0.002	−0.003
70	−0.001	−0.002	−0.003
75	−0.002	−0.003	−0.004
80	−0.002	−0.003	−0.005

Reference: as for Table II.2

Table II.4 Temperature variation of quotient of container volume, V, and apparent mass of water, $m(H_2O)$, or mercury, $m(Hg)$, filling it.
Assumptions: density of weights used, 8000 kg m^{-3}; air pressure, 101.325 kPa.

T/K	$(V/m(H_2O))$/cm^3 g^{-1}	$(V/m(Hg))$/cm^3 g^{-1}
283	1.001 39	0.073 685
284	1.001 48	0.073 698
285	1.001 58	0.073 712
286	1.001 70	0.073 725
287	1.001 83	0.073 738
288	1.001 97	0.073 752
289	1.002 13	0.073 765
290	1.002 29	0.073 778

T/K	$(V/m(H_2O))$/cm^3 g^{-1}	$(V/m(Hg))$/cm^3 g^{-1}
291	1.002 47	0.073 792
292	1.002 65	0.073 805
293	1.002 85	0.073 819
294	1.003 06	0.073 832
295	1.003 28	0.073 845
296	1.003 51	0.073 859
297	1.003 75	0.073 872
298	1.004	0.073 886

Reference: British Standards, BS 1797, 1968

Table II.5 Changes in the quotient given in Table II.4, $\Delta (V/m(H_2O))$, for water at other air pressures, p.

p/kPa	97	98	99	100	102	103	104	105
$\Delta(V/m(H_2O))/10^{-5}$ cm^3 g^{-1}	−5	−4	−3	−2	1	2	3	4

Reference: as for Table II.4

Table II.6 Values of reference pressures associated with phase transitions of pure substances at a temperature of 298 K, unless otherwise stated.
Transitions identified by change in electric resistance and/or volume.

Substance	transition	pressure/GPa
Ba	I–II	†5.53 ± 0.12
Ba	II–III	11.8 – 12.2
Bi	I–II	†2.55 ± 0.006
Bi	III–V	†7.67 ± 0.18
GaP		22
Hg	solid–liquid equilibrium	†0.7569 ± 0.0002 (at 273 K)

Substance	transition	pressure/GPa
NaCl		29.6
Pb	I–II	13.4
Sn	I–II	9.4
Tl	II–III	†3.67 ± 0.05
ZnS		15.4

† indicates a basic reference pressure
References: *NBS special publication 326*, 1971
Deeker et al., *J. Phys. Chem. Ref. Data*, *1*, 773 (NBS), 1972

Table II.7 1981 International Practical Pressure Scale: calculated values of pressure for various linear and volume compressions for NaCl at 298 K.

| Compression | | pressure/ | Compression | | pressure/ |
linear	volume	GPa	linear	volume	GPa
0.001	0.0030	0.071	0.068	0.1904	8.379
0.002	0.0059	0.144	0.070	0.1956	8.772
0.004	0.0119	0.293	0.072	0.2008	9.175
0.006	0.0178	0.446	0.074	0.2059	9.590
0.008	0.0238	0.604	0.076	0.2111	10.017
0.010	0.0297	0.767	0.078	0.2162	10.457
0.012	0.0355	0.936	0.080	0.2213	10.908
0.014	0.0414	1.109	0.082	0.2263	11.372
0.016	0.0472	1.288	0.084	0.2314	11.850
0.018	0.0530	1.472	0.086	0.2364	12.340
0.020	0.0588	1.662	0.088	0.2414	12.845
0.022	0.0645	1.858	0.090	0.2464	13.364
0.024	0.0702	2.060	0.092	0.2513	13.897
0.026	0.0759	2.268	0.094	0.2563	14.445
0.028	0.0816	2.482	0.096	0.2612	15.085
0.030	0.0873	2.703	0.098	0.2661	15.508
0.032	0.0929	2.930	0.100	0.2710	16.183
0.034	0.0985	3.164	0.102	0.2758	16.795
0.036	0.1041	3.405	0.104	0.2806	17.424
0.038	0.1097	3.653	0.106	0.2854	18.070
0.040	0.1152	3.909	0.108	0.2902	18.735
0.042	0.1207	4.172	0.110	0.2950	19.417
0.044	0.1262	4.443	0.112	0.2997	20.119
0.046	0.1317	4.722	0.114	0.3044	20.840
0.048	0.1371	5.009	0.116	0.3091	21.581
0.050	0.1426	5.304	0.118	0.3138	22.342
0.052	0.1480	5.608	0.120	0.3185	23.125
0.054	0.1534	5.921	0.122	0.3231	23.929
0.056	0.1587	6.243	0.124	0.3277	24.755
0.058	0.1641	6.574	0.126	0.3323	25.603
0.060	0.1694	6.915	0.128	0.3369	26.476
0.062	0.1747	7.266	0.130	0.3414	27.372
0.064	0.1799	7.626	0.132	0.3460	28.292
0.066	0.1852	7.997	0.134	0.3505	29.238

Reference: *Proceedings of 8th AIRAPT Conference*, Arkitektkopia, Uppsala, Sweden, vol. 1, 1982

Table II.8 Variation with pressure, p, of melting point for Ag, Au and Cu.

| | melting point/K | | | | melting point/K | | |
p/GPa	Ag	Au	Cu	p/GPa	Ag	Au	Cu
1	1292	1393	1392	11	1759	1902	1731
2	1350	1451	1432	12	1794	1947	1759
3	1405	1508	1469	13	1831	1991	1785
4	1457	1563	1505	14	1865	2035	1813
5	1507	1613	1539	15	1901	2077	1837
6	1556	1658	1570	16	1933	2118	1861
7	1601	1714	1605	17	1965	2157	1885
8	1645	1761	1637	18	1995	2195	1905
9	1685	1809	1669	19	2025	2229	1927
10	1723	1857	1701	20	2051	2261	1949

Reference: Akella and Kennedy, *J. Geophys. Res.*, 76, 4969, 1971

Table II.9 Variation with temperature in mass of H_2O in 1 m³ of saturated air at standard pressure, as determined by chemical hygrometer.

temperature/K	273	274	275	276	277	278	279	280	281	282
mass/g	4.85	5.2	5.55	5.95	6.35	6.8	7.25	7.75	8.25	8.8
temperature/K	283	284	285	286	287	288	289	290	291	292
mass/g	9.4	10	10.65	11.35	12.05	12.8	13.6	14.45	15.35	16.3
temperature/K	293	294	295	296	297	298	299	300	301	302
mass/g	17.3	18.35	19.4	20.55	21.75	23.05	24.35	25.75	27.2	28.75
temperature/K	303	304	305	306	307	308	309	310	311	312
mass/g	30.35	32.05	33.8	35.6	37.55	39.55	41.65	43.9	46.2	48.6

Reference: Kaye and Laby Tables of Physical and Chemical Constants, 15 ed., Longmans, 1986

Table II.10 Relative humidity values, determined by forced ventilated wet- and dry-bulb hygrometer, corresponding to various wet- and dry-bulb temperatures.

dry-bulb temp/K	wet-bulb depression/K																					
	0.5	1.0	1.5	2.0	2.5	3.0	3.5	4.0	4.5	5.0	5.5	6.0	6.5	7.0	7.5	8.0	8.5	9.0	9.5	10.0	10.5	11.0
	relative humidity/%																					
283	94	88	82	76	71	65	60	54	49	44	39	34	29	24	19	14	9	5	—	—	—	—
285	94	89	83	78	73	68	63	57	53	48	43	38	34	29	24	20	16	11	7	3	—	—
287	95	90	84	79	74	70	65	60	56	51	47	42	38	33	29	25	24	17	13	9	5	2
289	95	90	85	81	76	71	67	62	58	54	50	46	41	37	34	30	26	22	18	15	11	8
291	95	91	86	82	77	73	69	65	60	56	52	49	45	41	37	34	30	27	23	20	16	13
293	96	91	86	83	78	74	70	66	62	59	55	51	48	44	41	37	34	30	27	24	21	18
295	96	92	87	83	79	76	72	68	64	61	57	54	50	47	44	40	37	34	31	28	25	22
297	96	92	88	84	80	77	73	69	66	62	59	56	52	49	46	43	40	37	34	31	28	26

Reference: *Hygrometric Tables for Aspirated Psychrometer Readings MO 2656*, HMSO, 1961

Table II.11 Variation with temperature of relative humidity for air above various saturated salt solutions sited in a uniform constant-temperature enclosure.

Salt	temperature/K										
	273	278	283	288	293	298	303	308	313	323	333
	relative humidity/%										
KCl	89	88	88	87	86	85	84	83	82	81	80
K_2CO_3			47	44	44	43	43	43	42		
$KC_2H_3O_2$	25	24	24	23	23	22	22	21	20		
KNO_3	97	96	95	94	93	92	91	89	88	85	82
KOH		14	13	10	9	8	7	6	6	6	5
K_2SO_4	99	98	98	97	97	97	96	96	96	96	96
LiCl	15	14	13	13	12	12	12	12	11	11	11
$MgCl_2$	35	34	34	34	33	33	33	32	32	31	30
$Mg(NO_3)_2$	60	58	57	56	55	53	52	50	49	46	43
NaCl	76	76	76	76	76	75	75	75	75	75	75
$Na_2Cr_2O_7$	60	59	58	56	55	54	52	51	50	47	
$NaNO_2$					66	65	63	62	62	59	59
NH_4NO_3	77	74	72	69	65	62	59	55	53	47	42
$(NH_4)_2SO_4$	83	82	82	81	81	80	80	80	79	79	78

Reference: British Standards, BS 3718, 1964

Table II.12 Variation with temperature of density of pure air-free H_2O at standard pressure.

temperature/K	273	275	277	279	281	283	295	287	289	291
density/kg m^{-3}	999.84	999.94	999.97	999.94	999.85	999.7	999.5	999.24	998.94	998.59
temperature/K	293	295	297	299	301	303	305	307	309	311
density/kg m^{-3}	998.2	997.77	997.3	996.78	996.23	995.65	995.03	994.37	993.69	992.97
temperature/K	313	315	317	319	321	323	325	327	329	331
density/kg m^{-3}	992.22	991.44	990.63	989.79	988.93	988.04	987.12	986.18	985.21	984.22
temperature/K	333	335	337	339	341	343	345	347	349	351
density/kg m^{-3}	983.2	982.16	981.1	980.01	978.9	977.77	976.62	975.44	974.25	973.03
temperature/K	353	355	357	359	361	363	365	367	369	371
density/kg m^{-3}	971.8	970.54	969.27	967.97	966.65	965.32	963.97	962.59	961.2	959.79

Reference: Kell, *J. Chem. Engng. Data*, 20, 97, 1975

Table II.13 Variation with temperature of density of heavy H_2O at standard pressure.

temperature/K	278	283	288	293	298	303	308
density/kg m^{-3}	1105.62	1105.99	1105.87	1105.34	1104.45	1103.23	1101.73
temperature/K	313	318	323	328	333	338	343
density/kg m^{-3}	1099.96	1097.94	1095.7	1093.25	1090.6	1087.77	1084.75
temperature/K	348	353	358	363	368	373	
density/kg m^{-3}	1081.58	1078.24	1074.75	1071.12	1067.36	1063.46	

Reference: Kell, *J. Chem. Engng. Data*, 12, 66, 1967

Table II.14 Variation with temperature of density of Hg at standard pressure.

temperature/K	253	255	257	259	261	263
density/kg m^{-3}	136 44.59	136 39.62	136 34.66	136 29.7	136 24.75	136 19.8
temperature/K	265	267	269	271	273	275
density/kg m^{-3}	136 14.85	136 09.9	136 04.96	136 00.02	135 95.08	135 90.05
temperature/K	277	279	281	283	285	287
density/kg m^{-3}	135 85.21	135 80.29	135 75.36	135 70.44	135 65.52	135 60.6
temperature/K	289	291	293	295	297	299
density/kg m^{-3}	135 55.69	135 50.78	135 45.87	135 40.96	135 36.06	135 31.16
temperature/K	301	303	305	307	309	311
density/kg m^{-3}	135 26.26	135 21.36	135 16.47	135 11.58	135 06.69	135 01.81
temperature/K	313	315	317	319	321	323
density/kg m^{-3}	134 96.92	134 92.04	134 87.17	134 82.29	134 77.42	134 72.55
temperature/K	325	327	329	331	333	335
density/kg m^{-3}	134 67.68	134 62.81	134 57.95	134 53.09	134 48.23	134 43.37
temperature/K	337	339	341	343	345	347
density/kg m^{-3}	134 38.52	134 33.66	134 28.81	134 23.96	134 19.12	134 14.27
temperature/K	349	351	353	355	357	359
density/kg m^{-3}	134 09.43	134 04.59	133 99.75	133 94.92	133 90.08	133 85.25
temperature/K	361	363	365	367	369	371
density/kg m^{-3}	133 80.42	133 75.59	133 70.77	133 65.94	133 361.12	133 56.3
temperature/K	373	393	413	433	453	473
density/kg m^{-3}	133 51.48	133 03.4	132 55.4	132 07.6	131 59.8	131 12.1
temperature/K	493	513	533	553	573	593
density/kg m^{-3}	130 64.5	130 16.9	129 69.2	129 21.5	128 73.7	128 15.8

Reference: Cook, *Br. J. Appl. Phys.*, 7, 285, 1956

Table II.15 Typical density for some everyday substances at about STP
For densities of some other substances, see Tables XI.1 and XI.2.

Substance	density/ 10^3 kg m^{-3}	Substance	density/ 10^3 kg m^{-3}	Substance	density/ 10^3 kg m^{-3}
agate	2.6	gas carbon	1.9	olive oil	0.9
Alni	6.9	gelatine	1.3	paraffin oil	0.8
Alnico	7.1	German silver	8.4	paraffin wax	0.9
aluminium-bronze (8% Al)	7.7	glass (soda)	2.5	Permalloy C	8.6
amber	1.1	glass (Pyrex)	2.23	petroleum	0.8
asbestos	2.4	glass (lead)	3–4	phosphor bronze	8.9
ash (timber)	0.75	glycerine	1.3	pine (white)	0.5
asphalt	1.4	gold (22 carat)	17.5	pitch	1.1
balsa wood	0.2	gold (9 carat)	11.3	plaster of Paris	1.8
bamboo	0.4	granite	2.7	platinum iridium (90/10)	21.5
bearing metal (80% Sn)	7.3	gunmetal	8.2	porcelain	2.3
beech	0.75	'heavy alloy'†	16.8–18.0	quartz (crystal)	2.6
beeswax	0.95	ice	0.92	resin	1.1
beryllium-copper	8.2	Inconel	8.5	sand (dry)	1.6
bone	1.9	Invar	8.0	sealing wax	1.8
borax	1.7	ivory	1.8	seawater	1.03
boxwood	1.0	Keramot	1.6	silica (fused)	
brass (60/40)	8.4	lard	0.9	translucent	2.1
brass (70/30)	8.5	lignum vitae	1.3	transparent	2.2
brightray	8.4	linseed oil	0.95	silicon iron	6.9
butter	0.9	Lo Ex	2.7	silver sand	2.6
carbon steel (< 1% C)	7.8	Magnalium	2.6	slate	2.8
cast iron	7.0–7.4	mahogany	0.8	soft solder (70% Sn 30% Pb)	8.3
castor oil	0.95	manganin	8.5	stainless iron (12% Cr)	7.7
cedarwood	0.55	marble	2.7	stainless steel	7.8
charcoal	0.4	Mazak (No. 2)	6.7	Supermalloy	8.9
china clay (kaolin)	2.6	methylated spirit	0.8	tar	1.0
coal (anthracite)	1.6	mica	2.8	teak	0.85
coal (bituminous)	1.4	mild steel	7.9	Thiokol	1.4
Constantan	8.9	milk	1.03	tungsten carbide (6% Co)	15.0
cork	0.25	Monel	8.8	tungsten carbide (12% Co)	14.2
corundum	4.0	Mumetal	8.8	turpentine	0.85
cronite	8.1	Mycalex	2.4	wax (soft red)	1.0
duralumin	2.8	naphtha	0.8	white spirit	0.85
ebonite	1.2	nickel chromium	8.4	wrought iron	7.8
ebony	1.2	nickel silver	8.8	Y-alloy	2.8
Elinvar	8.1	Nimonic	8.2		
emery	4.0	oak	0.7		

† Tungsten with metallic additives.
Reference: Kaye and Laby Tables of Physical and Chemical Constants, 15 ed., Longmans, 1986

Table II.16 Values of Young's, shear and bulk elastic moduli, the Poisson ratio and tensile strength for some metals at temperature 293 K.

Metal	moduli/GPa			Poisson ratio	tensile strength/MPa	
	Young's	shear	bulk		usual	as wire
Ag	82.7	30.3	103.6	0.367		290
Al	70.3	26.1	75.5	0.345	90–100 (cast) 90–150 (rolled)	200–450
As			22			
Au	78	27	217	0.44		200–250
Bi	31.9	12	31.3	0.33		
Ca			17.2		42–60	
Cd	49.9	19.2	41.6	0.3		
Co					260–750	
Cr	279.1	115.4	160.1	0.21		
Cs			1.6			
Cu	129.8	48.3	137.8	0.343	120–170 (cast) 200–400 (rolled)	400–460 (hard-drawn) 280–310 (annealed)
Fe	152.3 (cast) 211.4 (soft)	60 (cast) 81.6 (soft)	109.5 (cast) 169.8 (soft)	0.27 (cast) 0.293 (soft)	100–230 (cast) 290–450 (wrought)	460 (annealed) 540–620 (charcoal hard-drawn)
Hg			25			
K			3.1			
Li			11.1			
Mg	44.7	17.3	35.6	0.291	60–80 (cast) 170–190 (extruded)	
Mn			118			
Mo			231			1100–3000
Na			6.3			
Nb	104.9	37.5	170.3	0.397		
Ni (unmagne- tised)	219.2 (hard) 199.5 (soft)	83.9 (hard) 76 (soft)	187.6 (hard) 177.3 (soft)	0.306 (hard) 0.312 (soft)		500–900
Pb	16.1	5.59	45.8	0.44	12–17 (cast)	
Pd			182			350–450
Pt	168	61	228	0.377		330–370
Rb			2.5			
Sn	49.9	18.4	58.2	0.357	20–35 (cast)	
Ta	185.7	69.2	196.3	0.342		800–1100
Ti	115.7	43.8	107.7	0.321		
Tl			43			
W	411	160.6	311	0.28		1500–3500
V	127.6	46.7	158	0.365		
Zn	108.4	43.4	72	0.249	110–150 (rolled)	
Zr						260–390 (annealed) 1000 (hard-drawn)

Reference: Bradfield, *Notes on Applied Science, No. 30* (HMSO) 1964
Poynting and Thomson, *Properties of Matter*, Griffin, 1929

Table II.17 Values of Young's, shear and bulk elastic moduli, the Poisson ratio and tensile strength for some alloys at temperature 293 K.

Alloy	moduli/GPa			Poisson ratio	tensile strength/MPa	
	Young's	shear	bulk		usual	as wire
brass (30% Cu, 70% Zn)	100.6	37.3	111.8	0.35	150–190 (cast) 230–270 (rolled)	350–550
Constantan	162.4	61.2	156.4	0.327		
duralumin						400–550
German silver						460
gunmetal (90% Cu, 10% Sn)					190–260	
Hidurax special (Cu/Ni and traces Al/Fe/Mn)	144.5	54.4	144.1	0.333		
Invar (0.2% C, 63.8% Fe, 36% Ni)	144	57.2	99.4	0.259		
nickel silver (c. 55% Cu, 18% Ni, 27% Zn)	132.5	49.7	132	0.333		
phosphor bronze					180–280 (cast)	690–1080 (hard-drawn)
Pt + 10% Rh						630
soft solder					55–75	
steel (castings)					400–600	
steel (high C spring)					700–770 (annealed) 800–1000 (5% Ni) 1000–1500 (Ni/ Cr) 930–1000 (tempered)	
steel (mild)	211.9	82.2	169.2	0.291	430–490 (0.2% C)	c 1100
steel (0.75% C)	210	81.1	168.7	0.293		
steel (0.75% C hardened)	201.4	77.8	165	0.296		
steel (stainless)	215.3	83.9	166	0.283		
steel (tool)	211.6	82.2	165.3	0.287		
steel (tool hardened)	203.2	78.5	165.2	0.295		
steel						1860–2330 (piano, hard-drawn)
tungsten carbide	534.4	219	319	0.22		

Reference: as for Table II.16

Table II.18 Values of Young's, shear and bulk elastic moduli, the Poisson ratio and tensile strength for glass and quartz at temperature 293 K.

Substance	moduli/GPa			Poisson ratio	tensile strength /MPa
	Young's	shear	bulk		
crown glass	71.3	29.2	41.2	0.22	30–90
fused quartz	73.1	31.2	36.9	0.17	c 1000 (fibre)
heavy flint glass	80.1	31.5	57.6	0.27	30–90

Reference: as for Table II.16

Table II.19 Values of Young's modulus of elasticity and tensile strength for some plastics at temperature 293 K.

Plastic	Young's modulus/GPa	Tensile strength/MPa	Plastic	Young's modulus/GPa	Tensile strength/MPa
nylon 6 (cast)	2.4–3.1	76–97	polyethylene	0.4–1.3	21–35
nylon 6 (moulded)	0.8–3.1	76–97	polyimide	c 3.1	69–104
nylon 66	1.2–2.9	62–83	polymethyl methacrylate	2.4–3.4	50–76
polyacetal		c 69	polypropylene	1.1–1.6	30–40
polybenzoxazole	c 3.5	82–117	polystyrene	2.7–4.2	34–52
polycarbonate	2.4	55–65			

Reference: as for Table II.16

Table II.20 Values of Young's modulus of elasticity, along the grain and in radial and tangential directions, and tensile strength for some woods.

Wood	Young's modulus/GPa			tensile strength/MPa
	along	radial	tangential	
ash	16	1.6	0.9	60–110
balsa	6	0.3	0.1	
beech	14	2.2	1.1	60–110
birch	16	1.1	0.6	
deal				30–70
fir	16	1.1	0.8	40–80
mahogany	12	1.1	0.6	60–110
oak	11			60–110
pine	16	1.1	0.6	20–50
walnut	11	1.2	0.6	
teak	13			60–110
spruce	10–16	0.4–0.9	0.4–0.6	

Reference: Hearmon, *Forest Products Research Special Report No. 7*, HMSO, 1948

Table II.21 Values of temperature coefficient of Young's modulus and shear moduli of elasticity of some substances at temperature 288 K.

Substance	temperature coefficient/10^{-4} K^{-1}		Substance	temperature coefficient/10^{-4} K^{-1}	
	Young's	shear		Young's	shear
Ag	7.5	4.5	German silver		6.5
Al	4.8	5.2	phosphor bronze		3
Au	4.8	3.3	Pt	0.98	1
brass	3.7	4.6	quartz fibre	−1.5	−1.1
Cu	3	3.1	Sn		5.9
Fe	2.3	2.8	steel	2.4	2.6

Reference: Kaye and Laby Tables of Physical and Chemical Constants, 15 ed., Longmans, 1986

Table II.22 Values of bulk modulus of elasticity of some non-metallic elements.

Element	Br$_2$	C (diamond)	C (graphite)	Cl$_2$ (liquid)	I$_2$	P (red)	P (white)	S	Se	Si
Modulus/GPa	1.9	542	33	1.1	7.7	10.9	4.9	7.7	8.3	100

Reference: as for Table II.16

Table II.23 Values of bulk modulus of elasticity of some liquids at stated temperature and pressure.

Liquid†	pressure/ 10^5 Pa	temperature/ K	modulus/ GPa	Liquid†	pressure/ 10^5 Pa	temperature/ K	modulus/ GPa
CCl_4		293	1.12	C_2H_5Cl	8–37	288.2	0.662
$CHCl_3$	100–200	293	1.1	C_2H_5OH	1–500	273	1.32
CH_3COOCH_3	8–37	287.3	1.04		150–200	583	0.024
olive oil		293.5	1.6	n-C_3H_7OH	8	290.7	1.04
paraffin oil		287.8	1.62	iso-C_3H_7OH	8	290.8	0.983
$CH_3COOC_2H_5$	8–37	286	0.974	turpentine		292.7	1.28
glycerine		293.5	4.03	n-C_4H_9OH	8	290.4	1.13
H_2O	1–25	288	2.05	iso-C_4H_9OH	8	290.9	1.03
	900–1000	288	2.75	$C_4H_{10}O$ (ether)	1–50	273	0.689
	900–1000	471	1.81		900–1000	273	1.56
	2500–3000	287.2	3.88		900–1000	471	0.703
H_2O (sea)			2.32	n-$C_4H_{11}OH$	8	290.7	1.12
Hg	8–37	293	26.2	C_5H_{12}		293	0.318
	100–200	288	30	petroleum		289.5	1.46
CH_3COOH	1–16	293	1.45	C_6H_6	8	290.9	1.1
CH_3OH	37	287.7	0.97	CS_2	8–37	288.6	1.16
C_2H_5Br	8–37	372.3	0.343				

† For compound names see Table XI.2
Reference: as for Table II.16

Table II.24 Values of tensile strength of some miscellaneous substances.

Substance	catgut	hemp rope	leather belt	silk fibre	spider thread
tensile strength/MPa	420	60–100	30–50	260	180

Reference: Poynting and Thomson, *Properties of Matter*, Griffin, 1929

Table II.25 Values of coefficient of viscosity for various gases at about standard pressure at stated temperatures.

Gas	temperature/ K	viscosity/ µPa s	Gas	temperature/ K	viscosity/ µPa s
air	78.8	5.5	HBr	291.7	18.2
	169	11.3		373.2	23.4
	273	17.3	HCl	285.5	13.9
	373	22		373.2	18.2
	573	29.8	HI	293	16.6
	873	39.4		373	23.2
	1083	44.2		523	31.9
	1407	52.1	He	15.6	2.7
Ar	273	21		81.4	8.7
	373	27.3		273	18.7
	573	37.7		373	23.2
	873	50.4		573	31.2
	1100	56.3		873	41.8
Br_2 vapour	285.8	15.1		1090	47.1
	372.7	22.7	H_2O vapour	273	9.2
	493.3	24.8		373	12.4
CH_4	91.4	3.48		573	20.3
	273	10.3		873	32.6
	373	13.5	H_2S	273	11.7
	573	18.8		373	15.9
	873	25.3	I_2 vapour	397	18.4
$CH_3CH=CH_2$	289.7	8.3		520	24
	373	10.7	Kr	273	23.4
	473	13.4		373	31.2
$CH_3CH_2CH_3$	291	8		573	44.2
	373.4	10.1		873	60.2
	472.3	14.3	N_2	251.5	15.6
CH_3CH_3	194.5	6.3		323	18.9
	273	8.5		473	25.1
	373.4	11.4		673	31.9
	473.3	14.1		873	37.8
C_2H_2	273	9.4		1098	41.9
C_2H_4	197.5	7	Ne	273	29.8
	273	9.7		373	37
	373	12.8		573	48.9
	473	15.4		873	64.4
	573	17.9		1100	72.1
Cl_2	273	12.3	NH_3	194.5	6.7
	373	16.9		273	9.2
	473	21		373	13
	573	25		573	20.6
CO	84.5	5.6		873	31.9
	273	16.6	NO	273	17.8
	323	18.8		373	22.7
	373	21		473	26.8
	573	29	N_2O	273	13.7
	873	38.6		373	18.4
CO_2	175.2	9		573	27
	273	13.7		873	37
	373	18.5	O_2	273	19.5
	573	27.1		373	24.4
	873	37.4		573	33.7
	1325	47.9		873	44.7
H_2	15.3	0.6		1102	50.1
	89.6	3.9	SO_2	198	8.6
	273	8.4		273	11.6
	373	10.4		373	16.4
	573	13.7		573	25.1
	773	16.9		873	36.1
	1098	21.4			

Table II.25 (Continued)

Gas	temperature/K	viscosity/µPa s
Xe	273	21.2
	373	28.8
	573	42
	873	58.6

Reference: Handbook of Chemistry and Physics, CRC Press, 1991/2
Kaye and Laby Tables of Physical and Chemical Constants, 15 ed., Longmans, 1986

Table II.26 Values of coefficient of viscosity for some gases at various high pressures at temperature 300 K, unless otherwise stated.

Gas	Pressure/MPa					
	2	5	10	20	30	50
	viscosity/µPa s					
air	18.7	19.3	20.5	23.7	27.5	
Ar	23.3	24	25.7	30.5	36.4	
CH$_4$	11.6	12.3	14	19.2	24.7	
H$_2$	8.98	9.01	9.09	9.31	9.59	
He	19.9	19.9	20	20.1	20.3	
N$_2$	18.3	18.9	20.1	23.2	26.8	34.4
at 200 K		14.6	17.6	26.4	34.9	48.9
at 350 K		20.9	21.9	24.2	26.9	32.7
at 500 K		26.5	27.1	28.4	29.9	33.2
O$_2$	20.9	21.5	22.9	27.1	32.2	

Reference: Kaye and Laby Tables of Physical and Chemical Constants, 15 ed., Longmans, 1986

Table II.27 Values of coefficient of viscosity for water at about standard pressure at stated temperatures.

temperature/K	273	278	283	288	293	303
viscosity/mPa s	1.792	1.519	1.307	1.138	1.002	0.7973
temperature/K	313	323	333	343	353	363
viscosity/mPa s	0.6526	0.5471	0.4670	0.4046	0.3551	0.3150

Reference: Kestin et al, *J. Phys. Chem. Ref. Data*, **7**(3) 941, 1978

Table II.28 Values of coefficient of viscosity for various glycerol/H$_2$O solutions at about standard pressure at stated temperatures.
Assumption: viscosity coefficient for H$_2$O as in Table II.27

mass glycerol/%	temperature/K			mass glycerol/%	temperature/K		
	293	303	313		293	303	313
	viscosity/Pa s				viscosity/Pa s		
10	0.00131	0.00103	0.000823	96	0.622	0.28	0.142
20	0.00175	0.00135	0.00107	97	0.763	0.339	0.165
50	0.00598	0.0042	0.00309	98	0.936	0.408	0.195
80	0.0599	0.0338	0.0207	99	1.146	0.498	0.234
95	0.521	0.236	0.121	100	1.408	0.61	0.283

Reference: Segur and Oberstar, *Ind. Eng. Chem* vol. 43, 2117, 1951

Table II.29 Values of coefficient of viscosity for various sucrose/H_2O solutions at about standard pressure at stated temperatures.

mass sucrose/%	temperature/K			mass sucrose/%	temperature/K		
	288	293	298		288	293	298
	viscosity/Pa s				viscosity/Pa s		
30	0.003 757	0.003 187	0.002 735	65	0.211 3	0.147 2	0.105 4
40	0.007 463	0.006 167	0.005 164	70	0.746 9	0.481 6	0.321 6
50	0.019 53	0.015 43	0.012 40	75	4.039	2.328	1.405
60	0.079 49	0.058 49	0.040 03				

Reference: Bingham and Jackson, *NBS Bulletin*, vol. 14, 1918
(note corrected to H_2O viscosities from Table II.27)

Table II.30 Values of coefficient of viscosity for some liquid elements, molten salts, glasses and minerals at about standard pressure at stated temperatures.

Substance	temperature/K	viscosity/mPa s	Substance	temperature/K	viscosity/mPa s
Al	973	2.96	KCl	1073	1.096
	1073	2.66	KNO_3	673	2.09
Au	1373	5.13		873	1.07
	1473	4.64	Na	373	0.68
Br_2	268.7	1.31		673	0.286
	302	0.91		1073	0.174
diorite	1473	$10^6 \times 1.26$	NaCl	1089	1.5
	1673	$10^4 \times 6.31$	$NaNO_3$	673	1.91
diopside	1573	$10^3 \times 3.31$	olivine	1473	$10^5 \times 3.16$
	1673	$10^3 \times 2.69$		1673	$10^4 \times 1.59$
flint glass (medium)	1173	$10^6 \times 7.94$	Pb	673	2.32
	1273	$10^5 \times 6.31$		873	1.55
	1473	$10^4 \times 2.51$		1073	1.24
	1673	$10^3 \times 5.01$	plate glass	1173	10^7
H_2SO_4	273	48.4		1273	$10^6 \times 1.07$
	303	20.1		1473	$10^4 \times 7.41$
	348	6.6		1673	$10^4 \times 1.18$
	373	4.1	SiO_2	1373	$10^{17} \times 3.98$
Hg	253	1.86		1573	$10^{14} \times 6.31$
	273	1.616		1873	$10^{11} \times 1.59$
	303	1.497		2273	$10^6 \times 2.51$
	348	1.322	Sn	673	1.33
	373	1.255		873	1.04
	473	1.052		1073	0.89
	613	0.921		1373	0.78
K	373	0.458		1473	0.77
	673	0.224			
	1073	0.141			

Reference: Janz, *NSRDS-NBS 15*, 1968
Smithell, *Metal Reference Book*, 6th edn., Butterworths, 1983.

Table II.31 Values of coefficient of viscosity for various organic liquids at about standard pressure at stated temperatures.

Liquid†	temperature/K	viscosity(mPa s)	Liquid†	temperature/K	viscosity(mPa s)
CCl_4	273	1.341	CH_3OH	174.7	13.9
	298	0.912		228.5	1.98
	323	0.662		273	0.82
	373	0.395		298	0.547
$CHBr_3$	288	2.152		323	0.403
	303	1.741	C_2H_5Br	153	5.6
$CHCl_3$	223	1.514		193	1.81
	273	0.709		273	0.487
	312	0.5		303	0.348
$CH_2=CHCH_2OH$	273	2.145	C_2H_5OH	173	98.96
	293	1.363		223	8.318
	313	0.914		273	1.873
	343	0.553		303	0.983
CH_2Cl_2	288	0.449		348	0.459
	303	0.393	$n-C_4H_9OH$	222	36.1
CH_3CCl_3	293	1.2		273	5.186
$CH_3CH_2CH_2COOH$	273	2.286		303	2.3
	313	1.12		373	0.54
	373	0.551	$2-C_4H_9OH$	288	4.21
$CH_3CH_2CH_2OH$	273	3.883	$(C_2H_5)_2O$	173	1.545
	293	2.256		223	0.544
	313	1.405		273	0.288
	343	0.76		303	0.214
$CH_3CH_2CH_3$	173	0.421		373	0.119
	223	0.215	C_5H_5N	293	0.974
	273	0.127	cyclo C_5H_{10}	286.5	0.493
$CH_3(CH_2)_3CH_3$	173	1.3	C_6H_5Br	273	1.592
	223	0.498		303	0.995
	273	0.273		348	0.605
	298	0.214	$C_6H_5CH_3$	223	2.124
CH_3CH_2COOH	283	1.289		273	0.768
	313	0.845		303	0.52
CH_3CH_3	293	19.9		373	0.272
	313	9.13	$C_6H_5C_2H_5$	290	0.691
	353	3.02	C_6H_5Cl	288	0.9
	373	1.99		313	0.631
CH_3CHO	273	0.28		373	0.367
	293	0.22	$C_6H_5COCH_3$	285	2.28
CH_3Cl	293	0.183		298	1.617
CH_3CN	273	0.442		353	0.734
	298	0.345	$C_6H_5NH_2$	267	13.8
CH_3COCH_3	180.5	2.148		273	9.45
	243	0.575		323	1.982
	273	0.402		373	0.808
	298	0.31	$C_6H_5NO_2$	276	2.9
	314	0.28		283	2.47
CH_3COOCH_3	273	0.484		298	1.842
	313	0.32		323	1.244
$CH_3COOC_2H_5$	223	1.284	C_6H_5OH	291.3	12.7
	273	0.575		323	3.49
	303	0.406		363	1.26
	348	0.264	C_6H_6	273	0.912
$CH_3COOC_4H_9$	273	1.004		298	0.603
	313	0.563		323	0.436
CH_3COOH	288	1.31		348	0.332
	303	1.037	$n-C_6H_{11}OH$	293	68
	323	0.792	cyclo C_6H_{12}	290	1.02
	373	0.457	$n-C_6H_{14}$	223	0.782
				273	0.387
				298	0.296
				248	0.19

(Continued)

Table II.31 (Continued)

Liquid†	temperature/K	viscosity(mPa s)
n-C_8H_{18}	223	1.837
	273	0.719
	298	0.516
	323	0.39
CO_2	223	0.227
	273	0.098
	303	0.053
CS_2	260	0.514
	273	0.436
	313	0.33
n-decane	293	0.92
isopentane	273	0.273
	293	0.223

Liquid†	temperature/K	viscosity(mPa s)
l-octanol	288	10.6
oil, castor	283	2420
	313	231
	373	16.9
oil, cottonseed	293	70.4
oil, linseed	303	33.1
	363	7.1
oil, machine light	310.8	34.2
oil, machine heavy	310.8	127.4
oil, olive	303	54
oil, rape	293	163
oil, soya bean	303	40.6
turpentine	273	2.248
	343	0.728

† For compound names, see Table XI.2
References: Handbook of Chemistry and Physics, CRC Press, 1991/2
Kaye and Laby Tables of Physical and Chemical Constants, 15 ed., Longmans, 1986

Table II.32 Variation of the coefficient of viscosity of water with pressure and temperature.

temperature/K	283	293	303	323	373
pressure/MPa			viscosity/mPa s		
49	1.266	0.992	0.796	0.559	0.294
98	1.251	0.992	0.804	0.572	0.306
196	1.283	1.025	0.84	0.604	0.33
392	1.549	1.165	0.953	0.686	0.381
588	1.738	1.37	1.105	0.787	0.436
784		1.632	1.309	0.91	0.496
981			1.555	1.059	0.56

Reference: Kaye and Laby Tables of Physical and Chemical Constants, 15 ed., Longmans, 1986
Partington, *Advanced Treatise on Physical Chemistry*, vol. 2, Longmans, 1951

Table II.33 Variation of the coefficient of viscosity of various liquids with pressure at stated temperatures.

Liquid[†]	temperature/K	pressure/mPa					
		98	100	392	400	784	1177
		viscosity/mPa s					
$CH_3(CH_2)_2CH_3$	303	0.424		1.441		4.695	14.391
	348	0.315		1.026		2.842	6.734
CH_3COCH_3	303	0.496		1.189		2.862	
	348	0.33		0.71		1.472	2.74
CH_3OH	303	0.833		1.678		3.187	5.642
	348	0.429	0.806			1.402	2.261
CS_2	303	0.497		1.108		2.374	5.317
C_2H_5OH	303	1.563		4.07		10.322	24.084
	348	0.753		1.965		4.351	8.4
$C_4H_{10}O$ (ether)	303	0.452		1.327		3.895	10.015
	348	0.273		0.771		1.869	3.957
$C_6H_5CH_3$	303	1.322	4.103			26	
	348	0.621		2.114		8.216	36.406
C_6H_6	303	1.248					
	323		0.898				
	348	0.687	0.664				
	373		0.521		1.946		
n-C_6H_{14}	298		0.622				
	303	0.606		2.312		9.221	
	348	0.443	0.42	1.501	1.408	4.712	13.243
	373		0.35		1.119		
n-C_8H_{18}	298		1.192		6.656		
	303	1.03		5.978			
	348	0.673	0.685	2.745	2.812	10.92	
	373		0.556		2.114		
$C_{16}H_{34}$	323		5.352				
	348		3.349				
	373		2.339		13.44		

[†] For compound names, see Table XI.2
Reference: as for Table II.32

Table II.34 Values of STP molecular mean speed, mean free path and mean time between collisions, for various gases.

gas	mean speed/ m s^{-1}	mean free path/nm	mean time/ ps	gas	mean speed/ m s^{-1}	mean free path/nm	mean time/ ps
Ar	380	62.6	165	H$_2$	1694	110.6	65
CHCl$_3$	220	161	732	He	1202	173.6	144
CH$_4$	600	48.1	80	N$_2$	454	58.8	130
C$_2$H$_4$	454	34.3	75	N$_2$O	362	38.7	107
C$_6$H$_6$	272	148.2	545	Ne	535	124	232
Cl$_2$	285	27.4	96	O$_2$	425	63.3	149
CO	454	58.6	129	SO$_2$	300	27.4	91
CO$_2$	362	39	108				

Reference: Hirschfelder et al., *Molecular Theory of Gases and Liquids*, Wiley, 1954

Table II.35 Variation with temperature of surface tension, γ, of water against air at about standard pressure.

temperature/K	273	283	288	293	298	303
γ/(mN m^{-1})	75.7	74.2	73.5	72.75	72	71.2
temperature/K	313	323	333	343	353	
γ/(mN m^{-1})	69.6	67.9	66.2	64.4	62.6	

Reference: as for Table II.32

Table II.36 Variation with percentage solute of surface tension, γ, of various aqueous solutions at stated temperatures, at about standard pressure.

CH$_3$CH$_2$CH$_2$OH at temperature 298 K							
mass/%	0.1	0.5	1	50	60	80	90
γ/(mN m^{-1})	67.1	56.18	49.3	24.34	24.15	23.66	23.41
C$_4$H$_8$O$_2$(dioxane) at temperature 299 K							
mass/%	0.44	4.7	20.17	55	76.45	91.9	97.77
γ/(mN m^{-1})	69.83	62.45	51.57	39.27	35.8	33.95	33.1
CH$_3$COOH at temperature 303 K							
mass/%	1	2.475	5.001	10.01	30.09	49.96	69.91
γ/(mN m^{-1})	68	64.4	60.1	54.6	43.6	38.4	34.3
CH$_3$OH at temperature 293 K							
volume/%	7.5	10	25	50	60	80	100
γ/(mN m^{-1})	60.90	59.04	46.38	35.31	32.95	27.26	22.63
C$_2$H$_5$OH at temperature 293 K							
volume/%	34	48	60	72	80	96	
γ/(mN m^{-1})	33.24	30.1	27.56	26.28	24.91	23.04	
n-C$_4$H$_9$OH at temperature 303 K							
mass/%	0.04	0.41	9.53	80.44	86.05	94.2	97.4
γ/(mN m^{-1})	69.33	60.38	26.97	23.69	23.47	23.29	22.25
C$_6$H$_5$OH at temperature 293 K							
mass/%	0.024	0.047	0.118	0.417	0.941	3.76	5.62
γ/(mN m^{-1})	72.6	72.2	71.3	66.5	61.1	46	42.3
glycerol at temperature 291 K							
mass/%	5	10	20	30	50	85	100
γ/(mN m^{-1})	72.9	72.9	72.4	72	70	66	63

(Continued)

Table II.36 (Continued)

HCl at temperature 293 K							
mass/%	1.78	3.52	6.78	12.81	16.97	23.74	35.29
γ/(mN m^{-1})	72.55	72.45	72.25	71.85	71.75	70.55	65.75
HCOOH at temperature 303 K							
mass/%	1	5	10	25	50	75	100
γ/(mN m^{-1})	70.07	66.2	62.78	56.29	49.5	43.4	36.51
H_2SO_4 at temperature 298 K							
mass/%	4.11	8.26	12.18	17.66	21.88	29.07	33.63
γ/(mN m^{-1})	72.21	72.55	72.8	73.36	73.91	74.8	75.29
KCl at temperature 293 K							
mass/%	0.74	3.6	6.93	13.88	18.77	22.97	24.7
γ/(mN m^{-1})	72.99	73.45	74.15	75.55	76.95	78.25	78.75
KOH at temperature 291 K							
mass/%		2.73	5.31	10.08	17.57		
γ/(mN m^{-1})		73.95	74.85	76.55	79.75		
NaCl at temperature 293 K							
mass/%	0.58	2.84	5.43	10.46	14.92	22.62	25.92
γ/(mN m^{-1})	72.92	73.75	74.39	76.05	77.65	80.95	82.55
$NaNO_3$ at temperature 293 K							
mass/%	0.85	4.08	7.84	14.53	29.82	37.3	47.06
γ/(mN m^{-1})	72.87	73.75	73.95	75.15	78.35	80.25	87.05
NaOH at temperature 291 K							
mass/%		2.72	5.66	16.66	30.56	35.9	
γ/(mN m^{-1})		74.35	75.85	83.05	96.05	101.05	
sucrose at temperature 298 K							
mass/%		10	20	30	40	55	
γ/(mN m^{-1})		72.5	73	73.4	74.1	75.7	

Reference: Handbook of Chemistry and Physics, CRC Press, 1991/2

Table II.37 Surface tension, γ, at about standard pressure, for some liquid elements and molten salts at stated temperatures, T, together with some values of A and B in the formula: $\gamma = A - BT$, and temperature range of formula applicability.

Substance	T/K	γ/mN m^{-1}	A/mN m^{-1}	B/mN m^{-1} K^{-1}	range/K
Al	973	900	1240.55	0.35	933–1073
Au	1373	1120	1833.96	0.52	1338–1473
Hg	298	485.5	546.54	0.2049	278–573
K	338	110.9	132.06	0.0625	338–773
KCl	1053	100.3	181.42	0.077	1043–1243
KNO_3	613	111	156.98	0.075	613–773
N_2	90	5.99	26.38	0.2266	78–90
O_2	89	13.4	36.2	0.2561	71–89
Na	373	209.9	250.18	0.108	373–773
NaCl	1083	113.3	191.13	0.0719	1078–1243
NaF	1273	185.2	289.59	0.082	1273–1353
$NaNO_3$	593	119.2	155.53	0.0613	593–873
Na_2SO_4	1173	194.5	253.28	0.0501	1173–1353
Pb	623	444.5	485.72	0.066	617–923
$PbCl_2$	793	135.3	233.65	0.124	783–853

Reference: Janz, *NSRDS–NBS 28*, 1969
Smithell, *Metal Reference Book*, 6th edn. Butterworths, 1983.

Table II.38 Surface tension, γ, at about standard pressure, for some liquids at temperature 293 K, unless otherwise stated, together with some values of A and B in the formula: $\gamma = A - BT$, and temperature range of formula applicability.

Liquid[†]	γ/mN m^{-1}	A/mN m^{-1}	B/mN m^{-1} K^{-1}	Range/K
CCl_4	27.04	62.91	0.122 4	288–378
$CHBr_3$	41.53			
$CHCl_3$	27.32	65.26	0.129 5	288–348
$CH_2=CHCH_2OH$	25.8			
CH_2Cl_2	26.52			
CH_3CCl_3 (at 387 K)	22			
$CH_3CH_2CH_2COOH$	26.8			
$CH_3CH_2CH_2OH$	23.71	46.47	0.077 7	283–363
CH_3CH_2COOH	26.7			
CH_3CHO	21.2			
CH_3Cl	16.2			
CH_3CN	29.3			
CH_3COCH_3	23.7	56.84	0.112	293–323
CH_3COOCH_3	25.37	63.14	0.128 9	283–333
$CH_3COOC_2H_5$	23.97	57.99	0.116 1	283–373
CH_3COOH	27.59	56.72	0.099 4	293–363
CH_3NO_2	36.82			
CH_3OH	22.5	45.1	0.077 3	283–333
C_2H_5Br	24.15			
C_2H_5OH	22.39	46.76	0.083 2	283–343
n-C_4H_9OH	25.39	51.7	0.089 83	283–373
$C_4H_{10}O$ (ether)	17.1	43.71	0.090 8	288–303
C_5H_5N	38			
C_6H_5Br	36.5			
$C_6H_5CH_3$	28.52	63.36	0.118 9	283–373
$C_6H_5C_2H_5$	29.2			
C_6H_5Cl	33.56			
$C_6H_5COCH_3$	39.8			
$C_6H_5NH_2$	42.67	74.45	0.108 5	288–363
$C_6H_5NO_2$	43.9			
C_6H_5OH (at 313 K)	39.27	72.7	0.106 8	313–413
C_6H_6	28.88	66.64	0.128 7	283–353
n-C_6H_{12}	25.5			
n-C_6H_{14}	18.4	48.34	0.102 2	283–333
n-C_8H_{18}	21.62	49.48	0.095 09	283–293
CO_2 (at 248 K)	9.13			
CS_2	32.32	75.8	0.148 4	283–323
glycerol	63.4	89.32	0.088 45	293–423
isopentane	13.72			
n-octanol	27.53			

[†] For compound names, see Table XI.2
Reference: Jasper *J. Phys. Chem. Ref. Data*, vol. 1, 841, 1972

Table II.39 Interfacial tension, γ, at about standard pressure, for some liquids at temperature 293 K.

Liquids	γ/mN m^{-1}	Liquids	γ/mN m^{-1}	Liquids	γ/mN m^{-1}
H_2O against		H_2O against		Hg against	
CCl_4	45	n-C_6H_{14}	51.1	$CHCl_3$	357
$CHCl_3$	28	n-C_8H_{18}	51	$CH_3(CH_2)_5CH_3$	379
$CH_3(CH_2)_5CH_3$	51	Hg	375	CH_3COCH_3	390
$CH_3(CH_2)_5COOH$	7	n-octanol	8.5	$C_4H_{10}O$	379
$C_4H_{10}O$ (ether)	10	olive oil	20	C_6H_6	357
C_6H_6	35	paraffin oil	48	$C_{17}H_{33}COOH$	322

Reference: as for Table II.31.

Section III
Properties of the Elements

		Page
III.1	32-column form of the Periodic Table (IUPAC 1989)	35
III.2	Standard Atomic Weights of the Elements (1989)	36
	1 Alphabetical order	36
	2 Atomic Number order	37
III.3	Electronic configurations of the elements (ground states)	38
III.4	Physical properties of the elements	40
III.5	Crystal structure, atomic radii and ionic radii of the elements	42
III.6	van der Waals radii of the elements	44
III.7	Ionization energies of the elements	44
III.8	Atomic electron affinities at 0 K	46
III.9	Proton affinities of atoms	46
III.10	Electronegativities of the elements	47
III.11	Abundances of the elements in the Earth's crust and in sea water	48
III.12	Average composition of the atmosphere (exclusive of water vapour)	48
III.13	Elemental composition of the adult human body	49

Section III
Properties of the Elements

Table III.1 32-Column Form of the Periodic Table (IUPAC, 1989).

1	2																	3	4	5	6	7	8	9	10	11	12	13	14	15	16	17	18
1																																	2
H																																	He
3	4																											5	6	7	8	9	10
Li	Be																											B	C	N	O	F	Ne
11	12																											13	14	15	16	17	18
Na	Mg																											Al	Si	P	S	Cl	Ar
19	20																	21	22	23	24	25	26	27	28	29	30	31	32	33	34	35	36
K	Ca																	Sc	Ti	V	Cr	Mn	Fe	Co	Ni	Cu	Zn	Ga	Ge	As	Se	Br	Kr
37	38																	39	40	41	42	43	44	45	46	47	48	49	50	51	52	53	54
Rb	Sr																	Y	Zr	Nb	Mo	Te	Ru	Rh	Pd	Ag	Cd	In	Sn	Sb	Te	I	Xe
55	56	57	58	59	60	61	62	63	64	65	66	67	68	69	70	71	72	73	74	75	76	77	78	79	80	81	82	83	84	85	86
Cs	Ba	La	Ce	Pr	Nd	Pm	Sm	Eu	Gd	Tb	Dy	Ho	Er	Tm	Yb	Lu	Hf	Ta	W	Re	Os	Ir	Pt	Au	Hg	Tl	Pb	Bi	Po	At	Rn
87	88	89	90	91	92	93	94	95	96	97	98	99	100	101	102	103															
Fr	Ra	Ac	Th	Pa	U	Np	Pu	Am	Cm	Bk	Cf	Es	Fm	Md	No	Lr															

Key to following two tables

* Element has no stable nuclides. One or more well-known isotopes are given in Table IX.39 with the appropriate relative atomic mass and half-life. However, three such elements (Th, Pa and U) do have a characteristic terrestrial isotopic composition, and for these an atomic weight is tabulated.
g geological specimens are known in which the element has an isotopic composition outside the limits for normal material. The difference between the atomic weight of the element in such specimens and that given in the Table may exceed the implied uncertainty.
m modified isotopic compositions may be found in commercially available material because it has been subjected to an undisclosed or inadvertent isotopic separation. Substantial deviations in atomic weight of the element from that given in the Table can occur.
r range in isotopic composition of normal terrestrial material prevents a more precise $A_r(E)$ being given; the tabulated $A_r(E)$ value should be applicable to any normal material.
† revised values – see Chemistry International (1992), *14*, 11.

Table III.2.1 Standard Atomic Weights of the Elements (1989).

For more details see: Pure and Applied Chemistry (1991), *63*, 975.
(Scaled to $A_r(^{12}C) = 12$)
The atomic weights of many elements are not invariant but depend on the origin and treatment of the material. The footnotes to this Table elaborate the types of variation to be expected for individual elements. The values of $A_r(E)$ and uncertainties, $U_r(E)$, in parantheses follow the last significant figure to which they are attributed, apply to elements as they exist on earth.
Alphabetical order in English

Name	Symbol	Atomic number	Atomic weight	Footnotes	Name	Symbol	Atomic number	Atomic weight	Footnotes
Actinium*	Ac	89			Neodymium	Nd	60	144.24(3)	g
Aluminium	Al	13	26.981539(5)		Neon	Ne	10	20.1797(6)	g m
Americium*	Am	95			Neptunium*	Np	93		
Antimony (Stibium)	Sb	51	121.757(3)	g	Nickel	Ni	28	58.6934(2)	
Argon	Ar	18	39.948(1)	g r	Niobium	Nb	41	92.90638(2)	
Arsenic	As	33	74.92159(2)		Nitrogen	N	7	14.00674(7)	g r
Astatine*	At	85			Nobelium*	No	102		
Barium	Ba	56	137.327(7)		Osmium	Os	76	190.23(3)†	g
Berkelium*	Bk	97			Oxygen	O	8	15.9994(3)	g r
Beryllium	Be	4	9.012182(3)		Palladium	Pd	46	106.42(1)	g
Bismuth	Bi	83	208.98037(3)		Phosphorus	P	15	30.973762(4)	
Boron	B	5	10.811(5)	g m r	Platinum	Pt	78	195.08(3)	
Bromine	Br	35	79.904(1)		Plutonium*	Pu	94		
Cadmium	Cd	48	112.411(8)	g	Polonium*	Po	84		
Caesium	Cs	55	132.90543(5)		Potassium (Kalium)	K	19	39.0983(1)	
Calcium	Ca	20	40.078(4)	g	Praseodymium	Pr	59	140.90765(3)	
Californium*	Cf	98			Promethium*	Pm	61		
Carbon	C	6	12.011(1)	r	Protactinium*	Pa	91	231.03588(2)	
Cerium	Ce	58	140.115(4)	g	Radium*	Ra	88		
Chlorine	Cl	17	35.4527(9)	m	Radon*	Rn	86		
Chromium	Cr	24	51.9961(6)		Rhenium	Re	75	186.207(1)	
Cobalt	Co	27	58.93320(1)		Rhodium	Rh	45	102.90550(3)	
Copper	Cu	29	63.546(3)	r	Rubidium	Rb	37	85.4678(3)	g
Curium*	Cm	96			Ruthenium	Ru	44	101.07(2)	g
Dysprosium	Dy	66	162.50(3)	g	Samarium	Sm	62	150.36(3)	g
Einsteinium*	Es	99			Scandium	Sc	21	44.955910(9)	
Erbium	Er	68	167.26(3)	g	Selenium	Se	34	78.96(3)	
Europium	Eu	63	151.965(9)	g	Silicon	Si	14	28.0855(3)	r
Fermium*	Fm	100			Silver	Ag	47	107.8682(2)	g
Fluorine	F	9	18.9984032(9)		Sodium (Natrium)	Na	11	22.989768(6)	
Francium*	Fr	87			Strontium	Sr	38	87.62(1)	g r
Gadolinium	Gd	64	157.25(3)	g	Sulfur	S	16	32.066(6)	g r
Gallium	Ga	31	69.723(1)		Tantalum	Ta	73	180.9479(1)	
Germanium	Ge	32	72.61(2)		Technetium*	Tc	43		
Gold	Au	79	196.96654(3)		Tellurium	Te	52	127.60(3)	g
Hafnium	Hf	72	178.49(2)		Terbium	Tb	65	158.92534(3)	
Helium	He	2	4.002602(2)	g r	Thallium	Tl	81	204.3833(2)	
Holmium	Ho	67	164.93032(3)		Thorium	Th	90	232.0381(1)	g
Hydrogen	H	1	1.00794(7)	g m r	Thulium*	Tm	69	168.93421(3)	
Indium	In	49	114.818(3)†		Tin	Sn	50	118.710(7)	g
Iodine	I	53	126.90447(3)		Titanium	Ti	22	47.88(3)	
Iridium	Ir	77	192.22(3)		Tungsten (Wolfram)	W	74	183.84(1)†	
Iron	Fe	26	55.847(3)		Unnilhexium*	Unh	106		
Krypton	Kr	36	83.80(1)	g m	Unnilpentium*	Unp	105		
Lanthanum	La	57	138.9055(2)	g	Unnilquadium*	Unq	104		
Lawrencium*	Lr	103			Unnilseptium*	Uns	107		
Lead	Pb	82	207.2(1)	g r	Uranium*	U	92	238.0289(1)	g m
Lithium	Li	3	6.941(2)	g m r	Vanadium	V	23	50.9415(1)	
Lutetium	Lu	71	174.967(1)	g	Xenon	Xe	54	131.29(2)	g m
Magnesium	Mg	12	24.3050(6)		Ytterbium	Yb	70	173.04(3)	g
Manganese	Mn	25	54.93805(1)		Yttrium	Y	39	88.90585(2)	
Mendelevium*	Md	101			Zinc	Zn	30	65.39(2)	
Mercury	Hg	80	200.59(2)		Zirconium	Zr	40	91.224(2)	g
Molybdenum	Mo	42	95.94(1)	g					

Table III.2.2 Standard Atomic Weights of the Elements (1989).
(Scaled to $A_r(^{12}C) = 12$)
The atomic weights of many elements are not invariant but depend on the origin and treatment of the material. The footnotes to this Table elaborate the types of variation to be expected for individual elements. The values of $A_r(E)$ and uncertainties, $U_r(E)$, in parantheses follow the last significant figure to which they are attributed, apply to elements as they exist on earth.
Order of Atomic Number

Atomic number	Name	Symbol	Atomic weight	Footnotes			Atomic number	Name	Symbol	Atomic weight	Footnotes	
1	Hydrogen	H	1.00794(7)	g	m	r	55	Caesium	Cs	132.90543(5)		
2	Helium	He	4.002602(2)	g		r	56	Barium	Ba	137.327(7)		
3	Lithium	Li	6.941(2)	g	m	r	57	Lanthanum	La	138.9055(2)	g	
4	Beryllium	Be	9.012182(3)				58	Cerium	Ce	140.115(4)	g	
5	Boron	B	10.811(5)	g	m	r	59	Praseodymium	Pr	140.90765(3)		
6	Carbon	C	12.011(1)			r	60	Neodymium	Nd	144.24(3)	g	
7	Nitrogen	N	14.00674(7)	g		r	61	Promethium*	Pm			
8	Oxygen	O	15.9994(3)	g		r	62	Samarium	Sm	150.36(3)	g	
9	Fluorine	F	18.9984032(9)				63	Europium	Eu	151.965(9)	g	
10	Neon	Ne	20.1797(6)	g	m		64	Gadolinium	Gd	157.25(3)	g	
11	Sodium (Natrium)	Na	22.989768(6)				65	Terbium	Tb	158.92534(3)		
12	Magnesium	Mg	24.3050(6)				66	Dysprosium	Dy	162.50(3)	g	
13	Aluminium	Al	26.981539(5)				67	Holmium	Ho	164.93032(3)		
14	Silicon	Si	28.0855(3)			r	68	Erbium	Er	167.26(3)	g	
15	Phosphorus	P	30.973762(4)				69	Thulium	Tm	168.93421(3)		
16	Sulfur	S	35.066(6)	g		r	70	Ytterbium	Yb	173.04(3)	g	
17	Chlorine	Cl	35.4527(9)		m		71	Lutetium	Lu	174.967(1)	g	
18	Argon	Ar	39.948(1)	g		r	72	Hafnium	Hf	178.49(2)		
19	Potassium (Kalium)	K	39.0983(1)				73	Tantalum	Ta	180.9479(1)		
20	Calcium	Ca	40.078(4)	g			74	Tungsten (Wolfram)	W	183.84(1)†		
21	Scandium	Sc	44.955910(9)				75	Rhenium	Re	186.207(1)		
22	Titanium	Ti	47.88(3)				76	Osmium	Os	190.23(3)†	g	
23	Vanadium	V	50.9415(1)				77	Iridium	Ir	192.22(3)		
24	Chromium	Cr	51.9961(6)				78	Platinum	Pt	195.08(3)		
25	Manganese	Mn	54.93805(1)				79	Gold	Au	196.96654(3)		
26	Iron	Fe	55.847(3)				80	Mercury	Hg	200.59(2)		
27	Cobalt	Co	58.93320(1)				81	Thallium	Tl	204.3833(2)		
28	Nickel	Ni	58.6934(2)				82	Lead	Pb	207.2(1)	g	r
29	Copper	Cu	63.546(3)			r	83	Bismuth	Bi	208.98037(3)		
30	Zinc	Zn	65.39(2)				84	Polonium*	Po			
31	Gallium	Ga	69.723(1)				85	Astatine*	At			
32	Germanium	Ge	72.61(2)				86	Radon*	Rn			
33	Arsenic	As	74.92159(2)				87	Francium*	Fr			
34	Selenium	Se	78.96(3)				88	Radium*	Ra			
35	Bromine	Br	79.904(1)				89	Actinium*	Ac			
36	Krypton	Kr	83.80(1)	g	m		90	Thorium*	Th	232.0381(1)	g	
37	Rubidium	Rb	85.4678(3)	g			91	Protactinium*	Pa	231.03588(2)		
38	Strontium	Sr	87.62(1)	g		r	92	Uranium*	U	238.0289(1)	g	m
39	Yttrium	Y	88.90585(2)				93	Neptunium*	Np			
40	Zirconium	Zr	91.224(2)	g			94	Plutonium*	Pu			
41	Niobium	Nb	92.90638(2)				95	Americium*	Am			
42	Molybdenum	Mo	95.94(1)	g			96	Curium*	Cm			
43	Technetium*	Tc					97	Berkelium*	Bk			
44	Ruthenium	Ru	101.07(2)	g			98	Californium*	Cf			
45	Rhodium	Rh	102.90550(3)				99	Einsteinium*	Es			
46	Palladium	Pd	106.42(1)	g			100	Fermium*	Fm			
47	Silver	Ag	107.8682(2)	g			101	Mendelevium*	Md			
48	Cadmium	Cd	112.411(8)	g			102	Nobelium*	No			
49	Indium	In	114.818(3)†				103	Lawrencium*	Lr			
50	Tin	Sn	118.710(7)	g			104	Unnilquadium*	Unq			
51	Antimony (Stibium)	Sb	121.757(3)	g			105	Unnilpentium*	Unp			
52	Tellurium	Te	127.60(3)	g			106	Unnilhexium*	Unh			
53	Iodine	I	126.90447(3)				107	Unnilseptium*	Uns			
54	Xenon	Xe	131.29(2)	g	m							

Table III.3 Electronic Configurations of the Elements (ground states)

Atomic number	Element	Shell K $n=1$ 1s	L 2 2s	2p	M 3 3s	3p	3d	N 4 4s	4p	4d	4f
1	Hydrogen	1									
2	Helium	2									
3	Lithium	2	1								
4	Beryllium	2	2								
5	Boron	2	2	1							
6	Carbon	2	2	2							
7	Nitrogen	2	2	3							
8	Oxygen	2	2	4							
9	Fluorine	2	2	5							
10	Neon	2	2	6							
11	Sodium	2	2	6	1						
12	Magnesium	2	2	6	2						
13	Aluminium	2	2	6	2	1					
14	Silicon	2	2	6	2	2					
15	Phosphorus	2	2	6	2	3					
16	Sulphur	2	2	6	2	4					
17	Chlorine	2	2	6	2	5					
18	Argon	2	2	6	2	6					
19	Potassium	2	2	6	2	6		1			
20	Calcium	2	2	6	2	6		2			
21	Scandium	2	2	6	2	6	1	2			
22	Titanium	2	2	6	2	6	2	2			
23	Vanadium	2	2	6	2	6	3	2			
24	Chromium	2	2	6	2	6	5	1			
25	Manganese	2	2	6	2	6	5	2			
26	Iron	2	2	6	2	6	6	2			
27	Cobalt	2	2	6	2	6	7	2			
28	Nickel	2	2	6	2	6	8	2			
29	Copper	2	2	6	2	6	10	1			
30	Zinc	2	2	6	2	6	10	2			
31	Gallium	2	2	6	2	6	10	2	1		
32	Germanium	2	2	6	2	6	10	2	2		
33	Arsenic	2	2	6	2	6	10	2	3		
34	Selenium	2	2	6	2	6	10	2	4		
35	Bromine	2	2	6	2	6	10	2	5		
36	Krypton	2	2	6	2	6	10	2	6		

Electronic Configurations of the Elements

Atomic number	Element	Shell K $n=1$	L 2	M 3	N 4 4s	4p	4d	4f	O 5 5s	5p	5d	5f	P 6 6s	6p	6d
37	Rubidium	2	8	18	2	6			1						
38	Strontium	2	8	18	2	6			2						
39	Yttrium	2	8	18	2	6	1		2						
40	Zirconium	2	8	18	2	6	2		2						
41	Niobium	2	8	18	2	6	4		1						
42	Molybdenum	2	8	18	2	6	5		1						
43	Technetium	2	8	18	2	6	6		1						
44	Ruthenium	2	8	18	2	6	7		1						
45	Rhodium	2	8	18	2	6	8		1						
46	Palladium	2	8	18	2	6	10								
47	Silver	2	8	18	2	6	10		1						

(Continued)

Table III.3 (Continued)

Atomic number	Shell Element	K n = 1	L 2	M 3	N 4				O 5				P 6		
					4s	4p	4d	4f	5s	5p	5d	5f	6s	6p	6d
48	Cadmium	2	8	18	2	6	10		2						
49	Indium	2	8	18	2	6	10		2	1					
50	Tin	2	8	18	2	6	10		2	2					
51	Antimony	2	8	18	2	6	10		2	3					
52	Tellurium	2	8	18	2	6	10		2	4					
53	Iodine	2	8	18	2	6	10		2	5					
54	Xenon	2	8	18	2	6	10		2	6					
55	Caesium	2	8	18	2	6	10		2	6			1		
56	Barium	2	8	18	2	6	10		2	6			2		
57	Lanthanum	2	8	18	2	6	10		2	6	1		2		
58	Cerium	2	8	18	2	6	10	2	2	6			2		
59	Praseodymium	2	8	18	2	6	10	3	2	6			2		
60	Neodymium	2	8	18	2	6	10	4	2	6			2		
61	Promethium	2	8	18	2	6	10	5	2	6			2		
62	Samarium	2	8	18	2	6	10	6	2	6			2		
63	Europium	2	8	18	2	6	10	7	2	6			2		
64	Gadolinium	2	8	18	2	6	10	7	2	6	1		2		
65	Terbium	2	8	18	2	6	10	9	2	6			2		
66	Dysprosium	2	8	18	2	6	10	10	2	6			2		
67	Holmium	2	8	18	2	6	10	11	2	6			2		
68	Erbium	2	8	18	2	6	10	12	2	6			2		
69	Thulium	2	8	18	2	6	10	13	2	6			2		
70	Ytterbium	2	8	18	2	6	10	14	2	6			2		
71	Lutetium	2	8	18	2	6	10	14	2	6	1		2		
72	Hafnium	2	8	18	2	6	10	14	2	6	2		2		
73	Tantalum	2	8	18	2	6	10	14	2	6	3		2		
74	Tungsten	2	8	18	2	6	10	14	2	6	4		2		
75	Rhenium	2	8	18	2	6	10	14	2	6	5		2		
76	Osmium	2	8	18	2	6	10	14	2	6	6		2		
77	Iridium	2	8	18	2	6	10	14	2	6	9				
78	Platinum	2	8	18	2	6	10	14	2	6	9		1		
79	Gold	2	8	18	2	6	10	14	2	6	10		1		
80	Mercury	2	8	18	2	6	10	14	2	6	10		2		
81	Thallium	2	8	18	2	6	10	14	2	6	10		2	1	
82	Lead	2	8	18	2	6	10	14	2	6	10		2	2	
83	Bismuth	2	8	18	2	6	10	14	2	6	10		2	3	
84	Polonium	2	8	18	2	6	10	14	2	6	10		2	4	
85	Astatine	2	8	18	2	6	10	14	2	6	10		2	5	
86	Radon	2	8	18	2	6	10	14	2	6	10		2	6	

Electronic Configurations of the Elements

Atomic number	Shell Element	K n = 1	L 2	M 3	N 4	O 5				P 6			Q 7
						5s	5p	5d	5f	6s	6p	6d	7s
87	Francium	2	8	18	32	2	6	10		2	6		1
88	Radium	2	8	18	32	2	6	10		2	6		2
89	Actinium	2	8	18	32	2	6	10		2	6	1	2
90	Thorium	2	8	18	32	2	6	10		2	6	2	2
91	Protoactinium	2	8	18	32	2	6	10	2	2	6	1	2
92	Uranium	2	8	18	32	2	6	10	3	2	6	1	2
93	Neptunium	2	8	18	32	2	6	10	4	2	6	1	2
94	Plutonium	2	8	18	32	2	6	10	6	2	6		2
95	Americium	2	8	18	32	2	6	10	7	2	6		2
96	Curium	2	8	18	32	2	6	10	7	2	6	1	2

(Continued)

Table III.3 (Continued)

Atomic number	Element	Shell K n = 1	L 2	M 3	N 4	O 5 5s	5p	5d	5f	P 6 6s	6p	6d	Q 7 7s
97	Berkelium	2	8	18	32	2	6	10	8	2	6	1	2
98	Californium	2	8	18	32	2	6	10	10	2	6		2
99	Einsteinium	2	8	18	32	2	6	10	11	2	6		2
100	Fermium	2	8	18	32	2	6	10	12	2	6		2
101	Mendelevium	2	8	18	32	2	6	10	13	2	6		2
102	Nobelium	2	8	18	32	2	6	10	14	2	6		2
103	Lawrencium	2	8	18	32	2	6	10	14	2	6	1	2

Table III.4 Physical Properties of the Elements

m.p.: melting point. s: sublimes. (p): melts under pressure. Values in parenthesis are estimated.
b.p.: boiling point under atmospheric pressure.
^{a}: primary and secondary fixed points on the International Practical Temperature Scale of 1968 as amended 1975.
Values of the density, ρ, specific heat capacity, C_p, and electrical conductivity, κ, are at 25 °C. Values at 0 °C are indicated by ^{b}.
^{c}: this property refers to ^{147}Pm
For more extensive data see: *Dictionary of Inorganic Compounds* (1992), Chapman and Hall, London
Handbook of Chemistry and Physics (1991–2), CRC.

Element	m.p. /°C	b.p. /°C	ρ kg m^{-3}	C_p J K^{-1} kg^{-1}	κ MS m^{-1}	Oxidation states
actinium	1 050	3 200	10 070	119.8		3
aluminium	660.46^a	2 467	2 698.9	904.3	37.7	3
americium	994	2 607	13 670	(140)	0.7	3, 4, 5, 6
antimony	630.755^a	1 635	6 691	207.0	2.6^b	3, 5
argon	−189.2	−185.856	1.783 7	520.7		
arsenic	817 (p)	613 (s)	5 727	343.1	3.8	3, 5
astatine	(302)	377		(140)		
barium	729	1 637	3 500	284.5	2.8	2
berkelium			14 790			3, 4
beryllium	1 278	2 970	1 848	1 882.8	25	2
bismuth	271.442^a	1 551	9 747	122.0	0.9^b	3, 5
boron	2 027	3 802 (s)	2 340	1 293.8	5×10^{-12}	3
bromine	−7.2	58.78	3 120	473.7	10^{-16}	1, 3, 4, 5, 6
cadmium	321.108^a	765	8 650	231.3	14.7^b	2
caesium	28.40	666	1 873	217.6	5.3	1
calcium	839	1 492	1 540	623.4	31.3	2
carbon	3 827	4 200	2 250 (graph) 3 510 (diamond)	719.6 523.0	0.07 10^{-17}	2, 4
cerium	799	3 257	6 773	192.3	1.4	3, 4
chlorine	−100.98	−34.6	3.214^b	478.1		1, 3, 4, 5, 6, 7
chromium	1 875	2 482	7 194	448.1	7.9	2, 3, 6
cobalt	1 495^a	2 956	8 900	441.8	17.9^b	2, 3
copper	1 084.88^a	2 582	8 960	384.9	60.7	1, 2
dysprosium	1 410	2 335	8 550	173.3	1.1	3
erbium	1 520	2 510	9 066	168.1^b	1.2	3
europium	826	1 440	5 243	175.7	1.1	2, 3
fluorine	−219.6	−188.1	1.696^b	827.6		1
gadolinium	1 313	3 270	7 886	297.1^b	0.8	3
gallium	29.8	2 237	5 905	330.5	1.8	3
germanium	937	2 830	5 323	309.6	3×10^{-6}	4
gold	1 064.43^a	2 967	19 320	129.0	48.8^b	1, 3
hafnium	2 227	4 606	13 276	147.3	3.4	4
helium	−269.65	−268.93	0.178 5^b	5 196.6		
holmium	1 474	2 720	8 797	164.6	1.1	3
hydrogen	−258.14	−252.87^a	0.089 88	14 393.0		1
indium	156.634^a	2 006	7 290	242.7	3.4	1, 3
iodine	113.5	184.35	4 953	217.6	10^{-11}	1, 3, 5, 7
iridium	2 447^a	4 547	22 550	130.6	21.3^b	2, 3, 4, 6
iron	1 535	2 887	7 874	449.4	11.2	2, 3

(Continued)

Table III.4 (Continued)

Element	m.p. /°C	b.p. /°C	ρ kg m⁻³	C_p J K⁻¹ kg⁻¹	κ MS m⁻¹	Oxidation states
krypton	−156.6	−152.9	3.733[b]	248.2		2
lanthanum	920	3 457	6 145	200.8	1.9	3
lead	327.502[a]	1 749	11 343	133.9	4.8	2, 4
lithium	180.5	1 317	534	3 305.4	11.7[b]	1
lutetium	1 652	3 315	9 842	153.5[b]	1.5	3
magnesium	648.8	1 100	1 738	1 025.0	22.4	2
manganese	1 244	2 100	7 473	478.7	0.5	2, 3, 4, 6, 7
mercury	−38.836[a]	356.66[a]	13 546	139.6	1.0	1, 2
molybdenum	2 623[a]	5 560	10 222	276.1	17.3	2, 3, 4, 5, 6
neodymium	1 021	3 027	7 003	209.2[b]	1.6	3
neon	−248.67	−246.048[a]	0.899 9[b]	1 030.0		
neptunium	640	(3 900)	20 450		0.8	3, 4, 5, 6
nickel	1 455[a]	2 782	8 902	456.0	14.6	2, 3
niobium	2 477[a]	4 540	8 578	264.8[b]	6.6[b]	3, 4
nitrogen	−209.86	−195.806[a]	1.240 9	1 038.8		1, 2, 3, 4, 5
osmium	3 045	5 027	22 570	130.0	12.3	2, 3, 4, 6, 8
oxygen	−218.4	−182.962[a]	1.429[b]	918.8		2
palladium	1 554[a]	2 927	11 995	244.4	10.0[b]	2, 4
phosphorus (white)	44.1	280.4	1 820	790.8	10^{-16}	3, 5
(red)	590 (p)	417 (s)	2 340			
platinum	1 769[a]	3 827	21 447	132.5[b]	9.4	2, 4, 6
plutonium	640	3 454	19 814	138.1	0.7	3, 4, 5, 6
polonium	254	960	9 320	(125)	0.7	2, 4
potassium	63.65	754	862	736.4	16.4[b]	1
praseodymium	931	3 343	6 773	188.3	1.5	3, 4
promethium	1 080[c]	(3 300[c])	7 220[c]	(167)	2	3
protactinium	1 425	4 410	15 370	(121)	5.6	4, 5
radium	700	1 630	(5 000)	119.5	1.0	2
radon	−71	−61.8	9.73	93.7		
rhenium	3 160	5 762	21 020	136.9	5.8	2, 4, 5, 6, 7
rhodium	1 963[a]	3 687	12 410	246.4	23.0[b]	2, 3, 4
rubidium	38.89	688	1 532	334.7	47.8	1
ruthenium	2 310	4 200	12 450	230.1[b]	14.9	3, 4, 5, 6, 8
samarium	1 077	1 870	7 520	175.7	1.1	2, 3
scandium	1 541	2 832	2 989	556.5	1.5	3
selenium	217	684.9	4 810	338.9	8[b]	2, 4, 6
silicon	1 410	2 480	2 330	677.8	4×10^{-4}	4
silver	961.93[a]	2 195	10 500	235.5[b]	62.9	1
sodium	97.81	883	971	1 235.3	20.1	1
strontium	769	1 384	2 583	286.5	5.0	2
sulphur (α)	112.8	444.674[a]	2 070	732.2	5×10^{-16}	2, 4, 6
(β)	119.0		1 957			
tantalum	2 996	5 487	16 654	142.3	8.1	5
technetium	2 172	5 030	11 500	(243)	10^{-3}	7
tellurium	449.5	989.8	6 240	196.6	2×10^{-4}	2, 4, 6
terbium	1 356	3 041	8 267	171.5	0.9	3, 4
thallium	303.5	1 457	11 850	129.7	5.6[b]	1, 3
thorium	1 750	4 227	11 720	142.2	7.1	3, 4
thulium	1 545	1 947	9 318	160.0	1.3	2, 3
tin	231.968 1[a]	2 270	7 280 (white)	225.9	8.7	2, 4
			5 770 (grey)	206.3	0.3[b]	
titanium	1 660	3 313	4 508	521.9	2.6	2, 3, 4
tungsten	3 422[a]	5 727	19 254	134.3	18.2	2, 4, 5, 6
uranium	1 132.3	3 677	18 950	117.1	3.6	3, 4, 5, 6
vanadium	1 890	3 000	6 110	502.1	4.0	2, 3, 4, 5
xenon	−111.9	−108.1	5.887	158.4		2, 4, 6, 8
ytterbium	819	1 193	6 965	146.4	3.7	2, 3
yttrium	1 502	3 397	4 472	297.1	1.8	3
zinc	419.58[a]	907	7 135	382.8	16.9	2
zirconium	1 852	4 377	6 506	275.7	2.3	2, 3, 4

Table III.5 Crystal Structure, Atomic Radii and Ionic Radii of the Elements

Structure of solid: bcc: body-centred cubic; c: cubic; com: complex; d: diamond; fcc: face-centred cubic; hex: hexagonal; hcp: hexagonal close packing; monoclin: monoclinic; rhomb: rhombic; tetr: tetragonal.

Atomic radius: half the distance of closest approach of the atoms in the structure of the element.

Ionic radius: effective size of an ion in a crystal, as determined by Pauling, for coordination number 6.

Figures in parentheses indicate the charge on the ion.

LS: low spin state; HS: high spin state.

For more extensive data see: L. Pauling, *The Chemical Bond*, Cornell University Press.
                      M.F.C. Ladd, *Structure and Bonding in Solid State Chemistry* (1979), Ellis-Horwood, Chichester.
                      F.A. Cotton and G. Wilkinson, *Advanced Inorganic Chemistry*, 5th ed. (1989), Wiley, London.

Element	Structure	Atomic radius /pm	Ionic radius /pm
actinium	fcc	203	111 (3+)
aluminium	fcc	143	50 (3+)
americium	hcp	182	107 (3+)
antimony	rhomb	182	76 (3+); 62 (5+)
argon	fcc	191	
arsenic	rhomb	116	69 (3+); 50 (5+)
astatine			62 (7+); 230 (1−)
barium	bcc	210	135 (2+)
beryllium	hcp	105	31 (2+)
bismuth	rhomb	146	102 (3+); 74 (5+)
boron	hcp	97	23 (3+)
bromine	rhomb (s)	114	196 (1−)
cadmium	hcp	149	95 (2+)
caesium	bcc	265	169 (1+)
calcium	fcc	197	99 (2+)
carbon	d	77	15 (4+)
cerium	fcc	183	111 (3+); 101 (4+)
chlorine	Cl_2	99	181 (1−)
chromium	bcc	125	61 (3+); 52 (6+)
cobalt	hcp	125	57 (2+, LS); 74 (2+, HS) 53 (3+, LS); 61 (3+, HS)
copper	fcc	128	96 (1+); 72 (2+)
dysprosium	hcp	177	91 (3+)
erbium	hcp	176	104 (3+)
europium	bcc	204	113 (3+)
fluorine	F_2	67	133 (1−)
gadolinium	hcp	179	111 (3+)
gallium	com	133	62 (3+)
germanium	d	122	53 (4+)
gold	fcc	144	137 (1+)
hafnium	hcp	154	82 (4+)
helium	hcp		
holmium	hcp	177	105 (3+)
hydrogen	hcp	37	208 (1−)
indium	tetr	151	80 (3+)
iodine	rhomb	138	218 (1−)
iridium	fcc	136	64 (4+)
iron	bcc	125	61 (2+, LS); 78 (2+, HS); 55 (3+, LS); 65 (3+, HS)
krypton	fcc	201	
lanthanum	hex	187	106 (3+)
lead	fcc	175	132 (2+); 78 (4+)
lithium	bcc	156	60 (1+)
lutetium	hcp	174	99 (3+)
magnesium	hcp	160	65 (2+)
manganese	bcc	118	67 (2+, LS); 82 (2+, HS)
mercury	rhomb	147	110 (2+)
molybdenum	bcc	136	92 (3+); 68 (4+); 60 (6+)

(Continued)

Table III.5 (Continued)

Element	Structure	Atomic radius /pm	Ionic radius /pm
neodymium	hex	181	115 (3+)
neon	fcc		
neptunium	rhomb	150	102 (3+)
nickel	fcc	124	70 (2+)
niobium	bcc	147	70 (5+)
nitrogen	hcp	71	12 (5+)
osmium	hcp	134	68 (4+)
oxygen	c	60	138 (2−)
palladium	fcc	138	66 (4+)
phosphorus	c (white)	110	35 (5+)
platinum	fcc	139	64 (4+)
plutonium	monoclin	162	100 (3+); 86 (4+)
polonium	c	167	102 (4+); 67 (6+)
potassium	bcc	231	133 (1+)
praesodymium	hex	183	116 (3+); 92 (4+)
promethium	hex		98 (3+)
protactinium	tetr	161	106 (3+)
radium		235	144 (2+)
radon	fcc	132	
rhenium	hcp	138	56 (7+)
rhodium	fcc	134	68 (3+)
rubidium	bcc	243	148 (1+)
ruthenium	hcp	132	65 (4+)
samarium	rhomb	180	113 (3+)
scandium	hcp	161	81 (3+)
selenium	hex	115	198 (2−); 42 (6+)
silicon	c	117	41 (4+)
silver	fcc	144	113 (1+)
sodium	bcc	186	95 (1+)
strontium	fcc	215	113 (2+)
sulphur	rhomb	104	184 (2−); 29 (6+)
tantalum	bcc	143	68 (5+)
technetium	hcp	135	58 (7+)
tellurium	hex	137	56 (6+)
terbium	hcp	178	109 (3+)
thallium	hcp	199	150 (1+); 95 (3+)
thorium	fcc	181	102 (4+)
thulium	hcp	175	104 (3+)
tin	c (grey) tetr (white)	140	71 (4+)
titanium	hcp	154	90 (2+); 68 (4+)
tungsten	bcc	137	66 (4+)
uranium	rhomb	138	97 (4+); 75 (6+)
vanadium	bcc	131	59 (5+)
xenon	fcc		
ytterbium	fcc	194	113 (2+); 100 (3+)
yttrium	hcp	180	106 (3+)
zinc	hcp	133	74 (2+)
zirconium	hcp	161	80 (4+)

Table III.6 Van der Waals Radii of the Elements
For further information see: M.F.C. Ladd, *Structure and Chemical Bonding in Solid State Chemistry* (1979), Ellis Horwood, Chichester.

Element	H				
Radius/pm	120				

Element	N	O	F	Ne
Radius/pm	150	140	135	160

Element	P	S	Cl	Ar
Radius/pm	190	185	180	192

Element	As	Se	Br	Kr
Radius/pm	200	200	195	197

Element	Sb	Te	I	Xe
Radius/pm	220	220	215	217

Methyl group, CH_3, and methylene group, CH_2: 200 pm
Half-thickness of aromatic nucleus: 185 pm

Table III.7 Ionization Energies of the Elements
(a) The first ionization energy is the minimum energy required to remove an electron from an isolated atom in the gaseous state; ie. it is the increase in energy for the process:
$$M(g) \rightarrow M^+(g) + e$$
(b) The second ionization energy is the minimum energy required to remove an electron from a unipositive ion in the gaseous state:
$$M^+(g) \rightarrow M^{2+}(g) + e$$
Higher ionization energies are defined in a similar manner.
The tabulated values of ΔU/kJ mol^{-1} strictly relate to 0 K; conversion to ΔH/kJ mol^{-1} at 298 K requires the addition of (5RT/2) kJ mol^{-1}, approximately 6 kJ mol^{-1}. The corresponding ionization potential (eV) can be calculated by multiplying the ionization energy by 1.025×10^{-2}.
Reference: *Handbook of Chemistry and Physics* (1988–89), CRC.

Atomic number	Element	ΔU/kJ mol^{-1}							
		1st	2nd	3rd	4th	5th	6th	7th	8th
1	H	1 310							
2	He	2 370	5 250						
3	Li	519	7 300	11 800					
4	Be	900	1 760	14 800	21 000				
5	B	799	2 420	3 660	25 000	32 800			
6	C	1 090	2 350	4 610	6 220	37 800	47 000		
7	N	1 400	2 860	4 590	7 480	9 440	53 200	64 300	
8	O	1 310	3 390	5 320	7 450	11 000	13 300	71 000	84 100
9	F	1 680	3 370	6 040	8 410	11 000	15 100	17 900	91 600
10	Ne	2 080	3 950	6 150	9 290	12 100	15 200	20 000	23 000
11	Na	494	4 560	6 940	9 540	13 400	16 600	20 100	25 500
12	Mg	736	1 450	7 740	10 500	13 600	18 000	21 700	25 600
13	Al	577	1 820	2 740	11 600	14 800	18 400	23 400	27 500
14	Si	786	1 580	3 230	4 360	16 000	20 000	23 600	29 100
15	P	1 060	1 900	2 920	4 960	6 280	21 200	25 900	30 500
16	S	1 000	2 260	3 390	4 540	6 990	8 490	27 100	31 700
17	Cl	1 260	2 300	3 850	5 150	6 540	9 330	11 000	33 600
18	Ar	1 520	2 660	3 950	5 770	7 240	8 790	12 000	13 800
19	K	418	3 070	4 600	5 860	7 990	9 620	11 400	14 900
20	Ca	590	1 150	4 940	6 480	8 120	10 700	12 300	14 600
21	Sc	632	1 240	2 390	7 110	8 870	10 700	13 600	15 300
22	Ti	661	1 310	2 720	4 170	9 620	11 600	13 600	17 000
23	V	648	1 370	2 870	4 600	6 280	12 400	14 600	16 700
24	Cr	653	1 590	2 990	4 770	7 070	8 700	16 600	17 700
25	Mn	716	1 510	3 250	5 190	7 360	9 750	11 500	18 800
26	Fe	762	1 560	2 960	5 400	7 620	10 100	12 800	14 600

(Continued)

Table III.7 (Continued)

Atomic number	Element	ΔU/kJ mol⁻¹							
		1st	2nd	3rd	4th	5th	6th	7th	8th
27	Co	757	1 640	3 230	5 100	7 910	10 500	13 300	16 400
28	Ni	736	1 750	3 390	5 400	7 620	10 900	13 800	17 000
29	Cu	745	1 960	3 550	5 690	7 990	10 500	14 300	17 500
30	Zn	908	1 730	3 828	5 980	8 280	11 000	13 900	18 100
31	Ga	577	1 980	2 960	6 190	8 700	11 400	14 400	17 700
32	Ge	762	1 540	3 300	4 390	8 950	11 900	14 900	18 200
33	As	966	1 950	2 730	4 850	6 020	12 300	15 400	18 900
34	Se	941	2 080	3 090	4 140	7 030	7 870	16 000	19 500
35	Br	1 140	2 080	3 460	4 850	5 770	8 370	10 000	20 300
36	Kr	1 350	2 370	3 560	5 020	6 370	7 570	10 700	12 200
37	Rb	402	2 650	3 850	5 110	6 850	8 300	9 800	12 900
38	Sr	548	1 060	4 120	5 440	6 940	9 000	10 500	12 200
39	Y	636	1 180	1 980	6 000	7 400	9 100	11 300	13 000
40	Zr	669	1 270	2 220	3 310	8 000	9 600	11 000	14 000
41	Nb	653	1 380	2 430	3 690	4 850	10 000	12 000	14 000
42	Mo	694	1 560	2 620	4 480	5 400	7 100	12 000	15 000
43	Tc	699	1 470	2 800	4 100	5 900	7 500	9 200	15 000
44	Ru	724	1 620	2 740	4 500	6 300	7 900	9 600	11 000
45	Rh	745	1 740	3 000	4 400	6 300	8 400	10 000	12 000
46	Pd	803	1 870	3 180	4 730	6 300	8 800	10 000	13 000
47	Ag	732	2 070	3 360	5 000	6 700	8 400	11 000	13 000
48	Cd	866	1 630	3 620	5 300	7 000	9 100	11 100	14 100
49	In	556	1 820	2 700	5 230	7 400	9 500	11 700	13 900
50	Sn	707	1 410	2 940	3 930	7 780	9 900	12 200	14 600
51	Sb	833	1 590	2 440	4 270	5 360	10 400	12 700	15 200
52	Te	870	1 800	3 010	3 680	5 860	7 000	13 200	15 800
53	I	1 010	1 840	3 000	4 030	5 000	7 400	8 700	16 400
54	Xe	1 170	2 050	3 100	4 300	5 800	8 000	9 800	12 200
55	Cs	376	2 420	3 300	4 400	6 000	7 100	8 300	11 300
56	Ba	502	966	3 390	4 700	6 000	7 700	9 000	10 200
57	La	540	1 100	1 850	5 000	6 400	7 700	9 600	11 000
72	Hf	531	1 440	2 010	3 010				
73	Ta	760	1 560	2 150	3 190	4 350			
74	W	770	1 710	2 330	3 420	4 600	5 900		
75	Re	762	1 600	2 500	3 600	5 000	6 300	7 500	
76	Os	841	1 630	2 400	3 800	5 000	6 700	8 000	9 600
77	Ir	887	1 550	2 600	3 800	5 400	7 100	8 400	10 000
78	Pt	866	1 870	2 750	3 970	5 400	7 200	8 800	10 500
79	Au	891	1 980	2 940	4 200	5 400	7 100	9 200	11 000
80	Hg	1 010	1 810	3 300	7 000	7 900			
81	Tl	590	1 970	2 870	4 900	6 200	7 800	9 500	11 300
82	Pb	716	1 450	3 080	4 080	6 700	8 100	9 900	11 800
83	Bi	703	1 610	2 460	4 350	5 400	8 500	10 300	12 300
84	Po	812							
85	At	920							
86	Rn	1 040	1 930	2 890	4 250	5 310			
87	Fr	381							
88	Ra	510	979						
89	Ac	669	1 170						
90	Th	674	1 110	1 930	2 760				
91	Pa								
92	U	385							

Table III.8 Atomic Electron Affinities at 0 K

The electron affinity of an element is the energy released when an electron is added to an isolated atom in the gaseous state:

$$X(g) + e = X^-(g)$$

The energy values, ΔU, strictly relate to 0 K and conversion to enthalpy value, ΔH, at 298 K requires the addition of $(5RT/2)$ kJ mol^{-1}, i.e. approximately 6 kJ mol^{-1}. The data are limited to those negative ions, which by virtue of their positive electron affinity, are stable.
For more extensive data see: H. Hotop and W. C. Lineberger, *J. Phys. Chem. Ref. Data*, (1975), **4**, 539; B. M. Smirnov, *Negative Ions* (1981), McGraw-Hill, New York.

Element	Electron affinity at 0 K /eV	ΔU/kJ mol^{-1}	Element	Electron affinity at 0 K /eV	ΔU/kJ mol^{-1}
aluminium	0.44	42.5	mercury	negative ion does not exist	
antimony	1.1	106	molybdenum	0.75	72
argon	negative ion does not exist		neon	negative ion does not exist	
arsenic	0.80	77	nickel	1.15	111
astatine	2.9	280	niobium	0.89	86
barium	negative ion does not exist		nitrogen	negative ion does not exist	
beryllium	negative ion does not exist		osmium	1.4	135
bismuth	0.95	92	oxygen	1.46	141
boron	0.28	27	palladium	0.56	54
bromine	3.364	324.6	phosphorus	0.746	72
cadmium	negative ion does not exist		platinum	2.13	205
caesium	0.471 6	46	polonium	1.9	183
calcium	negative ion does not exist		potassium	0.501 5	48
carbon	1.263	122	rare earths (estimate)	<~0.5	
chlorine	3.615	348.8	rhenium	0.15	15
chromium	0.67	65	rhodium	1.14	110
cobalt	0.66	64	rubidium	0.485 9	47
copper	1.23	118	ruthenium	1.1	106
fluorine	3.399	328.0	scandium	0.19	18
francium	(0.46)	(44)	selenium	2.021	195
gallium	0.3	29	silicon	1.385	134
germanuim	1.2	116	silver	1.30	126
gold	2.308 6	223	sodium	0.547 9	53
hafnium	negative ion does not exist		strontium	negative ion does not exist	
helium	0.077	7.4	sulphur	2.077 12	210.4
hydrogen	0.754 2	72.8	tantalum	0.32	31
indium	0.3	29	technetium	0.68	58
iodine	3.059	295	tellurium	1.971	190
iridium	1.57	151	thallium	0.3	31
iron	0.16	15	tin	1.2	116
krypton	negative ion does not exist		titanium	0.08	8
lanthanum	0.5	48	tungsten	0.82	79
lead	0.37	36	vanadium	0.53	51
lithium	0.618	60	xenon	negative ion does not exist	
magnesium	negative ion does not exist		yttrium	0.31	30
manganese	negative ion does not exist		zinc	negative ion does not exist	
			zirconium	0.43	41

Table III.9 Proton Affinities of Atoms

The proton affinity of an atom is defined as the energy released in the protonation reaction: $A + H^+ \rightarrow AH^+$
Reference: S.G. Lias, J.F. Liebman and R.D. Levin, *J.Phys. Chem. Ref. Data* (1984), *13*, 695.

Atom	Proton affinity/eV	Uncertainty (+/–) /eV	Atom	Proton affinity/eV	Uncertainty (+/–) /eV
Ar	3.87	0.04	I	6.3	0.2
Br	5.7	0.2	Kr	4.4	0.15
Cl	5.3	0.2	N	3.4	0.3
Cs	7.6	0.8	Ne	2.08	0.06
F	3.42	0.03	O	5.1	0.2
H	2.650	0.02	S	6.9	0.7
He	1.845	0.02	Xe	5.1	0.5

Table III.10 Electronegativities of the Elements

Allred-Rochow values from *J. Inorg. Nuclear Chem.* (1958), **5**, 264; Pauling-type values from A.I. Allred, *J. Inorg. Nuclear Chem.* (1961), **17**, 215; Mulliken-type values from H.O. Pritchard and H.A. Skinner, *Chem. Rev.* (1955), **55**, 745. (Figures in parentheses are oxidation states used for the Pauling-type values.)

Element	Allred-Rochow formula	Pauling method	Mulliken method	Element	Allred-Rochow formula	Pauling method	Mulliken method
H	2.20			Cd	1.46	1.69 (2)	1.4
He				In	1.49	1.78 (3)	1.80
Li	0.97	0.98 (1)	0.94	Sn	1.72	1.96 (4)	
Be	1.47	1.57 (2)	1.46	Sb	1.82	2.05 (3)	1.65
B	2.01	2.04 (3)	2.01	Te	2.01	2.1 (2)	2.10
C	2.50	2.55 (4)	2.63	I	2.21	2.66 (1)	2.56
N	3.07	3.04 (3)	2.33	Xe			
O	3.50	3.44 (2)	3.17	Cs	0.86	0.79 (1)	
F	4.10	3.98 (1)	3.91	Ba	0.97	0.89 (2)	
Ne				La	1.08		1.10
Na	1.10	0.93 (1)	0.93	Ce	1.06		1.12
Mg	1.23	1.31 (2)	1.32	Pr	1.07		1.13
Al	1.47	1.61 (3)	1.81	Nd	1.07		1.14
Si	1.74	1.90 (4)	2.44	Pm	1.07	1.2 (2)	
P	2.06	2.19 (3)	1.81	Sm	1.07		1.17
S	2.44	2.58 (2)	2.41	Eu	1.01	1.2 (2)	
Cl	2.83	3.16 (1)	3.00	Gd	1.11		1.20
Ar				Tb	1.10	1.2 (2)	
K	0.91	0.82 (1)	0.80	Dy	1.10		1.22
Ca	1.04	1.00 (2)		Ho	1.10		1.23
Sc	1.20	1.36 (3)		Er	1.11		1.24
Ti	1.32	1.54 (4)		Tm	1.11		1.25
V	1.45	1.63 (2)		Yb	1.06	1.1 (2)	
Cr	1.56	1.66 (2)		Lu	1.14		1.27
Mn	1.60	1.55 (2)		Hf	1.23	1.3 (4)	
Fe	1.64	1.83 (2)		Ta	1.33	1.5 (2)	
Co	1.70	1.88 (2)		W	1.40	2.36 (2)	
Ni	1.75	1.91 (2)		Re	1.46	1.9 (2)	
Cu	1.75	1.90 (2)	1.36	Os	1.52	2.2 (2)	
Zn	1.66	1.65 (2)	1.49	Ir	1.55	2.20 (2)	
Ga	1.82	1.81 (3)	1.95	Pt	1.44	2.28 (2)	
Ge	2.02	2.01 (4)		Au	1.42	2.54 (2)	
As	2.20	2.18 (3)	1.75	Hg	1.44	2.00 (2)	
Se	2.48	2.55 (2)	2.23	Tl	1.44	2.04 (3)	
Br	2.74	2.96 (1)	2.76	Pb	1.55	2.33 (4)	
Kr				Bi	1.67	2.02 (3)	
Rb	0.89	0.82 (1)		Po	1.76	2.0 (2)	
Sr	0.99	0.95 (2)		At	1.96	2.2 (1)	
Y	1.11	1.22 (3)		Rn			
Zr	1.22	1.33 (4)		Fr	0.86	0.7 (1)	
Nb	1.23	1.6 (2)		Ra	0.97	0.9 (2)	
Mo	1.30	2.16 (2)		Ac	1.00		
Tc	1.36	1.9 (2)		Th	1.11		
Ru	1.42	2.2 (2)		Pa	1.14		
Rh	1.45	2.28 (2)		U	1.22		1.38
Pd	1.35	2.20 (2)		Np	1.22		1.36
Ag	1.42	1.93 (2)	1.36	Pu	1.22		1.28

Table III.11 Abundances of the Elements in the Earth's Crust and in Sea Water.
Values are grams per tonne or parts per million (ppm)
For more extensive data see: *McGraw-Hill Encyclopedea of Science and Technology* (1970), 4, 627.

Element	Earth's crust	Sea water	Element	Earth's crust	Sea water
hydrogen	1 400	107 000	rhodium	0.005	
lithium	20	0.17	palladium	0.01	
beryllium	2	6×10^{-7}	silver	0.05	3×10^{-4}
boron	10	4.5	cadmium	0.2	1×10^{-4}
carbon	250	28	indium	0.05	
nitrogen	20	0.7	tin	3	8×10^{-4}
oxygen	466 000	857 000	antimony	0.2	3×10^{-4}
fluorine	650	1.3	tellurium	0.01	
sodium	28 000	10 800	iodine	0.5	0.06
magnesium	21 000	1 300	caesium	2	3×10^{-4}
aluminium	81 000	0.01	barium	400	0.02
silicon	277 000	2.9	lanthanum	30	3×10^{-6}
phosphorus	400	0.09	cerium	60	1×10^{-6}
sulphur	260	904	praseodymium	8	6×10^{-7}
chlorine	130	19 400	neodymium	30	3×10^{-6}
potassium	26 000	392	samarium	6	1×10^{-7}
calcium	36 000	411	europium	1	1×10^{-7}
scandium	20	4×10^{-5}	gadolinium	5	7×10^{-7}
titanium	4 400	1×10^{-3}	terbium	1	1×10^{-7}
vanadium	140	2×10^{-3}	dysprosium	3	9×10^{-7}
chromium	100	2×10^{-4}	holmium	1	2×10^{-7}
manganese	950	4×10^{-4}	erbium	3	9×10^{-7}
iron	50 000	4×10^{-3}	thulium	0.5	2×10^{-7}
cobalt	25	4×10^{-4}	ytterbium	3	8×10^{-7}
nickel	70	7×10^{-3}	lutetium	0.5	2×10^{-7}
copper	50	1×10^{-3}	hafnium	3	$< 8 \times 10^{-6}$
zinc	70	5×10^{3}	tantalum	2	$< 3 \times 10^{-6}$
gallium	15	3×10^{-3}	tungsten	1	$< 1 \times 10^{-6}$
germanium	1.5	6×10^{-5}	rhenium	0.001	8×10^{-6}
arsenic	1.8	3×10^{-3}	osmium	0.005	
selenium	0.1	1×10^{-4}	iridium	0.001	
bromine	2.5	67	platinum	0.01	
rubidium	90	0.12	gold	0.005	1×10^{-5}
strontium	390	8.1	mercury	0.08	2×10^{-4}
yttrium	30	1×10^{-5}	thallium	0.5	1×10^{-5}
zirconium	170	3×10^{-5}	lead	13	3×10^{-5}
niobium	20	2×10^{-5}	bismuth	0.2	2×10^{-5}
molybdenum	1.5	0.01	thorium	7	4×10^{-8}
ruthenium	0.01	7×10^{-7}	uranium	2	3×10^{-3}

Table III.12 Average Composition of the Atmosphere (exclusive of water vapour).
Reference: *Handbook of Chemistry and Physics* (1988–89), CRC.

Constituent	Content/ppm		Constituent	Content/ppm	
	By volume	By weight		By volume	By weight
N_2	780 900	755 100	CH_4	1.5	0.94
O_2	209 500	231 500	Kr	1.1	2.9
Ar	9 300	12 800	N_2O	0.5	0.8
CO_2	300	460	H_2	0.5	0.035
Ne	18	12.5	O_3[a]	0.4	0.7
He	5.24	0.72	Xe	0.08	0.36

[a] Variable, increases with height.

Table III.13 Elemental Composition of the Adult Human Body.
Reference: as for Table III.12.

Element	Percentage of total body mass	Element	Percentage of total body mass	Element	Percentage of total body mass
oxygen	61	silicon	0.026	barium	3×10^{-5}
carbon	23	iron	6.0×10^{-3}	tin	2×10^{-5}
hydrogen	10	fluorine	3.7×10^{-3}	manganese	2×10^{-5}
nitrogen	2.6	zinc	3.3×10^{-3}	iodine	2×10^{-5}
calcium	1.4	rubidium	4.6×10^{-4}	nickel	1×10^{-5}
phosphorus	1.1	strontium	4.6×10^{-4}	gold	1×10^{-5}
sulphur	0.2	bromine	2.9×10^{-4}	molybdenum	1×10^{-5}
potassium	0.2	lead	1.7×10^{-4}	chromium	3×10^{-6}
sodium	0.14	copper	1.0×10^{-4}	caesium	2×10^{-6}
chlorine	0.12	aluminium	9×10^{-5}	cobalt	2×10^{-6}
magnesium	0.027	cadmium	7×10^{-5}	uranium	1×10^{-7}
		boron	7×10^{-5}		

Section IV
General Properties of Molecules

		Page
IV.1	Hybrid orbitals	53
IV.2	Average bond lengths of diatomic molecules	53
IV.3	Bond lengths of elements	54
IV.4	Bond lengths and angles in polyatomic molecules	54
	Inorganic compounds	54
	Organic compounds	55
IV.5	Bond lengths and angles for some representative polymers	56
IV.6	Shapes of molecules and ions	57
IV.7	Electric dipole moments of organic compounds	58
IV.8	Electric dipole moments of inorganic compounds	59
IV.9	Electric dipole moments, polarizabilities and polarizability volumes of some inorganic and organic molecules	59
IV.10	Electron affinities of some diatomic molecules and radicals	60
IV.11	Lennard-Jones (12–6) potential parameters	60
IV.12	Madelung constants	60
IV.13	Character tables	61
IV.14	Bravais lattices	64
IV.15	Characteristics of seven crystal systems	65

Section IV
General Properties of Molecules

Table IV.1 Hybrid orbitals
Important forms are indicated by boldface.

Coordination number	Hybridization	Shape
2	**sp**, dp	linear
	p^2, ds, d^2	bent
3	**sp^2**, dp^2, ds^2, d^3	trigonal planar
	dsp	unsymmetrical planar
	p^3, d^{2}p	trigonal pyramidal
4	**sp^3**, d^{3}s	tetrahedral
	d^{2}sp, dp^3, d^{3}p	irregular tetrahedral
	d^4	tetragonal pyramidal
5	**dsp^3**, d^{3}sp	bipyramidal
	d^{2}sp^2, d^{4}s, d^{2}p^3, d^{4}p	tetragonal pyramidal
	d^{3}p^2	pentagonal planar
	d^5	pentagonal pyramidal
6	**d^{2}sp^3**	octahedral
	d^{4}sp, d^{5}p	trigonal prismatic
	d^{3}p^3	trigonal antiprismatic

Table IV.2 Average bond lengths of diatomic molecules
Tabulated values are for the internuclear separation r_e(pm) in the absence of vibration.
For more extensive data see: M.F.C. Ladd, *Structure and Bonding in Solid State Chemistry* (1979), Ellis Horwood, Chichester; *Handbook of Chemistry and Physics* (1988–89), CRC.

Single Bonds

	Br	C	Cl	F	H	I	N	O	P	S	Si
Br	228.4	193.7	214	176	140.8		214		218	227	216
C		154.1	176.7	138.1	109.1	213.5	147.9	143	187	181	186.5
Cl			198.8	162.8	127.5	232	197	170	204.3	199	198.9
F				141.7	91.7	257	136	142	153.5	158	156.0
H					74.13	160	100.8	97.1	142	134.6	148
I						266.2			247		243
N							109.7	115.0			157
O								120.8	165	143.2	163
P									218		
S										188.7	215
Si											235.1

Multiple Bonds

C : C 133.7 ± 0.6	N : N	125
C . C 139.5 ± 0.3 (in benzene)		
C :: C 120.4 ± 0.2	N :: N	110
C : N 115.8 ± 0.2	O : O	121
C : O 123 ± 1	S : S	188

Table IV.3 Bond Lengths of Elements

Bond lengths measured at temperatures in the range 18–25°C unless otherwise stated. bcc: body-centred cubic; fcc: face-centred cubic; hcp: hexagonal close packing.

For more extensive data see: *Handbook of Chemistry and Physics* (1988–89), CRC.

Bond	Length/pm	Bond	Length/pm
Ag–Ag	288.94	Mn–Mn (1095, γ)	273.11 (fcc)
Al–Al	286.3	(1134, δ)	266.79 (bcc)
As–As	249	Mo–Mo	272.51
Au–Au	288.41	N–N	109.758
B–B	158.9	Na–Na	371.57
Ba–Ba	434.7	Ni–Ni	249.16
Be–Be	222.60	O–O	120.8
Bi–Bi	309	O–O (O_3)	127.8
Br–Br	229.0	P–P	218
Ca–Ca (α)	394.7 (fcc)	Pb–Pb	350.03
(500, β)	387.7 (bcc)	Pd–Pd	275.11
Cd–Cd	297.88	Pt–Pt	274.6
Cl–Cl	198.8	Rb–Rb	495
Ce–Ce	365.0	Re–Re	274.1
Co–Co	250.61	Rh–Rh	269.01
Cr–Cr	249.80	Ru–Ru	265.02
Cs–Cs (−10)	530.9	S–S (S_2)	188.7
Cu–Cu	255.60	S–S (S_8)	207
F–F	141.7	Sb–Sb	290
Fe–Fe (α)	248.23 (bcc)	Se–Se	232.1
(916, γ)	257.8 (fcc)	Si–Si	235.17
(1394, δ)	253.9 (bcc)	Sn–Sn (α)	280.99
Ge–Ge	244.98	(β)	302.2
H–H	74.611	Sr–Sr	430.2
H–D	74.136	Th–Th (α)	359.5 (fcc)
D–D	74.164	(1450, β)	356 (bcc)
He–He	108.0	Ti–Ti (α)	289.56 (hcp)
Hg–Hg (−46)	300.5	(900, β)	286.36 (bcc)
I–I	266.2	U–U	277
K–K	454.4	V–V	262.24
Li–Li	303.90	W–W	274.09
Mg–Mg	319.71	Zn–Zn	266.94
		Zr–Zr	317.9

Table IV.4 Bond Lengths and Angles in Polyatomic Molecules

For more extensive data see: *Tables of Interatomic Distances and Configuration of Molecules and Ions*, Special Publication No. 18, Chemical Society, London (1958, 1965).

Figures in parentheses are uncertainties.

1. Inorganic Compounds

Compound	Formula	Bond	Length/pm	Bond	Angle /°
ammonia	NH_3	N–H	100.8(0.4)	H–N–H	107.3 (0.2)
antimony trichloride	$SbCl_3$	Sb–Cl	235.2(0.5)	Cl–Sb–Cl	99.5(1.5)
arsenic trichloride	$AsCl_3$	As–Cl	216.1(0.4)	Cl–As–Cl	98.4(0.5)
arsenic trioxide	As_4O_6	As–O	178(2)	O–As–O	99.0(2)
				As–O–As	128.0(2)
arsine	AsH_3	As–H	151.9(0.2)	H–As–H	91.83(0.3)
bismuth trichloride	$BiCl_3$	Bi–Cl	248(2)	Cl–Bi–Cl	100.0(6)
carbon dioxide	CO_2	C–O	115.98	O–C–O	180
carbon disulphide	CS_2	C–S	155.30	S–C–S	180
chlorine dioxide	ClO_2	Cl–O	149	O–Cl–O	118.5
chlorogermane	GeH_3Cl	Ge–H	152(3)	H–Ge–H	110.9(1.5)
chlorosilane	SiH_3Cl	Si–H	148.3(0.1)	Cl–Si–Cl	110.0(0.03)
		Si–Cl	204.79(0.07)		
dichlorosilane	SiH_2Cl_2	Si–H	146		
		Si–Cl	202(3)	Cl–Si–Cl	110(1)

(Continued)

Table IV.4 (Continued)

Compound	Formula	Bond	Length/pm	Bond	Angle /°
hydrogen phosphide	PH_3	P–H	141.5(0.3)	H–P–H	93.3(0.2)
hydrogen sulphide	H_2S	S–H	134.55	H–S–H	93.3
hydrogen peroxide	H_2O_2	O–H	96.0(0.5)	O–O–H	100(2)
		O–O	148(1)		
nitrosyl bromide	NOBr	O–N	115(4)	Br–N=O	117(3)
		N–Br	214(2)		
nitrosyl chloride	NOCl	O–N	114(2)	Cl–N=O	113(2)
		N–Cl	197(1)		
trans-nitrous acid	HNO_2	H–O	98	O–N=O'	118(2)
		O–N	146		
		N–O'	120		
oxygen chloride	OCl_2	O–Cl	170(2)	Cl–O–Cl	110.8(1)
ozone	O_3	O–O	127.8(3)	O–O–O	116.8(0.5)
phosphorus oxycloride	$POCl_3$	P–Cl	199(2)	Cl–P–Cl	103.5(1)
phosphorus trichloride	PCl_3	P–Cl	204.3(0.3)	Cl–P–Cl	100.1(0.3)
sulphur dioxide	SO_2	S–O	143.21	O–S–O	119.536
sulphur trioxide	SO_3	S–O	143	O–S–O	120
sulphuryl chloride	SO_2Cl_2	S–O	134(2)	O–S–O	119.75(5)
		S–Cl	199(2)	Cl–S–Cl	111.20(2)
				Cl–O–O	106.5(2)
water	H_2O	O–H	95.84	H–O–H	104.45

2. Organic Compounds

Compound	Formula	Bond	Length/pm	Bond	Angle /°
acetaldehyde	CH_3CHO	C–H	109	C–C=O	121(2)
		C–C	150(2)		
		C–O	122(2)		
acetone	$(CH_3)_2CO$	C–C	151.5	C–C–C	116.22
		C–O	121.5	C–C–O	121.9
acetylene	C_2H_2	C–H	105.9	H–C–C	180
		C–C	120.4(0.2)		
benzene	C_6H_6	C–H	108.4	H–C–C	120
		C–C	142.10(0.01)	C–C–C	120
carbon tetrachloride	CCl_4	C–Cl	176.6(0.3)	Cl–C–Cl	109.5
		Cl–Cl	288.7(0.4)		
chloromethane	CH_3Cl	C–H	111(1)	H–C–H	110(2)
		C–Cl	178.4(0.3)		
dichloromethane	CH_2Cl_2	C–H	106.8(0.5)	H–C–H	112(0.3)
		C–Cl	177.24(0.05)	Cl–C–Cl	111.8
dimethylether	$(CH_3)_2O$	C–O	143(3)	C–O–C	110(3)
ethane	C_2H_6	C–H	110.7	H–C–H	109.3
		C–C	153.6		
ethylene	C_2H_4	C–H	108.4	H–C–H	115.5
		C–C	133.7(0.6)		
formaldehyde	HCHO	C–H	106.0(3.8)	H–C–H	125.8(7)
		C–O	123.0(1.7)		
glycine	H_2NCH_2COOH	C–C	152	C–C–O	119.0
		C–O	127	O–C–O	122.0
		C–N	139	C–C–N	112.0
methane	CH_4	C–H	109.1	H–C–H	109.3
methanethiol	CH_3SH	C–S	181.77(0.2)	C–S–H	100.3(0.2)
		C–H	110.39(0.2)	H–C–H	110.3(0.2)
		S–H	132.9(0.4)		
methanol	CH_3OH	C–H	109.6(1)	H–C–H	109.3(0.75)
		C–O	142.7(0.7)	C–O–H	108.8(2)
		O–H	95.6(1.5)		

(Continued)

Table IV.4 (Continued)

Compound	Formula	Bond	Length/pm	Bond	Angle /°
trimethylamine	$(CH_3)_3N$	C–N	147(1)	C–N–C	108(4)
		C–H	106	H–C–H	109.5
trimethylarsine	$(CH_3)_3As$	C–H	109	C–As–C	96(5)
		C–As	198(2)		

Table IV.5 Bond Lengths and Angles for Some Representative Polymers
For further information see: F. Khoury and B.M. Fanconi, *A Physicist's Desk Reference*, Physics Vade Mecum, 2nd ed. (1985), ed. H.L. Anderson, p. 215, Amer. Inst. Phys., New York.

Polymer	Bond	Length/pm	Bond	Angle /°
polyethene (polyethylene)	C–C	153.2	C–C–C	112.01
	C–H	105.8	H–C–H	109.29
			C–C–H	108.88
polytetrafluoroethene (PTFE)	C–C	155.3	C–C–C	113.85
	C–F	136	F–C–F	108
			C–C–F	108.7
polyamides (C* denotes amide carbon)	C–N	147	N–C*–C	109.7
	N–C*	132	N–C*–C	115.4
	C*–O	124	C*–N–C	120.9
	N–H	100	O–C*–C	121.0
			O–C*–N	123.6
			C*–N–H	123.0
			C–N–H	116.1
polystyrene C–C$_1$–C* $\quad\;$ C$_2$ $\quad\;\;\;$ C$_3$	C–C$_1$	154	C–C$_1$–C*	116
	C$_1$–C$_2$	154	C$_2$–C$_1$–C	108
	C$_2$–C$_3$	140	C$_2$–C$_1$–C*	111
polypropene (polypropylene) C–C$_1$–C* $\quad\;$ C$_2$	C–C$_1$	154	C–C$_1$–C*	114
	C$_1$–C$_2$	154	C–C$_1$–C$_2$	110
poly(ethylene oxide)	C–C	154	C–C–O	110
	C–O	143	C–O–C	112
	C–H	109	H–C–H	109.5
polybutadiene C–C$_1$–C$_2$=C$_3$	C–C$_1$	153	C–C$_1$–C$_2$	121
	C$_1$–C$_2$	154	C$_1$–C$_2$–C$_3$	142
	C$_2$=C$_3$	115		

Table IV.6 Shapes of Molecules and Ions
For more information see: A.F. Wells, *Structural Inorganic Chemistry*, 3rd ed. (1962), Oxford University Press; *Tables of Interatomic Distances and Configuration of Molecules and Ions*, Special Publication No. 18, Chemical Society (1958, 1965).

Shape	Coordination Number	Examples
Tetrahedral	4	CX_4, SiX_4 (X = Br, Cl, F, H, I), NH_4^+, BF_4^-, MnO_4^-, PO_4^{3-}, SO_4^{2-}, $Ni(CO)_4$
Square planar (90°)	4	$AuCl_4^-$, $Pt(NH_3)_4^{2+}$, ICl_4^-, XeF_4
Trigonal bipyramidal (90° and 120°)	5	AsF_5, PCl_5, $MoCl_5$, PF_5
Octahedral (90°)	6	SF_6, UF_6, PCl_6^- (six-coordinated complexes of many metals)
Trigonal prismatic	6	MoS_2, WS_2
Dodecahedral	8	$Mo(CN)_8^{4-}$
Icosahedral	12	All forms of element boron, $B_{12}H_{12}^{2-}$, carboranes of $B_{10}C_2H_{12}$-type

Tables IV.7 Electric Dipole Moments of Organic Compounds

For more extensive data see: R.D. Nelson, D.R. Lide and A.A. Maryott, *Selected Values of Electric Dipole Moments for Molecules in the Gas Phase*, National Reference Data System NSRDS — NBS 10.

Values quoted are for molecules in the gas phase; values in parentheses are for molecules in solution in benzene, B or 1,4-dioxane, D. (The c.g.s unit is the Debye; 1 D = 3.335 64 × 10^{-30} C m)

Compound	10^{30} p/(C m)	Compound	10^{30} p/(C m)
acetaldehyde	8.97 (8.30, B)	dimethylether	4.34
acetamide	12.54 (11.47, B; 13.00, D)	1, 2-dimethylbenzene	2.07
		1, 4-dimethylbenzene	0
acetic acid	5.80	dimethylsulphoxide	13.21
acetic anhydride	9.34	1, 2-dinitrobenzene	20.01
acetone	9.61	1, 3-dinitrobenzene	12.98
acetonitrile	13.08	1, 4-dinitrobenzene	0
acetophenone	10.07	1, 4-dioxane	0
acetyl chloride	9.07	diphenyl	0
acetylene	0	diphenyl ether	4.10
alkanes (unbranched)	0	ethane-1, 2-diol	7.61
aniline	5.10 (5.02, B; 5.83, D)	ethanol	5.64 (5.67, B)
		ethyl acetate	5.94
anisole	4.60	ethylamine	4.07
benzamide	12.00 (12.17, B)	ethylbenzene	1.97
benzaldehyde	9.21 (9.24, B)	ethyl benzoate	6.67
benzene	0	ethylene	0
benzonitrile	13.94	ethylene oxide	6.30
benzophenone	9.84	ethyl nitrite	8.00 (7.34, B)
benzyl alcohol	5.70	formaldehyde	7.77
bromobenzene	5.17	formamide	12.43 (11.23, B; 12.87, D)
bromoethane	6.74 (6.30, B)		
bromomethane	6.04	furan	2.20
1-butene	1.13	hydrogen cyanide	9.94
carbon tetrahalides	0	iodobenzene	5.70
carbonyl chloride	3.90	iodoethane	6.34 (5.94, B)
chlorobenzene	5.64 (5.34, B)	iodomethane	5.40 (4.70, B)
chloroethane	6.84	isoquinoline	9.11
chloro-fluoro-methane	6.07	methanethiol	5.07
-difluoro- (freon 22)	4.74	methanol	5.70 (5.53, B)
-trifluoro- (freon 13)	1.7	methyl acetate	5.74
chloroform	3.40	methylamine	4.37
chloromethane	6.24	methylethyl ether	4.10
1-chloro-2-nitrobenzene	15.48	methyl isonitrile	12.84
1-chloro-3-nitrobenzene	12.44	nitrobenzene	14.07 (13.21, B)
1-chloro-4-nitrobenzene	9.44	nitroethane	12.17
4-chlorophenol	7.00	nitromethane	11.54
2-chlorotoluene	5.20	phenol	4.84
4-chlorotoluene	7.37	1-propanol	5.60
cyanamide	14.24	2-propanol	5.54
cyclopropane	0	propene	1.22
1, 2-diaminobenzene	5.10	propionaldehyde	8.47
1, 3-diaminobenzene	6.04	propionic acid	5.84
1, 4-diaminobenzene	4.54	propylene	1.22
diazomethane	5.00	propyne	2.61
cis-1, 2-dichloroethylene	6.34	pyrazine	0
trans-1, 2-dichloroethylene	0	pyridine	7.30
dichloro-fluoro-methane-(freon 21)	4.30	pyrrole	6.14
-difluoro- (freon 12)	1.7	quinoline	7.64
dichloromethane	5.34	thiazole	5.40
1, 2-dichlorobenzene	8.34	thiophene	1.83
1, 3-dichlorobenzene	5.74	thiourea	16.31
1, 4-dichlorobenzene	0	toluene	1.20
1,1-dichloroethane	6.87	1, 1, 1-trichloroethane	5.94
1, 2-dichloroethane	3.97	triethylamine	2.20
diethylamine	3.07	trimethylamine	2.04
diethylether	3.84	urea	15.21
dimethylamine	3.44	vinyl chloride	4.87
		2-xylene	2.07

Table IV.8 Electric Dipole Moments of Inorganic Compounds

Values are in the gas phase, except where stated. B: solution in benzene; C: solution in carbon tetrachloride, D: solution in 1, 4-dioxane D; H: solution in n-hexane.

For more extensive data see: references to Table IV.7, and G.D. Moody and J.D.R. Thomas, *Dipole Moments in Inorganic Chemistry* (1971), Arnold, London.

Molecule	$10^{30}\ p/(C\ m)$	Molecule	$10^{30}\ p/(C\ m)$
aluminium tribromide	17.1 (B)	hydrogen cyanide	9.94
ammonia	4.90	hydrogen fluoride	6.40
arsenic trichloride	5.30	hydrogen iodide	1.47
arsine	0.67	hydrogen peroxide	7.44
boron trichloride	0	hydrogen sulphide	3.07
carbon dioxide	0	lithium chloride	23.78
carbon monoxide	0.37	mercury (II) bromide	3.17 (B)
carbonyl sulphide	2.17	mercury (II) chloride	4.10 (B)
chlorine dioxide	5.64 (C)	nitrogen monoxide	0.53
chlorine monofluoride	2.94	nitrogen dioxide	0.97
chlorine monoxide	2.60 (C)	oxygen	0
cyanogen bromide	9.80	ozone	2.2
deuterium oxide	6.17	pentacarbonyl iron	2.10 (B)
dicyclopentadienyl, iron (II)	0 (B)	phosphine	1.93
dicyclopentadienyl, lead (II)	5.44 (B)	phosphorus trichloride	2.60
dicyclopentadienyl, tin (II)	3.40 (B)	phosphoryl chloride	8.00 (B)
diethyl beryllium	3.33 (H)	potassium chloride	34.26
diethyl cadmium	1.00 (H)	silver chloride	19.1
diethyl magnesium	16.01 (D)	sodium chloride	30.02
diethyl mercury	0 (H)	stibine	0.40
diethyl zinc	0 (H)	sulphur monochloride	5.34 (B)
dinitrogen oxide	0.56	sulphur dioxide	5.44
dinitrogen tetroxide	1.23	sulphur trioxide	0
ethyl lithium	2.90 (B)	sulphuryl chloride	5.47 (B)
hydrazine	6.14	thionyl chloride	4.60
hydrogen	0	tin (IV) chloride	3.17 (B)
hydrogen bromide	2.67	tin (IV) iodide	0 (B)
hydrogen chloride	3.62	water	6.17

Table IV.9 Electric Dipole Moments p, Polarizabilities (α) and Polarizability Volumes (α'') of Some Inorganic and Organic Molecules

For more extensive data see: references for Table IV.8.

Molecule	$10^{30}\ p/(C\ m)$	$10^{24}\ \alpha''/cm^3$	$10^{40}\ \alpha/J^{-1}\ C^2\ m^2$
ammonia	4.90	2.22	2.47
argon	0	1.66	1.85
benzene	0	10.4	11.6
carbon dioxide	0	2.63	2.93
carbon monoxide	0.37	1.98	2.20
carbon tetrachloride	0	10.5	11.7
chloroform	3.40	8.50	9.46
chloromethane	6.24	4.53	5.04
dichloromethane	5.34	6.80	7.57
helium	0	0.20	0.22
hydrogen	0	0.819	0.911
hydrogen bromide	2.67	3.61	4.01
hydrogen chloride	3.62	2.63	2.93
hydrogen fluoride	6.40	0.51	5.67
hydrogen iodide	1.47	5.45	6.06
methane	0	2.60	2.89
methanol	5.70	3.23	3.59
nitrogen	0	1.77	1.97
water	6.17	1.48	1.65

Table IV.10 Electron Affinities of some Diatomic Molecules and Radicals

The electron affinity of a neutral diatomic molecule is the lowest energy required to remove an electron from the molecular negative ion. For more extensive data see: B.M. Smirnov *Negative Ions* (1981), McGraw-Hill, New York.

Molecule or radical	Electron affinity/eV	Molecule or Radical	Electron affinity/eV
Al_2	2.42	NH	0.38
Br_2	2.6	NS	1.19
C_2	3.39	NaCl	0.66
CH	1.24	O_2	0.44
CN	3.82	OH	1.8277
CS	0.21	OD	1.8255
Cl_2	2.44	PH	1.03
F_2	2.96	PO	1.11
I_2	2.51	S_2	1.66
		SH	2.31
		SiH	1.28

Table IV.11 Lennard–Jones (12,6)-Potential Parameters

The Lennard–Jones (12,6)-potential $V(r)$ is written in the form:

$$V(r) = 4\varepsilon\{(\sigma/r)^{12} - (\sigma/r)^6\}$$

where the parameter ε is the depth of the minimum of the curve of potential energy against r, the distance between two molecules, which occurs at a separation of $r_e = 2^{1/6}\sigma$.

For more extensive treatment see: J.O. Hirschfelder, C.F. Curtiss and R.B. Bird, *Molecular Theory of Gases*, (1954), Wiley, New York.

Substance	$(\varepsilon/k)/K$	σ/pm
He	10.22	258
Ne	35.7	279
Ar	124	342
Kr	190	361
Xe	229	406
H_2	33.3	297
O_2	113	343
N_2	91.5	368
Cl_2	357	412
Br_2	520	427
CO_2	190	400
CH_4	137	382
CCl_4	327	588
C_2H_4	205	423
C_6H_6	440	527

Table IV.12 Madelung Constants

Reference: T.C. Waddington, *Adv. Inorg. Chem. Radiochem.* (1959) *1*, 157.

Lattice	Crystal type	$\mathcal{M}$
rock salt, NaCl	face-centred cubic	1.747 56
caesium chloride, CsCl	body-centred cubic	1.772 67
zinc blende, ZnS	face-centred cubic	1.638 06
wurtzite, ZnS	hexagonal	1.641 32
calcium chloride, $CaCl_2$	cubic	2.365
fluorite, CaF_2	cubic	2.519 39
rutile, TiO_2	tetragonal	2.408
cuprite, Cu_2O	cubic	2.221 24
corundum, Al_2O_3	rhombohedral	4.171 9
β-quartz, SiO_2	hexagonal	2.219 7

NB. For hexagonal and tetragonal crystals, the value of $\mathcal{M}$ depends on the details of the lattice parameters.

Table IV.13 Character Tables
For more extensive tables see: P. W. Artkins, M. S. Child, Tables for Group Theory (1970), Oxford University Press, Oxford.

1. The Groups C_1, C_s, C_i

C_1 (1)	E
A	1

C_s (m)	E	σ_h		
A'	1	1	x, y, R_z	x^2, y^2, z^2, xy
A''	1	−1	z, R_x, R_y	yz, xz

C_i ($\bar{1}$)	E	i		
A_g	1	1	R_x, R_y, R_z	$x^2, y^2, z^2, xy, xz, yz$
A_u	1	−1	x, y, z	

2. The Groups C_n ($n = 2, 3, \ldots, 8$)

C_2 (2)	E	C_2		
A	1	1	z, R_z	x^2, y^2, z^2, xy
B	1	−1	x, y, R_x, R_y	yz, xz

C_4 (4)	E	C_4	C_2	C_4^3		
A	1	1	1	1	z, R_z	$x^2 + y^2, z^2$
B	1	−1	1	−1		$x^2 − y^2, xy$
E	$\begin{Bmatrix}1\\1\end{Bmatrix}$	$\begin{matrix}i\\-i\end{matrix}$	$\begin{matrix}-1\\-1\end{matrix}$	$\begin{matrix}-i\\i\end{matrix}$	$(x, y) (R_x, R_y)$	(yz, xz)

C_3 (3)	E	C_3	C_3^2			$\epsilon = \exp(2\pi i/3)$
A	1	1	1	z, R_z	$x^2 + y^2, z^2$	
E	$\begin{Bmatrix}1\\1\end{Bmatrix}$	$\begin{matrix}\epsilon\\\epsilon^*\end{matrix}$	$\begin{matrix}\epsilon^*\\\epsilon\end{matrix}$	$(x, y) (R_x, R_y)$	$(x^2 − y^2, xy) (yz, xz)$	

C_5	E	C_5	C_5^2	C_5^3	C_5^4			$\epsilon = \exp(2\pi i/5)$
A	1	1	1	1	1	z, R_z	$x^2 + y^2, z^2$	
E_1	$\begin{Bmatrix}1\\1\end{Bmatrix}$	$\begin{matrix}\epsilon\\\epsilon^*\end{matrix}$	$\begin{matrix}\epsilon^2\\\epsilon^{2*}\end{matrix}$	$\begin{matrix}\epsilon^{2*}\\\epsilon^2\end{matrix}$	$\begin{matrix}\epsilon^*\\\epsilon\end{matrix}$	$(x, y)(R_x, R_y)$	(yz, xz)	
E_2	$\begin{Bmatrix}1\\1\end{Bmatrix}$	$\begin{matrix}\epsilon^2\\\epsilon^{2*}\end{matrix}$	$\begin{matrix}\epsilon^*\\\epsilon\end{matrix}$	$\begin{matrix}\epsilon\\\epsilon^*\end{matrix}$	$\begin{matrix}\epsilon^{2*}\\\epsilon^2\end{matrix}$		$(x^2 − y^2, xy)$	

C_6 (6)	E	C_6	C_3	C_2	C_3^2	C_6^5			$\epsilon = \exp(2\pi i/6)$
A	1	1	1	1	1	1	z, R_z	$x^2 + y^2, z^2$	
B	1	−1	1	−1	1	−1			
E_1	$\begin{Bmatrix}1\\1\end{Bmatrix}$	$\begin{matrix}\epsilon\\\epsilon^*\end{matrix}$	$\begin{matrix}-\epsilon^*\\-\epsilon\end{matrix}$	$\begin{matrix}-1\\-1\end{matrix}$	$\begin{matrix}-\epsilon\\-\epsilon^*\end{matrix}$	$\begin{matrix}\epsilon^*\\\epsilon\end{matrix}$	(x, y) (R_z, R_y)	(xz, yz)	
E_2	$\begin{Bmatrix}1\\1\end{Bmatrix}$	$\begin{matrix}-\epsilon^*\\-\epsilon\end{matrix}$	$\begin{matrix}-\epsilon\\-\epsilon^*\end{matrix}$	$\begin{matrix}1\\1\end{matrix}$	$\begin{matrix}-\epsilon^*\\-\epsilon\end{matrix}$	$\begin{matrix}-\epsilon\\-\epsilon^*\end{matrix}$		$(x^2 − y^2, xy)$	

C_7	E	C_7	C_7^2	C_7^3	C_7^4	C_7^5	C_7^6			$\epsilon = \exp(2\pi i/7)$
A	1	1	1	1	1	1	1	z, R_z	$x^2 + y^2, z^2$	
E_1	$\begin{Bmatrix}1\\1\end{Bmatrix}$	$\begin{matrix}\epsilon\\\epsilon^*\end{matrix}$	$\begin{matrix}\epsilon^2\\\epsilon^{2*}\end{matrix}$	$\begin{matrix}\epsilon^3\\\epsilon^{3*}\end{matrix}$	$\begin{matrix}\epsilon^{3*}\\\epsilon^3\end{matrix}$	$\begin{matrix}\epsilon^{2*}\\\epsilon^2\end{matrix}$	$\begin{matrix}\epsilon^*\\\epsilon\end{matrix}$	(x, y) (R_x, R_y)	(xz, yz)	
E_2	$\begin{Bmatrix}1\\1\end{Bmatrix}$	$\begin{matrix}\epsilon^2\\\epsilon^{2*}\end{matrix}$	$\begin{matrix}\epsilon^{3*}\\\epsilon^3\end{matrix}$	$\begin{matrix}\epsilon^*\\\epsilon\end{matrix}$	$\begin{matrix}\epsilon\\\epsilon^*\end{matrix}$	$\begin{matrix}\epsilon^3\\\epsilon^{3*}\end{matrix}$	$\begin{matrix}\epsilon^{2*}\\\epsilon_2\end{matrix}$		$(x^2 − y^2, xy)$	
E_3	$\begin{Bmatrix}1\\1\end{Bmatrix}$	$\begin{matrix}\epsilon^3\\\epsilon^{3*}\end{matrix}$	$\begin{matrix}\epsilon^*\\\epsilon\end{matrix}$	$\begin{matrix}\epsilon^2\\\epsilon^{2*}\end{matrix}$	$\begin{matrix}\epsilon^{2*}\\\epsilon^2\end{matrix}$	$\begin{matrix}\epsilon\\\epsilon^*\end{matrix}$	$\begin{matrix}\epsilon^{3*}\\\epsilon^3\end{matrix}$			

Table IV.13 (contd)

C_8	E	C_8	C_4	C_2	C_4^3	C_8^3	C_8^5	C_8^7	$\epsilon = \exp(2\pi i/8)$	
A	1	1	1	1	1	1	1	1	z, R_z	x^2+y^2, z^2
B	1	-1	1	1	1	-1	-1	-1		
E_1	$\begin{cases}1\\1\end{cases}$	$\begin{matrix}\epsilon\\\epsilon^*\end{matrix}$	$\begin{matrix}i\\-i\end{matrix}$	$\begin{matrix}-1\\-1\end{matrix}$	$\begin{matrix}-i\\i\end{matrix}$	$\begin{matrix}\epsilon^*\\-\epsilon\end{matrix}$	$\begin{matrix}\epsilon\\-\epsilon^*\end{matrix}$	$\begin{matrix}\epsilon^*\\\epsilon\end{matrix}\}$	(x, y) (R_x, R_y)	(xz, yz)
E_2	$\begin{cases}1\\1\end{cases}$	$\begin{matrix}i\\-i\end{matrix}$	$\begin{matrix}-1\\-1\end{matrix}$	$\begin{matrix}1\\1\end{matrix}$	$\begin{matrix}-1\\-1\end{matrix}$	$\begin{matrix}-i\\i\end{matrix}$	$\begin{matrix}i\\-i\end{matrix}$	$\begin{matrix}-i\\i\end{matrix}\}$		(x^2-y^2, xy)
E_3	$\begin{cases}1\\1\end{cases}$	$\begin{matrix}-\epsilon\\-\epsilon^*\end{matrix}$	$\begin{matrix}i\\-i\end{matrix}$	$\begin{matrix}-1\\-1\end{matrix}$	$\begin{matrix}-i\\i\end{matrix}$	$\begin{matrix}\epsilon^*\\\epsilon\end{matrix}$	$\begin{matrix}\epsilon\\\epsilon^*\end{matrix}$	$\begin{matrix}-\epsilon^*\\-\epsilon\end{matrix}\}$		

3. The Groups D_n ($n = 2, 3, 4, 5, 6$)

D_2 (222)	E	$C_2(z)$	$C_2(y)$	$C_2(x)$		
A	1	1	1	1		x^2, y^2, z^2
B_1	1	1	-1	-1	z, R_z	xy
B_2	1	-1	1	-1	y, R_y	xz
B_3	1	-1	-1	1	x, R_x	yz

D_3 (32)	E	$2C_3$	$3C_2$		
A_1	1	1	1		x^2+y^2, z^2
A_2	1	1	-1	z, R_z	
E	2	-1	0	$(x, y) (R_x, R_y)$	$(x^2-y^2, xy)(xz, yz)$

D_4 (422)	E	$2C_4$	$C_2(=C_4^2)$	$2C_2'$	$2C_2''$		
A_1	1	1	1	1	1		x^2+y^2, z^2
A_2	1	1	1	-1	-1	z, R_z	
B_1	1	-1	1	1	-1		x^2-y^2
B_2	1	-1	1	-1	1		xy
E	2	0	-2	0	0	$(x, y) (R_x, R_y)$	(xz, yz)

D_5	E	$2C_5$	$2C_5^2$	$5C_2$		
A_1	1	1	1	1		x^2+y^2, z^2
A_2	1	1	1	-1	z, R_z	
E_1	2	2 cos 72°	2 cos 144°	0	$(x, y) (R_x, R_y)$	(xz, yz)
E_2	2	2 cos 144°	2 cos 72°	0		(x^2-y^2, xy)

D_6 (622)	E	$2C_6$	$2C_3$	C_2	$3C_2'$	$3C_2''$		
A_1	1	1	1	1	1	1		x^2+y^2, z^2
A_2	1	1	1	1	-1	-1	z, R_z	
B_1	1	-1	1	-1	1	-1		
B_2	1	-1	1	-1	-1	1		
E_1	2	1	-1	-2	0	0	$(x, y) (R_x, R_y)$	(xz, yz)
E_2	2	-1	-1	2	0	0		(x^2-y^2, xy)

4. The Groups C_{nv} ($n = 2, 3, 4, 5, 6$)

C_{2v} (2 mm)	E	C_2	$\sigma_v(xz)$	$\sigma_v'(yz)$		
A_1	1	1	1	1	z	x^2, y^2, z^2
A_2	1	1	-1	-1	R_z	xy
B_1	1	-1	1	-1	x, R_y	xz
B_2	1	-1	-1	1	y, R_x	yz

C_{3v} (3 m)	E	$2C_3$	$3\sigma_v$		
A_1	1	1	1	z	x^2+y^2, z^2
A_2	1	1	-1	R_z	
E	2	-1	0	$(x, y) (R_x, R_y)$	$(x^2-y^2, xy) (xz, yz)$

C_{4v} (4 mm)	E	$2C_4$	C_2	$2\sigma_v$	$2\sigma_d$		
A_1	1	1	1	1	1	z	x^2+y^2, z^2
A_2	1	1	1	-1	-1	R_z	
B_1	1	-1	1	1	-1		x^2-y^2
B_2	1	-1	1	-1	1		xy
E	2	0	-2	0	0	$(x, y) (R_x, R_y)$	(xz, yz)

C_{5v}	E	$2C_5$	$2C_5^2$	$5\sigma_v$		
A_1	1	1	1	1	z	x^2+y^2, z^2
A_2	1	1	1	-1	R_z	
E_1	2	2 cos 72°	2 cos 144°	0	$(x, y) (R_x, R_y)$	(xz, yz)
E_2	2	2 cos 144°	2 cos 72°	0		(x^2-y^2, xy)

C_{6v} (6 mm)	E	$2C_6$	$2C_3$	C_2	$3\sigma_v$	$3\sigma_d$		
A_1	1	1	1	1	1	1	z	x^2+y^2, z^2
A_2	1	1	1	1	-1	-1	R_z	
B_1	1	-1	1	-1	1	-1		
B_2	1	-1	1	-1	-1	1		
E_1	2	1	-1	-2	0	0	$(x, y) (R_x, R_y)$	(xz, yz)
E_2	2	-1	-1	2	0	0		(x^2-y^2, xy)

5. The Groups C_{nh} ($n = 2, 3, 4, 5, 6$)

C_{2h} (2/m)	E	C_2	i	σ_h		
A_g	1	1	1	1	R_z	x^2, y^2, z^2, xy
B_g	1	−1	1	−1	R_x, R_y	xz, yz
A_u	1	1	−1	−1	z	
B_u	1	−1	−1	1	x, y	

C_{3h} ($\bar{6}$)	E	C_3	C_3^2	σ_h	S_3	S_3^5			$\epsilon = \exp(2\pi i/3)$
A'	1	1	1	1	1	1	R_z	$x^2 + y^2, z^2$	
E'	$\begin{cases}1\\1\end{cases}$	$\begin{matrix}\epsilon\\\epsilon^*\end{matrix}$	$\begin{matrix}\epsilon^*\\\epsilon\end{matrix}$	$\begin{matrix}1\\1\end{matrix}$	$\begin{matrix}\epsilon\\\epsilon^*\end{matrix}$	$\begin{matrix}\epsilon^*\\\epsilon\end{matrix}$	(x, y)	$(x^2 - y^2, xy)$	
A''	1	1	1	−1	−1	−1	z		
E''	$\begin{cases}1\\1\end{cases}$	$\begin{matrix}\epsilon\\\epsilon^*\end{matrix}$	$\begin{matrix}\epsilon^*\\\epsilon\end{matrix}$	$\begin{matrix}-1\\-1\end{matrix}$	$\begin{matrix}-\epsilon\\-\epsilon^*\end{matrix}$	$\begin{matrix}-\epsilon^*\\-\epsilon\end{matrix}$	(R_x, R_y)	(xz, yz)	

C_{4h} (4/m)	E	C_4	C_2	C_4^3	i	S_4^3	σ_h	S_4		
A_g	1	1	1	1	1	1	1	1	R_z	x^2+y^2, z^2
B_g	1	−1	1	−1	1	−1	1	−1		$x^2 - y^2, xy$
E_g	$\begin{cases}1\\1\end{cases}$	$\begin{matrix}i\\-i\end{matrix}$	$\begin{matrix}-1\\-1\end{matrix}$	$\begin{matrix}-i\\i\end{matrix}$	$\begin{matrix}1\\1\end{matrix}$	$\begin{matrix}i\\-i\end{matrix}$	$\begin{matrix}-1\\-1\end{matrix}$	$\begin{matrix}-i\\i\end{matrix}$	(R_x, R_y)	(xz, yz)
A_u	1	1	1	1	−1	−1	−1	−1	z	
B_u	1	−1	1	−1	−1	1	−1	1		
E_u	$\begin{cases}1\\1\end{cases}$	$\begin{matrix}i\\-i\end{matrix}$	$\begin{matrix}-1\\-1\end{matrix}$	$\begin{matrix}-i\\i\end{matrix}$	$\begin{matrix}-1\\-1\end{matrix}$	$\begin{matrix}-i\\i\end{matrix}$	$\begin{matrix}1\\1\end{matrix}$	$\begin{matrix}i\\-i\end{matrix}$	(x, y)	

C_{5h}	E	C_5	C_5^2	C_5^3	C_5^4	σ_h	S_5	S_5^7	S_5^3	S_5^9			$\epsilon = \exp(2\pi i/5)$
A'	1	1	1	1	1	1	1	1	1	1	R_z	x^2+y^2, z^2	
E'_1	$\begin{cases}1\\1\end{cases}$	$\begin{matrix}\epsilon\\\epsilon^*\end{matrix}$	$\begin{matrix}\epsilon^2\\\epsilon^{2*}\end{matrix}$	$\begin{matrix}\epsilon^{2*}\\\epsilon^2\end{matrix}$	$\begin{matrix}\epsilon^*\\\epsilon\end{matrix}$	$\begin{matrix}1\\1\end{matrix}$	$\begin{matrix}\epsilon\\\epsilon^*\end{matrix}$	$\begin{matrix}\epsilon^2\\\epsilon^{2*}\end{matrix}$	$\begin{matrix}\epsilon^{2*}\\\epsilon^2\end{matrix}$	$\begin{matrix}\epsilon^*\\\epsilon\end{matrix}$	(x, y)		
E'_2	$\begin{cases}1\\1\end{cases}$	$\begin{matrix}\epsilon^2\\\epsilon^{2*}\end{matrix}$	$\begin{matrix}\epsilon^*\\\epsilon\end{matrix}$	$\begin{matrix}\epsilon\\\epsilon^*\end{matrix}$	$\begin{matrix}\epsilon^{2*}\\\epsilon^2\end{matrix}$	$\begin{matrix}1\\1\end{matrix}$	$\begin{matrix}\epsilon^2\\\epsilon^{2*}\end{matrix}$	$\begin{matrix}\epsilon^*\\\epsilon\end{matrix}$	$\begin{matrix}\epsilon\\\epsilon^*\end{matrix}$	$\begin{matrix}\epsilon^{2*}\\\epsilon^2\end{matrix}$		$(x^2 - y^2, xy)$	
A''	1	1	1	1	1	−1	−1	−1	−1	−1	z		
E''_1	$\begin{cases}1\\1\end{cases}$	$\begin{matrix}\epsilon\\\epsilon^*\end{matrix}$	$\begin{matrix}\epsilon^2\\\epsilon^{2*}\end{matrix}$	$\begin{matrix}\epsilon^{2*}\\\epsilon^2\end{matrix}$	$\begin{matrix}\epsilon^*\\\epsilon\end{matrix}$	$\begin{matrix}-1\\-1\end{matrix}$	$\begin{matrix}-\epsilon\\-\epsilon^*\end{matrix}$	$\begin{matrix}-\epsilon^2\\-\epsilon^{2*}\end{matrix}$	$\begin{matrix}-\epsilon^{2*}\\-\epsilon^2\end{matrix}$	$\begin{matrix}-\epsilon^*\\-\epsilon\end{matrix}$	(R_x, R_y)	(xz, yz)	
E''_2	$\begin{cases}1\\1\end{cases}$	$\begin{matrix}\epsilon^2\\\epsilon^{2*}\end{matrix}$	$\begin{matrix}\epsilon^*\\\epsilon\end{matrix}$	$\begin{matrix}\epsilon\\\epsilon^*\end{matrix}$	$\begin{matrix}\epsilon^{2*}\\\epsilon^2\end{matrix}$	$\begin{matrix}-1\\-1\end{matrix}$	$\begin{matrix}-\epsilon^2\\-\epsilon^{2*}\end{matrix}$	$\begin{matrix}-\epsilon^*\\-\epsilon\end{matrix}$	$\begin{matrix}-\epsilon\\-\epsilon^*\end{matrix}$	$\begin{matrix}-\epsilon^{2*}\\-\epsilon^2\end{matrix}$			

C_{6h} (6/m)	E	C_6	C_3	C_2	C_3^2	C_6^5	i	S_3^5	S_6^5	σ_h	S_6	S_3			$\epsilon = \exp(2\pi i/6)$
A_g	1	1	1	1	1	1	1	1	1	1	1	1	R_z	x^2+y^2, z^2	
B_g	1	−1	1	−1	1	−1	1	−1	1	−1	1	−1			
E_{1g}	$\begin{cases}1\\1\end{cases}$	$\begin{matrix}\epsilon\\\epsilon^*\end{matrix}$	$\begin{matrix}-\epsilon^*\\-\epsilon\end{matrix}$	$\begin{matrix}-1\\-1\end{matrix}$	$\begin{matrix}-\epsilon\\-\epsilon^*\end{matrix}$	$\begin{matrix}\epsilon^*\\\epsilon\end{matrix}$	$\begin{matrix}1\\1\end{matrix}$	$\begin{matrix}\epsilon\\\epsilon^*\end{matrix}$	$\begin{matrix}-\epsilon^*\\-\epsilon\end{matrix}$	$\begin{matrix}-1\\-1\end{matrix}$	$\begin{matrix}-\epsilon\\-\epsilon^*\end{matrix}$	$\begin{matrix}\epsilon^*\\\epsilon\end{matrix}$	(R_x, R_y)	(xz, yz)	
E_{2g}	$\begin{cases}1\\1\end{cases}$	$\begin{matrix}-\epsilon^*\\-\epsilon\end{matrix}$	$\begin{matrix}-\epsilon\\-\epsilon^*\end{matrix}$	$\begin{matrix}1\\1\end{matrix}$	$\begin{matrix}-\epsilon^*\\-\epsilon\end{matrix}$	$\begin{matrix}-\epsilon\\-\epsilon^*\end{matrix}$	$\begin{matrix}1\\1\end{matrix}$	$\begin{matrix}-\epsilon^*\\-\epsilon\end{matrix}$	$\begin{matrix}-\epsilon\\-\epsilon^*\end{matrix}$	$\begin{matrix}1\\1\end{matrix}$	$\begin{matrix}-\epsilon^*\\-\epsilon\end{matrix}$	$\begin{matrix}-\epsilon\\-\epsilon^*\end{matrix}$		$(x^2 - y^2, xy)$	
A_u	1	1	1	1	1	1	−1	−1	−1	−1	−1	−1	z		
B_u	1	−1	1	−1	1	−1	−1	1	−1	1	−1	1			
E_{1u}	$\begin{cases}1\\1\end{cases}$	$\begin{matrix}\epsilon\\\epsilon^*\end{matrix}$	$\begin{matrix}-\epsilon^*\\-\epsilon\end{matrix}$	$\begin{matrix}-1\\-1\end{matrix}$	$\begin{matrix}-\epsilon\\-\epsilon^*\end{matrix}$	$\begin{matrix}\epsilon^*\\\epsilon\end{matrix}$	$\begin{matrix}-1\\-1\end{matrix}$	$\begin{matrix}-\epsilon\\-\epsilon^*\end{matrix}$	$\begin{matrix}\epsilon^*\\\epsilon\end{matrix}$	$\begin{matrix}1\\1\end{matrix}$	$\begin{matrix}\epsilon\\\epsilon^*\end{matrix}$	$\begin{matrix}-\epsilon^*\\-\epsilon\end{matrix}$	(x, y)		
E_{2u}	$\begin{cases}1\\1\end{cases}$	$\begin{matrix}-\epsilon^*\\-\epsilon\end{matrix}$	$\begin{matrix}-\epsilon\\-\epsilon^*\end{matrix}$	$\begin{matrix}1\\1\end{matrix}$	$\begin{matrix}-\epsilon^*\\-\epsilon\end{matrix}$	$\begin{matrix}-\epsilon\\-\epsilon^*\end{matrix}$	$\begin{matrix}-1\\-1\end{matrix}$	$\begin{matrix}\epsilon^*\\\epsilon\end{matrix}$	$\begin{matrix}\epsilon\\\epsilon^*\end{matrix}$	$\begin{matrix}-1\\-1\end{matrix}$	$\begin{matrix}\epsilon\\\epsilon^*\end{matrix}$	$\begin{matrix}\epsilon^*\\\epsilon\end{matrix}$			

Table IV.14 Bravais Lattices
P: primitive unit. I (Innenzentrierte): body-centred; F: face-centred. C: 001 face-centred without symmetry demanding also centring of the prism faces. Interaxial angles are 90° unless otherwise indicated.

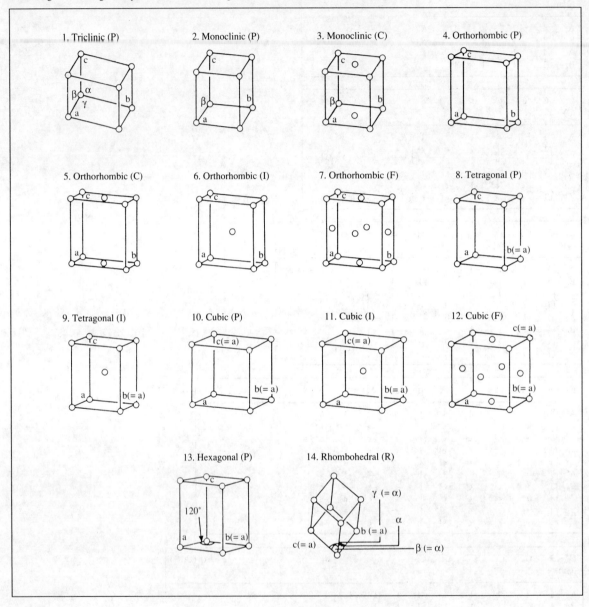

Table IV.15 Characteristics of Seven Crystal Systems

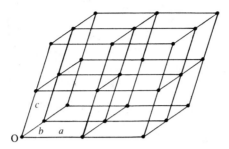

Crystal lattice

System	Axial relationships	Essential symmetry	Number of point groups	Examples
Triclinic	$a \neq b \neq c$ $\alpha \neq \beta \neq \gamma \neq 90°, 120°$	No axes or planes	2	$CuSO_4.5H_2O$, $K_2Cr_2O_7$
Monoclinic	$a \neq b \neq c$ $\alpha = \gamma = 90°$ $\beta \neq 90°, 120°$	1 2-fold axis (C_2) or 1 plane	3	$CaSO_4.2H_2O$, $Na_2CO_3.10H_2O$, sucrose
Orthorhombic (rhombic)	$a \neq b \neq c$ $\alpha = \beta = \gamma = 90°$	3 2-fold axes (C_2) or 1 2-fold axis and 2 perpendicular planes intersecting in a 2-fold axis	3	$BaSO_4$, KNO_3, $PbCO_3$, α-S
Tetragonal	$a = b \neq c$ $\alpha = \beta = \gamma = 90°$	1 4-fold axis (C_4)	7	KH_2PO_4, SnO_2, TiO_2 (rutile)
Cubic (regular)	$a = b = c$ $\alpha = \beta = \gamma = 90°$	4 3-fold axes (C_3), 3 2-fold axes (C_2) or 3 4-fold axes (C_4)	5	NaCl, C (diamond), CaF_2, ZnS, $NaClO_3$, Pb, Hg, Ag, Au
Hexagonal	$a = b \neq c$ $\alpha = \beta = 90°, \gamma = 120°$	1 6-fold axis (C_6)	7	C (graphite), H_2O (ice), NiAs, Mg, Zn, Cd
Trigonal (rhombohedral)	$a = b = c$ $\alpha = \beta = \gamma \neq 90°, < 120°$	1 3-fold axis (C_3)	5	$CaCO_3$, $NaNO_3$, quartz, As, Sb, Bi

Section V
Electricity and Magnetism

		Page
V.1	Resistivity values for some liquid metallic elements at stated temperatures	69
V.2	Resistivity values for single crystals of some non-cubic metallic elements along a- and c- axes at standard temperature, unless otherwise stated	69
V.3	Variation with temperature of resistivity of some polycrystalline solid metallic elements	70
V.4	Variation with temperature of resistivity of some alloys	71
V.5	Temperature coefficient of resistivity for resistance-wire alloys	71
V.6	Resistances per metre length of various standard wire gauge resistance wires for materials of Table V.5, together with copper	72
V.7	Approximate values of electric current required to maintain temperature rises, ΔT, of 500 K and 1000 K in straight horizontal Ni/Cr resistance wires, of various standard wire gauges, free to radiate in air	72
V.8	Resistivity values for semiconducting elements at standard temperature	72
V.9	Resistivity values one minute after measurement-voltage application for some ceramics at standard temperature and 50–60% relative humidity	73
V.10	Resistivity values one minute after measurement-voltage application for some plastics at standard temperature and 50–60% relative humidity	73
V.11	Resistivity values one minute after measurement-voltage application for some other insulators at standard temperature and 50–60% relative humidity	73
V.12	Relative permittivity of some gases and vapours for frequencies less than infrared at stated temperatures and, except where otherwise stated, standard pressure	74
V.13	Relative permittivity and loss tangent, at stated temperatures and frequencies, for some liquids of controlled purity used for electrical purposes	74
V.14	Relative permittivity and its temperature rate of change for various inorganic liquids at stated temperatures and standard pressure, unless otherwise stated	75
V.15	Values of relative permittivity and its temperature rate of change for various organic liquids at stated temperatures and standard pressure	76
V.16	Values of relative permittivity and loss tangent for some inorganic solids at stated temperatures and frequencies	77
V.17	Values of relative permittivity and loss tengent for some organic solids at stated temperatures and frequencies	78
V.18	Representative variation of magnetic flux density with magnetic field strength for various low-coercivity magnetic substances, together with initial and maximum relative permeability, saturation polarization, coercivity, remanence, Curie temperature, resistivity and specific total power loss	80
V.19	Representative values for some core materials of initial relative permeability, frequency range, loss factor at maximum frequency, Curie temperature and resistivity, together with some other information	82
V.20	Representative values of remanence, maximum of product of magnetic field strength and magnetic flux density (with field strength given in brackets), induction and magnetisation coercive forces, resistivity, Curie temperature and maximum operating temperature, for some permanent-magnet materials	83
V.21	Representative values of relative permeability and resistivity for some feebly magnetic steels and cast irons for various conditions	84
V.22	Mass susceptibility per kilogram at temperature 293 K for some paramagnetic and diamagnetic materials	84
V.23	Values of critical temperature and critical magnetic field(s) at temperature 0 K for some superconducting elements	86
V.24	Values of critical temperature and critical magnetic field at temperature of 4.2 K for some superconducting compounds and alloys	86

Section V
Electricity and Magnetism

For comprehensive cover of classical data see: Landolt-Bornstein, Springer-Verlag, 6th ed., vol II, parts 6 (1959) and 7 (1960).
Note: In order to improve worldwide harmonization of electrical measurement standards, the values of the national reference standards of electromotive force and resistance maintained at the National Physical Laboratory were respectively decreased by 8.06 and 1.61 ppm from 1.1.90.

Reference: Jones, Division of Electrical Science, NPL, Teddington, TW11 0LW from whom further information may be obtained.

Table V.1 Resistivity values for some liquid metallic elements at stated temperatures

Element	temperature/ K	resistivity / $10^{-8}\,\Omega\,m$	Element	temperature/ K	resistivity/ $10^{-8}\,\Omega\,m$
Ag	1473	19.4	K	373	17.5
Al	973	24.7		573	28.2
	1473	32.1		973	66.4
Au	1336	31		1473	160
Bi	573	129	Li	573	30
	973	155		973	40.5
	1273	172		1473	53
Cd	873	36.3	Mg	973	27.7
Cs	373	43.5		1173	28.7
	573	67	Na	373	9.7
	973	134		573	16.8
	1473	295		973	39.2
Cu	1356	21.3		1473	89
Fe	1823	139	Ni	1773	109
Ga	373	27.2	Pb	973	108
	573	31		1273	126
Hg	273	94.1	Rb	373	27.5
	373	103.5		573	48
	573	128		973	99
	973	214		1473	260
	1473	630	Sb	973	114
In	573	36.7		1273	123.5
	973	47	Sn	573	50
	1273	55		973	60
				1473	72
			Tl	973	85
				1073	88
			Zn	773	37

Reference: Kaye and Laby, *Tables of Physical and Chemical Constants*, 15 ed., Longmans, 1986

Table V.2 Resistivity values for single crystals of some non-cubic metallic elements along a- and c-axes at standard temperature, unless otherwise stated

Element	resistivity/$10^{-8}\,\Omega\,m$ a-axis	c-axis	Element	resistivity/$10^{-8}\,\Omega\,m$ a-axis	c-axis
Be	3.13	3.58	Ho	101.2	59.9
Bi	109	138	Lu	75.6	34
Cd	6.54	7.79	Mg	4.53	3.78
Dy	98.2	77.4	Sb	42.6	35.6
Er	87.6	47	Sn	10	13.1
Ga	17.3	55.5	Tb	122	101
Gd	135	122	Tm	87	46
Hg (at 227.7 K)	23.5	17.7	Zn	5.83	6.05

Reference: as for Table V.1

Table V.3 Variation with temperature of resistivity of some polycrystalline solid metallic elements

The highest temperature given is 1473 K unless otherwise stated.

Element	temperature/K					
	78	273	373	573	973	1473
	\multicolumn{6}{c}{resistivity/ $10^{-8}\,\Omega\,\mathrm{m}$}					
Ag	0.3	1.47	2.08	3.34	6.1	
Al	0.21	2.5	3.55	5.9		
As	5.5	26				
Au	0.5	2.05	2.88	4.63	8.6	
Ba	7	36				
Be		2.8	5.3	11.1	26	
Bi	35	107	156			
Ca	0.7	3.2	4.75	7.8	20	
Cd	1.6	6.8	9.8			
Ce		73	80	92	110	123 (at 1053 K)
Co	0.9	5.6	9.5	19.7	48	88.5
Cr	0.5	12.7	16.1	25.2	47.2	80
Cs	4.5	18.8				
Cu	0.2	1.55	2.23	3.6	6.7	
Dy	26	89	103	124	156.5	184
Er	41	81	103	135	183	216
Eu	60	89				
Fe	0.7	8.9	14.7	31.5	85.5	122
Ga	2.75	13.6				
Gd	20	126				
Hf		29.6	42.6			
Hg	5.8					
Ho	34	90	105	134	175	203
In	1.8	8	12.1			
Ir	0.9	4.7	6.8	10.8	22	33.5
K	1.38	6.1				
La		54	66	83	105	126 (at 1173 K)
Li	1.04	8.55	12.4			
Lu	16	54				
Mg	0.62	3.94	5.6	10		
Mn		138				
Mo	0.7	5	7.6	12.7	23.3	37.2
Na	0.8	4.2				
Nb	3	15.2	19.2	27.1	43	59
Nd		61	74	93	120	138 (at 1183 K)
Ni	0.55	6.2	10.3	22.5	40	
Np			119.3			121.3 (at 811 K)
Os		8.1	11.4	17.8	30.4	46
Pa	6.1	17.7				
Pb	4.7	19.2	27	50		
Pd	1.73	10	13.8	21	33	42
Pm		50	64	89	126	
Po		c 40				
Pr		65	78	96	118	134 (at 1123 K)
Pt	1.96	9.81	13.6	21	34.3	48.3
Pu	c 150	146	142	109		
Rb	2.2	11				
Re	2.62	17.2	24.9	39.7	63.5	84.4
Rh	0.46	4.3	6.2	10.2	20	33
Ru	1.34	7.1	10	15.6	27.8	44.4
Sb	8	39	59			
Sc		50.5	75	115	167	198
Sm	66	91.4				
Sn	2.1	11.5	15.8			
Sr	5	20	30	52.5	94.5	
Ta	2.5	12.3	16.7	25.5	43	61.5
Tb	27	113				
Te			22.6	33.3	51	65

(Continued)

Table V.3 (Continued)

Element	temperature/K					
	78	273	373	573	973	1473
	resistivity/ $10^{-8}\Omega$ m					
Th	3.9	14.7	20.8	32.5	53.6	68
Ti	4.6	39	58	90	142	
Tl	3.7	15	22.8	38		
Tm	31	67				
U	11	28	35	47		
Vb	2.6	18.2	25.3	38		54 (at 873 K)
W	0.6	4.9	7.3	12.4	24	39
Y	15.5	55				
Yb	13	27.7				
Zn	1.1	5.5	7.8	13		
Zr	7.3	40	58	88	125	110 (at 1273 K)

Reference: as for Table V.1

Table V.4 Variation with temperature of resistivity of some alloys

Alloy mass % age composition†	temperature/K				
	273	373	573	973	1473
	resistivity/10^{-8} Ω m				
2Al, 2Mn, 96Ti	110	123.1	149.5		
5Al, 2.5Sn, 92.5Ti	155.6	165.6	179.4		
Alumel	28.1	34.8	43.8	53.2	65.1
brass	6.3				
bronze	13.6				
0.1C, 2.3Sn, 97.6 Zr	91.5	105.2	123.9		
0.1C, 6.7Sn, 93.2 Zr	132.8	139.5	148.3		
Chromel P	70	72.8	79.3	89.3	100.1
Constantan	49				
German silver	40				
10Ir, 90Pt	24.8	28			
manganin	41.5				
Monel	42.9	50.1	52.5	63.3	
Nichrome	107.3	108.3	110	110.3	
10Rh, 90Pt	18.7	21.8			
steel, carbon	17	23.2	39.8	93.5	123.1
18/8	66.3	74.3	89.1	109.4	124
Era ATV	98	102.7	111	122	
Ni-Cr	27.7	33.7	48.7	99.4	122.2
silicon	41.9	47	60.1	105.7	127.1
stainless	55	63.4	80.2	114.1	

† More compositional details are given in Table XVII.3
Reference: as for Table V.1

Table V.5 Temperature coefficient of resistivity for resistance–wire alloys

Alloy	coefficient/K^{-1}	temperature range/K
Cu/Mn	c 3×10^{-6} minimum	288–293
Cu/Ni	0.000 04	
Ni/Cr	0.000 06	293–773
Ni/Cr/Al/one of Co, Cu, Fe, Mn, Mo	0.000 002 minimum	

Reference: as for Table V.1

Table V.6 Resistances per metre length of various standard wire gauge resistance wires for materials of Table V.5, together with copper

Gauge	diameter/ mm	resistance/ Ω m^{-1}				
		Cu/Mn	Cu/Ni	Ni/Cr	Ni/Cr/Al/ other	Cu at 293 K
12	2.642	0.076	0.09	0.197	0.243	0.00312
14	2.032	0.128	0.151	0.333	0.41	0.00532
16	1.626	0.2	0.235	0.52	0.64	0.00831
18	1.219	0.355	0.42	0.92	1.14	0.0148
20	0.914	0.63	0.745	1.65	2.03	0.0263
22	0.711	1.05	1.23	2.72	3.35	0.0434
24	0.559	1.69	2	4.4	5.4	0.0703
26	0.457	2.53	3	6.6	8.1	0.105
28	0.376	3.75	4.4	9.7	12	0.155
30	0.315	5.3	6.3	13.9	17.1	0.221
32	0.274	7	8.3	18.3	22.5	0.292
34	0.234	9.7	11.4	25.2	31	0.402
36	0.193	14.2	16.7	37	45.5	0.589
38	0.152	22.7	27	59	73	0.945
40	0.122	35.5	42	92	114	1.48
42	0.102	51	60.5	133	164	2.13
44	0.0813	80	94	208	255	3.32
46	0.061	142	168	370	455	5.91
48	0.0406	320	380	835	1030	13.3
50	0.0254	820	970	2130	2620	34

The wire-wound-resistor colour code is given in Table XVII.4
Reference: as for Table V.1

Table V.7 Approximate values of electric current required to maintain temperature rises, ΔT, of 500 K and 1000 K in straight horizontal Ni/Cr resistance wires, of various standard wire gauges, free to radiate in air

Gauge	current/A		Gauge	current/A	
	$\Delta T500$/K	$\Delta T1000$/K		$\Delta T500$/K	$\Delta T1000$/K
12	38	78	28	2.7	5.5
14	26	53	30	2.2	4.5
16	19	40	32	1.9	3.5
18	13	27	34	1.6	3
20	8.5	18	36	1.3	2.3
22	6.3	13	38	1	1.7
24	4.5	9.5	40	0.8	1.4
26	3.5	7	42	0.65	1.1

Reference: as for Table V.1

Table V.8 Resistivity values for semiconducting elements at standard temperature

Element	resistivity/Ω m	Element	resistivity/Ω m
C (amorphous)	$c\ 6 \times 10^{-5}$	Ge	10^{-3}–0.5
C (graphite)	3×10^{-6}–6×10^{-5}	Se	$c\ 0.1$
C (pyrolytic graphite, along planes)	$c\ 5 \times 10^{-6}$	Si	0.1–60
C (pyrolytic graphite, normal to planes)	$c\ 5 \times 10^{-3}$	Te	$c\ 3 \times 10^{-3}$

Reference: as for Table V.1

Table V.9 Resistivity values one minute after measurement–voltage application for some ceramics at standard temperature and 50–60% relative humidity

Ceramic	resistivity/Ω m	Ceramic	resistivity/Ω m
alumina	10^9–10^{12}	steatite	10^{11}–10^{13}
porcelain	10^{10}–10^{12}	titanates	10^6–10^{13}
pyrophyllite	10^{10}–10^{13}		

Reference: as for Table V.1

Table V.10 Resistivity values one minute after measurement–voltage application for some plastics at standard temperature and 50–60% relative humidity

Plastic	resistivity/Ω m	Plastic	resistivity/Ω m
acrylic (e.g. Perspex)	$> 10^{13}$	polystyrene (general purpose)	10^{15}–10^{19}
alkydes or polyester (no filler)	10^{12}–10^{13}	polystyrene (toughened)	10^{10}–10^{15}
aminos, melamines (cellulose)	$> 10^9$	polytetrafluoroethylene	10^{15}–10^{19}
aminos, melamines (mineral)	10^9	polythene (high density)	10^{14}–10^{15}
casein	10^7–10^8	polythene (low density)	10^{14}–10^{18}
cellulose acetate	10^8–10^{11}	polyurethanes	10^9–10^{12}
epoxy case resin (no filler)	10^{12}–10^{13}	polyvinyl chloride (rigid)	5×10^{12}–10^{13}
phenolics	10^6–10^{12}	polyvinyl chloride (flexible)	5×10^6–5×10^{12}
polyamides (nylon)	10^8–10^{13}	silicone (glass)	10^8–10^{12}
polychlorotrifluoroethylene	10^{16}		
polyethylene terephthalate	10^{15}–10^{17}		
polyformaldehyde	$c\ 6 \times 10^{12}$		
polypropylene	10^{13}–10^{15}		

Reference: as for Table V.1

Table V.11 Resistivity values one minute after measurement–voltage application for some other insulators at standard temperature and 50–60% relative humidity

Insulator	resistivity/Ω m	Insulator	resistivity/Ω m
B	10^{10}	mineral insulating oil	10^{11}–10^{15}
beeswax	10^{12}–10^{13}	paper (dry)	$c\ 10^{10}$
diamond	10^{10}–10^{11}	paraffin (kerosine coloured)	10^{11}–10^{12}
glass (soda-lime)	10^9–10^{11}	paraffin wax	10^{13}–10^{17}
glass (borosilicate: Pyrex)	10^{12}	quartz (fused)	$> 10^{16}$
glass (plate)	2×10^{11}	quartz (parallel optic axis)	10^{12}
guttapercha	10^7	quartz (perpendicular optic axis)	10^{14}
hard rubber (ebonite, etc.)	10^{13}–10^{15}	S	10^{14}–10^{15}
H$_2$O (distilled)	10^2–10^5	shellac	10^7
I$_2$	10^{13}	silicone oils	10^{12}
ivory	10^6	silicone rubber	10^9
marble	10^7–10^9	slate	10^5–10^6
mica (sheet)	10^{11}–10^{15}	soil	10^2–10^4
mica (moulded)	10^{13}	wood (paraffined)	10^8–10^{11}
mineral oil	$> 10^{10}$		

Reference: as for Table V.1

Table V.12 Relative permittivity of some gases and vapours for frequencies less than infrared at stated temperatures and, except where otherwise stated, standard pressure

Substance	temperature/ K	relative permittivity	Substance	temperature/ K	relative permittivity
air (dry)	293	1.000 536	CO_2	293	1.000 922
Ar	293	1.000 517		373	1.000 985
Br_2	273	1.012 8	CS_2	302	1.002 9
CH_4	273	1.000 944	D_2	273	1.000 27
CH_3CH_3	273	1.001 5	H_2	273	1.000 272
				373	1.000 264
CH_3OH	373	1.005 7	H_2O	373	1.006
C_2H_2	273	1.001 34	at 1333 Pa	293	1.000 121
C_2H_4	298	1.001 32	He	273	1.000 07
	383	1.001 44	N_2	293	1.000 547
C_2H_5OH	373	1.007 8	NH_3	274	1.007 1
$C_6H_5CH_3$	373	1.004 3	N_2O	298	1.001 03
C_6H_6	373	1.003 27	Ne	273	1.000 13
CO	298	1.000 64	O_2	293	1.000 494
				373	1.000 523
			SO_2	295	1.008 2
				373	1.000 75

Reference: *Handbook of Chemistry and Physics*, CRC Press, 1991/2
Kaye and Laby, *Tables of Physical and Chemical Constants*, 15 ed., Longmans, 1986

Table V.13 Relative permittivity and loss tangent, at stated temperatures and frequencies, for some liquids of controlled purity used for electrical purposes

Liquid	relative permittivity	frequency/ Hz	temperature/ K	loss tangent
castor oil	4.5	1000	293	
paraffin oil	2.2	1000	293	0.0001
pentachlorobiphenyl	5.2–4.3	50	273–373	0.07–0.0003
silicone fluid	2.2	$5 \times 10^{10} - 3 \times 10^9$	293	0.0002–0.0019
transformer oil	2.2	$5 \times 10^7 - 10^8 - 10^{10}$	293	0.0001–0.0042–0.0008
trichlorobiphenyl	7–5	$5 \times 10^4 - 2 \times 10^4$	263–373	0.2–0.0002

Reference: *Dielectric Constants of Pure Liquids*, NBS Circular no. 514, 1951
Kaye and Laby, *Tables of Physical and Chemical Constants*, 15 ed., Longmans, 1986
Handbook of Chemistry and Physics, CRC Press, 1991/2

Table V.14 Relative permittivity and its temperature rate of change for various inorganic liquids at stated temperatures and standard pressure, unless otherwise stated

Liquid	relative permittivity	temperature/ K	rate/ K^{-1}
$AlBr_3$	3.38	373	−0.0033
Ar	1.538	82	−0.0034
Br_2	3.09	293	−0.007
Cl_2	2.1	223	−0.0031
CO_2 (at 5 MPa)	1.6	293	
D_2	1.277	20	−0.004
D_2O	78.25	298	−0.0046
*H_2	1.228	20.4	−0.0034
HBr	7	188	−0.042
HCl	12	160	−0.08
HCN	114.9	293	−1.67
HF	17	200	
HI	3.39	223	−0.008
H_2NNH_2	52.9	293	−0.26
H_2O	78.54	298	−0.36
H_2O_2	84.2	273	
H_2S	9.26	184.5	
He	1.056	2.6	
N_2	1.454	70	−0.0029
NH_3	25	196	
	22.4	239.6	
N_2O_4	2.5	288	
*O_2	1.507	80	−0.0024
$POCl_3$	13.3	295	
S	3.52	391	
SO_2	17.6	253	−0.35

* Recommended as standard for calibrations
Reference: as for Table V.13

Table V.15 Values of relative permittivity and its temperature rate of change for various organic liquids at stated temperatures and standard pressure

For relative permittivities of some other substances, see Tables XI.8 and XI.9.

Liquid[†]	relative permittivity	temperature/ K	rate/ K^{-1}
CCl_4	2.238	293	−0.002 4
$CHCl_3$	4.806	293	−1.77
CH_2Cl_2	9.08	293	
$CH_3CH_2CH_2OH$	20.1	298	−13.56
$CH_3(CH_2)_3CH_3$	1.844	293	−0.001 6
$CH_3(CH_2)_4CH_2OH$	13.3	298	−0.107
$CH_3(CH_2)_5CH_3$	1.924	293	−0.001 4
$CH_3(CH_2)_6CH_3$	1.948	293	−0.001 31
$CH_3(CH_2)_7CH_3$	1.972	293	−0.001 34
$CH_3(CH_2)_8CH_3$	1.991	293	−0.001 29
$CH_3(CH_2)_9CH_3$	2.005	293	−0.001 24
$CH_3(CH_2)_{10}CH_2$	2.014	293	−0.001 21
$1,2-(CH_3)_2C_6H_4$	2.568	293	−0.002 66
$1,3-(CH_3)_2C_6H_4$	2.374	293	−0.001 95
$1,4-(CH_3)_2C_6H_4$	2.27	293	−0.001 6
CH_3CHO	21.8	283	
CH_3Cl	12.6	253	
CH_3CN	37.5	293	
CH_3COCH_3	20.7	298	−0.977
$CH_3CONHCH_3$	165	298	
$CH_3COOC_2H_5$	6.02	298	−0.208
CH_3COOH	6.15	293	
CH_3NO_2	36.7	298	
CH_3OH	32.63	298	−0.197
$(CH_3)_2SO$	46.7	298	
$C_2H_5COC_2H_5$	17	293	−0.088 4
C_2H_5OH	24.3	298	−0.151
$C_4H_3SO_2$	43.3	303	
C_4H_4S	2.76	289	
$l\text{-}C_4H_9OH$	17.1	293	−0.127
$C_4H_{10}O$	4.34	293	−0.021 7
C_5H_5N	12.3	298	
C_5H_{10}	1.965	293	
$C_5H_{11}OH$	13.9	298	−0.073 7
C_6H_5CHO	17.8	293	
$C_6H_5CH_2OH$	13.1	273	
$C_6H_5CH_3$	2.379	298	−0.002 43
$C_6H_5C_2H_5$	2.412	293	
*C_6H_5Cl	5.621	298	
C_6H_5N	12.3	298	
$C_6H_5NH_2$	6.89	293	−0.023 5
$C_6H_5NO_2$	34.82	298	−0.18
C_6H_5OH	9.78	333	−0.072 1
*C_6H_6	2.284	293	−0.002
$C_6H_{10}O$	18.3	293	
$C_6H_{11}OH$	15	298	−0.151
*C_6H_{12}	2.023	293	−0.001 6
C_6H_{14}	1.89	293	−0.001 55
CS_2	2.641	293	−0.002 68
1,4-dioxane	2.209	298	−0.001 7
furan	2.95	298	
glycerol	42.5	298	−0.203 5
ethylene glycol	37.7	298	−0.194
$HCON(CH_3)_2$	36.7	298	
$HCONCH_3$	182.4	298	
$HCONH_2$	109	293	
$HCOOH$	58.5	289	

[†] For compound names, see Table XI.2
* Recommended as standard for calibration; C_6H_{12} is preferred at normal temperatures
Reference: as for Table V.13

Table V.16 Values of relative permittivity and loss tangent for some inorganic solids at stated temperatures and frequencies

The range of relative permittivity and loss tangent is intended to indicate variation within stated ranges of temperature and/or frequency; however, values are very sensitive to composition and impurities.

Solid Al_2O_3	relative permittivity	frequency/ Hz	temperature/ K	loss tangent/ 10^{-3}
CERAMIC:				
Al_2O_3	8.5	$50-10^6$	293–373	2–0.5
$CaTiO_3$	150	10^6	293	0.3
$MgTiO_3$	14	$50-10^6$	293–423	0.1–0.4
$PbZrO_3$	110	10^6	293	3
porcelain	5.5	$50-10^6$	293–373	30–8
$SrTiO_3$	200	10^6	293	0.5
$SrZrO_3$	38	10^6	293	0.3
steatite	6	10^6-10^9	293	2
GLASS:				
borosilicate, normal	5.3	10^3-10^6	293	5–4
low alkali	5	10^6	293	3
v.low alkali	4	$50-10^8$	293	1.5–0.5
Corning 0010	6.32		293	
0080	6.75		293	
0120	6.65		293	
8870	9.5		293	
fused quartz	3.8	$50-10^8$	293–423	1–0.1
Pb	6.9	10^3-10^6	293	1.7–1.3
Pyrex 1710	6		293	
3320	4.71		293	
7040	4.65		293	
7050	4.77		293	
7052	5.07		293	
7060	4.7		293	
7720	4.5		293	
7740	5		293	
silica	3.81		293	
soda	7.5	10^6-10^8	293	10–8
MINERAL:				
Amber	2.8–2.6	$10^6-3 \times 10^9$	293	0.2–9
Asbestos board	3	10^6	293	220
Bitumen	3.5	10^3	293	30
granite	8	10^6	293	
gypsum	5.7	10^4	293	
marble	8	10^6	293	40
sand, dry	2.5	10^6	293	
sandstone	10	10^6	293	
soil, dry	3	10^6	293	
SINGLE CRYSTAL:				
AgCl	11.2	10^8	290–295	
apatite$^\perp$	9.5	3×10^8	290–295	
apatite$^\parallel$	7.41	3×10^8	290–295	
$CaCO_3^\perp$	8.5	10^3-10^4	293	
$CaCO_3^\parallel$	8	10^3-10^4	293	
CaF_2	7.4–6.8	$10^4-2 \times 10^6$	293	
diamond	5.7–5.5	$5 \times 10^2-10^8$	293	
GaAs	12	10^3	293	
Ge	16.3	10^3	293	
I_2	4	10^8	290–295	
KBr	5–4.9	10^3-10^{10}	293–298	0.2–0.7
KCl	4.9–4.8	10^3-10^6	293	
KF	5.3–6	10^3-10^6	293	
KI	5.1–5	10^3-10^6	293	
LiBr	13.2–12.1	10^3-10^6	293	
LiCl	11.8–11	10^3-10^6	293	

(Continued)

Table V.16 (Continued)

Solid Al$_2$O$_3$	relative permittivity	frequency/ Hz	temperature/ K	loss tangent/ 10^{-3}
LiF	8.9–9.1	10^3–10^{10}	293–298	0.2
LiI	16.8–11	10^3–10^6	293	
mica	7	50–10^8	293–373	1–0.2
NaBr	6.5–6	10^3–10^6	293	
NaCl	6.1–5.9	10^3–10^{10}	273–278	0.5–0.1
NaF	5.1–6	10^3–10^6	293	
NaI	7.3–6.6	10^3–10^6	293	
periclase	9.7	10^2–10^8	298	0.3
quartz$\perp$,$\parallel$	4.5–4.3	10^3–3×10^7	293	0.2
RbBr	4.9	10^3	293	
RbCl	4.9	10^3	293	
RbF	6.5	10^3	293	
RbI	4.9	10^3	293	
ruby$\perp$	13.3	10^4	290–295	
ruby$\parallel$	11.3	10^4	290–295	
rutile$\perp$	86	50–10^8	293	10–0.2
rutile$\parallel$	170†	10^8	290–295	
S (rhombic) 100	3.8	10^3	298	0.5
010	4	10^3	298	0.5
001	4.4	10^3	298	0.5
sapphire$\perp$	9.4	50–10^9	293	0.2
sapphire$\parallel$	11.6	50–10^9	293	0.2
Se	6.6	10^8	290–295	
Si	11.7	10^3	293	
zircon$\perp$,$\parallel$	12	10^8	290–295	

$\perp$ Indicates field perpendicular to optic axis
$\parallel$ Indicates field parallel to optic axis
† Value critically dependent on stoichiometric ratio
Reference: von Hippel, *Dielectric Materials and Applications*, Chapman and Hall, 1954
 Handbook of Chemistry and Physics, CRC Press, 1991/2
 Kaye and Laby, *Tables of Physical and Chemical Constants*, 15 ed., Longmans, 1986

Table V.17 Values of relative permittivity and loss tangent for some organic solids at stated temperatures and frequencies
The range of relative permittivity and loss tangent is intended to indicate variation within stated ranges of temperature and/or frequency; however, values are very sensitive to composition and impurities.

Solid	relative permittivity	frequency/ Hz	temperature/ K	loss tangent/ 10^{-3}
CELLULOSE & PAPER:				
cellophane	7.6–6.7	50–10^6	293	10–65
cotton rag*	1.7	50–5×10^4	293–363	0.8–6.5
	(3.5)	(50)	(293)	(1.8)
fibre	4.5	10^6	293	50
kraft	2.9	10^3	293	45
kraft, cable*	2.1	50	293–363	1.5–2
	(3.2)	(50)	(293–363)	(1.8–2)
kraft, tissue*	1.8	10^3	293–363	1–1.5
	(3.6)	(50)	(293)	(2.2)
pressboard	3.2	50	293	8
PLASTIC:				
Aniline resin	3.5	3×10^9	293	50
cellulose acetate	3.5	10^6–10^9	293	30–40
cellulose nitrate	6.6	10^6	300	
cellulose triacetate	3.8–3.2	50–10^8	293	10–30
ebonite	3–2.7	10^3–10^9	293	9–3
epoxy resin	3.6–3.5	10^3–10^8	298	20

(Continued)

Table V.17 (continued)

Solid	relative permittivity	frequency/ Hz	temperature/ K	loss tangent/ 10^{-3}
ethyl cellulose	3.01	10^3	298	
melamine resin	4.7	3×10^9	293	40
methyl cellulose	5.7	10^3	295	
phenolic resin (fabric filled)	5.5	10^6	293	50
phenolic resin (paper filled)	5	10^6–10^9	293	30–80
polyacrylate	2.76	10^3	300	
polyamide	4–3	50–10^8	293	20
polycarbonate	3.2–3	50–10^6	293	1–10
poly(dimethyl)phenyloxide	2.6	10^2–10^6	298	0.4–0.7
polyesters	3.12–4	10^3	298	
polyethylene	2.3	50–10^9	293	0.2–0.3
polyethylene terephthalate	3.2–2.9	50–10^8	293	2–1.5
polyimide	3.4	10^6	293	
polyisobutylene	2.2	50–3×10^9	293	0.2–0.5
polymethylmethacrylate	3.4–2.6	50–10^8	293	60–6
poly-4-methylpentene	2.1	10^2–10^4	293	0.2–0.1
polypropylene	2.2	50–10^6	293	0.5
polystyrene	2.6	50–10^9	293	0.2–0.5
polytetrafluororethylene	2.1	50–3×10^9	293	0.2
polyvinyl acetate	4	10^6–10^7	293	50
polyvinyl carbazole	2.8	50–10^8	293	0.5–1
polyvinyl chloride (plasticized)	4	10^6–10^7	293	60
polyvinyl chloride (unplasticized)	3.2–2.8	50–10^8	293	20–10
silicone resins	3.79–3.82	10^3	298	
urea resin (paper filled)	6	10^6	293	30
RUBBER:				
butadiene/styrene (unfilled)	2.5	50–10^8	293–353	0.5–7
butyl (unfilled)	2.4	50–10^8	293	3.5–1
chloroprene	6.5–5.7	10^3–10^6	293	30–90
natural (crepe)	2.4	10^6–10^7	293–253	1.5–10
natural (vulcan,soft)	3.2	10^6–10^7	293	28–20
silicone (67% TiO_2 filled)	8.6–8.5	50–10^8	293	5–1
silicone (unfilled)	3.2–3.1	10^3–10^8		
WAX:				
beeswax	2.75–3	10^6	293	
carnuba	2.75–3	10^6	293	
ozokerite	2.3	50–10^8	293	0.5–1
paraffin	2.2	10^6–10^9	293	0.2
petroleum jelly	2.1–1.9	50	293–333	0.1–0.5
rosin	2.4	3×10^9	293	0.6
tetrachloronapthalene	5.4–4.2	50–10^8	293	0.7–270
trichloronapthalene	5.4–4.2	50–10^8	293	0.7–270
WOOD[†]:				
balsa (0)	1.4–1.2	50–3×10^9	293	4–14
beech (16)	9.4–8.5	10^6–10^8	293	58–83
birch (10)	3.1	10^6–10^8	293	40–80
fir (11)	3.2	10^6–10^7	288	52–81
mahogany	2	3×10^9	293	34
pine (15)	8.2–7.3	10^6–10^8	293	59–94
walnut (0)	2	10^7	293	35
walnut (17)	5	10^7	293	140
whitewood (10)	3	10^6–10^8	293	40–75

* Values given first (second) are for dry unimpregnated (mineral-oil impregnated) material
† Bracketed numbers give percentage H_2O
Reference: as for Table V.16

Table V.18 Representative variation of magnetic flux density, B, with magnetic field strength, H, for various low-coercivity magnetic substances, together with initial and maximum relative permeability, μ_r, saturation polarization, J_s, coercivity, H_c, remanence, B_r, Curie temperature, T_C, resistivity, ρ, and specific total power loss, f_{tot}

Substance†				H/A m^{-1}			μ_r/ 10^3		J_s/ T
	1	10	50	10^3	5×10^3	5×10^4	init	max	
				B/T					
Co				0.21	0.7	1.22			1.76
Fe, high purity	1.5	1.99	2.01					1500	2.16
Fe, Armco		0.005	0.06	1.55	1.72	2.2	0.25	7	2.16
Fe, cast annealed				0.6	0.86	1.45			1.7
Fe, electrolytic		0.02	0.9	1.68	1.74	2.2			2.16
Fe, Swedish annealed		0.01	0.15	1.52	1.72	2.2			2.16
Fe-B ALLOYS:									
Metglas 260553									1.58
Metglas 260556									1.61
Fe-Co ALLOYS:									
Permendur 24 (24 Co, 76 Fe)		0.002	0.02	1.45	1.85	2.3	0.25	2	2.35
Permendur 49 (49 Co, 51 Fe)		0.01	0.13	1.85	2.3	2.4	1	7	2.35
Supermendur (49 Co, 49 Fe, 2V)			2.05	2.3	2.35	2.4		70	2.35
Fe-Ni ALLOYS □:									
HCR (50 Fe, 50 Ni)		0.3	1.46	1.55			0.5–1	50–100	1.6
Hyperm 36 (64 Fe, 36 Ni)							1.75	6	
Mumetal° (20–30 Fe, 70–80 Ni)	0.18	0.6	0.72				60	240	0.77
Mumetal plus° (20–30 Fe, 70–80 Ni)	0.3	0.7	0.75				80	300	0.77
Nilomag 641° (c. 35 Fe, 65 Ni)							0.4–1	200–400	1.4
Radiometal 4550 (c. 50 Fe, 50 Ni)	0.01	0.48	1.05	1.62			3–6	20–50	1.6
Radiometal 36 (65 Fe, 35 Ni)		0.15	0.72	1.2			2	15	1.3
Satmumetal (c. 50 Fe, 50 Ni)	0.2	1.15	1.3				65	240	1.5
Supermumetal° (20–30 Fe, 70–80 Ni)	0.45	0.72	0.76				140	350	0.77
Super Radiometal° (c. 50 Fe, 50 Ni)	0.06	1.02	1.25	1.62			2–11	50–120	1.6
Heusler alloy (13 Al, 61 Cu, 26 Mn)				0.25	0.38	0.45			0.48
Isoperm (11 Cu, 59 Fe, 30 Ni)							0.06	0.065	
Ni				0.45	0.55	0.68			0.615
Ni-Cu alloy (30 Cu, 70 Ni)				0.07	0.1				
Ni (wrought)							0.25	2	0.6
Perminvar (25 Co, 35 Fe, 40 Ni)							0.3	1.5	1.55
SHEET STEELS:									
Losil 400–50 (97.6 Fe, 2.4 Si)			0.06	0.4	1.48	1.69		7	2.03
Losil 800–65 (98.4 Fe, 1.6 Si)				0.4	1.53	1.73		5	2.08
Newcor 1000–65 (no Si)				0.38	1.59	1.75		8	2.15
Transil 300 35 (97.1 Fe, 2.9 Si)			0.1	0.54	1.46	1.65		8	2
Unisil*:	(96.9 Fe, 3.1 Si)								
27M4			0.95	1.56	1.86	1.96		75	2
30M5			0.75	1.56	1.86	1.96		59	2
35M6			0.7	1.53	1.84	1.94		58	2
Unisil H, 30M$_2$H* (97.1 Fe, 2.9 Si)			1.21	1.72	1.93	2	2.04	93	2
SOLID STEELS:									
C(annealed) (1 C, 99 Fe)				0.75	1.54	2			2
cast		0.004	0.05	1.38	1.68	2.12			2.1
constructional (0.3 C, 98.7 Fe, 1 Ni)				1.32	1.68	2.12			2.1
constructional (0.4 C, 1.5 Cr, 95.1 Fe, 3 Ni)				0.75	1.67	2.08			2.05
mild (0.1 C, 99.9 Fe)		0.004	0.055	1.46	1.74	2.15			2.15

† Mass % age composition of magnetic substance is given in brackets where appropriate.
□ names given by Telcon Metals, except for Nilomag 641 from Wiggin & Co.
° other elements also present in small amounts.
* grain oriented with preferred magnetic properties in direction of rolling of parent strip.

H_c/ A m^{-1}	B_r/ T	T_C/ K	ρ/ $10^{-8}\Omega$ m	f_{tot}/ W k g^{-1}
950		1388	9	
12		1043	10	
80	1.3	1043	11	
400				
30		1043	10	
70		1043		
8	0.7	678	125	0.15(1.7 T)
5	1.1	643	125	
950	1.65	1253	20	0.33(0.5 T)
140	1.5	1253	47	
20	2.1	1253	40	
10	1.5	798	40	
1	0.45	623	55	0.01(0.5 T)
0.8	0.45	623	55	
4	1.35	863	48	
12–24	0.4–1	803	50	0.12(0.5 T)
12	0.35	453–543	80	0.12(0.5 T)
2	0.7	823	60	
0.55	0.5	623	55	0.11(400 Hz, 0.5 T)
4–12	0.4–1.2	803	43	0.55(400 Hz, 0.5 T)
550		603		
400		631	9	
		283–373		
120	0.3	631	7	
100		988	19	
40		1021	44	3.6
70		1031	34	6.5
50		1043	12	8
40		1018	48	2.95
7		1018	48	0.84
7		1018	48	0.89
7		1018	48	1
6		1018	45	1.12 (1.7 T)
600				
250				
250				
500				
150			10	

f_{tot} is given for saturation polarization of 1.5 T and frequency 50 Hz, unless otherwise indicated.
μr initial is given for field strength of 0.4 A m^{-1}, except for Hyperm 36, for which the value is 0.8 A m^{-1}.
Reference: Kaye and Laby, *Tables of Physical and Chemical Constants*, 15 ed., Longmans, 1986

Table V.19 Representative values for some core materials of initial relative permeability, $\mu_{r,0}$, frequency range, $\Delta \nu$, loss factor f_L at maximum frequency, Curie temperature, T_C, and resistivity, ρ, together with some other information.

Core[†]		$\mu_{r,0}$	$\Delta\nu$/MHz	$f_L/10^{-6}$	T_C/K	$\rho/\Omega\,\text{m}$	other information
CARBONYL Fe POWDER:							
	type 100	30	0.1–2	700			permeability temperature factor/ 10^{-6} K^{-1} = 20
	type 500	12	1–10	250			permeability temperature factor/ 10^{-6} K^{-1} = 12
	type 900	10	1–50	600			permeability temperature factor/ 10^{-6} K^{-1} = 12
	type 901	5	10–100	1500			permeability temperature factor/ 10^{-6} K^{-1} = 12
Fe flake							$\mu_{r,0}$ at 1 kHz = 90; $\mu_{r,0}$ at 150 kHz = 65
FERRITE:							
Mn/Zn	type F5	2000			473	1	°flux density/ T = 0.48; *power loss density/mW cm^{-3} = 75
	type F6	1500			453	1	°flux density/ T = 0.45; *power loss density/mW cm^{-3} = 150
	type F8	1500	0.05–0.5	80	453	1	
	type F10	5000	0.01–0.1	12	453	1	
	type F11	600	0.1–1	50	493	5	
Mn Zn	type P10	2000			423	1	permeability temperature factor/ 10^{-6} K^{-1} = 0–2; 100 kHz loss factor/10^{-6} = 12
	type P11	2200			423	1	permeability temperature factor/ 10^{-6} K^{-1} = 0.5–1.5; 100 kHz loss factor/10^{-6} = 5
	type P12	2200			423	1	permeability temperature factor/ 10^{-6} K^{-1} = 0.4–1; 100 kHz loss factor/10^{-6} = 3
Ni/Zn	type F13	650	0.05–1	130	453	300	
	type F14	220	0.1–2	50	543	1000	
	type F16	125	1–10	100	543	1000	
	type F22	19	5–40	500	773	1000	
Perminvar	type F25	50	5–40	300	723	1000	
	type F29	12	10–200	1000	773	1000	

[†] Supplier: Neosid Ltd.
° for magnetic field strength of 800 Am^{-1}
* for flux density of 0.2 T at 16 kHz frequency
Reference: Snelling: *Soft Ferrites, Properties and Applications*, 2nd edn, Butterworth-Heinemann 1988

Table V.20 **Representative values of remanence, B_r, maximum of product $H.B$ of magnetic field strength and magnetic flux density (with field strength, in A m^{-1}, given in brackets), induction and magnetization coercive forces, $H_{c,B}$ and $H_{c,M}$, resistivity, ρ, Curie temperature, T_C, and maximum operating temperature, T_{max}, for some permanent-magnet materials**

Material[†]	B_r/ T	$H.B$/ kJ m^{-3}	$H_{c,B}$/ kA m^{-1}	$H_{c,M}$/ kA m^{-1}	ρ/ $\mu\Omega$ m	T_C/ K	T_{max}/ K
Al-Ni-Co ALLOYS:							
a Alcomax III (8 Al, 24 Co, 3Cu, 51.5 Fe, 13.5 Ni)	1.3	43 (42)	52	53	0.55	1123	823
cp Alcomax III (8 Al, 24 Co, 3Cu, 51.5 Fe, 13.5 Ni)	1.32	49 (45)	56	57	0.55	1133	823
i Alni (13 Al, 4 Cu, 58 Fe, 25 Ni)	0.56	10 (28)	46	49	0.63	1033	823
i Alnico (10 Al, 12 Co, 6 Cu, 53 Fe, 19 Ni)	0.73	13.5 (29)	45	48	0.65	1073	823
c Columax (8 Al, 24 Co, 3 Cu, 51.5 Fe, 13.5 Ni)	1.35	60 (51)	59	60	0.55	1133	823
a Hycomax II (7 Al, 29 Co, 4.5 Cu, 40 Fe, 14.5 Ni, 5 Ti)	0.85	32 (60)	95	97	0.5	1123	823
a Hycomax III (7.3 Al, 34 Co, 3 Cu, 36.45 Fe, 14 Ni, 5.25 Ti)	0.9	44 (85)	127	129	0.5	1123	823
CO-RARE EARTH ALLOYS:							
a, s Supermagloy (66 Co, 34 Sm)	0.8–0.93	120–175	610–700	1200–1400	0.7	998	473
bp Supermagloy (66 Co, 34 Sm)	0.3–0.59	16–64	220–420	480–800	10 000		333
FERRITES:							
a, b Feroba	0.25	11.2 (96)	175	240	100		393
i, s Feroba 1 (5.9 BaO, 94.1 Fe$_2$O$_3$)	0.22	8 (72)	135	220	10	723	453
a, s Feroba 2 (5.9 BaO, 94.1 Fe$_2$O$_3$)	0.39	28.5 (136)	150	160	10	723	453
a, s Feroba 3 (5.9 SrO, 94.1 Fe$_2$O$_3$)	0.37	26 (138)	240	300	10	723	453
b, i Feroba bonded	0.15	3.2 (50)	85	175	100		393

[†] mass % age composition given in brackets; a, b, bp, c, cp, i and s respectively indicate anisotropic, bonded, polymer-bonded, columnar, semicolumnar, isotropic and sintered materials

Reference: Kaye and Laby, *Tables of Physical and Chemical Constants*, 15 ed., Longmans, 1986

Table V.21 Representative values of relative permeability, μ_r, and resistivity, ρ, for some feebly magnetic steels and cast irons for various conditions

Substance	approx mass %age composition	condition	μ_r	ρ/Ω m
AUSTENITIC STAINLESS STEELS:				
AISI type 301	17.6 Cr, 74.6 Fe, 7.6 Ni	austenized	1.003	0.68
		19.5% cold reduction	1.15	
		55% cold reduction	14.8	
type 302	18.4 Cr, 72.6 Fe, 9 Ni	austenized	1.003	0.7
		20% cold reduction	1.008	
		44% cold reduction	1.05	
		68% cold reduction	1.59	
type 304	19 Cr, 70.3 Fe, 10.7 Ni	austenized	1.004	0.72
		13.8% cold reduction	1.005	
		32% cold reduction	1.04	
		65% cold reduction	1.55	
type 305	17.9 Cr, 70.4 Fe, 11.7 Ni	austenized	1.003	
		18.5% cold reduction	1.004	
		52.5% cold reduction	1.05	
type 310	24.3 Cr, 55 Fe, 20.7 Ni	austenized	1.002	0.94
		64.2% cold reduction	1.002	
type 316	17.5 Cr, 69.1 Fe, 13.4 Ni	austenized	1.003	0.74
		81% cold reduction	1.007	
type 321	18.3 Cr, 70.72 Fe, 10.3 Ni, 0.68 Ti	austenized	1.003	0.72
		16.5% cold reduction	1.018	
		41.5% cold reduction	1.4	
type 347	0.95 Co, 18.4 Cr, 69.95 Fe, 10.7 Ni	austenized	1.004	0.73
		13.5% cold reduction	1.007	
		40% cold reduction	1.06	
		60% cold reduction	1.25	
CAST IRONS:				
ductile Ni-resists, types D2, D2B, D2C, D4	0–5 Cr, 57.6–82 Fe, 0–2.4 Mn, 18–32 Ni	as cast	1.02–1.1	
Ni resist type 1	2 Cr, 6 Cu, 78 Fe, 14 Ni	as cast	1.03	
Nomag	83 Fe, 6 Mn, 11 Ni	as cast	1.03–1.05	

μ_r is given for magnetic field strength of 5 k A m^{-1}.
Reference: Post and Eberley, *Trans. Am. Soc. Metals*, 39, 868, 1947

Table V.22 Mass susceptibility per kilogram, χ_{mass}, at temperature 293 K for some paramagnetic and diamagnetic materials
The SI values quoted may be converted to cgs units per gram by multiplying by $10^3/4\pi$

Element	Ag	Al	Ar	As	Au
$\chi_{mass}/10^8$	−0.25	0.82	−0.6	−0.39	−0.19
Element	Hg	In	K	Kr	N
$\chi_{mass}/10^8$	−0.21	−0.14	0.65	−0.41	−0.54
Element	S	Sb	Si	Te	U
$\chi_{mass}/10^8$	−0.62	−1.09	−0.16	−0.39	2.19
Inorganic compound	CO_2	$CuSO_4 \cdot 5H_2O$	$FeSO_4 (NH_4)_2 SO_4 \cdot 6H_2O$	HCl	H_2O
$\chi_{mass}/10^8$	−0.59	7.7	40.6	−0.75	−0.9
Inorganic compound	NH_3	NO	NaCl	anhyd.$NiCl_2$	$NiSO_4 \cdot 7H_2O$
$\chi_{mass}/10^8$	−1.38	59.3	−0.64	78.5	20.1
Organic compound	araldite	CCl_4	$CHCl_3$	CH_3CN	CH_3COCH_3
$\chi_{mass}/10^8$	0.63	0.54	0.62	0.86	0.73
Organic compound	CH_3COOCH_3	CH_3OH	C_2H_5OH	$(C_2H_5)_2O$	C_5H_5N
$\chi_{mass}/10^8$	0.72	0.84	0.91	0.93	0.78
Organic compound	C_6H_5Cl	$C_6H_5NH_2$	C_6H_6	$C_6H_{10}O$	$C_6H_{11}OH$
$\chi_{mass}/10^8$	0.78	0.85	0.88(at 305 K)	0.79	0.92
Organic compound	glycerol	glycol	perspex	polyethylene	PVC
$\chi_{mass}/10^8$	0.78	0.78	0.5	−0.2	0.75

Reference: *Handbook of Chemistry and Physics*, CRC Press, 1991/2
Kaye and Laby, *Tables of Physical and Chemical Constants*, 15 ed., Longmans, 1986.

Bi	Cu	Ga	Ge	H	He
−1.7	−0.107	−0.3	−0.15	−2.49	−0.59
Na	Ne	O	P	Pb	Pt
0.75	−0.41	133.6	−1.13	−0.15	1.22
Xe					
−0.4					
H_2SO_4	$MnSO_4 \cdot 4H_2O$				
−0.5	81.2				
$NiSO_4 \cdot K_2SO_4 \cdot 7H_2O$					
13.9					
$CH_3COC_2H_5$	CH_3CONH_2				
0.79	0.73				
$C_6H_5CH_3$	C_6H_5CH				
0.9	0.72 (at 283 K)				
C_6H_{12}	C_8H_{18}	$C_{10}H_{18}$			
1.02 (at 300 K)	1.06	0.9			

Table V.23 Values of critical temperature,† T_c, and critical magnetic field(s), H_c at temperature of 0 K for some superconducting elements

Element	T_c/K	H_c/mT	Element	T_c/K	H_c/mT	Element	T_c/K		H_c/mT
Al	1.75	10.5	Hg(β)	3.949	33.9	Pb	7.2		80.3
Am(α)	0.6		In	3.41	28.2	Re	1.7		20
Am(β)	1		Ir	0.11	1.6	Ru	0.49		6.9
Be	0.026		La(α)	4.88	80	Sn	3.72		30.5
Cd	0.517	2.8	La(β)	6	160	Th	1.38		16
Ga(α)	1.083	5.8	Lu	0.1	35	Ti	0.4		5.6
Ga(β)	5.9, 6.2	56	Mo	0.92	9.6	W	0.015		0.115
Hf	0.128	1.27	Os	0.66	7	Zn	0.85		5.4
Hg(α)	4.154	41.1	Pa	1.4		Zr	0.61		4.7
Ga(γ)	7.62	300	Nb	9.25	173 405	Tc	7.8	116	312
Ga(δ)	7.85		Ta	4.47	45 200	V	5.4	26	268

† The temperature at which resistivity falls sharply to zero, in the absence of any external magnetic field and with no current flowing (contrast with meaning in table XI 12)
Reference: Kaye and Laby, *Tables of Physical and Chemical Constants*, 15 ed., Longmans, 1986

Table V.24 Values of critical temperature,† T_c, and critical magnetic field, H_c, at temperature of 4.2 K for some superconducting compounds and alloys

Compound	T_c/K	H_c/mT	Compound	T_c/K	H_c/mT	Compound	T_c/K	H_c/mT
*$Bi_2Sr_2CaCu_2O_x$	92		$NbC_{0.3}N_{0.7}$	17.8	12.5	TaC	10.35	0.5
La_2C_3	11		Nb_3Ga	20.7	33	TaN	14.3	
$(LaSr)_2CuO_4$	35		Nb_3Ge	23.6	37	*$Tl_2Ba_2Ca_2Cu_3O_z$	92	
$(LaTh)_2C_3$	14.3		NbN	17.3	15	V_3Ga	15.3	25
$LiTi_2O_4$	13.7	20	Nb_3Si	18.6	14.5	V_2Hf	9.2	20
$Li_{0.3}Ti_{1.1}S_2$	13		Nb_3Sn	18.3	26	V($Hf_{0.5}Z_{0.5}$)	10.1	24
MoC	14.3	5.2	$Nb_{0.15}Ti_{0.25}$	9.8	10	$V_{1.83}Nb_{0.13}Hf_{0.15}$	10.4	25
MoN	14.8		$Nb_{0.44}Ti_{0.56}$	9	14.1	V_3Si	17.1	23
$Mo_{0.66}Re_{0.34}$	11.8	1.2	(NbTi)N	18	13.5	$V_{1.81}Ta_{0.19}Hf_{0.11}$	10	26
Nb_3Al	19.1	29.5	$Nb_{0.75}Zr_{0.25}$	10.9	8.3	WC	10	
$Nb_3(Al_{0.7}Ge_{0.3})$	20.7	41	$Nb_{0.67}Zr_{0.33}$	11	8.3	*$YBa_2Cu_3O_y$	92	
NbC	11.1	1.7	$PbMo_{0.51}S_6$	14.4	51	Y_2C_3	11.5	
			$(Pd_{0.55}Cu_{0.45})H_{0.7}$	16.6		$(YTh)_2C_3$	17	

† See Table V. 23 footnote†
* According to the first reference, the broad spectrum of electrical engineering applications originally envisaged for these compounds may be difficult or impossible to achieve. However, they may be ideally suited for use as sensors for detecting small magnetic fields, temperature changes and strains. Moreover, the Y- and Bi-based materials may act as selective filters for gas molecules at 77 K and so have applications in gas liquefaction and purification. The situation is updated in the second and third references.
Reference: Taylor, *Physics World*, vol 2, 7/22 IOP, 1989
           Gough, *Physics World*, vol 4, 12/26 IOP, 1991
           Gough, *Physics World*, vol 5, 1/5 IOP, 1992
           Kaye and Laby, *Tables of Physical and Chemical Constants*, 15 ed., Longmans, 1986

Section VI
Electrochemistry

		Page
VI.1	Conductivity of potassium chloride solutions	89
VI.2	Molar conductivity of electrolytes in aqueous solution at 25 °C	89
VI.3	Limiting molar ionic conductivities in aqueous solution at 25 °C	90
VI.4	Limiting molar ionic conductivities in aqueous solution at various temperatures	91
VI.5	Limiting molar ionic conductivities at 25 °C in a range of solvents	92
VI.6	Electrical properties of fused salts	92
	1 Equivalent conductivities of single salt melts	92
	2 Cation transport numbers for fused salts	93
VI.7	Debye-Hückel-Onsager coefficients for 1:1 electrolytes at 25 °C	93
VI.8	Ionic strength as a function of concentration for various valence type electrolytes	93
VI.9	Relationship between molality, mean ionic molality, activity and mean ionic activity coefficient for various valence type electrolytes	93
VI.10	Mean stoichiometric ionic activity coefficients at 25 °C for electrolytes in aqueous solution	94
VI.11	Mean ionic activity coefficients for hydrochloric acid in water	95
VI.12	Mean ionic activity coefficients for hydrochloric acid at 25 °C in dioxane-water mixtures	95
VI.13	Activity coefficients calculated using the Debye-Hückel equation	96
VI.14	Transport numbers of cations in aqueous solution at 25 °C	97
VI.15	Transport numbers of cations of 2:1 electrolytes in aqueous solution at 25 °C, showing the effect of auto-complex formation	97
VI.16	Cation transport numbers of hydrochloric acid at 25 °C in dioxane-water mixtures	97
VI.17	Cation transport numbers in aqueous solution at various temperatures	98
VI.18	Diffusion coefficients of aqueous solutions of selected electrolytes at 25 °C	98
VI.19	Osmotic and activity coefficients of sodium, potassium and calcium chloride and sulphuric acid solutions at 25°C	99
VI.20	Osmotic and activity coefficients of sucrose solutions at 25 °C	99
VI.21	pH values of operational reference standard solutions	100
VI.22	pH values of useful buffer solutions	100
VI.23	Biological buffer solutions	102
VI.24	pH values of some common buffer solutions	102
VI.25	Approximate pH values of some common solutions, fluids and foodstuffs	105
VI.26	Acid-base indicators	107
VI.27	Standard (reduction) potentials at 298 K in aqueous solution	107
VI.28	Formal electrode potentials at 298 K	109
VI.29	Standard redox potentials for some biological half reactions at 298 K and pH = 7.0	109
VI.30	Standard (reduction) potentials in non-aqueous solution at 298 K (except liquid ammonia 265 K)	109
VI.31	Redox indicators for volumetric use	110
VI.32	Biological redox indicators	110
VI.33	Dissociation constants of inorganic acids in aqueous solution at 298 K	110
VI.34	Dissociation constants of organic acids in aqueous solution at 298 K	111
VI.35	Dissociation constants of inorganic bases in aqueous solution at 298 K	113
VI.36	Dissociation constants of organic bases in aqueous solution at 298 K	113
VI.37	Ionization constants for water and deuterium oxide	114
VI.38	Autoprotolysis constants for some solvents at 298 K	115
VI.39	Solubility products in water at 298 K	115
VI.40	Stepwise formation (stability) constants	116
VI.41	Formation constants for EDTA complexes at 293 K and an ionic strength of 0.1 mol dm^{-3}	117
VI.42	Hammett acidity function	117
VI.43	Half-wave potentials of selected inorganic species solvents	118
	1 Half-wave potentials of metals	118
	2 Half-wave potentials of some other inorganic species	119
	3 Half-wave potentials of some cations in non-aqueous solvents	120
VI.44	Half-wave potentials of selected organic compounds	122
VI.45	Isoelectric points (or range) for several proteins	123

Section VI
Electrochemistry

Table VI.1 Conductivity of Potassium Chloride Solutions
Standard solutions for calibrating conductivity cells, from: G. Jones and B.C. Bradshaw, *J. Amer. Chem. Soc.* (1933), **55**, 1780. Converted from (int. ohm)$^{-1}$ cm^{-1}.

Concentration	g KCl per 1 kg of solution in vacuo	κ/ohm^{-1} m^{-1}		
		0°C	18°C	25°C
1 D (demal)	71.135 2	6.514$_4$	9.779$_0$	11.128$_7$
0.1 D	7.419 13	0.7134$_4$	1.1161$_2$	1.2849$_7$
0.01 D	0.745 263	0.07732$_6$	0.12199$_2$	0.14080$_8$

The following equations may be used to calculate the molar conductivity of solutions of potassium chloride at 25 °C, and hence the conductivity ($\kappa = \Lambda c$):
For concentrations, c mol m^{-3}, in the range 0 to 20 mol m^{-3}:

$$\Lambda/\text{ohm}^{-1}\,\text{m}^2\,\text{mol}^{-1} = 149.93 \times 10^{-4} - 2.994 \times 10^{-4} c^{1/2} + 58.74 \times 10^{-7} c \lg c + 22.18 \times 10^{-7} c$$

(J.E. Lind, J.J. Zwolenik and R.M. Fuoss, *J. Amer. Chem. Soc.* (1959), **81**, 1557)
For concentrations, c mol dm^{-3}, in the range 0.01 to 0.1 mol dm^{-3}:

$$\Lambda/\text{ohm}^{-1}\,\text{cm}^2\,\text{mol}^{-1} = 149.86 - 94.83c + 25.47c \lg c + 229c^{3/2}$$

(Ying-Chech Chiu and R.M. Fuoss, *J. Phys. Chem.* (1968), **72** 4123)

Table VI.2 Molar Conductivity Λ of Electrolytes in Aqueous Solution at 25 °C
Prior to about 1968, the literature employed the quantity equivalent conductance $\Lambda_{\text{equiv}} = 1000 \kappa / c_{\text{equiv}}$ (units: ohm^{-1} cm^2 g equiv^{-1}) in which the concentration was expressed as equivalents per litre. To convert Λ_{equiv} to Λ, divide by 10^4 and multiply by the number of charges associated with the ionized molecule (i.e. zv); thus for aluminium sulphate $zv = 6$.
Reference: R.A. Robinson and R.H. Stokes, *Electrolyte Solutions*, 2nd ed. (1968), Butterworths, London.

1:1 Electrolytes

c/mol dm^{-3}	0	0.0005	0.0010	0.005	0.01	0.02	0.05	0.10
LiCl	115.03	113.15	112.40	109.40	107.32	104.65	100.11	95.86
NaCl	126.45	124.50	123.74	120.65	118.51	115.76	111.06	106.74
NaI	126.94	125.36	124.25	121.25	119.24	116.70	112.79	108.78
NaOH	247.80	245.60	244.70	240.80	238.00			
NaOOCCH$_3$	91.00	89.20	88.50	85.72	83.76	81.24	76.92	72.80
NaClO$_4$	117.48	115.64	114.87	111.75	109.59	106.96	102.40	98.43
KCl	149.86	147.81	146.95	143.55	141.27	138.34	133.37	128.96
KBr	151.90		146.09	143.43	140.48	135.68	131.39	
KI	150.38		144.37	142.18	139.45	134.97	131.11	
KNO$_3$	144.96	142.77	141.84	138.48	132.82	132.41	126.31	120.40
KHCO$_3$	118.00	116.10	115.34	112.24	110.08	107.22		
KClO$_4$	140.04	138.76	137.87	134.16	131.46	127.92	121.62	115.20
NH$_4$Cl	149.70		146.80	143.50	141.28	138.33	133.29	128.75
AgNO$_3$	133.36	131.36	130.51	127.20	124.76	121.41	115.24	109.14
CH$_3$COOH	390.72		48.63	22.80	16.20	11.57	7.36	5.20

2:1 Electrolytes

c/mol dm^{-3}	0	0.00025	0.0005	0.0025	0.005	0.01	0.025	0.05
			$10^4 \times \Lambda$/ohm^{-1} m^2 mol^{-1}					
CaCl$_2$	271.68	263.86	260.72	248.50	240.72	231.30	216.94	204.92
SrCl$_2$	271.60	263.80	260.66	248.48	240.58	231.08	216.50	204.38
BaCl$_2$	279.38	271.92	268.68	256.04	247.88	238.18	222.96	210.38
MgCl$_2$	258.80	251.22	248.22	236.62	229.10	220.08	206.16	194.20
Na$_2$SO$_4$	259.80	251.48	248.30	234.30	224.88	213.56	195.50	179.96

2:2 Electrolytes

c/mol dm^{-3}	0	0.00025	0.0005	0.0025	0.005	0.01	0.025	0.05
			$10^4 \times \Lambda$/ohm^{-1} m^2 mol^{-1}					
CuSO$_4$	534.4	486.40	461.04	376.28	332.48	288.80	236.20	202.32
ZnSO$_4$	531.2	485.60	462.12	381.96	339.64	296.96	244.80	210.56
NiSO$_4$		474.80	452.40	372.80	330.80	289.20	236.80	203.20

3:1 Electrolytes

c/mol dm^{-3}	0	0.00017	0.00033	0.0017	0.0033	0.0066	0.017	0.033
			$10^4 \times \Lambda$/ohm^{-1} m^2 mol^{-1}					
LaCl$_3$	437.4	418.8	411.0	382.5	365.4	345.9	318.6	297.3
K$_3$Fe(CN)$_6$	523.5	499.2	489.3	452.1				

4:1 Electrolytes

c/mol dm^{-3}	0	0.000 125	0.000 25	0.001 25	0.0025	0.005	0.0125	0.025
			$10^4 \times \Lambda$/ohm^{-1} m^2 mol^{-1}					
K$_4$Fe(CN)$_6$	738.0		668.96	584.36	539.32	491.28	430.80	391.48

Table VI.3 Limiting Molar Ionic Conductivities λ_i° in Aqueous Solution at 25 °C

The ionic mobility u_i may be calculated from the relationship $u_i = \lambda_i^\circ / F$; e.g. for the sodium ion at 25 °C, u° (Na$^+$) = 50.13 × 10^{-4}/964 85 = 5.20 × 10^{-8} m^2 V^{-1} s^{-1}.

For more extensive data see: R.A. Robinson and R.H. Stokes, *Electrolyte Solutions* (2nd edn, 1968), Butterworths, London.

Anions	$10^4 \times \lambda_i^\circ$/ohm^{-1} m^2 mol^{-1}	Cations	$10^4 \times \lambda_i^\circ$/ohm^{-1} m^2 mol^{-1}
Ag$^+$	61.92	Br$^-$	78.17
Al^{3+}	189	CO$_3^{2-}$	138.6
Ba^{2+}	127.3	Cl$^-$	76.34
Cd^{2+}	108.0	ClO$_3^-$	64.58
Cs$^+$	77.26	ClO$_4^-$	67.31
Ca^{2+}	119.0	CN$^-$	82
Co^{2+}	98	CNS$^-$	66.0
Cr^{3+}	201	F$^-$	55.41
Cu^{2+}	113.2	Fe(CN)$_6^{3-}$	297.3
Fe^{2+}	107.0	Fe(CN)$_6^{4-}$	444
Fe^{3+}	204	HCO$_3^-$	44.5
H$_3$O$^+$	349.85	HSO$_4^-$	52
Hg$_2^{2+}$	137.2	I$^-$	76.85
Hg^{2+}	127.2	IO$_3^-$	40.75
K$^+$	73.50	IO$_4^-$	54.53
La^{3+}	209.25	MnO$_4^-$	61.3
Li$^+$	38.68	NO$_3^-$	71.44
Mg^{2+}	106.12	OH$^-$	197.6

(Continued)

Table VI.3 (Continued)

Anions	$10^4 \times \lambda_i^\circ$/ohm^{-1} m^2 mol^{-1}	Cations	$10^4 \times \lambda_i^\circ$/ohm^{-1} m^2 mol^{-1}
Mn^{2+}	107.0	PO$_4^{3-}$	281.4
Na$^+$	50.15	SO$_4^{2-}$	159.6
NH$_4^+$	73.55	S$_2$O$_3^{2-}$	174.8
(CH$_3$)$_4$N$^+$	44.92	HCOO$^-$	54.6
(C$_2$H$_5$)$_4$N$^+$	32.66	CH$_3$COO$^-$	40.90
(C$_3$H$_7$)$_4$N$^+$	23.42	C$_2$H$_5$COO$^-$	35.8
(C$_4$H$_9$)$_4$N$^+$	19.47	C$_2$O$_4^{2-}$	148.2
(C$_5$H$_{11}$)$_4$N$^+$	17.47	CH$_2$(COO$^-$)$_2$	26.8
Ni^{2+}	108	PhCOO$^-$	32.38
Pb^{2+}	140	Pic$^-$	31.4
Rb$^+$	77.81		
Sr^{2+}	59.45		
Tl$^+$	74.7		
Zn^{2+}	105.6		

Table VI.4 Limiting Molar Ionic Conductivities in Aqueous Solution at Various Temperatures

The ionic mobility, u_i, may be calculated from the relationship: $u_i = \lambda_i/F$; e.g. for the sodium ion at 25 °C, u° (Na$^+$) = 50.13 × 10^{-4}/96485 = 5.20 × 10^{-8} m^2 V^{-1} s^{-1}.

^{a}: Data at 50 °C.
^{b}: Data at 75 °C.
For more extensive data see: R.A. Robinson and R.H. Stokes, *Electrolyte Solutions* (2nd edn, 1968), Butterworths, London.

Ion				$10^4 \times \lambda_i^\circ$/ohm^{-1} m^2 mol^{-1}			
	t/°C	0	18	25	35	45	100
Ag$^+$		32.9	54.3	61.92			175
Ba^{2+}		68.0	109.2	127.3	208^a	298^b	390
Ca^{2+}		62.4	101.4	119.0	146.6	176.4	360
H$_3$O$^+$		225	315	349.85	397.0	441.4	630
K$^+$		40.7	63.9	73.5	88.2	103.5	195
La^{3+}		103.2	178.5	209.25	357^a	519^b	645
Li$^+$		19.4	32.8	38.68	48.0	58.0	115
NH$_4^+$		40.2	63.9	73.55	88.7		180
Na$^+$		26.5	42.8	50.15	61.54	73.7	145
Rb$^+$		43.9	66.5	77.81	92.9	108.6	
Cl$^-$		41.0	66.0	76.34	92.2	108.9	212
Br$^-$		42.6	68.0	78.17	94.0	110.7	
Fe(CN)$_6^{4-}$		232	380	444.0	692^a	976^b	1284
I$^-$		41.4	66.5	76.85	92.3	108.6	
NO$_3^-$		40.0	62.3	71.44	85.5		195
OH$^-$		105	175.8	197.6	233.0	267.2	450
SO$_4^{2-}$		82	136.8	159.6			520
CH$_3$COO$^-$		20.1	34.6	40.90			
C$_2$O$_4^{2-}$		78	126	146	230^a	326^b	426

Table VI.5 Limiting Molar Ionic Conductivities ($10^4 \lambda_i^\circ / \Omega^{-1}$ m^2 mol^{-1}) at 25 °C in a Range of Solvents

NMe$_4^+$: tetramethylammonium ion: NEt$_4^+$: tetraethylammonium ion. NPr$_4^+$: tetra-iso-propylammonium ion: NBu$_4^+$ tetra-t-butylammonium ion: NAm$_4^+$: tetra-amylammonium ion. Pic$^-$: picrate ion.
Reference: as for Table VI.4.

	Water	Methanol	Ethanol	Ethylene glycol	Formamide	N-Methyl formamide	Dimethyl formamide	Acetonitrile	Nitromethane	Dimethyl sulphoxide	Sulpholene (303 K)	N-Methyl acetamide
ε_r	80.4	33.7	24.5	41.2	109	182.4	36.7	37	36.7	46.7	43.3	165
H$^+$	349.82	146.2	62.8	27.7	10.8		34.7	9.9		15.0		8.95
Li$^+$	38.68	39.6	17.09	2.11	9.03		25.0	69.3			4.34	6.45
Na$^+$	50.11	45.2	20.32	3.11	10.10	21.6	29.9	76.9		14.2	3.61	8.1
K$^+$	73.52	52.5	23.57	4.62	12.75	22.2	30.8	83.6		15.0	4.04	8.3
Rb$^+$	77.81	55.9	24.76		12.8		32.4	85.6			4.20	
Cs$^+$	77.26	60.9	26.58		13.9	24.3	34.5	87.3		16.6	4.34	
Ag$^+$	61.92	50.1	19.23				35.2	86.2		16.4	4.81	
NH$_4^+$	73.55				15.6		38.6				4.98	9.55
NMe$_4^+$	44.92	68.7	29.8	2.97	13.38		39.1	94.5	54.5	19.0	4.28	11.9
NEt$_4^+$	32.66	60.5	29.4	2.20	11.02	26.2	35.4	84.8	47.7	17.5	3.94	11.5
NPr$_4^+$	23.45	46.1	23.05	1.74	8.12		29.0	70.3	39.2	13.8	3.23	9.0
NBu$_4^+$	19.47	39.05	19.72	1.51	6.83		25.4	61.4	34.1	11.8	2.76	7.7
NAm$_4^+$	17.47	34.85			5.8		22.9	56.0		10.8	2.51	7.25
OH$^-$	198.30											
F$^-$	55.41			3.26								
Cl$^-$	76.34	52.35	21.85	5.07	17.12	19.8	55.1	98.7	62.5	24.0	9.29	11.7
Br$^-$	78.14	56.45	23.78	4.98	17.17	21.5	53.6	100.7	62.9	23.6	8.91	12.95
I$^-$	76.84	62.75	26.92	4.61	16.73	22.8	52.3	102.4		23.4	7.22	14.75
NO$_3^-$	71.46	61.1	25.65	4.81	17.4		57.3	106.4		26.5		14.65
CH$_3$COO$^-$	40.90	53.1			11.9							
Pic$^-$	31.4	47.1	25.0	2.21	9.1	13.1	37.4	77.7		16.8	5.32	11.95
SCN$^-$	66.0	62.3	27.4		17.2		59.7	113.4		28.6	9.63	16.1

Table VI.6 Electrical Properties of Fused Salts

For more extensive data (including density, surface tension and viscosity of fused salts) see: G.J. Janz et al., *J. Phys. Chem. Ref. Data* (1972) **1**, 581; (1974) **3**, 1; (1975) **4**, 871; (1977) **6**, 409; (1979) **8**, 125; (1980) **9**, 791, 831; (1983) **12**, 591.

1. Equivalent Conductivities of Single Salt Melts
$\Lambda = A_\Lambda \exp(-E_\Lambda/RT)$.

Fused salt	$10^4 \times A_\Lambda$/ohm^{-1} m^2 eq^{-1}	E_Λ/kJ mol^{-1}	Fused salt	$10^4 \times A_\Lambda$/ohm^{-1} m^2 eq^{-1}	E_Λ/kJ mol^{-1}
BaCl$_2$	772.5	25.12	KI	541.2	14.40
CdCl$_2$	224.4	10.46	KNO$_3$	657.4	14.97
CaBr$_2$	506.7	20.51	KOH	520.2	10.32
CaCl$_2$	675.2	22.11	RbBr	611.1	17.45
PbBr$_2$	700.8	19.33	RbCl	754.1	18.41
LiBr	585.3	8.86	RbI	568.1	16.73
LiCl	508.2	8.43	AgBr	210.2	4.62
LiF	312.6	7.46	AgCl	261.5	5.13
LiI	377.6	5.42	NaBr	622.7	13.51
MgBr$_2$	385.5	22.61	NaCl	544.6	12.51
MgCl$_2$	263.7	18.25	NaF	344.7	12.40
MgI$_2$	751.1	28.25	NaI	694.5	13.48
KBr	591.1	15.68	NaOH	668.2	13.05
KCl	548.0	14.29	ZnBr$_2$	35 684	61.1
KF	402.7	10.44			

Table VI.6 (Continued)
2. Cation Transport Numbers for Fused Salts

Fused Salt	$t/°C$	t_+
$BaCl_2$	960–1100	0.23
$CaCl_2$	780–1100	0.42
$CsCl$	705	0.565
$PbCl_2$	565–635	0.242
$LiNO_3$	350	0.84
$MgCl_2$	730–920	0.48
KCl	850	0.77

Fused Salt	$t/°C$	t_+
$RbCl$	785	0.58
$AgCl$	500	0.54
$AgNO_3$	350	0.72
$NaCl$	850	0.87
$NaNO_3$	350	0.71
$SrCl_2$	880–1165	0.26
$ZnCl_2$	420	0.685

Table VI.7 Debye–Hückel–Onsager Coefficients for (1:1) Electrolytes at 25 °C

The Debye–Hückel–Onsager equation for molar conductivities has the form

$$\Lambda_m = \Lambda_m^° - (A + B\Lambda_m^°) \, c^{1/2}$$

where the units are as follows: Λ_m and $\Lambda_m^°$ ohm^{-1}, cm^2 mol^{-1}, c mol dm^{-3}, A [ohm^{-1} cm^2 mol^{-1}/ (mol dm^{-3})$^{1/2}$] and B (mol dm^{-3})$^{-1/2}$.
For more details see: J. O'M. Bockris and A.K.N. Reddy, *Modern Electrochemistry* (1973), Plenum Press, New York.

Solvent	A	B
acetone	32.8	1.63
acetonitrile	22.9	0.716
ethanol	89.7	1.83
methanol	156.1	0.923
nitrobenzene	44.2	0.776
nitromethane	125.1	0.708
water	60.20	0.229

Table VI.8 Ionic Strength as a Function of Concentration for Various Valence Type Electrolytes

In terms of molalities, $\quad I = \frac{1}{2}\sum_i m_i z_i^2 = km.$

In terms of volume concentration, $\quad I = \frac{1}{2}\sum_i c_i z_i^2 = kc.$

where the summation extends over all ions in solution, m is the molality and c the concentration of the electrolyte.
Values of k, for different valence type electrolytes are tabulated:

	M^+	M^{2+}	M^{3+}	M^{4+}
A^-	1	3	6	10
A^{2-}	3	4	15	12
A^{3-}	6	15	9	42
A^{4-}	10	12	42	16

Table VI.9 Relationship between Molality, Mean Ionic Molality, Activity and Mean Ionic Activity Coefficient for Various Valence Type Electrolytes

For an electrolyte which dissociates into $\nu+$ cations and $\nu-$ anions, the mean ionic molality $m_\pm$ is given by

$$m_\pm = (m_+^{\nu+} m_-^{\nu-})^{1/\nu}$$

where $\nu = \nu+ + \nu-$. Similar relationships exist for $a_\pm$ and $\gamma_\pm$. The activity of the solute, a_2, is defined as follows:

$$a_2 = a_\pm^\nu = (m_\pm \gamma_\pm)^\nu$$

Reference: as for Table VI.4.

Type	Example	$\gamma_\pm$	$m_\pm$	a_2
Non-electrolyte	glucose			$m\gamma$
1:1, 2:2, 3:3	$KCl, MgSO_4, LaFe(CN)_6$	$(\gamma_+\gamma_-)^{1/2}$	m	$m^2\gamma_\pm^2$
2:1	$CaCl_2$	$(\gamma_+\gamma_-^2)^{1/3}$	$4^{1/3}m$	$4m^3\gamma_\pm^3$
1:2	Na_2SO_4	$(\gamma_+^2\gamma_-)^{1/3}$	$4^{1/3}m$	$4m^3\gamma_\pm^3$
3:1	$LaCl_3$	$(\gamma_+\gamma_-^3)^{1/4}$	$27^{1/4}m$	$27m^4\gamma_\pm^4$
1:3	$K_3Fe(CN)_6$	$(\gamma_+^3\gamma_-)^{1/4}$	$27^{1/4}m$	$27m^4\gamma_\pm^4$
4:1	$Th(NO_3)_4$	$(\gamma_+\gamma_-^4)^{1/5}$	$256^{1/5}m$	$256m^5\gamma_\pm^5$
1:4	$K_4Fe(CN)_6$	$(\gamma_+^4\gamma_-)^{1/5}$	$256^{1/5}m$	$256m^5\gamma_\pm^5$
3:2	$Al_2(SO_4)_3$	$(\gamma_+^2\gamma_-^3)^{1/5}$	$108^{1/5}m$	$108m^5\gamma_\pm^5$

Table VI.10 Mean Stoichiometric Ionic Activity Coefficients ($\gamma_\pm$) at 25 °C for Electrolytes in Aqueous Solution

For more extensive data see: R.A. Robinson and R.H. Stokes, *Electrolyte Solutions* (2nd edn, 1968), Butterworths, London.

$m =$	0.001	0.002	0.005	0.01	0.02	0.05	0.10	0.20	0.50	1.0	2.0	4.0
$AgNO_3$			0.925	0.897	0.86	0.793	0.734	0.657	0.536	0.429	0.28	0.21
$AlCl_3$							(0.337)	0.305	0.331	0.539		
$Al_2(SO_4)_3$							0.035	0.023	0.0143	0.0175		
$BaCl_2$							0.508	0.450	0.403	0.401		
$Ba(OH)_2$			0.773	0.712		0.526	0.443	0.370				
$CaCl_2$	0.89	0.85	0.785	0.731	0.66	0.583	0.515	0.472	0.448	0.500	0.792	2.93
$CdCl_2$			0.623	0.524		0.304	0.228	0.164	0.101	0.067	0.044	0.031
$CdSO_4$							0.150	0.103	0.061	0.041	0.032	
$CrCl_3$							0.331	0.298	0.314	0.481		
$Cr_2(SO_4)_3$							0.046	0.030	0.019	0.021		
CsBr							0.754	0.694	0.603	0.538	0.486	0.457
CsCl							0.756	0.694	0.606	0.544	0.496	0.474
CsI							0.754	0.692	0.599	0.533	0.470	
$CuCl_2$			0.783	0.723		0.577	0.508	0.457	0.413	0.419	0.468	0.575
$CuSO_4$	0.74		0.573	0.438	0.31	0.217	0.154	0.104	0.062	0.043		
HBr							0.805	0.782	0.789	0.871	1.168	
HCl	0.966	0.952	0.928	0.904	0.875	0.830	0.796	0.767	0.757	0.809	1.01	1.76
$HClO_4$							0.803	0.778	0.769	0.823	1.055	2.08
HI							0.818	0.807	0.839	0.963	1.356	
HNO_3	0.965	0.951	0.927	0.902	0.871	0.823	0.791	0.754	0.720	0.724	0.793	
H_2SO_4	0.830	0.757	0.639	0.544	0.453	0.340	0.265	0.209	0.156	0.132	0.124	0.171
KBr	0.965	0.952	0.927	0.903	0.872	0.822	0.772	0.722	0.657	0.617	0.593	0.608
KCl	0.965	0.952	0.927	0.902		0.816	0.770	0.718	0.649	0.604	0.573	0.577
KF							0.775	0.727	0.670	0.645	0.658	0.779
KI	0.965	0.951	0.927	0.905	0.88	0.84	0.778	0.733	0.676	0.645	0.637	0.673
KNO_3			0.926	0.898		0.799	0.739	0.663	0.545	0.443	0.333	
K_2SO_4			0.777	0.711		0.525	0.441	0.356	0.261			
KOH						0.824	0.798	0.760	0.732	0.756	0.863	1.314
$KOOC \cdot CH_3$							0.796	0.766	0.751	0.783	0.910	
LiBr							0.796	0.766	0.753	0.803	1.015	1.897
LiCl	0.963	0.948	0.921	0.89	0.86	0.82	0.790	0.757	0.739	0.774	0.921	1.510
LiI							0.815	0.802	0.824	0.910	1.198	
$MgCl_2$							0.528	0.488	0.480	0.569	1.051	5.53
$MgSO_4$							0.150	0.108	0.069	0.049	0.042	
$MnSO_4$							0.150	0.105	0.064	0.044	0.035	0.047
NH_4Cl							0.770	0.718	0.649	0.603	0.570	0.560
NH_4NO_3			0.925	0.897		0.799	0.740	0.677	0.582	0.504	0.419	0.331
NaBr							0.782	0.741	0.697	0.687	0.731	0.929
NaCl	0.966	0.953	0.929	0.904	0.875	0.823	0.778	0.735	0.681	0.657	0.668	0.783
$NaClO_4$							0.775	0.729	0.668	0.629	0.609	0.626
NaF							0.765	0.710	0.632	0.573		
NaH_2PO_4							0.744	0.675	0.563	0.468	0.371	0.293
NaI							0.787	0.751	0.723	0.736	0.820	
$NaNO_3$			0.929	0.905		0.821	0.762	0.703	0.617	0.548	0.478	0.408
NaOH				0.905		0.818	0.766	0.727	0.690	0.678	0.707	0.901
Na_2SO_4			0.778	0.714		0.536	0.452	0.371	0.270	0.204	0.154	0.138
$NaOOC \cdot CH_3$							0.791	0.757	0.735	0.757	0.851	
$NiCl_2$							0.523	0.479	0.464	0.536	0.906	2.96
$NiSO_4$							0.150	0.105	0.063	0.043	0.034	
$Th(NO_3)_4$							0.279	0.225	0.189	0.207	0.326	0.647
UO_2Cl_2							0.539	0.505	0.512	0.614	0.968	
UO_2SO_4							0.150	0.102	0.061	0.044	0.037	0.043
$ZnCl_2$			0.77	0.71		0.56	0.518	0.465	0.396	0.341	0.291	0.309
$ZnSO_4$			0.477	0.387		0.202	0.150	0.104	0.063	0.043	0.036	

Table VI.11 Mean Ionic Activity Coefficients ($\gamma_\pm$) for Hydrochloric Acid in Water
For more extensive data see: H.S. Harned and B.B. Owen, *The Physical Chemistry of Electrolytic Solutions* (1950), Reinhold, New York.

m \ $t/°C$	0	10	20	25	30	40	50	60
0.0001	0.9890	0.9890	0.9892	0.9891	0.9890	0.9885	0.9879	0.9879
0.0002	0.9848	0.9846	0.9844	0.9842	0.9835	0.9833	0.9831	0.9831
0.0005	0.9756	0.9756	0.9759	0.9752	0.9747	0.9741	0.9738	0.9734
0.001	0.9668	0.9666	0.9661	0.9656	0.9650	0.9643	0.9639	0.9632
0.002	0.9541	0.9544	0.9527	0.9521	0.9515	0.9505	0.9500	0.9491
0.005	0.9303	0.9300	0.9294	0.9285	0.9275	0.9265	0.9250	0.9235
0.01	0.9065	0.9055	0.9052	0.9048	0.9034	0.9016	0.9000	0.8987
0.02	0.8774	0.8773	0.8768	0.8755	0.8741	0.8715	0.8690	0.8666
0.05	0.8346	0.8338	0.8317	0.8304	0.8285	0.8246	0.8211	0.8168
0.10	0.8027	0.8016	0.7985	0.7964	0.7940	0.7891	0.7850	0.7813
0.20	0.7756	0.7740	0.7694	0.7667	0.7630	0.7569	0.7508	0.7437
0.50	0.7761	0.7694	0.7616	0.7571	0.7526	0.7432	0.7344	0.7237
1.0	0.8419	0.8295	0.8162	0.8090	0.8018	0.7865	0.7697	0.7541
1.5	0.9452	0.9270	0.9065	0.8962	0.8849	0.8601	0.8404	0.8178
2.0	1.078	1.053	1.024	1.009	0.9929	0.9602	0.9327	0.9072
3.0	1.452	1.401	1.345	1.316				
4.0	2.006	1.911	1.812	1.762				

Data at concentrations < 0.002 m were obtained from plots used for extrapolation.

Table VI.12 Mean Ionic Activity Coefficients ($\gamma_\pm$) for Hydrochloric Acid at 25 °C in Dioxane–Water Mixtures
For more extensive data see: H.S. Harned and B.B. Owen, *The Physical Chemistry of Electrolytic Solutions* (1950), Reinhold, New York.

m \ wt% dioxane	0	20	45	70	82
0.001	0.9656			0.700	0.3979
0.002	0.9521			0.618	0.3147
0.005	0.9285	0.896	0.808	0.505	0.2181
0.01	0.9048	0.862	0.753	0.418	0.1629
0.02	0.8755	0.821	0.692	0.342	0.1213
0.05	0.8304	0.763	0.607	0.258	0.0826
0.10	0.7964	0.720	0.547	0.212	0.0634
0.20	0.7667	0.676	0.496	0.180	0.0521
0.50	0.7571	0.660	0.465	0.163	0.0504
1.0	0.8090	0.704	0.497	0.187	
1.5	0.8962	0.786	0.570	0.240	
2.0	1.009	0.898	0.676		
3.0	1.316	1.219	1.001		

Table VI.13 Activity Coefficients Calculated Using the Debye–Hückel Equation

The limiting Debye-Hückel equation for an individual ion is

$$\lg \gamma_i = -Az^2 \sqrt{I}$$

and the extended form is

$$\lg \gamma_i = \frac{-Az^2 \sqrt{I}}{1 + Bå\sqrt{I}}$$

where I/mol dm^{-3} is the ionic strength, z, the charge of the ion

$$A = \frac{1.825 \times 10^6}{(\epsilon_r T)^{3/2}} = 0.509 \text{ dm}^{3/2} \text{ mol}^{-1/2}$$

at 25 °C in water, and

$$B = \frac{50.29 \times 10^8}{(\epsilon_r T)^{1/2}} = 0.33 \text{ dm}^{1/2} \text{ mol}^{-1/2}$$

at 25°C, in water and $å$, which may be regarded as the effective diameter of the hydrated ion, is expressed in Angström units, ϵ_r is the relative permittivity. The mean ionic activity coefficient (γ) is related to the activity coefficients of the individual ions (γ_+, γ_-) by the relationship: $\gamma = (\gamma_+^{\nu_+} \gamma_-^{\nu_-})^{1/\nu}$ for an electrolyte with ν ions, ν_+ cations and ν_- anions.
The table lists the values of activity coefficients of selected individual ions at different ionic strengths (a total of 130 ions are listed by J. Kielland, *J. Am. Chem. Soc.* (1937), **59**, 1675).

$å$	\multicolumn{8}{c}{Ionic strength I/mol dm^{-3}}
	0.0005	0.001	0.0025	0.005	0.01	0.0025	0.05	0.1
\multicolumn{9}{c}{Ion Charge 1}
9	0.975	0.967	0.950	0.933	0.914	0.88	0.86	0.83
6	0.975	0.965	0.948	0.929	0.907	0.87	0.835	0.80
5	0.975	0.964	0.947	0.928	0.904	0.865	0.83	0.79
4.5	0.975	0.964	0.947	0.928	0.902	0.86	0.82	0.775
4	0.975	0.964	0.947	0.927	0.901	0.855	0.815	0.77
3.5	0.975	0.964	0.946	0.926	0.900	0.855	0.81	0.760
3	0.975	0.964	0.945	0.925	0.899	0.85	0.805	0.755
2.5	0.975	0.964	0.945	0.924	0.898	0.85	0.80	0.75
\multicolumn{9}{c}{Ion Charge 2}
8	0.906	0.872	0.813	0.755	0.69	0.595	0.52	0.45
6	0.905	0.870	0.809	0.749	0.675	0.57	0.485	0.405
5	0.903	0.868	0.805	0.744	0.67	0.555	0.465	0.38
4.5	0.903	0.868	0.805	0.742	0.665	0.55	0.455	0.37
4	0.903	0.867	0.803	0.740	0.66	0.545	0.445	0.355
\multicolumn{9}{c}{Ion Charge 3}
9	0.802	0.738	0.632	0.54	0.445	0.325	0.245	0.18
5	0.796	0.728	0.616	0.51	0.405	0.27	0.18	0.115
4	0.796	0.725	0.612	0.505	0.395	0.25	0.16	0.095
\multicolumn{9}{c}{Ion Charge 4}
11	0.678	0.588	0.455	0.35	0.255	0.155	0.10	0.065
5	0.668	0.57	0.425	0.31	0.20	0.10	0.048	0.021

Values of $å$ for selected ions

$å$	Charge 1
9	H$^+$
6	Li$^+$
5	CHCl$_2$COO$^-$, CCl$_3$COO$^-$,
4–4.5	Na$^+$, IO$_3^-$, HCO$_3^-$, H$_2$PO$_4^-$, CH$_3$COO$^-$, NH$_2$CH$_2$COO$^-$, C$_2$H$_5$NH$_3^+$
3.5	OH$^-$, F$^-$, CNS$^-$, CNO$^-$, BrO$_3^-$, MnO$_4^-$, HCOO$^-$, H$_2$ citrate$^-$, CH$_3$NH$_3^+$
3	K$^+$, Cl$^-$, Br$^-$, I$^-$, CN$^-$, NO$_2^-$, NO$_3^-$
2.5	NH$_4^+$, Ag$^+$

$å$	Charge 2
8	Mg^{2+}
6	Ca^{2+}, Cu^{2+}, Zn^{2+}, Sn^{2+}, Mn^{2+}, Fe^{2+}, Ni^{2+}, Co^{2+}
5	Sr^{2+}, Ba^{2+}, Cd^{2+}, Hg^{2+}
4.5	Pb^{2+}, CO$_3^{2-}$, SO$_3^{2-}$, (COO$^-$)$_2$, H citrate^{2-}
4	Hg$_2^{2+}$, SO$_4^{2-}$, S$_2$O$_3^{2-}$, CrO$_4^{2-}$, HPO$_4^{2-}$

$å$	Charge 3
9	Al^{3+}, Fe^{3+}, Cr^{3+}, Ce^{3+}
5	Citrate^{3-}
4	PO$_4^{3-}$, Fe(CN)$_6^{3-}$, Cr(NH$_3$)$_6^{3+}$

$å$	Charge 4
11	Ce^{4+}, Sn^{4+}
5	Fe(CN)$_6^{4-}$

Table VI.14 Transport Numbers of Cations in Aqueous Solution at 25 °C

Limiting values of the transport numbers can be calculated from the data of Table VI.3, $t_+ = \lambda_+ / (\lambda_+ + \lambda_-)$.
[a]: molality = 0.005 [b]: molality = 0.025 [c]: data at 18 °C.
Reference: as for Table VI.12.

				t_+				
m	0	0.01	0.02	0.05	0.10	0.20	0.50
AgNO$_3$	0.4643	0.4648		0.4664	0.4682		
BaCl$_2$	0.4547	0.440[a]	0.4317[b]	0.4253		0.3986	0.3792
CdSO$_4$[c]	0.396	0.384			0.353		0.295
HCl	0.8209	0.8251	0.8270	0.8292	0.8314	0.835	0.838
KCl	0.4905	0.4902		0.4899	0.4898		0.4888
KNO$_3$	0.5071	0.5084		0.5093	0.5103		
LiCl	0.3362	0.3289		0.3211	0.3168		0.300
NaCl	0.3965	0.3918		0.3876	0.3854		

Table VI.15 Transport Numbers of Cations of 2:1 Electrolytes in Aqueous Solution at 25 °C, showing the Effect of Autocomplex Formation

Reference: as for Table VI.4.

m	Zn(ClO$_4$)$_2$	ZnI$_2$	ZnBr$_2$	ZnCl$_2$	CdCl$_2$	CdBr$_2$
0		0.440	0.407	0.403	0.409	
0.05		0.382	0.366	0.365		0.424
0.10	0.409	0.363	0.349	0.350	0.487	0.408
0.20	0.389	0.345	0.331	0.335	0.498	0.371
0.50	0.361	0.320	0.306	0.331	0.471	
0.54						0.222
1.0	0.335	0.291	0.286	0.171	0.386	0.066
1.90						−0.068
2.0	0.303	0.178	0.181	0.000	0.241	
3.0	0.281	0.056	−0.059	−0.137		
3.1						−0.185
4.0	0.271	−0.050	−0.151	−0.256	0.002	
5.0		−0.190	−0.233	−0.364		
5.1					−0.104	
6.1					−0.192	
8.0		−0.444	−0.445	−0.562		
10.0		−0.550	−0.563	−0.559		

Table VI.16 Cation Transport Numbers of Hydrochloric Acid at 25 °C in Dioxane–Water Mixtures

For more extensive data see: H.S. Harned and B.B. Owen, *The Physical Chemistry of Electrolytic Solutions* (1950), Reinhold, New York.

				t_+		
m	wt% dioxane	0	20	45	70	82
0		0.821	0.831	0.806	0.755	0.670
0.005		0.824	0.835	0.811	0.766	
0.01		0.825	0.836	0.813	0.769	
0.02		0.827	0.838	0.816	0.772	
0.05		0.830	0.841	0.820	0.774	0.722
0.10		0.832	0.843	0.823	0.775	0.756
0.20		0.835	0.844	0.826	0.776	0.740
0.50		0.838	0.843	0.829	0.774	0.648
1.0		0.841	0.840	0.828	0.772	
2.0		0.843	0.836	0.827		
3.0		0.843	0.833	0.825		

Table VI.17 Cation Transport Numbers in Aqueous Solution at Various Temperatures
For more extensive data see: H.S. Harned and B.B. Owen, *The Physical Chemistry of Electrolytic Solutions* (1950), Reinhold, New York.
$m = 0.1$.

	t_+					
$t/°C =$	5	15	25	35	45	50
HCl	0.850	0.840	0.832	0.823	0.814	0.810
KCl		0.492	0.490	0.489	0.487	
NaCl		0.382	0.385	0.398	0.393	

$m = 0.05$.

	t_+					
$t/°C =$	5	15	25	35	45	50
H_2SO_4		0.830	0.819	0.807	0.793	
$CaCl_2$		0.402	0.407	0.412		

Table VI.18 Diffusion Coefficients of Aqueous Solutions of Selected Electrolytes at 25 °C
The diffusion coefficient D may be defined by either of the equations
$$J = -D\, \partial c/\partial x \quad \text{or} \quad \partial c/\partial t = D\, \partial^2 c/\partial x^2$$
when diffusion occurs in the x-direction only. J is the diffusion flux across unit area normal to the x-direction, $\partial c/\partial x$ is the concentration gradient at a fixed time and $\partial c/\partial t$ is the rate of change of concentration with time at a fixed distance. Units of concentration: mol dm^{-3}.
For more extensive data see: R.A. Robinson and R.H. Stokes, *Electrolyte Solutions* (1968), Butterworths, London; A.L. Horvath, *Handbook of Aqueous Electrolyte Solutions* (1985), Ellis Horwood, Chichester.
[a] Values in this column are Nernst limiting values derived from the limiting mobilities of the ions:
[b] $c = 0.026$; [c] $c = 0.006$ mol dm^{-3}

					$10^9 \times D/m^2\,s^{-1}$					
Salt/$c =$	0[a]	0.001	0.005	0.01	0.05	0.10	0.50	1.0	2.0	3.0
$AgNO_3$	1.765		1.708							
$BaCl_2$	1.385	1.320	1.265		1.179	1.159	1.160	1.179		
CsCl	2.044	2.013	1.978	1.958		1.871	1.860	1.902	2.029	2.175
$CaCl_2$	1.335	1.249	1.179	1.188	1.121	1.110	1.140	1.203	1.307	1.265
HBr	3.400				3.156	3.146	3.388	3.87		
HCl	3.336				3.073	3.050	3.184	3.436	4.046	4.658
KBr	2.016				1.892	1.874	1.885	1.975	2.132	2.280
KCl	1.993	1.964	1.934	1.917	1.864	1.844	1.850	1.892	1.999	2.112
$KClO_4$	1.871	1.845	1.829	1.790						
KI	1.999				1.891	1.865	1.955	2.065	2.254	2.440
KNO_3	1.928	1.899	1.866	1.846						
$LaCl_3$	1.293	1.175	1.105	1.021[b]						
LiCl	1.366	1.345	1.323	1.312	1.280	1.269	1.278	1.302	1.362	1.430
$LiNO_3$	1.336		1.289	1.276		1.240	1.260	1.293	1.332	1.332
$MgSO_4$	0.849	0.768	0.710	0.704[c]						
NH_4Cl	1.994					1.838	1.861	1.921	2.051	2.164
NaBr	1.625				1.533	1.517	1.542	1.596	1.668	
NaCl	1.610	1.585	1.560	1.545	1.507	1.483	1.474	1.484	1.516	1.565
NaI	1.614				1.527	1.520	1.580	1.662	1.846	1.992
$NaNO_3$	1.568		1.516	1.503						

Table VI.19 Osmotic ($\emptyset_m$) and Activity ($\gamma_\pm$) Coefficients of Sodium, Potassium and Calcium Chloride and Sulphuric Acid Solutions at 25 °C

For more extensive data see: R.A. Robinson and R.H. Stokes, *Electrolyte Solutions* (1968), Butterworths, London; A.L. Horvath, *Handbook of Aqueous Electrolyte Solutions* (1985), Ellis Horwood, Chichester; R.N. Goldberg *J. Phys. Chem. Ref. Data* (1981), **10**, 1, 671; B.R. Staples *J. Phys. Chem. Ref. Data* (1981), **10**, 765, 778; K.S. Pitzer, J.C. Peiper and R.H. Busey *J. Phys. Chem. Ref. Data* (1984), **13**, 1.

m	NaCl		KCl		CaCl$_2$		H$_2$SO$_4$	
	$\emptyset_m$	$\gamma_\pm$	$\emptyset_m$	$\gamma_\pm$	$\emptyset_m$	$\gamma_\pm$	$\emptyset_m$	$\gamma_\pm$
0.1	0.9324	0.778	0.9266	0.770	0.854	0.518	0.680	0.2655
0.2	0.9245	0.735	0.9130	0.718	0.862	0.472	0.668	0.2090
0.4	0.9203	0.693	0.9017	0.666	0.894	0.448		
0.5							0.676	0.1557
0.6	0.9230	0.673	0.8976	0.637	0.940	0.453		
0.8	0.9288	0.662	0.8970	0.618	0.988	0.470		
1.0	0.9355	0.657	0.8974	0.604	1.046	0.500	0.721	0.1316
1.2	0.9428	0.654	0.8986	0.593	1.107	0.539		
1.4	0.9513	0.654	0.9010	0.586	1.171	0.587		
1.5							0.780	0.1263
2.0	0.9833	0.668	0.9124	0.573	1.376	0.792	0.846	0.1276
2.4	1.0068	0.683	0.9214	0.569				
2.5					1.568	1.063	0.916	0.1331
3.0	1.0453	0.714	0.9367	0.569	1.779	1.483	0.991	0.1422
3.4	1.0725	0.739	0.9477	0.571				
3.5					1.981	2.078	1.071	0.1547
4.0	1.1158	0.783	0.9647	0.577	2.182	2.934	1.150	0.1700
4.4	1.1456	0.817	0.9766	0.582				
4.5					2.383	4.17	1.226	0.1875
5.0	1.1916	0.874			2.574	5.89	1.303	0.2081
5.4	1.2229	0.916						
5.5					2.743	8.18	1.376	0.2312
6.0	1.2706	0.986			2.891	11.11	1.445	0.2567
7.0					3.081	18.28	1.576	0.3166
8.0					3.151	26.02	1.691	0.386
9.0					3.171	34.2	1.793	0.467
10.0					3.169	43.0	1.884	0.559
15.0							2.187	1.154
20.0							2.341	1.940
30.0							2.441	3.792
40.0							2.386	5.406
50.0							2.265	6.443
60.0							2.127	6.982
70.0							1.990	7.171

Table VI.20 Osmotic and Activity Coefficients of Sucrose Solutions at 25 °C

Reference: as for Table VI.19.

m	$\emptyset_m$	γ	m	$\emptyset_m$	γ
0.1	1.008	1.017	2.0	1.189	1.442
0.2	1.017	1.034	2.5	1.240	1.590
0.4	1.033	1.068	3.0	1.288	1.751
0.6	1.050	1.105	3.5	1.334	1.924
0.8	1.068	1.144	4.0	1.375	2.101
1.0	1.088	1.188	4.5	1.414	2.310
1.2	1.108	1.233	5.0	1.450	2.481
1.4	1.129	1.283	5.5	1.482	2.680
1.6	1.150	1.335	6.0	1.511	2.878
1.8	1.169	1.387			

Table VI.21 pH Values of Operational Reference Standard Solutions (British Standard, BS. 1647, 1984)

[a]: The last decimal figure is not significant but is included to facilitate smooth interpolation.
[b]: Tris = tris(hydroxymethyl) aminomethane.
[c]: Ratio of 3.5:1.
[d]: Ratio of 4:1.

Operational reference standard solution	Temperature/°C						
	0	10	20	25	30	37	40
0.1 mol kg^{-1} potassium tetraoxalate			1.475	1.479	1.483	1.490	1.493
0.05 mol kg^{-1} potassium tetraoxalate		1.638	1.644	1.646	1.648	1.649	1.650
0.05 mol kg^{-1} sodium hydrogen diglycolate		3.470	3.484	3.492	3.502	3.519	3.527
Saturated potassium hydrogen tartrate				3.556	3.549	3.544	3.542
0.05 mol kg^{-1} potassium hydrogen phthalate	4.000	3.997	4.000	4.005	4.011	4.022	4.027
0.1 mol dm^{-3} acetic acid + 0.1 mol dm^{-3} sodium acetate	4.664	4.652	4.645	4.644	4.643	4.647	4.650
0.01 mol dm^{-3} acetic acid + 0.01 mol dm^{-3} sodium acetate	4.729	4.717	4.712	4.713	4.715	4.722	4.726
0.02 mol kg^{-1} piperazine orthophosphate		6.419	6.310	6.259	6.209	6.143	6.116
0.025 mol kg^{-1} disodium hydrogen orthophosphate + 0.025 mol kg^{-1} potassium dihydrogen orthophosphate	6.961	6.912	6.873	6.857	6.843	6.828	6.823
0.03043 mol kg^{-1} disodium hydrogen orthophosphate + 0.08695 mol kg^{-1} potassium dihydrogen orthophosphate[c]	7.506	7.460	7.423	7.406	7.390	7.369	
0.04 mol kg^{-1} disodium hydrogen orthophosphate + 0.01 mol kg^{-1} potassium dihydrogen orthophosphate[d]		7.488	7.445	7.428	7.414	7.404	
0.05 mol kg^{-1} Tris hydrochloride + 0.01667 mol/kg Tris[b]	8.399	8.083	7.788	7.648	7.513	7.332	7.257
0.05 mol kg^{-1} disodium tetraborate (borax)	9.475	9.347	9.233	9.182	9.134	9.074	9.051
0.01 mol kg^{-1} disodium tetraborate	9.451	9.329	9.225	9.179	9.138	9.088	9.066
0.025 mol kg^{-1} sodium hydrogen carbonate + 0.025 mol kg^{-1} sodium carbonate	10.27	10.154	10.045	9.995	9.948	9.889	9.866
Saturated calcium hydroxide	13.360	12.965	12.602	12.431	12.267	12.049	11.959

Table VI.22 pH Values of Useful Buffer Solutions (British Standard, BS. 1647, 1984)

[a]: Bis (hydroxymethyl) aminotris (hydroxyethyl) methane; dry at 80 °C.
[b]: Tris (hydroxymethyl) methylglycine.

Buffer Solutions	Temperature/°C						
	0	5	10	15	20	25	30
0.01 mol dm^{-3} HCl + 0.09 mol dm^{-3} KCl						2.07	
0.05 mol dm^{-3} potassium dihydrogen citrate	3.86	3.84	3.82	3.80	3.79	3.78	3.78
0.01 mol dm^{-3} phenylacetic acid + 0.01 mol dm^{-3} sodium phenylacetate	4.27		4.27		4.28	4.29	4.30
0.02 mol dm^{-3} bis-tris + 0.02 mol dm^{-3} bis-tris HCl[a]	7.00	6.91	6.81	6.72	6.64	6.55	6.47
0.06 mol dm^{-3} tricine + 0.02 mol dm^{-3} sodium tricinate[b]		8.02	7.92	7.81	7.71	7.62	7.53
0.05 mol dm^{-3} tricine + 0.05 mol dm^{-3} sodium tricinate[b]		8.49	8.38	8.27	8.18	8.08	7.99

	Temperature/°C			
50	60	70	80	90
1.503	1.513	1.52	1.53	1.53
1.653	1.660	1.671[a]	1.689[a]	1.72
3.558	3.595			
3.544	3.553	3.570[a]	3.596[a]	3.627
4.050	4.080	4.116	4.159	4.208
4.663	4.684	4.713	4.75	4.80
4.743	4.768	4.800	4.839[a]	4.88
6.030	5.952			
6.814	6.817	6.830	6.85	6.90
7.018	6.794			
8.983	8.932	8.898	8.88	8.84
9.009	8.965	8.932	8.91	8.90
9.800	9.753	9.728[a]	9.725[a]	9.75
11.678	11.423	11.192[a]	10.984[a]	10.80

	Temperature/°C			
35	37	40	45	50
	3.08			
3.76	3.76	3.75	3.75	3.75
	4.32	4.33		4.38
6.39	6.36	6.31	6.24	6.17
7.43	7.41	7.36	7.28	7.20
7.90	7.86	7.82	7.74	7.66

Table VI.23 Biological Buffer Solutions

These buffer solutions conform to the following criteria: (a) pK_a values mainly in the range 6 to 8.4; (b) good solubility in water; (c) good stability; (d) minimum absorbance in the visible and u.v. regions. lg K: metal-buffer binding constant. *: negligible binding.
For further information see: N. Good et al, *Biochemistry* (1968) 5, 467.

Name		Formula	pK_a (20 °C)	lg K Mg^{2+}	lg K Ca^{2+}
2-(N-morpholino)-ethanesulphonic acid	MES	$CH_2CH_2OCH_2CH_2NCH_2CH_2SO_3H$	6.15	0.8	0.7
N-(2-acetamido)-iminodiacetic acid	ADA	$NH_2COCH_2N(CH_2COOH)_2$	6.62	2.5	4.0
piperazine-NN'-bis-2-ethanesulphonic acid	PIPES	$C_8H_{18}N_2O_6S_2$	6.80	*	*
N-(2-acetamido)-2-aminoethane-sulphonic acid	ACES	$NH_2COCH_2NHCH_2CH_2SO_3H$	6.88	0.4	0.4
2-aminoethyl-trimethylammonium chloride hydrochloride		$[(CH_3)_3NCH_2CH_2NH_2 \cdot HCl]Cl$	7.11	*	*
NN-bis-(2-hydroxyethyl)-2-aminoethane-sulphonic acid	BES	$(HOCH_2CH_2)_2NCH_2CH_2SO_3H$	7.15	*	*
3-(N-morpholino)propane-sulphonic acid	MOPS	$CH_2CH_2OCH_2CH_2N(CH_2)_3SO_3H$	7.20	*	*
N-tris (hydroxymethyl)-methyl-2-aminoethane-sulphonic acid	TES	$(HOCH_2)_3CNHCH_2CH_2SO_3H$	7.50	*	*
N-2-hydroxyethyl-piperazine-N'-2-ethane-sulphonic acid	HEPES	$C_8H_{18}N_2O_4S$	7.55	*	*
N-2-hydroxyethyl-piperazine-N'-3-propane-sulphonic acid	HEPPS	$C_9H_{20}N_2O_4S$	8.00	*	*
N-tris (hydroxymethyl)-methylglycine	Tricine	$(HOCH_2)_3CNHCH_2COOH$	8.15	1.2	2.4
tris(hydroxymethyl)-methylamine	Tris	$(HOCH_2)_3CNH_2$	8.3	*	*
NN-bis (2-hydroxyethyl)-glycine	Bicine	$(HOCH_2CH_2)_2NCH_2COOH$	8.35	1.5	2.8
N-glycylglycine		$NH_2CH_2CONHCH_2COOH$	8.4	0.8	0.8
2-(cyclohexylamino)-ethane-sulphonic acid	CHES	$C_8H_{17}NO_3S$	9.55		

Table VI.24 pH Values of Some Common Buffer Solutions

1. Sodium Acetate – Hydrochloric Acid (Walpole), pH Range 0.65–5.20

50 cm³ of 1.0 mol dm⁻³ sodium acetate (136.09 g $CH_3COONa \cdot 3H_2O$ in 1000 cm³ solution) + x cm³ 1 mol dm⁻³ HCl made up to final volume of 250 cm³.
See also: A.M. James and F.E. Prichard, *Practical Physical Chemistry*, 3rd ed. (1974), Longmans, London, and reference to Table VI.4.

x	pH	x	pH	x	pH
100	0.65	52.5	1.99	42.5	3.79
80	0.91	50.0	2.64	35.0	4.19
70	1.09	49.75	2.72	30.0	4.39
65	1.24	48.5	3.09	25.0	4.58
60	1.42	47.5	3.29	20.0	4.76
55	1.71	46.25	3.49	15.0	4.92
53.5	1.85	45.0	3.61	10.0	5.20

Table VI.24 (Continued)

2. Potassium Hydrogen Phthalate – Hydrochloric Acid, pH Range 2.2–4.0

50 cm^3 of 0.1 mol dm^{-3} potassium hydrogen phthalate (20.42 g in 1000 cm^3 solution) + x cm^3 0.1 mol dm^{-3} HCl made up to final volume of 100 cm^3.

x	pH	x	pH	x	pH
49.5	2.20	22.3	3.00	6.3	3.60
42.2	2.40	15.7	3.20	2.9	3.80
35.4	2.60	10.4	3.40	0.1	4.00
28.9	2.80				

3. Potassium Hydrogen Phthalate – Sodium Hydroxide, pH Range 4.1–5.9

50 cm^3 of 0.1 mol dm^{-3} potassium hydrogen phthalate (20.42 g in 1000 cm^3 solution) + x cm^3 0.1 mol dm^{-3} NaOH made up to final volume of 100 cm^3.

x	pH	x	pH	x	pH
1.3	4.10	19.4	4.90	36.6	5.50
4.7	4.30	25.5	5.10	40.6	5.70
8.7	4.50	31.6	5.30	43.7	5.90
13.6	4.70				

4. Universal buffer solution for use in u.v. spectrophotometry (Davies), pH Range 2.0–12.0

Solution A: 21.01 g citric acid, 16.61 g potassium dihydrogen phosphate, 19.07 g sodium tetraborate, 12.11 g tris(hydroxymethyl)-methylamine and 7.46 g potassium chloride made up to 1000 cm^3.
Solution B: 0.4 mol dm^{-3} hydrochloric acid.
Solution C: 0.4 mol dm^{-3} sodium hydroxide solution.
50 cm^3 of solution A + specified amount of solution B or C made up to final volume of 200 cm^3.

Vol. of B /cm^3	Vol. of C /cm^3	pH	Vol. of B /cm^3	Vol. of C /cm^3	pH
34.8	0	2.0	0	33.2	8.0
19.6	0	3.0	0	46.2	9.0
10.0	0	4.0	0	59.0	10.0
0	0.4	5.0	0	65.6	11.0
0	11.4	6.0	0	77.2	12.0
0	22.4	7.0			

5. Disodium Hydrogen Phosphate-Citric Acid (McIlvaine), pH Range 2.20–8.0

0.2 mol dm^{-3} disodium hydrogen phosphate (71.64 g Na$_2$HPO$_4$.12H$_2$O in 1000 cm^3 solution) and 0.1 mol dm^{-3} citric acid (21.01 g of hydrate in 1000 cm^3 solution).

Vol. of Na$_2$HPO$_4$ /cm^3	Vol. of citric acid /cm^3	pH	Vol. of Na$_2$HPO$_4$ /cm^3	Vol. of citric acid /cm^3	pH
0.40	19.60	2.2	10.72	9.28	5.2
2.18	17.82	2.6	11.60	8.40	5.6
4.11	15.89	3.0	12.63	7.37	6.0
5.70	14.30	3.4	13.85	6.15	6.4
7.10	12.90	3.8	15.45	4.55	6.8
8.28	11.72	4.2	17.39	2.61	7.2
9.35	10.65	4.6	18.73	1.27	7.6
10.30	9.70	5.0	19.45	0.55	8.0

Table VI.24 (Continued)

6. Acetate-veronal-hydrochloric Acid (Michaelis), pH Range 2.6–9.6
Solution A: 9.714 g $CH_3COONa \cdot 3H_2O$ and 14.714 g sodium diethylbarbiturate made up to 500 cm^3.
Solution B: 8.5 g sodium chloride in 100 cm^3.
Solution C: 0.1 mol dm^{-3} HCl.
To 5.0 cm^3 of solution A add 2.0 cm^3 of solution B and x cm^3 of C and make up to a final volume of 25 cm^3.
All these solutions have a constant ionic strength (0.173 mol dm^{-3}) and are isotonic with blood.

x	pH	x	pH	x	pH
0	9.64	4.0	7.66	11.0	4.33
0.25	9.16	5.0	7.42	12.0	4.13
0.50	8.90	6.0	6.99	13.0	3.88
0.75	8.68	7.0	6.75	14.0	3.62
1.0	8.55	8.0	6.12	15.0	3.20
2.0	8.18	9.0	5.32	16.0	2.62
3.0	7.90	10.0	4.66		

7. Acetic Acid–Sodium Acetate (Walpole), pH Range 3.42–5.89
0.2 mol dm^{-3} sodium acetate (27.22 g $CH_3COONa \cdot 3H_2O$ in 1000 cm^3 solution); 0.2 mol dm^{-3} acetic acid.

Vol. of CH_3COOH /cm^3	Vol. of CH_3COONa /cm^3	pH	Vol. of CH_3COOH /cm^3	Vol. of CH_3COONa /cm^3	pH
9.5	0.5	3.42	4.0	6.0	4.80
9.0	1.0	3.72	3.0	7.0	4.99
8.0	2.0	4.05	2.0	8.0	5.23
7.0	3.0	4.27	1.5	8.5	5.37
6.0	4.0	4.45	1.0	9.0	5.57
5.0	5.0	4.63	0.5	9.5	5.89

8. Phosphate Buffer Solution (Sørensen), pH range 5.29–8.04
0.0667 mol dm^{-3} disodium hydrogen phosphate (23.88 g $Na_2HPO_4 \cdot 12H_2O$ in 1000 cm^3 solution), 0.0667 mol dm^{-3} potassium dihydrogen phosphate (9.077 g KH_2PO_4 in 1000 cm^3 solution).

Vol. of Na_2HPO_4 /cm^3	Vol. of KH_2PO_4 /cm^3	pH	Vol. of Na_2HPO_4 /cm^3	Vol. of KH_2PO_4 /cm^3	pH
0.25	9.75	5.29	5.0	5.0	6.81
0.50	9.50	5.59	6.0	4.0	6.98
1.0	9.0	5.91	7.0	3.0	7.17
2.0	8.0	6.24	8.0	2.0	7.38
3.0	7.0	6.47	9.0	1.0	7.73
4.0	6.0	6.64	9.5	0.5	8.04

9. Tris-hydrochloric Acid, pH Range 7.0–9.0
50 cm^3 0.1 mol dm^{-3} tris(hydroxymethyl)methylamine (12.11 g in 1000 cm^3 solution) + x cm^3 of 0.1 mol dm^{-3} HCl made up to a final volume of 100 cm^3.

x	pH	x	pH	x	pH
46.6	7.00	34.5	7.80	12.2	8.60
44.7	7.20	29.2	8.00	8.5	8.80
42.0	7.40	22.9	8.20	5.7	9.00
38.5	7.60	17.2	8.40		

Table VI.24 (Continued)
10. Sodium diethylbarbiturate-hydrochloric acid (Michaelis), pH Range 6.8–9.6
0.1 mol dm^{-3} sodium diethylbarbiturate (20.618 g in 1000 cm^3 solution), 0.1 mol dm^{-3} HCl. These solutions are all of constant ionic strength (0.1 mol dm^{-3}).

Vol. of barbiturate /cm^3	Vol. of HCl /cm^3	pH	Vol. of barbiturate /cm^3	Vol. of HCl /cm^3	pH
5.22	4.78	6.80	8.23	1.77	8.40
5.36	4.64	7.00	8.71	1.29	8.60
5.54	4.46	7.20	9.08	0.92	8.80
5.81	4.19	7.40	9.36	0.64	9.00
6.15	3.85	7.60	9.52	0.48	9.20
6.62	3.38	7.80	9.74	0.26	9.40
7.16	2.84	8.00	9.85	0.15	9.60

11. Borax – Sodium Hydroxide (Nalgeli and Tyabji), pH Range 9.24–12.39
10 cm^3 0.05 mol dm^{-3} borax (19.07 g Na$_2$B$_4$O$_7$. 10H$_2$O in 1000 cm^3 solution) + x cm^3 0.1 mol dm^{-3} NaOH.

x	pH	x	pH	x	pH	x	pH	x	pH
0	9.24		9.96	10.85		12.68			12.12
6.66	9.94		10.54	11.34		13.97			12.28
8.24	10.21		11.20	11.72		15.03			12.39
9.25	10.49		11.96	11.98					

12. Sodium Bicarbonate–Sodium Hydroxide, pH Range 9.6–11.0
50 cm^3 0.05 mol dm^{-3} sodium bicarbonate (4.20 g NaHCO$_3$ in 1000 cm^3 solution) + x cm^3 0.1 mol dm^{-3} NaOH made up to a final volume of 100 cm^3.

x	pH	x	pH
5.0	9.60	16.5	10.40
7.6	9.80	19.1	10.60
10.7	10.00	21.2	10.80
13.8	10.20	22.7	11.00

13. Disodium Hydrogen Phosphate–Sodium Hydroxide (Ringer), pH Range 10.97–12.06.
50 cm^3 0.15 mol dm^{-3} disodium hydrogen phosphate (53.72 g Na$_2$HPO$_4$. 12H$_2$O) + x cm^3 0.1 mol dm^{-3} NaOH.

x	pH	x	pH
15.0	10.97	50.0	11.77
25.0	11.29	75.0	12.06

Table VI.25 Approximate pH Values of Some Common Solutions, Fluids and Foodstuffs
Acids
For further information see: E.S. West, W.R. Todd, H.S. Mason and J.T. van Bruggen, *Textbook of Biochemistry*, 4th ed. (1966), Macmillan, New York; *Handbook of Chemistry and Physics* (1989–90), CRC.

Acid	Approx. conc. /mol dm^{-3}	Approximate pH
acetic	1	2.4
	0.1	2.9
	0.01	3.4
arsenious	saturated	5.0
benzoic	0.01	3.1
boric	0.03	5.2
carbonic	saturated	3.8
citric	0.03	2.2
formic	0.1	2.3
hydrochloric	1	0.1
	0.1	1.1
	0.01	2.0
hydrogen sulphide	0.1	4.1
lactic	0.1	2.4
malic	0.05	2.2
oxalic	0.05	1.6
phosphoric	0.03	1.5
sulphuric	0.5	0.3
	0.05	1.2
	0.005	2.1
sulphurous	0.05	1.5
tartaric	0.05	2.2

Table VI.25 (Continued)
Bases

Base	Approx. conc. /mol dm^{-3}	Approximate pH
ammonia	1	11.6
	0.1	11.1
	0.01	10.6
ammonium acetate	1.0–0.001	7.0
calcium hydroxide	saturated	12.4
iron(II) hydroxide	saturated	9.5
methylamine	0.01	11.3
potassium cyanide	0.1	11.0
potassium hydroxide	1	14.0
	0.1	13.0
pyridine	0.1	8.6
sodium acetate	0.1	8.8
sodium bicarbonate	0.1	8.4
sodium carbonate	0.05	11.6
sodium hydroxide	1.0	14.0
	0.1	13.0
	0.01	12.0
trisodium phosphate	0.03	12.0

Human fluids

Fluid	pH	Fluid	pH
bile	6.8–7.0	milk	6.6–7.6
blood plasma	7.3–7.5	saliva	6.5–7.5
duodenal contents	4.8–8.2	spinal fluid	7.3–7.5
faeces	4.6–8.4	urine	4.8–8.4
gastric contents	1.0–3.0		

Foodstuffs

Food	pH	Food	pH
apples	2.9–3.3	limes	1.8–2.0
apricots	3.6–4.0	milk (cows)	6.3–6.6
asparagus	5.4–5.8	olives	3.6–3.8
bananas	4.5–4.7	oranges	3.0–4.0
beans	5.0–6.0	oysters	6.1–6.6
beer	4.0–5.0	peaches	3.4–3.6
blackberries	3.2–3.6	pears	3.6–4.0
bread (white)	5.0–6.0	peas	5.8–6.4
butter	6.1–6.4	pickles	3.0–3.6
cabbage	5.2–5.4	plums	2.8–3.0
carrots	4.9–5.3	potatoes	5.6–6.0
cheese	4.8–6.4	raspberries	3.2–3.6
cherries	3.2–4.0	rhubarb	3.1–3.2
cider	2.9–3.3	salmon	6.1–6.3
eggs	7.6–8.0	sauerkraut	3.4–3.6
flour (wheat)	5.5–6.5	soft drinks	2.0–4.0
gooseberries	2.8–3.0	spinach	5.1–5.7
grapefruit	3.0–3.3	strawberries	3.0–3.5
grapes	3.5–4.5	tomatoes	4.0–4.4
jams	3.5–4.0	turnips	5.2–5.6
jellies	2.8–3.4	vinegar	2.4–3.4
lemons	2.2–2.4	water	6.5–8.0
		wines	2.8–3.8

Table VI.26 Acid–Base Indicators

Colours: B blue; C colourless; P purple; R red; V violet; Y yellow. 0.01 M NaOH is 0.01 mol dm^{-3} NaOH.
For more extensive data see: A.I. Vogel, *A Textbook of Quantitative Inorganic Analysis*, 2nd ed. (1951), Longmans.

Indicator	pK	pH range	Colour change	Preparation
cresol red		0.2–1.8	R–Y	0.1 g in 26.2 cm^3 0.01 M NaOH to 250 cm^3 with water
cresol purple	1.5	1.2–2.8	R–Y	0.1 g in 26.2 cm^3 0.01 M NaOH to 250 cm^3 with water
thymol blue	1.7	1.2–2.8	R–Y	0.1 g in 21.5 cm^3 0.01 M NaOH to 250 cm^3 with water
2, 4-dinitrophenol	3.9	2.8–4.0	C–Y	saturated aqueous solution
methyl yellow	3.3	2.9–4.0	R–Y	0.1% in 90% aq. ethanol
methyl orange	3.7	3.1–4.4	R–Y	0.1% in water
bromophenol blue	4.0	3.0–4.6	Y–BV	0.1 g in 14.9 cm^3 0.01 M NaOH to 250 cm^3 with water
bromocresol green	4.7	3.8–5.4	Y–B	0.1 g in 14.3 cm^3 0.01 M NaOH to 250 cm^3 with water
methyl red	5.1	4.4–6.2	R–Y	0.02% in 60% aq. ethanol
chlorophenol red	6.0	5.4–6.8	Y–R	0.1 g in 23.6 cm^3 0.01 M NaOH to 250 cm^3 with water
bromocresol purple	6.3	5.2–6.8	Y–P	0.1 g in 18.5 cm^3 0.01 M NaOH to 250 cm^3 with water
litmus		5.0–8.0	R–B	0.5% in water
bromothymol blue	7.1	6.0–7.6	Y–B	0.1 g in 16 cm^3 0.01 M NaOH to 250 cm^3 with water
4-nitrophenol	7.2	5.3–7.6	C–Y	0.1% in water
neutral red	7.4	6.8–8.0	R–Y	0.01% in 50% aq. ethanol
phenol red	7.9	6.4–8.0	Y–R	0.1 g in 28.2 cm^3 0.01 M NaOH to 250 cm^3 with water
3-nitrophenol	8.3	6.8–8.6	C–Y	0.3% in water
cresol red	8.3	7.2–8.8	Y–R	0.1 g in 26.2 cm^3 0.01 M NaOH to 250 cm^3 with water
metacresol purple	8.3	7.4–9.2	Y–P	0.1 g in 26.2 cm^3 0.01 M NaOH to 250 cm^3 with water
thymol blue	8.9	8.0–9.6	Y–B	0.1 g in 21.5 cm^3 0.01 M NaOH to 250 cm^3 with water
phenolphthalein	9.4	8.2–10.0	C–R	0.05% in 50% aq. ethanol
thymolphthalein	10.0	9.4–10.6	C–B	0.04% in 50% aq. ethanol
alizarin yellow R	11.2	10.1–12.0	Y–V	0.01% in water

Table VI.27 Standard (reduction) potentials at 298 K in aqueous solution

For more extensive data see: S.G. Bratsch, *J. Phys. Chem. Ref. Data* (1989), **18**, 1.

Electrode	Electrode reaction	$E^\ominus$/V	(d$E^\ominus$/dT)/mV K^{-1}
Li$^+$ | Li	Li$^+$ + e $\to$ Li	−3.045	−0.59
Rb$^+$ | Rb	Rb$^+$ + e $\to$ Rb	−2.925	
K$^+$ | K	K$^+$ + e $\to$ K	−2.924	−1.07
Cs$^+$ | Cs	Cs$^+$ + e $\to$ Cs	−2.923	
Ba^{2+} | Ba	Ba^{2+} + 2e $\to$ Ba	−2.90	−0.40
Sr^{2+} | Sr	Sr^{2+} + 2e $\to$ Sr	−2.89	−0.23
Ca^{2+} | Ca	Ca^{2+} + 2e $\to$ Ca	−2.87	−0.21
Na$^+$ | Na	Na$^+$ + e $\to$ Na	−2.7109	−0.75
Mg^{2+} | Mg	Mg^{2+} + 2e $\to$ Mg	−2.375	+0.18
La^{3+} | La	La^{3+} + 3e $\to$ La	−2.37	
Ce^{3+} | Ce	Ce^{3+} + 3e $\to$ Ce	−2.335	
Th^{4+} | Th	Th^{4+} + 4e $\to$ Th	−1.90	
Be^{3+} | Be	Be^{3+} + 3e $\to$ Be	−1.85	
U^{3+} | U	U^{3+} + 3e $\to$ U	−1.8	
Al^{3+} | Al	Al^{3+} + 3e $\to$ Al (0.1 M NaOH)	−1.66	+0.53
Ti^{2+} | Ti	Ti^{2+} + 2e $\to$ Ti	−1.63	
V^{2+} | V	V^{2+} + 2e $\to$ V	−1.18	
Mn^{2+} | Mn	Mn^{2+} + 2e $\to$ Mn	−1.18	
Cr^{2+} | Cr	Cr^{2+} + 2e $\to$ Cr	−0.91	
OH$^-$ | H$_2$ (g), Pt	2H$_2$O + 2e $\to$ H$_2$ (g) + 2OH$^-$	−0.8277	−0.80
Zn^{2+} | Zn	Zn^{2+} + 2e $\to$ Zn	−0.7628	+0.10
Cr^{3+} | Cr	Cr^{3+} + 3e $\to$ Cr	−0.74	
U^{4+} | U^{3+}, Pt	U^{4+} + e $\to$ U^{3+}	−0.61	

(Continued)

Table VI.27 (Continued)

Electrode	Electrode reaction	$E°/V$	$(dE°/dT)/mV\ K^{-1}$
$Ga^{3+}\|Ga$	$Ga^{3+} + 3e \rightarrow Ga$	−0.560	
$S\|S^{2-}, Pt$	$S + 2e \rightarrow S^{2-}$	−0.508	
$Fe^{2+}\|Fe$	$Fe^{2+} + 2e \rightarrow Fe$	−0.44	+0.05
$Cr^{3+}, Cr^{2+}\|Pt$	$Cr^{3+} + e \rightarrow Cr^{2+}$	−0.41	
$Cd^{2+}\|Cd$	$Cd^{2+} + 2e \rightarrow Cd$	−0.4026	−0.09
$PbSO_4\|SO_4^{2-}$	$PbSO_4 + 2e \rightarrow Pb + SO_4^{2-}$	−0.3553	−0.99
$In^{3+}\|In$	$In^{3+} + 3e \rightarrow In$	−0.338	
$Tl^+\|Tl$	$Tl^+ + e \rightarrow Tl$	−0.3363	
$Co^{2+}\|Co$	$Co^{2+} + 2e \rightarrow Co$	−0.28	
$V^{3+}, V^{2+}\|Pt$	$V^{3+} + e \rightarrow V^{2+}$	−0.255	
$Ni^{2+}\|Ni$	$Ni^{2+} + 2e \rightarrow Ni$	−0.25	
$AgI\|I^-$	$AgI + e \rightarrow Ag + I^-$	−0.1519	−0.33
$Sn^{2+}\|Sn$	$Sn^{2+} + 2e \rightarrow Sn$	−0.1364	
$Pb^{2+}\|Pb$	$Pb^{2+} + 2e \rightarrow Pb$	−0.1263	−0.38
$Hg_2I_2\|I^-$	$Hg_2I_2 + 2e \rightarrow 2Hg + 2I^-$	−0.0405	
$Fe^{3+}\|Fe$	$Fe^{3+} + 3e \rightarrow Fe$	−0.036	
$D^+\|D_2(g), Pt$	$2D^+ + 2e \rightarrow D_2(g)$	−0.0034	
$H_3O^+\|H_2(g), Pt$	$2H_3O^+ + 2e \rightarrow 2H_2O + H_2(g)$	0	0
$AgBr\|Br^-$	$AgBr + e \rightarrow Ag + Br^-$	0.0713	−0.49
$HgO\|OH^-$	$HgO + H_2O + 2e \rightarrow Hg + 2OH^-$	0.0984	
$Hg_2Br_2\|Br^-$	$Hg_2Br_2 + 2e \rightarrow 2Hg + 2Br^-$	0.1396	
$Sb_2O_3\|Sb$	$Sb_2O_3 + 6H^+ + 6e \rightarrow 2Sb + 3H_2O$	0.1445	
$Sn^{4+}, Sn^{2+}\|Pt$	$Sn^{4+} + 2e \rightarrow Sn^{2+}$	0.154	
$Cu^{2+}, Cu^+\|Pt$	$Cu^{2+} + e \rightarrow Cu^+$	0.158	
$AgCl\|Cl^-$	$AgCl + e \rightarrow Ag + Cl^-$	0.2224	−0.66
$Hg_2Cl_2\|Cl^-$	$Hg_2Cl_2 + 2e \rightarrow 2Hg + 2Cl^-$	0.2682	−0.31
*Saturated calomel		0.2415	
*Normal calomel (1 mol dm^{-3} KCl)		0.2807	
*0.1 normal calomel (0.1 mol dm^{-3} KCl)		0.3337	
$Cu^{2+}\|Cu$	$Cu^{2+} + 2e \rightarrow Cu$	0.3402	+0.01
$Ag_2O\|Ag$	$Ag_2O + H_2O + 2e \rightarrow 2Ag + 2OH^-$	0.342	
$Fe(CN)_6^{3-}, Fe(CN)_6^{4-}\|Pt$	$Fe(CN)_6^{3-} + e \rightarrow Fe(CN)_6^{4-}$	0.36	
$O_2(g)\|OH^-, Pt$	$O_2(g) + 2H_2O + 4e \rightarrow 4OH^-$	0.401	
$Cu^+\|Cu$	$Cu^+ + e \rightarrow Cu$	0.522	
$I_3^-, I^-\|Pt$	$I_3^- + 2e \rightarrow 3I^-$	0.5338	
$I_2(s)\|I^-, Pt$	$I_2(s) + 2e \rightarrow 2I^-$	0.535	
$MnO_4^-\|MnO_2$	$MnO_4^- + 2H_2O + 3e \rightarrow MnO_2 + 4OH^-$	0.588	+0.01
$Hg_2SO_4\|SO_4^{2-}$	$Hg_2SO_4 + 2e \rightarrow 2Hg + SO_4^{2-}$	0.6158	
Quinhydrone electrode ($a_{H^+} = 1$)			
$Q\|QH_2, Pt$	$Q + 2H^+ + 2e \rightarrow QH_2$	0.6995	
$Fe^{3+}, Fe^{2+}\|Pt$	$Fe^{3+} + e \rightarrow Fe^{2+}$	0.771	+1.19
$Hg_2^{2+}\|Hg$	$Hg_2^{2+} + 2e \rightarrow 2Hg$	0.7961	
$Ag^+\|Ag$	$Ag^+ + e \rightarrow Ag$	0.7996	−1.00
$O_2(g)\|H_2O, Pt$	$O_2(g) + 4H^+ 4e \rightarrow 2H_2O$ (pH 7)	0.815	
$Pd^{2+}\|Pd$	$Pd^{2+} + 2e \rightarrow Pd$	0.83	
$Hg^{2+}\|Hg$	$Hg^{2+} + 2e \rightarrow Hg$	0.851	
$Hg^{2+}\|Hg_2^{2+}, Pt$	$2Hg^{2+} + 2e \rightarrow Hg_2^{2+}$	0.905	
$Br_2(l)\|Br^-, Pt$	$Br_2(l) + 2e \rightarrow 2Br^-$	1.065	−0.61
$O_2(g)\|H_2O, Pt$	$O_2(g) + 4H^+ + 4e \rightarrow 2H_2O$	1.229	−0.85
$Cr_2O_7^{2-}, Cr^{3+}\|Pt$	$Cr_2O_7^{2-} + 14H^+ + 6e \rightarrow 2Cr^{3+} + 7H_2O$	1.33	
$Cl_2(g)\|Cl^-, Pt$	$Cl_2(g) + 2e \rightarrow 2Cl^-$	1.3583	−1.25
$Au^{3+}\|Au$	$Au^{3+} + 3e \rightarrow Au$	1.42	
$MnO_4^-, Mn^{2+}\|Pt$	$MnO_4^- + 8H^+ + 5e \rightarrow Mn^{2+} + 4H_2O$	1.491	
$Ce^{4+}, Ce^{3+}\|Pt$	$Ce^{4+} + e \rightarrow Ce^{3+}$	1.61	
$Co^{3+}, Co^{2+}\|Pt$	$Co^{3+} + e \rightarrow Co^{2+}$	1.842	
$F_2(g)\|F^-, Pt$	$F_2(g) + 2e \rightarrow 2F^-$	2.87	

Table VI.28 Formal Electrode Potentials at 298 K

The potentials listed are for solutions 1.0 mol dm^{-3} in H$^+$ (except for the calomel electrodes).
For more extensive data see: E.H. Swift and E.A. Butler, *Quantitative Measurements and Chemical Equilibria* (1972), Freeman, San Francisco.

Electrode reaction	E /V	Electrolyte
Pb^{2+} + 2e → Pb	−0.29	H$_2$SO$_4$
	−0.14	HClO$_4$
Sn^{2+} + 2e → Sn	−0.16	HClO$_4$
2H$_3$O$^+$ + 2e → 2H$_2$O + H$_2$	−0.005	HCl, HClO$_4$
Sn^{4+} + 2e → Sn^{2+}	0.14	HCl
Hg$_2$Cl$_2$ + 2e → 2Hg + 2Cl$^-$	0.2415	Saturated KCl
	0.2807	1.0 mol dm^{-3} KCl
	0.3337	0.1 mol dm^{-3} KCl
Q (quinone) + 2H$^+$ + 2e → QH$_2$	0.696	HCl, H$_2$SO$_4$, HClO$_4$
Fe(CN)$_6^{3-}$ + e → Fe(CN)$_6^{4-}$	0.71	HCl
	0.72	HClO$_4$
Fe^{3+} + e → Fe^{2+}	0.68	H$_2$SO$_4$
	0.732	HClO$_4$
Tl^{3+} + 2e → Tl$^+$	0.77	HClO$_4$
Ag$^+$ + e → Ag	0.77	H$_2$SO$_4$
	0.792	HClO$_4$
2Hg^{2+} + 2e → 2Hg$^+$	0.907	HClO$_4$
MnO$_2$ + 4H$^+$ + 2e → Mn^{2+} + 2H$_2$O	1.24	HClO$_4$
Ce^{4+} + e → Ce^{3+}	1.44	H$_2$SO$_4$
	1.61	HNO$_3$
	1.70	HClO$_4$

Table VI.29 Standard Redox Potentials for Some Biological Half-reactions at 298 K and pH = 7.0

MB and MBH$_2$ represent the oxidized and reduced forms respectively of methylene blue.
For more information see: G. Milazzo and M. Blank, eds, *Biochemistry I, Biological Redox Reactions* (1983), Plenum Press, New York.

System	$E^{\circ\prime}$/V	System	$E^{\circ\prime}$/V
acetate/pyruvate	−0.70	MB/MBH$_2^*$	+0.011
acetate/acetaldehyde	−0.581	fumarate/succinate	+0.031
Fe^{3+}/Fe^{2+} (ferredoxin)	−0.432	Fe^{3+}/Fe^{2+} (myoglobin)	+0.046
H$^+$/H$_2$	−0.421	Fe^{3+}/Fe^{2+} (cytochrome b)	+0.050
CO$_2$/formate	−0.42	ubiquinone/ubihydroquinone	+0.10
NADP$^+$/NADPH	−0.324	Fe^{3+}/Fe^{2+} (hemoglobin)	+0.144
NAD$^+$/NADH	−0.320	cytc^{3+}/cytc^{2+} (Fe^{3+}/Fe^{2+})	+0.254
Fe^{3+}/Fe^{2+} peroxidase (horse radish)	−0.271	quinone/hydroquinone	+0.28
FAD/FADH$_2$	−0.219	cyta^{3+}/cyta^{2+} (Fe^{3+}/Fe^{2+})	+0.29
acetaldehyde/ethanol	−0.197	cytf^{3+}/cytf^{2+} (Fe^{3+}/Fe^{2+})	+0.365
pyruvate/lactate	−0.185	Cu^{2+}/Cu$^+$ (haemocyanin)	+0.540
oxaloacetate/malate	−0.166	O$_2$/H$_2$O	+0.816

Table VI.30 Standard (reduction) potentials in non-aqueous solution at 298 K (except liquid ammonia 265 K) $E^{\circ}_{H^+ + H_2O} = 0$ V

Reference: G.J. Janz and R.P.T. Tomkins, eds, *Non-aqueous Electrolytes Handbook*, vol 1 (1972), vol 2 (1973), Academic Press, New York.

Electrode	Water	Methanol	Acetonitrile	Formamide	Hydrazine	Ammonia
Li$^+$ | Li	−3.04	−3.13	−3.09		−3.11	−3.23
Ca^{2+} | Ca	−2.87		−2.61		−2.82	−2.73
Na$^+$ | Na	−2.71	−2.76	−2.73		−2.74	−2.84
Zn^{2+} | Zn	−0.76	−0.77	−0.60	−0.83	−1.32	−1.52
Cd^{2+} | Cd	−0.40	−0.46	−0.33	−0.48	−1.01	−1.19
I$^-$ | AgI, Ag	−0.152	−0.318				
Pb^{2+} | Pb	−0.13	−0.23	+0.02	−0.26	−0.56	−0.67
H$^+$ | H$_2$, Pt	0	−0.03	+0.14	−0.07	−0.91	−0.99
Br$^-$ | AgBr, Ag	+0.071	−0.14				
Cl$^-$ | AgCl, Ag	+0.2224	−0.010		+0.204		
Cu^{2+} | Cu	+0.34	+0.31	−0.24	+0.21		−0.56
Ag$^+$ | Ag	+0.799	+0.73	+0.37		−0.14	−0.16
Cl$^-$ | Cl$_2$, Pt	+1.36	+1.09	+0.72			+1.104

Tables VI.31 Redox Indicators for Volumetric Use
Reference: as for Table VI.26.

Indicator	$E^\ominus/V$ (at pH = 0)	Colour change	
		Oxidized form	Reduced form
phenosafranine	0.28	red	colourless
indigo tetrasulphonate	0.36	blue	colourless
methylene blue	0.53	blue	colourless
p-ethoxychrysoidine	0.76	yellow	red
diphenylamine	0.76	violet	colourless
barium diphenylamine sulphonate	0.84	violet	colourless
sodium diphenylamine sulphonate	0.845	violet	colourless
diphenylamine sulphonic acid	0.86	red–violet	colourless
lissamine green	0.99	orange	green
erioglaucine	1.00	orange	green–yellow
xylene cyanol FF	1.05		
N-phenylanthranilic acid	1.08	purple–red	colourless
1,10-phenanthroline hydrate	1.08	blue	red
1,10-phenanthroline–iron (II) complex	1.11	blue	red
5-nitro-1,10-phenanthroline–iron (II) complex	1.25	pale blue	red–violet

Table VI.32 Biological Redox Indicators
Reference: as for Table VI.24.

Indicator	$E^{\ominus\prime}/V$ (at pH = 7.0)	Colour change	
		Oxidized form	Reduced form
methylviologen	−0.440	violet	colourless
benzylviologen	−0.359	violet	colourless
neutral red	−0.325	red–violet	colourless
phenosafranine	−0.252	violet–blue	colourless
Janus green	−0.225	green	colourless
methylene blue	0.011	blue	colourless
phenolindo-2,6-dichlorophenol	0.217	blue	colourless
phenolindophenol	0.227	blue	colourless

Table VI.33 Dissociation Constants of Inorganic Acids in Aqueous Solution at 298 K
For more extensive data see: D.D. Perrin, *Ionization Constants of Inorganic Acids and Bases in Aqueous Solution*, IUPAC Chemical Data Series No. 29 (1982), Pergamon Press, Oxford.

K_a, equilibrium constant for the reaction $HA \rightleftharpoons H^+ + A^-$, where HA is an acid, it may be expressed as $pK_a = -\lg K_a$

Compound	$pK_{a,1}$;	$pK_{a,2}$;	$pK_{a,3}$
aluminium ion $[Al(H_2O)_6]^{3+}$	4.9		
ammonium ion (NH_4^+)	9.24		
arsenic(III) acid (arsenious acid)	9.29		
arsenic(V) acid	2.25,	6.77,	11.60
azoic acid (HN_3)	4.72		
boric acid	9.14,	12.74,	13.8 (293 K)
bromic(I) acid (hypobromous acid)	8.69		
carbonic acid	6.37,	10.25	
chloric(I) acid (hypochlorous acid)	7.53	(291 K)	
chromic acid	−0.61,	6.49	
chromium(III) ion $[Cr(H_2O)_6]^{3+}$	3.95		
hydrazinium ion ($N_2H_5^+$)	7.94		
hydrocyanic acid	9.31		
hydrofluoric acid	3.25		
hydrogen peroxide	11.62		

(Continued)

Table VI.33 (Continued)

Compound	$pK_{a,1}$;	$pK_{a,2}$;	$pK_{a,3}$
hydrogen sulphide	7.04,	11.96	(291 K)
iodic(V)acid	0.77		
iron(III) ion [Fe(H$_2$O)$_6$]$^{3+}$	2.22		
lead(II) ion [Pb(H$_2$O)$_n$]$^{2+}$	7.8		
nitric(III) acid (nitrous acid)	3.37	(285.5 K)	
periodic acid	1.64		
hypophosphorous acid	1.2		
phosphorous acid	2.00,	6.59	(291 K)
o-phosphoric acid	2.12,	7.21,	12.67
pyrophosphoric acid	0.85,	1.49,	5.77, 8.22 (291 K)
selenic(VI) acid	<0,	1.92	
selenic(IV) acid (selenous acid)	2.46,	7.31	
m-silic acid	9.70,	12.00	
o-silic acid	9.66,	11.70,	12.00,12.00 (303 K)
sulphuric(IV) acid (sulphurous acid)	1.81,	6.91	(291 K)
sulphuric(VI) acid	<0,	1.92	
telluric(IV) acid	2.48,	7.70	
telluric(VI) acid	7.68,	11.29	(291 K)

Table VI.34 Dissociation Constants of Organic Acids in Aqueous Solution at 298 K
For more extensive data see: E.P. Serjeant and B. Dempsey, *Ionization Constants of Organic Acids in Aqueous Solution*, IUPAC Chemical Data Series No. 23 (1979), Pergamon Press, Oxford.

Compound	$pK_{a,1}$;	$pK_{a,2}$;	$pK_{a,3}$
acetic acid	4.75		
aminoacetic acid (glycine)	2.35,	9.78	
bromoacetic acid	2.69		
chloroacetic acid	2.85		
dichloroacetic acid	1.48		
deuteroacetic acid (in D$_2$O)	5.25		
trichloroacetic acid	0.70		
flouroacetic acid	2.23		
hydroxyacetic acid (glycolic acid)	3.83		
iodoacetic acid	3.12		
phenylacetic acid	4.31		
thioacetic acid	3.33		
acetylenedicarboxylic acid	1.75,	4.40	
acrylic acid	4.25		
adipic acid	4.43,	5.41	
ascorbic acid	4.30,	11.82	
barbituric acid	4.01		
benzene-1,2-dicarboxylic acid (phthalic acid)	2.89,	5.51	
benzene-1,3-dicarboxylic acid (isophthalic acid)	3.54,	4.60	
benzene-1,4-dicarboxylic acid (terephthalic acid)	3.51,	4.82	
benzene sulphonic acid	0.70		
benzoic acid	4.19		
2-aminobenzoic acid (anthranilic acid)	6.97		
3-aminobenzoic acid	4.78		
4-aminobenzoic acid	4.92		
2-hydroxybenzoic acid (salicylic acid)	3.00,	13.40	
3-hydroxybenzoic acid	4.08,	9.93	
2-hydroxybenzoic acid	4.58,	9.23	
2-methylbenzoic acid (o-toluic acid)	3.91		
3-methylbenzoic acid (m-toluic acid)	4.27		
4-methylbenzoic acid (p-toluic acid)	4.36		
2-nitrobenzoic acid	2.17		
3-nitrobenzoic acid	3.49		
4-nitrobenzoic acid	3.44		

(Continued)

Table VI.34 (Continued)

Compound	$pK_{a,1}$	$pK_{a,2}$	$pK_{a,3}$
butanoic acid	4.83		
2-chlorobutanoic acid	2.86		
3-chlorobutanoic acid	4.05		
4-chlorobutanoic acid	4.52		
cis-cinnamic acid	3.89		
trans-cinnamic acid	4.44		
citric acid	3.13,	4.76,	6.40
α-crotonic acid	4.69		
isocrotonic acid	4.44		
cyclohexane-1,1-dicarboxylic acid	3.45,	6.11	
1,2-dihydroxybenzene (catechol)	9.12	12.08 (303 K)	
1,3-dihydroxybenzene (resorcinol)	9.15,	11.32 (303 K)	
1,4-dihydroxybenzene (hydroquinone)	9.91,	12.04 (303 K)	
formic acid	3.75		
fumaric acid	3.03,	4.44	
furoic acid	3.17		
glutaric acid	4.31,	5.41	
glycerol	14.15		
glycol	14.22		
hexanoic acid (n-caproic acid)	4.88		
1-hydroxynaphthalene (α-naphthol)	9.34		
2-hydroxynaphthalene (β-naphthol)	9.51		
lactic acid (D, L, DL)	3.86		
maleic acid	1.83,	6.07	
malic acid	3.40,	5.11	
malonic acid	2.83,	5.69	
1-naphthalenesulphonic acid	0.17		
1-naphthalenecarboxylic acid (α-naphthoic acid)	3.70		
2-naphthalenecarboxylic acid (β-naphthoic acid)	4.17		
octanoic acid (caprylic acid)	4.89		
oxalic acid	1.23,	4.19	
phenol	9.99		
2-chlorophenol	8.49		
3-chlorophenol	8.85		
4-chlorophenol	9.18		
2,4-dinitrophenol	4.09		
3,6-dinitrophenol	5.22		
2,4,6-trinitrophenol (picric acid)	0.29		
2-methylphenol (o-cresol)	10.20		
3-methylphenol (m-cresol)	10.01		
4-methylphenol (p-cresol)	10.17		
pimelic acid	4.48,	5.42	
propanoic acid (proprionic acid)	4.87		
2,2-dimethylpropanoic acid (trimethylacetic acid)	5.03 (291 K)		
2-methylpropanoic acid (isobutyric acid)	4.84 (293 K)		
pyruvic acid	2.49		
quinolinic acid	2.52		
2-pyridinecarboxylic acid	5.52		
3-pyridinecarboxylic acid	4.85		
4-pyridinecarboxylic acid	4.96		
succinic acid	4.16,	5.61	
tartaric acid (D, L and DL)	2.98,	4.34	
tartaric acid (meso-)	3.22,	4.82	
valeric acid	4.84		
iso-valeric acid	4.78		

Table VI.35 Dissociation Constants of Inorganic bases in Aqueous Solution at 298 K
For more extensive data see: D.D. Perrin, *Ionization Constants of Inorganic Acids and Bases in Aqueous Solution*, IUPAC Chemical Data Series No. 29 (1982), Pergamon Press, Oxford.

Compound	$pK_{b,1}$	$pK_{b,2}$
ammonium hydroxide	4.75	
barium hydroxide		0.64
beryllium hydroxide	10.30	
calcium hydroxide	2.43	1.40
deuteroammonium hydroxide	4.24	
hydrazine	6.06	
hydroxylamine	8.05	
lead hydroxide	3.02	
silver hydroxide	3.96	
strontium hydroxide	0.82	
zinc hydroxide	3.02	

Table VI.36 Dissociation Constants of Organic bases in Aqueous Solution at 298 K
The dissociation constant of a base B is given in terms of the pK_a value of its conjugate acid, BH^+:
$$BH^+(aq) \rightleftharpoons B(aq) + H^+(aq).$$
The pK_b for a base may be calculated from the pK_a value of its conjugate acid from the expression:
$$pK_b = pK_w - pK_a;$$ at 298 K this becomes $pK_b = 14.00 - pK_a$.
For more extensive data see: D. Perrin, *Dissociation Constants of Organic Bases in Aqueous Solution: Supplement* (1972), IUPAC, Butterworths, London.

Compound	$pK_{a,1}$	$pK_{a,2}$
acetamide	0.63	
acridine	5.58 (293 K)	
1-aminonaphthalene (α-naphthylamine)	3.92	
2-aminonaphthalene (β-naphthylamine)	4.16	
aniline	4.63	
2-chloroaniline	2.65	
3-chloroaniline	3.52	
4-chloroaniline	4.15	
N,N-diethylaniline	2.61 (295 K)	
N,N-dimethylaniline	5.15	
N-ethylaniline	5.12 (297 K)	
N-methylaniline	4.848	
2-hydroxyaniline (2-aminophenol)	4.74	9.66
2-methylaniline (o-toluidine)	4.44	
3-methylaniline (m-toluidine)	4.73	
4-methylaniline (p-toluidine)	5.08	
3-nitroaniline	2.466	
4-nitroaniline	1.00	
N-phenylaniline (diphenylamine)	0.79	
benzidine	4.66	3.57 (303 K)
benzylamine	9.33	
brucine	8.28	
n-butylamine	10.61	
iso-butylamine	10.41	
sec-butylamine	10.56	
tert-butylamine	10.83	
cyclohexylamine	10.64	
n-decylamine	10.64	
1,2-diaminobenzene	0.6	4.74
1,3-diaminobenzene	2.3	5.0
1,4-diaminobenzene	2.7	6.2
1,2-diaminoethane (ethylenediamine)	9.928	6.848
1,2-diaminopropane	9.82	6.61
1,3-diaminopropane	10.94	9.03 (283 K)

(Continued)

Table VI.36 (Continued)

Compound	$pK_{a,1}$	$pK_{a,2}$
ethanolamine	9.498	
diethanolamine	8.883	
triethanolamine	7.762	
ethylamine	10.807	
diethylamine	11.04	
triethylamine	11.01 (291 K)	
n-hexylamine	10.56	
methylamine	10.657	
dimethylamine	10.732	
trimethylamine	9.81	
nicotine	8.02	3.12
1,10-phenanthroline	4.84	
piperazine	9.83	5.56 (296 K)
piperidine	11.123	
n-propylamine	10.708	
isopropylamine	10.6	
pteridine	4.05 (293 K)	
purine	2.30	8.96 (293 K)
6-aminopurine (adenine)	4.12	9.83
pyrazine	0.65 (300 K)	
pyrazole	2.48	
pyridazine	2.24 (293 K)	
pyridine	5.25	
2-aminopyridine	6.82 (293 K)	
4-aminopyridine	9.11	
2,3-dimethylpyridine (2,3-lutidine)	6.57	
2,4-dimethylpyridine (2,4-lutidine)	6.99	
3,5-dimethylpyridine (3,5-lutidine)	6.15	
2-methylpyridine (α-picoline)	5.97	
3-methylpyridine (β-picoline)	5.68	
4-methylpyridine (γ-picoline)	6.02	
pyrrolidine	11.27	
quinine	8.52	4.13
quinoline	4.88	
isoquinoline	5.42 (293 K)	
thiazole	2.44 (293 K)	
tris-(hydroxymethyl)methylamine (Tris)	8.075	
urea	0.10	

Table VI.37 Ionization Constants for Water and Deuterium Oxide
The subscript m indicates a value on the molal scale.
Reference: H.S. Harned and R.A. Robinson, *Trans. Faraday Soc.* (1940) 36, 973; *Handbook of Chemistry and Physics* (1989–90), CRC.

Temperature/°C	pK_w for water	$^*pK_m$ for D_2O
0	14.9435	
5	14.7338	
10	14.5346	15.526
15	14.3463	
20	14.1669	15.136
24	14.0000	
25	13.9965	14.955
30	13.8330	14.784
35	13.6801	
40	13.5348	14.468
45	13.3960	
50	13.2617	14.182
55	13.1369	
60	13.0171	

Table VI.38 Autoprotolysis Constants for Some Solvents at 298 K

For the solvent SH capable of being both a proton donor and acceptor: $SH + SH \rightleftharpoons SH_2^+ + S^-$
for which $K_s = a_{SH_2^+} + a_{S^-}$. Since solvents have different molalities, a better comparison can be obtained by diving K_s by the square of the molality of the pure solvent: $K_s^1 = K_s/m_{SH}^2 = K_s \times M^2$ (M kg is the molar mass of SH).
Reference: F.T. Gucker and R.L. Seifert, *Physical Chemistry* (1967), English University Press.

Solvent	pK_s	pK_s^1
sulphuric acid	3.57	5.59
ethanolamine	5.7	8.1
water	14.00	17.49
acetic acid	14.5	16.9
methanol	16.6	19.6
ethanol	18.8	21.5
ammonia (240 K)	29.0	32.0
acetonitrile	>32	>35

Table VI.39 Solubility Products in water at 298 K

The values of K_s are tabulated in the forms a and b, where b is the exponent of 10 which multiplies a. Thus for 2.5, –11, K_s (for a 1 : 1 salt) = 2.5×10^{-11} mol^2 dm^{-6}. The units of K_s for a salt M_nX_m are mol$^{(m+n)}$ dm$^{-3(m+n)}$.
For more extensive data see: L.G. Sillen and A.E. Martell, *Stability Constants of Metal–Ion Complexes*, Chemical Society, London (1964).

Substance	a	b
aluminium hydroxide	1.0	−33
barium carbonate	5.1	−9
barium chromate (301 K)	5.0	−10
barium hydroxide	5.0	−3
barium oxalate	1.7	−7
barium sulphate	1.08	−10
cadmium sulphide (291 K)	2	−28
calcium carbonate (calcite)	4.8	−9
calcium fluoride	3.95	−11
calcium oxalate	2.57	−9
calcium sulphate	2.45	−5
calcium tartrate (291 K)	0.77	−6
cobalt sulphide (291 K)	3.0	−26
chromium (III) hydroxide	2.9	−29
copper (II) sulphide (291 K)	6	−36
copper (I) bromide (293 K)	4.15	−8
copper (I) chloride (293 K)	1.02	−6
copper (I) iodide	1.1	−12
copper (I) thiocyanate (291 K)	1.6	−11
iron (III) hydroxide (291 K)	1.1	−36
iron (II) hydroxide	8	−16
iron (II) sulphide	6	−18
lead bromide	7.9	−5
lead carbonate (291 K)	3.3	−14
lead chloride	2.0	−5
lead chromate (291 K)	1.77	−14
lead fluoride (291 K)	3.2	−8
lead iodide	1	−9
lead oxalate (291 K)	2.74	−11
lead sulphate	1.58	−8
lead sulphide (291 K)	3.4	−28
lithium carbonate	1.7	−3
magnesium ammonium phosphate	2.5	−13
magnesium carbonate	1	−5
magnesium fluoride (300 K)	6.4	−9
magnesium hydroxide	1.9	−13
magnesium oxalate (291 K)	8.57	−5
manganese hydroxide	1.9	−13
manganese sulphide (pink)	2.5	−10
(green)	2.5	−13

(Continued)

Table VI.39 (Continued)

Substance	a	b
mercury (II) sulphide (black)	1.6	−52
(red)	1.4	−53
mercury (I) bromide	5.8	−23
mercury (I) chloride	1.3	−18
mercury (I) iodide	4.5	−29
nickel sulphide (α)	3.0	−19
(β)	1.0	−24
(γ)	2.0	−26
potassium hydrogen tartrate (291 K)	3.0	−4
silver bromate	5.77	−5
silver bromide	5.2	−13
silver chloride	1.82	−10
silver chromate	1.1	−12
silver iodide	8.3	−17
silver phosphate	1.8	−18
silver sulphide	6	−50
silver thiocyanate	1.16	−12
strontium carbonate	1.1	−10
strontium fluoride (291 K)	2.8	−9
strontium oxalate (291 K)	5.61	−8
strontium sulphate (290 K)	3.8	−7
thallium chloride	1.9	−4
tin (II) hydroxide	3.0	−27
tin (II) sulphide	1.0	−25
zinc hydroxide (293 K)	2.0	−17
zinc sulphide	4.5	−24

Table VI.40 Stepwise Formation (Stability) Constants

K_n (mol dm^{-3}) is the stability constant for the reaction:

$$ML_{n-1} + L \rightleftharpoons ML_n \ (n = 1, 2, 3, \ldots).$$

The higher the stability constant the more stable the complex ion.

For more extensive data see: A.E. Martell and R.M. Smith *Critical Stability Constants*, 1 (1974) Amino Acids, 2 (1975) Amines, 3 (1977) Other Organic Ligands, 4 (1976) Inorganic Complexes, 5 (1982) First Supplement. Plenum, New York.

Ligand	Metal ion	I /mol dm^{-3}	lg K_1	lg K_2	lg K_3	lg K_4
NH$_3$	Ag$^+$	0	3.30	3.84		
	Cd^{2+}	0	2.51	1.96	1.30	0.79
	Cu^{2+}	0	3.99	3.34	2.73	1.97
	Co^{2+}	0	1.99	1.51	0.93	0.64
			lg K_5 = 0.04	lg K_6 = −0.74		
	Hg^{2+}	2	8.78	8.70	1.00	0.78
	Ni^{2+}	0	2.67	2.11	1.61	1.08
			lg K_5 = 0.63	lg K_6 = −0.96		
	Zn^{2+}	0	2.28	2.34	2.40	2.04
Cl$^-$	Cd^{2+}		2.00	0.70	−0.58	
	Hg^{2+}	0.5	6.74	6.48	0.95	1.04
	Fe^{3+}	0	1.48	0.65	−1.0	
	Sn^{2+}	0	1.51	0.73	−0.21	−0.55
CN$^-$	Ag$^+$	0	lg K_1K_2 = 20			
	Cd^{2+}	3	5.70	5.11	4.63	3.54
	Hg^{2+}	0.1	18.0	16.7	3.85	3.00
	Ni^{2+}		lg $K_1K_2K_3K_4$ = 22			
F$^-$	Al^{3+}	0.5	6.11	5.00	3.85	2.70
			lg K_5 = 1.60	lg K_6 = 0.70		
	Fe^{3+}	0.5	5.26	4.00	3.00	

(Continued)

Table VI.40 (Continued)

Ligand	Metal ion	I /mol dm^{-3}	lg K_1	lg K_2	lg K_3	lg K_4
EDTA (see Table VI. 41)						
I$^-$	Cd^{2+}		2.28	1.64	1.08	1.11
	Hg^{2+}	0.5	12.87	10.95	3.78	2.32
	I$_2$	2	2.9			
C$_2$O$_4^{2-}$	Fe^{3+}	0	9.40	6.78	4.00	
	Mg^{2+}	0	3.82	1.43		
	Mn^{3+}	2	10.0	6.59		
SO$_4^{2-}$	Al^{3+}	0	3.20	1.90		
	Cd^{2+}	0	2.30			
	Cu^{2+}	0	2.20			
	Fe^{3+}	0	4.04			
SCN$^-$	Fe^{3+}	0	2.15	1.30	0	
	Hg^{2+}	0	lg K_1K_2 = 17.25		2.71	1.78

Table VI.41 Formation Constants for EDTA Complexes at 293 K and an Ionic Strength of 0.1 mol dm^{-3}

The constant K refers to the equilibrium involving the species Y^{4-} (i.e. the ethylenediaminetetraacetate ion):

$$M^{n+} + Y^{4-} \rightleftharpoons MY^{(n-4)+} \qquad K = [MY^{(n-4)+}]/[M^{n+}][Y^{4-}]$$

For more information see: G. Schwarzenbach, *Complexometric Titrations* (1957), Interscience, New York.

Cation	lg K	Cation	lg K
Ag$^+$	7.32	La^{3+}	15.1
Al^{3+}	16.13	Li$^+$	2.8
Ba^{2+}	7.76	Mg^{2+}	8.69
Ca^{2+}	10.70	Mn^{2+}	13.79
Cd^{2+}	16.46	Na$^+$	1.7
Co^{2+}	16.31	Ni^{2+}	18.62
Cr^{2+}	13.0	Pb^{2+}	18.04
Cr^{3+}	24.0	Sr^{2+}	8.63
Cu^{2+}	18.80	Th^{4+}	23.2
Fe^{2+}	14.33	V^{3+}	25.9
Fe^{3+}	25.11	Zn^{2+}	16.50
Hg^{2+}	21.80		

Table VI.42 Hammett Acidity Function, H_0

For further information see: F.A. Cotton and G. Wilkinson, *Advanced Inorganic Chemistry*, 5th ed. (1989), Wiley, London.

Acid	H_0
HSO$_3$F + SbF$_5$ + SO$_3$	>16
HF + SbF$_5$ (3 mol dm^{-3})	15.2
HSO$_3$F	12.6
H$_2$S$_2$O$_7$	12.2
H$_2$SO$_4$	11.0
HF	10.2
HF + NaF (1 mol dm^{-3})	8.4
H$_3$PO$_4$	5.0
H$_2$SO$_4$ (63% in H$_2$O)	4.9
HCOOH	2.2

Table VI.43 Half-wave Potentials of Selected Inorganic Species

Half-wave potentials are with reference to the saturated calomel electrode (SCE) at 25 °C for which $E^\circ = -0.242$ V.
For more extensive data see: G.J. Janz and R.P.T. Tomkins, *Non-aqueous Electrolytes Handbook II* (1973), Academic Press, New York.

1. Half-wave Potentials of Metals

Metal	Solvent	Supporting electrolyte (concentration/mol dm^{-3})	$E_{\frac{1}{2}}$ /V	Comments
Al	water	BaCl$_2$ (0.025)	−1.75	
Ba	80% EtOH	Me$_4$NI	−1.82	
Bi	water	tartrate buffer (0.5)	−0.29	pH = 4.5
Ca	80% EtOH	Me$_4$NI	−2.13	
Cd	water	KCl or HCl (0.1)	−0.599	
	water	tartrate buffer (0.5)	−0.64	pH = 4.5
	water	NH$_4$OH (1) + NH$_4$Cl (1)	−0.81	
Co (Co^{2+})	water	KCl (0.1)	−1.4	
(Co^{3+})	water	NH$_4$OH (1) + NH$_4$Cl (1)	−0.5	Co$^{3+} \to$ Co^{2+}
			−1.3	Co$^{2+} \to$ Co
(Co^{2+})	water	KSCN (1)	−1.03	
(Co^{2+})	water	pyridine (0.1) + pyridinium chloride (0.1)	−1.07	
(Co^{2+})	DMSO	NaClO$_4$ (0.1)	−1.4	
(Co^{2+})	30% formamide	NaClO$_4$ (1)	−1.270	
Cr (Cr^{3+})	water	KCl (0.1)	−0.91	Cr$^{3+} \to$ Cr^{2+}
			−1.47	Cr$^{2+} \to$ Cr
(CrO$_4^{2-}$)	water	NaOH (1)	−0.85	Cr$^{6+} \to$ Cr^{3+}
(CrO$_4^{2-}$)	water	KCl (0.1)	−0.3	
			−1.0	Cr$^{6+} \to$ Cr^{3+}
			−1.5	Cr$^{3+} \to$ Cr^{2+}
			−1.7	Cr$^{2+} \to$ Cr
Cu (Cu^{2+})	water	KCl or HCl (0.1)	+0.4	Cu$^{2+} \to$ Cu
	water	NH$_4$OH (1) + NH$_4$Cl (1)	−0.24	Cu$^{2+} \to$ Cu$^+$
			−0.5	Cu$^+ \to$ Cu
(Cu^{2+})	water	KSCN (0.1)	−0.02	Cu$^{2+} \to$ Cu$^+$
			−0.39	Cu$^+ \to$ Cu
Fe (Fe^{2+})	water	KCl (0.1) or BaCl$_2$ (0.05)	−1.32	Fe$^{2+} \to$ Fe
(Fe^{3+})	water	KOH (1) + 8% mannitol	−0.9	Fe$^{3+} \to$ Fe^{2+}
			−1.5	Fe$^{2+} \to$ Fe
K	50% EtOH	Et$_4$NOH	−2.10	
Li	50% EtOH	Et$_4$NOH	−2.31	
Mn (Mn^{2+})	water	KCl (1)	−1.51	Mn$^{2+} \to$ Mn
	water	NH$_4$OH (1) + NH$_4$Cl (1)	−1.65	
	water	KSCN (1)	−1.55	
	30% formamide	NaClO$_4$ (1)	−1.508	
Na	50% EtOH	Et$_4$NOH	−2.07	
Ni	water	KCl (1)	−1.1	
	water	NH$_4$OH (1) + NH$_4$Cl (0.2)	−1.06	
	water	KSCN (1)	−0.70	
	water	KCl (1) + pyridine (0.5)	−0.78	
	DMSO	NaClO$_4$ (0.1)	−1.0	
	30% formamide	NaClO$_4$ (1)	−0.980	
Pb	water	KNO$_3$ (1)	−0.405	
	water	KCl or HCl (0.1)	−0.396	
	water	tartrate buffer (0.5)	−0.48	pH = 4.5
Sb (Sb^{3+})	water	HCl (1)	−0.15	Sb$^{3+} \to$ Sb
(Sb^{3+})	water	NaOH (1)	−1.26	
Sn (Sn^{2+})	water	HCl (1)	−0.47	
(SnO$_2$)	water	NaOH (1)	−1.22	SnO$_2 \to$ Sn
Ti (Ti^{4+})	water	HCl or HNO$_3$ (0.1)	−0.81	Ti$^{4+} \to$ Ti^{3+}
Tl	water	KCl, HCl or NaOH (0.1)	−0.46	
U (UO$_2^{2-}$)	water	HCl (0.02)	−0.18	U$^{6+} \to$ U^{5+}
			−0.92	U$^{5+} \to$ U^{3+}
V (V^{4+})	water	HCl (0.01)	−0.85	V$^{4+} \to$ V^{2+}

(Continued)

Table VI.43 (Continued)

Metal	Solvent	Supporting electrolyte (concentration/mol dm^{-3})	$E_{\frac{1}{2}}$ /V	Comments
Zn	water	KCl or KNO$_3$ (0.1)	−0.995	
	water	KSCN (0.1)	−1.01	
	water	tartrate buffer (0.5)	−1.23	pH = 4.5
	water	NaOH (1) [ZnO$_2$]	−1.53	
	30% formamide	NaClO$_4$ (1)	−1.024	

2. Half-wave Potentials of Some Other Inorganic Species

Species	Solvent	Supporting Electrolyte (conc./mol dm^{-3})	$E_{\frac{1}{2}}$ /V	Comments
NH$_3$	DMSO	Et$_4$NClO$_4$ (0.1)	+0.2	
NH$_4$Cl	DMSO	Et$_4$NClO$_4$ (0.1)	−0.17	
			−2.04	
Cl−	MeNO$_2$	Et$_4$NClO$_4$ (0.1)	+1.38	v. Ag/AgCl
CrO$_4^{2-}$	30% MeOH	NaOH (0.1)	−1.095	
	30% PrOH	NaOH (0.1)	−1.235	
HCl	DMSO	Et$_4$NClO$_4$ (0.1)	+0.01	
			−0.17	
			−1.14	
HClO$_4$	DMSO	Et$_4$NClO$_4$ (0.1)	−1.14	
I−	MeOH	Et$_4$NClO$_4$ (0.1)	+0.4	($E_{\frac{1}{4}}$)
	PrOH	LiClO$_4$ (0.2)	+0.42	($E_{\frac{1}{4}}$)
I$_2$	MeCN	Et$_4$NClO$_4$ (0.1)	+0.30	v. Ag/Ag$^+$
			+0.96	v. Ag/Ag$^+$
KCl	DMSO	Et$_4$NClO$_4$ (0.1)	+0.01	
			−0.17	
			−2.22	
KClO$_4$	DMSO	Et$_4$NClO$_4$ (0.1)	−2.11	
O$_2$	DMSO	NaClO$_4$ (0.1)	−0.65	
			−1.20	

Table VI.43 (Continued)
3. Half-wave Potentials of Some Cations in Non-aqueous Solutions

Ion	Solvent methanol	ethanol	ethane-1, 2-diol	acetone	tetrahydrofuran	formamide	N-methyl-formamide	N,N-dimethyl-formamide	N,N-dimethyl-acetamide
Li^+	−1.49	−1.46[a]		−1.40[a]	−1.435		−1.661[a]	−1.623	−1.690
Na^+	−1.22	−1.17[a]		−1.224[a]	−1.251		−1.321[a]	−1.349	−1.380
K^+	−1.24	−1.18[d]		−1.281[a]	−1.202		−1.334[a]	−1.371	−1.404
Rb^+	−1.23	−1.18[d]		−1.296[a]	−1.204		−1.335[a]	−1.358	−1.349
Cs^+	−1.20	−1.18[d]		−1.270[a]	−1.297		−1.294[a]	−1.335	−1.344
Cu^+				1.020[a]		0.718[a]			
Tl^+	0.422	0.449	0.366[b]	0.410[a]	0.408	0.375	0.286	0.261	0.259
Cd^{2+}	0.285	0.224[a]	0.271[b]	0.510[a]	0.320	0.156	0.132	0.126	0.129
Ba^{2+}	−1.06[a]	−0.991		−1.075[a]	−1.102		−1.33[a]	−1.305	−1.354
Pb^{2+}	0.467[e]	0.525	0.413[b]	0.708	0.511	0.387	0.282	0.270	0.259
Mn^{2+}			−0.644	−0.410[a]	−0.594			−0.86	−0.88
Co^{2+}			−0.614	0.00[a]	−0.420			−0.55	−0.35
Ni^{2+}			−0.763		−0.453			−0.22	−0.14
Cu^{2+}	0.963	0.936[d]	0.651	1.23[a]	0.817	0.858[a]	0.723	0.706	0.725
Zn^{2+}	−0.397	−0.180	−0.361	0.130[a]	−0.049		−0.390	−0.291	−0.233

Tabulated values of $E_{\frac{1}{2}}$ in volts against bis(biphenyl)chromium (I)/(0) for the reduction of metal ions in 0.1 mol dm^{-3} solutions of tetrabutylammonium perchlorate, Bu_4NClO_4, in non-aqueous solutions at 25 °C unless otherwise stated.
[a] 0.1 mol dm^{-3} Et_4NClO_4; [b] 0.05 mol dm^{-3} Bu_4NClO_4; [c] 30 °C; [d] iodides in 0.1 mol dm^{-3} solutions of Bu_4NI; [e] 0.05 mol dm^{-3} Bu_4NI;
[f] tetraphenylborates in 0.05 mol dm^{-3} tetrabutylammonium tetraphenylborate.
For more extensive data see: G. Gritzner, *Pure App. Chem.* (1990), **62**, 1839.

nitromethane	acetonitrile	benzonitrile	pyridine	dimethyl sulphoxide	tetramethylene sulphone	trimethyl phosphate
	−1.200[a]	−1.1115	−1.428	−1.86	−1.26[a,c]	−1.721
	−1.118[a]	−1.044	−1.201	−1.37[a]	−1.15[a,c]	−1.37[a]
	−1.223[a]	−1.131[f]	−1.231	−1.40[a]	−1.25[a,c]	−1.36[a]
	−1.224[a]	−1.174[f]	−1.232	−1.37[a]	−1.26[a,c]	−1.350
	−1.207[a]	−1.183[f]	−1.215	−1.361	−1.25[a,c]	−1.309
				0.595		
				0.18	0.41[a,c]	0.31
0.824	0.460	0.543[a]	0.031			
	−0.883	−0.88[a]	−1.036			
0.850	0.686	0.705	0.337			
	−0.34[a]					
	0.12[a]					
	0.44[a]					

Table VI.44 Half-wave Potentials of Selected Organic Compounds.

Solvents: % (w/v) with water. DMF: dimethylformamide. DMSO: dimethylsulphoxide. MeCN: acetonitrile. [a] gelatine present.
Electrodes DME:dropping mercury electrode. HDE:hanging drop electrode: MP: mercury pool. NHE: normal hydrogen electrode. SCE: saturated calomel electrode. NCE: normal calomel electrode. AgBr, AgCl, AgI, AgClO$_4$: silver–silver bromide,–silver chloride,–silver iodide, –silver perchlorate electrodes respectively. R$_4$NCE: mercury, mercury (I) chloride – tetraalkylammonium chloride electrode.
Temperature   In range 20 to 25 °C.
For more extensive data see: G.J. Janz and R.P.T. Tomkins, *Non-aqueous Electrolytes Handbook II* (1973), Academic Press, New York; L. Meites and P. Zuman, *CRC Handbook Series in Organic Electrochemistry*, Vol. II, (1978), CRC Press, Cleveland.

Compound	Solvent	Supporting Electrolyte (c/mol dm^{-3})	$E_{\frac{1}{2}}$ /V	versus	Electrodes
acetaldehyde	30% EtOH	Et$_4$NOH (0.05)	−2.03	SCE	DME/SCE
acetic acid	DMSO	Et$_4$NClO$_4$ (0.1)	−1.91 −2.22	SCE	DME/SCE
	pyridine	LiClO$_4$ (0.1)	−1.12 −1.36	Ag/Ag$^+$	DME/Ag, Ag$^+$
acetone	water	H$_2$SO$_4$ (1)	−0.23	MP	DME/MP
acrylonitrile	MeCN	Et$_4$NClO$_4$ (0.1)	−2.17	SCE	DME/SCE
adenine	water	HCl, KCl (0.5, pH = 1.2)	−1.08	SCE	DME/SCE
aniline	water	Bu$_4$NBr (0.1), HCl	−1.36	NHE	DME/R$_4$NCE
anthracene	DMF	Bu$_4$NClO$_4$ (0.1)	−2.405	AgClO$_4$	DME/AgClO$_4$
L-arabinose	water	H$_3$PO$_4$, (Me$_4$N)$_3$PO$_4$ (pH) = 8)	−2.12	SCE	DME/SCE
asparagine	DMSO	Et$_4$NClO$_4$ (0.1)	−2.38	SCE	DME/SCE
aspartic acid	DMSO	Et$_4$NClO$_4$ (0.1)	−2.14 −2.618	SCE	DME/SCE
benzaldehyde	water	CH$_3$COOH (0.1), KCl	−1.15	SCE	DME/SCE
benzil	50%MeOH	buffer (0.001), pH = 8.5)	−0.688	SCE	DME/SCE
	50% MeOH	buffer (0.001, pH = 11)	−0.780	SCE	DME/SCE
benzoic acid	DMF	Et$_4$NI (0.04)	−2.24	SCE	DME/AgI
1,4-benzoquinone	water	NH$_4$OH (0.5, pH = 7.1)	−0.18	SCE	DME/SCE
bromobenzene	MeCN	Et$_4$NBr (0.02)	−1.91	AgBr	DME/AgBr
bromoethane	DMF	Et$_4$NBr (0.01)	−2.167	SCE	DME/SCE
chlorobenzene	DMF	Et$_4$NBr	−2.13	AgBr	DME/AgBr
cyclohexanone	water	HCl (0.1)	−1.213	SCE	DME/AgCl
L-cysteine	water	buffer (pH = 11.5)	−0.58	SCE	DME/SCE
1,2-dinitrobenzene	20% MeOH	NaOH (0.1), LiCl (1)[a]	−0.39	SCE	DME/SCE
	DMF	Et$_4$NClO$_4$ (0.1)	−0.72 −1.08 −1.66 −2.24	SCE	HDE/SCE
1,3-dinitrobenzene	water	buffer (pH = 1.1)[a]	−0.10	SCE	DME/SCE
1,4-dinitrobenzene	20% MeOH	NaOH (0.1), LiCl (1), CaCl$_2$ (0.2)[a]	−0.331	SCE	DME/SCE
	DMF	Et$_4$NClO$_4$ (0.1)	−0.535 −0.84 −2.24	SCE	HDE/SCE
formic acid	pyridine		−1.48	SCE	DME/Ag, Ag$^+$
fumaric acid	pyridine	LiClO$_4$ (0.1)	−1.12 −1.33	MP	DME/MP
furfural	water	buffer (0.25, pH = 2.47)	−1.04 −1.38	SCE	DME/SCE
glutamic acid	DMSO	Et$_4$NClO$_4$ (0.1)	−2.123 −2.604	SCE	DME/SCE
hydroxyproline	DMSO	Et$_4$NClO$_4$ (0.1)	−2.317	SCE	DME/SCE
iodoethane	water	LiClO$_4$ (0.1)	−1.73	SCE	DME/SCE
isoleucine	DMSO	Et$_4$NClO$_4$ (0.1)	−2.435	SCE	DME/SCE
maleic acid	water	borate, KCl (0.3, pH = 8)	−1.50	SCE	DME/SCE
malic acid	DMSO	Et$_4$NClO$_4$ (0.1)	−2.25	SCE	DME/SCE
methionine	DMSO	Et$_4$NClO$_4$ (0.1)	−2.435	SCE	DME/SCE
methyl bromide	water	KBr (1)	−1.66	SCE	DME/SCE
methyl iodide	50% EtOH	Et$_4$NI (0.1)	−1.328	SCE	DME/SCE
naphthalene	DMF	Bu$_4$NClO$_4$ (0.1)	−2.917	AgClO$_4$	DME/AgClO$_4$
2-nitroaniline	44% EtOH	H$_2$SO$_4$ (0.1)	−0.335	SCE	DME/SCE
3-nitroaniline	44% EtOH	H$_2$SO$_4$ (0.1)	−0.270	SCE	DME/SCE
4-nitroaniline	80% MeOH	NaOH (0.1), LiCl (1)[a]	−0.918	SCE	DME/SCE

(Continued)

Table VI.44 (Continued)

Compound	Solvent	Supporting Electrolyte (c/mol dm^{-3})	$E_{\frac{1}{2}}$ /V	versus	Electrodes
nitrobenzene	30% MeOH	buffer (pH = 6.0)	–0.5	SCE	DME/SCE
nitroethane	water	H$_2$SO$_4$, NaCl (pH =0.2)	–0.58	SCE	DME/SCE
2-nitropropane	water	ClCH$_2$COOH, NaCl (0.1, pH = 1.9)	–0.49	SCE	DME/SCE
	DMF	LiCl (0.1)	–1.11	MP	DME/MP
2-nitrophenol	1% MeOH	KOH (0.1)	–0.72	SCE	DME/NCE
	DMF	Et$_4$NClO$_4$ (0.1)	–0.85 –1.74	SCE	HDE/SCE
3-nitrophenol	1% MeOH	KOH (0.1)	–0.80	SCE	DME/NCE
4-nitrophenol	water	OH$^-$ (pH = 13)	–0.94 –1.58	SCE	DME/SCE
oxalic acid	pyridine	Et$_4$NClO$_4$ (0.1)	–1.28 –1.69	SCE	DME/Ag, Ag$^+$
phenol	pyridine	Et$_4$NClO$_4$ (0.1)	–2.10	SCE	DME/Ag, Ag$^+$
phenylalanine	DMSO	Et$_4$NClO$_4$ (0.1)	–2.398	SCE	DME/SCE
piperidine	water	Na$_2$SO$_4$ (0.5, pH = 10.29)	–0.094	SCE	DME/MP
proline	DMSO	Et$_4$NClO$_4$ (0.1)	–2.307	SCE	DME/SCE
propanoic acid	DMSO	Et$_4$NClO$_4$ (0.1)	–2.36	SCE	DME/SCE
purine	water	(pH = 1.0)	–0.73 –1.00	SCE	DME/SCE
pyridine	water	buffer, KCl (0.5, pH = 7)	–1.75	SCE	DME/SCE
	DMF	Bu$_4$NI (0.1)	–2.76	AgCl	DME/AgCl
pyrimidine	water	HCl, KCl (pH = 0.3)	–0.632	SCE	DME/SCE
	DMF	Bu$_4$NI	–2.35	Ag, Ag$^+$	DME/Ag, Ag$^+$
quinhydrone	DMSO	Et$_4$NClO$_4$ (0.1)	+0.98 –0.31	SCE	Pt/SCE
quinoline	DMF	Bu$_4$NI (0.1)	–2.175	AgCl	DME/AgCl
succinic acid	DMSO	Et$_4$NClO$_4$ (0.1)	–2.29	SCE	DME/SCE
tartaric acid	pyridine	Et$_4$NClO$_4$ (0.1)	–1.33 –1.56	SCE	DME/Ag, Ag$^+$
threonine	DMSO	Et$_4$NClO$_4$ (0.1)	–2.371	SCE	DME/SCE
tryptophan	DMSO	Et$_4$NClO$_4$ (0.1)	–2.388	SCE	DME/SCE
tyrosine	DMSO	Et$_4$NClO$_4$ (0.1)	–2.379	SCE	DME/SCE
xanthine	pyridine	Et$_4$NClO$_4$ (0.1)	–1.61 –2.02	SCE	DME/Ag, Ag$^+$

Table VI.45 Isoelectric points (or range) for several proteins
Reference: *Handbook of Biochemistry* (1970), CRC.

Protein	pI*	Protein	pI*
pepsin	c. 1.0	fibrinogen	5.5
fetuin	3.4	catalase	5.6
keratins	3.7–5.0	carboxypepidase	6.0
α-casein	4.0	collagen	6.6
ovalbumin	4.6	myoglobin	6.99
hemocyanins	4.6–6.4	hemoglobin (horse)	6.9
gelatin	4.8	hemoglobin (human)	7.1
serum albumin (human)	4.8	hemoglobin (chick)	7.2
urease	5.1	ribonuclease	9.5
β-lactoglobulin	5.2	cytochrome c	10.65
insulin	5.35	lysozyme	11.1

* The precise value depends on the temperature and the ionic strength of the solution.

Section VII
Optics

		Page
VII.1	Luminance and correlated colour temperature for some light sources	127
VII.2	1931 Commission International de l'Eclairage (CIE) specification of colorimetric standard observer, for 0.5° to 4° field subtenses, in terms of variation of spectral tristimulus values with wavelength	128
VII.3	1951 CIE scotopic relative luminous efficiency function	129
VII.4	CIE standard illuminants for colorimetry	129
VII.5	Relative spectral power distribution of the CIE standard illuminants A, C, and D_{65}	130
VII.6	Hue discrimination for the normal observer	131
VII.7	Types of colour deficiency characteristics and inherited percentage incidence in the population	131
VII.8	Some substances which may produce significant red/green colour perception disturbance	132
VII.9	Some substances which may produce significant blue/yellow colour perception disturbance	132
VII.10	Variation of visual depth of focus with pupil diameter of eye	132
VII.11	Range and mean of main geometrical optical properties for 1000 eyes in vivo	132
VII.12	General specification of the Gullstrand-Emsley schematic eye	132
VII.13	Chromatic aberration of the Gullstrand-Emsley schematic eye	133
VII.14	International Commission for Optics recommended wavelengths for refractometry	133
VII.15	Some commonly used wavelengths in refractometry	133
VII.16	Variation of refractive index of standard air with wavelength	134
VII.17	Refractive indices, relative to a vacuum, of various gases for wavelength 589.3 nm and for radio wavelengths at standard pressure and temperature, unless otherwise stated	134
VII.18	Variation with wavelength of refractive index and rate of change of index with temperature for the refractometer-calibrating liquids distilled water and toluence 'Ultra' grade	134
VII.19	Refractive indices for various liquids for wavelength 589.3 nm at or near temperature 293 K	135
VII.20	Variation with wavelength of refractive index of some Chance Pilkington optical glasses at temperature 293 K and standard pressure	135
VII.21	Variation with wavelength of light transmittance of 25 mm thickness of some Chance Pilkington optical glasses at temperature 293 K and standard pressure	136
VII.22	Variation of reflectance for normally incident light in air with refractive index of a nonabsorbing reflecting substance	136
VII.23	Refractive and absorption indices and reflectance for normally incident light in air for various metals and semiconductors at stated wavelengths	136
VII.24	Variation with wavelength of refractive index of synthetic fused silica, calcium fluoride and lithium fluoride at temperature 293 K	137
VII.25	Variation with wavelength of refractive index of commonly used optical plastics	137
VII.26	Refractive indices at or near temperature 293 K of infrared optical materials at stated wavelengths, together with transmission range	138
VII.27	Refractive indices of some optical cements at wavelength 589.3 nm and temperature 293 K	138
VII.28	Variation with wavelength of ordinary and extraordinary refractive indices of calcite at temperature 291 K (293 K) for wavelength range 200–706 nm (801–3324 nm)	139
VII.29	Variation with wavelength of ordinary and extraordinary refractive indices of crystal quartz at temperature 293 K	139
VII.30	Ordinary and extraordinary refractive indices at or near temperature 293 K of various birefringent optical materials at stated wavelengths, together with transmission range	139
VII.31	Variation with angle of incidence of light of the percentage reflectance, at interface between air and glass (refractive index 1.52) for electric vector respectively parallel and perpendicular to the plane of incidence	140
VII.32	Refractive indices for common thin film materials and their percentage reflectances in air for light of wavelength 550 nm and for thickness to form a quarter-wave film on glass of refractive index 1.52	140
VII.33	Percentage reflectance in air for light of wavelength 550 nm of quarter-wave multilayer stacks of zinc sulphide and magnesium fluoride alternately for different numbers of layers on glass of refractive index 1.52	140
VII.34	Variation with wavelength of specific rotation for quartz at temperature 293 K	140
VII.35	Values of Verdet constants for various gases at wavelength 589.3 nm and STP	140
VII.36	Variation with wavelength of Verdet constant for water at 293 K	140
VII.37	Verdet constants for various liquids at stated temperature for wavelength 589.3 nm	141
VII.38	Verdet constants for various solids at stated temperatures and wavelengths	141
VII.39	Refractive indices, half-wave voltage and transmission range for some electro-optic substances	141
VII.40	Values of wavelengths of maximum emission and cut-off, decay constant and light conversion efficiency, relative to that of NaI (Tl) for some solid scintillators	142
VII.41	Operational mode, vacuum wavelength, average and peak powers, rate of change of output power relative to input power, and pulse duration for some representative lasers	142

Section VII
Optics

Table VII.1 Luminance and correlated colour temperature for some light sources
Correlated colour temperature is the black-body temperature of radiation whose CIE coordinates are closest to those of the radiation under investigation.

Source	luminance/ 10^4 lm sr^{-1} m^{-2}	correlated colour temperature/K
arc crater (solid C)	1.72×10^4	3780
arc crater (high intensity, 150 A)	8×10^4	5000–5500
C_2H_2 (Kodak burner)	10.8	2360
candle	0.5	1930
clear blue sky	c 0.4	12 000–24 000
Hg vapour (high pressure, compact, 250 W)	2×10^4	
Hg vapour (high pressure, compact, 1000 W)	4×10^4	
Hg vapour (low pressure in glass)	2.3	
Hg vapour (tubular fluorescent, colour matching 40 W)	1.2	6050
Hg vapour (tubular fluorescent, warm white 40 W)	1.9	3000
moon (through atmosphere)	c 0.4	
paraffin flame (flat wick)	1.25	2055
perfect diffuser in sunlight	c 4	
perfect diffuser in moonlight	c 10^{-5}	
starlit sky	c 5×10^{-9}	
W coil lamp (quartz halogen)	c 2800	3300
W strip lamp (gas filled)	540	2800
W strip lamp (gas filled)	990	3000
W strip lamp (vacuum)	124	2400
Xe arc (compact, 250 W)	2.5×10^4	6000
Xe arc (compact, 2 kW)	1.2×10^5	6000
zenith sun (through atmosphere)	c 1.6×10^5	5400

Reference: Kaye and Laby, *Tables of Physical and Chemical Constants*, 15 ed., Longman, 1986

Table VII.2 1931 Commission International de l'Eclairage (CIE) specification of colorimetric standard observer, for 0.5° to 4° field subtenses, in terms of variation of spectral tristimulus values with wavelength, λ

wave length/nm	spectral tristimulus values			wave length/nm	spectral tristimulus values		
	$\bar{x}(\lambda)$	$\bar{y}(\lambda)$†	$\bar{z}(\lambda)$		$\bar{x}(\lambda)$	$\bar{y}(\lambda)$†	$\bar{z}(\lambda)$
380	0.0014	0.0000	0.0065	580	0.9163	0.8700	0.0017
385	0.0022	0.0001	0.0105	585	0.9786	0.8163	0.0014
390	0.0042	0.0001	0.0201	590	1.0263	0.7570	0.0011
395	0.0076	0.0002	0.0362	595	1.0567	0.6949	0.0010
400	0.0143	0.0004	0.0679	600	1.0622	0.6310	0.0008
405	0.0232	0.0006	0.1102	605	1.0456	0.5668	0.0006
410	0.0435	0.0012	0.2074	610	1.0026	0.5030	0.0003
415	0.0776	0.0022	0.3713	615	0.9384	0.4412	0.0002
420	0.1344	0.0040	0.6456	620	0.8544	0.3810	0.0002
425	0.2148	0.0073	1.0391	625	0.7514	0.3210	0.0001
430	0.2839	0.0116	1.3856	630	0.6424	0.2650	0.0000
435	0.3285	0.0168	1.6230	635	0.5419	0.2170	0.0000
440	0.3483	0.0230	1.7471	640	0.4479	0.1750	0.0000
445	0.3481	0.0298	1.7826	645	0.3608	0.1382	0.0000
450	0.3362	0.0380	1.7721	650	0.2835	0.1070	0.0000
455	0.3187	0.0480	1.7441	655	0.2187	0.0816	0.0000
460	0.2908	0.0600	1.6692	660	0.1649	0.0610	0.0000
465	0.2511	0.0739	1.5281	665	0.1212	0.0446	0.0000
470	0.1954	0.0910	1.2876	670	0.0874	0.0320	0.0000
475	0.1421	0.1126	1.0419	675	0.0636	0.0232	0.0000
480	0.0956	0.1390	0.8130	680	0.0468	0.0170	0.0000
485	0.0580	0.1693	0.6162	685	0.0329	0.0119	0.0000
490	0.0320	0.2080	0.4652	690	0.0227	0.0082	0.0000
495	0.0147	0.2586	0.3533	695	0.0158	0.0057	0.0000
500	0.0049	0.3230	0.2720	700	0.0114	0.0041	0.0000
505	0.0024	0.4073	0.2123	705	0.0081	0.0029	0.0000
510	0.0093	0.5030	0.1582	710	0.0058	0.0021	0.0000
515	0.0291	0.6082	0.1117	715	0.0041	0.0015	0.0000
520	0.0633	0.7100	0.0782	720	0.0029	0.0010	0.0000
525	0.1096	0.7932	0.0573	725	0.0020	0.0007	0.0000
530	0.1655	0.8620	0.0422	730	0.0014	0.0005	0.0000
535	0.2257	0.9149	0.0298	735	0.0010	0.0004	0.0000
540	0.2904	0.9540	0.0203	740	0.0007	0.0002	0.0000
545	0.3597	0.9803	0.0134	745	0.0005	0.0002	0.0000
550	0.4334	0.9950	0.0087	750	0.0003	0.0001	0.0000
555	0.5121	1.0000	0.0057	755	0.0002	0.0001	0.0000
560	0.5945	0.9950	0.0039	760	0.0002	0.0001	0.0000
565	0.6784	0.9786	0.0027	765	0.0001	0.0000	0.0000
570	0.7621	0.9520	0.0021	770	0.0001	0.0000	0.0000
575	0.8425	0.9154	0.0018	775	0.0001	0.0000	0.0000
580	0.9163	0.8700	0.0017	780	0.0000	0.0000	0.0000

†$\bar{y}(\lambda)$ gives the photopic relative luminous efficiency function, $V(\lambda)$
Reference: *Colorimetry*, CIE Publication No. 15, 1971

Table VII.3 1951 CIE scotopic relative luminous efficiency function, $V'(\lambda)$

Note: for the region between scotopic and photopic (i.e. mesopic) there is at present no agreed data.

Wavelength/nm	$V'(\lambda)$	wavelength/nm	$V'(\lambda)$	wavelength/nm	$V'(\lambda)$
370	0	520	0.935	670	1.48×10^{-4}
380	5.89×10^{-4}	530	0.811	680	7.15×10^{-5}
390	2.209×10^{-3}	540	0.65	690	3.533×10^{-5}
400	0.009 29	550	0.481	700	1.78×10^{-5}
410	0.034 84	560	0.328 8	710	9.14×10^{-6}
420	0.096 6	570	0.207 6	720	4.78×10^{-6}
430	0.199 8	580	0.121 2	730	2.546×10^{-6}
440	0.328 1	590	0.065 5	740	1.379×10^{-6}
450	0.455	600	0.033 15	750	7.6×10^{-7}
460	0.567	610	0.015 93	760	4.25×10^{-7}
470	0.676	620	0.007 37	770	2.413×10^{-7}
480	0.793	630	3.335×10^{-3}	780	1.39×10^{-7}
490	0.904	640	1.497×10^{-3}	790	0
500	0.982	650	6.77×10^{-4}		
510	0.997	660	3.129×10^{-4}		

Reference: as for Table VII.2

Table VII.4 CIE standard illuminants for colorimetry.

Illuminant	Nature	chromaticity coordinates	
		x	y
A	black-body radiation at temperature 2856 K	0.4476	0.4074
C	radiation from illuminant A after passing through two solutions C_1 and C_2 as specified†, to yield a correlated colour temperature of 6774 K	0.3101	0.3162
D_{65}	radiation of correlated colour temperature of 6504 K; at present no recommended source specification	0.3127	0.3290

†solution C_1: $CuSO_4.5H_2O$ 3.412 g   solution C_2: $CoSO_4.(NH_4)_2SO_4.6H_2O$ 30.58 g
           $C_6H_8(OH)_6$ 3.412 g $CuSO_4.5H_2O$ 22.52 g
           C_5H_5N 30 ml H_2SO_4 (density 1.835 g ml^{-1}) 10 ml
           distilled H_2O to make 1 litre   distilled H_2O to make 1 litre
           A 10 mm thick layer of each solution is contained in a double cell of colourless optical glass

Reference: as for Table VII.2

Table VII.5 Relative spectral power distribution of the CIE standard illuminants A, C, and D_{65}

wavelength/nm	A	distribution C	D_{65}	wavelength/nm	A	distribution C	D_{65}
375	8.77	27.50	51.0	575	110.80	100.15	96.1
380	9.80	33.00	50.0	580	114.44	97.80	95.8
385	10.90	39.92	52.3	585	118.08	95.43	92.2
390	12.09	47.40	54.6	590	121.73	93.20	88.7
395	13.35	55.17	68.7	595	125.39	91.22	89.3
400	14.71	63.30	82.8	600	129.04	89.70	90.0
405	16.15	71.81	87.1	605	132.70	88.83	89.8
410	17.68	80.60	91.5	610	136.35	88.40	89.6
415	19.29	89.53	92.5	615	139.99	88.19	88.6
420	20.99	98.10	93.4	620	143.62	88.10	87.7
425	22.79	105.80	90.1	625	147.24	88.06	85.5
430	24.67	112.40	86.7	630	150.84	88.00	83.3
435	26.64	117.75	95.8	635	154.42	87.86	83.5
440	28.70	121.50	104.9	640	157.98	87.80	83.7
445	30.85	123.45	110.9	645	161.52	87.99	81.9
450	33.09	124.00	117.0	650	165.03	88.20	80.0
455	35.41	123.60	117.4	655	168.51	88.20	80.1
460	37.81	123.10	117.8	660	171.96	87.90	80.2
465	40.30	123.30	116.3	665	175.38	87.22	81.2
470	42.87	123.80	114.9	670	178.77	86.30	82.3
475	45.52	124.09	115.4	675	182.12	85.30	80.3
480	48.24	123.90	115.9	680	185.43	84.00	78.3
485	51.04	122.92	112.4	685	188.70	82.21	74.0
490	53.91	120.70	108.8	690	191.93	80.20	69.7
495	56.85	116.90	109.1	695	195.12	78.24	70.7
500	59.86	112.10	109.4	700	198.26	76.30	71.6
505	62.93	106.98	108.6	705	201.36	74.36	73.0
510	66.06	102.30	107.8	710	204.41	72.40	74.3
515	69.25	98.81	106.3	715	207.40	70.40	68.0
520	72.50	96.90	104.80	720	210.36	68.30	61.6
525	75.79	96.78	106.2	725	213.27	66.30	65.7
530	79.13	98.00	107.7	730	216.12	64.40	69.9
535	82.52	99.94	106.0	735	218.92	62.80	72.5
540	85.95	102.10	104.4	740	221.67	61.50	75.1
545	89.41	103.95	104.2	745	224.36	60.20	69.3
550	92.91	105.20	104.0	750	227.00	59.20	63.6
555	96.44	105.67	102.0	755	229.59	58.50	55.0
560	100.00	105.30	100.0	760	232.12	58.10	46.4
565	103.58	104.11	98.2	765	234.59	58.00	56.6
570	107.18	102.30	96.3	770	237.01	58.20	66.8

Reference: as for Table VII.2

Table VII.6 Hue discrimination for the normal observer

wavelength/nm	minimum perceptible wavelength change/nm	wavelength/nm	minimum perceptible wavelength change/nm
430	6	537	2.15
440	3	540	2.1
442	2.7	550	1.8
450	3.3	560	1.55
452	3.4	570	1.25
460	3	580	1.05
470	2.1	590	0.9
480	1.5	595	0.95
490	1.2	600	0.9
495	1.1	610	1.05
500	1.2	620	1.3
510	1.3	630	1.6
520	1.6	640	2.4
530	2	650	4.2

Reference: Wright, *Researches on Normal and Defective Colour Vision*, Kimpton, 1946

Table VII.7 Types of colour deficiency characteristics and inherited percentage incidence in the population

| Type | Characteristics | | | % incidence | |
	colour matching (additive)	hue discrimination	luminosity	male	female
Deuteranomaly	3 stimuli required, but greater than normal amount of green required in red/green mixture matching yellow	reduced in yellow	similar to normal	5	0.4
deuteranopia	2 stimuli required, only one of which is from red/yellow/green range	none in yellow	similar to normal	1	0.01
monochromatism	1 stimulus only required	none	usually as scotopic luminosity curve	extremely small	
protanomaly	3 stimuli required but greater than normal amount of red required in red/green mixture matching yellow	reduced in yellow	reduced in red, otherwise normal	1.5	0.03
protanopia	2 stimuli required, only one of which is from red/yellow/green range	none in yellow	reduced in red, otherwise normal	1	0.02
Tritanomaly	3 stimuli required but matches differ from normal	reduced in blue/green	reduced in blue, otherwise normal	very small	
tritanopia	2 stimuli required, only one of which is from blue/green range	none in blue/green	reduced in blue, otherwise normal	very small	

Reference: Fletcher and Voke, *Defective Colour Vision*, Hilger, 1985
Wright, *Researches on Normal and Defective Colour Vision*, Kimpton, 1946

Table VII.8 Some substances which may produce significant red/green colour perception disturbance

aspirin	ethopropazine	pheniprazine
CCl_4	ethylene glycol	piperacetazine
	fluphenazine	promazine
CH_3Br	hydrazines	promethazine
$CH_3C_6H_3(NO_2)_2$	hydroxychloroquine	rifampin
	ibuprofen	salicylates
CH_3OH	isocarboxazid	streptomycin
C_2H_5OH	isoniazid	thiethylperazine
	mesoridazine	thiopropazate
$C_6H_4(NO_2)_2$	methdilazine	thioridazine
$C_{19}H_{20}O_2N_2$	methotrimeprazine	Tl
	Mn	tobacco
$C_{20}H_{24}N_2O_2 \cdot 3H_2O$	nardelzine	tolbutamide
	nialamide	tranylcypromine
$C_{41}H_{64}O_{13}$	opium	trifluoperazine
CO	paramethadione	triflupromazine
CS_2	pargyline	trimeprazine
chlorodinitrobenzene	Pb	trimethadione
chloropropamide	Pb compounds	
chloropromazine	perazine	
chloroquine	pericyazine	
digitalis	perphenazine	
ethambutol	phenelzine	

Reference: Fletcher and Voke, *Defective Colour Vision*, Hilger, 1985

Table VII.9 Some substances which may produce significant blue/yellow colour perception disturbance

acetyldigitoxin	digitalis	lanadoside C
amodiaquine	erythromycin	lyndiol
anovlar	ethambutol	novaquine
$C_{27}H_{44}O_{12} \cdot 8H_2O$	gitalin	ovulen
	hydroxychloroquine	paramethadione
$C_{41}H_{64}O_{13}$	indomethacin	plaquenyl
chloroquine	isoniazid	streptomycin
		trimethadione

Reference: as for Table VII.8

Table VII.10 Variation of visual depth of focus with pupil diameter of eye

pupil diameter/ mm	depth of focus at ∞ /m	depth of focus at 25 cm/cm
4	∞ to 32	0.8
3	∞ to 24	1.1
2	∞ to 16	1.6
1	∞ to 8	3.2

Reference: Duke-Elder, *Text-book of Ophthalmology vol. 1*, Kimpton, 1932

Table VII.11 Range and mean of main geometrical optical properties for 1000 eyes in vivo

Property	Range	Mean
corneal power/D	+38 to +48	+42.74
corneal radius/ mm	7 to 8.8	7.86
depth of anterior chamber/mm	2.8 to 4.6	3.68
length (axial)/mm	21 to 26	
equivalent power of eye/D	+51 to +70	+60.1
equivalent power of lens/D	+15 to +29	+21.3

Reference: Woolf *Am. J. Optom.* vol. 25, 218, 1948 (translation of Stenstrom's paper)

Table VII.12 General specification of the Gullstrand–Emsley schematic eye

Property†	Relaxed eye	Accommodated eye
RADII OF CURVATURE		
cornea	+7.80	+7.80
crystalline: first surface	+10.00	+5.00
crystalline: second surface	−6.00	−5.00
AXIAL SEPARATIONS		
Depth of anterior chamber	3.60	3.20
thickness of crystalline	3.60	4.00
depth of vitreous body	16.69	16.69
Overall axial length	23.89	23.89
MEAN REFRACTIVE INDICES		
air	1	1
aqueous humour	1.3333	1.3333
crystalline	1.4160	1.4160
vitreous humour	1.3333	1.3333
SURFACE POWERS		
cornea	+42.73	+42.73
crystalline: first surface	+8.27	+16.54
crystalline: second surface	+13.78	+16.54
EQUIVALENT POWERS		
crystalline	+21.76	+32.31
eye	+60.49	+69.73
EQUIVALENT FOCAL LENGTHS OF EYE		
first (PF)	−16.53	−14.34
second (P'F')	+22.04	+19.12
DISTANCES FROM CORNEAL VERTEX		
first principal point	+1.55	+1.78
second principal point	+1.85	+2.13
first nodal point	+7.06	+6.56
second nodal point	+7.36	+6.91
entrance pupil	+3.05	+2.67
exit pupil	+3.69	+3.25
first principal focus	−14.98	−12.56
second principal focus	+23.89	+21.25
Refractive state	0	−8.47
Distance of near point from corneal vertex		−116.21

†All linear dimensions in millimetres and powers in dioptre
Reference: Bennett and Rabbetts, *Clinical Visual Optics*, Butterworths, 1984

Table VII.13 Chromatic aberration of the Gullstrand–Emsley schematic eye

Property†	Relaxed eye wavelength/nm					Accommodated eye wavelength/nm				
	380	480	587.6	643.8	780	380	480	587.6	643.8	780
equivalent power	63.09	61.36	60.49	60.19	59.68	66.62	64.74	63.81	63.49	62.96
DISTANCES FROM CORNEAL VERTEX:										
first principal pt	1.57	1.55	1.55	1.55	1.56	1.66	1.65	1.65	1.66	1.66
second principal pt	1.89	1.86	1.85	1.85	1.86	2.01	1.98	1.97	1.98	1.98
entrance pupil	3.04	3.05	3.05	3.05	3.06	2.89	2.9	2.91	2.91	2.92
exit pupil	3.69	3.69	3.69	3.69	3.69	3.53	3.53	3.53	3.53	3.52
CHROMATIC DIFFERENCES:										
power	2.6	0.87	0	−0.3	−0.81	2.81	0.93	0	−0.32	−0.85
refraction	−1.97	−0.65	0	0.22	0.6	−2.17	−0.71	0	0.24	0.64
magnification	0.9924	0.9973	1	1.001	1.0026	0.9922	0.9972	1	1.0011	1.0028

†All linear dimensions in millimetres and powers in dioptre
Reference as for Table VII.12

Table VII.14 1962 International Commission for Optics recommended wavelengths for refractometry

Source	wavelength†/nm	designating letter	Source	wavelength†/nm	designating letter
		Standard			
Cd	467.8		Hg	366.3	
Cd	508.6		Hg	404.7	h
Cd	643.8	C'	Hg	435.8	g
Cs	852.1	s	Hg	546.1	e
Cs	894.4		Hg	1014	t
He	587.6	d	Hg	1128.7	
He	667.8		Hg	1395.1	
He	706.5	r	Hg	1529.5	
He	2058.1		Hg	1813.1	
Hg	237.8		Hg	1970.1	
Hg	269.9		Hg	2325.4	
Hg	289.4		Rb	780	
Hg	296.7		Zn	213.9	
Hg	334.2				
		Secondary			
Cd	480	F'	Hg	292.5	
Hg	239.9		Hg	390.6	
Hg	275.3		Rb	794.7	

†Wavelength in standard air, i.e. dry air (0.03% CO_2 by volume) at temperature 288 K and standard pressure
Reference: *Optica Acta*, vol. 10, 217, 1963

Table VII.15 Some commonly used wavelengths in refractometry

Source	wavelength†/nm	designating letter	Source	wavelength†/nm	designating letter
H	434	G'	Hg	1357	
H	486.1	F	Hg	1367.3	
H	656.3	C	K	766.5	
Hg	365	i	K	769.9	
Hg	365.5		Na	589	
Hg	577		Na	589.6	D
Hg	579				

†Wavelength in standard air, i.e. dry air (0.03% CO_2 by volume) at temperature 288 K and standard pressure
Note: Laser radiations marked st in Table VII.41 are also commonly used, as are 488.122 nm Ar^+, 568.348 nm and 647.27 nm Kr^+, and 3392–231 nm He–Ne laser radiations.
Reference: Kaye and Laby, *Tables of Physical and Chemical Constants*, 15ed., Longmans, 1986

Table VII.16 Variation of refractive index of standard air with wavelength

Standard air is dry air (0.03% CO_2 by volume) at temperature 288 K and standard pressure

wavelength/nm	500	550	630	740	950
refractive index	1.000 278	1.000 277	1.000 276	1.000 275	1.000 274

Reference: *Handbook of Chemistry and Physics*, CRC Press, 1991/2

Table VII.17 Refractive indices, relative to a vacuum, of various gases for wavelength 589.3 nm and for radio wavelengths at standard pressure and temperature, unless otherwise stated

Gas	refractive index at 589.3 nm wavelength	radio wavelengths	Gas	refractive index at 589.3 nm wavelength	radio wavelengths
air	1.000 292	(CO_2 free) 1.000 288	Cl_2	1.000 773	
Ar	1.000 281		D_2		1.000 135
Br_2	1.001 132		H_2	1.000 132	1.000 136
$CHCl_3$	1.001 45		HCl	1.000 447	
CH_3COCH_3	1.001 09		H_2O(vapour)	1.000 254	1.000 061°
CH_3OCH_3	1.000 891		H_2S	1.000 634	
CH_3OH	1.000 586		He	1.000 036	1.000 035
CH_4	1.000 444		N_2	1.000 297	1.000 294
C_2H_5OH	1.000 878		NH_3	1.000 376	
$(C_2H_5)_2O$	1.001 533		NO	1.000 297	
C_5H_{12}	1.001 711		N_2O	1.000 516	
C_6H_6	1.001 762		O_2	1.000 272	1.000 266
CO	1.000 338*		SO_2	1.000 686	
CO_2	1.000 451	1.000 494			
CS_2	1.001 481				

*white light value
°at 1333 Pa pressure, temperature 293 K

Reference: Kaye and Laby, *Tables of Physical and Chemical Constants*, 15ed., Longmans, 1986
Handbook of Chemistry and Physics, CRC Press, 1991/2

Table VII.18 Variation with wavelength of refractive index and rate of change of index with temperature for the refractometer-calibrating liquids distilled water and toluene 'Ultra' grade

Refractive index is given for temperature 293 K; rate of change of index for toluene is over the range 293–298 K.

wavelength/nm	distilled H_2O		$C_6H_5CH_3$	
	index	rate/K^{-1}	index	rate/K^{-1}
404.66	1.526 120	−0.000 597	1.342 742	−0.000 101
435.84	1.517 830	−0.000 588	1.340 210	−0.000 100
479.99	1.509 285	−0.000 577		
486.13	1.508 315	−0.000 576	1.337 123	−0.000 099
546.07	1.500 715	−0.000 565	1.334 466	−0.000 098
587.56	1.496 920	−0.000 560	1.333 041	−0.000 097
589.00	1.496 800	−0.000 560		
589.59	1.496 755	−0.000 560	1.332 988	−0.000 097
632.80	1.493 680	−0.000 556	1.331 745	−0.000 096
643.85	1.493 005	−0.000 555		
656.28	1.492 285	−0.000 554	1.331 151	−0.000 096
706.52	1.489 795	−0.000 551	1.330 020	−0.000 095

Reference: Kaye and Laby, *Tables of Physical and Chemical Constants*, 15ed., Longmans, 1986

Table VII.19 Refractive indices for various liquids for wavelength 589.3 nm at or near temperature 293 K

Liquid[†]	index	Liquid[†]	index
benzyl benzoate	1.568	$KHgI_3$	1.717
CCl_4	1.46	KI and HgI_2 (aqueous)	1.82*
CH_2I_2	1.737	methyl salicylate	1.538
CH_2I_2 and S(saturated)	1.78	α-monobromonaphthalene	1.66
CH_3I_2 (35), SnI_4 (31), AsI_3 (16), S (10), SbI_3 (8)*	1.868	oil, cedar	1.516
		oil, cinnamon	1.601
C_6H_5Cl	1.525	oil, cloves	1.532
$C_6H_5CH_3$	1.497	oil, olive	1.46
$C_6H_5NH_2$	1.586	oil, paraffin	1.44
C_6H_6	1.501	oil, turpentine	1.47
$C_6H_{12}O_3$	1.405	paraffin, medicinal	1.48
C_9H_7N	1.627	P in CS_2	1.95*
ethyl cinnamate	1.559	P (yellow) (80), S (10), CH_2I_2 (10)	2.06
ethyl salicylate	1.523		
glycerol	1.47		
H_2O	1.333		
H_2S	1.885		
$HgI_2.BaI_2.5H_2O$ (aqueous)	1.793		
HgI_2 in $C_6H_5NH_2$ or C_9H_7N	2.2*		

[†]Figures in brackets are percentage mass composition
*maximum value obtainable
Reference: Kaye and Laby, *Tables of Physical and Chemical Constants*, 15ed., Longmans, 1986

Table VII.20 Variation with wavelength of refractive index of some Chance Pilkington optical glasses at temperature 293 K and standard pressure

Glass type[†]	365.02	404.66	486.13	587.56	656.28	706.52	1014
				refractive index			
BF 606439	1.641 81	1.629 90	1.615 31	1.605 62	1.601 52	1.599 24	1.591 29
Bor F614439	1.650 13	1.638 33	1.623 79	1.614 00	1.609 80	1.607 44	1.598 74
BSC 510644	1.528 82	1.522 90	1.515 18	1.509 70	1.507 27	1.505 86	1.500 41
DBC 610573	1.636 71	1.628 42	1.617 72	1.610 29	1.607 06	1.605 23	1.598 48
DEDF 805254	1.896 45	1.864 26	1.827 74	1.805 18	1.796 08	1.791 16	1.775 12
DF 620364	1.666 28	1.650 64	1.632 08	1.620 04	1.615 03	1.612 27	1.602 83
EDF 706300	1.773 38	1.749 38	1.722 56	1.705 85	1.699 03	1.695 30	1.682 73
ELF 548456	1.578 85	1.568 67	1.556 11	1.547 69	1.544 10	1.542 10	1.534 90
HC 519604	1.540 06	1.533 47	1.524 96	1.518 99	1.516 37	1.514 87	1.509 24
LAC 652585	1.678 98	1.670 42	1.659 34	1.651 60	1.648 21	1.646 28	1.639 04
LAF 717479	1.755 05	1.742 81	1.727 46	1.717 00	1.712 49	1.709 95	1.700 63
LF 579411	1.615 95	1.603 55	1.588 51	1.578 60	1.574 44	1.572 14	1.564 19
MBC 569561	1.594 03	1.586 07	1.575 90	1.568 83	1.565 76	1.564 02	1.557 61
TF 530512	1.556 61	1.548 18	1.537 57	1.530 33	1.527 21	1.525 44	1.519 00
ZC 508612	1.527 76	1.521 49	1.513 34	1.507 59	1.505 04	1.503 58	1.497 90

[†]Appearance of the letter C (F) indicates a crown (flint) glass; the other letter(s) denote the type or use e.g. ED extra dense, T telescope. The first 3 numbers are the first 3 significant figures of the difference of the 587.56 nm wavelength refractive index and 1; the second 3 numbers are the first 3 significant figures of the reciprocal dispersive power for the 656.28 nm and 486.13 nm wavelengths.
Reference: *Chance Pilkington optical glass catalogue*

Table VII.21 Variation with wavelength of light transmittance of 25 mm thickness of some Chance Pilkington optical glasses at temperature 293 K and standard pressure

Glass type[†]	wavelength/nm						
	360	380	400	420	440	460	500
				transmittance			
BSC 510644	0.915	0.941	0.980	0.982	0.982	0.987	0.989
DBC 610573	0.799	0.930	0.971	0.983	0.985	0.989	0.995
DEDF 805254		0.203	0.609	0.837	0.930	0.968	0.991
DF 620364	0.749	0.889	0.958	0.972	0.977	0.986	0.992
EDF 706300		0.342	0.768	0.886	0.922	0.939	0.966
HC 519604	0.897	0.930	0.978	0.978	0.976	0.975	0.987
LF 579411	0.816	0.908	0.969	0.974	0.973	0.978	0.984
MBC 569561	0.875	0.945	0.981	0.987	0.989	0.990	0.995
ZC 508612	0.747	0.872	0.954	0.968	0.974	0.982	0.990

[†] See footnote, Table VII.20
Reference: as for Table VII.20

Table VII.22 Variation of reflectance for normally incident light in air with refractive index of a nonabsorbing reflecting substance

index	1.46	1.52	1.6	1.8	2	3	4
reflectance/%	3.5	4.3	5.3	8.2	11.1	25	36

Reference: Kaye and Laby, *Tables of Physical and Chemical Constants*, 15ed., Longmans, 1986

Table VII.23 Refractive and absorption indices and reflectance for normally incident light in air for various metals and semiconductors at stated wavelengths

Substance	refractive index	absorption index	reflectance/%	wavelength/nm
Ag	0.055	3.32	98.2	550
	10.7	69	99.1	10 000
Al	0.14	2.35	91.8	220
	0.82	5.99	91.6	546
Au	0.33	2.32	81.5	550
	7.4	53.4	99	10 000
Cr	2.51	2.66	48.2	546
Cu	0.76	2.46	66.8	550
Ge	5.15	2.15	51.5	545
	4.003	0	36	10 000
Ni	1.85	3.27	60.7	540
Rh	1.62	4.63	77.1	546
Si	4.05	0.03	36.5	546
steel	2.4	3.4	58.5	550

Reference: Kaye and Laby, *Tables of Physical and Chemical Constants*, 15ed., Longmans, 1986

Table VII.24 Variation with wavelength of refractive index of synthetic fused silica, calcium fluoride and lithium fluoride at temperature 293 K

wavelength/nm	refractive index			wavelength/nm	refractive index		
	SiO_2	CaF_2	LiF		SiO_2	CaF_2	LiF
213.86	1.534 30	1.484 96	1.432 29	780.02	1.453 67	1.430 79	1.389 24
237.83	1.514 72	1.472 15	1.422 56	794.76	1.453 41	1.430 63	1.389 07
239.94	1.513 36	1.471 25	1.421 86	852.11	1.452 47	1.430 07	1.388 49
269.89	1.498 04	1.460 98	1.413 93	894.35	1.451 84	1.429 71	1.388 09
275.28	1.495 91	1.459 54	1.412 80	1014.0	1.450 24	1.428 84	1.387 07
289.36	1.490 99	1.456 20	1.410 18	1060.0	1.449 68		1.386 70
292.54	1.489 99	1.455 52	1.409 65	1128.7	1.448 87	1.428 15	1.386 17
296.73	1.488 73	1.454 67	1.408 97	1395.1	1.445 83	1.426 80	1.384 10
334.15	1.479 77	1.448 52	1.404 10	1529.5	1.444 27	1.426 17	1.383 02
366.33	1.474 35	1.444 79	1.401 11	1813.1	1.440 71	1.424 83	1.380 56
404.66	1.469 62	1.441 52	1.398 47	1970.1	1.438 53	1.424 06	1.379 09
435.84	1.466 69	1.439 50	1.396 83	2058.1	1.437 23	1.423 61	1.378 22
467.82	1.464 29	1.437 85	1.395 47	2325.4	1.432 93	1.422 16	1.375 30
479.99	1.463 50	1.437 31	1.395 02	2800.0		1.419 23	1.369 37
486.13	1.463 12	1.437 05	1.394 80	3400.0		1.414 87	1.360 32
508.58	1.461 86	1.436 18	1.394 08	4000.0		1.409 63	1.349 37
546.07	1.460 08	1.434 97	1.393 05	4600.0		1.403 57	1.336 40
587.56	1.458 46	1.433 89	1.392 11	5000.0		1.399 08	1.326 56
632.80	1.457 02		1.391 26	5600.0			1.309 88
643.85	1.456 70	1.432 72	1.391 07	5893.2		1.387 12	
656.28	1.456 37	1.432 50	1.390 87	6000.0			1.297 40
667.82	1.456 07	1.432 31	1.390 69	7071.8		1.368 05	
706.52	1.455 15	1.431 71	1.390 14	8250.5		1.344 40	
769.90	1.453 86	1.430 91	1.389 35	9429.1		1.316 05	

Reference: Kaye and Laby, *Tables of Physical and Chemical Constants*, 15ed., Longmans, 1986

Table VII.25 Variation with wavelength of refractive index of commonly used optical plastics

Plastic	wavelength/nm		
	486.1	589.3	656.3
	refractive index		
methyl methacrylate (acrylic)	1.498	1.492	1.489
methyl methacrylate styrene copolymer	1.574	1.563	1.558
polycarbonate	1.598	1.584	1.578
polystyrene	1.604	1.59	1.585

Reference: Kaye and Laby, *Tables of Physical and Chemical Constants*, 15ed., Longmans, 1986

Table VII.26 Refractive indices at or near temperature 293 K of infrared optical materials at stated wavelengths, together with transmission range

Material	index	wavelength/nm	range/nm
AgCl	2.0224	1000	400– c 28 000
	1.9069	20 000	
As_2S_3	2.4777	1000	600–13 000
	2.3816	10 000	
BaF_2	1.4759	546	c 250– c 15 000
	1.3964	10 346	
CdTe(Irtran 6)	2.838	1000	900– c 16 000
	2.672	10 000	
CsBr	1.6779	1000	c 300– c 55 000
	1.5624	39 000	
CsI	1.7572	1000	250– c 80 000
	1.6366	50 000	
diamond	2.4235	546	c 250–> 80 000
GaAs	3.135	10 000	1000– c 15 000
Ge	4.0032	10 000	1800–23 000
KBr	1.5639	546	c 250–40 000
	1.4866	21 180	
KCl	1.4932	546	210– c 30 000
	1.389	20 400	
KI	1.6731	546	c 250–45 000
	1.5964	20 000	
MgO (Irtran 5)	1.7227	1000	250–8500
	1.4824	8000	
NaCl	1.5518	500	210–26 000
	1.3831	20 000	
NaF	1.3264	546	< 190–15 000
	1.233	10 300	
PbF_2	1.7722	550	250– c 16 000
	1.6367	10 000	
Si	3.417	10 000	1200–> 15 000
$SrTiO_3$	2.4254	560	390–6800
	2.1221	5000	
thallium bromo-iodide	2.6806	540	600–> 40 000
	2.2887	30 000	
ZnS (Irtran 2)	2.2907	1000	1000–14 500
	2.1688	12 000	
(Cleartan)	2.3884	546	400–14 500
	2.171	12 000	
ZnSe (Irtran 4)	2.485	1000	450– c 21 500
	2.31	15 000	

Reference: Kaye and Laby, *Tables of Physical and Chemical Constants*, 15ed., Longmans, 1986

Table VII.27 Refractive indices of some optical cements at wavelength 589.3 nm and temperature 293 K

Cement	index	Cement	index
Beetle 8128	1.55	Epikote 817	c 1.57
BS No.8	1.59	HT	1.45
cellulose caprate	1.47	Loctite 357 (UV curing)	1.47
cellulose caprate (plasticized)	1.49	soft Canada balsam	1.53
Crystic 191 LV	1.56	solid Canada balsam	1.54

Reference: *Optica Acta* vol.14, 401, 1967

Table VII.28 Variation with wavelength of ordinary and extraordinary refractive indices, n_o and n_e, of calcite at temperature 291 K (293 K) for wavelength range 200–706 nm (801–3324 nm).

wavelength/nm	refractive index n_o	refractive index n_e	wavelength/nm	refractive index n_o	refractive index n_e
200	1.902 84	1.576 49	1422	1.635 90	
303	1.719 59	1.513 65	1497	1.634 57	1.477 44
410	1.680 14	1.496 40	1609	1.632 61	
508	1.665 27	1.489 56	1682	1.631 27	
643	1.655 04	1.484 90	1749		1.476 38
706	1.652 07	1.483 53	1761	1.629 74	
801	1.648 69	1.482 16	1849	1.628 00	
905	1.645 78	1.480 98	1909		1.475 73
1042	1.642 76	1.479 85	1946	1.626 02	
1159	1.640 51	1.479 10	2053	1.623 72	
1229	1.639 26	1.478 70	2100		1.474 92
1307	1.637 89	1.478 31	2172	1.620 99	
1396	1.636 37	1.477 89	3324		1.473 92

Reference: Kaye and Laby, *Tables of Physical and Chemical Constants*, 15ed., Longmans, 1986

Table VII.29 Variation with wavelength of ordinary and extraordinary refractive indices, n_o and n_e, of crystal quartz at temperature 293 K.

wavelength/nm	refractive index n_o	refractive index n_e	wavelength/nm	refractive index n_o	refractive index n_e
185	1.675 78	1.689 88	1541.4	1.527 81	1.536 30
198	1.650 87	1.663 94	1681.5	1.525 83	1.534 22
231	1.613 95	1.625 55	1761.4	1.524 68	1.533 01
340	1.567 47	1.577 37	1945.7	1.521 84	1.530 04
394	1.558 46	1.568 05	2053.1	1.520 05	1.528 23
434	1.553 96	1.563 39	2300	1.515 61	
508	1.548 22	1.557 46	2600	1.509 86	
589.3	1.544 24	1.553 35	3000	1.499 53	
768	1.539 03	1.547 94	3500	1.484 51	
832.5	1.537 73	1.546 61	4000	1.466 17	
991.4	1.535 14	1.543 92	4200	1.456 9	
1159.2	1.532 83	1.541 52	5000	1.417	
1307.0	1.530 90	1.539 51	6450	1.274	
1395.8	1.529 77	1.538 32	7000	1.167	
1479.2	1.528 65	1.537 16			

Reference: Kaye and Laby, *Tables of Physical and Chemical Constants*, 15ed., Longmans, 1986
Handbook of Chemistry and Physics, CRC Press, 1991/2

Table VII.30 Ordinary and extraordinary refractive indices, n_o and n_e, at or near temperature 293 K of various birefringent optical materials at stated wavelengths, together with transmission range

material	refractive index n_o	refractive index n_e	wavelength/nm	range/nm
Al_2O_3 (sapphire)	1.7708		546	140–6500
	1.7555	1.7479	1014	
KH_2PO_4 (KDP)	1.5115	1.4698	546	250–c 1700
	1.4954	1.4604	1014	
MgF_2	1.3786	1.3904	546	110–7500
	1.3731	1.3846	1083	
$(NH_4)H_2PO_4$ (ADP)	1.5266	1.4808	546	130–c 1700
	1.5084	1.469	1014	
TiO_2	2.652	2.958	546	c 430–6200
	2.484	2.747	1014	

Reference: Kaye and Laby, *Tables of Physical and Chemical Constants*, 15ed., Longmans, 1986

140 Optics

Table VII.31 Variation with angle of incidence of light of the percentage reflectance, at interface between air and glass (refractive index 1.52) for electric vector respectively parallel ($R_\parallel$) and perpendicular ($R_\perp$) to the plane of incidence

angle of incidence/°	$R_\parallel$/%	$R_\perp$/%	angle of incidence/°	$R_\parallel$/%	$R_\perp$/%
0	4.3	4.3	56.7	0	15.7
15	3.9	4.7	60	0.2	18.3
30	2.7	6.1	75	10.6	40.8
45	0.9	9.7	85	49.2	73.8

Reference: Kaye and Laby, *Tables of Physical and Chemical Constants*, 15ed., Longmans, 1986

Table VII.32 Refractive indices for common thin-film materials and their percentage reflectances in air for light of wavelength 550 nm and for thickness to form a quarter-wave film on glass of refractive index 1.52

material	index	reflectance/%	material	index	reflectance/%
Al_2O_3	1.63	7.4	Ta_2O_5	2.15	25.5
MgF_2	1.38	1.3	TiO_2	2.2–2.7	27.3–42.9
Na_3AlF_6 (cryolite)	1.35	0.8	ZnS	2.35	32.5
PbF_2	1.75	11.3	ZrO_2	2.05	22
SiO_2	1.46	2.8			

Reference: Kaye and Laby, *Tables of Physical and Chemical Constants*, 15ed., Longmans, 1986

Table VII.33 Percentage reflectance in air for light of wavelength 550 nm of quarter-wave multilayer stacks of zinc sulphide and magnesium fluoride alternately for different numbers of layers on glass of refractive index 1.52

number of layers	3	5	7	9	11
reflectance/%	68.3	87.7	95.6	98.5	99.5

Reference: Kaye and Laby, *Tables of Physical and Chemical Constants*, 15ed., Longmans, 1986

Table VII.34 Variation with wavelength of specific rotation for quartz at temperature 293 K

For specific rotations for some other substances, see Tables XI.4, XI.5 and XI.6.

wavelength/nm	specific rotation/°	wavelength/nm	specific rotation/°	wavelength/nm	specific rotation/°
214.3	235.972	404.7	48.112	589.3	21.726
257.1	143.266	435.8	41.546	656.2	17.318
318	84.972	480	33.674	718.4	14.304
358.2	64.459	546.1	25.535	760.4	12.668

Reference: *Handbook of Chemistry and Physics*, CRC Press, 1991/2

Table VII.35 Values of Verdet constants for various gases at wavelength 589.3 nm and STP

Gas	$CH_3CH_2CH_3$	CH_4	iso-C_4H_{10}	CO_2	H_2	He	N_2	O_2
constant/10^{-6} min A^{-1}	50.05	24.15	68.12	13.22	8.867	0.667	8.861	7.598

Reference: Kaye and Laby, *Tables of Physical and Chemical Constants*, 15ed., Longmans, 1986

Table VII.36 Variation with wavelength of Verdet constant for water at 293 K

wavelength/nm	248.2	280.4	302.2	334.2	435.8	546.1	589.3
constant/10^{-2} min A^{-1}	13.657	9.418	7.677	5.931	3.168	1.935	1.645

Reference: Kaye and Laby, *Tables of Physical and Chemical Constants*, 15ed., Longmans, 1986

Table VII.37 Verdet constants for various liquids at stated temperatures for wavelength 589.3 nm

Liquid	temperature/K	constant/10^{-2} min A^{-1}	liquid	temperature/K	constant/10^{-2} min A^{-1}
CCl_4	288	4.03	C_2H_5OH	293	1.41
$CHCl_3$	293	2.06	$C_6H_5CH_3$	301	3.38
$CH_3(CH_2)_3CH_3$	288	1.48	C_6H_{14}	288	1.57
CH_3COCH_3	293	1.42	CS_2	291	5.28
CH_3OH	293	1.17			

Reference: Kaye and Laby, *Tables of Physical and Chemical Constants*, 15ed., Longmans, 1986

Table VII.38 Verdet constants for various solids at stated temperatures and wavelengths

Solid	temperature/K		wavelength/nm	constant/10^{-2} min A^{-1}
CaF_2 (fluorite)			253.7	7.526
			589.3	1.127
crystalline quartz	293		257.3	13.56
(along axis)	293	435.8	3.997	
	293		589.3	2.091
GLASS:				
dense barium crown	293		589.3	2.4
dense flint	293		589.3	4.85
extra dense flint	293		589.3	6.55
Hoya Faraday rotator FR4			632.8	−13.07
FR5			632.8	−31.54
light flint	293		589.3	3.9
medium barium crown	293		589.3	2.5
plate	293		589.3	2.3
Schott low stress optical coefficient (SR57)	293		589.3	10.3
NaCl (rock salt)			259.9	34.03
			404.7	9.74
			670.8	3.08
SiO_2 (fused)	298		284.8	9.59
	298		334.2	6.48
	298		546.1	2.175

Reference: Kaye and Laby, *Tables of Physical and Chemical Constants*, 15ed., Longmans, 1986

Table VII.39 Refractive indices, half-wave voltage and transmission range for some electro-optic substances
Ordinary and extraordinary refractive indices, n_o and n_e, are given for some substances.

Substance	index		$\frac{1}{2}$-wave voltage/kV	range/nm
$BaTiO_3$	2.4		0.66	430–7000
CS_2	1.63		370	400–3000
CuCl	1.93		6.2	400–20000
hexamine	1.59		91	300–2000
H_2O	1.331		310	300–1300
$KTa_{0.65}Nb_{0.35}O_3$	2.29		c 0.6	400–5500
$KTaO_3$ at temperature 4.2 K	2.24		0.95	
nitrobenzene	1.55		34	460–1500
$SrTiO_3$	2.38		c 10	400–5500
ZnS	2.36		10.2	400–1300
	n_o	n_e		
$Ba_{0.25}Sr_{0.75}Nb_2O_6$	2.31	2.3	c 0.1	400–6000
KD_2PO_4 (KDDP)	1.51	1.47	3.5	250–1800
KH_2PO_4 (KDP)	1.507	1.467	8.7	250–1600
$LiNbO_3$	2.3	2.21	2.8	400–4500
$LiTaO_3$	2.176	2.18	2.4	
$(NH_4)H_2PO_4$ (ADP)	1.522	1.477	10.5	250–1300

Reference: Nye, *Physical Properties of Crystals*, OUP, 1964
Harvey, *Coherent Light*, Wiley, 1970

Table VII.40 Values of wavelengths of maximum emission and cut-off, decay constant and light conversion efficiency, relative to that of NaI (Tl) for some solid scintillators

Scintillator	wavelength/nm max	cut-off	constant/ μs	efficiency	Scintillator	wavelength/nm max	cut-off	constant/ μs	efficiency
CaF$_2$(Eu)	435	405	0.94	50	CsI(Na)	420	300	0.63	85
CaWO$_4$	430	300	†0.9–20	†14–18	CsI(Tl)	565	330	1	45
CdWO$_4$	530	450	†0.9–20	†17–20	KI(Tl)	426	325	0.24/2.5	24
BaF$_2$	325	134	0.63	10	LiI(Eu)	470–485	450	1.4	35
Bi$_4$Ge$_3$O$_{12}$	480	350	0.3	8	NaI(Tl)	410	320	0.23	100
CsF	390	220	0.005	5	TlCl(Be, I)	465	390	0.2	2.5

†More than one value is due to the presence of more than one decay component
Reference: Anderson (ed.), *A Physicist's Desk Reference*, American Institute of Physics, 1989

Table VII.41 Operational mode, vacuum wavelength, λ, average and peak powers, P_{av} and P_{max}, rate of change, dP_{out}/dP_{in}, of output power relative to input power, and pulse duration, Δt for some representative lasers

Note: According to Hanna (*Microlasers going macro*, Physics World, vol.2, 9/35, 1989), semiconductor-diode lasers powering solid-state lasers have superior characteristics to most other types and so will largely supersede them in the near future. Parts of this table may thus shortly become of historical interest only. (See also Macfarlane, Physics World, vol. 4, 10/8, IOP, 1991)

Laser†	Mode*	λ/nm	P_{av}/W	P_{max}/kW	dP_{out}/dP_{in}	Δt/ns
■ st Ar†	cw	514.673	10–150		< 0.001	
■ st CO$_2$	cw	10 600	(1–50) × 10^3		0.1–0.2	
°CO$_2$ (TEA)	p	10 600	10^3	10^4	0.1	100–500
Cu	p	510.696	40	10^2	0.01–0.02	20–40
°FEL	cw	c 400 to 10^6	very high potentially		0.01–0.1	
GaAs (diode)	cw	850	1		0.4	
HF	cw	2600–3300	10^4–10^6			
HF	p	2600–3300		10^3		
He–Cd	cw	325.1	0.1			
		441.69				
■ st He-Ne	cw	632.991	10^{-3}–10^{-2}		< 0.001	
KrF	p	248	5 × 10^2	5 × 10^3	0.01	10
N$_2$	p	337.1	0.1	10^3	< 0.001	10
°Nd: YAG (solid state)	cw	1064	2 × 10^2		0.01–0.03	
°Nd: YAG (solid state)	p	1064	10^3	10	0.01–0.03	10^6–5 × 10^6
°Nd: YAG (solid state)	p	1064	10	2 × 10^4	0.01–0.03	10–20
rhodamine 6G (dye)	p	590	10^2	10^2	0.005	10^4
rhodamine 6G (dye)	cw	590	5			
■ st ruby (solid state)	p	694.5	1	10–10^4	< 0.001	10^6–10

† material gaseous, unless otherwise stated
* cw represents continuous wave, p represents pulsed or repetitively pulsed
■ st indicates standard: see note to Table VII.15.
° TEA, FEL and YAG respectively denote transversely excited atmospheric, free electron laser and yttrium aluminium garnet
Reference: Svelto, *Principles of Lasers, 3rd edn*. Plenum, 1989
        Kaye and Laby, *Tables of Physical and Chemical Constants*, 15ed., Longmans, 1986
        Physics World Buyers Guide, IOP, 1991

Section VIII
Spectroscopy

		Page
VIII.1	Electromagnetic spectrum	145
VIII.2	Types of optical spectra	146
VIII.3	The effective cut-off wavelengths of some common solvents in the UV and visible regions	146
VIII.4	Cut-off wavelengths of materials for use in the UV and visible regions	147
VIII.5	Wavelength calibration standards in UV and visible regions	147
	1 Rare earth filters	147
	2 Discharge tube emission lines	147
VIII.6	Photometric calibration standards for use in the ultraviolet and visible regions	148
VIII.7	Spectroscopic solvents for use in the infrared region	150
VIII.8	Transmission regions of various optical materials for use as cell windows in the infrared region	152
VIII.9	Wavelength calibration standards in infrared region	152
VIII.10	Wavelength calibration standards for spectroscopes	153
VIII.11	Spectral series of hydrogen	153
VIII.12	Wavelengths of the Fraunhofer lines (sun's spectrum)	153
VIII.13	Absorption characteristics of some common organic chromophores	154
VIII.14	Molecular parameters of gases	155
VIII.15	Spectroscopic data on the properties of diatomic molecules in their ground state	156
VIII.16	Spectroscopic data on the properties of triatomic molecules	156
VIII.17	Rotational constants and moments of inertia of symmetric top molecules in their ground states	157
VIII.18	Rotational constants and moments of inertia of asymmetric rotor molecules	157
VIII.19	Fundamental vibrational frequencies of polyatomic molecules	158
VIII.20	Potential energy barriers to internal rotation about single bonds	158
VIII.21	Nuclear spins, moments, resonant frequencies and receptivities of naturally occurring magnetically active isotopes	159
VIII.22	Chemical shifts of some common NMR solvents	160
VIII.23	Ranges of ^{1}H chemical shifts of some common functional groups	161
VIII.24	Ranges of ^{13}C chemical shifts in functional groups	162
VIII.25	Ranges of ^{15}N chemical shifts in some common functional groups	163
VIII.26	Some chemical shifts of ^{17}O, ^{19}F, ^{27}Al, ^{31}P and ^{35}Cl for substances indicated	164
VIII.27	Mass-composition of most useful fragments to mass 99	164
VIII.28	Common losses from molecular ions	166
VIII.29	Recommended fluorescent quantum yield references in various emission ranges	167

Section VIII
Spectroscopy

Table VIII. 1 Electromagnetic Spectrum

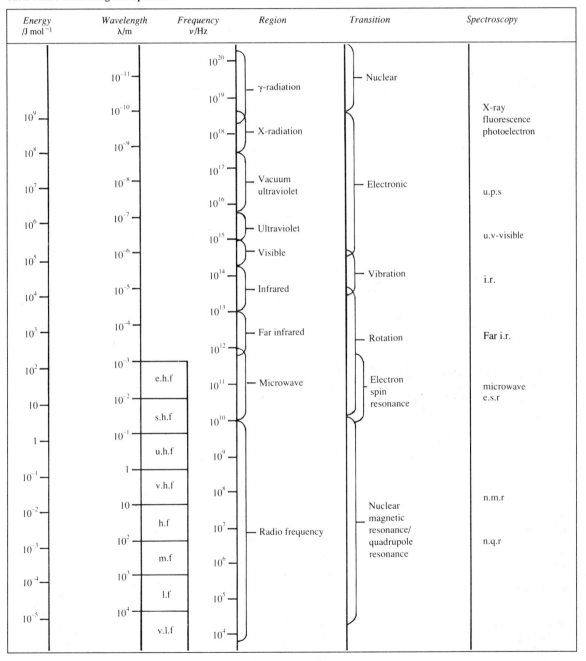

145

Table VIII.2 Types of Optical Spectra

Average thermal kinetic energy = $RT/2$ = 1.3 kJ mol^{-1} or 100 cm^{-1}.

Description	Frequency range	Wavenumber /cm^{-1}	Energy /kJ mol^{-1}	Type of molecular energy	Information obtained
radio	10^3 10^9	3.3×10^{-10} 0.033	4×10^{-8} 4×10^{-4}		
microwave	10^9 10^{11}	0.033 3.3	4×10^{-4} 4×10^{-2}	rotation of heavy molecules	interatomic distances, dipole moments, nuclear interactions
far infrared	10^{11} 10^{13}	3.3 333	4×10^{-2} 4	rotation of light molecules, vibration of heavy molecules	interatomic distances, force constants
infrared	10^{13} 10^{14}	333 3330	4 40	vibration of light molecules, vibration–rotation	interatomic distances, force constants, charge distributions
Raman	10^{11} 10^{14}	3.3 3330	4×10^{-2} 40	pure rotation, vibration–rotation	interatomic distances, force constants, charge distributions
visible, ultraviolet	10^{14} 10^{16}	3330 3.3×10^5	40 4000	electronic transitions	all above and bond dissociation energies
X-rays and γ-rays	10^{16} 10^{21}	3.3×10^5 3.3×10^{10}	4000 4×10^8	ionization	all above and crystal structures

Table VIII.3 The Effective Cut-off Wavelengths of Some Common Solvents in the UV and Visible Regions

Wavelengths at which the transmittance falls to 25% (A = 0.60) for the specified path length, measured against water as the reference
Reference: D.A. Skoog and D.M. West, *Fundamentals of Analytical Chemistry*, 3rd ed. (1976), Holt Saunders, New York.

Solvent	Path length	Wavelength/nm	
		10 mm	1 mm
n-hexane		199.2	190.5
n-heptane		200.0	188.7
iso-octane		202.0	188.7
diethyl ether		205.3	199.2
ethanol		207.5	196.1
iso-propanol		209.2	
methanol		210.5	196.1
cyclohexane		211.9	196.1
acetonitrile		213.2	200.8
1,4-dioxane		216.0	208.3
dichloromethane		232.6	
tetrahydrofuran		238.1	
chloroform		246.9	
carbon tetrachloride		257.1	
dimethyl sulphoxide		270.3	
dimethyl formamide		271.0	
benzene		280.1	
pyridine		305.8	
acetone		331.1	

Table VIII.4 Cut-off Wavelengths of Materials for Use in the Ultraviolet and Visible Regions
Reference: A.M. James and F.E. Prichard, *Practical Physical Chemistry*, 3rd ed. (1974), Longmans, London.

Material	*Cut-off wavelength*/nm	*General properties*
fused silica	120	unaffected by most solvents
crystal quartz	120	unaffected by most solvents
LiF	120	slightly soluble in water
NaCl	200	hygroscopic
KBr	250	hygroscopic
silicate glass	350	unaffected by most solvents

Table VIII. 5 Wavelength Calibration Standards in Ultraviolet and Visible Regions of the Spectrum
1. Rare-earth Filters.
H: holmium glass
D: didymium glass
[a] for filter with extra red density.

Filter	*Wavelength of absorption peak* /nm	*Uncertainty* (+/−) /nm
H	190.2	0.2
H	192.3	0.2
H	241.5	0.2
H	279.4	0.3
H	287.5	0.35
H	333.7	0.55
H	360.9	0.75
H	418.4	1.1
H	453.2	1.4
H	536.2	2.3
D	573.0	3.0
D	586.0	3.0
H	637.5	3.8
D	685.0	4.5
D[a]	741.0	5.5
D[a]	803.0	6.3

2. Discharge Tube Emission Lines

Neon emission/nm	*Mercury emission*/nm
533.1	
534.1	253.7
540.1	364.9
585.3	404.5
594.5	435.8
614.3	546.1
633.4	576.9
640.2	579.0
667.8	
693.0	
717.4	
724.5	

See also Tables VII.14 and VII.15.

Table VIII.6 Photometric Calibration Standards for Use in the Ultraviolet and Visible Regions

Solution A: cobalt ammonium sulphate, 14.481 g of $CoSO_4(NH_4)_2SO_4 \cdot 6H_2O$ dissolved in 1 dm^3 of 1% sulphuric acid (10 cm^3 of concentrated sulphuric acid, specific gravity 1.835, to 1 dm^3 with water).
Solution B: copper sulphate, 20.000 g of $CuSO_4 \cdot 5H_2O$ dissolved in 1 dm^3 of 1% sulphuric acid.
Absorbance Values A in 10 mm Cell. All measurements at 25 °C.
[a]: most reliable values for both solutions. [b]: values adopted by IUPAC.
For further information see: C. Burgess and A. Knowles, *Standards in Absorption Spectroscopy* (1981), Chapman and Hall, London.

Wavelength λ/nm	Solution A ϵ /m^2 mol^{-1}	Solution A A	Solution B ϵ /m^2 mol^{-1}	Solution B A
350	0.01037	0.0038	0.01124	0.0090
360	0.01092	0.0040	0.00786	0.0063
370	0.01365	0.0050	0.00574	0.0046
380	0.01774	0.0065	0.00437	0.0035
390	0.02402	0.0088	0.00349	0.0028
400[a]	0.03412	0.0125	0.00287	0.0023
410	0.04586	0.0168	0.00237	0.0019
420	0.06114	0.0224	0.00200	0.0016
430	0.09281	0.0340	0.00175	0.0014
440	0.14249	0.0522	0.00150	0.0012
450[a]	0.21101	0.0773	0.00137	0.0011
460	0.28145	0.1031	0.00137	0.0011
470	0.33113	0.1213	0.00150	0.0012
480	0.36821	0.1349	0.00175	0.0014
490	0.40179	0.1472	0.00225	0.0018
500[a]	0.44623	0.1635	0.00324	0.0026
510	0.47544	0.1742	0.00474	0.0038
520	0.46100	0.1689	0.00687	0.0055
530	0.39637	0.1452	0.00986	0.0079
540	0.30381	0.1113	0.01386	0.0111
550[a]	0.21154	0.0775	0.01935	0.0155
560	0.13539	0.0496	0.02697	0.0216
570	0.08408	0.0308	0.03646	0.0292
580	0.05651	0.0207	0.04870	0.0390
590	0.04313	0.0158	0.06467	0.0518
600[a]	0.03739	0.0137	0.08490	0.0680
610	0.03384	0.0124	0.11048	0.0885
620	0.03139	0.0115	0.14048	0.1125
630	0.03057	0.0112	0.17853	0.143
640	0.03003	0.0110	0.22475	0.180
650[a]	0.02866	0.0105	0.27964	0.224
660	0.02647	0.0097	0.34214	0.274
670	0.02374	0.0087	0.41448	0.332
680	0.02074	0.0076	0.48944	0.392
690	0.01801	0.0066	0.57306	0.459
700[a]	0.01474	0.0054	0.65796	0.527
710	0.01255	0.0046	0.73909	0.592
720	0.01037	0.0038	0.81903	0.656
730	0.00873	0.0032	0.89261	0.715
740	0.00819	0.0030	0.95896	0.768
750[a]	0.00764	0.0028	1.02000	0.817
Hg404.7[a]	0.03931[b]	0.0144	0.00262[b]	0.0021
Hg435.8[a]	0.11929[b]	0.0437	0.00162[b]	0.0013
Hg491.6[a]	0.40860[b]	0.1497	0.00237[b]	0.0019
He501.6[a]	0.45342[b]	0.1661	0.00350[b]	0.0028
Hg546.1[a]	0.24592[b]	0.0901	0.01685[b]	0.0135
Hg578.0[a]	0.05977[b]	0.0219	0.04594[b]	0.0368
He587.6[a]	0.04558[b]	0.0167	0.06080[b]	0.0487
He667.8[a]	0.02429[b]	0.00089	0.39829[b]	0.319

Table VIII.6 (Continued)
Molar Extinction Coefficients of Potassium Chromate and Potassium Dichromate Solutions Used as Transmission Standards

Solution	λ/nm	ϵ/m^{2}mol^{-1}	A (10 mm)
50 mg K$_2$Cr$_2$O$_7$ in 1 dm^3 0.02 mol dm^{-3} H$_2$SO$_4$	235	368.3	0.626
(reference 0.02 mol dm^{-3} H$_2$SO$_4$)	257	427.7	0.727
	313	143.6	0.244
	350	315.2	0.536
0.1 g K$_2$Cr$_2$O$_7$ in 1 dm^3 0.02 mol dm^{-3} H$_2$SO$_4$	235	368.3	1.251
(reference 0.02 mol dm^{-3} H$_2$SO$_4$)	257	427.7	1.454
	313	143.6	0.488
	350	315.2	1.071
40 mg K$_2$CrO$_4$ in 1 dm^3 0.05 mol dm^{-3} KOH	273	369.0	0.760
(reference 0.05 mol dm^{-3} KOH)	373	480.0	0.989

150 Spectroscopy

Table VIII.7 Spectroscopic Solvents for use in the Infrared Region
The white windows in the chart below show the regions in which the solvents listed have at least 25% transmittance at the path length indicated. The diagrams also indicate the regions over which these solvents have been examined.
Reference: B.D.H. Chemicals Ltd (1983), Poole.

Solvent	Path length
Acetonitrile	0.2 mm
Benzene	0.05 mm
Bromoform	0.2 mm
Carbon disulphide	0.2 mm
Carbon tetrachloride	0.2 mm
Chloroform	0.2 mm
Deuterochloroform-d	0.2 mm
Dichloromethane	0.2 mm
Dimethylformamide	0.05 mm
Hexachlorobuta-1,3-diene	a
Paraffin liquid	a
Pyridine	0.2 mm
Tetrachloroethylene	0.2 mm

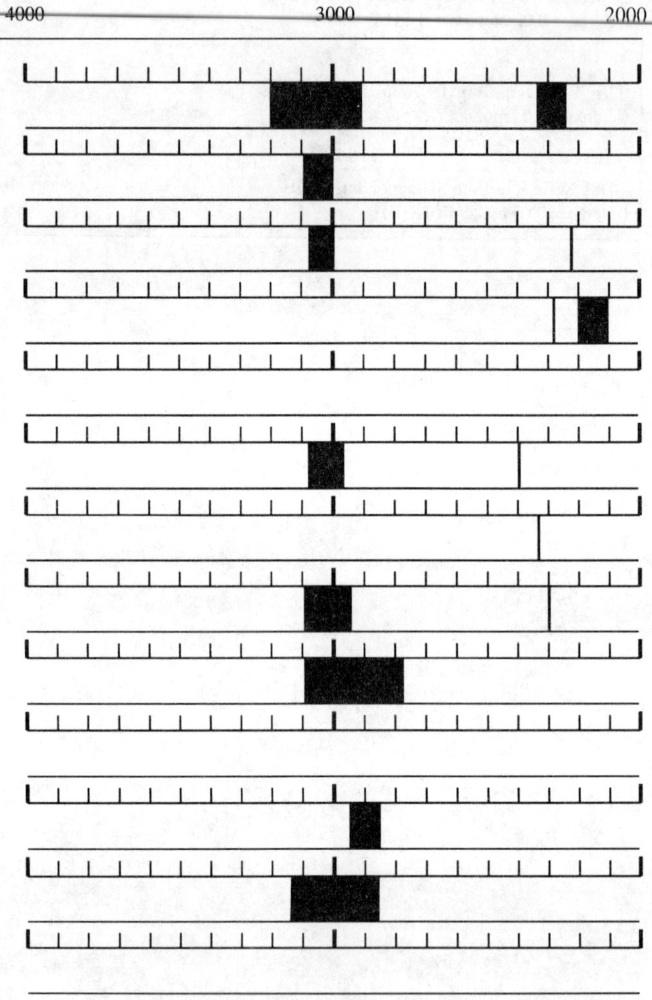

a: Liquid film between caesium iodide.

Solvents for IR Region 151

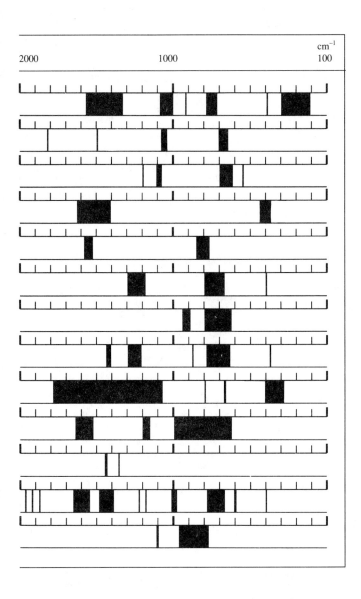

Table VIII.8 Transmission Regions of Various Optical Materials for Use as Cell Windows in the Infrared Region (limiting wavelengths at which a sample 2 mm thick has 10% transmission)

Reference: R.G.J. Miller and B.C. Stace, *Laboratory Methods in Infrared Spectroscopy*, 2nd ed. (1972), Heydon, London.

Material	Range		General properties
	Wavelength $\lambda/\mu m$	Wavenumber ω/cm^{-1}	
fused silica	0.12–4.5	83 300–2200	unaffected by most solvents
crystal quartz	0.12–4.5	83 300–2200	unaffected by most solvents
calcite	0.2–5.5	50 000–1670	slightly soluble in water, soluble in acids
sapphire	0.14–6.5	71 500–1540	unaffected by most solvents
Irtran-1 (MgF_2)	0.11–7.5	91 000–1330	insoluble in water, soluble in acids
LiF	0.12–9	83 000–1100	slightly soluble in water
Irtran-3 (CaF_2)	0.13–12	76 900–830	insoluble in water, resistant to acids and bases
Irtran-2 (ZnS)	1.6–14	10 000–700	insoluble in water, acids, bases, and organic solvents; high refractive index
BaF_2	0.2–15	50 000–700	insoluble in water, soluble in acids; resistant to fluorine and fluorides
NaCl	0.2–26	50 000–400	hygroscopic
AgCl	0.4–28	25 000–370	insoluble in water, soluble in acids and NH_4Cl
KCl	0.21–30	47 600–330	hygroscopic
AgBr	0.45–35	22 200–280	insoluble in water, acetone and alcohols
KBr	0.25–40	40 000–250	hygroscopic
KRS-5 (TlBrI)	0.5–40	20 000–250	not hygroscopic, poisonous, high refractive index, toxic
KI	0.25–45	40 000–220	hygroscopic
TlBr	0.42–48	23 800–200	slightly soluble in water, alcohol
CsBr	0.3–55	33 000–180	hygroscopic
CsI	0.24–80	41 700–125	hygroscopic, soluble in alcohol
diamond	0.25–80	40 000–125	unaffected by most solvents
polyethylene (polythene)	16–1000	625–10	inexpensive far-i.r. material, insoluble in water, swells in organic solvents

Table VIII.9 Wavelength Calibrations Standards in the Infrared Region

For more extensive data for calibration in the range 3950–600 cm^{-1} using ammonia, carbon dioxide, methane, water vapour and hydrogen chloride see: IUPAC, *Calibration Standards for IR. Tables of Wavenumbers for Calibration of IR Spectrometers* (1961), Butterworths, London: *Pure and Appl. Chem.* (1960/1), **1**, 603.

The tabulated values of the principal absorption bands of polystyrene are recommended reference standards (uncertainty, +/–)

Wavenumber/cm^{-1}	Wavelength/μm	Wavenumber/cm^{-1}	Wavelength/μm
3062	3.266	1494	6.693
3027.1 (0.3)	3.303	1181.4 (0.3)	8.465
2924 (2)	3.420	1154.3 (0.3)	8.663
2850.7 (0.3)	3.508	1069.1 (0.3)	9.354
1944.0 (1)	5.144	1028.0 (0.3)	9.728
1871.0 (0.3)	5.345	906.7 (0.3)	11.03
1801.6 (0.3)	5.550	698.9 (0.5)	14.31
1601.4 (0.3)	6.244		

© (1961) IUPAC.

Table VIII.10 Wavelength Calibration Standards for Spectroscopes
See also: *Handbook of Chemistry and Physics* (1978–79), CRC.

Source	Wavelength/nm	Source	Wavelength/nm
K, solar	393.4	mercury arc	546.1
H_1, solar	396.9	mercury arc	577.0
mercury arc	404.7	mercury arc	579.1
hydrogen tube	434.0	sodium flame	589.0
mercury arc	435.8	sodium flame	589.6
strontium flame	460.8	mercury arc	623.4
hydrogen tube	486.1	hydrogen tube	656.3
mercury arc	491.6	lithium flame	670.8
mercury arc	496.0	B, solar	686.9
magnesium flame	517.3	mercury arc	690.7
magnesium flame	518.4	potassium flame	766.5
E, solar	527.0	potassium flame	769.9
thallium flame	535.1		

Table VIII.11 Spectral Series of Hydrogen
The lines in the various series can be fitted to the expression:

$$1/\lambda = R_H (1/n_1^2 - 1/n_2^2)$$

with $n_1 = 1$ (Lyman), $n_1 = 2$ (Balmer), $n_1 = 3$ (Paschen); in each case $n_2 = n_1 + 1$, $n_1 + 2 \ldots$
For further information see: W.L. Weise and G.A. Martin, Atomic Spectroscopy, in *A Physicist's Desk Reference*, *Physics Vade Mecum*, 2nd ed., ed. H.L. Anderson (1989), p. 92, Amer, Inst. Phys.

Name		Wavelength/nm	Name		Wavelength/nm
Lyman	α	121.567	Balmer	δ	410.173
	β	102.573		ϵ	397.007
	γ	97.253 7	Paschen	α	1875.10
	δ	94.974 3		β	1281.81
	ϵ	93.780 3		γ	1093.81
Balmer	α	656.280		δ	1004.94
	β	486.132		ϵ	954.597
	γ	434.046			

Table VIII.12 Wavelengths of the Fraunhöfer Lines (Sun's Spectrum)
At 15 °C and 1 atmosphere pressure.
Reference: as for VIII.10.

Line	Due to	Wavelength/nm	Line	Due to	Wavelength/nm
U	Fe	294.79	g	Ca	422.6742
t	Fe	299.44	G	Fe	430.7914
T	Fe	302.1067		Ca	430.7749
s	Fe	304.7623	G'	H	434.0477
	Fe	310.0683	F	H	486.1344
S_1	Fe	310.0326		Fe	516.7510
S_2	Fe	309.9943	b_4	Mg	516.7330
	Ca	318.1277	b_2	Mg	517.2700
R	Ca	317.9343	b_1	Mg	518.3621
Q	Fe	328.6773	E_2	Fe	526.9557
P	Ti	336.1194	D_2	Na	588.9977
O	Fe	344.1020	D_1	Na	589.5944
N	Fe	358.1210	C	H	656.2816
M	Fe	372.7636	B	O	686.9955
L	Fe	382.0438		O	762.1
K	Ca	393.3684	A	O	759.4
H	Ca	396.8494	Z		822.85
h	H	410.1750	Y		899.00

Table VIII.13 Absorption Characteristics of Some Common Organic Chromophores
Reference: as for Table VIII.3.

Chromophore	Example	Solvent	λ_{max}/nm	ϵ_{max}/m^2 mol^{-1}
ALIPHATIC COMPOUNDS				
alkene	$C_6H_{13}CH{:}CH_2$	n-heptane	177	1 260
			210	40
conjugated alkene	$CH_2{:}CHCH{:}CH_2$	n-heptane	217	2 100
	$t{-}CH_2{:}CHCH{:}CHCH{:}CH_2$	n-hexane	247	3 390
			258	4 365
			268	3 630
alkyne	$C_5H_{11}C{:}CCH_3$	n-heptane	177.5	1 000
			222.5	16
carbonyl	$(CH_3)_2CO$	n-hexane	186	100
			280	1.6
	$C_2H_5COCH_3$	ethanol	273.5	1.7
	CH_3CHO	gas	178	302
			181.5	1 120
carboxyl	CH_3COOH	ethanol	208	3.2
amido	CH_3CONH_2	methanol	205	16.2
azo	$CH_3N{:}NCH_3$	water	345	2
imino	$CH_3C({:}NH)NH_2$	water (pH=13)	219	110
nitro	CH_3NO_2	ethanol	260	3.9
nitroso	C_4H_9NO	ether	300	10
			665	2
nitrate	$C_2H_5ONO_2$	dioxane	270	1.2
AROMATIC COMPOUNDS				
anthracene		ethanol	218	1 100
			324	280
			357	770
benzaldehyde	C_6H_5CHO	cyclohexane	241	1 410
			277.5	120
benzene	C_6H_6	n-hexane	204	790
			256	20
benzenesulphonamide	$C_6H_5SO_2NH_2$	ethanol	219	890
			259	32
			264	50
benzoic acid	C_6H_5COOH	ethanol	227	1 150
benzonitrile	C_6H_5CN	ethanol	222	1 290
			230	1 100
chrysene		ethanol	222	3 550
			268	15 850
			320	1 320
diphenyl	$C_6H_5 . C_6H_5$	ethanol	247	1 740
naphthalene		ethanol	221	10 965
			275.5	575
			286	390
perylene		ethanol	251	5 010
			387	1 200
			434	3 630
phenanthrene		heptane	187.5	3 020
			211.5	3 715
phenol	C_6H_5OH	ethanol	218.5	602
			271	190
stilbene (*trans*)	$C_6H_5CH{:}CHC_6H_5$	ethanol	244	2 455
styrene	$C_6H_5CH{:}CH_2$	ethanol	245.3	1 510
toluene	$C_6H_5CH_3$	ethanol	207	933
			260	30

Table VIII.14 Molecular Parameters of Gases

For more extensive data see: D.R. Stull and H. Prophet, *JANAF Thermochemical Tables* (3rd edn, 1985, NSRDS–NBS 37).
Dissociation energies D_o are with respect to the corresponding atoms in the ground state; σ is the symmetry number; θ_r, θ_v and θ_e the characteristic temperatures of rotation, vibration and electronic excitation (values of θ_e in excess of 10 000 K are not given); g_o is the degeneracy of the ground electronic state.

[a]: Values of θ_r for non-linear polyatomic molecules are values of $\theta_A \theta_B \theta_C$.   [b]: Values in parentheses are degeneracies.

Species	D_o/kJ mol^{-1}	σ	θ_r/K[a]	θ_v/K[b]	g_o	θ_e/K[b]
H	0				2	
C	0				1	23.6(3)
						62.6(5)
N	0				4	
O	0				5	228.1(3)
						325.9
Cl	0				2	1 269.53(2)
I	0				4	10 939.3(2)
H$_2$	432.073	2	87.547	6 338.3	1	
HD	435.44	1	64.27	5 226.0	1	
D$_2$	439.59	2	43.03	4 307.0	1	
N$_2$	941.1	2	2.875 05	3 392.01	1	
O$_2$	491.888	2	2.079	2 273.64	3	
Cl$_2$	239.216	2	0.345 6	807.3	1	
Br$_2$	190.17	2	0.116 5	462.0	1	
I$_2$	148.81	2	0.053 76	308.65	1	
HCl	427.772	1	15.234 4	4 301.38	1	
HBr	362.20	1	12.29	3 680.0	1	
HI	294.67	1	9.369	3 322.24	1	
CO	1 070.11	1	2.777 1	3 121.48	1	
NO	627.7	1	2.452 0	2 738.87	2	174.2(2)
CO$_2$	1 596.23	2	0.561 67	960.10(2)	1	
				1 932.09		
				3 380.14		
N$_2$O		1	0.69	847.0(2)		
				1 850		
				3 200		
C$_2$H$_2$		2	1.593	880(2)		
				1 050(2)		
				2 840		
				4 730		
				4 850		
NO$_2$	928.3	2	4.243 01	1 088.9	2	
				1 953.6		
				2 396.3		
H$_2$O	917.773	2	11 331.5	2 294.27	1	
				5 261.71		
				5 403.78		
NH$_3$	1 157.77	3	1 876.0	1 367	1	
				2 341(2)		
				4 800		
				4 955(2)		
CH$_4$	1 640.57	12	435.6	1 957(3)		
				2 207.1(2)		
				4 196.2		
				4 343.3(3)		
N$_2$O$_4$	1 909.82	4	6.5793 × 10^{-3}	72	1	
				374		
				554		
				619		
				691		
				971		
				1 079		
				1 184		
				1 814		
				1 975		
				2 460		
				2 515		

Table VIII.15 Spectroscopic Data on the Properties of Diatomic Molecules in their Ground State

For more extensive data see: G. Herzberg, *Molecular Spectra and Molecular Structure*, Vol. I (1950), Van Nostrand, Princeton; A.A. Radzig and B.M. Smirnov, *Reference Data on Atoms, Molecules and Ions* (1985), Springer Series in Chemical Physics 31, Springer, Berlin.

Molecule	Bond length r_e/nm	Dissociation energy D_e/eV	Fundamental vibration ω_e/cm^{-1}	Rotational constant B/cm^{-1}	Force constant f/N m^{-1}	Moment of inertia $10^{47} I$/kg m^2
$^1H_2^+$	0.106	2.648$_1$	2297	29.8	160.0	
$^1H_2$	0.074 16$_6$	4.476$_3$	4395.2$_4$	60.864	547.9	0.457
$^1H^2H$ (HD)	0.074 13	4.513	3817.09	45.655	572.1	0.608
$^2H_2$ (D$_2$)	0.074 17	4.556	3118.4$_6$	30.442	577.0	0.918
$^{19}F_2$	0.141 2	1.60	916.6	0.890 2	470.0	31.45
$^{35}Cl_2$	0.198 8	2.475	564.9	0.244 1	322.7	114.84
$^{79}Br^{81}Br$	0.228 3$_6$	1.971	323.2	0.080 91	246.1	342.02
$^{127}I_2$	0.266 6$_6$	1.5417	214.57	0.037 35	172.2	749.74
$^{127}I^{35}Cl$	0.232 070	2.152	384.18	0.114 162	238.5	245.37
$^1H^{19}F$	0.091 68	5.85	4138.52	20.955 7	965.7	1.326
$^1H^{35}Cl$	0.127 460	4.430	2989.74	10.590 9	516.3	2.623
$^1H^{81}Br$	0.141 38	3.75$_4$	2649.67	8.464 88	411.5	3.278
$^1H^{127}I$	0.160 4$_1$	3.056$_4$	2309.5$_3$	6.510 8	313.8	4.239
$^{14}N_2$	0.109 76	9.756	2359.61	1.998 7	2293.8	13.91
$^{14}N^1H$	0.103 8	3.4	3300	16.65	598.7	1.67
$^{16}O_2$	0.120 739$_8$	5.115	1580.361$_3$	1.445 67	1176.8	19.36
$^{16}O^1H$	0.097 06	4.35	3735.21	18.871	773.4	1.472
$^{12}C^{16}O$	0.112 82	11.108	2170.21	1.931 3$_9$	1902.2	14.492
$^{12}C^1H$	0.111 98	3.47	2861.6	14.457	445.2	1.922
$^{14}N^{16}O$	0.115 08	6.487	1904.03	1.704 6	1594.4	16.42
$^7Li_2$	0.267 2$_5$	1.14	351.43$_5$	0.672 7	25.5	41.49
$^{23}Na_2$	0.307 86	0.73	159.23	0.154 71	17.2	180.98
$^{39}K_2$	0.392 5	0.51	92.40	0.056 2	9.80	402.32
$^{23}Na^{35}Cl$	0.252	4.25	360	0.218 1	118.0	146.35
$^{39}K^{35}Cl$	0.279	4.42	281	0.128 6	85.2	238.41
$^7Li^1H$	0.159 53$_5$	2.429	1405.649	7.513 1	101.8	3.292

Table VIII.16 Spectroscopic Data on the Properties of Triatomic Molecules

For more extensive data see: G. Herzberg, *Molecular Spectra and Molecular Structure* Vol. I (1950), van Nostrand, Princeton; A.A Radzig and B.M. Smirnov, *Reference Data on Atoms, Molecules and Ions*, (1985), Springer Series in Chemical Physics 31, Springer, Berlin.

Molecule	Geometric Parameters Bond angle	Geometric Parameters Bond length r_e/nm	Moments of inertia $10^{47} I$/kg m^2	Fundamental vibration ω/cm^{-1}	Rotational constants /cm^{-1}
CO$_2$	180	0.1160 (CO)	71.67	1388 667 2349	B_0 0.39
CS$_2$	180	0.155 (CS)	256.4	658 397 1532	B_0 0.109
N$_2$O	180	0.113 (NN) 0.118 (NO)	66.94	2224 1285 589	B_0 0.419
HCN	180	0.1065 (HC) 0.1153 (CN)	18.935	3311 2097 712	B_0 1.48
^{2}HCN (DCN)	180	0.1065 (DC) 0.1153 (CN)	23.16	1906 569 2629	B_0 1.209
OCS	180	0.115 (OC) 0.156 (CS)	227.3	2062 859 520	B_0 0.0203
NO$_2$	134.25	0.120 (NO)		1320 750	A_0 8.0 B_0 0.43

(Continued)

Table VIII.16 (Continued)

Molecule	Geometric Parameters		Moments of inertia $10^{47} I$/kg m^2	Fundamental vibration ω/cm^{-1}	Rotational constants /cm^{-1}
	Bond angle	Bond length r_e/nm			
SO$_2$	119	0.143 (SO)	12.3 73.2 85.5	1618 1151 518	C_0 0.41 A_0 2.03 B_0 0.34
O$_3$	116.8	0.1272 (OO)		1362 1740 701	C_0 0.29 A_0 3.55 B_0 0.445
^{2}H$_2$O (D$_2$O)	105	0.096 (DO)	1.790 3.812 5.752	1042 2666 1179 2784	C_0 0.395
H$_2$O	104.5	0.0958 (HO)	1.024 1.921 2.947	3657 1595 3756	A_0 27.88 B_0 14.51 C_0 9.287
H$_2$S	92.1	0.134 (HS)		2615 1183 2625	A_0 10.4 B_0 8.99 C_0 4.73

Table VIII.17 Rotational Constants and Moments of Inertia of Symmetric Top Molecules in their Ground States

For more extensive data see: G. Herzberg, *Molecular Spectra and Molecular Structure*, Vol. II (1960), van Nostrand, Princeton.

Molecule	B_0 /cm^{-1}	A_0 /cm^{-1}	Moments of inertia $10^{47} \times I$/kg m^2		
			I_A	I_B	I_C
NH$_3$	9.941	6.30$_9$	4.43$_7$	2.816	2.816
ND$_3$	5.138	3.15$_7$	8.86$_8$	5.448	5.448
CH$_3$D	3.878$_5$	5.245	5.33	7.218	7.218
CH$_3$F	0.8496	5.100	5.49	32.95	32.95
CH$_3$Cl	0.49	5.09$_7$	5.49$_2$	57.$_1$	57.$_1$
CH$_3$Br	0.31	5.08$_2$	5.50$_8$	90	90
CH$_3$I	0.28	5.07$_7$	5.51$_4$	100	100
C$_2$H$_6$	0.6621	2.538	11.03	42.28	42.28
BF$_3$	0.35$_5$	0.178	158	78.9	78.9
C$_2$H$_4$	0.9116*	4.867	5.752	28.08	33.85
C$_2$D$_4$	0.6522*	2.437	11.48$_7$	37.93	49.42
HCHO	1.215*	9.404	2.976$_8$	21.65	24.62
HCOOH	0.348*	2.55$_4$	10.96	75.3	86.3

*: Values of $\tilde{B}$

Table VIII.18 Rotational Constants and Moments of Inertia of Asymmetric Rotor Molecules

For further information see: M.D. Harmony, Molecular Spectroscopy and Structure, in *A Physicist's Desk Reference*, *Physics Vade Mecum*, 2nd ed., ed. H.L. Anderson (1989), p. 238, Amer. Inst. Phys.

Molecule	Rotational constants/cm^{-1}			Moments of inertia $10^{47} \times I$/kg m^2		
	A	B	C	I_A	I_B	I_C
CH$_3$CHO	1.900	0.339	0.304	14.7	82.6	92.2
(CH$_3$)$_2$CO	0.339	0.284	0.164	82.6	98.5	170.9
C$_2$H$_5$OH	1.111	0.304	0.267	25.2	92.1	104.6
CH$_3$COOH	0.378	0.316	0.178	74.0	88.5	157.6
CH$_3$OH	4.256	0.822	0.793	6.58	34.1	35.3
CH$_3$NH$_2$	0.755	0.724	0.344	37.1	38.7	81.4
CH$_2$F$_2$	1.639	0.354	0.309	17.1	79.1	90.7
C$_6$H$_5$F	0.189	0.086	0.059	148.2	326.4	474.6
CH$_2$CH$_2$O	0.850	0.738	0.470	32.9	37.9	59.5
C$_4$H$_5$N	0.305	0.300	0.151	91.9	93.2	185.2

Table VIII.19 Fundamental Vibrational Frequencies of Polyatomic Molecules
For more extensive data see: T. Shimanouchi. *J. Phys. Chem. Ref. Data* (1977), **6**, 993.

Molecule	Fundamental vibrational frequencies/cm^{-1}								
	ω_1	ω_2	ω_3	ω_4	ω_5	ω_6	ω_7	ω_8	ω_9
4-atoms									
BF_3	888	719	1503	482					
SO_3	1065	498	1391	530					
NH_3	3337	950	3444	1627					
NF_3	1032	647	905	493					
PH_3	2323	992	2328	1118					
PF_3	892	487	860	344					
HCHO	2766	1746	1501	2843	1247	1164			
C_2H_2	3374	1974	3289	612	730				
C_2N_2	2330	846	2158	503	234				
5-atoms									
CH_4	2917	1534	3019	1306					
CCl_4	458	218	776	314					
CH_3Cl	2937	1355	732	3042	1452	1017			
CH_3F	2930	1464	1049	3006	1467	1182			
CH_2Cl_2	2999	1467	717	282	1153	3040	898	1266	758
SiH_2Cl_2	2224	954	527	188	710	2237	602	876	590
$VOCl_3$	1042	408	163	502	246	125			
6-atoms									
PF_5	816	648	947	575	1024	533	174	520	
PCl_5	395	370	465	299	592	273	100	261	
VF_5	718	608	784	331	810	282	110	336	
CH_3CN	2965	2267	1400	920	3009	1454	1041	361	
7-atoms									
SF_6	772	642	932	613	522	344			
TeF_6	697	670	751	327	314	197			
$H_3CC\!:\!CH$	3334	2918	2142	1382	931	3008	1452	1053	633
							ω_{10} 328		
$F_3CC\!:\!CH$	3327	2165	1253	812	536	1179	686	612	453
							ω_{10} 171		

Table VIII.20 Potential Energy Barriers to Internal Rotation about Single Bonds
Reference: J. Dale, *Tetrahedron* (1966), 22, 3373.

Molecule	Bond	Barrier/kJ mol^{-1}
CH_3–CH_3	C–C	11.5
CH_3–CCl_3	C–C	11.3
CH_3–CF_3	C–C	14.4
CH_3CH_2–CH_3	C–C	13.8
$(CH_3)_3CH$	C–C	16.2
$(CH_3)_4C$	C–C	17.6
CH_3–$CH\!=\!CH_2$	C–C	8.8
$(CH_3)_2C\!=\!CH_2$	C–C	9.2
$CH_3CH\!=\!C\!=\!CH_2$	C–C	6.7
$CH_3C\!\equiv\!CCH_3$	C–C	0
CH_3–CHO	C–C	4.8
CH_3–OH	C–O	4.5

Molecule	Bond	Barrier/kJ mol^{-1}
$(CH_3)_2O$	C–O	11.4
$(CH_3)_2S$	C–S	8.4
CH_3SH	C–S	5.3
CH_3NH_2	C–N	8.1
$(CH_3)_2NH$	C–N	13.7
$(CH_3)_3N$	C–N	18.5
$(CH_3)_4Si$	C–Si	5.4
$(CH_3)_3SiH$	C–Si	7.7
$(CH_3)_2SiH_2$	C–Si	7.0
CH_3SiH_3	C–Si	7.0

Table VIII. 21 Nuclear Spins, Moments, Resonant Frequencies and Receptivities of Naturally Occurring Magnetically Active Isotopes

The table shows one nucleus for each element; many elements have several magnetically active isotopes. Some of importance are ^{6}Li, ^{10}B, ^{37}Cl, ^{79}Br, ^{115}Sn, ^{131}Xe, ^{201}Hg and ^{204}Tl.

μ is the magnetic moment, given in units of the nuclear magneton, μ_N; I the spin quantum number; Q the quadrupole moment (spin $\frac{1}{2}$ nuclei have no quadrupole moment); ν the resonant frequency; D^p the receptivity (the relative sensitivity figure, taking ^{1}H as unity, used to compare signal areas theoretically obtainable at a given magnetic field for different nuclei taking into account both the different magnetic moments and the different natural abundances).

For more information see: R.K. Harris and B.E. Mann, eds, *NMR and the Periodic Table* (1978), Academic Press, London.

Nuclide	I	μ/μ_N	$Q/10^{-28}$ m^2	ν/MHz	D^p	Natural abundance /%
^{0}n	$\frac{1}{2}$	−1.913 12		29.167		
^{1}H	$\frac{1}{2}$	2.792 68		42.575 9	1	99.985
^{2}H	1	0.857 387	0.002 73	6.535 66	1.5 × 10^{-6}	0.015
^{3}He	$\frac{1}{2}$	−2.127 4		32.433 8	5.8 × 10^{-7}	0.000 14
^{7}Li	$\frac{3}{2}$	3.256 36	−0.03	16.546	0.27	92.5
^{9}Be	$\frac{3}{2}$	−1.177 45	0.052	5.983 4	0.014	100
^{11}B	$\frac{3}{2}$	2.668 0	0.036	13.659 5	0.133	80.2
^{13}C	$\frac{1}{2}$	0.702 2		10.705 4	1.8 × 10^{-4}	1.11
^{14}N	1	0.403 75	0.016	3.075 6	1.0 × 10^{-3}	99.63
^{17}O	$\frac{5}{2}$	−1.893 7	−0.026	5.771 9	1.1 × 10^{-5}	0.038
^{19}F	$\frac{1}{2}$	2.627 27		40.054 3	0.83	100
^{21}Ne	$\frac{3}{2}$	−0.661 76	0.09	3.361 1	6.3 × 10^{-6}	0.27
^{23}Na	$\frac{3}{2}$	2.217 40	0.15	11.262 1	9.3 × 10^{-2}	100
^{25}Mg	$\frac{5}{2}$	−0.854 49	0.22	2.605 4	2.7 × 10^{-4}	10
^{27}Al	$\frac{5}{2}$	3.638 5	0.149	11.094 0	0.21	100
^{29}Si	$\frac{1}{2}$	−0.555 26		8.457 8	3.7 × 10^{-4}	4.67
^{31}P	$\frac{1}{2}$	1.131 7		17.235	6.6 × 10^{-2}	100
^{33}S	$\frac{3}{2}$	0.642 57	−0.064	3.265 4	1.7 × 10^{-5}	0.75
^{35}Cl	$\frac{3}{2}$	0.821 81	−0.079	4.171 7	3.6 × 10^{-3}	75.77
^{39}K	$\frac{3}{2}$	0.391 47	0.11	1.986 8	4.7 × 10^{-4}	93.26
^{43}Ca	$\frac{7}{2}$	−1.317 2	0.2	2.864 6	9.3 × 10^{-6}	0.135
^{45}Sc	$\frac{7}{2}$	4.755 9	−0.22	10.343 4	0.301	100
^{47}Ti	$\frac{5}{2}$	−0.788 46	0.29	2.399 7	1.5 × 10^{-4}	7.4
^{51}V	$\frac{7}{2}$	5.148 5	0.04	11.192 2	0.381	99.75
^{53}Cr	$\frac{3}{2}$	−0.473 54	−0.03	2.406 5	8.6 × 10^{-5}	9.5
^{55}Mn	$\frac{5}{2}$	3.468 0	0.55	10.554 2	0.175	100
^{57}Fe	$\frac{1}{2}$	0.090 60		1.375 8	7.4 × 10^{-7}	2.15
^{59}Co	$\frac{7}{2}$	4.616 3	0.40	10.072	0.277	100
^{61}Ni	$\frac{3}{2}$	−0.749 8	0.16	3.804 7	6.3 × 10^{-4}	1.13
^{63}Cu	$\frac{3}{2}$	2.226 2	−0.16	11.285	6.4 × 10^{-2}	69.2
^{67}Zn	$\frac{5}{2}$	0.875 6	0.15	2.663	1.2 × 10^{-4}	4.10
^{71}Ga	$\frac{3}{2}$	2.561 7	0.112	12.984 0	5.6 × 10^{-2}	39.9
^{73}Ge	$\frac{9}{2}$	−0.879 18	−0.2	1.485 2	1.1 × 10^{-4}	7.8
^{75}As	$\frac{3}{2}$	1.439	0.3	7.291 9	2.5 × 10^{-2}	100
^{77}Se	$\frac{1}{2}$	0.534		8.118	5.3 × 10^{-4}	7.6
^{81}Br	$\frac{3}{2}$	2.269 6	0.28	11.498	4.9 × 10^{-2}	49.31
^{83}Kr	$\frac{9}{2}$	−0.970 3	0.15	1.638 0	2.2 × 10^{-4}	11.5
^{87}Rb	$\frac{3}{2}$	2.750 6	0.13	13.931	4.9 × 10^{-2}	27.83
^{87}Sr	$\frac{9}{2}$	−1.089 3	0.2	1.845 2	1.9 × 10^{-4}	7.0
^{89}Y	$\frac{1}{2}$	−0.137 33		2.086 0	1.2 × 10^{-4}	100
^{91}Zr	$\frac{5}{2}$	−1.302 84	?	3.957 8	1.0 × 10^{-3}	11.2
^{93}Nb	$\frac{9}{2}$	6.167	−0.2	10.404 8	0.482	100
^{95}Mo	$\frac{5}{2}$	−0.913 5	0.12	2.774	5.1 × 10^{-4}	15.9
^{101}Ru	$\frac{5}{2}$	−0.72	?	2.1	1.7 × 10^{-4}	17.0
^{103}Rh	$\frac{1}{2}$	−0.088 3		1.343 4	3.1 × 10^{-5}	100
^{105}Pd	$\frac{5}{2}$	−0.639	?	1.892 1	6.2 × 10^{-5}	22.2
^{109}Ag	$\frac{1}{2}$	−0.130 5		1.980 8	4.9 × 10^{-5}	48.17
^{111}Cd	$\frac{1}{2}$	−0.595 00		9.028 3	1.2 × 10^{-3}	12.8
^{115}In	$\frac{9}{2}$	5.534 8	1.16	9.331 0	0.332	95.7
^{119}Sn	$\frac{1}{2}$	−1.046 1		15.869	4.4 × 10^{-3}	8.6
^{121}Sb	$\frac{5}{2}$	3.359 2	−0.5	10.189	9.2 × 10^{-2}	57.3
^{125}Te	$\frac{1}{2}$	−0.887 2		13.453	2.2 × 10^{-3}	7.0
^{127}I	$\frac{5}{2}$	2.809 1	−0.7	8.518 3	9.3 × 10^{-2}	100

(Continued)

Table VIII.21 (Continued)

Nuclide	I	μ/μ_N	$Q/10^{-28}\text{m}^2$	ν/MHz	D^p	Natural abundance /%
^{129}Xe	1/2	−0.776 8		11.777	5.6×10^{-3}	26.4
^{133}Cs	7/2	2.578 8	−0.003	5.584 69	4.7×10^{-2}	100
^{137}Ba	3/2	0.936 5	0.2	4.731 5	8.0×10^{-4}	11.2
^{139}La	7/2	2.778	0.21	6.014 4	5.9×10^{-2}	99.911
^{141}Pr	5/2	4.16	−0.059	12.5	0.26	100
^{143}Nd	7/2	−1.063	−0.48	2.303	3.2×10^{-4}	12.2
^{147}Sm	7/2	−0.813	−0.208	1.72	1.3×10^{-4}	15.1
^{151}Eu	5/2	3.466	1.16	10.35	8.7×10^{-2}	47.9
^{157}Gd	3/2	−0.338 8	2	1.9	5.0×10^{-5}	15.7
^{159}Tb	3/2	2.008	1.3	9.7	3.0×10^{-2}	100
^{163}Dy	5/2	0.676	1.6	1.94	1.6×10^{-4}	24.9
^{165}Ho	7/2	4.12	2.82	8.73	0.10	100
^{167}Er	7/2	−0.569	2.83	1.23	7.1×10^{-5}	22.9
^{169}Tm	1/2	−0.231		3.47	5.7×10^{-4}	100
^{171}Yb	1/2	0.493 0		7.456	7.8×10^{-4}	14.4
^{175}Lu	7/2	2.230	5.7	4.818 9	4.8×10^{-2}	97.39
^{177}Hf	7/2	0.790 2	3	1.3	1.2×10^{-4}	18.6
^{181}Ta	7/2	2.35	3	5.096	3.6×10^{-2}	99.99
^{183}W	1/2	0.116 9		1.771 6	1.0×10^{-5}	14.3
^{187}Re	5/2	3.204	2.6	9.683 7	8.6×10^{-2}	62.6
^{189}Os	3/2	0.656 5	0.8	3.303 4	3.8×10^{-4}	16.1
^{193}Ir	3/2	0.158 3	1.5	0.9	2.6×10^{-5}	62.7
^{195}Pt	1/2	0.602 2		9.152 3	3.4×10^{-3}	33.8
^{197}Au	3/2	0.144 86	0.59	0.741 2	2.5×10^{-5}	100
^{199}Hg	1/2	0.504 15		7.612	9.5×10^{-4}	16.8
^{205}Tl	1/2	1.627		24.567	0.14	70.5
^{207}Pb	1/2	0.588 1		8.898	2.1×10^{-3}	22.1
^{209}Bi	9/2	4.080	−0.4	6.842	0.137	100
^{235}U	7/2	−0.43	4.1	0.75	8.5×10^{-7}	0.72

Table VIII.22 Chemical Shifts of Some Common NMR Solvents

[a]: residual protons; [b]: $H_2O = 0$; [c]: $NO_3^- = -2.6$.
Reference: as for Table VIII.21.

Solvent	δ/ppm			
	^{1}H[a]	^{13}C	^{14}N, ^{15}N	^{17}O
CCl$_4$		96.7		
CD$_3$COOD	2.1	21.1		
	4.76	177.3		
(CD$_3$)$_2$CO	2.2	30.2		523[b]
		205.1		
CD$_3$CN	2.0	0.3	−137.6[c]	
		117.2		
C$_6$D$_6$	7.4	128.7		
CD$_3$Cl	7.3	77.6		
CD$_2$Cl$_2$	5.3	54.2		
(CD$_3$)$_2$SO	2.6	39.5		
CD$_3$OD	3.5	49.3		−34[b]
CD$_3$NO$_2$	4.3	57.3	0[c]	

Table VIII.23 Ranges of ¹H Chemical Shifts of Some Common Functional Groups

For more detailed information see: N.F. Chamberlain, *The Practice of NMR Spectroscopy—with Spectra-structure Correlations for Hydrogen-1* (1974), Plenum Press, New York.

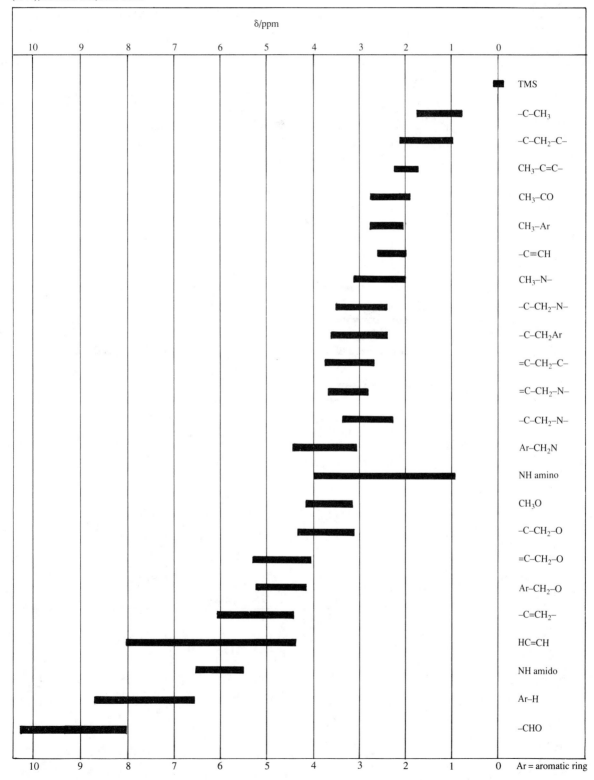

Ar = aromatic ring

Table VIII.24 Ranges of ^{13}C Chemical Shifts in Functional Groups
For more details see: W. Bremser, B. Franke and H. Wagner, *Chemical Shift Ranges in ^{13}C NMR Spectroscopy* (1982), VCH, Cambridge, UK.

$\delta(Me_4Si) = 0$.

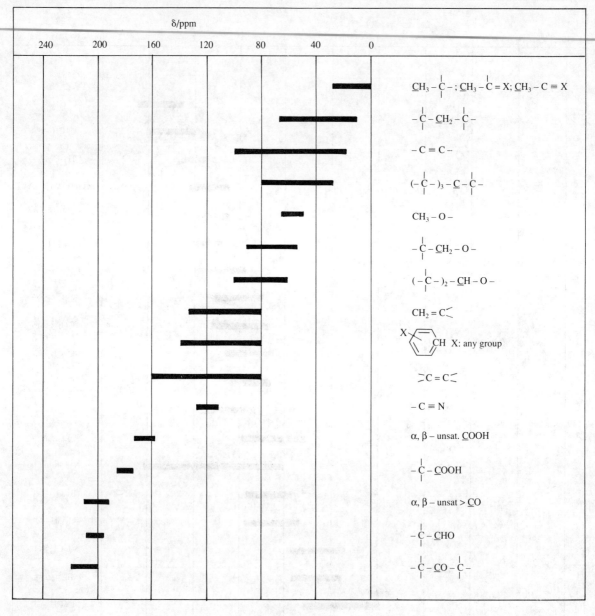

Table VIII.25 Ranges of ^{15}N Chemical Shifts in Some Common Functional Groups
For additional information see: G.C. Levy and R.L. Lichter, *^{15}N nmr Spectroscopy* (1979), Wiley, Chichester.

$\delta(MeNO_2) = 0$.

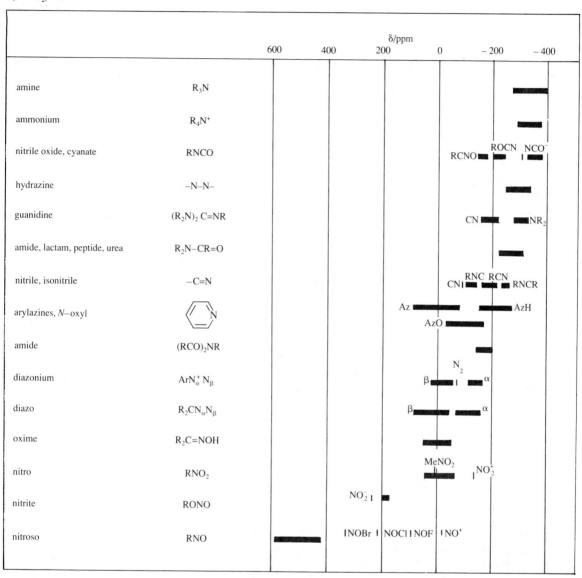

Table VIII.26 Some Chemical Shifts of ^{17}O, ^{19}F, ^{27}Al, ^{31}P, and ^{35}Cl for Substances Indicated
Reference: as for Table VIII.21.

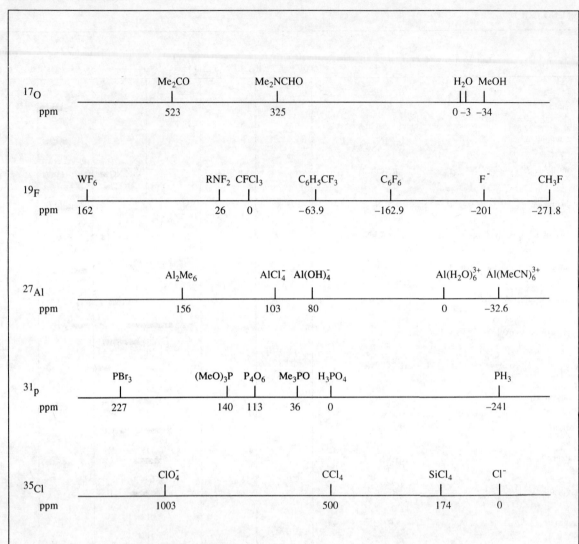

Table VIII.27 Mass composition of Most Useful Fragment Ions to Mass 99
Reference: V.G. Instruments Inc (1990), Danver, Mass.

m/z	Groups associated with loss	Possible inference
15	CH_3^+	
18	$H_2O^{+\cdot}$	
26	$C_2H_2^{+\cdot}$	hydrocarbon
27	$C_2H_3^+$	
28	CO^+; $C_2H_4^{+\cdot}$; $N_2^{+\cdot}$	carbonyl; ethyl; azo compounds
29	CHO^+; $C_2H_5^+$	aldehyde; ethyl compound
30	$CH2=NH_2^+$	primary amine
31	$CH_2=OH^+$	primary alcohol
35/37 (3:1)	Cl^+	chloro compounds
36/38 (3:1)	$HCl^{+\cdot}$	chloro compounds

(Continued)

Table VIII.27 (Continued)

m/z	Groups associated with loss	Possible inference
41	$C_3H_5^+$	hydrocarbon
42	$CH_2=C=O^{+\cdot}$; $C_3H_6^{+\cdot}$	acetates; hydrocarbon
43	CH_3CO^+	CH_3COX
43	$C_3H_7^+$	C_3H_7X
44	$CH_3CH=NH_2^+$	some aliphatic amines
44	$O=C=NH_2^+$	primary amides
44	$CO_2^{+\cdot}$; $C_3H_8^{+\cdot}$	
44	$CH_2=CH(OH)^{+\cdot}$	some aldehydes
45	$CH_2=OCH_3^+$	some ethers and alcohols
45	$CH_3CH=OH^+$	some ethers and alcohols
47	$CH_2=SH^+$	aliphatic thiol
49/51 (3 : 1)	CH_2Cl^+	chloromethyl compounds
50	$C_4H_2^{+\cdot}$	aromatic compound
51	$C_4H_3^+$	C_6H_5X
55	$C_4H_7^+$	unsaturated hydrocarbon
56	$C_4H_8^{+\cdot}$	hydrocarbon
57	$C_4H_9^+$	C_4H_9X
57	$CH_3CH_2CO^+$	ethyl ketone, propanoate ether
58	$CH_2=C(OH)CH_3^{+\cdot}$	some methyl ketones
58	$CH_2=C(OH)CH_3^{+\cdot}$	some dialkyl ketones
58	$(CH_3)_2N=CH_2^+$	some aliphatic amines
59	$COOCH_3^+$	methyl ester
59	$CH_2=C(OH)NH_2^{+\cdot}$	some primary amides
59	$CH_2=OC_2H_5^+$ and isomers	some ethers
61	$CH_3CO(OH_2)^+$	$CH_3COOC_nH_{2n+1}$ ($n>1$)
61	$CH_2CH_2SH^+$	aliphatic thiol
66	$H_2S_2^{+\cdot}$	dialkyl sulphide
68	$CH_2CH_2CH_2CN^+$	monosubstituted pyrrole
69	CF_3^+	
69	$C_5H_9^+$	hydrocarbon
71	$C_5H_{11}^+$	$C_5H_{11}X$
72	$CH_2=C(OH)C_2H_5^{+\cdot}$	some ethyl alkyl ketones
72	$C_3H_7CH=NH_2^+$ and isomers	some amines
73	$C_4H_9O^+$	
73	$(CH_3)_3Si^+$	$(CH_3)_3SiX$
74	$CH_2=C(OH)OCH_3^{+\cdot}$	some methyl esters
75	$(CH_3)_2Si=OH^+$	$(CH_3)_3SiOX$
75	$C_2H_5CO(OH_2)^+$	$C_2H_5COOC_nH_{2n+1}$ ($n>1$)
77	$C_6H_5^+$	C_6H_5X
78	$C_6H_6^{+\cdot}$	C_6H_5X
78	$C_5H_4N^+$	monosubstituted pyridines
80/82 (1 : 1)	$HBr^{+\cdot}$	bromo compounds
80	$C_5H_6N^+$	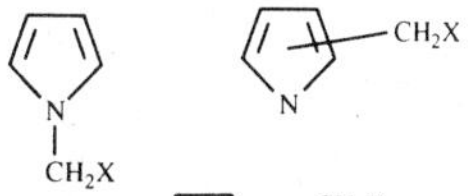
81	$C_5H_5O^+$	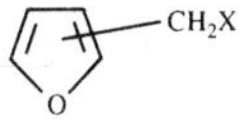
85	$C_6H_{13}^+$	$C_6H_{13}X$
85	$C_4H_9CO^+$	C_4H_9COX
85	$C_5H_9CO^+$	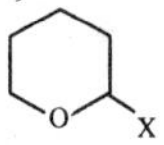
85	$C_4H_5O_2^+$	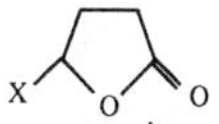
86	$C_4H_9CH=NH_2^+$ and isomers	some amines
87	$CH_2=CHC(=OH)OCH_3^+$	$XCH_2CH_2COOCH_3$
91	$C_7H_7^+$	$C_6H_5CH_2X$; $CH_3C_6H_4X$
92	$C_7H_8^{+\cdot}$	$C_6H_5CH_2$alkyl

(Continued)

Table VIII.27 (Continued)

m/z	Groups associated with loss	Possible inference
92	$C_6H_6N^+$	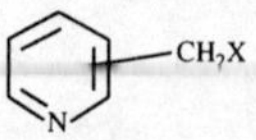 pyridine-CH_2X
91/93 (3 : 1)	$C_4H_8Cl^+$	n-alkyl chloride ($\geq$ hexyl)
93/95 (1 : 1)	$BrCH_2^+$; $BrCH_2^+$	$BrCH_2X$
94	$C_6H_6O^{+\cdot}$	C_6H_5O-alkyl (alkyl $\neq CH_3$)
95	$C_5H_3O_2^+$	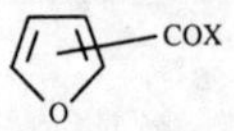 furan-COX
95	$C_6H_7O^+$	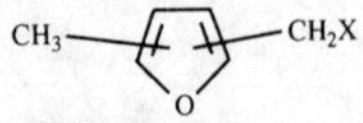 CH_3-furan-CH_2X
97	$C_5H_5S^+$	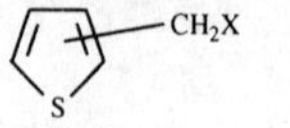 thiophene-CH_2X
99	$C_7H_{15}^+$	$C_7H_{15}X$
99	$C_5H_7O_2^+$	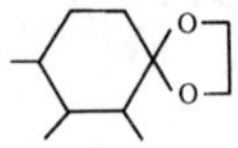 cyclohexane dioxolane
99	$C_5H_7O_2^+$	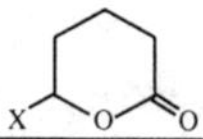 X-lactone

Table VIII.28 Common Losses from Molecular Ions
Reference: as for Table VIII.27.

Ion	Groups associated with loss	Possible inference
M − 1	H	
M − 2	H_2	
M − 15	CH_3	
M − 16	O	$ArNO_2$, $\geq NO$, sulphoxide
M − 16	NH_2	$ArSO_2NH_2$, -$CONH_2$
M − 17	OH	
M − 17	NH_3	
M − 18	H_2O	alcohol, aldehyde, ketone etc.
M − 19	F	fluorides
M − 20	HF	fluorides
M − 26	C_2H_2	aromatic hydorcarbon
M − 27	HCN	aromatic nitriles
M − 27	HCN	nitrogen heterocycles
M − 28	CO	quinones
M − 28	C_2H_4	aromatic ethyl ethers
M − 28	C_2H_4	ethyl esters, n-propyl ketones
M − 29	CHO	ethyl ketones
M − 29	C_2H_5	ethyl ketones, Ar-n-C_3H_7
M − 30	C_2H_6	
M − 30	CH_2O	aromatic methyl ether
M − 30	NO	Ar-NO_2
M − 31	OCH_3	methyl ester
M − 32	CH_3OH	methyl ester
M − 32	S	
M − 33	$H_2O + CH_3$	
M − 33	HS	thiols
M − 34	H_2S	thiols
M − 41	C_3H_5	propyl ester
M − 42	CH_2CO	methyl ketone
M − 42	CH_2CO	aromatic acetate, $ArNHCOCH_3$
M − 42	C_3H_6	n- or iso-butyl ketone
M − 42	C_3C_6	aromatic propyl ether, Ar-n-C_4H_9
M − 43	C_3H_7	propyl ketone, Ar-n-C_4H_9
M − 43	CH_3CO	methyl ketone

(Continued)

Table VIII.28 (Continued)

Ion	Groups associated with loss	Possible inference
M – 44	CO_2	ester, anhydride
M – 44	C_3H_8	
M – 45	COOH	carboxylic acid
M – 45	OC_2H_5	ethyl ester
M – 46	C_2H_5OH	ethyl ester
M – 46	NO_2	Ar-NO_2
M – 48	SO	aromatic sulphoxide
M – 55	C_4H_7	butyl ester
M – 56	C_4H_8	Ar-n-C_5H_{11}, ArO-n-C_4H_9
M – 56	C_4H_8	Ar-iso-C_5H_{11}, ArO-iso-C_4H_9
M – 56	C_4H_8	pentyl ketone
M – 57	C_4H_9	butyl ketone
M – 57	C_2H_5CO	ethyl ketone
M – 58	C_4H_{10}	
M – 60	CH_3COOH	acetate

Table VIII.29 Recommended Fluorescent Quantum Yield References in various Emission Ranges
For more extensive date see: D.F. Eaton, *Pure and App. Chem.* (1988), **60**, 1107.
All concentration less than 10^{-5} mol dm^{-5} (absorbance < 0.1), temperature 20 ± 1 °C (except for tryptophan 25 °C).

Region/nm	Compound	Solvent	ϕ_f	(Uncertainty)
270 – 300	benzene	cyclohexane	0.05	(0.02)
300 – 380	tryptophan	water (pH 7.2)	0.14	(0.02)
300 – 400	naphthalene	cyclohexane	0.23	(0.02)
315 – 480	2-aminopyridine	0.05 mol dm^{-3} H_2SO_4	0.60	(0.05)
360 – 480	anthracene	ethanol	0.27	(0.03)
400 – 500	9,10-diphenyl-anthracene	cyclohexane	0.90	(0.02)
400 – 600	quinine bisulphate	0.5 mol dm^{-3} H_2SO_4	0.546	
600 – 650	rhodamine 101	ethanol	1.0	(0.02)
600 – 650	cresyl violet	methanol	0.54	(0.02)

© (1988) IUPAC

Section IX
Atomic and Nuclear Physics

		Page
IX.1	Ionic mobilities of various atomic and molecular ions in some gases at standard pressure and, unless otherwise indicated, standard temperature	171
IX.2	Electronic drift speeds in some gases for various values of applied field strength per unit pressure	172
IX.3	Variation with pressure of the ionic recombination coefficient for air at temperature 291 K	172
IX.4	Variation with pressure and temperature of the ionic recombination coefficient for oxygen	172
IX.5	Ionic recombination coefficient for some gases at stated temperatures and pressures	173
IX.6	Variation of ionic recombination coefficient for air, at temperature 288 K and standard pressure, with number of ions per cubic metre	173
IX.7	Variation of the ratio of mean agitation energy of a free electron in a gas to that of a gas molecule with applied field strength per unit pressure, for various gases	173
IX.8	Variation with electron energy of ratio of total electron-atom (or -molecule) collision cross-section area to area of first Bohr orbit of hydrogen atom, for various gases	173
IX.9	Work function of various solid elements obtained from contact potential difference and/or photoelectric and/or thermionic measurements	174
IX.10	Auger-electron kinetic energies resulting from the most intense transitions of some common elements	175
IX.11	Kinetic energy of photoelectrons ejected from atomic energy levels 1s to 4 $d_{5/2}$ due to excitation with $AlK\alpha_{1,2}$ radiation, energy 1486.6 eV, for elements of atomic numbers 1–92	176
IX.12	Kinetic energy of photoelectrons ejected from atomic energy levels $4f_{5/2}$ to $6d_{5/2}$ due to excitation with $AlK\alpha_{1,2}$ radiation energy 1486.6 eV, for elements of atomic numbers 49–92	178
IX.13	Kinetic energy of photoelectrons ejected from atomic energy levels 1s to $4d_{5/2}$ due to excitation with $MgK\alpha_{1,2}$ radiation energy 1253.6 eV, for elements of atomic numbers 1–92	179
IX.14	Kinetic energy of photoelectrons ejected from atomic energy levels $4f_{5/2}$ to $6d_{5/2}$ due to excitation with $MgK\alpha_{1,2}$ radiation energy 1253.6 eV, for elements of atomic numbers 49–92	181
IX.15	Energy values of X-ray absorption edges and of characteristic X-radiation types for K series, for elements of atomic numbers 4–103	182
IX.16	Energy values of X-ray absorption edges and of characteristic X-radiation types for L series, for elements of atomic numbers 4–103	184
IX.17	Energy values of X-ray absorption edges and of characteristic X-radiation types for M series, for elements of atomic numbers 29–103	186
IX.18	Approximate intensities of some characteristic X-radiations relative to that of the most intense of its series	187
IX.19	Variation with photon energy of values of mass absorption coefficient for various elements and concrete for a collimated incident photon beam which emerges undeflected without photon energy loss	188
IX.20	Inelastic mean free path of electrons, usually of 1 keV energy, in various substances, and n values	189
IX.21	Range in aluminium of electrons for various electron energies	189
IX.22	Range in various substances of protons for various proton energies	190
IX.23	Range in various substances of alpha particles for various particle energies	190
IX.24	Range in various substances of pions for various pion energies	190
IX.25	Range in various substances of muons for various muon energies	191
IX.26	Range in aluminium of some heavy ions for various ion energies	191
IX.27	Range in carbon of some heavy ions for various ion energies	191
IX.28	Range in germanium of some heavy ions for various ion energies	192
IX.29	Range in gold of some heavy ions for various ion energies	192
IX.30	Range in silver of some heavy ions for various ion energies	192
IX.31	Range in titanium of some heavy ions for various ion energies	193
IX.32	Range in uranium of some heavy ions for various ion energies	193
IX.33	Mass stopping power in various substances for protons of various energies	193
IX.34	Mass stopping power in various substances for alpha particles of various energies	194
IX.35	Mass stopping power in various substances for pions of various energies	194
IX.36	Mass stopping power in various substances for muons of various energies	194
IX.37	Air kerma-rate constant for gamma radiation of energy greater than 50 keV for some common radionuclides	195
IX.38	International Commission on Radiation Protection risk and risk-weighting factors of fatal cancer and serious hereditary defect-inducing radiation in adult humans for a dose equivalent of 1 sievert	195
IX.39	Values of atomic number, mass number, half-life and/or natural abundance, decay scheme, mass excess, cross section, parity and spin for various nuclides	196
IX.40	Average values of the components of the total energy released by slow-neutron induced fission in some fission materials	263
IX.41	Half-life of principal activity, principal resonance energy, average activation cross section and resonance integral for some neutron foil resonance detectors	263

IX.42	Half-life of principal activity, mean response energy in fission spectrum, average activation cross section and resonance integral for some neutron foil threshold detectors	263
IX.43	Average values, for neutron thermal spectrum, of cross section for elastic scattering and for radioactive capture by cladding and structural elements in nuclear reactors	264
IX.44	Values, for neutron slowing down region, of cross section for elastic scattering by, and of capture resonance integral for, cladding and structural elements in nuclear reactors	264
IX.45	Average values, for neutron fission spectrum, of cross section for elastic scattering, for absorption, and for inelastic scattering by cladding and structural elements in nuclear reactors	264
IX.46	Average values, for neutron thermal spectrum, of cross section for elastic scattering and for absorption for coolant components in nuclear reactors	264
IX.47	Values, for neutron slowing down region, of cross section for elastic scattering by, and of absorption resonance integral for, coolant components in nuclear reactors	264
IX.48	Average values, for neutron fission spectrum, of cross section for elastic and inelastic scattering and for radiative capture by coolant components in nuclear reactors	265
IX.49	Average values, for neutron thermal spectrum, of cross section for elastic scattering and for absorption for nuclide or element poisons and absorbers in nuclear reactors	265
IX.50	Values, for neutron slowing down region, of cross section for elastic scattering by, and of absorption resonance integral for, nuclide or element poisons and absorbers in nuclear reactors	265
IX.51	Average values, for neutron thermal spectrum, of number of neutrons produced per fission, of cross section for elastic scattering, for radiative capture and for fission by fissile and fertile nuclides in nuclear reactors	265
IX.52	Values, for neutron slowing down region, of cross section of elastic scattering by, and of capture and fission resonance integrals for, fissile and fertile nuclides in nuclear reactors, together with average number of neutrons per fission	266
IX.53	Average values, for neutron fission spectrum, of number of neutrons produced per fission, of cross section for elastic and inelastic scattering, for radiative capture and for fission by fissile and fertile nuclides in nuclear reactors	266
IX.54	Average values, for core and blanket, of number of neutrons produced per fission and of neutron fission cross section for nuclides in a typical fast nuclear reactor fuelled by plutonium-uranium oxide and with liquid metal coolant	266
IX.55	Thermal diffusion length and average diffusion coefficient over thermal neutron spectrum for neutrons in some moderators and shielding materials at temperature 293 K, together with mean thermal-neutron lifetimes	266
IX.56	Some fusion reactions of possible significance for electrical energy generation	267
IX.57	Half-life of principal activity, threshold response energy and activation cross section for 14 MeV fusion neutrons for some neutron foil detectors	267
IX.58	The more important radionuclides produced in each chemical element likely to occur in the structural materials of a fusion-reactor vacuum vessel and breeding blanket	267
IX.59	Values of average reaction cross section over energy spectrum of fusion neutrons emitted from a deuterium-tritium plasma at 20 KeV ion temperature, for various target elements and reaction types	268
IX.60	Charge, mass, mean life and spin for various leptons together with their principal decay modes	268
IX.61	Baryon number, charge, mass and Z-component of isospin for the various quarks together with name and value of other quantum numbers	268
IX.62	Baryon number, mass, mean life, quark content and spin of non-strange hadrons together with their principal decay modes	269
IX.63	Baryon number, strangeness, mass, mean life, quark content and spin of strange hadrons together with their principal decay modes	269
IX.64	Baryon number, charm, strangeness, mass, mean life, quark content and spin of charmed hadrons together with their principal decay modes	269
IX.65	Mass, spin and width of beautyonium states together with their $b\bar{b}$ configuration and principal modes of decay	270
IX.66	Mass, spin and width of charmonium states together with their $c\bar{c}$ configuration and principal modes of decay	270
IX.67	Charge, mass and width of gauge bosons of spin 1 together with the force transmitted and principal decay modes	270

Section IX
Atomic and Nuclear Physics

Table IX.1 Ionic mobilities of various atomic and molecular positive ions in some gases at standard pressure and, unless otherwise indicated, standard temperature

Ion[†]	Gas										
	Ar	CH_4	CO	CO_2	H_2	He	Kr	N_2	Ne	O_2	Xe
	mobility/10^{-4} m^2 s^{-1} V^{-1}										
Ar (a)	1.53										
Ar (m)	2.3										
CH_4 (m)		3.5									
CO (m)			1.6								
*Cs (a)	2.23		1.98		13.4	19.1	1.42	2.25	6.5		0.99
H (a)					16						
H (m)					11.3						
He (a)						10.2					
He (m)						20.3					
*K (a)	2.81		2.32	1.45	14	22.7	1.98	2.7	8		1.44
Kr (a)							0.9				
Kr (m)							1.2				
*Li (a)	4.99		2.63		13.3	25.6	4	4.21	12		3.04
N (a)								3.3			
N (m)								1.8			
*Na (a)	3.23		2.44	1.63	15	23.4	2.34	3.4	8.7		1.8
Ne (a)									4.2		
Ne (m)									6.5		
O (m)										2.2	
*Rb (a)	2.4		2.08	1.23	13.4	21.2	1.59	2.39	7.18		1.1
Xe (a)											0.6
Xe (m)											0.8

† (a) and (m) signify atomic and molecular respectively
* Values at temperature 293 K
Reference: Ellis *et al*, *Atomic Data and Nuclear Data Tables,* vol. 17, 1976
   McDaniel, *Collision Phenomena in Ionized Gases,* Wiley, 1964

Table IX.2 Electronic drift speeds in some gases for various values of applied field strength per unit pressure

Gas	Field strength per unit pressure/V m N^{-1}										
	0.075	0.15	0.3	0.45	0.6	0.75	1.5	3	4.5	7.5	15
	Drift speed / 10^3 m s^{-1}										
air	3.5	5	8	9.3	11	12	17	26	34	51	88
Ar	2.3	2.7	3.2	3.5	3.9	4.1	6.2	13	30		
*Ar/CF$_4$	34	63	100	120	120	111	66				
*Ar/CH$_4$	49	55	45	36	32	30	26	24			
*Ar/C$_2$H$_2$	14.3	27.3	44.3	49.3	48.3	45.8	45.2				
*Ar/N$_2$	4	6	9	13	16.8	19.3	27.3				
BF$_3$	0.13	0.25	0.5	0.75	1	1.3	2.5	5	7.5	12.5	25
CH$_4$	8	24	60	80	95	100	100				
CO	4.8	6.5	9.4	11	13	14	17	20	25	29	
CO$_2$	0.56	1.1	2.3	3.4	4.6	5.7	11	33	66	109	137
H$_2$	3.1	5	6.8	8	9.2	10	15	23	30	42	70
He	2.7	4	6	7	8.5	9.6	15	36			
Kr	1.5	1.8	2.1	2.4	2.5	3.1					
N$_2$	3.1	3.9	5	6	7	8	14	22	30	43	80
Ne	4	5	7	10	12	16	26	52			
Xe	1	1.2	1.4	1.6	1.7	1.8					

*By volume, mixture contains 90% Ar
Reference: Pack *et al, Phys. Rev.* vol.127, 2084, 1962
Christophorou *et al, Nucl. Inst. and Methods*, vol. 163, 141, 1978.

Table IX.3 Variation with pressure of the ionic recombination coefficient for air at temperature 291 K

pressure/ 10^5 N m^{-2}	coefficient/ 10^{-12} m^3 s^{-1}	pressure/ 10^5 N m^{-2}	coefficient/ 10^{-12} m^3 s^{-1}	pressure/ 10^5 N m^{-2}	coefficient/ 10^{-12} m^3 s^{-1}
0.05	0.12	0.7	1.22	8	0.698
0.1	0.21	0.8	1.4	9.7	0.594
0.2	0.37	0.9	1.55	12.5	0.48
0.3	0.55	1	1.72	15.1	0.407
0.4	0.7	1.01	1.65	17.8	0.344
0.5	0.88	4.6	1.058	21.1	0.29
0.6	1.05	6.4	0.844	25.7	0.237

Reference: Loeb, *Basic Processes of Gaseous Electronics,* California University Press, 1955

Table IX.4 Variation with pressure and temperature of the ionic recombination coefficient for oxygen

Pressure/ 10^5 N m^{-2}	temperature/ K	coefficient/ 10^{-12} m^3 s^{-1}	pressure/ 10^5 N m^{-2}	temperature/ K	coefficient/ 10^{-12} m^3 s^{-1}
0.1	293	0.7	0.9	293	2.08
0.2	293	1.3	1.01	200	3.7
0.3	293	1.6	1.01	250	2.7
0.4	293	1.75	1.01	295	1.32
0.5	293	1.88	1.01	298	2.08
0.6	293	1.96	1.01	300	1.9
0.7	293	2.01	1.01	350	1.5
0.8	293	2.05	1.01	400	1.2

Reference: as for Table IX.3

Table IX.5 Ionic recombination coefficient for some gases at stated temperatures and pressures

Gas	pressure / 10^5 N m^{-2}	temperature/ K	coefficient/ 10^{-12} m^3 s^{-1}
Ar	1.01	295	1.06
CO_2	4.6	291	5.76
CO_2	7.9	291	3.58
CO_2	12.5	291	2.24
CO_2	17.4	291	1.56
CO_2	22	291	1.24
H_2	1.01	295	0.28
N_2	1.01	295	1.06

Reference: as for Table IX.3

Table IX.6 Variation of ionic recombination coefficient for air, at temperature 288 K and standard pressure, with number of ions per cubic metre

Number/10^{12}	1.5	2.7	3.1	4.1	4.4
coefficient/10^{-12} m^3 s^{-1}	2.65	2.5	2.5	2.3	2.3

Reference: as for Table IX.3

Table IX.7 Variation of the ratio of mean agitation energy of a free electron in a gas to that of a gas molecule with applied field strength per unit pressure, for various gases

Gas	field strength per unit pressure/V m N^{-1}									
	0.5	1	2	3	4	5	10	20	30	40
	Energy ratio									
air	7	15	26	34	39	42	50	68	90	106
Ar	220	320	316	312	310	314	324			
Br_2				40	52	59	74	82	89	91
CO	5	8	14	19	23	27	38	48	68	
CO_2	1.3	1.6	2	4	9	17	58	89	117	
Cl_2							53	71	77	82
H_2	6	10	18	23	28	32	55	100	130	150
He	37	74	126	152	176					
I_2							27	76	91	99
N_2	16	25	34	39	42	44	52	68	89	114
NO	7	8	10	11	12	14	38			
O_2	11	18	32	41	47	52	68	96	133	

Reference: Huxley et al, Aus. J. Phys. vol.12, 303, 1959

Table IX.8 Variation with electron energy of ratio of total electron-atom (or -molecule) collision cross-section area to area of first Bohr orbit of hydrogen atom, for various gases
Area of first Bohr orbit of hydrogen atom has the value 0.88×10^{-20} m^2.

Gas	Electron Energy/eV														
	0.25	1	2.25	4	9	16	25	36	49	64	81	100	200	300	400	500
	Ratio†															
Ar	0.6	0.6	4	8.5	21	22	15	11	8 (3.6)	7.5	7	6.5 (3.3)	(2.7)	(2.4)	(1.9)	(1.7)
CO	15	14	38	18	14	14	14		(2.9)			(3.5)	(3)	(2.4)	(2)	(1.8)
Cd				54	43	40	41.5	39		34	31	29				
Cs		470	675	350	300	230	160	135	100	80	67	46				
H						6	7.6	8.6	9							
H_2		15	17	16	10.5	7	5.5	4	3.5 (1.1)			(1.1)	(0.8)	(0.66)	(0.56)	(0.48)
He	6.3	6.3	6.2	5.6	4.6	3.5	2.6	2	1.7 (0.23)			(0.4)	(0.34)	(0.28)	(0.25)	(0.23)
Hg		80	70	55	23	18	18	20	20 (6.6)	18.5	18	17.5 (5.7)	(4)	(3.15)	(2.6)	
K		450	420	270	220	160	125	80	60	50	50	40				
Kr	2	1	6	15	30	29	20	14	11.5	9	8	7.5				
N_2	17	10	29	13	12	12.5	14	13	11.5 (2.5)			(3.3)	(2.9)	(2.4)	(2.1)	(1.8)
Na		280	380	180	130	110	85	65	50	48	46	35				
Ne	1	1.9	2.5	2.8	3.5	3.7	3.7	3.6	3.5 (0.46)	3.4	3	2.9 (0.85)	(0.89)	(0.77)	(0.68)	(0.61)
NO									(2.9)			(3.7)	(3.2)	(2.7)	(2.5)	(2)
O_2	6	7	7.5	8	11	12	12	12	(2.5)			(3.3)	(3)	(2.5)	(2.1)	(1.8)
Rb		374	390	329	220	170	142	85	68	51	45	43				
Tl		6	10	16	12.5	11	9.5	9	8.8	8	7	6.2				
Xe	7.5	2	15	33	43	32.5	22.5	16								
Zn				80	45	35	27	24	23		23	23	21.5			

†Figures in brackets refer to an ionizing collision in which the first electron is removed
Reference: Massey and Burhop, *Electronic and Ionic Impact Phenomena*, OUP, 1952

Table IX.9 Work function of various solid elements obtained from contact potential difference and /or photoelectric and/or thermionic measurements

Solid	Method			Solid	Method		
	contact p.d.	photo-electric	thermionic		contact p.d.	photo-electric	thermionic
	work function/eV				work function/eV		
Ag	4.29	4.26		Mn		4.08	
Al	4.19	4.28		Mo	4.21	4.49	4.33
As		4.79		Na	2.46	2.36	
Au	5.28	5.1		Nb	4.37		4.3
Ba	2.35	2.52		Ni	5.25	5.15	5.24
Be	3.91			Pb	3.83	4.25	
Bi		4.34		Pd		5.4	
C	4.65–5			Pt		5.63	5.36
Ca		2.87		Rb		2.05	
Cd	4.22			Re			4.72
Co		4.97		Rh			4.72
Cr		4.44	4.6	Ru	4.73	4.71	
Cs	1.82	1.95		Sb		4.56	
Cu	4.51	4.65		Si	4.75	4.95	
Fe	4.16	4.6		Sn	4.43	4.28	
Ga		4.35		Ta	4.22	4.3	4.33
Ge	4.83	5.15		Th	3.71		
Hf			3.65	Ti	4.2	4.33	4.1
In		4.08		U	3.63	3.47	3.47
Ir			4.57	W	4.55	4.55	4.55
K	2.01	2.3		Zn	4.11	3.63	
Li	2.32			Zr			4
Mg	3.61						

Reference: Rivière, *Solid State Surface Science*, vol.1, Dekker, 1969

Table IX.10 Auger-electron kinetic energies resulting from the most intense transitions of some common elements

Element	energy/eV	transition†	Element	energy/eV	transition†	Element	energy/eV	transition†
Ag	48	N_3VV	Fe	44	$M_{2,3}VV$		116	$N_7O_{4,5}V$
	56	N_2VV		594	$L_3M_{2,3}M_{2,3}$		246	$N_5N_7O_{4,5}$
	270	$M_{4,5}N_1V$		647	$L_3M_{2,3}V$		265	$N_4N_7O_{4,5}$
	308	$M_{4,5}N_{2,3}V$		700	L_3VV	Pt	13	N_7O_3V
	362	$M_{4,5}VV$	Ge	24	$M_{4,5}VV$		45	O_3VV
Al	66	$L_{2,3}VV$		44	$M_3M_{4,5}M_{4,5}$		68	N_7VV
	1380	$KL_{2,3}L_{2,3}$		48	$M_2M_{4,5}M_{4,5}$		160	$N_5N_{6,7}N_{6,7}$
As	36	$M_{4,5}VV$		84	$M_{2,3}M_{4,5}V$		171	$N_4N_{6,7}O_{6,7}$
	42	$M_3M_{4,5}M_{4,5}$		1140	$L_3M_{4,5}M_{4,5}$		238	$N_5N_{6,7}V$
	47	$M_2M_{4,5}M_{4,5}$		1173	$L_2M_{4,5}M_{4,5}$		254	$N_4N_{6,7}V$
	91	$M_{2,3}M_{4,5}V$	K	250	$L_{2,3}VV$	S	150	$L_{2,3}VV$
Au	42	O_3VV	Mg	47	$L_{2,3}VV$	Si	91	$L_{2,3}VV$
	71	N_7VV		1180	$KL_{2,3}L_{2,3}$		1610	$KL_{2,3}L_{2,3}$
	151	$N_5N_{6,7}N_{6,7}$	Mn	40	$M_{2,3}VV$	Sn	22	$N_{4,5}VV$
	165	$N_4N_{6,7}N_{6,7}$		540	$L_3M_{2,3}M_{2,3}$		64	$N_{2,3}N_{4,5}V$
	244	$N_5N_{6,7}V$		588	$L_3M_{2,3}V$		310	$M_5N_1N_{4,5}$
	261	$N_4N_{6,7}V$		637	L_3VV		360	$M_5N_{2,3}N_{4,5}$
Bi	60	$O_2O_{4,5}O_{4,5}$	Mo	27	$N_{2,3}VV$		423	$M_5N_{4,5}N_{4,5}$
	102	$N_7O_{4,5}O_{4,5}$		120	$M_{4,5}N_1N_{2,3}$	Ta	26	$N_{6,7}VV$
	128	$N_7O_{4,5}$		185	$M_{4,5}N_{2,3}N_{2,3}$		163	$N_5N_{6,7}N_{6,7}$
	250	$N_5N_7O_{4,5}$		220	$M_{4,5}N_{2,3}V$		173	$N_4N_{6,7}N_{6,7}$
	270	$N_4N_7O_{4,5}$	N	380	KVV		199	$N_5N_{6,7}V$
C	270	KVV	Na	30	$L_{2,3}VV$		209	$N_4N_{6,7}V$
Ca	290	$L_{2,3}M_{2,3}M_{2,3}$		990	$KL_{2,3}L_{2,3}$		341	$N_3N_{6,7}N_{6,7}$
Cl	180	$L_{2,3}VV$	Nb	22	$N_{2,3}VV$	W	20	$N_{6,7}VV$
Co	52	$M_{2,3}VV$		102	$M_{4,5}N_1N_{2,3}$		164	$N_5N_{6,7}N_{6,7}$
	651	$L_3M_{2,3}M_{2,3}$		164	$M_{4,5}N_{2,3}N_{2,3}$		176	$N_4N_{6,7}N_{6,7}$
	711	$L_3M_{2,3}V$		194	$M_{4,5}N_{2,3}V$		205	$N_5N_{6,7}V$
	771	L_3VV	Ni	60	$M_{2,3}VV$		216	$N_4N_{6,7}V$
Cr	35	$M_{2,3}VV$		716	$L_3M_{2,3}M_{2,3}$		347	$N_3N_{6,7}N_{6,7}$
	486	$L_3M_{2,3}M_{2,3}$		781	$L_3M_{2,3}V$	Zn	58	M_3VV
	527	$L_3M_{2,3}V$		847	L_3VV		830	$L_3M_{2,3}M_{2,3}$
Cu	60	M_3VV	O	510	KVV		910	$L_3M_{2,3}V$
	772	$L_3M_{2,3}M_{2,3}$	P	116	$L_{2,3}VV$		990	L_3VV
	844	$L_3M_{2,3}V$	Pb	46	$O_3O_{4,5}O_{4,5}$	Zr	20	$N_{2,3}VV$
	917	L_3VV		57	$O_2O_{4,5}O_{4,5}$		90	$M_{4,5}N_1N_{2,3}$
F	650	KVV		92	$N_7O_{4,5}O_{4,5}$		115	$M_{4,5}N_{2,3}N_{2,3}$
							145	$M_{4,5}N_{2,3}V$

†V refers to the valence band; K, L, M, N and O are the usual atomic energy level notation for X-rays
Reference: Rivière, *The Analyst*, vol. 108, 649, 1983

Table IX.11 Kinetic energy of photoelectrons ejected from atomic energy levels 1s to $4d_{5/2}$ due to excitation with Al K $\alpha_{1,2}$ radiation, energy 1486.6 eV, for elements of atomic numbers 1–92
See Table III.2 for atomic numbers.

Element	1s	2s	$2p_{1/2}$	$2p_{3/2}$	3s	$3p_{1/2}$	$3p_{3/2}$	$3d_{3/2}$	$3d_{5/2}$	4s	$4p_{1/2}$	$4p_{3/2}$	$4d_{3/2}$	$4d_{5/2}$
						†photoelectron kinetic energy/eV								
Ac										216	405	595	811	846
Ag					768	883	914	1113	**1119**	1390	1423	1430	1482	
Al		**1368**	1412	1413	1484	1483								
Ar		1165	**1238**	**1240**	1460	1473								
As			127	**162**	1282	1339	1345	1444			1483			
At										443	599	745	952	978
Au										727	842	940	1133	1152
B	**1297**			1481										
Ba					193	349	423	689	**705**	1232	1294	1306	1393	1396
Be	**1374**													
Bi										547	680	807	1022	1045
Br					1229	1296	1304	**1415**	**1416**	1458	1480	1481		
C	**1202**			1479										
Ca		1048	1135	**1139**	1442	1460				1480				
Cd					715	835	869	1075	**1082**	1378	1419		1476	
Ce					51	213	300	584	**602**	1196	1262	1278	1375	
Cl		1215	**1284**	**1285**	1468	1479								
Co		560	692	**707**	1385	1426		1483		1481				
Cr		791	902	**911**	1411	1443		1483		1482				
Cs					268	420	488	746	**760**	1255	1313	1324	1407	1409
Cu		389	534	**554**	1366	1412		1484		1483				
Dy								153	**191**	1069	1154	1193	1331	
Er								32	**76**	1036	1119	1165	1309	1318
Eu							5	325	**355**	1125	1202	1229	1352	
F	**800**	1454		1477										
Fe		639	764	**777**	1393	1431		1482		1482				
Fr										332	505	675	882	908
Ga		188	343	**370**	1327	1379	1383	1468			1484			
Gd								268	**300**	1110	1197	1215	1345	
Ge		71	238	**269**	1305	1358	1365	1457		1482				
H	**1472**													
He	**1461**													
Hf										947	1048	1105	1262	**1272**
Hg										685	809	914	1107	1126
Ho								94	**134**	1050	1142	1179	1324	
I					413	555	611	854	**866**	1299	1363		1436	
In					660	783	821	1035	**1042**	1364	1408		1469	
Ir										795	908	991	1174	1191
K		1108	1189	**1192**	1452	1468				1483				
Kr					1196	1263	1272	**1397**		1461	1475			
La					124	281	362	637	**654**	1215	1280	1294	1387	
Li	**1431**													
Lu										979	1075	1126	1281	**1290**
Mg	**180**	1396		1434	1483									
Mn		716	833	**845**	1402	1437		1482		1482				
Mo					981	1076	1093	1255	**1258**	1424	1451		1484	
N	**1087**			1476										
Na	**413**	1422		1454	1484									
Nb					1017	1107	1122	1278	**1281**	1427	1452		1482	
Nd							83	188	486	**508**	1170	1242	1261	1368
Ne	**619**	1440		1467										
Ni		477	614	**631**	1374	1417		1482		1481				
O	**953**	1462		1478										
Os										831	939	1017	1196	1213
P		1296	**1349**	**1350**	1469	1475								
Pa										98	213	441	705	748
Pb										592	722	841	1050	1073

(Continued)

Table IX.11 (Continued)

Element	1s	2s	2p$_{1/2}$	2p$_{3/2}$	3s	3p$_{1/2}$	3p$_{3/2}$	3d$_{3/2}$	3d$_{5/2}$	4s	4p$_{1/2}$	4p$_{3/2}$	4d$_{3/2}$	4d$_{5/2}$
						atomic energy level								
					†photoelectron kinetic energy/eV									
Pd					816	926	954	1145	**1151**	1399	1434		1484	
Pm						14	129	434	**459**	1154	1230	1248	1365	
Po										490	634	780	985	1012
Pr						148	243	534	**554**	1181	1249	1268	1372	
Pt										763	876	966	1155	1172
Ra										277	428	606	850	883
Rb					1163	1238	1247	**1374**	1375	1456	1471	1472		
Re										860	968	1041	1212	1225
Rh					858	964	989	1174	**1178**	1404	1438		1483	
Rn										388	556	717	919	944
Ru					900	1003	1025	1202	**1206**	1411	1442		1483	
S		1256	**1320**	1321	1469	1477								
Sb					542	674	720	949	**958**	1333	1387		1454	
Sc		985	1079	**1083**	1432	1453		1478		1483				
Se			9	**50**	1254	1317	1324	1429			1480			
Si		1337	**1385**	1386	1477	1482								
Sm							66	379	**405**	1140	1220	1238	1356	
Sn					602	729	771	992	**1001**	1349	1397		1462	
Sr					1128	1206	1216	**1350**	1352	1448	1466			
Ta										920	1021	1081	1244	1256
Tb							210	**244**		1088	1175	1200	1338	
Tc					941	1041	1060	1229	**1233**	1417	1447		1483	
Te					479	616	667	903	**913**	1317	1375		1446	
Th										156	317	518	771	809
Ti		922	1024	**1030**	1425	1451		1482		1482				
Tl										640	764	876	1079	1099
Tm								**18**		1014	1100	1149	1306	
U										45	213	441	705	748
V		857	965	**973**	1419	1448		1483		1482				
W										890	994	1060	1227	1240
Xe					340	486	548	800	**813**	1277	1339		1422	
Y					1092	1173	1185	1326	**1328**	1440	1460		1482	
Yb										998	1089	1142	1287	**1301**
Zn		292	443	**466**	1350	1399		1477		1481				
Zr					1055	1141	1155	1303	**1305**	1434	1457		1482	

†Values to nearest electron volt; they are most accurate when the element exists as a pure non-volatile solid. The values in bold-face type are those of maximum intensity for each element. The reference point for all values in Tables IX.11–IX.14 is a binding energy value of 84 eV for the Au 4f$_{7/2}$ atomic energy level.
Reference: Roberts, *Sci. Prog*, vol. 68, 65, Oxford, 1982

Table IX.12 Kinetic energy of photoelectrons ejected from atomic energy levels $4f_{5/2}$ to $6d_{5/2}$ due to excitation with Al K $\alpha_{1,2}$ radiation, energy 1486.6 eV, for elements of atomic numbers 49–92
See Table III.2 for atomic numbers

Element	\multicolumn{14}{c}{atomic energy level}
	$4f_{5/2}$	$4f_{7/2}$	5s	$5p_{1/2}$	$5p_{3/2}$	$5d_{3/2}$	$5d_{5/2}$	$5f_{5/2}$	$5f_{7/2}$	6s	$6p_{1/2}$	$6p_{3/2}$	$6d_{3/2}$	$6d_{5/2}$
	\multicolumn{14}{c}{†photoelectron kinetic energy/eV}
Ac		**1166**	1213	1270	1318	1405				1433	1453		1483	
At		**1275**	1290	1337	1370	1445				1467	1477			
Au	1399	**1403**	1378	1414	1432	1483								
Ba			1446	1469	1471									
Bi	1324	**1328**	1326	1369	1393	1459	1461			1477	1483			
Ce		1485	1448	1466										
Cs			1463	1472	1474									
Dy		1481	1423	1459										
Er		1481	1426	1456										
Eu		1485	1454	1463										
Fr		**1217**	1251	1303	1345	1427				1451	1470			
Gd		1485	1449	1465										
Hf	1467	1468	1421	1447	1455	1478								
Hg	1383	**1387**	1365	1405	1428	1479								
Ho		1482	1434	1465										
I			1472	1482										
In				1485										
Ir	1422	**1425**	1390	1422	1435	1482								
La			1453	1471										
Lu		1479	1429	1457										
Nd		1484	1448	1464										
Os	**1433**	**1435**	1402	1427	1440	1485								
Pa	1114	**1126**	1176	1262		1391		1452		1420	1443		1482	
Pb	1353	**1357**	1338	1381	1399	1464	1466			1482	1485			
Pm		1481	1447	1463										
Po	1299	**1304**	1308	1353	1381	1454				1473	1480			
Pr		1483	1448	1463										
Pt	1411	**1414**	1384	1420	1434	1483								
Ra		**1187**	1231	1285	1333	1418				1442	1467			
Re	**1443**	**1445**	1403	1440	1451	1481								
Rn		**1247**	1271	1312	1358	1437				1459	1474			
Sb			1479	1483										
Sm		1480	1448	1464										
Sn			1485	1484										
Ta	**1459**	**1460**	1414	1441	1449	1480								
Tb		1483	1446	1460										
Te			1474	1483										
Th	1141	**1150**	1195	1256	1304	1391	1398			1426	1436	1442	1483	
Tl	1363	**1367**	1349	1386	1410	1470	1472			1482				
Tm		1480	1432	1462										
U	1094	**1105**	1162	1226	1290	1380	1389	1448		1415	1443	1453	1481	
W	1449	**1452**	1408	1439	1450	1479								
Xe			1467	1478										
Yb		1479	1431	1462										

†See footnote Table IX.11
Reference: as for Table IX.11

Table IX.13. Kinetic energy of photoelectrons ejected from atomic energy levels 1s to 4d$_{5/2}$ due to excitation with Mg K $\alpha_{1,2}$ radiation, energy 1253.6 eV, for elements of atomic numbers 1–92

Element	1s	2s	2p$_{1/2}$	2p$_{3/2}$	3s	3p$_{1/2}$	3p$_{3/2}$	3d$_{3/2}$	3d$_{5/2}$	4s	4p$_{1/2}$	4p$_{3/2}$	4d$_{3/2}$	4d$_{5/2}$
					†*photoelectron kinetic energy*/eV									
Ac										172	362		578	613
Ag					535	650	681	880	**886**	1157	1190	1197	1249	
Al		**1135**	1179	1180	1251	1250								
Ar		932	**1005**	**1007**	1227	1240								
As					1049	1106	**1112**	1211		1250				
At										210	366	512	719	745
Au										494	609	707	900	919
B	**1064**			1248										
Ba						116	190	456	**472**	999	1061	1073	1160	1163
Be	**1141**													
Bi										314	447	574	789	812
Br					996	1063	1071	**1182**	**1183**	1225	1247	1248		
C	**969**			1246										
Ca		815	902	**906**	1209	1227				1247				
Cd					482	602	636	842	**849**	1145	1186		1243	
Ce							67	351	**369**	963	1029	1045	1142	
Cl		982	**1051**	**1052**	1235	1246								
Co		327	459	**474**	1152	1193		1250	1248					
Cr		558	669	**678**	1178	1210		1250	1249					
Cs					35	187	255	513	**527**	1022	1080	1091	1174	1176
Cu		156	301	**321**	1133	1179		1251		1250				
Dy										836	921	960	**1098**	
Er										803	886	932	1076	**1085**
Eu								92	**122**	892	969	996	1119	
F	**567**	1221		1244										
Fe		406	531	**544**	1160	1198		1249		1249				
Fr										99	272	442	649	675
Ga			110	**137**	1094	1146	1150	1235		1251				
Gd								35	**67**	877	964	982	1112	
Ge			5	**36**	1072	1125	1132	1224		1249				
H	**1239**													
He	**1228**													
Hf										714	815	872	1029	**1039**
Hg										452	576	681	874	893
Ho										817	909	946	**1091**	
I					180	322	378	621	**633**	1066	1130		1203	
In					427	550	588	802	**809**	1131	1175		1236	
Ir										562	675	758	941	958
K		875	956	**959**	1219	1235		1250						
Kr					963	1030	1039	**1164**		1228	1242			
La						48	129	404	**421**	982	1047	1061	1154	
Li	**1198**													
Lu										746	842	893	1048	**1057**
Mg		**1163**	1201		1250									
Mn		483	600	**612**	1169	1204		1249		1249				
Mo					748	843	860	1022	**1025**	1191	1218		1251	
N	**854**			1243										
Na	**180**	1189		1221	1251									
Nb					784	874	889	1045	**1048**	1194	1219		1249	
Nd								253	**275**	937	1009	1028	1135	
Ne	**386**	1207		1234										
Ni		244	381	**398**	1141	1184		1249		1248				
O	**720**	1229		1245										
Os										598	706	784	963	980
P		1063	**1116**	**1117**	1236	1242								
Pa										28	246		509	544
Pb										359	489	608	817	840
Pd					583	693	721	912	**918**	1166	1201		1251	

(Continued)

Table IX.13 (Continued)

Element	1s	2s	2p$_{1/2}$	2p$_{3/2}$	3s	3p$_{1/2}$	3p$_{3/2}$	3d$_{3/2}$	3d$_{5/2}$	4s	4p$_{1/2}$	4p$_{3/2}$	4d$_{3/2}$	4d$_{5/2}$
					†*photoelectron kinetic energy*/eV									
Pm								201	**226**	921	997	1015	1132	
Po								257		401	547	752	779	
Pr							10	301	**321**	948	1016	1035	1139	
Pt										530	643	733	922	939
Ra										44	195	373	617	650
Rb					930	1005	1014	**1141**	**1142**	1223	1238	1239		
Re										627	735	808	979	992
Rh					625	731	756	941	**945**	1171	1205		1250	
Rn										155	323	484	686	711
Ru					667	770	792	969	**973**	1178	1209		1250	
S		1023	**1087**	**1088**	1236		1244							
Sb					309	441	487	716	**725**	1100	1154		1221	
Sc		752	846	**850**	1199	1220		1245		1250				
Se					1021	1084	1091	**1196**		1247				
Si		1104	**1152**	**1153**	1244	1249								
Sm								146	**172**	907	987	1005	1123	
Sn					369	496	538	759	**768**	1116	1164		1229	
Sr					895	973	983	**1117**	**1119**	1215	1233			
Ta										687	788	848	1011	1023
Tb									**11**	855	942	967	1105	
Tc					708	808	827	996	**1000**	1184	1214		1250	
Te					246	383	434	670	**680**	1084	1142		1213	
Th										84	285	538		576
Ti		689	791	**797**	1192	1218		1249		1249				
Tl										407	531	643	846	866
Tm										781	867	916	**1073**	
U												208	472	515
V		624	732	**740**	1186	1215		1250		1249				
W										657	761	827	994	1007
Xe					107	253	315	567	**580**	1044	1106			1189
Y					859	940	952	1093	**1095**	1207	1227			1249
Yb										765	856	909	1054	**1068**
Zn		59	210	**233**	1117	1166			1244	1248				
Zr					822	908	922	1070	**1072**	1201	1224		1249	

†See footnote, Table IX.11
Reference: as for Table IX.11

Table IX.14 Kinetic energy of photoelectrons ejected from atomic energy levels $4f_{5/2}$ to $6d_{5/2}$ due to excitation with Mg K $\alpha_{1/2}$ radiation, energy 1253.6 eV, for elements of atomic numbers 49–92

See Table III.2 for atomic numbers

Element	$4f_{5/2}$	$4f_{7/2}$	5s	$5p_{1/2}$	$5p_{3/2}$	$5d_{3/2}$	$5d_{5/2}$	$5f_{5/2}$	$5f_{7/2}$	6s	$6p_{1/2}$	$6p_{3/2}$	$6d_{3/2}$	$6d_{5/2}$
					†photoelectron kinetic energy/eV									
Ac	**933**		980	1037	1085		1172			1200		1220	1250	
At	**1042**		1057	1104	1137		1212			1234		1244		
Au	1166	**1170**	1145	1181	1199		1250							
Ba			1213	1236	1238									
Bi	1091	**1095**	1093	1136	1160	1226	1228			1244		1250		
Ce		1252	1215	1233										
Cs			1230	1239	1241									
Dy		1248	1190	1226										
Er		1248	1193	1223										
Eu		1252	1221	1230										
Fr	**984**		1018	1070	1112		1194			1218		1237		
Gd		1252	1216	1232										
Hf	1234	1235	1188	1214	1222		1245							
Hg	1150	**1154**	1132	1172	1195		1246							
Ho		1249	1201	1232										
I			1239				1249							
In				1252										
Ir	1189	**1192**	1157	1189	1202		1249							
La			1220	1238										
Lu		1246	1196	1224										
Nd		1251	1215	1231										
Os	**1200**	**1202**	1169	1194	1207		1252							
Pa	881	893	943	1029			1158	1219		1187		1210	1249	
Pb	1110	**1114**	1105	1148	1166	1231	1233			1249		1252		
Pm		1248	1214	1230										
Po	1066	**1071**	1075	1120	1148		1221			1240		1247		
Pr		1250	1215	1230										
Pt	1178	**1181**	1151	1187	1201		1250							
Ra		**954**	998	1052	1100		1185			1209		1234		
Re	**1210**	**1212**	1170	1207	1218		1248							
Rn		**1014**	1038	1088	1125		1204			1226		1241		
Sb			1246	1250										
Sm		1247	1215	1231										
Sn			1252	1251										
Ta	**1226**	**1227**	1181	1208	1216		1247							
Tb		1250	1213	1227										
Te			1241	1250										
Th	908	917	962	1023	1071	1158	1165			1193	1203	1209	1250	
Tl	1130	**1134**	1116	1153	1177	1237	1239			1249				
Tm		1247	1199	1229										
U	861	872	929	993	1057	1147	1156	1215		1182		1210	1220	1248
W	1216	**1219**	1175	1206	1217		1246							
Xe			1234	1245										
Yb		1246	1198	1229										

†See footnote, Table IX.11
Reference: as for Table IX.11

Table IX.15. Energy values of X-ray absorption edges and of characteristic X-radiation types for K series, for elements of atomic numbers 4–103
See Table III.2 for atomic numbers.

Element	Characteristic X-radiation type†					
	Absorption edge	KN_{III} ($K\beta_2$)	KM_{III} ($K\beta_1$)	KM_{II} ($K\beta_3$)	KL_{III} ($K\alpha_1$)	KL_{II} ($K\alpha_2$)
	*energy/keV					
Ac	106.738	105.837	102.846	102.101	90.884	87.675
Ag	25.531	25.463	24.943	24.912	22.163	21.991
Al	1.562		1.557		1.487	1.486
Am	125.027	123.891	120.36	119.317	106.523	102.083
Ar	3.203		3.19		2.958	2.956
As	11.877	11.864	11.727	11.721	10.544	10.509
At	95.73	94.983	92.302	91.722	81.523	78.943
Au	80.729	80.182	77.985	77.58	68.804	66.99
B	0.188				0.183	
Ba	37.414	37.27	36.378	36.303	32.194	31.817
Be	0.115				0.109	
Bi	90.54	89.866	87.354	86.831	77.118	74.815
Bk	131.59	130.355	126.663	125.443	112.138	107.205
Br	13.483	13.470	13.292	13.285	11.924	11.878
C	0.282				0.277	
Ca	4.034		4.013		3.692	3.688
Cd	26.727	26.653	26.095	26.061	23.173	22.985
Ce	40.41	40.243	39.258	39.17	34.72	34.279
Cf	135.96	134.681	130.851	129.601	116.03	110.71
Cl	2.824		2.816		2.622	2.62
Cm	128.22	127.066	123.423	122.325	109.29	104.441
Co	7.712		7.649		6.93	6.915
Cr	5.987		5.947		5.415	5.405
Cs	35.966	35.833	34.985	34.918	30.973	30.625
Cu	8.993		8.905	8.903	8.048	8.028
Dy	53.761	53.491	52.119	51.956	45.999	45.208
Er	57.464	57.164	55.681	55.491	49.128	48.221
Es	139.49	138.169	134.238	132.916	119.08	113.47
Eu	48.486	48.256	47.036	46.902	41.542	40.902
F	0.692				0.677	
Fe	7.112		7.058		6.404	6.391
Fm	143.09	141.724	137.693	136.347	122.19	116.28
Fr	101.131	100.306	97.477	96.807	86.114	82.231
Ga	10.386	10.366	10.271	10.261	9.252	9.231
Gd	50.207	49.964	48.696	48.554	42.996	42.309
Ge	11.115	11.101	10.983	10.978	9.887	9.856
Hf	65.345	64.973	63.236	62.979	55.79	54.611
Hg	83.109	82.532	80.261	79.822	70.819	68.894
Ho	55.593	55.308	53.878	53.707	47.547	46.699
I	33.168	33.054	32.295	32.239	28.612	28.317
In	27.953	27.872	27.275	27.237	24.209	24.002
Ir	76.111	75.62	73.56	73.203	64.896	63.287
K	3.607		3.59		3.314	3.311
Kr	14.33	14.315	14.113	14.105	12.65	12.598
La	38.894	38.739	37.802	37.721	33.442	33.034
Lr	154.38	152.9	148.67	146.92	132.02	125.1
Lu	63.311	62.96	61.286	61.049	54.07	52.965
Md	146.78	145.37	141.234	139.761	125.39	119.17
Mg	1.309		1.302		1.253	
Mn	6.537		6.49		5.899	5.888
Mo	20.002	19.96	19.608	19.59	17.479	17.374
N	0.397				0.393	
Na	1.08		1.071		1.041	
Nb	18.985	18.947	18.623	18.606	16.615	16.521
Nd	43.538	43.345	42.272	42.166	37.361	36.847

(Continued)

Table IX.15 (Continued)

Element	Characteristic X-radiation type†					
	Absorption edge	KN_{III} ($K\beta_2$)	KM_{III} ($K\beta_1$)	KM_{II} ($K\beta_3$)	KL_{III} ($K\alpha_1$)	KL_{II} ($K\alpha_2$)
	*energy/keV					
Ne	0.874		0.858		0.848	
Ni	8.339		8.265		7.478	7.461
No	150.54	149.092	144.852	143.295	128.66	122.1
Np	118.678	117.591	114.243	113.312	101.068	97.077
O	0.533				0.525	
Os	73.869	73.404	71.416	71.077	63.001	61.487
P	2.143		2.139		2.014	2.013
Pa	112.599	111.606	108.435	107.606	95.883	92.287
Pb	88.008	87.367	84.936	84.45	74.969	72.804
Pd	24.365	24.303	23.819	23.792	21.178	21.021
Pm	45.152	44.947	43.825	43.713	38.725	38.171
Po	93.113	92.403	89.801	89.25	79.301	76.863
Pr	41.958	41.778	40.748	40.653	36.026	35.55
Pt	78.4	77.883	75.751	75.364	66.832	65.123
Pu	121.818	120.703	117.261	116.277	103.761	99.552
Ra	103.909	103.039	100.13	99.432	88.476	85.434
Rb	15.202	15.185	14.962	14.952	13.396	13.336
Re	71.67	71.23	69.309	68.994	61.14	59.718
Rh	23.229	23.173	22.723	22.698	20.216	20.073
Rn	98.402	97.617	94.866	94.246	83.793	81.065
Ru	22.123	22.072	21.656	21.637	19.279	19.15
S	2.471		2.464		2.308	2.307
Sb	30.499	30.402	29.725	29.677	26.359	26.11
Sc	4.486		4.461		4.09	4.086
Se	12.666	12.652	12.496	12.489	11.222	11.181
Si	1.84		1.836		1.74	1.739
Sm	46.801	46.584	45.413	45.289	40.118	39.523
Sn	29.211	29.122	28.491	28.439	25.272	25.044
Sr	16.106	16.085	15.836	15.826	14.166	14.098
Ta	67.405	67.011	65.221	64.946	57.533	56.277
Tb	51.965	51.709	50.382	50.228	44.481	43.744
Tc	21.048	21.002	20.619	20.599	18.367	18.251
Te	31.817	31.712	30.995	30.944	27.472	27.201
Th	109.641	108.69	105.611	104.831	93.358	89.952
Ti	4.965		4.932		4.511	4.505
Tl	85.532	84.924	82.575	82.384	72.872	70.832
Tm	59.374	59.059	57.513	57.303	50.742	49.773
U	115.606	114.561	111.303	110.424	98.44	94.659
V	5.463		5.427		4.952	4.944
W	69.517	69.1	67.244	66.951	59.318	57.982
Xe	34.551	34.428	33.625	33.562	29.779	29.459
Y	17.037	17.015	16.737	16.725	14.958	14.882
Yb	61.322	60.991	59.374	59.157	52.389	51.354
Zn	9.673	9.658	9.572	9.567	8.639	8.616
Zr	17.997	17.963	17.662	17.649	15.77	15.692

† Common designations are bracketed beneath the term labels for each type
* The smaller energy values may vary by a few eV depending on the elements' chemical state; satellite radiations are not included
Reference: Kaye and Laby, *Tables of Physical and Chemical Constants*, 15ed., Longmans, 1986
Handbook of Chemistry and Physics, CRC Press, 1991/2

Table IX.16. Energy values of X-ray absorption edges and of characteristic X-radiation types for L series, for elements of atomic numbers 4–103
See Table III.2 for atomic numbers

Element	Absorption edges			Characteristic X-radiation type†							
	L_I	L_{II}	L_{III}	L_IN_{III} ($L\gamma_3$)	L_IM_{III} ($L\beta_3$)	L_IM_{II} ($L\beta_4$)	$L_{II}N_{IV}$ ($L\gamma_1$)	$L_{II}M_{IV}$ ($L\beta_1$)	$L_{III}N_V$ ($L\beta_2$)	$L_{III}M_V$ ($L\alpha_1$)	$L_{III}M_{IV}$ ($L\alpha_2$)
						*energy/keV					
Ac	19.823	19.063	15.854	18.922	15.931	15.186	18.402	15.711	15.227	12.652	12.502
Ag	3.822	3.54	3.368	3.754	3.234	3.203	3.511	3.151	3.342	2.985	2.979
Al	0.087	0.076	0.075								
Am	23.773	22.944	18.504	22.637	19.106	18.063	22.065	18.852	17.676	14.617	14.412
Ar	0.286	0.247	0.245								
As	1.536	1.368	1.333		1.386	1.38		1.316		1.282	
At	17.495	16.787	14.207	16.748	14.067	13.487	16.262	13.882	13.708	11.427	11.302
Au	14.353	13.739	11.925	13.806	11.609	11.204	13.383	11.432	11.585	9.705	9.618
B			0.005								
Ba	5.964	5.597	5.22	5.82	4.928	4.853	5.529	4.827	5.158	4.47	4.45
Be			0.006								
Bi	16.391	15.725	13.422	15.717	13.203	12.682	15.261	13.031	12.981	10.836	10.728
Bk	25.275	24.385	19.452	24.04	20.348	19.128		20.019		15.32	15.086
Br	1.791	1.605	1.559		1.6	1.593		1.523		1.48	
C			0.005								
Ca	0.403	0.346	0.342								
Cd	4.034	3.742	3.554	3.96	3.402	3.368	3.71	3.319	3.525	3.134	3.131
Ce	6.516	6.131	5.69	6.349	5.364	5.276	6.051	5.261	5.617	4.839	4.82
Cf	26.11	25.25	19.93	24.831	21.001	19.751		20.763		15.677	15.443
Cl	0.237	0.204	0.202								
Cm	24.46	23.779	18.93	23.306	19.663	18.565		19.552		14.959	14.703
Co	0.929	0.797	0.782			0.87		0.791		0.776	
Cr	0.694	0.582	0.572			0.654		0.583		0.573	
Cs	5.7	5.341	4.993	5.567	4.719	4.652	5.278	4.619	4.936	4.289	4.271
Cu	1.109	0.965	0.945		1.023	1.019		0.95		0.93	
Dy	9.013	8.553	7.762	8.743	7.371	7.208	8.409	7.249	7.627	6.496	6.458
Er	9.725	9.243	8.336	9.425	7.942	7.752	9.078	7.813	8.18	6.951	6.906
Es	26.9	26.02	20.41	25.579	21.648	20.326		21.39		16.036	15.78
Eu	8.024	7.584	6.944	7.794	6.574	6.44	7.467	6.455	6.835	5.843	5.815
F			0.015								
Fe	0.846	0.721	0.708			0.792		0.719		0.705	
Fm	27.7	26.81	20.9	26.334	22.303	20.957		22.044		16.402	16.134
Fr	18.63	17.9	15.017	17.805	14.976	14.306	17.307	14.775	14.456	12.031	11.892
Ga	1.316	1.155	1.134		1.197	1.191		1.125		1.098	
Gd	8.343	7.898	7.211	8.1	6.832	6.69	7.772	6.713	7.034	6.058	6.026
Ge	1.426	1.259	1.228		1.294	1.289		1.218		1.188	
Hf	11.262	10.734	9.555	10.89	9.153	8.896	10.517	9.016	9.348	7.891	7.837
Hg	14.835	14.215	12.29	14.258	11.987	11.548	13.834	11.823	11.927	9.999	9.898
Ho	9.365	8.894	8.046	9.08	7.65	7.479	8.74	7.529	7.901	6.719	6.681
I	5.186	4.851	4.556	5.072	4.313	4.257	4.799	4.221	4.509	3.938	3.926
In	4.25	3.951	3.744	4.169	3.572	3.534	3.915	3.487	3.712	3.288	3.28
Ir	13.416	12.824	11.215	12.925	10.865	10.508	12.514	10.705	10.919	9.175	9.096
K	0.34	0.296	0.293								
Kr	1.923	1.732	1.68		1.706	1.698		1.637		1.586	
La	6.235	5.86	5.452	6.08	5.143	5.062	5.786	5.037	5.385	4.651	4.629
Lr	30.24	29.28	22.36	28.76	24.53	22.78		24.13		17.5	17.21
Lu	10.869	10.346	9.241	10.518	8.844	8.607	10.144	8.709	9.049	7.655	7.604
Md	28.53	27.61	21.39	27.12	22.984	21.511		22.707		16.768	16.487
Mg	0.062		0.056								
Mn	0.768	0.649	0.638			0.721		0.649		0.637	
Mo	2.867	2.628	2.523	2.825	2.473	2.455	2.611	2.396	2.508	2.295	2.291
N			0.004								
Na			0.039								
Nb	2.698	2.464	2.37	2.66	2.336	2.319	2.449	2.257	2.357	2.166	2.163

(Continued)

Table IX.16 (Continued)

Element	Absorption edges			Characteristic X-radiation type†							
	L_I	L_{II}	L_{III}	$L_I N_{III}$ ($L\gamma_3$)	$L_I M_{III}$ ($L\beta_3$)	$L_I M_{II}$ ($L\beta_4$)	$L_{II} N_{IV}$ ($L\gamma_1$)	$L_{II} M_{IV}$ ($L\beta_1$)	$L_{III} N_V$ ($L\beta_2$)	$L_{III} M_V$ ($L\alpha_1$)	$L_{III} M_{IV}$ ($L\alpha_2$)
				*energy/keV							
Nd	7.095	6.691	6.177	6.902	5.829	5.723	6.597	5.722	6.091	5.231	5.208
Ne			0.026								
Ni	1.016	0.878	0.861			0.941		0.869		0.852	
No	29.38	28.44	21.88	27.932	23.692	22.135		23.403		17.139	16.843
Np	22.427	21.601	17.61	21.34	17.992	17.061	20.785	17.751	16.84	13.944	13.76
O			0.008								
Os	12.968	12.382	10.868	12.503	10.515	10.176	12.092	10.349	10.592	8.905	8.835
P	0.153	0.13	0.129								
Pa	21.088	20.312	16.716	20.095	16.924	16.095	19.581	16.715	16.022	13.3	13.119
Pb	15.863	15.204	13.039	15.222	12.791	12.305	14.769	12.618	12.625	10.555	10.453
Pd	3.619	3.344	3.187	3.557	3.073	3.046	3.318	2.99	3.163	2.838	2.833
Pm	7.398	6.981	6.427	7.193	6.071	5.959	6.88	5.962	6.334	5.433	5.408
Po	16.94	16.25	13.812	16.23	13.628	13.077	15.756	13.452	13.342	11.131	11.014
Pr	6.802	6.408	5.932	6.622	5.592	5.497	6.321	5.485	5.853	5.034	5.009
Pt	13.88	13.277	11.568	13.363	11.231	10.844	12.944	11.073	11.251	9.439	9.364
Pu	23.097	22.266	18.057	21.982	18.54	17.556	21.417	18.293	17.256	14.279	14.084
Ra	19.222	18.475	15.433	18.352	15.443	14.745	17.848	15.238	14.839	12.34	12.196
Rb	2.067	1.866	1.806	2.051	1.827	1.817		1.752		1.694	1.692
Re	12.522	11.952	10.53	12.082	10.161	9.846	11.682	10.006	10.273	8.651	8.584
Rh	3.421	3.156	3.013	3.365	2.915	2.89	3.132	2.835	2.992	2.698	2.692
Rn	18.047	17.337	14.609	17.262	14.511	13.891	16.777	14.323	14.079	11.727	11.595
Ru	3.23	2.973	2.844	3.179	2.763	2.744	2.952	2.683	2.825	2.556	2.554
S	0.193	0.164	0.163								
Sb	4.706	4.389	4.14	4.609	3.932	3.884	4.345	3.843	4.101	3.604	3.594
Sc	0.462	0.4	0.396					0.4		0.395	
Se	1.662	1.485	1.444		1.492	1.485		1.419		1.379	
Si	0.118	0.101	0.1								
Sm	7.707	7.278	6.683	7.49	6.319	6.195	7.169	6.205	6.582	5.635	5.61
Sn	4.475	4.167	3.939	4.377	3.75	3.703	4.127	3.661	3.903	3.442	3.433
Sr	2.217	2.008	1.94	2.197	1.947	1.937		1.872		1.806	1.804
Ta	11.672	11.128	9.872	11.278	9.488	9.213	10.894	9.345	9.649	8.147	8.089
Tb	8.679	8.221	7.484	8.423	7.096	6.942	8.086	6.976	7.358	6.273	6.239
Tc	3.047	2.797	2.681	3.001	2.618	2.598	2.778	2.537	2.664	2.424	2.421
Te	4.942	4.616	4.345	4.837	4.12	4.069	4.568	4.03	4.302	3.77	3.759
Th	20.449	19.689	16.283	19.498	16.419	15.639	18.993	16.215	15.622	12.97	12.809
Ti	0.529	0.46	0.454					0.458		0.452	
Tl	15.344	14.7	12.66	14.736	12.387	12.196	14.293	12.217	12.272	10.271	10.177
Tm	10.097	9.601	8.632	9.782	8.236	8.026	9.426	8.103	8.465	7.181	7.134
U	21.757	20.947	17.166	20.712	17.454	16.575	20.167	17.219	16.429	13.614	13.438
V	0.626	0.519	0.511			0.585		0.519		0.511	
W	12.092	11.535	10.199	11.675	9.819	9.526	11.284	9.671	9.959	8.396	8.335
Xe	5.442	5.092	4.772	5.319	4.516	4.453	5.035	4.415	4.72	4.11	4.095
Y	2.372	2.155	2.079	2.347	2.072	2.06		1.996		1.923	1.92
Yb	10.479	9.968	8.933	10.148	8.531	8.314	9.781	8.402	8.755	7.415	7.367
Zn	1.208	1.057	1.034		1.107	1.102		1.035		1.012	
Zr	2.535	2.305	2.227	2.503	2.2	2.187	2.292	2.118	2.215	2.043	2.04

†,* See footnotes †, * Table IX.15
Reference: as for Table IX.15

Table IX.17 Energy values of X-ray absorption edges and of characteristic X-radiation types for M series, for elements of atomic numbers 29–103
See Table III.2 for atomic numbers.

Element	Absorption edges			Characteristic X-radiation type†			
	M_{III}	M_{IV}	M_V	$M_{IV}N_V$ (M γ)	$M_{IV}N_{VI}$ (M β)	$M_V N_{VII}$ (M α₁)	$M_V N_{VI}$ (M α₂)
				*energy/keV			
Ac	3.892	3.352	3.202	3.265	3.054	2.913	2.904
Ag	0.588	0.389	0.383	0.562			
Am	4.667	4.092	3.887	3.839			
As	0.15	0.052					
At	3.428	2.905	2.78	2.929	2.707	2.581	2.582
Au	2.744	2.307	2.22	2.404	2.22	2.142	2.133
Ba	1.036	0.77	0.75	0.974			
Bi	3.186	2.694	2.586	2.745	2.534	2.422	2.426
Bk	4.927	4.366	4.132				
Br	0.191	0.082					
Cd	0.632	0.423	0.42	0.603			
Ce	1.152	0.87	0.851	1.079	0.902	0.883	
Cf	5.109	4.487	4.253				
Cm	4.797	4.227	3.971				
Cs	0.981	0.722	0.704	0.924			
Cu		0.015					
Dy	1.642	1.304	1.266	1.507	1.325	1.293	
Er	1.783	1.43	1.385	1.627	1.443	1.406	
Es	5.252	4.63	4.374				
Eu	1.45	1.129	1.101	1.341	1.153	1.131	
Fm	5.397	4.766	4.498				
Fr	3.654	3.125	2.986	3.093	2.881	2.746	2.742
Ga	0.115	0.03					
Gd	1.511	1.185	1.153	1.334	1.209	1.185	
Ge	0.132	0.041					
Hf	2.109	1.718	1.664	1.902	1.7	1.646	
Hg	2.848	2.392	2.291	2.485	2.285	2.195	2.184
Ho	1.715	1.365	1.327	1.57	1.383	1.348	
I	0.873	0.63	0.618	0.826			
In	0.678	0.464	0.456	0.646			
Ir	2.551	2.119	2.04	2.255	2.062	1.988	1.983
Kr	0.217	0.095					
La	1.092	0.823	0.801	1.025	0.854	0.833	
Lr	5.71	5.15	4.86				
Lu	2.025	1.637	1.586	1.833	1.623	1.572	
Md	5.546	4.903	4.622				
Mo	0.394	0.232	0.228	0.379			
Nb	0.362	0.207	0.204	0.349			
Nd	1.266	0.969	0.946	1.18	0.997	0.978	
No	5.688	5.037	4.741				
Np	4.435	3.85	3.666	3.665	3.435	3.262	3.251
Os	2.453	2.033	1.963	2.177	1.988	1.921	1.918
Pa	4.164	3.597	3.416	3.47	3.251	3.083	3.07
Pb	3.072	2.586	2.484	2.658	2.442	2.345	2.34
Pd	0.546	0.354	0.349	0.522			
Pm	1.327	1.019	0.994	1.234			
Po	3.312	2.798	2.681	2.842	2.62	2.501	2.503
Pr	1.21	0.923	0.898	1.131	0.95	0.929	
Pt	2.649	2.204	2.219	2.332	2.134	2.065	2.059
Pu	4.557	3.973	3.778	3.756	3.527	3.346	3.332
Ra	3.779	3.237	3.093	3.185	2.967	2.829	2.823
Rb	0.24	0.114	0.112				
Re	2.361	1.946	1.879	2.104	1.910	1.845	1.843
Rh	0.506	0.321	0.315	0.485			
Rn	3.536	3.014	2.882	3.006	2.794	2.663	2.662

(Continued)

Table IX.17 (Continued)

Element	Absorption edges			Characteristic X-radiation type†			
	M_{III}	M_{IV}	M_V	$M_{IV}N_V$ (M γ)	$M_{IV}N_{VI}$ (M β)	$M_V N_{VII}$ (M α_1)	$M_V N_{VI}$ (M α_2)
				*energy/keV			
Ru	0.467	0.29	0.288	0.448			
Sb	0.774	0.546	0.536	0.735			
Se	0.17	0.066					
Sm	1.388	1.073	1.048	1.287	1.1	1.081	
Sn	0.72	0.506	0.497	0.684			
Sr	0.27	0.136	0.134				
Ta	2.184	1.783	1.725	1.961	1.76		1.702
Tb	1.583	1.245	1.211	1.457	1.266	1.24	
Tc	0.429	0.26	0.257	0.412			
Te	0.822	0.586	0.575	0.779			
Th	4.03	3.474	3.313	3.369	3.145	2.996	2.984
Tl	2.957	2.483	2.389	2.569	2.36	2.27	2.266
Tm	1.861	1.498	1.451	1.694	1.503	1.462	
U	4.303	3.728	3.552	3.566	3.337	3.171	3.161
W	2.273	1.864	1.803	2.033	1.835	1.776	1.774
Xe	0.926	0.677	0.662	0.874			
Y	0.3	0.159	0.156				
Yb	1.948	1.566	1.518	1.77	1.568	1.521	1.507
Zn		0.022					
Zr	0.335	0.187	0.184	0.323			

†,* See footnotes †, *, Table IX.15
Reference: as for Table IX. 15

Table IX.18 Approximate intensities of some characteristic X-radiations relative to that of the most intense of its series

	Characteristic X-radiation type†										
	KN_{III} (K β_2)	KM_{III} (K β_1)	KM_{II} (K β_3)	KL_{II} (K α_2)	$L_I N_{III}$ (L γ_3)	$L_I M_{III}$ (L β_3)	$L_I M_{II}$ (L β_4)	$L_{II} N_{IV}$ (L γ_1)	$L_{III} N_V$ (L β_2)	$L_{III} M_V$ (Lα_1)	$L_{III} M_{IV}$ (L α_2)
	intensity relative to that of KL_{III} (K α_1)				intensity relative to that of $L_{II} M_{IV}$ (L β_1)						
low atomic number elements	0.02	0.2	0.1	0.5	0.05	0.5	0.2	0.05	0.05	0.9	0.1
high atomic number elements	0.05	0.2	0.1	0.53	0.05	0.35	0.2	0.05	0.05	0.9	0.1

† See footnote †, Table IX.15
Reference: as for Table IX.15

Table IX.19 Variation with photon energy of values of mass absorption coefficient for various elements and concrete for a collimated incident photon beam which emerges undeflected without photon energy loss

Substance	photon energy/MeV											
	0.2	0.3	0.4	0.5	0.6	0.8	1	2	5	10	50	100
	mass absorption coefficient/10^{-2} m^2 kg^{-1}											
Ag	2.963	1.557	1.13	0.931	0.814	0.676	0.592	0.421	0.358	0.388	0.581	0.666
Al	1.223	1.042	0.928	0.845	0.78	0.684	0.615	0.432	0.284	0.232	0.231	0.252
Be	1.089	0.946	0.847	0.774	0.716	0.629	0.565	0.394	0.235	0.163	0.102	0.099
C	1.229	1.066	0.954	0.871	0.806	0.708	0.636	0.444	0.271	0.196	0.143	0.146
concrete[†]	1.27	1.08	0.96	0.88	0.81	0.71	0.64	0.45	0.29	0.23	0.22	0.23
Fe	1.458	1.098	0.94	0.841	0.77	0.67	0.599	0.426	0.315	0.299	0.383	0.433
Ge	1.658	1.13	0.933	0.821	0.745	0.643	0.573	0.409	0.316	0.316	0.43	0.489
H	2.429	2.112	1.893	1.729	1.599	1.405	1.263	0.877	0.505	0.325	0.142	0.119
I	3.65	1.768	1.215	0.969	0.83	0.674	0.584	0.412	0.361	0.4	0.613	0.704
Pb	9.985	4.026	2.323	1.613	1.248	0.887	0.71	0.522	0.427	0.497	0.806	0.931
U	12.98	5.191	2.922	1.976	1.49	1.016	0.789	0.488	0.446	0.519	0.849	0.983
W	7.844	3.238	1.925	1.378	1.093	0.806	0.662	0.443	0.41	0.475	0.762	0.88

[†] Approximate percentage composition by mass: Al 5, Ca 8, O 50, Si 31, unspecified 6
Reference: Hubbell, *Int. J. Appl. Radioat. Isot.*, vol. 33, 1269, 1982

Table IX.20 Inelastic mean free path of electrons, usually of 1 keV energy, in various substances, and n values

n is defined as follows: mean free path for electron energy E keV $= E^n \times$ mean free path for electron of energy 1 keV.

Substance	electron energy/keV	mean free path/nm	n	Substance	electron energy/keV	mean free path/nm	n
Ag	1	1.5	0.5	Mo	1	1.6	0.71
Al	1	1.9	0.74	Na	0.173	1.3	
Al_2O_3	1	1.4	0.54	NaCl	1	3.8	0.56
				NaF	1	4.2	0.56
Au	1	1.7	0.5	Ni	1.186	1.3	
Be	1	1.6	0.5				
barium stearate	1	5	0.62	NiO	0.511	1	
butylamine	1	4.2	0.48				
C (amorphous)	1	1.4		polybromoparaxylylene	1.065	1.9	
C (diamond)	1	2					
C (graphite)	1	3.7					
cadmium arachidate	1	3.4	0.5	polychloroparaxylyene	1.065	1.9	
Co	1.194	1.2		polyethylene	0.969	8.4	
Cr	1.211	1.1		polyethylene terephthalate	0.969	6.8	
Cr_2O_3	1	1.9	0.52				
Cs	1.26	2.6		polyparaxylene	1	1.5	2
				polystyrene	0.969	7.4	
Cu	1	1.4	0.75				
Fe	1.2	1.3		polytetrafluoroethylene	0.969	7.2	
Fe_3O_4	0.519	2.3		polyvinyl chloride	0.969	8.4	
Ge	1	2.3	0.6	polyvinylidene fluoride	1	1.4	1.4
GeO_2	0.25	0.6		Si	1	2.9	0.62
				SiO_2	1	2.2	0.7
graphite fluoride	0.969	4.4		Sn	0.52	1.8	
In	0.408	1		Ti	0.422	1	
α iodostearic-acid	0.867	14					
K	0.173	2.1		V	1.216	1.1	
				W	1	1.3	0.36
KI	1	6.3	0.59	WO_3	1.45	2.6	
				Zn	1.167	1.8	

Reference: Wagner et al, *Surface and Interface Analysis*, vol. 2, 53, 1980

Table IX.21 Range in aluminium of electrons for various electron energies

Range/ mg cm^{-2}	0.16	0.24	0.38	0.68	1.06	1.5	2.6	4	6.4	9.2	13.5	18.5	26.5
Energy/ MeV	0.01	0.012	0.015	0.02	0.025	0.03	0.04	0.05	0.065	0.08	0.1	0.12	0.15
Range/ mg cm^{-2}	42	59	78	120	165	235	310	420	520	680	950	1210	1480
Energy/ MeV	0.2	0.25	0.3	0.4	0.5	0.65	0.8	1	1.2	1.5	2	2.5	3
Range/ mg cm^{-2}	2020	2540	3300	4100	5200	6200	7800						
Energy/ MeV	4	5	6.5	8	10	12	15						

Reference: Katz and Penfold, *Rev. Mod. Phys.*, vol. 24, 28, 1952

Table IX.22 Range in various substances of protons for various proton energies

Substance	energy/MeV													
	0.005	0.01	0.05	0.1	0.5	1	2	5	10	20	50	100	200	500	1000
	range/kg m^{-2}													
	$\times 10^{-5}$	$\times 10^{-5}$	$\times 10^{-4}$	$\times 10^{-4}$	$\times 10^{-3}$	$\times 10^{-2}$	$\times 10^{-2}$	$\times 10^{-1}$	$\times 10^{-1}$					
air	17.9	31	9.7	16.5	10.1	2.95	9.2	4.36	14.6	5.02	26	89.7	300	1350	3740
Ag	33.8	62.3	25.8	49.3	28.4	7.26	20.3	8.39	25.8	8.23	39.7	132	429	1870	5120
Al	21.5	38.4	13.2	24.1	15	3.93	11.4	5.13	16.8	5.68	29	99	330	1470	4080
C	17.6	30.2	9.4	16.1	9.6	2.75	8.44	4.02	13.6	4.71	24.6	85.4	287	1300	3600
Cu	31.4	60	25.1	46.6	24.4	6.02	16.3	6.87	21.7	7.08	35	117	386	1700	4670
H	3.7	6.2	1.9	3.2	2.7	0.88	2.94	1.52	5.4	1.93	10.4	36.6	125	569	1590
H$_2$O	13.2	23	7.1	12.2	8	2.4	7.6	3.68	12.5	4.31	22.5	77.9	262	1180	3280
Pb	48.7	92.4	40.4	73.2	40.2	10.3	28.4	11.3	33.9	10.6	49.5	161	518	2230	6050
Ti	27.6	51.9	19.6	33.6	17.6	4.76	13.5	5.99	19.4	6.45	32.4	110	364	1610	4450
U	47.5	86.5	36.1	68.2	40.5	10.6	29.2	11.8	35.6	11.1	51.5	16.8	537	2310	6250

Reference: Janni, *Atomic Data and Nuclear Data Tables*, vol.27, nos. 2–5, 1982

Table IX.23 Range in various substances of alpha particles for various particle energies

Substance	energy/MeV													
	0.01	0.05	0.1	0.5	1	2	5	10	20	50	100	200	500	1000
	range/kg m^{-2}													
	$\times 10^{-5}$	$\times 10^{-4}$	$\times 10^{-4}$	$\times 10^{-3}$	$\times 10^{-3}$	$\times 10^{-3}$	$\times 10^{-2}$	$\times 10^{-2}$	$\times 10^{-1}$					
air	28.8	9.57	15.7	3.92	6.53	12.8	4.35	12.7	4.14	2.17	7.48	26	133	439
Ag	47.1	20.9	36.6	12	19.4	36.4	10.9	28.9	83.9	3.79	12.4	41.1	195	599
Al	35.1	10.9	16.7	4.9	8.77	17.8	5.85	16.3	4.97	2.37	8.1	28	142	464
C	18.8	6.6	11.1	3.6	5.95	12.3	4.23	12.4	3.97	2.03	7.18	25.8	133	428
Cu	47.4	20.6	34.9	9.2	17.9	32	9.08	23.6	6.84	3.15	10.6	36.1	179	570
H	9	3	4.3	1.12	1.71	3.4	1.28	4.35	1.51	0.8	2.9	10.3	54.2	183
H$_2$O	16.9	5.95	10	3.24	4.95	10.3	3.68	11.3	3.73	1.86	6.45	22.5	116	383
Pb	56.6	27.2	47.5	17	28.3	52.7	15.5	42.3	11.7	5	15.7	50	239	755
Ti	35.7	13.6	22.1	6.6	10.8	20.2	6.75	19.5	6	2.85	9.6	32.5	163	540
U	61	28.6	49.5	15.8	27.6	54.9	16.5	43.8	12.3	5.22	16.7	52.3	247	785

Reference: as for Table IX.22

Table IX.24 Range in various substances of pions for various pion energies

Substance	energy/MeV									
	1	2	5	10	20	50	100	200	500	1000
	range/kg m^{-2}									
	$\times 10^{-2}$	$\times 10^{-1}$	$\times 10^{-1}$							
air	10.8	3.67	18.9	6.57	22.5	107	315	829	2500	5170
Ag	20.1	6.28	29.8	9.86	32.5	150	434	1130	3370	6970
Al	12.6	4.2	21.2	7.29	24.7	117	343	903	2730	5700
C	10	3.43	17.9	6.24	21.4	102	302	798	2410	5020
Cu	16.6	5.33	26	8.73	29.1	135	395	1030	3080	6400
H	3.88	1.38	7.45	2.65	9.22	44.6	133	355	1080	2260
H$_2$O	9.19	3.15	16.4	5.7	19.5	93	275	728	2210	4640
Pb	26.7	8.15	37.5	12.2	39.6	179	515	1330	3920	8030
Ti	14.7	4.81	23.9	8.13	27.4	128	375	984	2960	6180
U	28	8.57	39.2	12.7	41.1	185	533	1370	4050	8280

Reference: as for Table IX.22

Table IX.25 Range in various substances of muons for various muon energies

Substance	energy/MeV									
	1	2	5	10	20	50	100	200	500	1000
	range/kg m^{-2}									
	$\times 10^{-2}$	$\times 10^{-1}$	$\times 10^{-1}$							
air	13.4	4.57	23.7	8.18	27.6	127	357	895	2560	5150
Ag	23.9	7.58	36.5	12.1	39.5	176	489	1210	3450	6950
Al	15.4	5.19	26.5	9.05	30.3	138	389	976	2810	5700
C	12.4	4.28	22.5	7.78	26.4	121	343	863	2480	5020
Cu	20	6.5	32.1	10.8	35.5	160	446	1110	3160	6390
H	4.89	1.74	9.44	3.32	11.4	53.2	152	385	1110	2270
H$_2$O	11.4	3.92	20.6	7.1	24	111	313	788	2280	4660
Pb	31.5	9.76	45.7	14.8	47.9	210	579	1430	4000	7980
Ti	17.8	5.9	29.7	10.1	33.5	152	425	1060	3040	6180
U	33.1	10.2	47.7	15.4	49.6	218	598	1470	4120	8220

Reference: as for Table IX.22

Table IX.26 Range in aluminium of some heavy ions for various ion energies

mass unit/ MeV amu^{-1}	ions											
	$^{107}_{47}$Ag	$^{27}_{13}$Al	$^{40}_{18}$Ar	$^{197}_{79}$Au	$^{12}_{6}$C	$^{153}_{63}$Eu	$^{74}_{32}$Ge	$^{4}_{2}$He	$^{58}_{28}$Ni	$^{181}_{73}$Ta	$^{238}_{92}$U	$^{90}_{40}$Zr
	range/10^{-2} kg m^{-2}											
0.0125	0.16	0.12	0.11	0.19	0.096	0.16	0.16	0.12	0.15	0.18	0.19	0.14
0.05	0.55	0.34	0.33	0.72	0.25	0.6	0.54	0.28	0.46	0.68	0.75	0.49
0.2	1.56	0.85	0.97	2.16	0.73	1.72	1.43	0.72	1.15	1.96	2.28	1.32
0.8	3.56	2.17	2.4	4.67	2.1	3.94	3.08	3.1	2.71	4.51	5.1	3.07
3.2	8.3	8.2	8	10.8	10.8	9.68	8.3	24.5	7.6	10.4	11.4	8.1
12	31.8	48	42	33.2	81	33.1	33.5	233	31.5	32	33.5	30.5

Reference: Littmark and Ziegler, *The Range and Energy Loss of Particles in Matter*, vol. 6, Pergamon Press, 1980
Northcliffe and Schilling, *Nuclear Data Tables*, nos. 3–4, vol. 7A, 1970

Table IX.27 Range in carbon of some heavy ions for various ion energies

mass unit/ MeV amu^{-1}	ions											
	$^{107}_{47}$Ag	$^{27}_{13}$Al	$^{40}_{18}$Ar	$^{197}_{79}$Au	$^{12}_{6}$C	$^{153}_{63}$Eu	$^{74}_{32}$Ge	$^{4}_{2}$He	$^{58}_{28}$Ni	$^{181}_{73}$Ta	$^{238}_{92}$U	$^{90}_{40}$Zr
	range/10^{-2} kg m^{-2}											
0.0125	0.12	0.087	0.09	0.15	0.065	0.13	0.13	0.084	0.11	0.14	0.16	0.11
0.05	0.42	0.24	0.28	0.54	0.17	0.45	0.43	0.2	0.36	0.53	0.6	0.36
0.2	1.12	0.59	0.67	1.47	0.52	1.23	1.02	0.5	0.88	1.42	1.68	0.93
0.8	2.44	1.53	1.68	3.37	1.48	2.81	2.18	2.2	1.8	3.26	3.67	2.16
3.2	6.18	6.33	5.99	8.15	8.3	7.25	6.15	18.4	5.6	7.81	8.65	6.1
12	26	38.3	34.6	27	65.5	25.8	27.2	189	27.2	26	27.3	25

Reference: as for Table IX.26

Table IX.28 Range in germanium of some heavy ions for various ion energies

energy per mass unit/ MeV amu^{-1}	$^{107}_{47}$Ag	$^{27}_{13}$Al	$^{40}_{18}$Ar	$^{197}_{79}$Au	$^{12}_{6}$C	$^{153}_{63}$Eu	$^{74}_{32}$Ge	$^{4}_{2}$He	$^{58}_{28}$Ni	$^{181}_{73}$Ta	$^{238}_{92}$U	$^{90}_{40}$Zr
					range/10^{-2} kg m^{-2}							
0.0125	0.24	0.21	0.21	0.27	0.19	0.24	0.25	0.25	0.23	0.28	0.27	0.21
0.05	0.91	0.62	0.63	1.09	0.47	0.96	0.88	0.56	0.78	1.13	1.12	0.79
0.2	2.6	1.47	1.68	3.47	1.15	2.86	2.39	1.35	2.1	3.43	3.64	2.23
0.8	5.75	3.65	4.05	7.62	3.55	6.55	5.09	5.2	4.55	7.32	8.27	5.1
3.2	12.7	12.2	12	16.3	16	14.8	12.7	35.2	11.6	15.8	17.5	12.5
12	41.9	66	55	45.4	109	43.7	46.5	305	43.5	43.2	46.3	41.5

Reference: as for Table IX.26

Table IX.29 Range in gold of some heavy ions for various ion energies

energy per mass unit/ MeV amu^{-1}	$^{107}_{47}$Ag	$^{27}_{13}$Al	$^{40}_{18}$Ar	$^{197}_{79}$Au	$^{12}_{6}$C	$^{153}_{63}$Eu	$^{74}_{32}$Ge	$^{4}_{2}$He	$^{58}_{28}$Ni	$^{181}_{73}$Ta	$^{238}_{92}$U	$^{90}_{40}$Zr
					range/10^{-2} kg m^{-2}							
0.0125	0.38	0.36	0.39	0.38	0.35	0.36	0.42	0.58	0.41	0.37	0.36	0.36
0.05	1.47	1	1.18	1.67	0.82	1.52	1.38	1.2	1.32	1.65	1.66	1.31
0.2	4.17	2.4	2.76	5.53	1.9	4.66	3.79	2.55	3.35	5.17	5.73	3.63
0.8	9.7	6.27	7	12.7	6.25	11.1	8.6	9.4	7.65	12.3	13.7	8.75
3.2	21.4	20.5	20.5	27.4	27	24.7	21.2	58.5	19.3	26.5	29	20.8
12	65	97	83	70.3	163	67.2	69.8	459	65.5	69	70.8	63

Reference: as for Table IX.26

Table IX.30 Range in silver of some heavy ions for various ion energies

energy per mass unit/ MeV amu^{-1}	$^{107}_{47}$Ag	$^{27}_{13}$Al	$^{40}_{18}$Ar	$^{197}_{79}$Au	$^{12}_{6}$C	$^{153}_{63}$Eu	$^{74}_{32}$Ge	$^{4}_{2}$He	$^{58}_{28}$Ni	$^{181}_{73}$Ta	$^{238}_{92}$U	$^{90}_{40}$Zr
					range/10^{-2} kg m^{-2}							
0.0125	0.27	0.24	0.26	0.29	0.24	0.26	0.28	0.3	0.25	0.28	0.29	0.23
0.05	1	0.64	0.73	1.18	0.58	1	0.96	0.68	0.84	1.12	1.21	0.87
0.2	2.82	1.58	1.83	3.76	1.32	3.11	2.57	1.6	2.27	3.54	3.97	2.42
0.8	6.37	4.02	4.4	8.34	3.95	7.2	5.65	6	5.05	8.08	9.11	5.64
3.2	14.8	14.5	14.3	19	19.2	17.1	14.8	42.2	13.2	18.4	20.2	14.5
12	49.1	72.2	62.8	53.4	128	51.9	53.5	357	49	53.1	53.9	49.8

Reference: as for Table IX.26

Table IX.31 Range in titanium of some heavy ions for various ion energies

energy per mass unit/ MeV amu^{-1}	$^{107}_{47}$Ag	$^{27}_{13}$Al	$^{40}_{18}$Ar	$^{197}_{79}$Au	$^{12}_{6}$C	$^{153}_{63}$Eu	$^{74}_{32}$Ge	$^{4}_{2}$He	$^{58}_{28}$Ni	$^{181}_{73}$Ta	$^{238}_{92}$U	$^{90}_{40}$Zr
					range/10^{-2} kg m^{-2}							
0.0125	0.15	0.15	0.13	0.2	0.11	0.16	0.15	0.15	0.14	0.17	0.2	0.15
0.05	0.6	0.41	0.43	0.79	0.27	0.65	0.56	0.36	0.49	0.7	0.8	0.53
0.2	1.73	0.94	1.08	2.39	0.9	1.92	1.57	0.85	1.38	2.16	2.53	1.46
0.8	3.86	2.43	2.65	5.2	2.35	4.39	3.44	3.46	3.05	5.02	5.68	3.4
3.2	9.7	9.51	9.2	12.4	12.9	11.2	9.7	28.7	8.8	12	13.3	9.5
12	33.8	54.1	47	36.5	91.2	35.4	38.3	262	36.5	36.9	36.6	36

Reference: as for Table IX.26

Table IX.32 Range in uranium of some heavy ions for various ion energies

energy per mass unit/ MeV amu^{-1}	$^{107}_{47}$Ag	$^{27}_{13}$Al	$^{40}_{18}$Ar	$^{197}_{79}$Au	$^{12}_{6}$C	$^{153}_{63}$Eu	$^{74}_{32}$Ge	$^{4}_{2}$He	$^{58}_{28}$Ni	$^{181}_{73}$Ta	$^{238}_{92}$U	$^{90}_{40}$Zr
					range/10^{-2} kg m^{-2}							
0.0125	0.38	0.34	0.38	0.36	0.36	0.34	0.38	0.47	0.39	0.36	0.35	0.35
0.05	1.55	0.99	1.05	1.64	0.8	1.45	1.35	1.05	1.25	1.59	1.65	1.24
0.2	4.04	2.28	2.63	5.4	1.8	4.49	3.65	2.35	3.22	5.12	5.67	3.52
0.8	9.6	6.12	6.8	12.8	6.15	11	8.4	9.3	7.5	12.2	13.7	8.65
3.2	20.3	21	21	28.3	28.1	25.5	21.7	61.5	19.9	27.3	29.9	21.6
12	69.5	104	89.5	76	179	72.5	74.5	490	70.5	75.2	76.5	67.5

Reference: as for Table IX.26

Table IX.33 Mass stopping power in various substances for protons of various energies

| Substance | energy/MeV | | | | | | | | | | | | | |
	0.005	0.01	0.05	0.1	0.5	1	2	5	10	20	50	100	200	500	1000
	stopping power/MeV kg^{-1} m^2										×10^{-1}	×10^{-1}	×10^{-1}	×10^{-1}	×10^{-1}
air	25.6	36.7	70.7	71.8	34	21	13	6.63	3.87	2.22	10.7	6.29	3.89	2.38	1.93
Ag	6.47	9.67	17.2	18.6	13.3	9.28	6.38	3.67	2.31	1.41	7.19	4.36	2.76	1.73	1.41
Al	17.3	25.4	44.8	43.1	25.1	17.2	11	5.74	3.4	1.98	9.66	5.72	3.55	2.19	1.77
C	27.1	40.1	71.4	73.4	36.4	22.9	14.3	7.14	4.13	2.35	11.2	6.58	4.06	2.48	2
Cu	7	10.3	19.1	21.6	16.5	11.8	8.03	4.42	2.71	1.62	8.1	4.86	3.05	1.9	1.55
H	171	220	389	341	110	66.3	38.5	18.1	10.1	5.64	26.3	15.2	9.28	5.6	4.48
H$_2$O	38	51.3	97.2	95.3	41.5	25.4	15.5	7.78	4.5	2.57	12.3	7.22	4.45	2.72	2.19
Pb	2.87	4.01	9.63	12.3	8.93	6.39	4.54	2.74	1.77	1.11	5.8	3.57	2.29	1.45	1.2
Ti	9.91	14.1	29.8	35	20.8	14.1	9.35	4.96	2.98	1.76	8.68	5.17	3.23	2	1.62
U	3.73	5.43	10.8	12.3	8.63	6.15	4.33	2.6	1.68	1.06	5.57	3.44	2.2	1.4	1.16

Reference: as for Table IX.22

Table IX.34 Mass stopping power in various substances for alpha particles of various energies

Substance	\multicolumn{13}{c}{energy/MeV}
	0.01	0.05	0.1	0.5	1	2	5	10	20	50	100	200	500	1000
	\multicolumn{14}{c}{stopping power/MeV kg^{-1} m^2}
air	34.8	74	104	197	192	139	76.2	46.9	28.1	13.4	7.62	4.36	2.14	1.38
Ag	10.6	21.7	29.9	57.5	62.5	52.9	34.3	22.8	14.2	7.87	4.75	2.87	1.48	0.96
Al	26.3	64.1	90.4	130	123	98.7	60.5	37.9	22.8	11.4	6.61	3.82	1.91	1.23
C	56.7	98	125	191	181	136	77.2	47.9	28.5	13.6	7.77	4.45	2.18	1.41
Cu	11.3	23.5	33.1	65.1	70.6	62.8	42.8	28.1	17.1	9.19	5.46	3.21	1.63	1.06
H	127	281	419	806	711	461	226	130	74.1	33.8	18.8	10.5	5.13	3.22
H$_2$O	50	105	163	251	222	160	88.1	52.7	31.1	15	8.59	4.92	2.45	1.55
Pb	6.4	14	19.7	38.1	40.1	34.9	24	16.1	11	6.1	3.81	2.29	1.23	0.8
Ti	20.9	44	83.5	110	106	82.8	50.8	32.1	19.2	10	5.9	3.45	1.76	1.12
U	5.4	13.1	19.3	42	44.5	35.7	22.7	15.4	9.9	5.82	3.62	2.25	1.19	0.78

Reference: as for Table IX.22

Table IX.35 Mass stopping power in various substances for pions of various energies

Substance	\multicolumn{10}{c}{energy/MeV}
	1	2	5	10	20	50	100	200	500	1000
	\multicolumn{10}{c}{stopping power/MeV kg^{-1} m^2}
			×10^{-1}	×10^{-1}	×10^{-1}	×10^{-1}	×10^{-1}	×10^{-1}	×10^{-1}	×10^{-1}
air	5.28	3.06	14.7	8.64	5.08	2.87	2.14	1.84	1.81	1.93
Ag	3.03	1.88	9.63	5.79	3.56	2.06	1.56	1.36	1.34	1.42
Al	4.61	2.71	13.2	7.69	4.63	2.63	1.96	1.69	1.64	1.72
C	5.66	3.25	15.5	8.91	5.33	2.99	2.22	1.91	1.87	1.96
Cu	3.61	2.18	10.9	6.5	3.96	2.28	1.71	1.49	1.47	1.54
H	14.1	7.88	36.5	20.7	12.2	6.8	5	4.26	4.12	4.34
H$_2$O	6.17	3.55	16.9	9.76	5.83	3.28	2.44	2.08	2.02	2.09
Pb	2.29	1.46	7.7	4.7	2.93	1.72	1.31	1.16	1.16	1.25
Ti	4	2.39	11.8	6.93	4.2	2.4	1.8	1.55	1.51	1.59
U	2.17	1.38	7.37	4.52	2.82	1.66	1.27	1.12	1.13	1.21

Reference: as for Table IX.22

Table IX.36 Mass stopping power in various substances for muons of various energies

Substance	\multicolumn{10}{c}{energy/MeV}
	1	2	5	10	20	50	100	200	500	1000
	\multicolumn{10}{c}{stopping power/MeV kg^{-1} m^2}
			×10^{-1}	×10^{-1}	×10^{-1}	×10^{-1}	×10^{-1}	×10^{-1}	×10^{-1}	×10^{-1}
air	4.24	2.44	11.7	6.86	4.2	2.5	1.98	1.8	1.85	2
Ag	2.51	1.54	7.83	4.74	2.97	1.81	1.45	1.33	1.37	1.47
Al	3.73	2.18	10.6	6.24	3.83	2.3	1.81	1.64	1.67	1.76
C	4.54	2.59	12.3	7.19	4.38	2.6	2.05	1.86	1.9	2.01
Cu	2.95	1.77	8.85	5.29	3.29	1.99	1.59	1.46	1.49	1.58
H	11.2	6.23	28.9	16.7	10	5.9	4.61	4.14	4.19	4.47
H$_2$O	4.94	2.83	13.5	7.88	4.81	2.86	2.25	2.03	2.04	2.13
Pb	1.92	1.21	6.3	3.87	2.46	1.52	1.23	1.14	1.19	1.29
Ti	3.26	1.93	9.48	5.63	3.48	2.1	1.66	1.51	1.54	1.63
U	1.81	1.15	6.04	3.72	2.37	1.47	1.18	1.1	1.16	1.26

Reference: as for Table IX.22

Table IX.37 Air kerma-rate constant for gamma radiation of energy greater than 50 keV from some common radionuclides
Values of the constant may in practice depend on source construction and on bremsstrahlung plus annihilation radiation contributions.

Nuclide constant/	$^{137}_{56}$Ba† + $^{137}_{55}$Cs	$^{60}_{27}$Co	$^{59}_{26}$Fe	$^{131}_{53}$I	$^{192}_{77}$Ir	$^{140}_{57}$La	$^{24}_{11}$Na	$^{226}_{88}$Ra††	$^{182}_{73}$Ta	$^{170}_{69}$Tm
10^{-17} m² Gy s⁻¹ Bq⁻¹	2.1	8.5	4.1	1.4	2.9	7.4	12	5.5	4.4	0.018

†Conventional isomer
††Including decay products in 0.5 mm Pt
Reference: Kaye and Laby, *Tables of Physical and Chemical Constants*, 15 ed., Longmans, 1986.

Table IX.38 International Commission on Radiation Protection risk and risk-weighting factors of fatal cancer and serious hereditary defect-inducing radiation in adult humans for a dose equivalent of 1 sievert

Site	Risk factor/10^{-4}	Risk-weighting factor
body (whole)	165	1
bone surfaces	5	0.03
breast	25	0.15
gonads (hereditary risk)	40	0.25
lung	20	0.12
red bone marrow	20	0.12
skin	1.7	0.01
thyroid	5	0.03
tissue (other)	10	0.06

Reference: *ICRP Recommendations*, Publication no. 26, 1977

Table IX.39 Values of atomic number, Z, and mass number, A, half-life and/or natural abundance, decay scheme, mass excess, cross section, parity and spin for various nuclides

The mass excess is the measured atomic mass in MeV less 931.478 times the mass number. Parity, P, is indicated by the superscript to the right of the value of the spin, J, the combination having the symbol J^P. Reference and significance of other symbols on p. 262.

Z	Nuclide□	A†	half-life and/or abundance	decay scheme⊕	mass excess/ MeV	cross section*/barn	J^P
89	Ac	209	0.1 s	α 7.59	9.1		
	Ac	210	0.35 s	α 7.46	8.9		
	Ac	211	0.25 s	α 7.48; β⁺; K	7.4		
	Ac	212	0.9 s	α 7.38	7.2		
	Ac	213	0.8 s	100% α 7.36	6.2		
	Ac	214	8.2 s	≥ 45% α 7.21; ≥ 38% α 7.08; K	6.1		
	Ac	215	0.17 s	α 7.60	6.0		
	Ac	216	0.0003 s	90% α 9.07; 10% α 8.99	8.0		
	Ac	216 m	0.0003 s	46% α 9.10; 50% α 9.02; 2% α 8.28; 2% α 8.19	8.0		
	Ac	217	1.1×10^{-7} s	100% α 9.65	8.70		
	Ac	218	2.7×10^{-7} s	100% α 9.20	10.84		
	Ac	219	7×10^{-6} s	100% α 8.66	11.56		
	Ac	220	0.026 s	24% α 7.85; 21% α 7.68; 23% α 7.61; 32% α (dist); γ	13.75		
	Ac	221	0.052 s	70% α 7.65; 20% α 7.44; 10% α 7.38	14.52		
	Ac	222	5.5 s	93% α 7.00; 6% α 6.96;	16.62		
	Ac	222 m	66 s	α (7.00); IT; K			
	Ac	223	2.2 min	32% α 6.66; 45% α 6.65; 14% α 6.56; 8% α (dist); 1% K	17.83		
	Ac	224	2.9 h	2% α 6.21; 3% α 6.14; 2% α 6.06; 3% α (dist); 90% K; 56% γ 0.217; 25% γ 0.133	20.22		
	Ac	225	10.0 d	51% α 5.83; 27% α 5.79 (double); 10% α 5.73; 12% α (dist); 2% γ 0.100; 1% γ 0.150	21.63		
	Ac	226	29 h	24% β⁻ 1.11; 10% β⁻ 1.04; 49% β⁻ 0.88; 17% K; 27% γ 0.230; 18% γ 0.158; 6% γ 0.254	24.301		
	Ac	227	21.77 yr	1% α 4.95 (double); 54% β⁻ 0.044; 35% β⁻ 0.035; 10% β⁻ 0.019	25.850	900 (γ) (s)	$\frac{3-}{2}$
	Ac	228	6.13 h	11% β⁻ 2.14; 32% β⁻ 1.17; 27% γ 0.911; 16% γ 0.969; 12% γ 0.339	28.90		
	Ac	229	63 min	β⁻; γ	30.7		
	Ac	230	122 s	β⁻; γ	33.8		
	Ac	231	7.5 min	β⁻; γ	35.9		
	Ac	232	35 s	β⁻	39.2		
47	Ag	100	2.3 min	β⁺; K; γ	−77.9		
	Ag	101	10.8 min	β⁺; K; 92% γ 0.263; 33% γ 0.668; 33% γ 1.164	−81.3		
	Ag	102	13.0 min	β⁺; K; 97% γ 0.557; 58% γ 0.719; 17% γ 1.745	−82.33		5⁺
	Ag	102 m	7.7 min	β⁺ 4.0; K; 49% IT; 42% γ 0.557; 10% γ 1.835; 7% γ 2.055	−82.32		2⁺
	Ag	103	1.10 h	58% K; 21% β⁺ 1.7; 22% γ 0.119; 20% γ 0.148; 9% γ 0.267	−84.80		$\frac{7+}{2}$
	Ag	103 m	5.7 s	100% IT; 21% γ 0.134	−84.67		
	Ag	104	69 min	β⁺; K; 92% γ 0.556; 66% γ 0.768; 25% γ 0.942	−85.15		5⁺

(Continued)

Table IX.39 (Continued)

Z	Nuclide □ A†	half-life and/or abundance	decay scheme⊕	mass excess/ MeV	cross section*/barn	J^P
	Ag	104 m	33 min	β^+ 2.7; K; 33% IT; 60% γ 0.556; 3% γ 1.239; 1% γ 1.782	−85.14		2^+
	Ag	105	41.3 d	100% K; 42% γ 0.345; 31% γ 0.280; 12% γ 0.443, 0.645	−87.08		$\frac{1-}{2}$
	Ag	105 m	7.2 min	100% IT	−87.05		
	Ag	106	23.96 min	K; β^+ 2.0; 17% γ 0.512	−86.93		1^+
	Ag	106 m	8.5 d	100% K; 88% γ 0.512; 29% γ 0.717; 28% γ 0.451	−86.84		6^+
	Ag	107	51.83%		−88.404	37 (γ); 0.3 (γm) (s)	$\frac{1-}{2}$
	Ag	107 m	44.3 s	100% IT; 5% γ 0.093	−88.311		$\frac{7+}{2}$
	Ag	108	2.37 min	96% β^- 1.6; 2% β^- 1.0; 2% K; 2% γ 0.633	−87.60		1^+
	Ag	108 m	127 yr	K; β^+; 9% IT; 91% γ 0.434; 91% γ 0.614; 91% γ 0.723	−87.49		6^+
	Ag	109	48.17%		−88.722	88 (γ); 4 (γm)	$\frac{1-}{2}$
	Ag	109 m	39.8 s	100% IT; 4% γ 0.088	−88.634		$\frac{7+}{2}$
	Ag	110	24.4 s	95% β^- 2.9; 5% γ 0.658	−87.456		1^+
	Ag	110 m	252.2 d	30% β^- 0.5; 67% β^- 0.09; 1% IT; 94% γ 0.658; 73% γ 0.885; 34% γ 0.937	−87.338	80 (γ)	6^+
	Ag	111	7.45 d	94% β^- 1.03; 5% γ 0.342; 1% γ 0.246	−88.226	3 (γ) (r)	$\frac{1-}{2}$
	Ag	111 m	65 s	100% IT; 1% γ 0.060	−88.166		
	Ag	112	3.14 h	54% β^- 4.0; 21% β^- 3.4; 42% γ 0.617; 5% γ 1.388; 3% γ 0.607, 1.614	−86.62		2
	Ag	113	1.15 min	β^-; γ	−86.8		
	Ag	113	5.37 h	85% β^- 2.0; (2% m); 9% γ 0.298; 2% γ 0.259; 1% γ 0.316	−87.04		$\frac{1}{2}$
	Ag	114	4.5 s	90% β^- 4.9; 10% γ 0.558; 1% γ 0.576	−85.2		1^+
	Ag	115	18 s	β^-; γ			
	Ag	115	20.0 min	32% β^- 3.2; 24% β^- 3.0; (9% m); 32% γ 0.230; 8% γ 0.214; 6% γ 0.473	−84.9		
	Ag	116	2.68 min	β^-; 84% γ 0.513; 13% γ 0.700; 9% γ 1.305	−82.6		
	Ag	116 m	10.5 s	β^-; 2% IT; 95% γ 0.513; 42% γ 0.706; 16% γ 1.029	−82.5		
	Ag	117	1.21 min	β^-; γ	−82.2		
	Ag	117	5.3 s	β^-	−82.2		
	Ag	118	3.7 s	β^-; 95% γ 0.488; 40% γ 0.677; 12% γ 3.226	−80.2		
	Ag	118 m	2.8 s	β^-; 41% IT; 59% γ 0.488; 58% γ 0.677; 32% γ 1.059	−80.1		
	Ag	119	2.1 s	3% β^- 4.2; 31% β^- 3.9; (22% m); 11% γ 0.626; 10% γ 0.366; 10% γ 0.399 (double)	−79.3		
	Ag	120	1.17 s	β^-; γ	−78		
	Ag	120 m	0.32 s	β^-; 37% IT; γ	−78		
	Ag	121	<3 s	β^-; γ			
	Ag	122	1.5 s	β^-; γ	−71		
	Ag	123	0.39 s	β^-; n			
13	Al	23	0.47 s	β^+; p	6.77		
	Al	24	2.05 s	7% β^+ 8.7; 48% β^+ 4.4; 39% β^+ 3.4; 96% γ 1.368; 43% γ 2.753; 41% γ 7.066	−0.05		4^+

(Continued)

Table IX.39 (Continued)

Z	Nuclide □	A†	half-life and/or abundance	decay scheme⊕	mass excess/ MeV	cross section*/barn	J^P
	Al	24 m	0.13 s	93% IT, 4% β⁺ 12.9, 93% γ 0.439	0.39		1^+
	Al	25	7.17 s	99% β⁺ 3.26; 1% γ 1.612	−8.913		$\frac{5}{2}^+$
	Al	26	7.2×10^5 yr	82% β⁺ 1.17; 18% K; 100% γ 1.809; 2% γ 1.130	−12.208		5^+
	Al	26 m	6.35 s	100% β⁺ 3.2	−11.979		0^+
	Al	27	100%		−17.194	0.23 (γ)	$\frac{5}{2}^+$
	Al	28	2.241 min	100% β⁻ 2.87; 100% γ 1.779	−16.848		3^+
	Al	29	6.6 min	89% β⁻ 2.41; 89% γ 1.273; 7% γ 2.426; 4% γ 2.028	−18.21		$\frac{5}{2}^+$
	Al	30	3.7 s	18% β⁻ 6.3; 67% β⁻ 5.0; 65% γ 2.235; 41% γ 1.263; 33% γ 3.498	−15.9		
	Al	31	0.64 s	β⁻; γ	−15.1		
95	Am	237	1.22 h	100% K; 47% γ 0.280; 8% γ 0.438; 4% γ 0.474	46.6		$\frac{5}{2}$
	Am	237 f	5×10^{-9} s	SF	48.7		
	Am	238	1.6 h	100% K; 29% γ 0.963; 24% γ 0.919; 11% γ 0.561	48.42		1^+
	Am	239	11.9 h	100% K; 15% γ 0.278; 11% γ 0.228; 3% γ 0.210, 0.226	49.39		$\frac{5}{2}^-$
	Am	239 f	1.6×10^{-7} s	SF	51.9		
	Am	240	50.8 h	100% K; 73% γ 0.988; 25% γ 0.889; 2% γ 0.099	51.44		
	Am	240 f	9×10^{-4} s	SF	54.0		
	□Am	241	432.0 yr	85% α 5.49; 13% α 5.44; 2% α (dist); 36% γ 0.060	52.932	560 (γ); 62 (γm); 3.2(f)	$\frac{5}{2}^-$
	Am	241 f	2×10^{-6} s	SF	55.1		
	Am	242	16.01 h	37% β⁻ 0.66; 46% β⁻ 0.62; 17% K	55.463	2100 (f) (r)	1^-
	Am	242 m	152 yr	100% IT	55.511	1650 (γ) (r); 7400 (f) (r)	5^-
	Am	242 f	0.0140 s	SF	57.8		
	Am	243	7370 yr	88% α 5.28; 11% α 5.23; 1% α 5.18; 66% γ 0.075; 6% γ 0.044	57.170	6 (γ) (r); 80 (γm) (r)	$\frac{5}{2}^-$
	Am	243 f	5×10^{-6} s	SF	59.2		
	Am	244	10.1 h	100% β⁻ 0.39; 66% γ 0.744; 28% γ 0.898; 19% γ 0.154	59.877	2200 (f) (r)	
	Am	244 m	26 min	80% β⁻ 1.50; 20% β⁻ 1.46	59.95	1600 (f) (r)	
	Am	244 f	0.0011 s	SF	61.5		
	Am	245	2.05 h	77% β⁻ 0.90; 6% γ 0.253	61.897		
	Am	245 f	6.4×10^{-7} s	SF			
	Am	246	39 min	β⁻; 52% γ 0.679; 35% γ 0.205; 25% γ 0.154			
	Am	246	25.0 min	7% β⁻ 2.2; 37% β⁻ 1.2; 29% γ 1.079; 26% γ 0.799; 18% γ 1.062	64.9		
	Am	246 f	7×10^{-5} s	SF			
	Am	247	24 min	76% β⁻ 1.4; 24 β⁻ 1.3; 23% γ 0.285; 6% γ 0.226	67.1		
18	Ar	33	0.18 s	18% β⁺ 10.6; 48% β⁺ 9.8; 27% β⁺ 5.1; 34% p; 48% γ 0.810	−9.39		$\frac{1}{2}^+$
	Ar	34	0.84 s	94% β⁺ 5.0; 3% γ 0.666; 1% γ 0.461; 1% γ 3.129	−18.379		0^+
	Ar	35	1.77 s	98% β⁺ 4.94; 1% γ 1.219	−23.049		$\frac{3}{2}^+$
	Ar	36	0.337%		−30.231	5 (γ) (r)	0^+

(Continued)

Table IX.39 (Continued)

Z	Nuclide	$A^\dagger$	half-life and/or abundance	decay scheme	mass excess/ MeV	cross section*/barn	J^P
	Ar	37	35.02 d	100% K	−30.948		$\frac{3+}{2}$
	Ar	38	0.063%		−34.715	0.8 (γ) (r)	0^+
	Ar	39	269 yr	100% β^- 0.57	−33.24	600 (γ) (r)	$\frac{7-}{2}$
	Ar	40	99.60%		−35.040	0.6 (γ)	0^+
	Ar	41	1.83 h	1% β^- 2.49; 99% β^- 1.20; 99% γ 1.294	−33.068		$\frac{7-}{2}$
	Ar	42	33 yr	100% β^- 0.60	−34.42		0^+
	Ar	43	5.4 min	β^-; γ	−32.0		
	Ar	44	11.9 min	β^-; 66% γ 0.182; 66% γ 1.705; 33% γ 1.887	−32.3		0^+
33	As	68	2.7 min	9% β^+ 7.2; 28% β^+ 6.2; 29% β^+ 4.8; 66% γ 1.016; 24% γ 0.651; 23% γ 0.763	−58.8		
	As	69	15 min	90% β^+ 2.95; 2% K; 5% γ 0.233; 2% γ 0.146; 1% γ 0.087	−63.12		
	As	70	53 min	β^+ 2.8; 16% K; 82% γ 1.040; 21% γ 0.668; 21% γ 1.114	−64.34		4
	As	71	61 h	68% K; β^+ 0.81; 84% γ 0.175; 4% γ 1.096; 3% γ 0.500	−67.893		$\frac{5-}{2}$
	As	72	26.0 h	β^+ 3.3; 23% K; 80% γ 0.834; 8% γ 0.630; 1% γ 1.051	−68.23		2^-
	As	73	80.3 d	100% K (100% m); 11% γ (m) 0.053	−70.949		$\frac{3-}{2}$
	As	74	17.79 d	17% β^- 1.35; 15% β^- 0.72; 31% β^+; 37% K; 60% γ 0.596; 15% γ 0.635	−70.860		2^-
	As	75	100%		−73.034	4.4 (γ)	$\frac{3-}{2}$
	As	76	26.3 h	51% β^- 2.97; 36% β^- 2.41; 45% γ 0.559; 6% γ 0.657; 3% γ 1.216	−72.291		2^-
	As	77	38.83 h	98% β^- 0.69; 2% γ 0.239	−73.916		$\frac{3-}{2}$
	As	78	90.7 min	34% β^- 4.3; 17% β^- 3.7; 54% γ 0.614; 18% γ 0.695; 11% γ 1.309	−72.7		
	As	79	9.0 min	95% β^- 2.1; (98% m); 9% γ (m) 0.096; 2% γ 0.365; 2% γ 0.432	−73.71		$\frac{3-}{2}$
	As	80	15.2 s	56% β^- 5.7; 27% β^- 5.0; 42% γ 0.662; 7% γ 1.645; 4% γ 1.207	−72.1		1
	As	81	33 s	67% β^- 3.8; (3% m); 20% γ 0.468; 8% γ 0.491; 1% γ 0.521	−72.6		
	As	82	14 s	15% β^- 5.5; 44% β^- 4.3; 19% β^- 3.7; 72% γ 0.654; 39% γ 1.895; 27% γ 0.819, 1.731			
	As	82	19 s	80% β^- 7.2; 15% γ 0.654; 4% γ 1.731; 2% γ 2.353	−70.4		
	As	83	13s	β^-; (70% m); γ	−69.9		
	As	84	0.7 s	β^-	−66.2		
	As	84	5.3 s	β^-; 49% γ 1.455; 21% γ 0.667; 5% γ 2.087	−66.2		
	As	85	2.03 s	β^-; 23% n; 16% γ 1.455; 7% γ 0.667; 2% γ 1.112	−63.5		
	As	86	0.9 s	β^-; 4% n; γ	−59.7		
	As	87	0.6 s	β^-	−56		

(Continued)

Table IX.39 (Continued)

Z	Nuclide	A†	half-life and/or abundance	decay scheme⊕	mass excess/ MeV	cross section*/barn	J^P
85	At	196	0.3 s	α 7.06	4.1		
	At	197	0.4 s	α 6.96	−6.0		
	At	198	5 s	α 6.75	−6.7		
	At	198 m	1.5 s	α 6.85			
	At	199	7 s	α 6.64	−8.5		
	At	200 (g)	42 s	32% α 6.47; 21% α 6.42; β$^+$; K; γ			
	At	200 (m)	4.3 s	α 6.54			
	At	201	1.5 min	71% α (6.34): β$^+$; K; γ	−10.5		
	At	202	3.0 min	9% α 6.23; 6% α 6.13; β$^+$; K; γ			
	At	203	7.3 min	31% α (6.09); β$^+$; K; γ	−12.0		
	At	204	9.3 min	4% α (5.95); β$^+$; K; 94% γ 0.683; 91% γ 0.515; 71% γ 0.425	−12.0		
	At	205	26 min	10% α 5.90; 87% K; β$^+$; 28% γ 0.719; 8% γ 0.669; 5% γ 0.629	−13.0		$\frac{9-}{2}$
	At	206	31 min	1% α 5.70; 82% K; 7% β$^+$ 3.3; 97% γ 0.700; 86% γ 0.477; 48% γ 0.396	−12.7		
	At	207	1.80 h	10% α 5.76; β$^+$; K; 49% γ 0.814; 22% γ 0.588; 14% γ 0.300	−13.3		$\frac{9-}{2}$
	At	208	1.63 h	1% α 5.64; β$^+$; K; 98% γ 0.686; 90% γ 0.660; 46% γ 0.177	−12.6		
	At	209	5.4 h	4% α 5.65; 96% K; 94% γ 0.545; 86% γ 0.782; 66% γ 0.790	−12.89		$\frac{9-}{2}$
	At	210	8.3 h	β$^+$; K; 99% γ 1.181; 79% γ 0.245; 47% γ 1.483	−11.98		5^+
	At	211	7.21 h	42% α 5.87; 58% K	−11.65		$\frac{9-}{2}$
	At	212	0.315 s	84% α 7.68; 15% α 7.62; 1% α 7.09	−8.63		
	At	212m	0.122 s	33% α 7.90; 65% α 7.84; 2% α (dist)	−8.40		
	At	213	1.1×10^{-7} s	100% α 9.08	−6.59		$\frac{9-}{2}$
	At	214	2×10^{-6} s	100% α 8.82	−3.39		
	At	215	1.0×10^{-4} s	100% α 8.03	−1.26		
	At	216	3.0×10^{-4} s	97% α 7.80; 2% α 7.70; 1% α 7.48	2.24		1
	At	217	0.032 s	100% α 7.07	4.38		
	At	218	1.7 s	4% α 6.76; 90% α 6.69; 6% α 6.65	8.10		
	At	219	0.9 min	97% α (6.28); β$^-$	10.5		
79	Au	175	0.1 s	α 6.44	−17.2		
	Au	176	1.3 s	20% α 6.29; 80% α 6.26	−18.4		
	Au	177	1.3 s	35% α 6.15; 65% α 6.11	−21.2		
	Au	178	3 s	α 5.92	−22.4		
	Au	179	7 s	α 5.85	−24.8		
	Au	181	11 s	1% α 5.62; β$^+$; K	−27.6		
	Au	182	22 s	β$^+$; K; 46% γ 0.155; 18% γ 0.265; 7% γ 0.855	−28.2		
	Au	183	42 s	β$^+$; K	−30.0		
	Au	184	53 s	β$^+$; K; γ	−30.2		
	Au	185	4.3 min	β$^+$; K; γ	−31.7		
	Au	185	6.8 min	β$^+$; K; γ	−31.7		
	Au	186	11 min	β$^+$; K; γ	−31.7		3
	Au	186	<2 min	β$^+$; K	−31.7		
	Au	187	8 min	β$^+$; K; γ	−32.9		$\frac{1}{2}$

(Continued)

Table IX.39 (Continued)

Z	Nuclide	$A^\dagger$	half-life and/or abundance	decay scheme	mass excess/ MeV	cross section*/barn	J^P
	Au	188	8.8 min	β^+; K; γ	−32.5		1
	Au	189	28.7 min	β^+; K; γ	−33.4		$\frac{1+}{2}$
	Au	189 m	4.6 min	β^+; K; γ	−33.2		$\frac{11-}{2}$
	Au	190	43 min	98% K; β^+ 3.4; 71% γ 0.296; 25% γ 0.302; 10% γ 0.598	−32.88		1^-
	Au	191	3.2 h	100% K; γ	−33.9		$\frac{3+}{2}$
	Au	191 m	0.9 s	100% IT; γ	−33.6		
	Au	192	5.0 h	99% K; β^+ 2.5; γ	−32.77		1^-
	Au	193	17.5 h	K; β^+; 11% γ 0.186; 7% γ 0.256; 4% γ 0.268	−33.4		$\frac{3+}{2}$
	Au	193 m	3.9 s	100% IT; 65% γ 0.258	−33.1		$\frac{11-}{2}$
	Au	194	39.5 h	97% K; β^+ 1.5; 61% γ 0.328; 11% γ 0.294; 6% γ 1.469	−32.26		1^-
	Au	195	183 d	100% K; 11% γ 0.099; 1% γ 0.130	−32.57		$\frac{3+}{2}$
	Au	195 m	30.6 s	100% IT; 67% γ 0.262; 1% γ 0.200	−32.25		$\frac{11-}{2}$
	Au	196	6.18 d	93% K; 7% β^- 0.25; 88% γ 0.356; 23% γ 0.333; 7% γ 0.426	−31.16		2^-
	Au	196 m$_1$	8.2 s	100% IT	−31.08		5^+
	Au	196 m$_2$	9.7 h	100% IT; (100%m$_1$); 43% γ 0.148; 38% γ 0.188; 8% γ 0.168	−30.57		12^-
	Au	197	100%		−31.15	98.8 (γ)	$\frac{3+}{2}$
	Au	197m	7.7 s	100% IT; 73% γ 0.279; 3% γ 0.131; 1% γ 0.202	−30.74		$\frac{11-}{2}$
	Au	198	2.70 d	99% β^- 0.96; 96% γ 0.412; 1% γ 0.676	−29.59	$2.5 \times 10^4 (\gamma)$	2^-
	Au	198 m	2.3 d	100% IT; 77% γ 0.215; 70% γ 0.097; 51% γ 0.180	−28.78		
	Au	199	3.14 d	6% β^- 0.45; 72% β^- 0.29; 42% γ 0.158; 9% γ 0.208	−29.10	30 (γ) (r)	$\frac{3+}{2}$
	Au	200	12.6 h	77% β^- 2.2; 21% γ 0.368; 15% γ 1.226; 4% γ 1.263	−27.3		1
	Au	200 m	19 h	β^-; 16% IT; 83% γ 0.368; 82% γ 0.498; 80% γ 0.579	−26		12
	Au	201	26 min	82% β^- 1.3; 2% γ 0.543; 1% γ 0.167, 0.517, 0.613	−26.4		
	Au	202	29 s	90% β^- 3.5; 10% γ 0.440; 3% γ 1.125; 2% γ 1.204, 1.307	−23.9		
	Au	203	53 s	β^-; γ	−23.0		
	Au	204	40 s	β^-; γ			
5	B	8	0.77 s	93% β^+ 13.7; 100% 2 α	22.922		2^+
	B	10	19.8%		12.052	3838 (α)	3^+
	B	11	80.2%		8.668	0.005 (γ) (r)	$\frac{3-}{2}$
	B	12	0.020 s	97% β^- 13.37; 2% α; 1% γ 4.439	13.370		1^+
	B	13	0.017 s	92% β^- 13.44; 8% γ 3.68	16.562		$\frac{3-}{2}$
	B	14	0.016 s	87% β^- 14.5; 90% γ 6.09; 9% γ 6.73	23.66		2^-
56	Ba	117	1.9 s	β^+; K; p			
	Ba	119	5.4 s	β^+; K; p	−64.5		
	Ba	120	32 s	β^+; K; γ	−69		0^+
	Ba	121	30 s	β^+; K	−70.6		
	Ba	122	2.0 min	β^+; K	−74.3		0^+

(Continued)

Table IX.39 (Continued)

Z	Nuclide	A	half-life and/or abundance	decay scheme	mass excess/ MeV	cross section*/barn	J^P
	Ba	123	2.7 min	β^+; K; γ	−75.7		
	Ba	124	10.5 min	β^+; K; γ	−78.8		0^+
	Ba	125	3.5 min	β^+; K; γ	−79.5		
	Ba	125	8 min	β^+; K			
	Ba	126	100 min	β^+; K; 20% γ 0.234; 8% γ 0.258; 6% γ 0.241	−82.6		0^+
	Ba	127	12.7 min	β^+; K; 11% γ 0.181; 8% γ 0.115; 2% γ 0.066	−82.8		
	Ba	128	2.4 d	100% K; 14% γ 0.273	−85.48		0^+
	Ba	129	2.2 h	K; β^+ 1.4; γ	−85.05		$\frac{1}{2}^+$
	Ba	129 m	2.1 h	β^+; K; γ	−84.77		
	Ba	130	0.106%		−87.30	8 (γ) (r); 2.5 (γm) (r)	0^+
	Ba	131	12.0 d	100%; K; 42% γ 0.496; 28% γ 0.124; 22% γ 0.216	−86.73		$\frac{1}{2}^+$
	Ba	131 m	14.6 min	100% IT; 56% γ 0.108; 1% γ 0.078	−86.54		$\frac{9}{2}^-$
	Ba	132	0.101%		−88.45	7 (γ) (r); 0.6 (γm) (r)	0^+
	Ba	133	10.7 yr	100% K; 62% γ 0.356; 34% γ 0.081; 18% γ 0.303	−87.57		$\frac{1}{2}^+$
	Ba	133 m	38.9 h	100% IT; 18% γ 0.276	−87.28		$\frac{11}{2}^-$
	Ba	134	2.42%		−88.97	2 (t)	0^+
	Ba	135	6.59%		−87.87	6 (γ) (r); 0.014 (γm) (r)	$\frac{3}{2}^+$
	Ba	135 m	28.7 h	100% IT; 16% γ 0.268	−87.60		$\frac{11}{2}^-$
	Ba	136	7.85%		−88.91	0.4 (t)	0^+
	Ba	136 m	0.308 s	100% IT; 100% γ 0.819; 100% γ 1.048	−86.88		7^-
	Ba	137	11.2%		−87.73	5.1 (γ) (r)	$\frac{3}{2}^+$
	Ba	137 m	2.551 min	100% IT; 90% γ 0.662	−87.07		$\frac{11}{2}^-$
	Ba	138	71.7%		−88.27	0.4 (γ) (r)	0^+
	Ba	139	82.9 min	72% β^- 2.3; 22% γ 0.166	−84.93	6 (γ) (r)	
	Ba	140	12.79 d	22% β^- 1.01; 40% β^- 0.99; 24% γ 0.537; 13% γ 0.030; 6% γ 0.163	−83.29	1.6 (γ)	0^+
	Ba	141	18.3 min	β^-; 46% γ 0.190; 25% γ 0.304; 23% γ 0.277	−80.0		
	Ba	142	10.6 min	β^- 2.1; 20% γ 0.255; 15% γ 1.204; 13% γ 0.895	−77.8		0^+
	Ba	143	13.6 s	β^-; γ	−74.0		
	Ba	144	11.9 s	β^-; γ	−72.0		0^+
	Ba	145	5 s	β^-; γ	−67.8		
	Ba	148	0.5 s	β^-			0^+
4	Be	7	53.3 d	100% K; 10% γ 0.477	15.770	49000 (γ) (r)	$\frac{3}{2}^-$
	Be	8	7×10^{-17} s	100% 2α	4.942		0^+
	Be	9	100%		11.348	0.0076 (γ)	$\frac{3}{2}^-$
	Be	10	1.6×10^6 yr	100% β^- 0.556	12.608		0^+
	Be	11	13.8 s	57% β^- 11.5; 33% γ 2.12; 5% γ 6.79; 2% γ 5.85	20.18		$\frac{1}{2}^+$
	Be	12	0.011 s	β^-; n	25.0		0^+
83	Bi	189	<1.5 s	α 6.67	−9.9		
	Bi	190	5 s	90% α (6.45); β^+; K	−10.9		
	Bi	191	13 s	40% α (6.31); β^+; K	−13.1		
	Bi	191 m	20 s	α 6.86			
	Bi	192	40 s	20% α (6.05); β^+; K	−13.7		

(Continued)

Table IX.39 (Continued)

Z	Nuclide	$A^\dagger$	half-life and/or abundance	decay scheme$^\oplus$	mass excess/ MeV	cross section*/barn	J^P
	Bi	193	64 s	60% α (5.89); β$^+$; K	−15.6		
	Bi	193 m	3.5 s	24% α 6.48; 1% α 6.18; β$^+$; K			
	Bi	194	1.7 min	β$^+$; K; 100% γ 0.965; 87% γ 0.575; 70% γ 0.280	−16.0		
	Bi	195	2.8 min	β$^+$; K	−17.7		
	Bi	195 m	90 s	4% α (6.10); β$^+$; K			
	Bi	196	4.5 min	β$^+$; K; 100% γ 1.049; 62% γ 0.688; 46% γ 0.372	−17.8		
	Bi	197 (m)	8 min	β$^+$; K			
	Bi	198	11.9 min	β$^+$; K; 100 γ 1.063; 80% γ 0.198; 79% γ 0.562	−19.3		
	Bi	198 m	8 s	100% IT; 39% γ 0.248	−19.1		
	Bi	199 (g)	27 min	1000 K	−20.6		$\frac{9-}{2}$
	Bi	199 (m)	24.7 min	α	−20.0		
	Bi	200	36 min	K; β$^+$; 100% γ 1.026; 90% γ 0.462; 85% γ 0.420	−20.5		7
	Bi	200 m	0.4 s	100% IT; γ	−20.0		10
	Bi	201	1.77 h	β$^+$; K; γ	−21.4		$\frac{9-}{2}$
	Bi	201 m	59 min	β$^+$; K; IT; γ	−20.6		
	Bi	202	1.67 h	99% K; β$^+$; 100% γ 0.961; 84% γ 0.422; 61% γ 0.657	−21.0		5
	Bi	203	11.8 h	100% K; (23%m$_1$); 29% γ 0.820; 16% γ (m) 0.825; 13% γ 0.897	−21.6		$\frac{9-}{2}$
	Bi	204	11.2 h	100% K; (13% m); 100% γ 0.899; 74% γ 0.375; 59% γ 0.984	−20.8		6$^+$
	Bi	205	15.31 d	100% K; 32% γ 1.764; 31% γ 0.703; 16% γ 0.988	−21.07		$\frac{9-}{2}$
	Bi	206	6.24 d	100% K; 99% γ 0.803; 67% γ 0.881; 41% γ 0.516	−20.03		6$^+$
	□Bi	207	38 yr	100% K; (83%m); 98% γ 0.570; 72% γ (m) 1.064; 7% γ 1.770	−20.06		$\frac{9-}{2}$
	Bi	208	3.7×10^5 yr	100% K; 100% γ 2.615	−18.88		
	Bi	209	100%		−18.27	0.019 (γ) (r); 0.014 (γm) (r)	$\frac{9-}{2}$
	Bi	210	5.01 d	100% β$^-$ 1.16	−14.80	0.050 (γ) (s)	1$^-$
	Bi	210 m	3.0×10^6 yr	55% α 4.94; 40% α 4.90; 4% α 4.57; 50% γ 0.266; 28% γ 0.305; 4% γ 0.650	−14.53		9$^-$
	Bi	211	2.15 min	84% α 6.62; 16% α 6.28; 13% γ 0.351	−11.87		
	Bi	212	60.6 min	10% α 6.09; 26% α 6.05; 55% β$^-$ 2.25; 6% γ 0.727; 1%γ 0.040, 0.785, 1.621	−8.14		1
	Bi	212 m$_1$	25 min	α 6.34; β$^-$	−7.88		
	Bi	212 m$_2$	9 min	β$^-$			
	Bi	213	45.6 min	2% α 5.87; 66% β$^-$ 1.42; 31% β$^-$ 0.98; 21% γ 0.440	−5.24		
	Bi	214	19.7 min	18% β$^-$ 3.27; 18%β$^-$ 1.54; 18% β$^-$ 1.51; 46% γ 0.60.9; 16% γ 1.765; 15% γ 1.120	−1.21		
	Bi	215	7 min	β$^-$	1.7		
97	Bk	242	7 min	100% K	57.8		
	Bk	242 f$_1$	6×10^{-7} s	SF			
	Bk	242 f$_2$	10^{-8} s	SF			
	Bk	243	4.5 h	100% K; γ	58.69		
	Bk	244	4.4 h	100% K; γ	60.65		
	Bk	244 f	8×10^{-7} s	SF			

(Continued)

Table IX.39 (Continued)

Z	Nuclide	$A^\dagger$	half-life and/or abundance	decay scheme	mass excess/ MeV	cross section*/barn	J^P
	Bk	245	4.90 d	100% K; 91% γ 0.253; 2% γ 0.381	61.81		$\frac{3}{2}^-$
	Bk	245 f	2×10^{-9} s	SF			
	Bk	246	1.80 d	100% K; 61% γ 0.799; 6% γ 1.081; 5% γ 0.834	64.0		
	Bk	247	1.4×10^3 yr	17% α 5.71; 45% α 5.53; 13% α 5.69; 25% α (dist); 40% γ 0.084; 30% γ 0.265	65–48		
	Bk	248	23.5 h	65% β⁻ 0.7; 30% K; 5% γ 0.551	68.0		
	Bk	248	>9 yr	?	68.0		
	Bk	249	325 d	100% β⁻ 0.13	69.848	100 (t) (r) <6 (f) (r)	$\frac{7}{2}^+$
	Bk	250	3.22 h	5% β⁻ 1.78; 83% β⁻ 0.75; 45% γ 0.989; 35% γ 1.032; 4% γ 1.029	72.95	1000 (f) (r)	2^-
	Bk	251	56 min	β⁻; γ	75.3		
35	Br	72	1.31 min	23% β⁺ 7.1; 20% β⁺ 6.6; 70% γ 0.862; 17% γ 1.317; 13% γ 0.455	−58.9		
	Br	73	3.3 min	β⁺ 3.5; K; γ	−63.7		
	Br	74	25.3 min	β⁺; K; 84% γ 0.635 (double); 19% γ 0.282; 8% γ 2.615	−65.30		
	Br	74m	42 min	β⁺; K; 117% γ 0.635 (double); 37% γ 0.728; 9% γ 1.269	−65.1		
	Br	75	96 min	β⁺ 1.7; 25% K; 92% γ 0.286; 7% γ 0.141; 4% γ 0.377, 0.428, 0.432	−69.16		
	Br	76	16.1 h	β⁺ 3.9; 43% K; 74% γ 0.559; 16% γ 0.657; 14% γ 1.854	−70.30		1^-
	Br	77	57.0 h	99% K; 1% β⁺ 0.34; 23% γ 0.239; 22% γ 0.521; 4% γ 0.297	−73.242		$\frac{3}{2}^-$
	Br	77m	4.3 min	100% IT; 14% γ 0.106	−73.136		$\frac{9}{2}^+$
	Br	78	6.46 min	β⁺ 2.5; 8% K; 14% γ 0.614	−73.458		1^+
	Br	79	50.69%		−76.070	10.8 (γ); 2.4 (γm)	$\frac{3}{2}^-$
	Br	79m	4.86 s	100% IT; 76% γ 0.207	−75.863		$\frac{9}{2}^+$
	Br	80	17.68 min	85% β⁻ 2.0; 6% β⁻ 1.4; 6% K; 3% β⁺ 0.86; 7% γ 0.616; 1% γ 0.666	−75.891		1^+
	Br	80m	4.42 h	100% IT; 39% γ 0.037	−75.805		5^-
	Br	81	49.31%		−77.976	2.7 (γ)	$\frac{3}{2}^-$
	Br	82	35.34 h	99% β⁻ 0.44; 83% γ 0.776; 70% γ 0.554; 43% γ 0.619	−77.498		5^-
	Br	82m	6.05 min	98% IT; 2% β⁻ 3.1	−77.452		2^-
	Br	83	2.39 h	99% β⁻ 0.92; (100% m); 5% γ (m) 0.009; 1% γ 0.530	−79.03		
	Br	84	31.8 min	33% β⁻ 4.7; 13% β⁻ 3.8; 42% γ 0.882; 15% γ 1.898; 7% γ 2.484, 3.928	−77.76		2^-
	Br	84m	6.0 min	100% β⁻ 2.2; 100% γ 0.424; 98% γ 0.882; 97% γ 1.463	−77.5		
	Br	85	2.87 min	96% β⁻ 2.5; (100% m); 14% γ (m) 0.305; 3% γ 0.802; 2% γ 0.925	−78.7		$\frac{3}{2}^-$
	Br	86	55.7 s	15% β⁻ 7.3; 51% β⁻ 3.0; 62% γ 1.565; 19% γ 2.751; 10% γ 1.362	−76.0		

(Continued)

Table IX.39 (Continued)

Z	Nuclide □	A†	half-life and/or abundance	decay scheme⊕	mass excess/ MeV	cross section*/barn	J^P
	Br	87	55.6 s	β^-; 2% n; 32% γ 1.420; 12% γ 1.476; 9% γ 1.578	−74.2		
	Br	88	16.7 s	β^-; 6% n; 77% γ 0.775; 16% γ 0.802; 5% γ 1.441, 3.932	−71.1		
	Br	89	4.37 s	β^-; 13% n; γ	−69.1		
	Br	90	1.96 s	β^-; 23% n; γ	−65.2		
	Br	91	0.541 s	β^-; 10% n			
	Br	92	0.37 s	β^-; 16% n	−58		
6	C	9	0.127 s	β^+; 100% p	28.91		
	C	10	19.15 s	99% β^+ 1.87; 99% γ 0.718; 1% γ 1.022	15.703		0^+
	C	11	20.38 min	100% β^+ 0.96	10.650		$\frac{3-}{2}$
	C	12	98.89%		0.0	0.0034 (γ)	0^+
	C	13	1.11%		3.1250	0.0009 (γ)	$\frac{1-}{2}$
	□C	14	5730 yr	100% β^- 0.156	3.0199		0^+
	C	15	2.45 s	32% β^- 9.77; 68% β^- 4.47; 68% γ 5.299	9.873		$\frac{1+}{2}$
	C	16	0.75 s	84% β^- 5.4; 100% n	13.69		0^+
20	Ca	37	0.173 s	16% β^+ 10.6; 47% β^+ 5.5; 76% p; γ	−13.16		$\frac{3+}{2}$
	Ca	38	0.44 s	74% β^+ 5.6; 25% β^+ 4.0; (100% m); 25% γ 1.568	−22.06		0^+
	Ca	39	0.860 s	100% β^+ 5.49	−27.282		$\frac{3+}{2}$
	Ca	40	96.94%		−34.847	0.4 (γ) (r)	0^+
	Ca	41	1.03×10^3 yr	100% K	−35.139		$\frac{7-}{2}$
	Ca	42	0.647%		−38.544	0.7 (γ) (r)	0^+
	Ca	43	0.135%		−38.405	6 (γ) (r)	$\frac{7-}{2}$
	Ca	44	2.09%		−41.466	0.88 (γ)	0^+
	Ca	45	165 d	100% β^- 0.26	−40.810		$\frac{7-}{2}$
	Ca	46	0.0035%		−43.138	0.7 (γ) (r)	0^+
	Ca	47	4.540 d	16% β^- 1.99; 84% β^- 0.68; 77% γ 1.297; 7% γ 0.489; 7% γ 0.808	−42.343		$\frac{7-}{2}$
	Ca	48	0.187%		−44.216	1.1 (γ) (r)	0^+
	Ca	49	8.72 min	92% β^- 2.2; 92% γ 3.084; 7% γ 4.072	−41.286		
	Ca	50	14 s	100% β^- 3.1; (100% m) 97% γ (m) 0.257; 58% γ 0.072; 58% γ 1.519	−39.57		0^+
48	Cd	100	1.1 min	β^+; K; γ	−73.4		0^+
	Cd	101	1.2 min	β^+; K; γ	−75.5		
	Cd	102	5.5 min	K; β^+; (95% m); 62% γ 0.481; 13% γ 1.037; 9% γ 0.505	−79.4		0^+
	Cd	103	7.3 min	β^+; K; γ	−80.6		
	Cd	104	58 min	99% K; 1% β^+ 0.4; (100% m) 47% γ 0.084; 20% γ 0.709; 6% γ 0.559	−83.6		0^+
	Cd	105	56 min	K; β^+ 1.7; (85% m); 5% γ 0.962; 4% γ 0.347; 4% γ 1.302	−84.34		$\frac{5+}{2}$
	Cd	106	1.25%		−87.13	1(γ) (r)	0^+
	Cd	107	6.50 h	100% K; (100% m); 5% γ (m) 0.093	−86.99		$\frac{5+}{2}$
	Cd	108	0.89%		−89.25	1.2 (γ) (r)	0^+
	□Cd	109	453 d	100% K; (100% m); 4% γ (m) 0.088	−88.540	700 (γ) (r)	$\frac{5+}{2}$

(Continued)

Table IX.39 (Continued)

Z	Nuclide	A	half-life and/or abundance	decay scheme	mass excess/ MeV	cross section*/barn	J^P
	Cd	110	12.5%		−90.349	11 (γ) (r); 0.10 (γm) (r)	0^+
	Cd	111	12.8%		−89.254	24 (γ) (r)	$\frac{1}{2}^+$
	Cd	111m	48.6 min	100% IT; 94% γ 0.245; 31% γ 0.150	−88.858		$\frac{11}{2}^-$
	Cd	112	24.1%		−90.578	2 (γ) (r)	0^+
	Cd	113	9×10^{15} yr 12.2%	100% β^- 0.32	−89.050	19 800 (γ)	$\frac{1}{2}^+$
	Cd	113m	14 yr	100% β^- 0.58	−88.787		$\frac{11}{2}^-$
	Cd	114	28.7%		−90.020	0.30 (γ) (s); 0.04 (γm) (s)	0^+
	Cd	115	53.38 h	63% β^- 1.1; 33% β^- 0.6; (100% m); 46% γ (m) 0.336; 34% γ 0.528; 10% γ 0.492	−88.09		$\frac{1}{2}^+$
	Cd	115m	44.8 d	97% β^- 1.6; 2% γ 0.934; 1% γ 1.291	−87.92		$\frac{11}{2}^-$
	Cd	116	7.5%		−88.718	0.05 (γ) (s); 0.025 (γm) (s)	0^+
	Cd	117	2.4 h	21% β^- 2.2; 32% β^- 0.6; (94% m); 29% γ 0.273; 18% γ 0.344; 18% γ 1.303	−86.42		$\frac{1}{2}^+$
	Cd	117m	3.3 h	52% β^- 0.6; (1% m); 25% γ 1.997; 23% γ 1.066; 15% γ 0.564, 1.434	−86.29		$\frac{11}{2}^-$
	Cd	118	50.3 min	100% β^- 0.7	−86.71		0^+
	Cd	119	2.7 min	β^- 3.2; (93% m); 25% γ 0.293; 13% γ 0.343; 8% γ 1.610, 1.764	−84.2		$\frac{1}{2}^+$
	Cd	119m	1.9 min	1% β^- 2.2; 30% β^- 1.6; 28% γ 1.025; 24% γ 2.021; 21% γ 0.721	−84.1		$\frac{11}{2}^-$
	Cd	120	50.8 s	β^- 1.7	−83.98		0^+
	Cd	121	12.8 s	β^-; γ	−81		
	Cd	121	4.8 s	β^-; γ	−81		
	Cd	122	5.8 s	β^-	−80		0^+
	Cd	124	0.9 s	β^-; 50% γ 0.180; 23% γ 0.063; 13% γ 0.143	−76		0^+
58	Ce	129	3.5 min	β^+; K; γ			
	Ce	130	25 min	β^+; K; γ			0^+
	Ce	131	5 min	β^+; K; γ	−79.5		
	Ce	131	10 min	89% K; β^+; 20% γ 0.169; 9% γ 0.396; 9% γ 1.440	−79.5		
	Ce	132	3.5 h	100% K; 79% γ 0.182; 11% γ 0.155; 5% γ 0.217	−82.3		0^+
	Ce	133	97 min	β^+; K; 45% γ 0.097; 16% γ 0.077; 11% γ 0.558	−82.2		$\frac{1}{2}$
	Ce	133	5.4 h	β^+; K; 39% γ 0.477; 20% γ 0.510; 19% γ 0.058	−82.2		$\frac{9}{2}^-$
	Ce	134	76 h	100% K	−84.8		0^+
	Ce	135	17.8 h	99% K; β^+; 42% γ 0.266; 23% γ 0.300; 19% γ 0.607	−84.6		$\frac{1}{2}$
	Ce	135m	20 s	100% IT; 74% γ 0.214; 21% γ 0.150; 20% γ 0.083, 0.296	−84.1		
	Ce	136	0.190%		−86.50	6 (γ) (r); 1.0 (γm) (r)	0^+
	Ce	137	9.0 h	100% K; 2% γ 0.447	−85.9		$\frac{3}{2}^+$
	Ce	137m	34.4 h	99% IT; 1% K; 11% γ 0.254	−85.7		$\frac{11}{2}^-$

(Continued)

Table IX.39 (Continued)

Z	Nuclide □	A†	half-life and/or abundance	decay scheme⊕	mass excess/ MeV	cross section*/barn	J^P
	Ce	138	0.254%		−87.57	1.0 (γ) (r); 0.015 (γm) (r)	0^+
	Ce	139	137.2 d	100% K; 80% γ 0.166	−86.97		$\frac{3+}{2}$
	Ce	139m	56.4 s	100% IT; 93% γ 0.754	−86.21		$\frac{11-}{2}$
	Ce	140	88.5%		−88.08	0.56 (γ) (r)	0^+
	Ce	141	32.55 d	31% β⁻ 0.58; 69% β⁻ 0.44; 48% γ 0.145	−85.44	29 (γ) (r)	$\frac{7-}{2}$
	Ce	142	11.1%		−84.54	0.95 (γ)	0^+
	Ce	143	33.0 h	39% β⁻ 1.4; 47% β⁻ 1.1; 42% γ 0.293; 12% γ 0.057; 5% γ 0.665, 0.722	−81.61	6.2 (γ) (r)	$\frac{3-}{2}$
	Ce	144	284 d	76% β⁻ 0.32 (2% m); 11% γ 0.134; 2% γ 0.080	−80.43	1.0 (γ)	0^+
	Ce	145	3.0 min	76% β⁻ 1.7; 68% γ 0.725; 14% γ 0.063; 10% γ 1.148	−77.1		
	Ce	146	14 min	β⁻ 0.7; 53% γ 0.317; 20% γ 0.218; 8% γ 0.133	−75.8		0^+
	Ce	147	56 s	β⁻; γ	−72.2		
	Ce	148	48 s	β⁻; γ	−70.8		0^+
	Ce	149	5.0 s	β⁻; γ	−67.5		
	Ce	150	4 s	β⁻; γ	−65		0^+
	Ce	151	1.0 s	β⁻; γ	−62.7		
98	Cf	240	1.1 min	α 7.59	58.0		0^+
	Cf	241	4 min	α 7.34	59.2		
	Cf	242	3.5 min	80% α 7.39; 20% α 7.35	59.33		0^+
	Cf	243	10 min	30% α (7.17); 70% K	60.9		
	Cf	244	19 min	75% α 7.22; 25% α 7.18	61.47		0^+
	Cf	245	44 min	α 7.14; 70% K	63.38		
	Cf	246	36 hr	78% α 6.76; 22% α 6.72	64.096		0^+
	Cf	247	3.15 h	100% K; 1% γ 0.294	66.2		
	Cf	248	334 d	82% α 6.26; 18% α 6.22	67.24		0^+
	Cf	249	351 yr	2% α 6.19; 83% α 5.81; 5% α 5.76; 10% α (dist); 66% γ 0.388; 15% γ 0.333; 3% γ 0.253	69.722	480 (γ) (s); 1630 (f)	$\frac{9-}{2}$
	Cf	250	13.1 yr	83% α 6.03; 17% α 5.99	71.170	2000 (γ) (s); <350 (f) (r)	0^+
	Cf	251	900 yr	12% α 6.01; 27% α 5.85; 35% α 5.68; 26% α (dist); 17% γ 0.177; 5% γ 0.227; 1% γ 0.285	74.13	2900 (γ) (s); 4500 (f) (s)	$\frac{1+}{2}$
	□Cf	252	2.64 yr	82% α 6.12; 15% α 6.08; 3% SF	76.03	20 (γ) 32 (f) (s)	0^+
	Cf	253	17.8 d	100% β⁻ 0.29	79.30	18 (γ) (r); 1300 (f) (r)	
	Cf	254	60.5 d	100% SF	81.34	100 (t) (r)	0^+
	Cf	256	12 min	100% SF			0^+
17	Cl	32	0.298 s	1% β⁺ 11.7; 60% β⁺ 9.5; 92% γ 2.230; 21% γ 4.770; 4% γ 2.464	−13.33		1^+
	Cl	33	2.51 s	98% β⁺ 4.56; 1% γ 0.841; 1% γ 1.966; 1% γ 2.867	−21.003		$\frac{3+}{2}$
	Cl	34	1.526 s	100% β⁺ 4.47	−24.438		0^+
	Cl	34m	32.0 min	28% β⁺ 2.5; 24% β⁺ 1.3; 47% IT; 42% γ 0.146; 42% γ 2.127; 13% γ 1.176	−24.292		3^+
	Cl	35	75.77%		−29.014	43 (γ)	$\frac{3+}{2}$
	Cl	36	3.00×10^5 yr	98% β⁻ 0.710; 2% K	−29.522		2^+
	Cl	37	24.23%		−31.762	0.43 (γ)	$\frac{3+}{2}$

(Continued)

Table IX.39 (Continued)

Z	Nuclide	$A^\dagger$	half-life and/or abundance	decay scheme	mass excess/ MeV	cross section*/barn	J^P
	Cl	38	37.3 min	58% β^- 4.92; 31% β^- 1.11; 42% γ 2.167; 31% γ 1.642	−29.798		2^-
	Cl	38m	0.715 s	100% IT; 100% γ 0.671	−29.127		5^-
	Cl	39	56.2 min	7% β^- 3.44; 83% β^- 1.92; 54% γ 1.267; 47% γ 0.250; 37% γ 1.517	−29.80		$\frac{3+}{2}$
	Cl	40	1.32 min	9% β^- 7.5; 28% β^- 3.2; 20% β^- 2.9; 77% γ 1.461; 30% γ 2.840; 15% γ 2.622	−27.5		2^-
	Cl	41	34 s	β^-; γ	−27.4		
96	Cm	238	2.3 h	α 6.52; <90% K	49.40		0^+
	Cm	239	2.9 h	100% K; γ	51.1		
	Cm	240	26.8 d	71% α 6.29; 29% α 6.25	51.71		0^+
	Cm	240f	10^{-11} s	SF			
	Cm	241	32.8 d	1% α 5.94; 99% K; 71% γ 0.472; 4% γ 0.132; 4% γ 0.431	53.70		$\frac{1+}{2}$
	Cm	241f	1.5×10^{-8} s	SF	55.7		
	Cm	242	162.8 d	74% α 6.11; 26% α 6.07	54.802	20 (γ)	0^+
	Cm	242f$_1$	4×10^{-11} s	SF	57.6		
	Cm	242f$_2$	1.8×10^{-7} s	SF			
	Cm	243	28.5 yr	2% α 6.07; 73% α 5.79; 11% α 5.74; 14% α (dist); 14% γ 0.278; 11% γ 0.228; 3% γ 0.210	57.177	1000 (t) (r); 130 (γ); 610 (f)	$\frac{5+}{2}$
	Cm	243f	4×10^{-8} s	SF	58.7		
	□Cm	244	18.10 yr	77% α 5.81; 23% α 5.76	58.450	14 (γ); 1 (f)	0^+
	Cm	244f	$>10^{-7}$ s	SF	61.5		
	Cm	245	8540 yr	1% α 5.49; 1% α 5.45; 93% α 5.36; 5% α 5.30; 5% γ 0.133; 5% γ 0.172	61.001	350 (γ); 2000 (f)	$\frac{7+}{2}$
	Cm	245f	1.3×10^{-8} s	SF	62.7		
	Cm	246	4710 yr	79% α 5.39; 21% α 5.34	62.616	1.3 (γ); 0.2 (f)	0^+
	Cm	247	1.6×10^7 yr	14% α 5.27; 6% α 5.21; 71% α 4.87; 9% α (dist); 72% γ 0.402; 3% γ 0.278; 2% γ 0.288	65.53	60 (γ); 100 (f) (s)	$\frac{9-}{2}$
	Cm	248	3.40×10^5 yr	75% α 5.08; 17% α 5.03; 8% SF	67.39	4 (γ) (s); 0.3 (f)	0^+
	Cm	249	65 min	96% β^- 0.90; 2% γ 0.634; 1% γ 0.560	70.75	2 (γ) (r)	$\frac{1+}{2}$
	Cm	250	1.1×10^4 yr	100% SF	72.99	80 (γ) (r)	0^+
	Cm	251	16.8 min	73% β^- 1.42; 12% γ 0.543; 2% γ 0.530; 1% γ 0.390, 0.438	76.7		
27	Co	53m	0.25 s	1% p; β^+	−39.45		
	Co	54	0.193 s	100% β^+ 7.22	−48.010		0^+
	Co	54m	1.43 min	100% β^+ 4.5; 100% γ 0.411; 100% γ 1.130; 100% γ 1.407	−47.811		
	Co	55	17.54 h	β^+ 1.5; 23% K; 75% γ 0.931; 20% γ 0.477; 16% γ 1.409	−54.024		$\frac{7-}{2}$
	□Co	56	78.8 d	81% K; 19% β^+ 1.5; 100% γ 0.847; 67% γ 1.238; 17% γ 2.599	−56.037		4^+
	Co	57	271.7 d	100% K; 86% γ 0.122; 11% γ 0.136; 10% γ 0.014	−59.342		$\frac{7-}{2}$
	□Co	58	70.8 d	85% K; 15% β^+ 0.47; 99% γ 0.811; 1% γ 0.864	−59.844	1900 (γ)	2^+
	Co	58m	9.2 h	100% IT	−59.819	1.4×10^5 (γ)	5^+

(Continued)

Table IX.39 (Continued)

Z	Nuclide	$A^\dagger$	half-life and/or abundance	decay scheme	mass excess/ MeV	cross section*/barn	J^P
	Co	59	100%		−62.226	18 (γ) 19 (γm)	$\frac{7-}{2}$
	□Co	60	5.272 yr	100% β⁻ 0.32; 100% γ 1.173; 100% γ 1.332	−61.647	2.0 (γ) (s)	5⁻
	Co	60m	10.47 min	100% IT; 2% γ 0.059	−61.588	58 (γ) (s)	2⁺
	Co	61	1.65 h	96% β⁻ 1.25; 86% γ 0.067; 3% γ 0.909	−62.897		$\frac{7-}{2}$
	Co	62(g)	1.50 min	64% β⁻ 4.1; 28% β⁻ 3.0; 83% γ 1.173; 16% γ 2.302; 13% γ 1.129	−61.43		
	Co	62(m)	13.9 min	62% β⁻ 3.0; 20% β⁻ 2.2; 98% γ 1.173; 69% γ 1.164; 19% γ 2.004	−61.41		
	Co	63	27.5 s	94% β⁻ 3.6; 49% γ 0.087; 3% γ 0.982; 2% γ 0.156	−61.85		
	Co	64	0.30 s	90% β⁻ 7.3; 10% γ 1.346; 5% γ 0.931	−59.79		
24	Cr	45	0.05 s	β⁺; p	−19.5		
	Cr	46	0.26 s	100% β⁺ 6.6	−29.46		0⁺
	Cr	48	21.56 h	100% K; 100% γ 0.116; 100% γ 0.308	−42.82		0⁺
	Cr	49	41.9 min	β⁺ 1.6; K; 51% γ 0.091; 29% γ 0.153; 17% γ 0.062	−45.329		$\frac{5-}{2}$
	Cr	50	4.35%		−50.258	15.9 (γ)	0⁺
	Cr	51	27.70 d	100% K; 10% γ 0.320	−51.448		$\frac{7-}{2}$
	Cr	52	83.79%		−55.415	0.8 (γ) (r)	0⁺
	Cr	53	9.50%		−55.284	18 (γ) (r)	$\frac{3-}{2}$
	Cr	54	2.36%		−56.931	0.38 (γ) (r)	0⁺
	Cr	55	3.52 min	100% β⁻ 2.60	−55.106		$\frac{3-}{2}$
	Cr	56	5.9 min	100% β⁻ 1.53; 100% γ 0.026; 100% γ 0.083	−55.27		0⁺
55	Cs	116	57 s	β⁺ 3.3; K	−73.3		
	Cs	117	8 s	β⁺; K	−66.9		
	Cs	118	16 s	β⁺; K	−67.9		
	Cs	119	38 s	β⁺; K	−72.5		
	Cs	120	61 s	β⁺; K; γ	−73.4		
	Cs	121	126 s	β⁺; K; γ	−77.1		
	Cs	122	4.5 min	β⁺; K; γ			
	Cs	122	21 s	β⁺; K; γ	−78.0		
	Cs	123	5.9 min	β⁺; K; 15% γ 0.098; 15% γ 0.598; 9% γ 0.177	−81.2		
	Cs	123 m	1.6 s	100% IT; γ			
	Cs	124	31 s	β⁺; K; γ	−81.5		
	Cs	125	45 min	61% K; 27% β⁺ 2.05; 25% γ 0.526; 9% γ 0.112; 5% γ 0.412	−84.04		$\frac{1+}{2}$
	Cs	126	1.64 min	β⁺ 3.8; 18% K; 42% γ 0.389; 5% γ 0.491; 5% γ 0.925	−84.3		1⁺
	Cs	127	6.2 h	97% K; 1% β⁺ 1.1; 62% γ 0.411; 12% γ 0.125; 5% γ 0.462	−86.23		$\frac{1+}{2}$
	Cs	128	3.62 min	β⁺ 2.9; 39% K; 26% γ 0.443; 2% γ 0.527; 1% γ 1.140	−85.94		1⁺
	Cs	129	32.3 h	100% K; 32% γ 0.372; 23% γ 0.411; 4% γ 0.549	−87.49		$\frac{1+}{2}$
	Cs	130	29.9 min	K; β⁺ 2.0; 2% β⁻ 0.44; 4% γ 0.536	−86.86		1⁺

(Continued)

Table IX.39 (Continued)

Z	Nuclide A†		half-life and/or abundance	decay scheme⊕	mass excess/ MeV	cross section*/barn	J^P
	Cs	131	9.688 d	100% K	−88.07		$\frac{5+}{2}$
	Cs	132	6.47 d	97% K; 2% β⁻ 0.8; 1% β⁺ 0.4; 98% γ 0.668; 2% γ 0.465; 1% γ 0.630	−87.18		2
	Cs	133	100%		−88.09	27 (γ); 2.5 (γm)	$\frac{7+}{2}$
	Cs	134	2.06 yr	70% β⁻ 0.66; 98% γ 0.605; 85% γ 0.796; 15% γ 0.569	−86.91	140 (γ) (r)	4⁺
	Cs	134 m	2.91 h	100% IT; 13% γ 0.127	−86.77		8⁻
	Cs	135	3.0 × 10⁶ yr	100% β⁻ 0.21	−87.67	9 (γ) (s)	$\frac{7+}{2}$
	Cs	135 m	53 min	100% IT; 100% γ 0.787; 96% γ 0.840	−86.04		
	Cs	136	13.1 d	2% β⁻ 0.68; 94% β⁻ 0.34; (15% m); 100% γ 0.818; 80% γ 1.048; 47% γ 0.341	−86.36		5⁺
	Cs	136 m	19 s	100% IT			
	□Cs	137	30.17 yr	5% β⁻ 1.17; 95% β⁻ 0.51; (95% m); 85% γ (m) 0.662	−86.56	0.11 (γ) (r)	$\frac{7+}{2}$
	Cs	138	32.2 min	9% β⁻ 3.9; 41% β⁻ 2.8; 75% γ 1.436; 28% γ 1.010; 27% γ 0.463	−83.0		3⁻
	Cs	138 m	2.9 min	75% IT; 20% β⁻ 3.3; 25% γ 0.463; 25% γ 1.436; 20% γ 0.192	−82.9		
	Cs	139	9.5 min	84% β⁻ 4.3; 7% γ 1.283; 2% γ 0.627; 1% γ 1.421	−80.6		
	Cs	140	65 s	14% β⁻ 6.0; 20% β⁻ 5.4; 72% γ 0.602; 12% γ 0.908; 6% γ 1.201	−77.2		
	Cs	141	24.9 s	50% β⁻ 5.0; γ	−75.0		
	Cs	142	1.69 s	β⁻; γ	−71.0		
	Cs	143	1.78 s	β⁻ 5.7; 2% n; γ	−68.4		
	Cs	144	1.00 s	β⁻; 3% n; γ	−63.9		
	Cs	145	0.58 s	β⁻; 12% n; γ	−61.7		
	Cs	146	0.34 s	β⁻; 14% n			
29	Cu	58	3.20 s	82% β⁺ 7.5; 16% γ 1.455; 12% γ 1.448; 5% γ 0.040	−51.662		1⁺
	Cu	59	81.5 s	58% β⁺ 3.8; 15% γ 1.302; 12% γ 0.878; 8% γ 0.339	−56.352		$\frac{3-}{2}$
	Cu	60	23.4 min	β⁺ 3.8; 7% K; 88% γ 1.332; 45% γ 1.792; 22% γ 0.826	−58.343		2⁺
	Cu	61	3.41 h	β⁺ 1.2; 38% K; 13% γ 0.284; 12% γ 0.656; 7% γ 0.067	−61.981		$\frac{3-}{2}$
	Cu	62	9.73 min	98% β⁺ 2.9; 2% K	−62.796		1⁺
	Cu	63	69.2%		−65.579	4.4 (γ)	$\frac{3-}{2}$
	Cu	64	12.70 h	41% K; 19% β⁺ 0.66; 40% β⁻ 0.58; 1% γ 1.346	−65.423		1⁺
	Cu	65	30.8%		−67.262	2.17 (γ)	$\frac{3-}{2}$
	Cu	66	5.10 min	92% β⁻ 2.6; 8% γ 1.039	−66.257	140 (γ) (s)	1⁺
	Cu	67	62.0 h	20% β⁻ 0.58; 23% β⁻ 0.49; 56% β⁻ 0.40; 47% γ 0.185; 17% γ 0.093; 7% γ 0.091	−67.31		$\frac{3-}{2}$
	Cu	68	30 s	31% β⁻ 4.6; 40% β⁻ 3.5; 17% β⁻ 2.3; γ	−65.39		1⁺
	Cu	68 m	3.75 min	86% IT; 2% β⁻ 2.9; 74% γ 0.526; 71% γ 0.084; 16% γ 0.111	−64.66		
	Cu	69	3.0 min	79% β⁻ 2.5; 10% γ 1.006; 6% γ 0.834; 3% γ 0.530	−65.9		

(Continued)

Table IX.39 (Continued)

Z	Nuclide	$A^\dagger$	half-life and/or abundance	decay scheme	mass excess/ MeV	cross section*/barn	J^P
	Cu	70	5 s	46% β^- 6.2; 54% β^- 5.3; 54% γ 0.885	−63.4		1^+
	Cu	70	42 s	30% β^- 4.5; 60% β^- 3.3; 100% γ 0.885; 87% γ 0.902; 57% γ 1.252	−63.3		
66	Dy	147 m	59 s	IT; γ	−63.5		
	Dy	148	3.1 min	β^+; K; 100% γ 0.620	−67.8		0^+
	Dy	149	4.1 min	β^+; K; γ	−67.5		
	Dy	150	7.17 min	31% α (4.23); β^+ 0.5; K; 68% γ 0.397	−69.1		0^+
	Dy	151	17 min	6% α (4.07); β^+; K; 11% γ 0.477; 10% γ 0.547; 9% γ 0.176	−68.60		$\frac{7^-}{2}$
	Dy	152	2.37 h	100% K; 98% γ 0.257	−70.12		0^+
	Dy	153	6.3 h	β^+; K; 7% γ 0.081; 7% γ 0.214; 6% γ 0.100	−69.16		$\frac{7^-}{2}$
	Dy	154	10^7 yr	α 2.87	−70.39		0^+
	Dy	155	10.0 h	97% K; β^+ 1.08; 68% γ 0.227; 4% γ 0.185; 3% γ 1.090	−69.16		$\frac{3^-}{2}$
	Dy	156	0.057%		−70.53	33 (γ) (s)	0^+
	Dy	157	8.1 h	100% K; 94% γ 0.326; 2% γ 0.182; 1% γ 0.083	−69.43		$\frac{3^-}{2}$
	Dy	158	0.100%		−70.41	70 (γ) (r)	0^+
	Dy	159	144.4 d	100% K; 2% γ 0.058	−69.17		$\frac{3^-}{2}$
	Dy	160	2.3%		−69.67	60 (γ)	0^+
	Dy	161	19.0%		−68.06	570 (γ)	$\frac{5^+}{2}$
	Dy	162	25.5%		−68.18	160 (γ)	0^+
	Dy	163	24.9%		−66.38	130 (γ)	$\frac{5^-}{2}$
	Dy	164	28.1%		−65.97	900 (γ); 1800 (γm)	0^+
	Dy	165	2.33 h	83% β^- 1.29; 4% γ 0.095	−63.61	4000 (γ) (r)	$\frac{7^+}{2}$
	Dy	165 m	1.26 min	98% IT; 2% β^- 0.9; 3% γ 0.018	−63.50	2100 (γ) (r)	$\frac{1^-}{2}$
	Dy	166	81.5 h	5% β^- 0.48; 92% β^- 0.40; 13% γ 0.082	−62.58		0^+
	Dy	167	6.2 min	7% β^- 2.0; 85% β^- 1.8; 48% γ 0.570; 28% γ 0.259; 25% γ 0.310	−60.0		
68	Er	151	23 s	β^+; K	−58.2		
	Er	152	9.8 s	90% α (4.80); β^+; K	−60.4		0^+
	Er	153	36 s	38% α (4.67); β^+; K	−60.3		
	Er	154	3.8 min	β^+; K	−62.4		0^+
	Er	155	5.3 min	β^+; K	−62.06		
	Er	156	20 min	β^+; K; 37% γ 0.035; 6% γ 0.030	−63.9		
	Er	157	24 min	β^+; K; γ	−63.1		$\frac{3^-}{2}$
	Er	158	2.4 h	K; β^+ 0.7; 11% γ 0.072; 5% γ 0.387; 2% γ 0.249	−65.0		0^+
	Er	159	36 min	β^+; K; (21% m); 37% γ 0.624; 28% γ 0.649; 8% γ(m) 0.206	−64.4		$\frac{3^-}{2}$
	Er	160	28.6 h	100% K	−66.05		0^+
	Er	161	3.24 h	100% K; (27% m); 61% γ 0.827; 12% γ (m) 0.211; 3% γ 0.592	−65.20		$\frac{3^-}{2}$
	Er	162	0.14%		−66.34	19 (γ)	0^+
	Er	163	75.1 min	100% K	−65.17		$\frac{5^-}{2}$
	Er	164	1.56%		−65.94	13 (γ)	0^+
	Er	165	10.4 h	100% K	−64.52		$\frac{5^-}{2}$

(Continued)

Table IX.39 (Continued)

Z	Nuclide	A†	half-life and/or abundance	decay scheme⊕	mass excess/ MeV	cross section*/barn	J^P
	Er	166	33.4%		−64.92	5 (γ) (r); 15 (γm) (r)	0^+
	Er	167	22.9%		−63.29	650 (γ) (r)	$\frac{7+}{2}$
	Er	167 m	2.28 s	100% IT; 42% γ 0.208	−63.08		$\frac{1-}{2}$
	Er	168	27.1%		−62.99	2.0 (γ)	0^+
	Er	169	9.40 d	55% β⁻ 0.35; 45% β⁻ 0.34	−60.92		$\frac{1-}{2}$
	Er	170	14.9%		−60.10	5.7 (γ)	0^+
	Er	171	7.52 h	2% β⁻ 1.49; 94% β⁻ 1.07; 64% γ 0.308; 29% γ 0.296; 21% γ 0.112	−57.71	300 (γ) (r)	$\frac{5-}{2}$
	Er	172	49 h	3% β⁻ 0.48; 46% β⁻ 0.35; 48% β⁻ 0.28; 45% γ 0.610; 43% γ 0.407; 3% γ 0.068, 0.446	−56.49		0^+
	Er	173	1.4 min	23% β⁻ 2.2; 64% β⁻ 1.3; 54% γ 0.895; 48% γ 0.199; 47% γ 0.193	−53.7		
99	Es	243	21 s	α 7.89	64.8		
	Es	244	37 s	4% α (7.57); β⁺; K	66.0		
	Es	245	1.3 min	40% α (7.73); 60% K	66.4		
	Es	246	8 min	10% α (7.36); β⁺; K	67.9		
	Es	247	4.7 min	7% α (7.31); 93% K	68.55		
	Es	248	28 min	100% K	70.2		
	Es	249	1.70 h	1% α 6.77; 99% K; 40% γ 0.379; 9% γ 0.813; 3% γ 0.375	71.12		
	Es	250	8.6 h	100% K; 74% γ 0.829; 22% γ 0.303; 20% γ 0.349	73.2		
	Es	250	2.1 h	100% K; 17% γ 0.989; 14% γ 1.032	73.2		
	Es	251	33 h	100% K; 2% γ 0.178	74.51		
	Es	252	472 d	62% α 6.63; 11% α 6.56; 5% α (dist); 22% K; 16% γ 0.785; 12% γ 0.139; 2% γ 0.924	77.2		
	Es	253	20.47 d	90% α 6.63; 7% α 6.59; 3% α (dist)	79.012	160 (γm); <60 (f) (r)	$\frac{7+}{2}$
	Es	254	276 d	93% α 6.43; 2% α 6.42; 3% α 6.36; 2% α (dist); 2% γ 0.063	81.99	2800 (f)	
	Es	254 m	39.3 h	β⁻; 29% γ 0.649; 25% γ 0.694; 13% γ 0.689	82.07	1800 (f) (r)	2^+
	Es	255	38.3 d	7% α 6.30; 1% α 6.26; β⁻	84.1	65 (γ) (r)	
	Es	256	7.6 h	β⁻; γ	87.3		
	Es	256	22 min	β⁻	87.3		
63	Eu	138	1.5 s	β⁺			
	Eu	138	35 s	β⁺			
	Eu	139	22 s	β⁺; K; γ			
	Eu	140	1.3 s	β⁺; K; γ			
	Eu	141	40 s	β⁺; K; (3% m); 14% γ 0.394; 9% γ 0.385; 5% γ 0.383, 0.593	−69.9		
	Eu	141 m	3.3 s	β⁺; K; (66% m); 33% IT; 1% γ 0.096; 1% γ 0.395 (double)	−69.8		
	Eu	142	2.4 s	β⁺ 7.0; K; 15% γ 0.768; 2% γ 0.890; 2% γ 1.287, 1.658	−71.5		1^+
	Eu	142	1.22 min	β⁺; K; 100% γ 0.768; 92% γ 1.023; 87% γ 0.557	−71.5		

(Continued)

Table IX.39 (Continued)

Z	Nuclide $A^†$	half-life and/or abundance	decay scheme$^⊕$	mass excess/ MeV	cross section*/barn	J^P
	Eu 143	2.61 min	$β^+$ 4.1; 28% K; 7% γ 1.107; 3% γ 1.537; 2% γ 0.108, 1.913	−74.41		
	Eu 144	10.2 s	$β^+$ 5.3; K; 9% γ 1.660; 2% γ 0.818; 1% γ 2.423	−75.64		1^+
	Eu 145	5.93 d	98% K; 2% $β^+$ 1.7; 65% γ 0.894; 16% γ 1.659; 15% γ 0.654	−77.94		$\frac{5+}{2}$
	Eu 146	4.59 d	96% K; $β^+$ 1.5; 99% γ 0.747; 44% γ 0.633; 38% γ 0.634	−77.11		4^-
	Eu 147	22 d	99% K; $β^+$ 0.7; 22% γ 0.197; 19% γ 0.121; 9% γ 0.678	−77.54		$\frac{5+}{2}$
	Eu 148	54 d	100% K; 99% γ 0.550; 71% γ 0.630; 19% γ 0.414, 0.611	−76.24		5^-
	Eu 149	93.1 d	100% K; 4% γ 0.328; 3% γ 0.277; 1% γ 0.255	−76.44		$\frac{5+}{2}$
	Eu 150	36 yr	100% K; 94% γ 0.334; 79% γ 0.439; 52% γ 0.584			
	Eu 150	12.6 h	89% $β^-$ 1.01; 11% K; 4% γ 0.334; 2% γ 0.407	−74.76		0
	Eu 151	47.9%		−74.65	5800 (γ); 3200 (γm_1); 4 (γm_2)	$\frac{5+}{2}$
	Eu 152	13.2 yr	73% K; 8% $β^-$ 1.48; 31% γ 0.122; 27% γ 0.344; 21% γ 1.408	−72.88		3^-
	Eu 152 m_1	9.3 h	73% $β^-$ 1.87; 24% K; 13% γ 0.842; 11% γ 0.963; 6% γ 0.122	−72.84		0^-
	Eu 152 m_2	96 min	100% IT; 72% γ 0.090; 1% γ 0.018, 0.077	−72.74		8^-
	Eu 153	52.1%		−73.36	380 (γ)	$\frac{5+}{2}$
	Eu 154	8.5 yr	14% $β^-$ 1.86; 34% $β^-$ 0.61; 40% γ 0.123; 36% γ 1.274; 20% γ 0.723	−71.73		3^-
	Eu 154 m	45.8 min	100% IT; 36% γ 0.068; 28% γ 0.101; 9% γ 0.036	−71.6		
	Eu 155	4.96 yr	13% $β^-$ 0.25; 49% $β^-$ 0.14; 34% γ 0.086 (double); 23% γ 0.105; 1% γ 0.045	−71.83	4000 (γ)	$\frac{5+}{2}$
	Eu 156	15.1 d	31% $β^-$ 2.5; 30% $β^-$ 0.5; 10% γ 0.811; 9% γ 0.089; 8% γ 1.231	−70.08		0^+
	Eu 157	15.1 h	41% $β^-$ 1.30; 23% γ 0.064; 19% γ 0.411; 11% γ 0.371	−69.47		
	Eu 158	45.9 min	5% $β^-$ 3.4; 23% $β^-$ 2.5; 27% γ 2.4; 32% γ 0.944; 17% γ 0.977; 14% γ 0.079	−67.2		
	Eu 159	18.1 min	$β^-$ 2.6; 20% γ 0.068; 10% γ 0.079; 7% γ 0.096	−65.9		
	Eu 160	50 s	$β^-$; γ	−63.5		
9	F 17	64.5 s	100% $β^+$ 1.74	1.9517		$\frac{5+}{2}$
	F 18	109.7 min	97% $β^+$ 0.64; 3% K	0.873		1^+
	F 19	100%		−1.4874	0.0096 (γ) (r)	$\frac{1+}{2}$
	F 20	11.00 s	100% $β^-$ 5.392; 100% γ 1.634	−0.017		2^+
	F 21	4.35 s	29% $β^-$ 5.69; 63% $β^-$ 5.34; 71% γ 0.351; 8% γ 1.395	−0.05		$\frac{5+}{2}$

(Continued)

Table IX.39 (Continued)

Z	Nuclide	A	half-life and/or abundance	decay scheme	mass excess/ MeV	cross section*/barn	J^P
	F	22	4.23 s	3% β^- 7.50; 54% β^- 5.33; 16% β^- 5.21; 100% γ 1.275; 82% γ 2.083; 62% γ 2.166	2.83		4^+
	F	23	2.2 s	14% β^- 6.7; 10% β^- 6.2; 22% β^- 5.1; 35% β^- 4.7; 48% γ 1.701; 33% γ 2.129; 23% γ 1.822	3.4		
26	Fe	49	0.08 s	β^+; p	−24.5		
	Fe	52	8.28 h	57% β^+ 0.80; 43% K; (100% m); 99% γ 0.169	−48.33		0^+
	Fe	53	8.51 min	β^+ 2.7; K; 42% γ 0.378	−50.944		$\frac{1}{2}^-$
	Fe	53 m	2.51 min	100% IT; 99% γ 0.701; 86% γ 1.012; 86% γ 1.328	−47.904		
	Fe	54	5.8%		−56.251	2.2 (γ) (r)	0^+
	□Fe	55	2.69 yr	100% K	−57.479		$\frac{3}{2}^-$
	Fe	56	91.8%		−60.604	2.6 (γ) (r)	0^+
	Fe	57	2.15%		−60.179	2.4 (γ) (r)	$\frac{1}{2}^-$
	Fe	58	0.29%		−62.152	1.14 (γ)	0^+
	Fe	59	44.56 d	53% β^- 0.47; 45% β^- 0.27; 57% γ 1.099; 43% γ 1.292; 3% γ 0.192	−60.661		$\frac{3}{2}^-$
	Fe	60	3×10^5 yr	100% β^- 0.15 (100% m); 2% γ (m) 0.059	−61.44		0^+
	Fe	61	5.98 min	15% β^- 2.8; 37% β^- 2.6; 27% β^- 2.5; 44% γ 1.205; 43% γ 1.027; 22% γ 0.298	−59.0		
	Fe	62	68 s	100% β^- 2.1; 100% γ 0.506	−58.86		0^+
100	Fm	245	4 s	α 8.15	70.0		
	Fm	246	1.2 s	92% α (8.24); 8% SF	70.13		0^+
	Fm	247	9 s	α 8.18	71.5		
	Fm	247	35 s	α 7.93; α 7.87; <50% K	71.5		
	Fm	248	36 s	80% α 7.87; 20% α 7.83	71.89		0^+
	Fm	249	3 min	α 7.53	73.5		
	Fm	250	30 min	α 7.44	74.07		0^+
	Fm	250 m	1.8 s	100% IT			
	Fm	251	5.3 h	2% α 6.83; 98% K; 2% γ 0.880; 1% γ 0.406; 1% γ 0.453	76.0		
	Fm	252	25.4 h	85% α 7.04; 15% α 7.00	76.82		0^+
	Fm	253	3.0 d	5% α 6.94; 3% α 6.67; 4% α (dist); 88% K; 3% γ 0.272	79.35		$\frac{1}{2}^+$
	Fm	254	3.240 h	85% α 7.19; 14% α 7.15; 1% α 7.05	80.90		0^+
	Fm	255	20.1 h	93% α 7.02; 5% α 6.96; 1% α 7.08; 1% α 6.89	83.79	26 (γ) (r); 3300 (f) (s)	$\frac{7}{2}^+$
	Fm	256	2.63 h	8% α (6.92); 92% SF	85.48		0^+
	Fm	257	100.5 d	1% α 6.76; 4% α 6.70; 93% α 6.52; 2% α 6.44; 25% γ 0.115; 11% γ 0.241; 9% γ 0.179	88.59	5800 (t) (s); 3000 (f) (r)	
	Fm	258	3.8×10^{-4} s	100% SF			0^+
	Fm	259	1.5 s	100% SF			
87	Fr	203	0.7 s	α 7.13; β^+; K	1.2		
	Fr	204	2.1 s	70% α 7.03; 30% α 6.97	0.9		
	Fr	205	3.7 s	α 6.91	−1.0		
	Fr	206	16.0 s	85% α (6.79); β^+; K	−1.2		
	Fr	207	14.8 s	93% α (6.77); β^+; K	−2.7		
	Fr	208	58.0 s	74% α (6.64); β^+; K	−2.8		

(Continued)

Table IX.39 (Continued)

Z	Nuclide	$A^\dagger$	half-life and/or abundance	decay scheme	mass excess/MeV	cross section*/barn	J^P
	Fr	209	50.0 s	89% α 6.65; β⁺; K	−3.8		$\frac{9^-}{2}$
	Fr	210	3.2 min	α 6.57; β⁺; K; γ	−3.6		
	Fr	211	3.1 min	α 6.53; β⁺; K; γ	−4.2		
	Fr	212	19 min	9% α 6.41; 10% α 6.38; 18% α 6.26; 7% α (dist); β⁺; K; γ	−3.7		
	Fr	213	34.7 s	99% α 6.78; 1% K	−3.56		
	Fr	214	0.0050 s	93% α 8.43; 5% α 8.36; 1% α 7.94; 1% α 7.61	−0.97		
	Fr	214m	0.0034 s	46% α 8.55; 51% α 8.48; 3% α (dist)	−0.84		
	Fr	215	1.2×10^{-7} s	100% α 9.36	0.31		$\frac{9^-}{2}$
	Fr	216	7.0×10^{-7} s	100% α 9.01	2.98		
	Fr	217	2×10^{-5} s	100% α 8.32	4.31		$\frac{9^-}{2}$
	Fr	218	0.0007 s	93% α 7.87; 5% α 7.56; 1% α 7.53; 1% α (dist)	7.05		
	Fr	219	0.020 s	99% α 7.31; 1% α 6.97	8.62		
	Fr	220	27.4 s	61% α 6.69; 12% α 6.64; 10% α 6.58; 17% α (dist); γ	11.47		
	Fr	221	4.8 min	83% α 6.34; 1% α 6.24; 15% α 6.13; 1% α 5.98, 11% γ 0.218	13.27		
	Fr	222	14.4 min	β⁻	16.34		
	Fr	223	21.8 min	63% β⁻ 1.09; 34% γ 0.050; 8% γ 0.080; 3% γ 0.235	18.38		
	Fr	224	2.7 min	β⁻; γ	21.7		
	Fr	225	3.9 min	β⁻	23.8		
	Fr	226	48 s	β⁻; γ	27.5		
	Fr	227	2.4 min	β⁻; γ	29.6		
	Fr	228	39 s	β⁻			
	Fr	229	0.8 min	β⁻			
31	Ga	63	32.4 s	55% β⁺ 4.5; 11% γ 0.637; 10% γ 0.627; 6% γ 0.193	−56.7		
	Ga	64	2.62 min	β⁺ 6.1; 43% γ 0.992; 14% γ 0.808; 13% γ 3.366	−58.84		0^+
	Ga	65	15.2 min	β⁺ 2.2; 14% K; 55% γ 0.115; 12% γ 0.061; 9% γ 0.153	−62.654		$\frac{3^-}{2}$
	Ga	66	9.45 h	51% β⁺ 4.2; 44% K; 38% γ 1.039; 23% γ 2.752; 6% γ 0.834	−63.723		0^+
	Ga	67	78.3 h	100% K; 38% γ 0.093; 24% γ 0.185; 19% γ 0.300	−66.879		$\frac{3^-}{2}$
	Ga	68	68.3 min	89% β⁺ 1.9; 10% K; 3% γ 1.077	−67.085		1^+
	Ga	69	60.1%		−69.322	1.7 (γ)	$\frac{3^-}{2}$
	Ga	70	21.1 min	99% β⁻ 1.66	−68.905		1^+
	Ga	71	39.9%		−70.142	4.6 (γ)	$\frac{3^-}{2}$
	Ga	72	14.12 h	11% β⁻ 3.2; 28% β⁻ 1.8; 21% β⁻ 0.7; 96% γ 0.834; 26% γ 2.202; 24% γ 0.630	−68.591		3^-
	Ga	73	4.86 h	44% β⁻ 1.5; 47% β⁻ 1.2; (99% m); 47% γ 0.297; 10% γ (m) 0.053; 7% γ 0.326	−69.73		
	Ga	74	8.25 min	5% β⁻ 4.8; 51% β⁻ 2.5; 91% γ 0.596; 45% γ 2.354; 15% γ 0.608	−68.0		
	Ga	74m	10 s	100% IT; 75% γ 0.057	−68.0		1^+
	Ga	75	2.10 min	β⁻; γ	−68.6		

Table IX.39 (Continued)

Z	Nuclide	A	half-life and/or abundance	decay scheme	mass excess/ MeV	cross section*/barn	J^P
	Ga	76	27.1 s	14% β^- 6.2; 10% β^- 5.7; 10% β^- 5.2; 66% γ 0.563; 26% γ 0.546; 16% γ 1.108	−66.4		
	Ga	77	13.0 s	β^-	−66.4		
	Ga	78	5.09 s	β^-	−63.7		
	Ga	79	3.0 s	β^-; γ	−62.8		
	Ga	80	1.66 s	β^-; n	−59.5		
	Ga	81	1.23 s	β^-; n; γ			
	Ga	83	0.31 s	β^-; n			
64	Gd	143	108 s	β^+; K; 95% γ 0.272; 18% γ 0.588; 12% γ 0.799	−68.5		
	Gd	144	4.5 min	β^+; K; γ	−71.9		0^+
	Gd	145	22 min	β^+; K; 35% γ 1.758; 34% γ 1.881; 10% γ 1.042	−72.9		$\frac{1}{2}^+$
	Gd	145m	85 s	95% IT; β^+; K; 83% γ 0.721; 4% γ 0.387	−72.2		$\frac{11}{2}^-$
	Gd	146	48.3 d	100% K; 50% γ 0.115; 50% γ 0.116; 50% γ 0.155	−75.36		0^+
	Gd	147	38.1 h	100% K; 60% γ 0.229; 32% γ 0.396; 20% γ 0.929	−75.21		$\frac{7}{2}^-$
	Gd	148	98 yr	α 3.18	−76.27		0^+
	Gd	149	9.3 d	100% K; 53% γ 0.150; 31% γ 0.299; 25% γ 0.347	−75.13		$\frac{7}{2}^-$
	Gd	150	1.8×10^6 yr	α 2.73	−75.77		0^+
	Gd	151	120 d	100% K; γ	−74.17		$\frac{7}{2}^-$
	Gd	152	1.1×10^{14} yr 0.20%	α 2.14	−74.70	1100 (γ)	0^+
	Gd	153	242 d	100% K; 27% γ 0.097; 19% γ 0.103; 2% γ 0.070	−73.12		$\frac{3}{2}^-$
	Gd	154	2.1%		−73.70	90 (γ)	0^+
	Gd	155	14.8%		−72.07	6.1×10^4 (γ)	$\frac{3}{2}^-$
	Gd	156	20.6%		−72.54	2 (γ)	0^+
	Gd	157	15.7%		−70.83	2.6×10^5 (γ)	$\frac{3}{2}^-$
	Gd	158	24.8%		−70.69	2.4 (γ)	0^+
	Gd	159	18.6 h	β^- 0.97; 10% γ 0.363; 2% γ 0.058	−68.56		$\frac{3}{2}^-$
	Gd	160	21.8%		−67.94	0.77 (γ)	0^+
	Gd	161	3.6 min	5% β^- 1.6; 85% β^- 1.5; 61% γ 0.361; 23% γ 0.315; 14% γ 0.102	−65.51	4×10^4 (γ) (r)	$\frac{5}{2}^-$
	Gd	162	9 min	100% β^- 1.0; 53% γ 0.442; 46% γ 0.403; 7% γ 0.039	−64.4		0^+
32	Ge	64	64 s	β^+; 37% γ 0.427; 17% γ 0.667; 11% γ 0.128	−54.4		0^+
	Ge	65	31 s	β^+; 33% γ 0.650; 27% γ 0.062; 21% γ 0.809	−56.4		
	Ge	66	2.27 h	72% K; β^+; 29% γ 0.044; 28% γ 0.382; 11% γ 0.109, 0.273	−61.62		0^+
	Ge	67	19.0 min	β^+ 3.2; 4% K; 84% γ 0.167; 5% γ 1.473; 3% γ 0.911	−62.45		
	Ge	68	288 d	100% K	−66.97		0^+
	Ge	69	39.1 h	64% K; β^+ 1.2; 27% γ 1.106; 12% γ 0.574; 10% γ 0.872	−67.096		$\frac{5}{2}^-$
	Ge	70	20.5%		−70.561	3.3 (γ) (r)	0^+
	Ge	71	11.2 d	100% K	−69.906		$\frac{1}{2}^-$
	Ge	72	27.4%		−72.583	1.0 (γ) (r)	0^+
	Ge	73	7.8%		−71.294	15 (γ) (r)	$\frac{9}{2}^+$
	Ge	73m	0.50 s	100% IT; 11% γ 0.053	−71.227		$\frac{1}{2}^-$

(Continued)

Table IX.39 (Continued)

Z	Nuclide	A	half-life and/or abundance	decay scheme	mass excess/ MeV	cross section*/barn	J^P
	Ge	74	36.5%		−73.442	0.36 (γ) (r); 0.16 (γm) (r)	0^+
	Ge	75	82.78 min	87% β⁻ 1.18; 11% γ 0.265; 1% γ 0.199	−71.856		$\frac{1}{2}^-$
	Ge	75m	47.7 s	100% IT; 39% γ 0.140	−71.716		$\frac{7}{2}^+$
	Ge	76	7.8%		−73.214	0.06 (γ) (r) 0.10 (γm) (r)	0^+
	Ge	77	11.30 h	3% β⁻ 2.5; 21% β⁻ 1.5; 53% γ 0.264; 30% γ 0.211; 28% γ 0.216	−71.214		$\frac{7}{2}$
	Ge	77m	53 s	58% β⁻ 2.8; 21% β⁻ 2.6; 20% IT; 21% γ 0.216; 11% γ 0.160	−71.055		$\frac{1}{2}^-$
	Ge	78	1.47 h	96% β⁻ 0.71; 96% γ 0.277; 4% γ 0.294	−71.8		0^+
	Ge	79	42 s	75% β⁻ 4.2; 25% γ 0.230; 15% γ 0.543	−69.6		
	Ge	80	29.5 s	β⁻; 25% γ 0.266; 6% γ 0.110; 5% γ 1.564	−69.4		0^+
	Ge	81	10 s	β⁻ 6.3; γ	−66.3		
	Ge	82	4.6 s	β⁻; γ	−66.0		0^+
	Ge	83	1.9 s	β⁻	−62.5		
	Ge	84	1.2 s	β⁻			0^+
1	H	1	99.985%		7.2890	0.332 (γ)	$\frac{1}{2}^+$
	H	2	0.015%		13.1358	0.0005 (γ)	1^+
	□H	3	12.35 yr	100% β⁻ 0.01862	14.9499		$\frac{1}{2}^+$
2	He	3	0.00014%		14.9313	5330 (p)	$\frac{1}{2}^+$
	He	4	99.9999%		2.4249		0^+
	He	6	0.81 s	100% β⁻ 3.51	17.597		0^+
	He	8	0.122 s	88% β⁻ 9.7; 12% n; 88% γ 0.981	31.61		0^+
72	Hf	157	0.12 s	α 5.68	−39.0		
	Hf	158	3.0 s	α 5.27	−42.2		0^+
	Hf	159	5.6 s	α 5.09	−42.8		
	Hf	160	12 s	α 4.77	−45.8		0^+
	Hf	161	17 s	α 4.60	−46.1		
	Hf	166	6.8 min	β⁺; K; (100%m); 41% γ 0.079; 4% γ 0.342; 4% γ 0.408	−53.5		0^+
	Hf	167	2.0 min	β⁺; K; γ	−53.2		
	Hf	168	26.0 min	98% K; β⁺; γ	−55.1		0^+
	Hf	169	3.26 min	86% K; 14% β⁺ 1.9; 89% γ 0.493; 10% γ 0.370; 4% γ 0.124	−54.5		
	Hf	170	15.9 h	100% K; (5% m); 33% γ 0.165; 23% γ 0.621; 19% γ 0.120	−56.1		0^+
	Hf	171	12.1 h	β⁺; K; γ	−55.3		$\frac{7}{2}^+$
	Hf	172	1.87 yr	100% K; (100% m); 22% γ 0.024; 10% γ 0.126; 5% γ 0.067	−56.3		0^+
	Hf	173	24.0 h	100% K; 83% γ 0.124; 34% γ 0.297; 13% γ 0.140	−55.3		$\frac{1}{2}^-$
	Hf	174	2 × 10¹⁵ yr 0.16%	α 2.50	−55.83	400 (γ)	0^+
	Hf	175	70 d	100% K; 87% γ 0.343; 2% γ 0.089; 1% γ 0.433	−54.55		$\frac{5}{2}^-$
	Hf	176	5.2%		−54.57	30 (γ)	0^+

(Continued)

Table IX.39 (Continued)

Z	Nuclide	A	half-life and/or abundance	decay scheme	mass excess/ MeV	cross section*/barn	J^P
	Hf	177	18.6%		−52.88	390 (γ); 1.0 (γm$_1$)	$\frac{7}{2}^-$
	Hf	177m$_1$	1.1 s	100% IT; 83% γ 0.208; 50% γ 0.229; 38% γ 0.379	−51.56		$\frac{23}{2}^+$
	Hf	177m$_2$	51 min	100% IT; (100% m$_1$); 75% γ 0.277; 68% γ 0.295; 65% γ 0.327	−50.14		$\frac{37}{2}^-$
	Hf	178	27.1%		−52.43	40 (γ); 50 (γm$_1$)	0^+
	Hf	178m$_1$	4.3 s	100% IT; 97% γ 0.426; 94% γ 0.326; 81% γ 0.213	−51.29		8^-
	Hf	178m$_2$	31 yr	100% IT; (100% m$_1$); 97% γ 0.426; 94% γ 0.326; 84% γ 0.574	−49.99		16^+
	Hf	179	13.7%		−50.46	50 (γ); 0.4 (γm) (s)	$\frac{9}{2}^+$
	Hf	179m$_1$	18.7 s	100% IT; 95% γ 0.214	−50.09		$\frac{1}{2}^-$
	Hf	179m$_2$	25 d	100% IT; 65% γ 0.453; 38% γ 0.362; 27% γ 0.123	−49.36		$\frac{25}{2}^-$
	Hf	180	35.2%		−49.78	14 (γ)	0^+
	Hf	180m	5.5 h	100% IT; 94% γ 0.332; 85% γ 0.443; 81% γ 0.215	−48.64		8^-
	Hf	181	42.4 d	93% β$^-$ 0.41; 81% γ 0.482; 40% γ 0.133; 13% γ 0.346	−47.40	30 (γ) (r)	$\frac{1}{2}^-$
	Hf	182	9 × 10^6 yr	100% β$^-$ 0.16; 80% γ 0.270; 7% γ 0.156; 3% γ 0.114	−46.0		0^+
	Hf	182m	62 min	10% β$^-$ 0.95; 43% β$^-$ 0.48; (10% m$_2$); 46% IT; 46% γ 0.344; 38% γ 0.224; 24% γ 0.507	−44.8		
	Hf	183	64 min	25% β$^-$ 1.55; 68% β$^-$ 1.15; 65% γ 0.784; 38% γ 0.073; 27% γ 0.459	−43.27		
	Hf	184	4.1 h	46% β$^-$ 1.1; 38% β$^-$ 0.7; 48% γ 0.139; 38% γ 0.345; 15% γ 0.181	−41.5		0^+
80	Hg	178	0.5 s	84% α (6.43); β$^+$; K	−15.9		0^+
	Hg	179	1.09 s	53% α (6.27); β$^+$; K	−16.8		
	Hg	180	2.9 s	α 6.12	−19.9		0^+
	Hg	181	3.6 s	23% α 6.00; 3% α 5.92; β$^+$; K	−20.8		$\frac{1}{2}^-$
	Hg	182	11 s	9% α 5.87; β$^+$; K; γ	−23.2		0^+
	Hg	183	9 s	11% α 5.91; 1% α 5.83; 61% K; β$^+$	−23.7		$\frac{1}{2}^-$
	Hg	184	30.6 s	1% α 5.54; β$^+$; K; γ	−26.0		0^+
	Hg	185	48 s	5% α 5.65; β$^+$; K; γ	−26.1		$\frac{1}{2}^-$
	Hg	185m	17 s	α 5.38; IT?; γ			
	Hg	186	1.4 min	96% K; β$^+$; γ	−28.4		0^+
	Hg	187	1.6 min	β$^+$; K; γ	−28.1		
	Hg	187	2.4 min	β$^+$; K; γ	−28.1		$\frac{3}{2}^-$
	Hg	188	3.3 min	β$^+$; K; γ	−29.9		0^+
	Hg	189	8.7 min	β$^+$; K; γ	−29.2		
	Hg	189	7.5 min	β$^+$; K; γ	−29.2		$\frac{3}{2}^-$
	Hg	190	20 min	100% K; γ	−31.0		0^+
	Hg	191	4.9 min	β$^+$; K; γ	−30.5		
	Hg	191m	51 min	β$^+$; K; (90% m); 55% γ 0.253; 18% γ 0.420; 17% γ 0.579	−30.3		
	Hg	192	4.9 h	100% K; 40% γ 0.275; 6% γ 0.157; 4% γ 0.307	−32.0		0^+
	Hg	193	4 h	K; β$^+$; γ	−31.0		$\frac{3}{2}^-$

(Continued)

Table IX.39 (Continued)

Z	Nuclide	A†	half-life and/or abundance	decay scheme⊕	mass excess/ MeV	cross section*/barn	J^P
	Hg	193m	11 h	92% K; (92% m); 8% IT; 60% γ (m) 0.258; 25% γ 0.408; 14% γ 0.573	−30.9		$\frac{13+}{2}$
	Hg	194	260 yr	100% L	−32.21		0^+
	Hg	195	10 h	100% K; (3% m); 8% γ 0.779; 7% γ 0.061; 2% γ 0.180, 0.585, 0.600	−31.05		$\frac{1-}{2}$
	Hg	195m	41 h	51% IT; 49% K; (49% m); 38% γ 0.262; 9% γ 0.560; 3% γ 0.388	−30.87		$\frac{13+}{2}$
	Hg	196	0.15%		−31.85	3000 (γ) (s); 120 (γ m) (s)	0^+
	Hg	197	64.1 h	100% K; 19% γ 0.077; 1% γ 0.191	−30.74		$\frac{1-}{2}$
	Hg	197m	23.8 h	94% IT; 6% K; (6% m); 34% γ 0.134; 5% γ (m) 0.279	−30.44		$\frac{13+}{2}$
	Hg	198	10.0%		−30.96	0.018 (γ m)	0^+
	Hg	199	16.8%		−29.56	2000 (γ)	$\frac{1-}{2}$
	Hg	199m	42.6 min	100% IT; 52% γ 0.158; 12% γ 0.374	−29.03		$\frac{13+}{2}$
	Hg	200	23.1%		−29.51	< 60 (γ)	0^+
	Hg	201	13.2%		−27.67	< 60 (γ)	$\frac{3-}{2}$
	Hg	202	29.8%		−27.36	5.0 (γ)	0^+
	Hg	203	46.8 d	100% β⁻ 0.21; 82% γ 0.279	−25.28		$\frac{5-}{2}$
	Hg	204	6.9%		−24.70	0.4 (γ) (r)	0^+
	Hg	205	5.2 min	96% β⁻ 1.54; 2% γ 0.204	−22.30		$\frac{1-}{2}$
	Hg	206	8.2 min	63% β⁻ 1.31; 26% γ 0.305; 2% γ 0.650; 1% γ 0.344	−20.96		0^+
67	Ho	150	40 s	β⁺; K; 100% γ 0.391; 100% γ 0.654; 100% γ 0.804	−62.0		
	Ho	151	47 s	10% α (4.61); β⁺; K; γ	−63.4		
	Ho	151	35.6 s	20% α (4.52); β⁺; K; γ	−63.4		
	Ho	152	52 s	6% α (4.46); β⁺; K; 97% γ 0.648; 95% γ 0.684; 93% γ 0.614			
	Ho	152	2.4 min	2% α (4.39); β⁺; K; 97% γ 0.614; 16% γ 0.648	−63.7		
	Ho	153	2.0 min	β⁺; K; γ			
	Ho	153	9.3 min	β⁺; K; γ			
	Ho	154	12 min	β⁺; K; γ	−64.64		1
	Ho	154	3.2 min	β⁺; K; 95% γ 0.335; 76% γ 0.412; 51% γ 0.477			
	Ho	155	49 min	α 3.94; β⁺ 2.1; K; 8% γ 0.240; 4% γ 0.136; 2% γ 0.185, 0.325	−66.06		$\frac{5}{2}$
	Ho	156m	56 min	β⁺; K; IT; γ			1
	Ho	157	12.6 min	β⁺; K; 20% γ 0.280; 15% γ 0.341; 7% γ 0.193	−66.9		$\frac{7-}{2}$
	Ho	158	11.5 min	β⁺ 2.9; K; γ	−66.43		5^+
	Ho	158m₁	27 min	65% IT; β⁺; K; γ	−66.37		2^-
	Ho	158m₂	21 min	β⁺; K; γ			
	Ho	159	33 min	100% K; 34% γ 0.121; 23% γ 0.132; 14% γ 0.253, 0.310	−67.32		$\frac{7-}{2}$
	Ho	159m	8.3 s	100% IT; 40% γ 0.206; 5% γ 0.166	−67.11		$\frac{1+}{2}$
	Ho	160	25.6 min	100% K; γ	−66.39		5^+
	Ho	160m	5.0 h	65% IT; β⁺ K; γ	−66.33		2^-
	Ho	160	3 s	?			

(Continued)

Table IX.39 (Continued)

Z	Nuclide	A†	half-life and/or abundance	decay scheme⊕	mass excess/ MeV	cross section*/barn	J^P
	Ho	161	2.5 h	100% K; 29% γ 0.026; 3% γ 0.078; 3% γ 0.103	−67.20		$\frac{7-}{2}$
	Ho	161m	6.7 s	100% IT; 45% γ 0.211	−66.99		$\frac{1+}{2}$
	Ho	162	15 min	95% K; β⁺ 1.1; 8% γ 0.080; 4% γ 1.319; 1% γ 1.373	−66.05		1^+
	Ho	162m	68 min	61% IT; β⁺; K; γ	−65.9		6^-
	Ho	163	33 yr	100% K	−66.38		$\frac{7-}{2}$
	Ho	163m	1.09 s	100% IT; 78% γ 0.298	−66.08		$\frac{1+}{2}$
	Ho	164	29 min	58% K; 29% β⁻ 1.00; 13% β⁻ 0.91; 3% γ 0.091	−64.94		1^+
	Ho	164m	37 min	100% IT; 11% γ 0.037; 7% γ 0.057	−64.80		6
	Ho	165	100%		−64.90	62 (γ); 3 (ym)	$\frac{7-}{2}$
	Ho	166	26.8 h	51% β⁻ 1.85; 48% β⁻ 1.77; 6% γ 0.081; 1% γ 1.379	−63.07		0^-
	Ho	166m	1200 yr	80% β⁻ 0.06; 73% γ 0.184; 63% γ 0.810; 59% γ 0.712	−63.06		
	Ho	167	3.1 h	15% β⁻ 0.97; 43% β⁻ 0.30; (13%m); 57% γ 0.347; 24% γ 0.321; 5% γ 0.208, 0.238	−62.32		
	Ho	168	3.0 min	71% β⁻ 1.9; 36% γ 0.741; 34% γ 0.821; 18% γ 0.816	−60.3		3^+
	Ho	169	4.8 min	β⁻; 22% γ 0.788; 12% γ 0.853; 11% γ 0.761, 0.778	−58.79		
	Ho	170	43 s	β⁻ 4.0; γ	−56.1		
	Ho	170	2.8 min	23% β⁻ 3.0; 28% β⁻ 2.1; 35% γ 0.932; 25% γ 0.182; 19% γ 0.890; 1.139	−56.1		
53	I	115	1.3 min	β⁺; K	−76.8		
	I	116	2.9 s	β⁺ 6.7; K; 8% γ 0.679; 1% γ 0.540	−77.6		1^+
	I	117	2.20 min	54% K; β⁺ 3.3; γ	−80.9		
	I	118	14.3 min	β⁺ 5.4; 46% K; 95% γ 0.605; 12% γ 0.545; 12% γ 1.338	−80.6		
	I	118m	8.5 min	β⁺; K; IT; γ	−80.5		
	I	119	19.3 min	β⁺ 2.4; 49% K; 90% γ 0.257; 3% γ 0.636; 2% γ 0.321	−83.8		
	I	120	1.35 h	54% K; 19% β⁺ 4.6; 73% γ 0.560; 11% γ 1.523; 9% γ 0.641	−83.79		2^-
	I	120m	53 min	β⁺ 3.7; K; 100% γ 0.560, 87% γ 0.601; 67% γ 0.615	−82.9		
	I	121	2.12 h	94% K; β⁺ 1.2; 84% γ 0.212; 6% γ 0.532; 2% γ 0.599	−86.12		$\frac{5+}{2}$
	I	122	3.6 min	β⁺ 3.1; 23% K; 18% γ 0.564; 1% γ 0.693; 1% γ 0.793	−86.16		1^+
	I	123	13.02 h	100% K; 83% γ 0.159; 1% γ 0.529	−88.0		$\frac{5+}{2}$
	I	124	4.15 d	75% K; 11% β⁺ 2.1; 61% γ 0.603; 10% γ 0.723; 10% γ 1.691	−87.361		2^-
	I	125	60.2 d	100% K; 7% γ 0.035	−88.841	900 (γ) (s)	$\frac{5+}{2}$
	I	126	13.0 d	53% K; 1% β⁺ 1.1; 10% β⁻ 1.25; 32% β⁻ 0.86; 4% β⁻ 0.37; 35% γ 0.389; 34% γ 0.666; 4% γ 0.754	−87.91	6000 (γ) (r)	2^-
	I	127	100%		−88.980	6.1 (γ)	$\frac{5+}{2}$

(Continued)

Table IX.39 (Continued)

Z	Nuclide A†	half-life and/or abundance	decay scheme⊕	mass excess/ MeV	cross section*/barn	J^P
	I 128	24.99 min	77% β⁻ 2.1; 15% β⁻ 1.7; 2% β⁻ 1.1; 6% K; 16% γ 0.443; 2% γ 0.527	−87.734		1^+
	I 129	1.57×10^7 yr	100% β⁻ 0.15; 8% γ 0.040	−88.505	9 (γ); 18 (ym)	$\frac{7^+}{2}$
	I 130	12.36 h	1% β⁻ 1.2; 48% β⁻ 1.0; 46% β⁻ 0.6; 99% γ 0.536; 96% γ 0.669; 82% γ 0.739	−86.90	18 (γ) (r)	5^+
	I 130m	9.16 min	83% IT; 15% β⁻ 2.5; 17% γ 0.536 1% γ 0.586	−86.85		2^+
	I 131	8.040 d	89% β⁻ 0.61; (1%m); 81% γ 0.364; 7% γ 0.637; 6% γ 0.284	−87.451	0.7 (γ)	$\frac{7^+}{2}$
	I 132	2.285 h	16% β⁻ 2.1; 19% β⁻ 1.2; 99% γ 0.668; 76% γ 0.773; 18% γ 0.955	−85.71		4^+
	I 132m	83 min	86% IT; 8% β⁻ 1.5; 13% γ 0.600; 13% γ 0.668; 13% γ 0.773	−85.6		
	I 133	20.9 h	1% β⁻ 1.5; 84% β⁻ 1.2; (3% m); 86% γ 0.530; 5% γ 0.875; 2% γ 1.298	−85.90		$\frac{7^+}{2}$
	I 133m	9 s	100% IT; 100% γ 0.647; 100% γ 0.912	−84.27		
	I 134	52.5 min	12% β⁻ 2.4; 32% β⁻ 1.3; 95% γ 0.847; 65% γ 0.884; 15% γ 1.073	−84.0		
	I 134m	3.50 min	98% IT; 2% β⁻ 2.5; (2%m); 79% γ 0.272; 10% γ 0.044	−83.7		
	I 135	6.61 h	2% β⁻ 2.2; 23% β⁻ 1.4; 22% β⁻ 1.0; (15%m); 29% γ 1.260; 23% γ 1.132; 10% γ 1.678	−83.80		$\frac{7^+}{2}$
	I 136	45 s	67% β⁻ 5.1; 100% γ 0.381; 100% γ 1.313; 78% γ 0.197			
	I 136	83.4 s	30% β⁻ 5.6; 34% β⁻ 4.4; 68% γ 1.313; 25% γ 1.321; 7% γ 2.415, 2.634	−79.4		
	I 137	24.5 s	β⁻; 6% n; 13% γ 1.219; 7% γ 0.601; 5% γ 1.303	−76.7		
	I 138	6.3 s	58% β⁻ 7.7; 5% n; 93% γ 0.589; 17% γ 0.875; 7% γ 0.484	−71.9		
	I 139	2.3 s	β⁻; 10% n; γ	−69		
	I 140	0.86 s	β⁻; 14% n; γ			
	I 141	0.48 s	β⁻; n; γ			
49	In 104	1.5 min	β⁺; K; 100% γ 0.658; 100% γ 0.833; 32% γ 0.879	−75.6		
	In 105	5.1 min	β⁺; K; γ	−79.3		
	In 106	5.32 min	β⁺ 4.9; K; 93% γ 0.633; 8% γ 0.861; 6% γ 1.621	−80.59		
	In 106	6.3 min	β⁺; K; 100% γ 0.633; 90% γ 0.861; 39% γ 0.998			
	In 107	32.4 min	65% K; β⁺ 2.3; 47% γ 0.205; 12% γ 0.506; 10% γ 0.321	−83.5		$\frac{9^+}{2}$
	In 107m	50 s	100% IT; 94% γ 0.678	−82.9		$\frac{1^-}{2}$
	In 108	40 min	K; β⁺ 3.5; 76% γ 0.633; 12% γ 1.986; 9% γ 3.452	−84.1		3^+
	In 108	58 min	K; β⁺; 100% γ 0.633; 95% γ 0.876; 39% γ 0.243	−84.1		

(Continued)

Table IX.39 (Continued)

Z	Nuclide A	half-life and/or abundance	decay scheme	mass excess/ MeV	cross section*/barn	J^P
In	109	4.2 h	94% K; β^+ 0.8; 74% γ 0.203; 5% γ 0.624; 1.149	−86.32		$\frac{9+}{2}$
In	109m$_1$	1.3 min	100% IT; 93% γ 0.650	−85.87		$\frac{1-}{2}$
In	109m$_2$	0.20 s	100% IT; 98% γ 0.678; 75% γ 1.428; 20% γ 0.404, 1.044	−84.41		
In	110	4.9 h	100% K; 99% γ 0.658; 95% γ 0.885; 69% γ 0.937			7^+
In	110	69 min	β^+ 2.3; K; 98% γ 0.658; 2% γ 2.129; 2% γ 2.211	−86.41		2^+
In	111	2.83 d	100% K; 94% γ 0.171; 94% γ 0.245	−88.41		$\frac{9+}{2}$
In	111m	7.6 min	100% IT; 87% γ 0.537	−87.87		$\frac{1-}{2}$
In	112	14.4 min	44% β^- 0.66; 34% K; 22% β^+ 1.6; 6% γ 0.617; 2% γ 0.607	−88.00		1^+
In	112m	20.9 m	100% IT; 13% γ 0.155	−87.85		4^+
In	113	4.3%		−89.372	3 (γ); 8 (γm)	$\frac{9+}{2}$
In	113m	99.5 min	100% IT; 64% γ 0.392	−88.980		$\frac{1-}{2}$
In	114	71.9 s	98% β^- 1.98; 2% K	−88.576		1^+
In	114m	49.51 d	97% IT; 3% K; 16% γ 0.190; 3% γ 0.558; 3% γ 0.725	−88.386		5^+
In	115	5×10^{14} yr 95.7%	100% β^- 0.49	−89.54	41 (γ); 70 (γm$_1$); 91 (γm$_2$)	$\frac{9+}{2}$
In	115m	4.486 h	95% IT; 5% β^- 0.83; 46% γ 0.336	−89.21		$\frac{1-}{2}$
In	116	14.10 s	99% β^- 3.3; 1% γ 1.294	−88.25		1^+
In	116m$_1$	54.1 min	51% β^- 1.1; 36% γ 0.9; 85% γ 1.294; 56% γ 1.097; 32% γ 0.417	−88.13		5^+
In	116m$_2$	2.16 s	100% IT; (100%m); 35% γ 0.162	−87.96		8^-
In	117	42 min	100% β^- 0.74; 100% γ 0.553; 87% γ 0.159	−88.94		$\frac{9+}{2}$
In	117 m	1.93 h	37% β^- 1.8; 16% β^- 1.6; 47% IT; 17% γ 0.315; 14% γ 0.159	−88.63		$\frac{1-}{2}$
In	118	5.0 s	85% β^- 4.2; 15% γ 1.230; 5% γ 0.528	−87.5		1^+
In	118	4.4 min	24% β^- 2.0; 64% β^- 1.3; 96% γ 1.230; 82% γ 1.051; 55% γ 0.683	−87.4		
In	118	8.5 s	99% IT; 1% β^- 1.9; 22% γ 0.138; 1% γ 0.254	−87.2		
In	119	2.1 min	100% β^- 1.6; 99% γ 0.763	−87.73		$\frac{9+}{2}$
In	119m	18 min	94% β^- 2.6; 5% IT; 2% γ 0.311	−87.42		$\frac{1-}{2}$
In	120	44 s	19% β^- 3.1; 42% β^- 2.2; 97% γ 1.172; 60% γ 1.023; 30% γ 0.864	−85.8		
In	120	3.0 s	81% β^- 5.6; 19% γ 1.172; 1% γ 0.704; 1% γ 1.186	−85.5		1^+
In	121	30.0 s	100% β^- 2.4; 87% γ 0.926; 8% γ 0.262; 7% γ 0.657	−85.84		$\frac{9+}{2}$
In	121m	3.8 min	β^- 3.7; 1% IT; 21% γ 0.060; 1% γ 1.041; 1% γ 1.102	−85.53		$\frac{1-}{2}$
In	122	9.2 s	100% β^- 4.3; 100% γ 1.141; 56% γ 1.003; 18% γ 1.194	−83.4		
In	122	1.5 s	71% β^- 6.4; 29% γ 1.141	−83.5		

(Continued)

Table IX.39 (Continued)

Z	Nuclide A†	half-life and/or abundance	decay scheme	mass excess/ MeV	cross section*/barn	J^P
	In	123	6.0 s	β^-; (100%m); 63% γ 1.130; 32% γ 1.020; 3% γ 0.619	−83.44	
	In	123m	48 s	β^- 4.6; (100% m); γ	−83.12	
	In	124	2.4 s	β^-; γ		
	In	124	3.2 s	β^-, γ	−81.1	
	In	125	2.32 s	1% β^- 4.8, 85% β^- 4.0; 76% γ 1.335; 10% γ 1.032; 8% γ 0.618	−80.5	
	In	125	12.2 s	β^-; γ		
	In	126	1.53 s	β^-	−77.9	
	In	127	1.3 s	β^-; γ	−77.4	
	In	127	3.7 s	β^-; n; γ	−77.4	
	In	129	2.5 s	β^-; n; γ		
	In	129	0.99 s	β^-; n; γ	−73.1	
	In	130	0.58 s	β^-; n; 100% γ 0.775; 100% γ 1.217; 80% γ 0.127	−70.1	
	In	131	0.29 s	β^-; n; γ	−70	
	In	132	0.12 s	β^-; n; γ	−65	
77	Ir	172	2 s	α 5.81	−27.3	
	Ir	173	3.0 s	α 5.67	−29.9	
	Ir	174	4 s	α 5.48	−30.9	
	Ir	175	5 s	α 5.39	−33.2	
	Ir	176	8 s	α 5.12	−33.8	
	Ir	177	21 s	α 5.01	−35.8	
	Ir	178	12 s	β^+; K; γ	−36.3	
	Ir	179	4 min	β^+; K	−37.9	
	Ir	180	1.5 min	β^+; K; γ	−37.9	
	Ir	181	5.0 min	β^+; K; γ	−39.3	
	Ir	182	15 min	β^+; K; 45% γ 0.273; 35% γ 0.127; 9% γ 0.236, 0.912	−39.0	
	Ir	183	55 min	β^+; K; γ	−40.1	
	Ir	184	3.0 h	K; β^+; 68% γ 0.264; 30% γ 0.120; 26% γ 0.390	−39.5		5
	Ir	185	14 h	β^+; K; γ	−40.3		$\frac{5}{2}$
	Ir	186	16 h	98% K; 2% β^+ 1.9; 62% γ 0.297; 41% γ 0.137; 34% γ 0.435	−39.16		5
	Ir	186	1.7 h	β^+; K; γ		
	Ir	187	10.5 h	100% K; 6% γ 0.913; 4% γ 0.401, 0.427, 0.611	−39.7		$\frac{3+}{2}$
	Ir	188	41.5 h	100% K; 30% γ 0.155, 19% γ 2.215; 18% γ 0.633	−38.32		2−
	Ir	189	13.1 d	100% K; (8%m); 6% γ 0.245; 4% γ 0.069; 1% γ 0.059	−38.5		$\frac{3+}{2}$
	Ir	190	11.8 d	100% K; 52% γ 0.186, 39% γ 0.605; 35% γ 0.519	−36.7		
	Ir	190m$_1$	1.2 h	100% IT	−36.7		
	Ir	190m$_2$	3.2 h	95% K; (95%m); 5% IT; 95% γ (m) 0.502; 94% γ (m) 0.616; 90% γ (m) 0.361	−36.5		
	Ir	191	37.3%		−36.70	540 (γ); 400 (γm$_1$); 0.1 (γm$_2$)	$\frac{3+}{2}$
	Ir	191m	4.9 s	100% IT, 26% γ 0.129	−36.53		$\frac{11-}{2}$
	Ir	192	74.2 d	48% β^- 0.67; 41% β^- 0.54; 5% K; 83% γ 0.316; 48% γ 0.468; 30% γ 0.308	−34.83	1000 (t) (r)	4
	Ir	192m$_1$	1.45 min	100% IT	−34.77		1
	Ir	192m$_2$	241 yr	100% IT; γ	−34.67		9
	Ir	193	62.7%		−34.52	110 (γ)	$\frac{3+}{2}$

(Continued)

Table IX.39 (Continued)

Z	Nuclide A†	half-life and/or abundance	decay scheme	mass excess/ MeV	cross section*/barn	J^P
	Ir 193m	10.6 d	100% IT	34.44	0.05 (γm)	$\frac{11-}{2}$
	Ir 194	19.2 h	86% β⁻ 2.25; 13% γ 0.328; 3% γ 0.294; 1% γ 0.645	−32.51		
	Ir 194m	171 d	β⁻; 97% γ 0.483; 93% γ 0.329; 62% γ 0.601			
	Ir 195	2.5 h	13% β⁻ 1.1; 60% β⁻ 1.0; γ	−31.69		
	Ir 195m	3.7 h	33% β⁻ 1.0; 32% β⁻ 0.4; (48%m); 14% γ 0.320; 14% γ 0.433; 14% γ 0.685	−31.57		
	Ir 196	52 s	80% β⁻ 3.2; 19% γ 0.356; 10% γ 0.779; 5% γ 0.447	−29.4		
	Ir 196m	1.40 h	80% β⁻ 1.1; 97% γ 0.394; 96% γ 0.521; 94% γ 0.356, 0.447	−29.0		
	Ir 198	8 s	β⁻; γ	−25.5		
19	K 36	0.344 s	44% β⁺ 9.8; 42% β⁺ 5.2; 82% γ 1.970; 32% γ 2.433; 30% γ 2.208	−17.43		2^+
	K 37	1.23 s	98% β⁺ 5.13; 2% γ 2.794	−24.799		$\frac{3+}{2}$
	K 38	7.61 min	100% β⁺ 2.72; 100% γ 2.168	−28.802		3^+
	K 38m	0.926 s	100% β⁺ 5.0	−28.671		0^+
	K 39	93.26%		−33.806	2.1 (γ) (r)	$\frac{3+}{2}$
	K 40	1.28 × 10⁹ yr 0.0117%	89% β⁻ 1.31; 11% K; 11% γ 1.461	−33.535	70 (γ) (r)	4^-
	K 41	6.73%		−35.560	1.5 (γ)	$\frac{3+}{2}$
	K 42	12.361 h	81% β⁻ 3.52; 19% γ 1.525	−35.023		2^-
	K 43	22.2 h	2% β⁻ 1.8; 89% β⁻ 0.8; 88% γ 0.373; 80% γ 0.618; 11% γ 0.397; 11% γ 0.594	−36.59		$\frac{3+}{2}$
	K 44	22.2 min	34% β⁻ 5.66; 29% β⁻ 2.36; 58% γ 1.157; 23% γ 2.151; 10% γ 2.519	−35.81		2^-
	K 45	20 min	β⁻; γ	−36.61		$\frac{3+}{2}$
	K 46	115 s	50% β⁻ 6.3; 28% β⁻ 2.7; 91% γ 1.347; 28% γ 3.700; 9% γ 3.015	−35.42		
	K 47	17.5 s	β⁻; 100% γ 2.013; 85% γ 0.586; 15% γ 0.565	−35.70		$\frac{1+}{2}$
	K 48	6.8 s	β⁻ 8.2; 80% γ 3.832; 32% γ 0.780; 17% γ 0.675	−32.2		
36	Kr 72	17.4 s	β⁺; K; 19% γ 0.415; 15% γ 0.310; 8% γ 0.163	−53.9		0^+
	Kr 73	26 s	44% β⁺ 5.5; 1% p; 66% γ 0.178; 13% γ 0.151; 11% γ 0.474	−57.0		
	Kr 74	11.5 min	β⁺ 2.2; K; 31% γ 0.090; 20% γ 0.203; 11% γ 0.297	−62.0		0^+
	Kr 75	4.2 min	β⁺; K; 75% γ 0.132; 22% γ 0.155; 6% γ 0.153	−64.2		
	Kr 76	14.82 h	100% K; 38% γ 0.316; 20% γ 0.270, 18% γ 0.046	−69.1		0^+
	Kr 77	75 min	β⁺ 1.9; 20% K; (8%m); 84% γ 0.130; 39% γ 0.147; 4% γ 0.312	−70.24		
	Kr 78	0.356%		−74.15	4.7 (γ); 0.21 (γm)	0^+
	Kr 79	35.0 h	93% K; 7% β⁺ 0.61; 13% γ 0.261; 10% γ 0.398; 8% γ 0.606	−74.44		$\frac{1-}{2}$
	Kr 79m	50 s	100% IT; 27% γ 0.130	−74.31		$\frac{7+}{2}$

(Continued)

Table IX.39 (Continued)

Z	Nuclide	$A^\dagger$	half-life and/or abundance	decay scheme	mass excess/ MeV	cross section*/barn	J^P
	Kr	80	2.27%		−77.90	11.5 (γ); 4.6 (γm)	0^+
	Kr	81	2.1×10^5 yr	100% K; 4% γ 0.276	−77.65		$\frac{7+}{2}$
	Kr	81m	13 s	100% IT; 67% γ 0.190	−77.46		$\frac{1-}{2}$
	Kr	82	11.6%		−80.59	23 (γ); 20 (γm)	0^+
	Kr	83	11.5%		−79.985	200 (γ)	$\frac{9+}{2}$
	Kr	83m	1.83 h	100% IT; 5% γ 0.009	−79.943		$\frac{1-}{2}$
	Kr	84	57.0%		−82.432	0.042 (γ); 0.09 (γm)	0^+
	Kr	85	10.70 yr	100% β⁻ 0.69	−81.472	1.7 (γ)	$\frac{9+}{2}$
	Kr	85m	4.48 h	79% β⁻ 0.84; 21% IT; 79% γ 0.151; 14% γ 0.305	−81.167		$\frac{1-}{2}$
	Kr	86	17.3%		−83.263	0.06 (γ) (r)	0^+
	Kr	87	76.3 min	31% β⁻ 3.9; 40% β⁻ 3.5; 50% γ 0.403; 9% γ 2.555; 8% γ 0.845	−80.707		
	Kr	88	2.86 h	14% β⁻ 2.9; 67% β⁻ 0.5; 35% γ 2.392; 26% γ 0.196; 13% γ 0.835, 2.196	−79.69		0^+
	Kr	89	3.18 min	23% β⁻ 4.9; 15% β⁻ 2.3; 20% γ 0.221; 17% γ 0.586; 7% γ 0.498; 0.904, 1.473	−76.8		
	Kr	90	32.3 s	29% β⁻ 4.4; 63% β⁻ 2.6; (12%m); 38% γ 1.119; 33% γ 0.122; 30% γ 0.539	−75.2		0^+
	Kr	91	8.57 s	10% β⁻ 6.1; 42% γ 0.109; 18% γ 0.507; 7% γ 0.613	−71.8		
	Kr	92	1.84 s	2% β⁻ 6.0; 88% β⁻ 4.6; 64% γ 0.142; 60% γ 1.219; 15% γ 0.813	−69.2		0^+
	Kr	93	1.29 s	5% β⁻ 7.3; 27% β⁻ 6.8; 2% n; 36% γ 0.254; 24% γ 0.252; 24% γ 0.324	−65.6		
	Kr	94	0.20 s	β⁻; 6% n; γ	−61.3		0^+
57	La	126	1.0 min	β⁺; K; γ			
	La	127	3.8 min	β⁺; K	−77.8		
	La	128	4.9 min	β⁺; K; γ	−78.7		
	La	129	10 min	β⁺; K; γ	−81.1		
	La	129m	0.56 s	100% IT; 24% γ 0.068; 4% γ 0.105	−80.9		
	La	130	8.7 min	β⁺; K; 81% γ 0.357; 27% γ 0.551; 18% γ 0.544, 0.908	−81.6		
	La	131	61 min	76% K; β⁺ 1.9; 24% γ 0.108; 19% γ 0.418; 17% γ 0.366, 1.910	−83.8		$\frac{3+}{2}$
	La	132	4.8 h	β⁺ 3.7; K; 76% γ 0.465; 16% γ 0.567; 9% γ 0.663	−83.74		2^-
	La	132m	24.3 min	76% IT; β⁺; K; 44% γ 0.135; 22% γ 0.465; 7% γ 0.285	−83.55		6^-
	La	133	3.91 h	β⁺; K; 2% γ 0.279; 1% γ 0.290; 1% γ 0.302	−85.6		$\frac{5+}{2}$
	La	134	6.67 min	61% β⁺ 2.7; 38% K; 5% γ 0.605	−85.27		1^+
	La	135	19.4 h	100% K; 2% γ 0.481	−86.67		$\frac{5+}{2}$
	La	136	9.87 min	64% K; β⁺ 1.9; 3% γ 0.818	−86.0		1^+
	La	137	6×10^4 yr	100% K	−87.1		$\frac{7+}{2}$
	La	138	1.12×10^{11} yr 0.089%	68% K; 32% β⁻ 0.25; 68% γ 1.436; 32% γ 0.788	−86.52	57 (γ)	5^+

Table IX.39 (Continued)

Z	Nuclide	$A^\dagger$	half-life and/or abundance	decay scheme	mass excess/ MeV	cross section*/barn	J^P
	La	139	99.911%		−87.23	9.2 (γ)	$\frac{7+}{2}$
	La	140	40.3 h	10% β⁻ 2.2; 41% β⁻ 1.4; 96% γ 1.596; 43% γ 0.487; 23% γ 0.816	−84.32	2.7 (γ) (s)	3⁻
	La	141	3.9 h	97% β⁻ 2.4; 3% γ 1.354	−83.01		
	La	142	93 min	13% β⁻ 4.5; 20% β⁻ 2.1; 49% γ 0.641; 15% γ 2.398; 10% γ 2.543	−80.02		2⁻
	La	143	14.0 min	96% β⁻ 3.3; 2% γ 0.620; 1% γ 0.644	−78.3		
	La	144	39.9 s	β⁻; γ	−74.9		
	La	145	29 s	β⁻; γ	−72.9		
	La	146	11 s	β⁻; γ	−69.5		
	La	148	1.3 s	β⁻; γ	−64.0		
3	Li	6	7.5%		14.086	942 (α)	1⁺
	Li	7	92.5%		14.908	0.045 (γ) (r)	$\frac{3-}{2}$
	Li	8	0.844 s	100% β⁻ 13.1, 100% 2α	20.947		2⁺
	Li	9	0.178 s	65% β⁻ 13.6; 35% n, 2α	24.955		
	Li	11	0.0085 s	39% β⁻ 20.4; 61% n	40.94		
103	Lr	255	22 s	40% α 8.43; 60% α 8.37	90.3		
	Lr	256	26 s	19% α 8.52; 38% α 8.43; 19% α 8.39; 24% α (dist)	91.8		
	Lr	257	0.65 s	85% α 8.86; 15% α 8.80	93.0		
	Lr	258	4.4 s	10% α 8.65; 35% α 8.61; 45% α 8.59; 10% α 8.54	94.8		
	Lr	259	5 s	100% α 8.46	96.0		
	Lr	260	3.0 min	100% α 8.04	98.1		
71	Lu	155	0.07 s	α 5.63	−42.6		
	Lu	156	0.23 s	α 5.54	−43.8		
	Lu	156	0.5 s	α 5.43	−43.8		
	Lu	164	3.17 min	β⁺; K; γ	−54.6		
	Lu	165	11.8 min	β⁺; K; 25% γ 0.121; 23% γ 0.132; 13% γ 0.174	−56.2		$\frac{1}{2}$
	Lu	166	2.7 min	K; β⁺; 77% γ 0.228; 41% γ 0.337; 32% γ 0.368	−56.1		
	Lu	166m₁	1.4 min	β⁺; 42% IT; 26% γ 0.228; 22% γ 0.102; 19% γ 0.285	−56.1		
	Lu	166m₂	2.1 min	β⁺; K; 23% γ 1.427; 16% γ 2.099; 15% γ 1.257	−56.1		
	Lu	167	52 min	98% K; β⁺ 2.1; 16% γ 0.030; 9% γ 0.239; 4% γ 0.213	−57.5		$\frac{7+}{2}$
	Lu	168	5.5 min	β⁺; K; γ	−57.1		
	Lu	168m	6.7 min	K; β⁺; γ	−56.9		3⁺
	Lu	169	34.1 h	99% K; β⁺ (51% m); 23% γ 0.960; 22% γ 0.191; 9% γ 1.450	−57.88		$\frac{7+}{2}$
	Lu	169m	2.7 min	100% IT	−57.85		$\frac{1-}{2}$
	Lu	170	2.02 d	β⁺, K; 9% γ 0.084; 8% γ 1.280; 6% γ 2.042	−57.32		0⁺
	Lu	170m	0.7 s	100% IT; 1% γ 0.045	−57.23		4⁻
	Lu	171	8.22 d	100% K; 51% γ 0.740; 12% γ 0.667; 6% γ 0.076	−57.82		$\frac{7+}{2}$
	Lu	171m	79 s	100% IT	−57.75		$\frac{1-}{2}$
	Lu	172	6.70 d	100% K; 63% γ 1.094; 28% γ 0.901; 19% γ 0.181	−56.73		4⁻
	Lu	172m	3.7 min	100% IT	−56.68		1⁻
	Lu	173	499 d	100% K; 13% γ 0.272; 8% γ 0.079; 3% γ 0.101	−56.87		$\frac{7+}{2}$

(Continued)

Table IX.39 (Continued)

Z	Nuclide	A	half-life and/or abundance	decay scheme	mass excess/ MeV	cross section*/barn	J^P
	Lu	174	3.3 yr	100%K; 6% γ 1.242; 4% γ 0.076	−55.56		
	Lu	174m	142 d	99% IT; 1% K; 14% γ 0.045; 7% γ 0.067; 1% γ 0.273, 0.992	−55.39		
	Lu	175	97.39%		−55.16	10 (γ); 16 (γm)	$\frac{7+}{2}$
	Lu	176	3.6×10^{10} yr 2.61%	99% β⁻ 0.59; 93% γ 0.307; 85% γ 0.202; 13% γ 0.088	−53.38	2000 (γ); 7 (γm)	7⁻
	Lu	176m	3.68 h	40% β⁻ 1.31; 60% β⁻ 1.22; 9% γ 0.088	−53.25		1⁻
	Lu	177	6.71 d	78% β⁻ 0.50; 11% γ 0.208; 7% γ 0.113	−52.38		$\frac{7+}{2}$
	Lu	177m	160 d	78% β⁻ 0.15; (78%m₁); 22% IT; 65% γ (m) 0.208; 39% γ (m) 0.229; 30% γ (m) 0.379	−51.41		$\frac{23-}{2}$
	Lu	178	28.4 min	62% β⁻ 2.13; 28% β⁻ 2.04; 7% γ 0.093; 5% γ 1.341; 2% γ 1.310	−50.30		1⁺
	Lu	178m	22.9 min	83% β⁻ 1.3; (100%m₁); 97% γ (m) 0.426; 94% γ (m) 0.326; 81% γ (m) 0.213	−50.0		
	Lu	179	4.6 h	87% β⁻ 1.35; 12% γ 0.214	−49.11		
	Lu	180	5.7 min	1% β⁻ 1.7; 91% β⁻ 1.5; 50% γ 0.408; 26% γ 1.200; 24% γ 1.106	−46.7		
101	Md	248	7 s	5% α 8.36; 15% α 8.32; β⁺; K	77.0		
	Md	249	24 s	> 20% α (8.03); β⁺; K	77.3		
	Md	250	52 s	2% α 7.82; 4% α 7.75; β⁺; K	78.6		
	Md	252	2 min	β⁺; K	80.5		
	Md	254	10 min	100% K	83.4		
	Md	254	28 min	100% K	83.4		
	Md	255	27 min	7% α 7.33; 93% K; 7% γ 0.430	84.8		
	Md	256	75 min	6% α 7.22; 2% α 7.15; 2% α (dist); 90% K	87.4		
	Md	257	5.0 hr	10% α (7.07); 90% K	89.0		
	Md	258	56 d	28% α 6.79 72% α 6.72	91.8		
	Md	259	95 min	100% SF			
12	Mg	21	0.123 s	16% β⁺ 12.08; 41% β⁺ 11.75; 11% β⁺ 10.36; 10% β⁺ 7.61; 32% p	10.91		$\frac{5+}{2}$
	Mg	22	3.86 s	40% β⁺ 3.19; 54% β⁺ 3.11; 100% γ 0.583; 60% γ 0.074; 5% γ 1.280	−0.394		0⁺
	Mg	23	11.33 s	91% β⁺ 3.04; 9% γ 0.440	−5.471		$\frac{3+}{2}$
	Mg	24	78.99%		−13.931	0.053 (y) (r)	0⁺
	Mg	25	10.00%		−13.191	0.18 (γ) (r)	$\frac{5+}{2}$
	Mg	26	11.01%		−16.212	0.038 (γ)	0⁺
	Mg	27	9.46 min	72% β⁻ 1.77; 28% β⁻ 1.60; 73% γ 0.844; 27% γ 1.014	−14.585	0.15 (γ) (r)	$\frac{1+}{2}$
	Mg	28	20.9 hr	95% β⁻ 0.46; 95% γ 0.031; 54% γ 1.342; 36% γ 0.401; 36% γ 0.942	−15.016		0⁺
	Mg	29	1.5 s	β⁻; (50%) γ 2.224; (25%) γ 0.960; (25%) γ 1.398 [Only relative (γ)-ray intensities given]	−10.8		
	Mg	30	1 s	β⁻	−9.8		

(Continued)

Table IX.39 (Continued)

Z	Nuclide	A	half-life and/or abundance	decay scheme	mass excess/ MeV	cross section*/barn	J^P
25	Mn	50	0.2827 s	100% β^+ 6.61	−42.626		0^+
	Mn	50m	1.76 min	8% β^+ 3.7; 69% β^+ 3.5; 28% β^+ 3.0; 100% γ 0.783; 100% γ 1.098; 69% γ 1.443	−42.40		5^+
	Mn	51	46.2 min	β^+ 2.2; K	−48.240		$\frac{5-}{2}$
	Mn	52	5.591 d	72% K; 28% β^+ 0.58; 100% γ 1.434; 95% γ 0.936; 90% γ 0.744	−50.704		6^+
	Mn	52m	21.1 min	98% β^+ 3.7; 2% IT; 98% γ 1.434; 2% γ 0.378	−50.327		2^+
	Mn	53	3.74×10^6 yr	100% K	−54.687	70 (γ) (r)	$\frac{7-}{2}$
	□Mn	54	312.20 d	100% K; 100% γ 0.835	−55.554		3^+
	Mn	55	100%		−57.710	13.2 (γ)	$\frac{5-}{2}$
	Mn	56	2.579 h	56% β^- 2.84; 28% β^- 1.04; 99% γ 0.847; 27% γ 1.811; 14% γ 2.113	−56.909		3^+
	Mn	57	1.54 min	81% β^- 2.68; 10% γ 0.122; 4% γ 0.692; 1% γ 0.352	−57.49		$\frac{5-}{2}$
	Mn	58	65 s	β^-; 82% γ 0.811; 55% γ 1.323; 20% γ 0.459	−55.80		3^+
	Mn	58	3.0 s	β^- 6.3	−55.83		
42	Mo	90	5.67 hr	75% K; 25% β^+ 1.1; (94% m); 78% γ 0.257; 64% γ 0.122; 6% γ 0.203	−80.17		0^+
	Mo	91	15.49 min	94% β^+ 3.4; 6% K	−82.20		$\frac{9+}{2}$
	Mo	91m	64 s	50% IT; β^+; K; (50%m); 48% γ 0.653; 24% γ 1.508; 19% γ 1.208	−81.55		$\frac{1-}{2}$
	Mo	92	14.8%		−86.807	0.3 (t) (r)	0^+
	Mo	93	4×10^3 yr	100% K; (96%m)	−86.803		$\frac{5+}{2}$
	Mo	93m	6.9 h	100% IT; 100% γ 0.685; 99% γ 1.477; 57% γ 0.263	−84.378		$\frac{21+}{2}$
	Mo	94	9.3%		−88.412		0^+
	Mo	95	15.9%		−87.712	14 (γ) (r)	$\frac{5+}{2}$
	Mo	96	16.7%		−88.795	1.0 (γ) (r)	0^+
	Mo	97	9.6%		−87.545	2 (γ) (r)	$\frac{5+}{2}$
	Mo	98	24.1%		−88.115	0.13 (γ)	0^+
	Mo	99	66.02 h	81% β^- 1.21; (87%m); 76% γ (m) 0.141; 13% γ 0.739; 6% γ 0.181	−85.970		$\frac{1+}{2}$
	Mo	100	9.6%		−86.19	0.20 (γ) (r)	0^+
	Mo	101	14.6 min	12% β^- 2.6; 21% β^- 0.8; 23% γ 0.591 (double); 20% γ 0.192; 13% γ 1.013	−83.52		$\frac{1+}{2}$
	Mo	102	11.0 min	β^-; γ	−83.56		0^+
	Mo	103	60s	β^-	−80.6		
	Mo	104	1.00 min	β^-; γ	−81.7		0^+
	Mo	105	36s	β^-; γ	−77.1		
	Mo	106	9.5s	β^-; γ	−76		0^+
	Mo	108	0.9s	β^-	−71		0^+
0	n	1	10.6 min	100% β^- 0.7824	8.0714		$\frac{1+}{2}$
7	N	12	0.0110 s	94% β^+ 16.3; 4% α; 2% γ 4.439	17.338		1^+
	N	13	9.96 min	100% β^+ 1.19	5.346		$\frac{1-}{2}$
	N	14	99.63%		2.8634	1.8 (p)	1^+
	N	15	0.366%		0.1015	4×10^{-5} (γ) (r)	$\frac{1-}{2}$

(Continued)

Table IX.39 (Continued)

Z	Nuclide □	A†	half-life and/or abundance	decay scheme⊕	mass excess/ MeV	cross section*/barn	J^P
	N	16	7.1 s	26% β⁻ 10.42; 68% β⁻ 4.29; 69% γ 6.129; 5% γ 7.115; 1% γ 2.75	5.682		2⁻
	N	17	4.17 s	2% β⁻ 8.7; 50% β⁻ 3.3; 95% n; 3% γ 0.871	7.87		$\frac{1}{2}^-$
	N	18	0.63 s	100% β⁻ 9.6; 100% γ 1.982; 72% γ 0.82; 72% γ 1.65	13.3		
11	Na	20	0.448 s	79% β⁺ 11.23; 20% α; 79% γ 1.634	6.84		2⁺
	Na	21	22.5 s	95% β⁺ 2.53; 5% γ 0.351	−2.186		$\frac{3}{2}^+$
	□Na	22	2.602 yr	90% β⁺ 0.545; 10% K; 100% γ 1.275	−5.184	0.0003 (γ)	3⁺
	Na	23	100%		−9.530	0.53 (γ)	$\frac{3}{2}^+$
	Na	24	15.03 h	100% β⁻ 1.39; 100% γ 1.368; 100% γ 2.754	−8.418		4⁺
	Na	24m	0.020 s	IT; β⁻ 5.99 (weak); ~100% γ 0.472	−7.945		1⁺
	Na	25	60 s	63% β⁻ 3.83; 27% β⁻ 2.86; 15% γ 0.975; 13% γ 0.390; 13% γ 0.585	−9.36		$\frac{5}{2}^+$
	Na	26	1.1 s	88% β⁻ 7.52; 99% γ 1.809; 6% γ 1.130; 3% γ 1.412	−6.89		3⁺
	Na	27	0.30 s	86% β⁻ 8.0; 14% β⁻ 7.3; 86% γ 0.985; 14% γ 1.699	−5.6		
	Na	28	0.031 s	69% β⁻ 13.9; 30% γ 1.474; 16% γ 2.380	−1.1		1⁺
	Na	29	0.043 s	β⁻; 15% n; 12% γ 2.570; 7% γ 1.510	2.66		
	Na	30	0.055 s	β⁻; 33% n	8.4		
	Na	31	0.017 s	β⁻; 30% n	10.6		
	Na	32	0.015 s	β⁻	16.4		
	Na	33	0.02 s	β⁻			
41	Nb	86	1.4 min	β⁺; γ	−69.3		
	Nb	87	2.6 min	β⁺; K; 26% γ 0.201; 19% γ 0.471; 10% γ 1.066	−74.4		
	Nb	87	3.9 min	β⁺; K; (100%m); 97% γ (m) 0.201; 26% γ (m) 0.135	−74.4		
	Nb	88	7.8 min	β⁺; K; 90% γ 1.057; 57% γ 1.083; 43% γ 0.400	−76.4		
	Nb	88	14.3 min	β⁺; K; 100% γ 1.057; 98% γ 1.083; 80% γ 0.503	−76.4		
	Nb	89	2.0 h	β⁺ 3.2; K; (1%m); 4% γ 1.627; 3% γ 1.833; 3% γ 2.572	−80.62		
	Nb	89	66 min	β⁺ 2.6; 26% K; (100%m); 90% γ (m) 0.588; 86% γ 0.507; 7% γ 0.770	−80.62		
	Nb	90	14.6 h	53% β⁺ 1.5; 47% K; (98% m); 93% γ 1.129; 83% γ (m) 2.319; 66% γ 0.141	−82.65		8⁺
	Nb	90m	18.8 s	100% IT; 64% γ 0.122	−82.53		4⁻
	Nb	91m	62 d	97% IT; 3% K; 3% γ 1.205; 1% γ 0.104	−86.532		$\frac{1}{2}^-$
	Nb	92	3 × 10⁷ yr	100% K; γ	−86.448		7⁺
	Nb	92m	10.14 d	100% K; 99% γ 0.934; 2% γ 0.913; 1% γ 1.848	−86.313		2⁺
	Nb	93	100%		−87.209	1.1 (γ)	$\frac{9}{2}^+$

(Continued)

Table IX.39 (Continued)

Z	Nuclide	A†	half-life and/or abundance	decay scheme⊕	mass excess/ MeV	cross section*/barn	J^P
	Nb	93m	13.6 yr	100% IT	−87.179		$\frac{1-}{2}$
	Nb	94	2.0×10^4 yr	100% β⁻ 0.47; 100% γ 0.871; 98% γ 0.703	−86.367	15 (γ) (s); 0.59 (γm) (r)	6⁺
	Nb	94m	6.26 min	100% IT	−86.326		3⁺
	Nb	95	34.97 d	100% β⁻ 0.16; 100% γ 0.766	−86.787		$\frac{9+}{2}$
	Nb	95m	87h	97% IT; 3% β⁻ 1.0; 26% γ 0.236; 2% γ 0.204	−86.552		$\frac{1-}{2}$
	Nb	96	23.4h	96% β⁻ 0.75; 97% γ 0.778; 55% γ 0.569; 49% γ 1.091	−85.608		
	Nb	97	72 min	98% β⁻ 1.27; 98% γ 658; 1% γ 1.025	−85.612		$\frac{9+}{2}$
	Nb	97m	60 s	100% IT; 98% γ 0.743	−84.868		$\frac{1-}{2}$
	Nb	98	2.9 s	90% β⁻ 4.59; 3% γ 0.787; 2% γ 1.024; 1% γ 0.645, 0.972, 1.432	−83.53		1⁺
	Nb	98m	51.3 min	5% β⁻ 3.2; 16% β⁻ 2.5; 28% β⁻ 2.0; 93% γ 0.787; 73% γ 0.723; 18% γ 1.169	−83.45		
	Nb	99	15.0 s	100% β⁻ 3.4; 90% γ 0.138; 45% γ 0.098	−82.35		
	Nb	99m	2.6 min	β⁻; γ	−81.98		
	Nb	100	1.5 s	β⁻; γ	−80.0		
	Nb	100	3.1 s	β⁻; γ			
	Nb	101	7.0 s	β⁻; γ	−79.0		
	Nb	102	4.3 s	β⁻; γ	−76.4		
	Nb	102	1.3 s	β⁻; γ	−76.4		
	Nb	103	1.5 s	β⁻; γ	−75.4		
	Nb	104	0.8 s	β⁻; γ	−72.7		
	Nb	104	4.8 s	β⁻; γ			
	Nb	106	1 s	β⁻; γ			
60	Nd	129	6 s	β⁺; K; p			
	Nd	130	28 s	β⁺; K			0⁺
	Nd	132	1.8 min	β⁺; K			0⁺
	Nd	133	1.2 min	β⁺; K; γ			
	Nd	134	9 min	β⁺; γ			0⁺
	Nd	135	12 min	β⁺; K; 51% γ 0.204; 23% γ 0.042; 15% γ 0.441	−76.3		$\frac{9}{2}$
	Nd	135	6 min	β⁺; K	−76.3		
	Nd	136	50.6 min	94% K; β⁺; 33% γ 0.109; 20% γ 0.040; 12% γ 0.575	−79.2		0⁺
	Nd	137	39 min	β⁺; K; 17% γ 0.076; 13% γ 0.581; 10% γ 0.307	−79.4		$\frac{1+}{2}$
	Nd	137m	1.6 s	100% IT; 64% γ 0.234; 58% γ 0.178; 35% γ 0.108	−78.9		$\frac{11-}{2}$
	Nd	138	5.1 h	100% K; 3% γ 0.326	−82.0		0⁺
	Nd	139	29.7 min	74% K; 25% β⁺ 1.8; 6% γ 0.405; 2% γ 1.074; 1% γ 0.669	−82.05		$\frac{3+}{2}$
	Nd	139m	5.5 h	87% K; 1% β⁺ 1.2; 12% IT; 40% γ 0.114; 35% γ 0.738; 26% γ 0.708, 0.982	−81.82		$\frac{11-}{2}$
	Nd	140	3.37 d	100% K	−84.22		0⁺
	Nd	141	2.5 h	97% K; β⁺ 0.8; 1% γ 1.127	−84.20		$\frac{3+}{2}$
	Nd	141m	61 s	100% IT; 91% γ 0.756	−83.45		$\frac{11-}{2}$
	Nd	142	27.2%		−85.95	19 (γ) (r)	0⁺
	Nd	143	12.2%		−84.00	320 (γ)	$\frac{7-}{2}$
	Nd	144	2.1×10^{15} yr 23.8%	α 1.83	−83.75	4 (γ) (r)	0⁺

(Continued)

Table IX.39 (Continued)

Z	Nuclide A†		half-life and/or abundance	decay scheme⊕	mass excess/ MeV	cross section*/barn	J^P
	Nd	145	8.3%		−81.43	41 (γ)	$\frac{7-}{2}$
	Nd	146	17.2%		−80.92	1.3 (γ)	0^+
	Nd	147	10.98 d	83% β⁻ 0.081; 27% γ 0.091; 12% γ 0.531; 2% γ 0.319	−78.14	440 (γ) (r)	$\frac{5-}{2}$
	Nd	148	5.7%		−77.41	2.5 (γ)	0^+
	Nd	149	1.73 h	1% β⁻ 1.58; 26% β⁻ 1.48; 24% β⁻ 1.15; 31% γ 0.211; 22% γ 0.114; 20% γ 0.030	−74.37		$\frac{5-}{2}$
	Nd	150	5.6%		−73.68	1.2 (γ)	0^+
	Nd	151	12.4 min	β⁻ 2.3; 47% γ 0.116; 17% γ 0.256; 15% γ 1.181	−70.95		
	Nd	152	11.4 min	β⁻; 31% γ 0.278; 21% γ 0.250; 8% γ 0.016	−70.15		0^+
	Nd	154	40 s	β⁻; γ			0^+
10	Ne	17	0.109 s	1% β⁺ 13.5; 54% β⁺ 8.0; 98% p	16.48		$\frac{1-}{2}$
	Ne	18	1.67 s	92% β⁺ 3.42; 8% γ 1.041	5.32		0^+
	Ne	19	17.22 s	99% β⁺ 2.22	1.751		$\frac{1+}{2}$
	Ne	20	90.51%		−7.043	0.038 (γ) (r)	0^+
	Ne	21	0.27%		−5.733	0.7 (γ) (r)	$\frac{3+}{2}$
	Ne	22	9.22%		−8.026	0.05 (γ) (r)	0^+
	Ne	23	37.6 s	67% β⁻ 4.37; 32% β⁻ 3.94; 33% γ 0.440	−5.155		$\frac{5+}{2}$
	Ne	24	3.38 min	92% β⁻ 1.99; (100% m); 100% γ (m) 0.472; 8% γ 0.874	−5.95		0^+
	Ne	25	0.60 s	77% β⁻ 7.1; 96% γ 0.090; 18% γ 0.980; 2% γ 1.069	−2.2		
28	Ni	53	0.05 s	β⁺; p	−29.4		
	Ni	56	6.10 d	100% K; 99% γ 0.158; 75% γ 0.812; 48% γ 0.750	−53.90		0^+
	Ni	57	36.0 h	60% K; 34% β⁺ 0.86; 78% γ 1.378; 15% γ 1.919; 13% γ 0.127	−56.077		$\frac{3-}{2}$
	Ni	58	68.3%		−60.224	4.6 (γ) (r)	0^+
	Ni	59	8 × 10⁴ yr	100% K	−61.153	92 (t)	$\frac{3-}{2}$
	Ni	60	26.1%		−64.470	2.8 (γ) (r)	0^+
	Ni	61	1.13%		−64.219	2.5 (γ) (r)	$\frac{3-}{2}$
	Ni	62	3.59%		−66.745	14.2 (γ)	0^+
	Ni	63	100 yr	100% β⁻ 0.066	−65.513	23 (γ) (r)	$\frac{1-}{2}$
	Ni	64	0.91%		−67.098	1.49 (γ)	0^+
	Ni	65	2.520 h	61% β⁻ 2.14; 28% β⁻ 0.66; 24% γ 1.482; 16% γ 1.115; 5% γ 0.366	−65.125	24 (γ) (s)	$\frac{5-}{2}$
	Ni	66	54.8 h	100% β⁻ 0.24	−66.02		0^+
	Ni	67	18 s	β⁻ 3.8; γ	−63.5		
102	No	252	2.3 s	55% α 8.42; 18% α 8.37; 27% SF	82.87		0^+
	No	253	1.8 min	α 8.01	84.3		
	No	254	55 s	α 8.10	84.73		0^+
	No	254m	0.28 s	100% IT			
	No	255	3.3 min	28% α 8.12; 7% α 8.08; 7% α 7.93; 20% α (dist); 38% K	86.9		
	No	256	3.2 s	α 8.43	87.8		0^+
	No	257	26 s	19% α 8.32; 26% α 8.27; 55% α 8.22	90.22		

(Continued)

Table IX.39 (Continued)

Z	Nuclide A†	half-life and/or abundance	decay scheme⊕	mass excess/ MeV	cross section*/barn	J^P
	No 259	58 min	11% α 7.59; 18% α 7.52; 30% α 7.49; 41% α (dist); 22% K	94.01		
93	Np 229	4.0 min	α 6.89; K	33.76		
	Np 230	4.6 min	α 6.66	35.23		
	Np 231	48.8 min	α 6.28; K; γ	35.63		
	Np 232	14.7 min	100% K; 52% γ 0.327; 33% γ 0.819; 24% γ 0.867	37.3		
	Np 233	36.2 min	100% K; 1% γ 0.312	38.0		
	Np 234	4.4 d	100% K; 19% γ 1.559; 12% γ 1.528; 10% γ 1.602	39.95	1000 (f) (r)	
	Np 235	396 d	100% K	41.04	160 (γ) (r)	$\frac{5+}{2}$
	Np 236	1.2×10^5 yr	91% K; β⁻; 27% γ 0.160; 7% γ 0.104		3000 (f)	
	Np 236	22.5 h	50% K; 38% β⁻ 0.54; 12% β⁻ 0.50; 1% γ 0.642	43.43		1
	Np 237	2.14×10^6 yr	47% α 4.79; 25% α 4.77; 8% α 4.76; 20% α (dist); 14% γ 0.029; 13% γ 0.086	44.869	180 (γ); 0.02 (f)	$\frac{5+}{2}$
	Np 237f	5×10^{-8} s	IT; SF	47.6		
	Np 238	2.117 d	35% β⁻ 1.25; 50% β⁻ 0.26; 28% γ 0.984; 21% γ 1.029; 9% γ 1.026	47.453	2100 (f) (s)	2^+
	Np 239	2.35 d	7% β⁻ 0.72; 45% β⁻ 0.43; 36% β⁻ 0.33; 28% γ 0.106; 14% γ 0.278; 11% γ 0.228	49.306	20 (γ) (r); 32 (γm) (r)	$\frac{5+}{2}$
	Np 240	67 min	100% β⁻ 0.78; 29% γ 0.566; 23% γ 0.974; 22% γ 0.601	52.2		
	Np 240m	7.5 min	β⁻ 2.2; 22% γ 0.555; 13% γ 0.597; 1% γ 0.263; 0.303			1
	Np 241	16.0 min	β⁻ 1.4; γ	54.3		
8	O 13	0.009 s	β⁺ 16.7; 12% p	23.11		$\frac{3-}{2}$
	O 14	70.6 s	1% β⁺ 4.12; 99% β⁺ 1.81; 99% γ 2.313	8.008		0^+
	O 15	122.1 s	100% β⁺ 1.73	2.855		$\frac{1-}{2}$
	O 16	99.76%		−4.7370	0.00018 (γ) (r)	0^+
	O 17	0.038%		−0.810	0.24 (α) (s)	$\frac{5+}{2}$
	O 18	0.202%		−0.783	0.00016 (γ)	0^+
	O 19	26.8 s	40% β⁻ 4.62; 60% β⁻ 3.27; 91% γ 0.197; 55% γ 1.357; 4% γ 0.110	3.331		$\frac{5+}{2}$
	O 20	13.5 s	100% β⁻ 2.75; 100% γ 1.057	3.80		0^+
76	Os 169	3 s	α 5.56	−30.6		
	Os 170	7 s	α 5.40	−33.5		
	Os 171	8 s	α 5.24	−34.2		
	Os 172	19 s	β⁺; K	−36.8		
	Os 173	16 s	β⁺; K; γ	−37.4		
	Os 174	45 s	β⁺; K; γ	−39.6		0^+
	Os 176	3.6 min	β⁺; K; γ	−41.8		0^+
	Os 178	5.0 min	β⁺; K; γ	−43.4		0^+
	Os 179	7 min	β⁺; K; γ	−42.9		
	Os 180	22 min	100% K; 18% γ 0.020	−44.2		0^+
	Os 181	105 min	β⁺; K; 46% γ 0.239; 21% γ 0.827; 13% γ 0.118			
	Os 181	2.7 min	β⁺; K; 87% γ 0.145; 24% γ 0.118; 3% γ 1.119	−43.4		
	Os 182	22.0 h	100% K; 52% γ 0.510; 33% γ 0.180; 7% γ 0.263	−44.6		0^+

(Continued)

Table IX.39 (Continued)

Z	Nuclide	A†	half-life and/or abundance	decay scheme⊕	mass excess/ MeV	cross section*/barn	J^P
	Os	183	13 h	100% K; 86% γ 0.382; 20% γ 0.114; 8% γ 0.168	−43.5		$\frac{9+}{2}$
	Os	183m	9.9 h	89% K; 11% IT; 55% γ 1.102; 25% γ 1.108; 7% γ 1.035	−43.3		$\frac{1-}{2}$
	Os	184	0.018%		−44.23	3000 (γ)	0^+
	Os	185	94 d	100% K; 81% γ 0.646; 7% γ 0.875; 5% γ 0.880	−42.79		$\frac{1-}{2}$
	Os	186	2 × 10¹⁵ yr 1.6%	α 2.75	−42.99	80 (γ)	0^+
	Os	187	1.6%		−41.21	330 (γ)	$\frac{1-}{2}$
	Os	188	13.3%		−41.13	<5 (γ)	0^+
	Os	189	16.1		38.98	20 (γ); 3 × 10⁻⁴ (γm)	$\frac{3-}{2}$
	Os	189m	5.7 h	100% IT	−38.95		$\frac{9-}{2}$
	Os	190	26.4%		−38.70	4 (γ); 9 (γm)	0^+
	Os	190m	9.9 min	100% IT; 99% γ 0.502; 99% γ 0.616; 95% γ 0.361	−36.99		10^-
	Os	191	15.4 d	100% β⁻ 0.14; (100% m); 26% γ (m) 0.129	−36.39		$\frac{9-}{2}$
	Os	191m	13.1 h	100% IT	−36.31		$\frac{3-}{2}$
	Os	192	41.0%		−35.88	2.0 (γ)	0^+
	Os	192m	6.1 s	100% IT; 70% γ 0.569; 66% γ 0.206; 59% γ 0.453	−33.86		
	Os	193	30.6 h	54% β⁻ 1.1; 4% γ 0.139; 4% γ 0.460; 3% γ 0.073	−33.39	1500 (γ) (r)	
	Os	194	6.0 yr	67% β⁻ 0.10; 33% β⁻ 0.06; 3% γ 0.043	−32.42		0^+
	Os	195	6.5 min	β⁻	−29.7		
	Os	196	35 min	β⁻; 6% γ 0.408; 5% γ 0.126; 3% γ 0.315			0^+
15	P	28	0.270 s	68% β⁺ 11.5; 98% γ 1.779; 12% γ 4.497; 9% γ 7.536	−7.16		3^+
	P	29	4.15 s	99% β⁺ 3.92; 1% γ 1.273	−16.949		$\frac{1+}{2}$
	P	30	2.50 min	β⁺ 3.21; K	−20.205		1^+
	P	31	100%		−24.440	0.18 (γ) (r)	$\frac{1+}{2}$
	P	32	14.28 d	100% β⁻ 1.71	−24.305		1^+
	P	33	25.3 d	100% β⁻ 0.249	−26.337		$\frac{1+}{2}$
	P	34	12.5 s	85% β⁻ 5.4; 15% γ 2.128	−24.6		1^+
	P	35	47 s	β⁻; 99% γ 1.572	−24.9		
91	Pa	216	0.20 s	α 7.92			
	Pa	222	0.006 s	30% α 8.54; 20% α 8.33; 50% α 8.18	21.96		
	Pa	223	0.007 s	45% α 8.20; 55% α 8.01	22.33		
	Pa	224	1.0 s	100% α 7.49	23.80		
	Pa	225	1.8 s	70% α 7.25; 30% α 7.20	24.32		
	Pa	226	1.8 min	38% α 6.86; 34% α 6.82; 2% α 6.73; 26% K	26.03		
	Pa	227	38.3 min	43% α 6.47; 23% α 6.42 (double); 8% α 6.40; 11% α (dist); 15% K; 5% γ 0.065	26.83		
	Pa	228	22 h	2% α (6.12); 98% K; 16% γ 0.911; 13% γ 0.463; 13% γ 0.969	28.87		3^+
	Pa	229	1.4 d	100% K			
	Pa	230	18 d	91% K; 9% β⁻ 0.51; 28% γ 0.952; 8% γ 0.919; 6% γ 0.455	32.166	1500 (f) (r)	

(Continued)

Table IX.39 (Continued)

Z	Nuclide	A	half-life and/or abundance	decay scheme	mass excess/ MeV	cross section*/barn	J^P
	Pa	231	3.28×10^4 yr	11% α 5.06; 25% α 5.01; 23% α 4.95; 41% α (dist); 10% γ 0.016, 0.027; 8% γ 0.013	33.423	200 (γ); 0.019 (f) (s)	$\frac{3}{2}^-$
	Pa	232	1.31 d	1% β⁻ 1.29; 71% β⁻ 0.32; 42% γ 0.969; 19% γ 0.894; 12% γ 0.150	35.93	800 (γ) (r); 700 (f) (r)	
	Pa	233	27.0 d	3% β⁻ 0.57; 36% β⁻ 0.23; 27% β⁻ 0.15; 37% γ 0.312; 7% γ 0.300; 4% γ 0.341	37.49	19 (γ); 20 (γm)	$\frac{3}{2}^-$
	Pa	234	6.75 h	β⁻; 30% γ 0.880 (complex); 23% γ 0.926 (complex); 20% γ 0.131	40.35		4
	Pa	234m	1.175 min	99% β⁻ 2.29	40.4		
	Pa	235	24.2 min	97% β⁻ 1.4; (100%m); γ	42.3		
	Pa	236	9.1 min	10% β⁻ 3.1; 44% β⁻ 2.4; 30% γ 0.642; 8% γ 0.687; 5% γ 1.763	45.5		
	Pa	237	8.7 min	19% β⁻ 2.3; 52% β⁻ 1.4; 34% γ 0.854; 16% γ 0.865; 15% γ 0.529	47.6		
	Pa	238	2.3 min	β⁻; γ	51.3		
82	Pb	186	8 s	2% α 6.32; β⁺; K	−14.3		0^+
	Pb	187	17 s	1% α 6.19; 1% α 6.08; β⁺; K	−14.9		
	Pb	188	25 s	3% α (5.98); β⁺; K	−17.5		0^+
	Pb	189	51 s	β⁺; K	−17.9		
	Pb	190	1.2 min	β⁺; K	−20.2		0^+
	Pb	191	1.3 min	β⁺; K	−20.2		
	Pb	192	2.3 min	β⁺; K	−22.3		0^+
	Pb	193	5.8 min	β⁺; K; γ	−22.1		
	Pb	194	11 min	β⁺; K; γ	−23.8		0^+
	Pb	195	16.4 min	β⁺; K; γ	−23.6		
	Pb	196	37 min	100% K; γ	−25.2		0^+
	Pb	197m	42 min	β⁺; K; (81%m) 19% IT; 137% γ 0.386 (double); 25% γ 0.222; 19% γ 0.773	−24.3		
	Pb	198	2.4 h	100% K; 24% γ 0.173; 18% γ 0.290; 8% γ 0.865	−25.9		0^+
	Pb	199	90 min	99% K; β⁺; 62% γ 0.367; 13% γ 0.353; 11% γ 1.135	−25.3		$\frac{5}{2}^-$
	Pb	199m	12.2 min	93% IT; β⁺; K; 18% γ 0.424; 4% γ 0.369; 4% γ 0.382	−24.9		$\frac{13}{2}^+$
	Pb	200	21.5 h	100% K; 38% γ 0.148; 4% γ 0.236; 4% γ 0.257	−26.2		0^+
	Pb	201	9.4 h	100% K; 78% γ 0.331; 10% γ 0.361; 7% γ 0.946	25.33		$\frac{5}{2}^-$
	Pb	201m	61 s	100% IT; 55% γ 0.628	−24.70		$\frac{13}{2}^+$
	Pb	202	3×10^5 yr	100% L	−25.94		0^+
	Pb	202m	3.62 h	91% IT; 9% K; 92% γ 0.961; 86% γ 0.422; 50% γ 0.787	−23.77		9^-
	Pb	203	52.0 h	100% K; 81% γ 0.279; 4% γ 0.401	−24.79		$\frac{5}{2}^-$
	Pb	203m₁	6.1 s	100% IT; 70% γ 0.825; 8% γ 0.820	−23.97		$\frac{13}{2}^+$
	Pb	203m₂	0.48 s	100% IT; γ	−21.84		$\frac{29}{2}^-$
	Pb	204	1.42%		−25.12	0.66 (γ)	0^+
	Pb	204m	66.9 min	100% IT; 99% γ 0.899; 96% γ 0.912; 89% γ 0.374	−22.93		9^-
	Pb	205	1.4×10^7 yr	100% L	−23.78	3.8 (γ) (r)	$\frac{5}{2}^-$

(Continued)

Table IX.39 (Continued)

Z	Nuclide	A†	half-life and/or abundance	decay scheme	mass excess/ MeV	cross section*/barn	J^P
	Pb	206	24.1%		−23.80	0.03 (γ)	0^+
	Pb	207	22.1%		−22.46	0.71 (γ)	$\frac{1}{2}-$
	Pb	207m	0.81 s	100% IT; 98% γ 0.570; 87% γ 1.064	−20.83		$\frac{13+}{2}$
	Pb	208	52.3%		−21.76	0.0005 (γ)	0^+
	Pb	209	3.25 h	100% β⁻ 0.64	−17.62		$\frac{9+}{2}$
	▫Pb	210	22.3 yr	19% β⁻ 0.063; 81% β⁻ 0.016; 4% γ 0.046	−14.74	0.5 (γ)	0^+
	Pb	211	36.1 min	92% β⁻ 1.37; 3% γ 0.405; 3% γ 0.832; 2% γ 0.427	−10.492		
	▫Pb	212	10.64 h	12% β⁻ 0.57; 83% β⁻ 0.33; 43% γ 0.239; 3% γ 0.300; 1% γ 0.115	−7.56		0^+
	Pb	213	10.2 min	β⁻	−3.1		
	Pb	214	26.8 min	6% β⁻ 1.02; 42% β⁻ 0.70; 48% β⁻ 0.65; 37% γ 0.352; 19% γ 0.295; 8% γ 0.242	−0.185		0^+
46	Pd	97	3.3 min	β⁺; K; γ	−77.8		
	Pd	98	17.5 min	K; β⁺; γ	−81.3		0^+
	Pd	99	21.4 min	β⁺ 2.2; K; (97% m); 73% γ 0.136; 15% γ 0.264; 7% γ 0.673	−86.11		
	Pd	100	3.6 d	100% K; γ	−85.23		0^+
	Pd	101	8.5 h	94% K; 6% β⁺ 0.8; (100%m); 18% γ 0.296; 11% γ 0.590; 6% γ 0.270	−85.43		
	Pd	102	1.0%		−87.93	5 (γ) (r)	0^+
	Pd	103	16.96 d	100% K; (100%m)	−87.48		$\frac{5+}{2}$
	Pd	104	11.0%		−89.400		0^+
	Pd	105	22.2%		−88.422		$\frac{5+}{2}$
	Pd	106	27.3%		−89.913	0.28 (γ); 0.013 (γm)	0^+
	Pd	107	6.5 × 10⁶ yr	100% β⁻ 0.03	−88.37		$\frac{5+}{2}$
	Pd	107m	21.3 s	100% IT; 68% γ 0.215	−88.16		$\frac{11-}{2}$
	Pd	108	26.7%		−89.523	11(γ) (r); 0.19 (γm) (r)	0^+
	Pd	109	13.43 h	100% β⁻ 1.0; (100% m); 4% γ (m) 0.088	−87.606		$\frac{5+}{2}$
	Pd	109m	4.69 min	100% IT; 56% γ 0.189	−87.417		$\frac{11-}{2}$
	Pd	110	11.8%		−88.34	0.36 (γ) (r); 0.02 (γm) (r)	0^+
	Pd	111	22 min	96% β⁻ 2.1; (99%m); 1% γ 0.071; 1% γ 0.580; 1% γ (m) 0.060	−86.03		
	Pd	111m	5.5 h	71% IT; β⁻; (23%m); 32% γ 0.172; 7% γ 0.071; 5% γ 0.391	−85.86		
	Pd	112	21.1 h	100% β⁻ 0.28; 4% γ 0.019	−86.33		0^+
	Pd	113	1.5 min	β⁻	−83.6		
	Pd	114	2.4 min	β⁻	−83.8		0^+
	Pd	115	37.4 s	β⁻			
	Pd	116	14 s	β⁻	−80		0^+
	Pd	117	5 s	β⁻			
	Pd	118	3.1 s	β⁻	−76.2		0^+
61	Pm	132	4 s	β⁺; K			
	Pm	133	12 s	β⁺; K			
	Pm	134	24 s	β⁺; K; γ			
	Pm	135	0.9 min	β⁺; K; γ			

(Continued)

Table IX.39 (Continued)

Z	Nuclide	A	half-life and/or abundance	decay scheme	mass excess/ MeV	cross section*/barn	J^P
	Pm	136	107 s	β^+; K; 89% γ 0.374; 49% γ 0.603; 31% γ 0.858	−71.4		
	Pm	137	2.4 min	β^+; K; γ	−74.2		
	Pm	138	3.5 min	β^+; K; 93% γ 0.521; 37% γ 0.729; 20% γ 0.493	−75.0		
	Pm	139	4.15 min	β^+; K; 12% γ 0.403; 3% γ 0.368; 3% γ 0.463	−77.6		
	Pm	140	9.2 s	β^+ 5.0; K; 5% γ 0.774; 1% γ 0.717; 1% γ 1.490	−78.2		1^+
	Pm	140m	5.9 min	58% K; β^+ 3.2; 100% γ 0.774; 100% γ 1.028; 92% γ 0.420	−77.8		
	Pm	141	20.9 min	β^+ 2.7; 43% K; 4% γ 1.223; 2% γ 0.886; 1% γ 0.194	−80.47		$\frac{5+}{2}$
	Pm	142	40.5 s	β^+ 3.9; 31% K; 3% γ 1.576	−81.1		1^+
	Pm	143	265 d	100% K; 39% γ 0.742	−82.96		$\frac{5+}{2}$
	Pm	144	350 d	100% K; 100% γ 0.696; 99% γ 0.618; 42% γ 0.477	−81.42		5^-
	Pm	145	17.7 yr	100% K; 2% γ 0.072; 1% γ 0.067	−81.27		$\frac{5+}{2}$
	Pm	146	5.5 yr	63% K; 34% β^- 0.8; 3% β^- 0.2; 63% γ 0.454; 37% γ 0.747; 23% γ 0.736	−79.44	8000 (γ) (r)	3^-
	Pm	147	2.623 yr	100% β^- 0.22	−79.04	97 (γ); 85 (γm)	$\frac{7+}{2}$
	Pm	148	5.37 d	54% β^- 2.46; 35% β^- 0.99; 23% γ 0.550; 22% γ 1.465; 13% γ 0.915	−76.87	<3000 (γ) (r)	1^-
	Pm	148m	41.3 d	1% β^- 1.0; 56% β^- 0.4; 5% IT; 93% γ 0.550; 89% γ 0.630; 32% γ 0.726	−76.73	1.1×10^4 (γ)	6^-
	Pm	149	53.1 h	96% β^- 1.0; 3% γ 0.286	−76.06	1400 (γ) (r)	$\frac{7+}{2}$
	Pm	150	2.68 h	8% β^- 3.2; 28% β^- 2.3; 73% γ 0.334; 19% γ 1.325; 17% γ 1.166	−73.6		
	Pm	151	28.4 h	11% β^- 1.2; 42% β^- 0.8; 22% γ 0.005; 22% γ 0.340; 8% γ 0.168	−73.39	<700 (γ) (r)	$\frac{5+}{2}$
	Pm	152	4.1 min	61% β^- 3.5; 16% γ 0.122; 5% γ 0.842; 5% γ 0.961, 0.963	−71.3		
	Pm	152	7.5 min	β^-; 54% γ 0.245; 44% γ 0.122; 27% γ 0.340			
	Pm	152	15 min	β^-; γ			
	Pm	153	5.3 min	55% β^- 1.7; 25% γ 0.036; 14% γ 0.127; 8% γ 0.028	−70.8		
	Pm	154	1.7 min	12% β^- 3.1; 24% β^- 2.0; 35% β^- 1.9; 19% γ 2.059; 12% γ 0.082; 12% γ 0.840, 1.394	−68.5		
	Pm	154	2.7 min	β^-; 39% γ 0.185; 25% γ 1.440 (double); 19% γ 0.082			
84	Po	194	0.6 s	α 6.85	−10.8		0^+
	Po	195	4.5 s	α 6.61	−11.1		
	Po	195m	2.0 s	α 6.70			
	Po	196	5.5 s	α 6.52	−13.2		0^+
	Po	197	58 s	90% α (6.28); β^+; K	−13.2		
	Po	197m	26 s	α 6.38			
	Po	198	1.8 min	70% α (6.18); β^+; K	−15.1		0^+
	Po	199	5.2 min	12% α (5.95); β^+ K; 48% γ 1.034; 25% γ 1.021; 23% γ 0.362	−15.1		

(Continued)

Table IX.39 (Continued)

Z	Nuclide □ A†	half-life and/or abundance	decay scheme⊕	mass excess/ MeV	cross section*/barn	J^P
	Po 199m	4.2 min	39% α (6.06); β⁺; K; 61% γ 1.002; 26% γ 0.500; 7% γ 0.274			
	Po 200	11.6 min	14% α (5.86); β⁺; K; γ	−16.7		0^+
	Po 201	15.2 min	2% α (5.68); β⁺; K; 53% γ 0.890; 29% γ 0.905; 13% γ 0.848	−16.4		$\frac{3}{2}$
	Po 201m	9.0 min	3% α 5.79; 53% IT; β⁺; K; 43% γ 0.967; 19% γ 0.412; 9% γ 0.419	−16.0		
	Po 202	44 min	2% α (5.59); β⁺; K; 47% γ 0.687; 13% γ 0.316; 10% γ 0.166	−17.8		0^+
	Po 203	33 min	β⁺; K; 57% γ 0.909; 20% γ 1.091; 19% γ 0.894	−17.4		$\frac{5-}{2}$
	Po 203m	1.2 min	96% IT; β⁺; K; 51% γ 0.641; 4% γ 0.905; 2% γ 0.577	−16.7		
	Po 204	3.57 h	1% α 5.38; 99% K; γ	−18.3		0^+
	Po 205	1.80 h	β⁺; K; 35% γ 0.872; 27 % γ 1.001; 24% γ 0.850	−17.58		$\frac{5-}{2}$
	Po 206	8.8 d	5% α (5.22); 95% K; 32% γ 1.032; 23% γ 0.286; 23% γ 0.511	−18.19		0^+
	Po 207	5.7 h	99% K; β⁺; 59% γ 0.992; 29% γ 0.743; 18% γ 0.912	−17.15		$\frac{5-}{2}$
	Po 207m	2.8 s	100% IT; 99% γ 0.815; 43% γ 0.268; 30% γ 0.301	−15.77		$\frac{19-}{2}$
	Po 208	2.898 yr	100% α 5.12	−17.48		0^+
	Po 209	102 yr	100% α 4.88	−16.37		$\frac{1-}{2}$
	Po 210	138.38 d	100% α 5.30	−15.96	<0.03 (γ) (r)	0^+
	Po 211	0.516 s	98% α 7.45; 1% 6.89; 1% α 6.57; 1% γ 0.570, 0.898	−12.44		
	Po 211m	25.5 s	7% α 8.88; 2% α 8.00; 91% α 7.28; (91%m); 80% γ (m) 0.570; 80% γ (m) 1.064	−10.98		
	Po 212	2.96 × 10⁻⁷ s	α 8.78	−10.38		0^+
	Po 212m	45 s	97% α 11.65; 1% α 9.10; 2% α 8.52; 3% γ 2.615; 2% γ 0.570	−7.48		
	Po 213	4 × 10⁻⁶ s	100% α 8.38	−6.66		$\frac{9+}{2}$
	Po 214	1.64 × 10⁻⁴ s	100% α 7.69	−4.48		0^+
	Po 215	0.00178 s	100% α 7.39	−0.541		
	Po 216	0.145 s	100% α 6.78	1.77		0^+
	Po 217	<10 s	α 6.54	6.0		
	Po 218	3.05 min	100% α 6.00	8.35		0^+
59	Pr 129	24 s	β⁺; K			
	Pr 130	28 s	β⁺; K			
	Pr 132	1.6 min	β⁺; K; γ	−75.3		
	Pr 133	6.5 min	β⁺; K; γ	−78.0		$\frac{5}{2}$
	Pr 134	17 min	β⁺; K; γ	−78.5		2^+
	Pr 134	11 min	β⁺; K; γ	−78.5		
	Pr 135	25 min	75% K; β⁺ 2.5; γ	−81.0		$\frac{3}{2}$
	Pr 136	13.1 min	β⁺; K; 76% γ 0.552; 52% γ 0.540; 18% γ 1.092	−81.4		2^+
	Pr 137	1.28 h	75% K; 25% β⁺ 1.7; 2% γ 0.837; 1% γ 0.434; 1% γ 0.514	−83.2		$\frac{5+}{2}$
	Pr 138	1.4 min	β⁺ 3.4; K; 2% γ 0.789; 1% γ 0.688	−83.13		1^+

(Continued)

Table IX.39 (Continued)

Z	Nuclide □ A†	half-life and/or abundance	decay scheme⊕	mass excess/ MeV	cross section*/barn	J^P
	Pr	138m	2.02 h	77% K; 23% β+ 1.6; 100% γ 0.789; 100% γ 1.038; 80% γ 0.302	−82.77		7−
	Pr	139	4.41 h	92% K; β+ 1.1	−84.85		$\frac{5+}{2}$
	Pr	140	3.39 min	51% K; 48% β+ 2.4	−84.69		1+
	Pr	141	100%		−86.02	7.6 (γ); 3.9 (γm)	$\frac{5+}{2}$
	Pr	142	19.2 h	96% β− 2.2; 4% γ 1.576	−83.79	20 (γ) (r)	2−
	Pr	142m	14.6 min	100% IT	−83.79		5−
	Pr	143	13.57 d	100% β− 0.93	−83.07	90 (γ)	$\frac{7+}{2}$
	Pr	144	17.3 min	98% β− 3.0; 1% γ 0.696	−80.75		0−
	Pr	144m	7.2 min	100% IT	−80.69		3−
	Pr	145	5.98 h	97% β− 1.8	−79.63		
	Pr	146	24.0 min	β− 4.1; 48% γ 0.454; 18% γ 1.525; 8% γ 0.736, 0.789	−76.8		
	Pr	147	13 min	β−; 22% γ 0.315; 17% γ 0.641; 15% γ 0.578	−75.4		
	Pr	148	2.30 min	42% β− 4.5; 91% γ 0.302; 16% γ 0.450; 12% γ 0.697	−72.6		
	Pr	149	2.3 min	55% β− 3.0; 7% γ 0.139; 6% γ 0.110; 6% γ 0.165	−71.4		
	Pr	150	6.2 s	β− 5.7; γ	−68.0		
	Pr	151	4.0 s	β−; γ	−67.4		
78	Pt	173	<1 s	α 6.19	−21.8		
	Pt	174	0.7 s	80% α (6.03); β+; K	−24.9		0+
	Pt	175	2.5 s	75% α (5.95); β+; K	−25.6		
	Pt	176	6.3 s	42% α 5.74; β+; K	−28.5		0+
	Pt	177	7 s	6% α 5.53; 3% α 5.49; β+; K	−29.4		
	Pt	178	21 s	7% α 5.46; β+; K	−31.6		0+
	Pt	179	33 s	β+; K	−32.0		$\frac{1}{2}$
	Pt	180	50 s	β+; K	−34.1		0+
	Pt	181	51 s	β+; K	−34.1		
	Pt	182	2.6 min	β+; K; γ	−36.0		0+
	Pt	183	7 min	β+; K	−35.6		
	Pt	184	17.3 min	β+; K; γ	−37.2		0+
	Pt	185	71 min	β+; K; γ	−36.5		
	Pt	185	33 min	β+; K; γ	−36.5		
	Pt	186	2.0 h	100% K; γ	−37.8		0+
	Pt	187	2.35 h	β+; K; γ	−36.8		$\frac{3}{2}$
	Pt	188	10.2 h	100% K; 19% γ 0.188; 18% γ 0.195; 7% γ 0.382	−37.79		0+
	Pt	189	10.9 h	K; β+ 0.9; 6% γ 0.721; 5% γ 0.094; 5% γ 0.608	−36.6		$\frac{3-}{2}$
	Pt	190	6 × 10¹¹ yr 0.013%	α 3.18	−37.32	800 (γ)	0+
	Pt	191	2.9 d	100% K; (1%m); 14% γ 0.539; 8% γ 0.409; 6% γ 0.360	−35.70		$\frac{3-}{2}$
	Pt	192	0.78%		−36.28	10 (γ); 2 (γm) (r)	0+
	Pt	193	50 yr	100% K	−34.46		
	Pt	193m	4.33 d	100% IT	−34.31		
	Pt	194	32.9%		−34.77	1.1 (γ) (r); 0.09 (γm) (r)	0+
	Pt	195	33.8%		−32.80	27 (γ) (r)	$\frac{1-}{2}$
	Pt	195m	4.02 d	100% IT; 11% γ 0.099; 3% γ 0.130; 2% γ 0.031	−32.54		$\frac{13+}{2}$
	Pt	196	25.3%		−32.65	0.7 (γ) (r); 0.05 (γm) (r)	0+
	Pt	197	18.3 h	11% β− 0.72; 81% β− 0.64; 17% γ 0.077; 3% γ 0.192	−30.43		$\frac{1-}{2}$

(Continued)

Table IX.39 (Continued)

Z	Nuclide	$A^\dagger$	half-life and/or abundance	decay scheme	mass excess/ MeV	cross section*/barn	J^P
	Pt	197m	94 min	97% IT; 3% β^- 0.7; (3%m); 11% γ 0.346; 2% γ (m) 0.279	−30.03		$\frac{13+}{2}$
	Pt	198	7.2%		−29.92	3.7 (γ); 0.03 (γm)	0^+
	Pt	199	30.8 min	63% β^- 1.7; 15% γ 0.543; 6% γ 0.494; 5% γ 0.317	−27.42	15 (γ) (r)	
	Pt	199m	14 s	100% IT; 85% γ 0.392	−27.00		
	Pt	200	12.6 h	β^-; γ	−26.6		0^+
	Pt	201	2.5 min	β^- 2.7; γ	−23.7		
94	Pu	232	34 min	α 6.60; >80% K	38.36		0^+
	Pu	233	20.9 min	100% K; γ	40.04		
	Pu	234	8.8 h	4% α 6.20; 2% α 6.15; 94% K	40.34		0^+
	Pu	235	25.6 min	100% K; 2% γ 0.049	42.2		
	Pu	236	2.85 yr	69% α 5.77; 31% α 5.72	42.89	150 (f) (r)	0^+
	Pu	236f$_1$	3×10^{-11} s	SF			
	Pu	236f$_2$	3×10^{-8} s	SF	46.4		
	Pu	237	45.3 d	100% K; 3% γ 0.059	45.09	2100 (f) (r)	$\frac{7-}{2}$
	Pu	237m	0.18 s	100% IT; 2% γ 0.146	45.23		$\frac{1+}{2}$
	Pu	237f$_1$	1.1×10^{-7} s	SF	47.4		
	Pu	237f$_2$	1.1×10^{-6} s	SF	47.7		
	Pu	238	87.7 yr	72% α 5.50; 28% α 5.45	46.161	500 (γ) (r); 17 (f) (r)	0^+
	Pu	238f$_1$	5×10^{-10} s	SF	48.6		
	Pu	238f$_2$	7×10^{-9} s	SF	49.9		
	Pu	239	2.413×10^4 yr	73% α 5.16; 15% α 5.14; 12% α 5.10; (100%m)	48.585	271 (γ); 742 (f)	$\frac{1+}{2}$
	Pu	239f$_1$	8×10^{-6} s	SF	50.8		
	Pu	240	6570 yr	73% α 5.16; 27% α 5.12	50.123	290 (γ)	0^+
	Pu	240f	3.8×10^{-9} s	SF	52.5		
	Pu	241	14.36 yr	100% β^- 0.02	52.953	370 (γ); 1010 (f)	$\frac{5+}{2}$
	Pu	241f	2.3×10^{-5} s	SF	55.0		
	Pu	242	3.76×10^5 yr	74% α 4.90; 26% α 4.86	54.715	19 (γ)	0^+
	Pu	242f$_1$	4×10^{-9} s	SF			
	Pu	242f$_2$	3×10^{-8} s	SF			
	Pu	243	4.955 h	60% β^- 0.58; 21% β^- 0.50; 23% γ 0.084; 1% γ 0.042	57.753	100 (γ) (r); 200 (f) (r)	$\frac{7+}{2}$
	Pu	243f	6×10^{-8} s	SF	59.6		
	Pu	244	8.1×10^7 yr	80% α 4.59; 20% α 4.55	59.80	1.7 (γ)	0^+
	Pu	244f	4×10^{-10} s	SF			
	Pu	245	10.5 h	β^-; 26% γ 0.327; 6% γ 0.560; 5% γ 0.308	63.16	150 (γ)	
	Pu	246	10.85 d	β^-; 26% γ 0.044; 24% γ 0.224; 10% γ 0.180	65.3		0^+
88	Ra	206	0.4 s	α 7.27; β^+; K	4.0		0^+
	Ra	207	1.3 s	α 7.13; β^+; K	3.7		
	Ra	209	4.7 s	α 7.01	2.0		
	Ra	210	3.8 s	α 7.02	0.6		0^+
	Ra	211	15 s	α 6.91; β^+; K	0.8		
	Ra	212	13.0 s	α 6.90; β^+; K	−0.1		0^+
	Ra	213	2.7 min	36% α 6.73; 39% α 6.62; 5% α 6.52; β^+; K; 6% γ 0.110; 1% γ 0.215	0.29		
	Ra	213m	0.0021 s	1% α 8.47; 99% IT; 99% γ 0.546; 99% γ 1.063; 46% γ 0.161	2.06		
	Ra	214	2.46 s	α 7.14	0.09		0^+

(Continued)

Table IX.39 (Continued)

Z	Nuclide	$A^\dagger$	half-life and/or abundance	decay scheme	mass excess/ MeV	cross section*/barn	J^P
	Ra	215	0.0016 s	96% α 8.70; 1% α 8.17, 3% α 7.88	2.53		
	Ra	216	1.8×10^{-7} s	α 9.35	3.29		0^+
	Ra	217	1.6×10^{-6} s	α 8.99	5.88		
	Ra	218	1.4×10^{-5} s	α 8.39	6.64		0^+
	Ra	219	0.010 s	35% α 7.98; 65% α 7.68	9.38		
	Ra	220	0.023 s	99% α 7.46; 1% α 6.99; 1% γ 0.465	10.26		0^+
	Ra	221	30 s	30% α 6.76; 20% α 6.67; 34% α 6.61; 16% α (dist); 15% γ 0.089; 13% γ 0.152; 2% γ 0.176	12.96		
	Ra	222	38 s	97% α 6.56; 3% α 6.24; 3% γ 0.324	14.31		0^+
	Ra	223	11.435 d	9% α 5.75; 54% α 5.72; 24% α 5.61; 13% α (dist); 14% γ 0.269; 6% γ 0.154; 4% γ 0.324	17.235	134 (γ) (r)	$\frac{1}{2}^+$
	Ra	224	3.67 d	95% α 5.69; 5% α 5.45; 4% γ 0.241	18.81	12 (γ) (r)	0^+
	Ra	225	14.8 d	33% β⁻ 0.36; 67% β⁻ 0.32; 29% γ 0.040	21.99		
	Ra	226	1600 yr	94% α 4.78; 6% α 4.60; 3% γ 0.186	23.67	20 (γ) (r)	0^+
	Ra	227	42 min	33% β⁻ 1.31; 24% β⁻ 1.29; 17% γ 0.027; 5% γ 0.300; 5% γ 0.303	27.19		
	Ra	228	5.77 yr	60% β⁻ 0.039; 40% β⁻ 0.015	28.94	36 (γ) (s)	0^+
	Ra	229	4.0 min	β⁻	32.7		
	Ra	230	93 min	β⁻; γ	34.6		0^+
37	Rb	74	0.065 s	100% β⁺ 9.6	−51.4		
	Rb	75	18 s	β⁺; γ	−57.5		
	Rb	76	39 s	β⁺; γ	−60.6		
	Rb	77	3.9 min	β⁺; K; 66% γ 0.066; 26% γ 0.179; 12% γ 0.393	−65.1		
	Rb	78	18 min	β⁺; K; γ	−68.8		
	Rb	78m	6.0 min	β⁺; K; IT; γ	−68.7		
	Rb	79	23.0 min	β⁺; 16% K; (37%m); 24% γ 0.688; 16% γ 0.183; 13% γ 0.505	−70.9		
	Rb	80	34 s	74% β⁺ 4.7; 22% β⁺ 4.1; 25% γ 0.616; 2% γ 0.640; 2% γ 0.704	−72.19		1^+
	Rb	81	4.58 h	73% K; β⁺ 1.1; (97%m); 66% γ (m) 0.190; 19% γ 0.446; 2% γ 0.457	−75.39		$\frac{3}{2}^-$
	Rb	81m	32 min	β⁺; γ	−75.31		$\frac{9}{2}^+$
	Rb	82	1.25 min	β⁺ 3.4; 4% K; 14% γ 0.776	−76.21		1^+
	Rb	82m	6.2 h	74% K; β⁺; 83% γ 0.776; 63% γ 0.554; 37% γ 0.619	−76.1		5^-
	Rb	83	86.2 d	100% K; (76%m); 46% γ 0.520; 30% γ 0.530; 14% γ 0.553	−78.91		$\frac{5}{2}^-$
	Rb	84	32.8 d	75% K; 11% β⁺ 1.7; 11% β⁺ 0.8; 3% β⁻ 0.9; 74% γ 0.882; 1% γ 1.900	−79.752	12(p) (r)	2^-
	Rb	84m	20.5 min	100% IT; 65% γ 0.248; 33% γ 0.216; 32% γ 0.464	−79.288		

(Continued)

Table IX.39 (Continued)

Z	Nuclide	A	half-life and/or abundance	decay scheme	mass excess/ MeV	cross section*/barn	J^P
	Rb	85	72.17%		−82.159	0.40 (γ); 0.047 (γm)	$\frac{5-}{2}$
	Rb	86	18.82 d	91% β⁻ 1.77; 9% γ 1.077	−82.738		2^-
	Rb	86m	1.020 min	100% IT; 98% γ 0.556	−82.182		6^-
	Rb	87	4.72 × 10¹⁰ yr 27.83%	100% β⁻ 0.27	−84.596	0.12 (γ) (r)	$\frac{3-}{2}$
	Rb	88	17.8 min	77% β⁻ 5.3; 21% γ 1.836; 14% γ 0.898; 2% γ 2.678	−82.60	1.0 (γ) (r)	2^-
	Rb	89	15.15 min	25% β⁻ 4.5; 33% β⁻ 2.2; 33% β⁻ 1.3; 58% γ 1.032; 43% γ 1.248; 13% γ 2.196	−81.72		
	Rb	90	153 s	37% β⁻ 6.4; γ	−79.6		
	Rb	90m	258 s	β⁻; IT; γ	−79.5		
	Rb	91	58.7 s	β⁻ 5.7; 32% γ 0.094; 12% γ 2.564; 10% γ 3.600	−77.97		
	Rb	92	4.54 s	94% β⁻ 7.8; 4% γ 0.815; 1% γ 0.570, 1.385, 2.821	−75.1		
	Rb	93	5.85 s	β⁻; 1% n; γ	−72.9		
	Rb	94	2.73 s	β⁻; 10% n; γ	−68.8		
	Rb	95	0.38 s	β⁻; 8% n; γ	−66.6		
	Rb	96	0.203 s	β⁻; 13% n; γ	−62.8		
	Rb	97	0.170 s	β⁻; 27% n; γ			
	Rb	98	0.12 s	β⁻; 13% n			
	Rb	99	0.076 s	β⁻			
75	Re	170	8 s	β⁺; K; γ	−38.9		
	Re	172	30 s	β⁺; K; γ	−41.5		
	Re	174	2.4 min	β⁺; K; γ	−43.6		
	Re	175	4.6 min	β⁺; K; γ	−45.2		
	Re	176	5.2 min	β⁺; K; γ	−45.0		
	Re	177	14 min	β⁺; K; γ	−46.1		
	Re	178	13.2 min	89% K; β⁺; 45% γ 0.237; 23% γ 0.106; 9% γ 0.939	−45.8		
	Re	179	19.7 min	99% K; β⁺; (27%m); 28% γ 0.430; 27% γ 0.290; 13% γ 1.680	−46.6		
	Re	180	2.4 min	92% K; 8% β⁺ 1.8; 98% γ 0.902; 23% γ 0.104; 12% γ 0.825	−45.83		
	Re	181	20 h	100% K; 57% γ 0.366; 12% γ 0.361; 6% γ 0.639	−46.4		$\frac{5+}{2}$
	Re	182	64 h	100% K; 27% γ 0.229; 21% γ 1.121; 17% γ 1.221	−45.4		
	Re	182	12.7 h	100% K; 32% γ 1.121; 25% γ 1.221; 15% γ 1.189	−45.4		2^+
	Re	183	71 d	100% K; 24% γ 0.162; 8% γ 0.046; 3% γ 0.209, 0.292	−45.79		
	Re	184	38 d	100% K; 38% γ 0.903; 37% γ 0.792; 17% γ 0.111	−44.19		3^-
	Re	184m	169 d	75% IT; 25% K; 14% γ 0.105; 11% γ 0.253; 10% γ 0.217	−44.00		8^+
	Re	185	37.40%		−43.80	110 (γ); 0.3 (γm) (s)	$\frac{5+}{2}$
	Re	186	90.6 h	71% β⁻ 1.08; 21% β⁻ 0.94; 8% K; 9% γ 0.137; 1% γ 0.123	−41.91		1^-
	Re	186m	2 × 10⁵ yr	100% IT; γ	−41.8		
	Re	187	4 × 10¹⁰ yr 62.60%	100% β⁻ 0.003	−41.21	74 (γ); 1.0 (γm)	$\frac{5+}{2}$

(Continued)

Table IX.39 (Continued)

Z	Nuclide	A	half-life and/or abundance	decay scheme	mass excess/ MeV	cross section*/barn	J^P
	Re	188	16.98 h	70% β⁻ 2.12; 15% γ 0.155; 1% γ 0.478; 1% γ 0.633	−39.01		1⁻
	Re	188m	18.7 min	100% IT; 16% γ 0.064; 11% γ 0.106; 5% γ 0.092	−38.83		
	Re	189	24.3 h	60% β⁻ 1.0; (8%m); 6% γ 0.217; 4% γ 0.219; 3% γ 0.245	−37.97		
	Re	190	3.1 min	89% β⁻ 1.8; 35% γ 0.558; 27% γ 0.829; 25% γ 0.569	−35.5		
	Re	190m	3.2 h	β⁻; 49% IT; 50% γ 0.187; 28% γ 0.558; 26% γ 0.569	−35.3		
	Re	192	16 s	β⁻; γ	−32		
45	Rh	95	5.0 min	β⁺ 3.2; K; 72% γ 0.942; 21% γ 1.352; 6% γ 0.678	−78.3		
	Rh	95m	1.96 min	88% IT; β⁺; K; 80% γ 0.543; 8% γ 0.784; 2% γ 3.407	−77.8		
	Rh	96	9.9 min	β⁺; K; 100% γ 0.833; 98% γ 0.685; 80% γ 0.635	−79.63		
	Rh	96m	1.51 min	60% IT; β⁺; K; 39% γ 0.833; 8% γ 1.099; 6% γ 1.692	−79.58		
	Rh	97	31 min	β⁺; K; 75% γ 0.421; 12% γ 0.840; 9% γ 0.879	−82.6		
	Rh	97m	44 min	β⁺; K; 5% IT; 51% γ 0.189; 13% γ 0.422; 13% γ 2.245	−82.3		
	Rh	98	8.7 min	β⁺ 3.4; K; 94% γ 0.652; 5% γ 0.745; 5% γ 1.817	−83.17		
	Rh	98m	3.5 min	β⁺; K; 96% γ 0.652; 96% γ 1.414; 78% γ 0.745	−83.16		
	Rh	99	15.0 d	97% K; β⁺ 1.1; 38% γ 0.528; 35% γ 0.353; 33% γ 0.090	−85.52		
	Rh	99m	4.7 h	90% K; β⁺; 66% γ 0.341; 16% γ 1.261; 13% γ 0.618	−85.45		$\frac{9}{2}^+$
	Rh	100	20.8 h	95% K; 5% β⁺ 2.6; 78% γ 0.540; 35% γ 2.376; 20% γ 0.823, 1.553	−85.59		1⁻
	Rh	100m	4.7 min	93% IT; K; β⁺; 70% γ 0.075; 31% γ 0.265; 11% γ 0.540	−85.25		
	Rh	101	3.3 yr	100% K; 70% γ 0.127; 68% γ 0.198; 13% γ 0.325	−87.41		$\frac{1}{2}^-$
	Rh	101m	4.34 d	93% K; 7% IT; 87% γ 0.307; 4% γ 0.545	−87.25		$\frac{9}{2}^+$
	Rh	102	2.9 yr	100% K; 95% γ 0.475; 56% γ 0.631; 44% γ 0.697			
	Rh	102m	206 d	62% K; 10% β⁺ 1.3; 4% β⁺ 0.8; 17% β⁻ 1.1; 2% β⁻ 0.5; 5% IT; 45% γ 0.475; 4% γ 0.628; 3% γ 0.469	−86.78		
	Rh	103	100%		−88.024	134 (γ); 11 (γm)	$\frac{1}{2}^-$
	Rh	103m	56.12 min	100% IT	−87.984		
	Rh	104	42.8 s	98% β⁻ 2.45; 2% γ 0.556	−86.952	40 (γ) (r)	1⁺
	Rh	104m	4.4 min	100% IT; 48% γ 0.051; 3% γ 0.097; 2% γ 0.078	−86.823	800 (γ) (r)	5⁺
	Rh	105	35.5 h	75% β⁻ 0.57; 19% γ 0.319; 5% γ 0.306	−87.86	11 000 (γ); 5000 (γ m)	
	Rh	105m	45 s	100% IT; 20% γ 0.130	−87.73		$\frac{1}{2}^-$
	Rh	106	29.8 s	80% β⁻ 3.5; 19% γ 0.512; 10% γ 0.622; 2% γ 1.051	−86.37		1⁺

(Continued)

Table IX.39 (Continued)

Z	Nuclide	$A^{\dagger}$	half-life and/or abundance	decay scheme	mass excess/ MeV	cross section*/barn	J^P
	Rh	106m	130 min	1% β^- 1.7; 84% β^- 0.9; 86% γ 0.512; 31% γ 1.047; 29% γ 0.717	−86.24		
	Rh	107	21.7 min	β^-; 66% γ 0.303; 9% γ 0.392; 5% γ 0.312	−86.86		
	Rh	108	16.8 s	54% β^- 4.5; 21% β^- 3.4; 43% γ 0.434; 21% γ 0.619; 11% γ 0.511	−85.0		1^+
	Rh	108	6.0 min	7% β^- 3.4; 68% β^- 1.6; 91% γ 0.434; 58% γ 0.581; 50% γ 0.947	−85.1		
	Rh	109	80 s	β^- 2.5; 62% γ 0.327; 22% γ 0.291; 14% γ 0.249	−85.1		
	Rh	110	3.0 s	β^- 5.5; γ	−82.8		
	Rh	110	29 s	19% β^- 4.5; 42% β^- 2.6; 92% γ 0.374; 41% γ 0.546; 28% γ 0.440	−82.9		
	Rh	112	4.7 s	β^-; γ	−80		
86	Rn	200	1.0 s	α 6.91	−3.7		0^+
	Rn	201	7.0 s	α 6.72	−4.0		
	Rn	201(m)	3.8 s	α 6.77			
	Rn	202	9.9 s	α 6.64	−5.9		0^+
	Rn	203	45 s	65% α (6.50); β^+; K	−6.0		
	Rn	203m	28 s	α 6.55	−6.0		
	Rn	204	75 s	72% α (6.42); β^+; K	−7.8		0^+
	Rn	205	170 s	23% α (6.26); β^+; K; γ	−7.6		
	Rn	206	5.7 min	64% α 6.26; β^+; K	−9.0		0^+
	Rn	207	9.3 min	23% α 6.13; β^+; K; 63% γ 0.344; 21% γ 0.747; 16% γ 0.403, 0.674	−8.7		$\frac{5-}{2}$
	Rn	208	24.4 min	52% α 6.14; β^+; K	−9.6		0^+
	Rn	209	29 min	17% α 6.04; 80% K; β^+; 47% γ 0.408; 21% γ 0.746; 14% γ 0.337	−8.99		$\frac{5-}{3}$
	Rn	210	2.4 h	96% α 6.04; 4% K; 2% γ 0.458; 1% γ 0.073, 0.233	−9.61		0^+
	Rn	211	14.6 h	9% α 5.85; 16% α 5.78; 1% α 5.62; β^+; K; 46% γ 0.674; 33% γ 1.363; 29% γ 0.678	−8.76		$\frac{1-}{2}$
	Rn	212	23 min	100% α 6.26	−8.67		0^+
	Rn	213	0.025 s	99% α 8.09; 1% α 7.55	−5.71		
	Rn	214	2.7×10^{-7} s	α 9.04	−4.33		0^+
	Rn	215	2.3×10^{-6} s	100% α 8.67	−1.18		
	Rn	216	4×10^{-5} s	α 8.05	0.25		0^+
	Rn	217	5×10^{-4} s	100% α 7.74	3.65		$\frac{9+}{2}$
	Rn	218	0.035 s	100% α 7.13	5.21		0^+
	Rn	219	3.96 s	81% α 6.82; 12% α 6.55; 7% α 6.42; 10% γ 0.271; 7% γ 0.402	8.83		
	Rn	220	55.6 s	100% α 6.29	10.60	<0.2 (γ) (r)	0^+
	Rn	221	25 min	20% α ; β^-	14.4		
	Rn	222	3.824 d	100% α 5.49	16.37	0.7 (γ) (r)	0^+
	Rn	223	43 min	β^-			
	Rn	224	1.8 h	β^-; γ	22.3		0^+
	Rn	225	4.5 min	β^-			
	Rn	226	6 min	β^-			0^+
44	Ru	92	3.7 min	β^+; K; 92% γ 0.214; 88% γ 0.259; 63% γ 0.135	−75		0^+

(Continued)

Table IX.39 (Continued)

Z	Nuclide	A†	half-life and/or abundance	decay scheme	mass excess/ MeV	cross section*/barn	J^P
	Ru	93	60 s	β+; K; 5% γ 0.681; 1% γ 1.435	−77.3		
	Ru	93m	10.8 s	β+; K; 21% IT; (79%m); 36% γ 1.396; 24% γ 1.111; 20% γ 0.734	−76.6		
	Ru	94	52 min	100% K; (100%m); 79% γ 0.367; 21% γ 0.892; 2% γ 0.525	−82.57		0^+
	Ru	95	1.65 h	85% K; β+ 1.2; (3%m); 71% γ 0.336; 21% γ 1.097; 18% γ 0.627	−83.45		$\frac{5+}{2}$
	Ru	96	5.5%		−86.08	0.25 (γ)	0^+
	Ru	97	2.88 d	100% K; 86% γ 0.216; 10% γ 0.325; 1% γ 0.569	−86.07		$\frac{5+}{2}$
	Ru	98	1.86%		−88.226		0^+
	Ru	99	12.7%		−87.620	4 (γ)	$\frac{5+}{2}$
	Ru	100	12.6%		−89.222	5.8 (γ)	0^+
	Ru	101	17.0%		−87.952	5 (γ)	$\frac{5+}{2}$
	Ru	102	31.6%		−89.101	1.3 (γ)	0^+
	Ru	103	39.4 d	6% β− 0.72; 87% β− 0.22; (100%m); 86% γ 0.497; 5% γ 0.610	−87.261		$\frac{5+}{2}$
	Ru	104	18.7%		−88.099	0.47 (γ)	0^+
	Ru	105	4.44 h	2% β− 1.8; 48% β− 1.2; (28% m); 48% γ 0.724; 18% γ 0.469; 16% γ 0.676	−85.94	0.30 (γ)	
	Ru	106	367 d	100% β− 0.039	−86.33		0^+
	Ru	107	4.2 min	75% β− 3.2; 14% γ 0.194; 7% γ 0.849; 6% γ 0.463	−83.7		
	Ru	108	4.5 min	70% β− 1.2; 30% β− 1.0; 30% γ 0.165	−83.8		0^+
	Ru	109	34 s	β−; γ	−80.8		
	Ru	109	13 s	β−	−80.8		
	Ru	110	16 s	β−	−80		0^+
	Ru	111	1.5 s	β−			
16	S	29	0.20 s	β+; p	−3.2		$\frac{5+}{2}$
	S	30	1.22 s	19% β+ 5.1; 78% β+ 4.4; 80% γ 0.678; 3% γ 2.342	−14.062		0^+
	S	31	2.61 s	99% β+ 4.4; 1% γ 1.266	−19.044		$\frac{1+}{2}$
	S	32	95.02%		−26.015	0.53 (γ) (r)	0^+
	S	33	0.75%		−26.586	0.09 (α) (r)	$\frac{3+}{2}$
	S	34	4.21%		−29.931	0.24 (γ) (r)	0^+
	S	35	87.4 d	100% β− 0.167	−28.846		$\frac{3+}{2}$
	S	36	0.017%		−30.666	0.15 (γ) (r)	0^+
	S	37	4.99 min	6% β− 4.8; 94% β− 1.7; 94% γ 3.103	−26.91		
	S	38	170 min	12% β− 2.94; 84% β− 1.00; 84% γ 1.942; 2% γ 1.746	−26.86		0^+
51	Sb	108	7.0 s	75% β+ 7.3; 25% β+ 6.4; 100% γ 1.206; 25% γ 0.905	−72.4		
	Sb	109	18.3 s	β+; K; γ	−76.1		
	Sb	110	23.0 s	β+; K; 92% γ 1.211; 31% γ 0.985; 13% γ 1.243	−76.8		
	Sb	111	75 s	β+ 3.3; K; 64% γ 0.154; 27% γ 0.489; 8% γ 1.032	−81.5		
	Sb	112	54 s	β+ 4.8; K; 95% γ 1.257; 14% γ 0.991; 3% γ 0.670	−81.6		

(Continued)

Table IX.39 (Continued)

Z	Nuclide A†	half-life and/or abundance	decay scheme⊕	mass excess/ MeV	cross section*/barn	J^P
	Sb	113	6.7 min	β^+ 2.4; K; (16%m); 80% γ 0.498; 10% γ 0.331; 5% γ 0.939	−84.44		
	Sb	114	3.5 min	β^+ 4.1; K; 100% γ 1.300; 18% γ 0.888; 6% γ 0.322	−84.1		
	Sb	115	31.8 min	33% β^+ 1.5; 67% K; 99% γ 0.497; 3% γ 0.491	−87.01		$\frac{5+}{2}$
	Sb	116	16 min	72% K; β^+ 2.3; 85% γ 1.294; 25% γ 0.932; 14% γ 2.225	−86.93		3^+
	Sb	116m	60.4 min	81% K; β^+ 1.3; 100% γ 1.294; 72% γ 0.973; 52% γ 0.543	−86.32		8^-
	Sb	117	2.80 h	98% K; 2% β^+ 0.57; 86% γ 0.159	−88.65		$\frac{5+}{2}$
	Sb	118	3.5 min	75% β^+ 2.66; K; 3% γ 1.230; 1% γ 1.267	−87.967		1^+
	Sb	118m	5.00 h	100% K; 100% γ 0.255; 100% γ 1.229; 95% γ 1.050	−87.75		8^-
	Sb	119	38.0 h	100% K; 16% γ 0.024	−89.48		$\frac{5+}{2}$
	Sb	120	15.8 min	56% K; 44% β^+ 1.7; 2% γ 1.172	−88.42		1^+
	Sb	120	5.76 d	100% K; 100% γ 0.197; 100% γ 1.172; 99% γ 1.023			8^-
	Sb	121	57.3%		−89.588	6.1(γ); 0.06 (γm)	$\frac{5+}{2}$
	Sb	122	2.68 d	26% β^- 1.98; 66% β^- 1.42; 5% β^- 0.72; 3% K; 70% γ 0.564; 4% γ 0.693; 1% γ 1.140, 1.257	−88.323		2^-
	Sb	122m	4.21 min	100% IT; 43% γ 0.061; 18% γ 0.076; 10% γ 0.025	−88.160		
	Sb	123	42.7%		−89.218	4.0 (γ); 0.04 (γm)	$\frac{7+}{2}$
	Sb	124	60.20 d	23% β^- 2.3; 52% β^- 0.6; 98% γ 0.603; 49% γ 1.691; 11% γ 0.723	−87.613	7 (γ) (r)	3^-
	Sb	124m$_1$	93 s	80% IT; 20% β^- 1.2; 20% γ 0.498; 20% γ 0.603; 20% γ 0.646	−87.603		
	Sb	124m$_2$	20.2 min	100% IT; (100% m); γ	−87.58		
	Sb	125	2.71 yr	14% β^- 0.62; 40% β^- 0.31; (23% m); 30% γ 0.428; 18% γ 0.601; 12% γ 0.636	−88.252		$\frac{7+}{2}$
	Sb	126	12.4 d	16% β^- 1.9; 32% β^- 0.5; 100% γ 0.666; 100% γ 0.695; 88% γ 0.415 (double)	−86.40		
	Sb	126m	19.0 min	82% β^- 1.9; 14% IT; 86% γ 0.415; 86% γ 0.661; 86% γ 0.695	−86.38		
	Sb	127	3.9 d	2% β^- 1.5; 35% β^- 0.9; (17% m); 36% γ 0.686; 25% γ 0.473; 15% γ 0.784	−86.70		$\frac{7+}{2}$
	Sb	128	9.10 h	20% β^- 2.0; 100% γ 0.743; 100% γ 0.754; 61% γ 0.314	−84.8		8^-
	Sb	128 m	10.0 min	78% β^- 2.4; 4% IT; 96% γ 0.743; 96 % γ 0.754; 91% γ 0.314	−84.7		5^+
	Sb	129	4.41 h	3% β^- 2.3; 27% β^- 0.6; 23% β^- 0.5; (17% m); 46% γ 0.812; 21% γ 0.915; 19% γ 0.545	−84.63		$\frac{7+}{2}$

(Continued)

Table IX.39 (Continued)

Z	Nuclide	A†	half-life and/or abundance	decay scheme	mass excess/ MeV	cross section*/barn	J^P
	Sb	130	40 min	18% β⁻ 2.9; γ 100% γ 0.793; 100% γ 0.839; 78% γ 0.331	82.4		
	Sb	130	6.5 min	3% β⁻ 3.4; 36% β⁻ 2.2; 100% γ 0.893; 86% γ 0.793; 41% γ 0.182			
	Sb	131	23.03 min	7% β⁻ 2.9; 29% β⁻ 1.2; (7% m); 46% γ 0.943; 26% γ 0.933; 23% γ 0.642	−82.1		
	Sb	132	2.8 min	20% β⁻ 3.9; 99% γ 0.974; 86% γ 0.697; 15% γ 0.990	−79.7		
	Sb	132	4.2 min	44% β⁻ 3.6; 100% γ 0.697; 100% γ 0.974; 66% γ 0.151			
	Sb	133	2.7 min	β⁻; γ	−79.0		
	Sb	134	10.4 s	43% β⁻ 7.1; 57% β⁻ 6.4; 100% γ 1.279; 97% γ 0.297; 57% γ 0.706	−73.9		
	Sb	134	0.8 s	β⁻ 8.8	−73.9		
	Sb	135	1.71 s	β⁻; 20% n; γ	−70.4		
	Sb	136	0.82 s	β⁻; 32% n			
21	Sc	40	0.183 s	19% β⁺ 9.6; 50% β⁺ 5.6; 100% γ 3.736; 41% γ 0.756; 25% γ 2.046	−20.527		4
	Sc	41	0.596 s	100% β⁺ 5.48	−28.644		$\frac{7}{2}^-$
	Sc	42	0.681 s	100% β⁺ 5.40	−32.121		0^+
	Sc	42m	62.0 s	100% β⁺ 2.8; 100% γ 0.438; 100% γ 1.226; 100% γ 1.525	−31.503		7^+
	Sc	43	3.89 h	β⁺ 1.2; K; 22% γ 0.373	−36.185		$\frac{7}{2}^-$
	Sc	44	3.93 h	95% β⁺ 1.4; 5%K; 100%γ 1.157; 1% γ 1.499	−37.811		2^+
	Sc	44m	2.442 d	99% IT; 1%K; 87% γ 0.271; 1% γ 1.002; 1% γ 1.126	−37.540		6^+
	Sc	45	100%		−41.067	17 (γ); 9 (γm) (r)	$\frac{7}{2}^-$
	Sc	45m	0.32 s	100% IT	−41.054		$\frac{3}{2}^+$
	Sc	46	83.80 d	100% β⁻ 0.36; 100% γ 0.889; 100% γ 1.121	−41.756	8 (γ) (s)	4^+
	Sc	46m	18.7 s	100% IT; 62% 0.143	−41.613		1^-
	Sc	47	3.422 d	31% β⁻ 0.60; 69% β⁻ 0.44; 69% γ 0.159	−44.331		$\frac{7}{2}^-$
	Sc	48	43.7 h	91% β⁻ 0.65; 100% γ 0.984; 100% γ 1.312; 98% γ 1.037	−44.50		6^+
	Sc	49	57.0 min	100% β⁻ 2.00	−46.555		
	Sc	50	1.71 min	14% β⁻ 4.2; 86% β⁻ 3.6; 100% γ 1.121; 100% γ 1.554; 86% γ 0.524	−44.54		
	Sc	50m	0.35 s	100% IT; 97% γ 0.257	−44.28		
	Sc	51	12.4 s	28% β⁻ 5.1; 34% β⁻ 4.4; 52% γ 1.437; 32% γ 2.144; 15% γ 1.568	−43.22		
34	Se	69	27.4 s	β⁺ 5.8; 63% γ 0.098; 27% γ 0.066; 14% γ 0.691	−56.30		
	Se	70	41.1 min	β⁺; K; 35% γ 0.050; 29% γ 0.426; 9% γ 0.377	−61.7		0^+
	Se	71	4.9 min	β⁺ 3.4; K; 47% γ 0.147; 13% γ 0.831; 10% γ 1.096	−63.5		
	Se	72	8.4 d	100% K; 57% γ 0.046	−67.89		0^+
	Se	73	7.2 h	β⁺ 1.7; 35% K; 97% γ 0.361; 71% γ 0.067	−68.21		$\frac{7}{2}^+$

(Continued)

Table IX.39 (Continued)

Z	Nuclide	A†	half-life and/or abundance	decay scheme⊕	mass excess/ MeV	cross section*/barn	J^P
	Se	73m	41 min	73% IT; β⁺ 1.7; K; 3% γ 0.254; 2% γ 0.085; 2% γ 0.393	−68.18		$\frac{1-}{2}$
	Se	74	0.87%		−72.213	52 (γ)	0^+
	Se	75	118.5 d	100% K; 58% γ 0.265; 54% γ 0.136; 24% γ 0.280	−72.169		$\frac{5+}{2}$
	Se	76	9.0%		−75.259	64 (γ); 21 (γm) (r)	0^+
	Se	77	7.6%		−74.606	42 (γ) (r)	$\frac{1-}{2}$
	Se	77m	17.4 s	100% IT; 53% γ 0.162	−74.444		$\frac{7+}{2}$
	Se	78	23.5%		−77.032	0.41 (γ); 0.3 (γm) (r)	0^+
	Se	79	≤6.5×10⁴ yr	100% β⁻ 0.16	−75.91		$\frac{7+}{2}$
	Se	79m	3.91 min	100% IT; 10% γ 0.096	−75.82		$\frac{1-}{2}$
	Se	80	49.8%		−77.761	0.6 (γ); 0.07 (γm)	0^+
	Se	81	18.2 min	99% β⁻ 1.6	−76.391		
	Se	81m	57.3 min	100% IT; 10% γ 0.103	−76.288		
	Se	82	9.2% 10²⁰ yr	β⁻ β⁻	−77.59	0.006 (γ); 0.04 (γm)	0^+
	Se	83	22.4 min	β⁻ 2.9; 69% γ 0.357; 44% 0.510; 32% γ 0.225	−75.33		
	Se	83m	70.4 s	31% β⁻ 3.9; 33% β⁻ 2.9; 21% γ 1.031; 17% γ 0.357; 15% γ 0.674, 0.988	−75.11		
	Se	84	3.5 min	100% β⁻ 1.4; 100% γ 0.409	−75.94		0^+
	Se	85	31 s	60% β⁻ 6.1; 22% γ 0.345; 4% γ 0.609; 3% γ 3.396	−72.6		
	Se	86	16.7 s	β⁻; γ	−70.9		0^+
	Se	87	5.9 s	β⁻; γ	−66		
	Se	88	1.5 s	β⁻; 1% n; γ	−64.1		0^+
	Se	91	0.27 s	β⁻; 21% n			
14	Si	25	0.22 s	β⁺; p	3.82		
	Si	26	2.21 s	75% β⁺ 3.83; 22% β⁺ 3.00; (100% m)	−7.143		0^+
	Si	27	4.11 s	100% β⁺ 3.79	−12.385		$\frac{5+}{2}$
	Si	28	92.23%		−21.491	0.17 (γ) (r)	0^+
	Si	29	4.67%		−21.894	0.10 (γ) (r)	$\frac{1+}{2}$
	Si	30	3.10%		−24.432	0.11 (γ)	0^+
	Si	31	2.62 h	100% β⁻ 1.49	−22.949	0.5 (γ) (r)	$\frac{3+}{2}$
	Si	32	650 yr	100% β⁻ 0.21	−24.09		0^+
	Si	33	6.2 s	β⁻; γ	−20.6		
	Si	34	2.8 s	β⁻; γ	−19.9		0^+
62	Sm	133	32.0 s	β⁺; K; p			
	Sm	134	12 s	β⁺; K			0^+
	Sm	135	10 s	β⁺; K; p			
	Sm	137	44 s	β⁺; K			
	Sm	138	3.0 min	β⁺; K; γ			0^+
	Sm	139	2.5 min	β⁺; K; γ	−72.4		
	Sm	139m	9.5 s	94% IT; β⁺; 37% γ 0.190; 34% γ 0.267; 33% γ 0.155	−71.9		
	Sm	140	14.8 min	β⁺; K; γ	−75.5		0^+
	Sm	141	10.2 min	53% K; β⁺ 3.2; 42% γ 0.404; 37% γ 0.438; 7% γ 1.293	−75.9		$\frac{1+}{2}$
	Sm	141m	22.5 min	β⁺; K; 74% γ 0.197; 40% γ 0.432; 20% γ 0.777	−75.7		$\frac{11-}{2}$
	Sm	142	72.5 min	90%K; β⁺1.0; γ	−78.98		0^+
	Sm	143	8.83 min	54% K; β⁺ 2.4; 2% γ 1.057; 1% γ 1.515	−79.51		$\frac{3+}{2}$

(Continued)

Table IX.39 (Continued)

Z	Nuclide A†	half-life and/or abundance	decay scheme	mass excess/ MeV	cross section*/barn	J^P
	Sm	143m	63 s	100% IT; 90% γ 0.754	78.76		$\frac{11-}{2}$
	Sm	144	3.1%		-81.96	0.7 (γ) (r)	0^+
	Sm	145	340 d	100% K; 12% γ 0.061	-80.66	110 (γ) (r)	$\frac{7-}{2}$
	Sm	146	1.03×10^8 yr	α 2.55	-80.98		0^+
	Sm	147	1.06×10^{11} yr 15.1%	α 2.23	-79.27	60 (γ)	$\frac{7-}{2}$
	Sm	148	8×10^{15} yr 11.3%	α 1.96	-79.34	4.7 (γ)	0^+
	Sm	149	13.9%		-77.14	4.2×10^4 (γ)	$\frac{7-}{2}$
	Sm	150	7.4%		-77.05	104 (γ)	0^+
	Sm	151	90 yr	99% β⁻ 0.076	-74.57	1.5×10^4 (γ)	$\frac{5-}{2}$
	Sm	152	26.6%		-74.76	204 (γ)	0^+
	Sm	153	46.8 h	21% β⁻ 0.81; 44% β⁻ 0.71; 28% γ 0.103; 4% γ 0.070; 1% γ 0.097	-72.56		$\frac{3+}{2}$
	Sm	154	22.6%		-72.45	5 (γ)	0^+
	Sm	155	22.4 min	93% β⁻ 1.53; 70% γ 0.104; 4% γ 0.246; 2% γ 0.141	-70.20		$\frac{3-}{2}$
	Sm	156	9.4 hr	52% β⁻ 0.69; 45% β⁻ 0.42; 24% γ 0.088; 21% γ 0.204; 11% γ 0.166	-69.37		0^+
	Sm	157	8.0 min	β⁻; γ	-66.9		
50	Sn	106	1.9 min	K; β⁺; γ	-77.0		0^+
	Sn	107	2.90 min	β⁺; K; γ	-78.4		
	Sn	108	10.5 min	100% K; 74% γ 0.397; 51% γ 0.273; 24% γ 0.169	-81.9		0^+
	Sn	109	18.0 min	β⁺; K; (25% m₁); 35% γ 1.098; 23% γ (m) 0.650; 14% γ 1.322	-82.6		$\frac{7+}{2}$
	Sn	110	4.1 h	100% K; 98% γ 0.283	-85.83		0^+
	Sn	111	35 min	71% K; 29% β⁺ 1.5; 1% γ 0.761, 1.152, 1.915	-85.94		$\frac{7+}{2}$
	Sn	112	1.01%		-88.658	0.4 (γ) (r); 0.3 (γm) (r)	0^+
	Sn	113	115.1 d	100% K; (100% m); 64% γ (m) 0.392; 2% γ 0.255	-88.332		$\frac{1+}{2}$
	Sn	113m	21.4 m	91% IT; 9% K; 1% γ 0.079	-88.253		$\frac{7+}{2}$
	Sn	114	0.67%		-90.560		0^+
	Sn	115	0.38%		-90.035	50 (γ) (r)	$\frac{1+}{2}$
	Sn	116	14.8%		-91.526	0.006 (γm) (r)	0^+
	Sn	117	7.75%		-90.399	3 (γ) (r)	$\frac{1+}{2}$
	Sn	117m	14.0 d	100% IT; 86% γ 0.159	-90.084		$\frac{11-}{2}$
	Sn	118	24.3%		-91.654	0.08 (γm) (r)	0^+
	Sn	119	8.6%		-90.067	2 (γ)	$\frac{1+}{2}$
	Sn	119m	250 d	100% IT; 16% γ 0.024	-89.977		$\frac{11-}{2}$
	Sn	120	32.4%		-91.102	0.16 (γ); 0.001 (γm)	0^+
	Sn	121	27.06 h	100% β⁻ 0.39	-89.202		$\frac{3+}{2}$
	Sn	122	4.56%		-89.946	0.001 (γ); 0.15 (γm)	0^+
	Sn	123	129 d	99% β⁻ 1.40; 1% γ 1.089	-87.821		$\frac{11-}{2}$
	Sn	123m	40.1 min	100% β⁻ 1.3; 86% γ 0.160	-87.796		
	Sn	124	5.64%		-88.240	0.005 (γ); 0.13 (γm)	0^+
	Sn	125	9.63 d	82% β⁻ 2.4; 9% γ 1.066; 4% γ 1.089; 3% γ 0.823	-85.903		$\frac{11-}{2}$

(Continued)

Table IX.39 (Continued)

Z	Nuclide □ A†	half-life and/or abundance	decay scheme⊕	mass excess/ MeV	cross section*/barn	J^P
	Sn 125m	9.52 min	98% β⁻ 2.0; 97% γ 0.332	−85.876		$\frac{3+}{2}$
	Sn 126	10⁵ yr	100% β⁻ 0.25; (100% m); 37% γ 0.088; 10% γ 0.064; 9% γ 0.087	−86.02		0^+
	Sn 127	2.16 h	22% β⁻ 2.9; 12% β⁻ 0.3; 38% γ 1.114; 19% γ 1.096; 11% γ 0.823	−83.79		
	Sn 127m	4.13 min	β⁻ 2.4; 99% γ 0.491; 5% γ 1.348; 4% γ 1.564	−83.78		
	Sn 128	59.3 min	84% β⁻ 0.6; (100% (m)); 58% γ 0.482; 27% γ 0.075; 17% γ 0.557	−83.4		0^+
	Sn 129	2.23 min	β⁻; γ	−80.6		
	Sn 129m	7.5 min	β⁻; γ	−80.6		
	Sn 130	3.7 min	15% β⁻ 1.3; 85% β⁻ 1.0; 71% γ 0.192; 59% γ 0.780; 36% γ 0.070	−80.4		0^+
	Sn 130m	1.7 min	β⁻; γ			
	Sn 131	63 s	β⁻; γ	−77.5		
	Sn 132	41 s	100% β⁻ 1.8; 49% γ 0.086; 43% γ 0.340; 42% γ 0.247, 0.899	−76.6		0^+
	Sn 133	1.47 s	β⁻ 7.5; n; γ	−71.5		
	Sn 134	1.04 s	β⁻; 17% n			0^+
38	Sr 77	9 s	β⁺; γ	−58.0		
	Sr 78	31 min	β⁺; K	−65.5		0^+
	Sr 79	8.1 min	β⁺; K; γ	−65.5		
	Sr 80	106 min	β⁺; K; 40% γ 0.589; 10% γ 0.175; 7% γ 0.553	−70.4		0^+
	Sr 81	26 min	β⁺; 13% K; 36% γ 0.153; 30% γ 0.148; 21% γ 0.188	−71.40		
	Sr 82	25.0 d	100% K	−76.00		0^+
	Sr 83	32.4 h	76% K; β⁺ 1.23; 30% γ 0.763; 20% γ 0.382 (double); 5% γ 0.418	−76.66		$\frac{7+}{2}$
	Sr 83m	5.0 s	100% IT; 88% γ 0.259	−76.41		$\frac{1-}{2}$
	Sr 84	0.56%		−80.641	0.3 (γ); 0.6 (γm) (s)	0^+
	Sr 85	64.85 d	100% K; 99% γ 0.514	−81.10		$\frac{9+}{2}$
	Sr 85m	68.0 min	87% IT; 13% K; 85% γ 0.232; 11% γ 0.151	−80.86		$\frac{1-}{2}$
	Sr 86	9.8%		−84.512	0.84 (γm) (s)	0^+
	Sr 87	7.0%		−84.869		$\frac{9+}{2}$
	Sr 87m	2.805 h	100% IT; 82% γ 0.388	−84.481		$\frac{1-}{2}$
	Sr 88	82.6%		−87.911	0.0057 (γ) (r)	0^+
	Sr 89	50.6 d	100% β⁻ 1.49	−86.203	0.42 (γ) (r)	$\frac{5+}{2}$
	Sr 90	28.8 yr	100% β⁻ 0.55	−85.935	0.8 (γ) (r)	0^+
	Sr 91	9.48 h	31% β⁻ 2.7; 24% β⁻ 1.4; 35% β⁻ 1.1; (57% m); 54% γ (m) 0.556; 33% γ 1.024; 23% γ 0.750; 8% γ 0.653	−83.666		$\frac{5+}{2}$
	Sr 92	2.71 h	3% β⁻ 1.93; 96% β⁻ 0.55; 90% γ 1.384; 4% γ 0.953; 3% γ 0.242, 0.431, 1.142	−82.89		0^+
	Sr 93	7.43 min	11% β⁻ 3.4; 23% β⁻ 2.4; (39% m); 73% γ 0.590; 25% γ 0.876; 24% γ 0.888	−80.3		

(Continued)

Table IX.39 (Continued)

Z	Nuclide	A†	half-life and/or abundance	decay scheme⊕	mass excess/ MeV	cross section*/barn	J^P
	Sr	94	74.1 s	99% β⁻ 2.0; 93% γ 1.428, 3% γ 0.724; 2% γ 0.622, 0.704	−79.0		0^+
	Sr	95	24.4 s	53% β⁻ 6.1; 24% γ 0.686; 5% γ 2.717; 4% γ 2.933	−75.1		
	Sr	96	1.06 s	β⁻; 75% γ 0.809; 66% γ 0.122; 15% γ 0.932	−73.1		0^+
	Sr	97	0.40 s	20% β⁻ 7.2; 42% β⁻ 5.0; 24% γ 1.905; 23% γ 0.954; 12% γ 0.307, 0.652	−69.1		
	Sr	98	0.65 s	65% β⁻ 5.8; 29% β⁻ 5.2; 29% γ 0.119; 10% γ 0.445; 9% γ 0.429	−67.4		0^+
	Sr	99	0.6 s	β⁻; 3% n			
73	Ta	166	32 s	β⁺; K; γ	−46.1		
	Ta	167	3 min	β⁺; K	−48.0		
	Ta	168	2.4 min	β⁺; K; 37% γ 0.124; 28% γ 0.262; 10% γ 0.750	−48.4		
	Ta	169	5 min	β⁺; K; γ	−50.0		
	Ta	170	6.8 min	β⁺; K; 21% γ 0.101; 16% γ 0.221; 9% γ 0.987 (double)	−50.1		
	Ta	171	23.3 min	β⁺; K; γ	−51.6		
	Ta	172	37 min	85% K; β⁺; 54% γ 0.214; 17% γ 0.095; 15% γ 1.109	−51.4		
	Ta	173	3.7 h	K; β⁺; 17% γ 0.172; 6% γ 0.070; 5% γ 0.090	−52.4		
	Ta	174	1.1 h	K; β⁺; 57% γ 0.207; 16% γ 0.091; 5% γ 1.206	−52.0		3
	Ta	175	10.5 h	K; β⁺; 14% γ 0.207; 12% γ 0.349; 11% γ 0.267 (double)	−52.4		$\frac{7^+}{2}$
	Ta	176	8.1 h	99% K; β⁺ 2.1; 23% γ 1.159; 11% γ 0.088; 5% γ 0.202; 0.711	−51.5		
	Ta	177	56.6 h	100% K; 7% γ 0.113; 1% γ 0.208	−51.72		$\frac{7^+}{2}$
	Ta	178	9.3 min	99% K; 1% β⁺ 0.89; 7% γ 0.093; 1% γ 1.341; 1% γ 1.351	−50.5		1^+
	Ta	178	2.4 h	100% K; (100% m₁); 97% γ (m) 0.426; 94% γ (m) 0.326; 81% γ (m) 0.213			
	Ta	179	665 d	100% K	−50.35		
	Ta	180(g)	> 10¹³ yr 0.0123%			700 (γ)	
	Ta	180(m)	8.1 h	87% K; 10% β⁻ 0.71; 3% β⁻ 0.61; 5% γ 0.093	−48.91		1
	Ta	181	99.9877%		−48.43	21 (γ); 0.010 (γm)	$\frac{7^+}{2}$
	Ta	182	115.0 d	5% β⁻ 0.59; 40% β⁻ 0.52; 29% β⁻ 0.26; 41% γ 0.068; 35% γ 1.121; 27% γ 1.221	−46.42	8200 (γ)	3^-
	Ta	182m₁	0.283 s	100% IT	−46.40		5^+
	Ta	182m₂	15.8 min	100% IT; 47% γ 0.172; 36% γ 0.147; 24% γ 0.185	−45.90		10^-
	Ta	183	5.1 d	1% β⁻ 0.66; 91% β⁻ 0.62; (5% m); 27% γ 0.246; 12% γ 0.108; 12% γ 0.354	−45.28		$\frac{7^+}{2}$
	Ta	184	8.7 h	β⁻; 74% γ 0.414; 45% γ 0.253; 33% γ 0.921	−42.82		

(Continued)

Table IX.39 (Continued)

Z	Nuclide	A	half-life and/or abundance	decay scheme	mass excess/ MeV	cross section*/barn	J^P
	Ta	185	49 min	β⁻; 26% γ 0.178; 22% γ 0.174; 4% γ 0.066, 0.244	−41.36		
	Ta	186	10.5 min	2% β⁻ 2.9; 63% β⁻ 2.2; 58% γ 0.198; 49% γ 0.215; 43% γ 0.511	−38.6		
65	Tb	146	23 s	β⁺; K; 90% γ 1.580; 46% γ 1.079; 15% γ 1.417	−67.3		
	Tb	147	1.6 h	95% K; β⁺; 75% γ 1.152; 32% γ 0.694; 20% γ 0.140	−70.5		$\frac{5+}{2}$
	Tb	147	1.8 min	β⁺; K; 85% γ 1.398; 14% γ 1.798; 1% γ 0.998	−70.5		$\frac{11-}{2}$
	Tb	148	2.2 min	β⁺; K; γ			
	Tb	148	60 min	80% K; β⁺ 4.6; 86% γ 0.784; 25% γ 0.489; 12% γ 1.077	−70.6		2⁻
	Tb	149	4.15 h	17% α 3.97; 79% K; β⁺; 33% γ 0.352; 28% γ 0.165; 21% γ 0.388	−71.43		
	Tb	149m	4.16 min	β⁺; K; 90% γ 0.796; 7% γ 0.165	−71.39		
	Tb	150	3.3 h	90% K; β⁺ 3.7; 72% γ 0.638; 15% γ 0.496; 5% γ 0.792	−71.10		
	Tb	150	6.0 min	β⁺; K; 100% γ 0.638; 70% γ 0.650; 42% γ 0.438			
	Tb	151	17.6 h	99% K; β⁺; 26% γ 0.252; 25% γ 0.108; 25% γ 0.287	−71.61		$\frac{1}{2}$
	Tb	152	17.5 h	87% K; 6% β⁺ 2.8; 66% γ 0.344; 10% γ 0.586; 9% γ 0.271 (double)	−70.85		2⁻
	Tb	152m	4.2 min	78% IT; 22% K; 62% γ 0.283; 20% γ 0.344; 20% γ 0.411	−70.35		
	Tb	153	2.30 d	100% K; 40% γ 0.212; 9% γ 0.170; 8% γ 0.110	−71.33		$\frac{5+}{2}$
	Tb	154	21 h	98% K; β⁺ 2.4; γ	−70.24		0
	Tb	154m₁	9 h	β⁺; K; 22% IT; γ			3
	Tb	154m₂	23 h	98% K; 2% IT; 87% γ 0.248; 76% γ 0.347; 49% γ 1.420			
	Tb	155	5.3 d	100% K; 29% γ 0.087; 23% γ 0.105; 7% γ 0.180	−71.26		$\frac{3+}{2}$
	Tb	156	5.4 d	100% K; 66% γ 0.534; 42% γ 0.199; 32% γ 1.222	−70.10		3⁻
	Tb	156	24 h	100% IT; γ			
	Tb	156m	5.0 h	IT; K; γ	−70.01		
	Tb	157	150 yr	100% K	−70.77		$\frac{3+}{2}$
	Tb	158	150 yr	82% K; 17% β⁻ 0.84; 43% γ 0.944; 20% γ 0.962; 11% γ 0.080	−69.48		3⁻
	Tb	158m	10.5 s	100% IT; 1% γ 0.110	−69.37		0⁻
	Tb	159	100%		−69.54	23 (γ)	$\frac{3+}{2}$
	Tb	160	72.1 d	27% β⁻ 0.87; 47% β⁻ 0.57; 30% γ 0.879; 27% γ 0.299; 25% γ 0.966	−67.84	500 (γ) (r)	3⁻
	Tb	161	6.90 d	10% β⁻ 0.59; 66% β⁻ 0.52; 16% γ 0.049; 10% γ 0.075; 2% γ 0.057	−67.47		$\frac{3+}{2}$
	Tb	162	7.8 min	99% β⁻ 1.3; 80% γ 0.260; 43% γ 0.807; 39% γ 0.888	−65.8		
	Tb	163	19.5 min	9% β⁻ 1.3; 35% β⁻ 0.8; 26% γ 0.351; 22% γ 0.495; 11% γ 0.422	−64.7		$\frac{3+}{2}$

(Continued)

Table IX.39 (Continued)

Z	Nuclide A	half-life and/or abundance	decay scheme	mass excess/ MeV	cross section*/barn	J^P
	Tb 164	3.0 min	β^-; 24% γ 0.169; 22% γ 0.755; 20% γ 0.215, 0.611	−62.1		
43	Tc 90	50 s	β^+; γ			
	Tc 90	7.9 s	β^+ 7.9; γ	−71		
	Tc 91	3.14 min	β^+ 5.2; K; (1% m); 14% γ 2.451; 8% γ 1.605; 7% γ 1.565	−76.0		
	Tc 91	3.3 min	β^+; K; (98% m); 55% γ 0.503; 47% γ (m) 0.653; 4% γ 0.928			
	Tc 92	4.4 min	β^+ 4.1; K; 100% γ 0.773; 100% γ 1.510; 80% γ 0.329	−78.94		
	Tc 93	2.75 hr	87% K; β^+ 0.8; 66% γ 1.363; 24% γ 1.520; 10% γ 1.477	−83.61		$\frac{9+}{2}$
	Tc 93m	44 min	80% IT; 20% K; 60% γ 0.392; 16% γ 2.645; 2% γ 3.129	−83.22		$\frac{1-}{2}$
	Tc 94	293 min	89% K; 11% β^+ 0.82; 100% γ 0.871; 100% γ 0.703; 98% γ 0.850	−84.16		7^+
	Tc 94m	53 min	β^+ 2.4; 28% K; 94% γ 0.871; 6% γ 1.869; 5% γ 1.522	−84.08		
	Tc 95	20.0 h	100% K; 93% γ 0.766; 4% γ 1.074; 2% γ 0.948	−86.01		$\frac{9+}{2}$
	Tc 95m	61 d	96% K; 4% IT; 66% γ 0.204; 32% γ 0.582; 28% γ 0.835	−85.97		$\frac{1-}{2}$
	Tc 96	4.35 d	100% K; 99% γ 0.778; 97% γ 0.850; 81% γ 0.813	−85.82		7^+
	Tc 96m	52 min	98% IT; 2% K; 2% γ 0.778; 1% γ 1.200	−85.79		4^+
	Tc 97	2.6×10^6 yr	100% K	−87.224		$\frac{9+}{2}$
	Tc 97m	90 d	100% IT	−87.128		$\frac{1-}{2}$
	Tc 98	4.2×10^6 yr	100% β^- 0.39; 100% γ 0.652; 100% γ 0.745	−86.43	3 (γm) (r)	
	Tc 99	2.1×10^5 yr	100% β^- 0.29	−87.326	19 (γ)	$\frac{9+}{2}$
	Tc 99m	6.01 h	100% IT; 89% γ 0.141	−87.184		$\frac{1-}{2}$
	Tc 100	15.8 s	93% β^- 3.20; 7% γ 0.540; 6% γ 0.591	−86.019		1^+
	Tc 101	14.3 min	89% β^- 1.3; 88% γ 0.307; 6% γ 0.545; 3% γ 0.127	−86.33		$\frac{9+}{2}$
	Tc 102	5.3 s	41% β^- 4.5; 30% β^- 4.0; 53% γ 0.475; 11% γ 1.105 (double); 10% γ 0.628	−84.6		1^+
	Tc 102m	4.4 min	3% β^- 3.7; 35% β^- 2.1; 2% IT; 85% γ 0.475; 26% γ 0.628; 16% γ 0.630			
	Tc 103	50 s	β^-; γ	−84.9		
	Tc 104	18.1 min	24% β^- 3.9; 89% γ 0.358; 15% γ 0.531; 14% γ 0.535	−83.9		
	Tc 105	7.6 min	41% β^- 3.4 (double); 11% γ 0.143; 10% γ 0.108; 8% γ 0.321	−82.5		
	Tc 106	36 s	β^-; γ	−80.0		
	Tc 107	21.2 s	β^-; γ	−79.5		
	Tc 108	5.1 s	β^-; γ	−76		
	Tc 109	1.4 s	β^-			
	Tc 110	0.82 s	β^-; γ			

(Continued)

Table IX.39 (Continued)

Z	Nuclide A†	half-life and/or abundance	decay scheme⊕	mass excess/ MeV	cross section*/barn	J^P
52	Te 107	2.2 s	α 3.28			
	Te 108	5.3 s	α 3.08	−65.3		0^+
	Te 109	4.2 s	α 3.08	−67.5		
	Te 111	19 s	β^+; K; p	−74.1		
	Te 113	2.0 min	β^+; K; γ	−79.0		
	Te 114	17 min	β^+; K	−81.5		0^+
	Te 115	6.0 min	β^+ 2.7; K; 32% γ 0.723; 25% γ 1.381; 24% γ 1.327	−82.58		
	Te 115	7.5 min	β^+; K; 61% γ 0.770; 24% γ 1.072; 15% γ 1.032			
	Te 116	2.50 h	K; β^+; 29% γ 0.094; 5% γ 0.103; 1% γ 0.630	−85.4		0^+
	Te 117	61 min	70% K; β^+ 1.8; 65% γ 0.720; 16% γ 1.716; 11% γ 2.300	−85.16		$\frac{1+}{2}$
	Te 118	6.00 d	100% K	−87.67		0^+
	Te 119	16.0 h	97% K; 3% β^+ 0.6; 84% γ 0.644; 10% γ 0.700; 4% γ 1.750	−87.19		$\frac{1+}{2}$
	Te 119m	4.68 d	100% K; 66% γ 0.153; 66% γ 1.213; 27% γ 0.270	−86.9		$\frac{11-}{2}$
	Te 120	0.091%		−89.40	2.0 (γ) (r); 0.3 (γm) (r)	0^+
	Te 121	16.8 d	100% K; 80% γ 0.573; 18% γ 0.508; 1% γ 0.470	−88.49		$\frac{1+}{2}$
	Te 121m	154 d	90% IT; 10% K; 83% γ 0.212; 2% γ 1.102	−88.19		$\frac{11-}{2}$
	Te 122	2.5%		−90.304	3 (γ) (r)	0^+
	Te 123	10^{14} yr 0.89%	100% K	−89.166	400 (γ) (r)	$\frac{1+}{2}$
	Te 123m	119.7 d	100% IT; 84% γ 0.159	−88.918		$\frac{11-}{2}$
	Te 124	4.6%		−90.518	6.7 (γ) (r); 0.05 (γm) (r)	0^+
	Te 125	7.0%		−89.019	1.6 (γ) (r)	$\frac{1+}{2}$
	Te 125m	58 d	100% IT	−88.874		$\frac{11-}{2}$
	Te 126	18.7%		−90.066	0.9 (γ) (r); 0.13 (γm) (r)	0^+
	Te 127	9.4 h	99% β^- 0.69; 1% γ 0.418	−88.285		$\frac{3+}{2}$
	Te 127m	109 d	98% IT; 2% β^- 0.7; 1% γ 0.058	−88.197		$\frac{11-}{2}$
	Te 128	31.7% 10^{24} yr	$\beta^- \beta^-$	−88.992	0.20 (γ); 0.016 (γm)	0^+
	Te 129	70 min	89% β^- 1.47; 16% γ 0.028; 7% γ 0.460; 1% γ 0.487	−87.007		$\frac{3+}{2}$
	Te 129m	33.5 d	63% IT; 33% β^- 1.6; γ	−86.901		$\frac{11-}{2}$
	Te 130	34.5% 2×10^{21} yr	$\beta^- \beta^-$	−87.348	0.2 (γ) (r); 0.03 (γm) (r)	0^+
	Te 131	25.0 min	60% β^- 2.10; 22% β^- 1.65; 68% γ 0.150; 18% γ 0.452; 5% γ 1.147	−85.201		$\frac{3+}{2}$
	Te 131m	30 h	β^- 2.4; 22% IT; 38% γ 0.774; 21% γ 0.852; 20% γ 0.150	85.019		$\frac{11-}{2}$
	Te 132	78 h	100% β^- 0.22; 88% γ 0.228; 14% γ 0.050; 2% γ 0.112, 0.116	−85.21		0^+
	Te 133	12.5 min	26% β^- 2.7; 34% β^- 2.2; 73% γ 0.312; 33% γ 0.408; 12% γ 1.333	−82.9		
	Te 133m	55.4 min	30% β^- 2.4; (10% m); 17% IT; 63% γ 0.913; 22% γ 0.648; 12% γ 0.915	−82.6		

(Continued)

Table IX.39 (Continued)

Z	Nuclide	A†	half-life and/or abundance	decay scheme⊕	mass excess/ MeV	cross section*/barn	J^P
	Te	134	42 min	42% β⁻ 0.5; 43% β⁻ 0.4; 30% γ 0.767; 22% γ 0.210; 20% γ 0.278	−82.7		0^+
	Te	135	19.2 s	β⁻; γ	−77.6		
	Te	136	17.5 s	β⁻; 1% n; γ	−74.8		0^+
	Te	137	3 s	β⁻; 3% n; γ			
	Te	138	1.4 s	β⁻; 6% n			
90	Th	215	1.2 s	40% α 7.52; 52% α 7.40; 8% α 7.33	10.9		
	Th	216	0.028 s	α 7.92	10.4		0^+
	Th	217	2.5×10^{-4} s	α 9.25	12.14		
	Th	218	10^{-7} s	100% α 9.66	12.36		0^+
	Th	219	1.0×10^{-6} s	α 9.34	14.47		
	Th	220	10^{-5} s	α 8.79	14.66		0^+
	Th	221	0.0017 s	38% α 8.47; 56% α 8.15; 6% α 7.73	16.93		
	Th	222	0.0028 s	α 7.98	17.20		0^+
	Th	223	0.66 s	40% α 7.32; 60% α 7.29	19.26		
	Th	224	1.0 s	79% α 7.17; 19% α 7.00; 1% α 6.76; 1% α 6.70; 9% γ 0.177; 1% γ 0.410	19.99		0^+
	Th	225	8.0 min	8% α 6.80; 13% α 6.50; 39% α 6.48; 30% α (dist); 10% K; 27% γ 0.323	22.30		
	Th	226	30.9 min	75% α 6.34; 23% α 6.23; 2% α 6.10; 3% γ 0.111; 1% γ 0.242	23.19		0^+
	Th	227	18.72 d	24% α 6.04; 23% α 5.98; 20% α 5.76; 33% α (dist); 11% γ 0.236; 7% γ 0.050; 6% γ 0.256	25.806	200 (γ) (r)	$\frac{3+}{2}$
	Th	228	1.913 yr	73% α 5.42; 27% α 5.34; 1% γ 0.084	26.76	100 (γ) (r)	0^+
	Th	229	7340 yr	11% α 4.90; 56% α 4.85; 8% α 4.81; 25% α (dist); 5% γ 0.194; 4% γ 0.031; 4% γ 0.125 (double)	29.581	30 (f)	$\frac{5+}{2}$
	Th	230	8.0×10^4 yr	76% α 4.69; 24% α 4.62	30.861	40 (γ)	0^+
	Th	231	25.52 h	37% β⁻ 0.31; 36% β⁻ 0.29; 15% γ 0.026; 75% γ 0.084; 1% γ 0.081, 0.090	33.812		$\frac{5+}{2}$
	Th	232	1.41×10^{10} yr 100%	77% α 4.02; 23% α 3.96	35.45	7.4 (γ)	0^+
	Th	233	22.3 min	85% β⁻ 1.25; 3% γ 0.029; 3% γ 0.086; 1% γ 0.459	38.73	1400 (γ) (s); 15 (f)	$\frac{1+}{2}$
	Th	234	24.10 d	72% β⁻ 0.18; (100% m); 5% γ 0.093 (double); 4% γ 0.063 (double)	40.61	2.0 (γ)	0^+
	Th	235	6.9 min	β⁻; γ	44.2		
	Th	236	38 min	46% β⁻ 1.1; 51% β⁻ 1.0; 5% γ 0.111; 1% γ 0.113, 0.132, 0.230	46.6		0^+
22	Ti	41	0.080 s	9% β⁺ 9.7; 15% β⁺ 7.6; 24% β⁺ 6.0; 100% p	−15.78		$\frac{3+}{2}$
	Ti	42	0.202 s	43% β⁺ 6.0; 56% β⁺ 5.4; 56% γ 0.611	−25.12		0^+

(Continued)

Table IX.39 (Continued)

Z	Nuclide	A	half-life and/or abundance	decay scheme	mass excess/ MeV	cross section*/barn	J^P
	Ti	43	0.49 s	100% β^+ 5.84	−29.32		$\frac{7-}{2}$
	Ti	44	48 yr	100% K; 94% γ 0.78; 88% γ 0.068	−37.546		0^+
	Ti	45	3.08 h	86% β^+ 1.04; 14% K	−39.004		$\frac{7-}{2}$
	Ti	46	8.2%		−44.123	0.6 (γ) (r)	0^+
	Ti	47	7.4%		−44.931	1.7 (γ) (r)	$\frac{5-}{2}$
	Ti	48	73.7%		−48.488	7.9 (γ) (r)	0^+
	Ti	49	5.4%		−48.559	2.1 (γ) (r)	$\frac{7-}{2}$
	Ti	50	5.2%		−51.432	0.18 (γ)	0^+
	Ti	51	5.80 min	92% β^- 2.15; 93% γ 0.320; 7% γ 0.929	−49.733		$\frac{3-}{2}$
	Ti	52	1.7 min	100% β^- 1.83; 100% γ 0.124	−49.47		0^+
	Ti	53	33 s	β^-; 45% γ 0.128; 39% γ 0.228; 25% γ 1.676	−46.8		
81	Tl	184	11 s	2% α 6.16; β^+; K; γ	−16.9		
	Tl	185m	1.7 s	α 5.98; IT; γ	−18.7		
	Tl	186	28 s	β^+; K; γ	−19.9		
	Tl	186m	3 s	100% IT; 80% γ 0.374	−19.5		
	Tl	187m	16 s	α 5.51; IT; γ	−21.6		
	Tl	188	71 s	β^+; K; γ	−22.3		
	Tl	189	1.4 min	β^+; K; γ	−24.0		
	Tl	189	2.3 min	β^+; K; γ	−24.0		
	Tl	190	2.6 min	β^+; K; γ	−24.2		
	Tl	190	3.7 min	β^+; K; γ			
	Tl	191	5.2 min	98% K; β^+; γ	−25.7		
	Tl	192	10.8 min	β^+; K; γ	−25.6		
	Tl	192	10.6 min	β^+; K; γ	−25.6		
	Tl	193	21 min	>96% K; β^+; γ	−27.0		$\frac{1+}{2}$
	Tl	193m	2.1 min	100% IT; γ			
	Tl	194	33 min	β^+; K; γ	−26.8		2^-
	Tl	194m	32.8 min	β^+; K; 100% γ 0.428; 100% γ 0.636; 77% γ 0.749	−26.5		
	Tl	195	1.16 h	99% K; β^+; 4% γ 0.563; 3% γ 0.884; 3% γ 1.364	−27.9		$\frac{1+}{2}$
	Tl	195m	3.6 s	100% IT; 91% γ 0.384	−27.4		$\frac{9-}{2}$
	Tl	196	1.84 h	β^+; K; γ	−27.4		2
	Tl	196m	1.41 h	β^+; K; 4% IT; γ	−27.0		
	Tl	197	2.84 h	99% K; β^+; 12% γ 0.426; 8% γ 0.152; 4% γ 0.309 (double)	−28.3		$\frac{1+}{2}$
	Tl	197m	0.54 s	100% IT; 90% γ 0.385; 30% γ 0.222	−27.7		$\frac{9-}{2}$
	Tl	198	5.3 h	99% K; β^+ 2.4; 78% γ 0.412; 10% γ 0.637; 10% γ 0.676	−27.5		2^-
	Tl	198m	1.87 h	β^+; K; 44% IT; 59% γ 0.412; 59% γ 0.637; 54% γ 0.587	−27.0		7^+
	Tl	199	7.4 h	100% K; 12% γ 0.208; 12% γ 0.455; 9% γ 0.247	−28.1		$\frac{1+}{2}$
	Tl	200	26.1 h	100% K; 89% γ 0.368; 30% γ 1.206; 14% γ 0.579	−27.06		2^-
	Tl	201	73 h	100% K; 9% γ 0.167 (double); 2% γ 0.135	−27.19		$\frac{1+}{2}$
	Tl	202	12.23 d	100% K; 91% γ 0.440	−25.99		2^-
	Tl	203	29.5%		−25.77	10 (γ)	$\frac{1+}{2}$
	Tl	204	3.77 yr	97% β^- 0.76; 3% K	−24.35	22 (γ) (r)	2^-
	Tl	205	70.5%		−23.84	0.10 (γ) (r)	$\frac{1+}{2}$
	Tl	206	4.20 min	100% β^- 1.53	−22.27		0^-
	Tl	206m	3.6 min	100% IT; γ	−19.63		

(Continued)

Table IX.39 (Continued)

Z	Nuclide A†	half-life and/or abundance	decay scheme⊕	mass excess/ MeV	cross section*/barn	J^P
	Tl	207	4.77 min	100% β⁻ 1.42	−21.04		$\frac{1+}{2}$
	Tl	207m	1.3 s	100% IT; γ	−19.70		$\frac{11-}{2}$
	Tl	208	3.053 min	51% β⁻ 1.8; 22% β⁻ 1.5; 23% β⁻ 1.3; 100% γ 2.615; 86% γ 0.583; 22% γ 0.511	−16.77		
	Tl	209	2.2 min	100% β⁻ 1.8; 100% γ 1.566; 72% γ 0.467; 67% γ 0.117	−13.65		
	Tl	210	1.30 min	β⁻; 99% γ 0.795; 79% γ 0.296; 21% γ 1.310	−9.25		
69	Tm	153	1.6 s	α 5.10	−53.9		
	Tm	154	5 s	α 4.96	−54.5		
	Tm	154	3.0 s	α 5.03	−54.5		
	Tm	155	39 s	α 4.45	−56.5		
	Tm	156	80 s	α 4.23; β⁺; K; γ	−56.9		
	Tm	156	19 s	α 4.46	−56.9		
	Tm	157	3.6 min	β⁺; K; γ	−58.5		
	Tm	158	4.0 min	β⁺; K; 69% γ 0.192; 19% γ 0.335; 8% γ 1.150	−58.4		
	Tm	159	9.0 min	β⁺; K; 8% γ 0.038; 8% γ 0.085; 7% γ 0.271	−60.2		$\frac{5}{2}$
	Tm	160	9.2 min	85% K; β⁺; 35% γ 0.126; 13% γ 0.729; 9% γ 0.264, 1.369	−60.1		1^-
	Tm	161	30 min	β⁺; K; 25% γ 0.046; 20% γ 1.648; 10% γ 0.084	−61.7		$\frac{7}{2}$
	Tm	162	21.8 min	93% K; β⁺ 3.8; 17% γ 0.102; 8% γ 0.799; 7% γ 0.227	−61.5		1^-
	Tm	162m	24 s	90% IT; β⁺; K; 7% γ 0.067; 7% γ 0.812; 7% γ 0.900 (double)			
	Tm	163	1.8 h	100% K; 20% γ 0.104; 20% γ 0.240 (double) 13% γ 1.434	−62.99		$\frac{1+}{2}$
	Tm	164	2.0 min	61% K; β⁺ 2.9; 7% γ 0.091; 2% γ 1.155; 1% γ 0.769	−61.98		1^+
	Tm	164m	5.1 min	80% IT; β⁺; K; 17% γ 0.208; 11% γ 0.315; 9% γ 0.240			6
	Tm	165	30.06 h	100% K; 35% γ 0.243; 25% γ 0.297 (double); 8% γ 0.807	−62.92		$\frac{1+}{2}$
	Tm	166	7.7 h	98% K; β⁺ 1.9; 20% γ 2.052; 19% γ 0.779; 16% γ 0.184	−61.87		2^+
	Tm	167	9.25 d	100% K; (98%m); 41% γ (m) 0.208; 2% γ 0.532	−62.54		$\frac{1+}{2}$
	Tm	168	93.1 d	98% K; β⁻; 50% γ 0.198; 46% γ 0.816; 22% γ 0.447	−61.31		3
	Tm	169	100%		−61.27	98 (γ)	$\frac{1+}{2}$
	Tm	170	128.6 d	76% β⁻ 0.97; 3% γ 0.084	−59.79	92 (γ) (s)	1^-
	Tm	171	1.92 yr	98% β⁻ 0.097	−59.21	4.5 (γ) (s)	$\frac{1+}{2}$
	Tm	172	63.6 h	29% β⁻ 1.87; 36% β⁻ 1.79; 7% γ 0.079; 6% γ 1.094; 6% γ 1.387	−57.38		2^-
	Tm	173	8.2 h	2% β⁻ 1.32; 77% β⁻ 0.92; 88% γ 0.399; 7% γ 0.460; 1% γ 0.063	−56.23		
	Tm	174	5.4 min	83% β⁻ 1.2; 100% γ 0.366; 90% γ 0.992; 87% γ 0.273	−53.9		
	Tm	175	15 min	23% β⁻ 1.9; 36% β⁻ 0.9; 65% γ 0.515; 15% γ 0.941; 13% γ 0.364	−52.3		

(Continued)

Table IX.39 (Continued)

Z	Nuclide	A	half-life and/or abundance	decay scheme	mass excess/ MeV	cross section*/barn	J^P
	Tm	176	1.9 min	β^-; 44% γ 0.190; 33% γ 1.069; 24% γ 0.382	−49.6		
92	U	226	0.5 s	α 7.43	27.19		0^+
	U	227	1.1 min	100% α 6.87	28.9		
	U	228	9.1 min	67% α 6.69; 28% α 6.60; ≤5% K	29.22		0^+
	U	229	58 min	13% α 6.36; 4% α 6.33; 2% α 6.30; 1% α 6.22; 80% K	31.20		
	U	230	20.8 d	68% α 5.89; 32% α 5.82; 1% γ 0.072	31.61	20 (f) (r)	0^+
	U	231	4.2 d	100% K; 12% γ 0.026; 7% γ 0.084; 1% γ 0.220	33.78	300 (f) (r)	
	U	232	72 yr	68% α 5.32; 32% α 5.26	34.60	74 (γ); 76 (f)	0^+
	U	233	1.59×10^5 yr	85% α 4.82; 13% α 4.78; 2% α 4.73	36.915	46 (γ); 530 (f)	$\frac{5+}{2}$
	U	234	2.45×10^5 yr 0.0054%	72% α 4.77; 28% α 4.72	38.143	100 (γ)	0^+
	U	235	7.04×10^8 yr 0.720%	5% α 4.60; 57% α 4.40; 18% α 4.37; 20% α (dist); 54% γ 0.186; 11% γ 0.144; 5% γ 0.163	40.916	98 (γ); 580 (f)	$\frac{7-}{2}$
	U	235m	26 min (variable)	100% IT	40.917		$\frac{1+}{2}$
	U	236	2.342×10^7 yr	74% α 4.49; 26% α 4.45	42.442	5 (γ)	0^+
	U	236 f	1.2×10^{-7} s	88% IT; 12% SF	44.8		
	U	237	6.75 d	43% β^- 0.25; 53% β^- 0.24; 36% γ 0.060; 23% γ 0.208; 2% γ 0.026, 0.165	45.389	400 (γ)	$\frac{1+}{2}$
	U	238	4.47×10^9 yr 99.275%	77% α 4.20; 23% α 4.15	47.307	2.7 (γ); 2.7 (f)	0^+
	U	238f	2.0×10^{-7} s	IT; SF; γ	49.866		
	U	239	23.5 min	18% β^- 1.27; 73% β^- 1.20; 50% γ 0.075; 5% γ 0.044	50.572	22 (γ) (r); 15 (f) (r)	$\frac{5+}{2}$
	U	240	14.1 h	100% β^- 0.36; (100% m); 2% γ 0.044	52.71		0^+
106	Unh	263	0.9 s	α 9.25			
105	Unp	260	1.5 s	17% α 9.12; 25% α 9.07; 48% α 9.04; 10% SF	104		
		261	2 s	75% α (8.93); 25% SF	104		
		262	40 s	α 8.66	106		
104	Unq	257	5 s	35% α 9.00; 30% α 8.95; 20% α 8.78; 15% α 8.70	96.0		
		259	3 s	α 8.86	98.5		
		261	65 s	α 8.29	101.3		
23	V	46	0.4223 s	100% β^+ 6.0	−37.071		0^+
	V	47	32.6 min	β^+ 1.9; K	−42.001		$\frac{3+}{2}$
	V	48	15.976 d	50% β^+ 0.70; 50% K, 100% γ 0.984; 98% γ 1.312; 8% γ 0.944	−44.473		4^+
	V	49	330 d	100% K	−47.957		$\frac{7-}{2}$
	V	50	0.250%		−49.219	50 (γ)	6^+
	V	51	99.75%		−52.199	4.9 (γ)	$\frac{7-}{2}$
	V	52	3.75 min	99% β^- 2.55; 100% γ 1.434	−51.439		3^+

(Continued)

Table IX.39 (Continued)

Z	Nuclide A†	half-life and/or abundance	decay scheme	mass excess/ MeV	cross section*/barn	J^P
	V 53	1.6 min	89% β⁻ 2.4; 90% γ 1.006; 10% γ 1.289	−51.86		$\frac{7-}{2}$
	V 54	43 s	11% β⁻ 5.2; 45% β⁻ 2.9; 100% γ 0.835; 82% γ 0.986; 50% γ 2.255	−49.9		
74	W 162	<0.25 s	α 5.53	−34.1		0^+
	W 163	2.5 s	α 5.39	−35.3		
	W 164	6 s	α 5.15	−38.0		0^+
	W 165	5 s	α 4.91	−38.7		
	W 166	16 s	α 4.74	−41.5		0^+
	W 172	7 min	K; β⁺; γ	−48.8		0^+
	W 173	16 min	β⁺; K; γ	−48.5		
	W 174	29 min	100% K; γ	−50.1		0^+
	W 175	34 min	β⁺; K; γ	−49.5		
	W 176	2.3 h	100% K; γ	−50.6		0^+
	W 177	135 min	β⁺; K; 58% γ 0.116 (double); 16% γ 0.186 (double); 13% γ 0.427	−49.7		
	W 178	21.5 d	100% K	−50.4		0^+
	W 179	38 min	100% K; 18% γ 0.031	−49.28		
	W 179 m	6.4 min	100% IT; 9% γ 0.222	−49.06		
	W 180	0.13%		−49.62	10 (γ) (r)	0^+
	W 181	121 d	100% K	−48.24		$\frac{9+}{2}$
	W 182	26.3%		−48.23	21 (γ)	0^+
	W 183	14.3%		−46.35	10.1 (γ)	$\frac{1-}{2}$
	W 183m	5.4 s	100% IT; γ	−46.04		
	W 184	30.7%		−45.69	1.8 (γ); 0.002 (γm)	0^+
	W 185	75.1 d	100% β⁻ 0.43	−43.37		$\frac{3-}{2}$
	W 185m	1.66 min	100% IT; 5% γ 0.066; 4% γ 0.132; 3% γ 0.174	−43.17		$\frac{11+}{2}$
	W 186	28.6%		−42.50	38 (γ)	0^+
	W 187	23.9 h	33% β⁻ 1.31; 53% β⁻ 0.63; 26% γ 0.686; 21% γ 0.480; 11% γ 0.072	−39.89	70 (γ)	$\frac{3-}{2}$
	W 188	69 d	99% β⁻ 0.35	−38.66		0^+
	W 189	11.5 min	β⁻ 2.5; γ	−35.5		
	W 190	30 min	β⁻ 1.0; 39% γ 0.158; 11% γ 0.162	−34.2		0^+
54	Xe 113	2.8 s	β⁺; K; p			
	Xe 115	18 s	β⁺; K; γ	−68.9		
	Xe 116	57 s	β⁺ 3.3; K; γ	−73.3		0^+
	Xe 117	61 s	65% K; β⁺; γ	−74.5		
	Xe 118	6 min	86% K; β⁺; γ	−77.3		0^+
	Xe 119	6 min	82% K; β⁺; γ	−78.8		
	Xe 120	40 min	97% K; 2% β⁺ 0.9; 29% γ 0.025; 9% γ 0.073; 7% γ 0.178	−81.8		0^+
	Xe 121	38.8 min	92% K; β⁺ 2.8; γ	−82.3		
	Xe 122	20.1 h	100% K; 8% γ 0.350; 3% γ 0.149; 2% γ 0.417	−85.2		0^+
	Xe 123	2.08 h	87% K; β⁺ 1.5; 49% γ 0.149; 15% γ 0.178; 9% γ 0.330	−85.3		
	Xe 124	0.096%		−87.5	100 (γ); 20 (γm)	0^+

(Continued)

Table IX.39 (Continued)

Z	Nuclide □ A†	half-life and/or abundance	decay scheme⊕	mass excess/ MeV	cross section*/barn	J^P
	Xe	125	17.3 h	100% K; 55% γ 0.188; 29% γ 0.243; 6% γ 0.055	−87.11		
	Xe	125m	57 s	100% IT; 58% γ 0.112; 19% γ 0.142	−86.86		
	Xe	126	0.090%		−89.16	3 (γ); 0.4 (γm)	0^+
	Xe	127	36.41 d	100% K; 68% γ 0.203; 25% γ 0.172; 17% γ 0.375	−88.32		
	Xe	127m	69 s	100% IT; 68% γ 0.125; 37% γ 0.173	−88.02		
	Xe	128	1.92%		−89.861	0.4 (γm)	0^+
	Xe	129	26.4%		−88.698	20 (γ) (r)	$\frac{1}{2}^+$
	Xe	129m	8.89 d	100% IT; 5% γ 0.197	−88.461		$\frac{11}{2}^-$
	Xe	130	4.1%		−89.881	0.4 (γm)	0^+
	Xe	131	21.2%		−88.421	90 (γ) (r)	$\frac{3}{2}^+$
	Xe	131m	11.77 d	100% IT; 2% γ 0.164	−88.257		$\frac{11}{2}^-$
	Xe	132	26.9%		−89.286	0.4 (γ); 0.03 (γm)	0^+
	Xe	133	5.25 d	99% β⁻ 0.35; 37% γ 0.081	−87.66	190 (γ) (r)	$\frac{3}{2}^+$
	Xe	133m	2.19 d	100% IT; 10% γ 0.233	−87.43		$\frac{11}{2}^-$
	Xe	134	10.4%		−88.13	0.25 (γ); 0.003 (γm)	0^+
	Xe	134m	0.29 s	100% IT; 100% γ 0.847; 100% γ 0.884; 68% γ 0.233	−86.16		
	Xe	135	9.10 h	96% β⁻ 0.91; 90% γ 0.250; 3% γ 0.608	−86.51	2.6×10⁶ (γ)	$\frac{3}{2}^+$
	Xe	135m	15.6 min	100% IT; 81% γ 0.527	−85.98		$\frac{11}{2}^-$
	Xe	136	8.9%		−86.43	0.16 (γ)	0^+
	Xe	137	3.82 min	67% β⁻ 4.3; 31% γ 0.455; 1% γ 0.849	−82.22		
	Xe	138	14.1 min	β⁻; 30% γ 0.258; 20% γ 0.434; 19% γ 1.768	−80.2		0^+
	Xe	139	39.7 s	22% β⁻ 4.9; 21% β⁻ 4.4; 50% γ 0.219; 19% γ 0.297; 18% γ 0.175	−75.8		
	Xe	140	13.6 s	β⁻; 8% γ 0.622; 5% γ 0.118; 4% γ 0.080	−73.2		0^+
	Xe	141	1.73 s	β⁻; γ	−69.0		
	Xe	142	1.24 s	β⁻; γ	−66.1		0^+
	Xe	143	0.30 s	β⁻			
	Xe	143	0.96 s	β⁻			
	Xe	144	1.2 s	β⁻			0^+
	Xe	145	0.9 s	β⁻			
39	Y	81	5.0 min	β⁺; K; γ			
	Y	83	7.1 min	β⁺ 3.3; 5% K; (2% m); 20% γ 0.036; 6% γ 0.490; 6% γ 0.882 (double)	−72.4		
	Y	83	2.85 min	β⁺; K; (100% m); 88% γ (m) 0.259; 33% γ 0.422; 13% γ 0.495	−72.4		
	Y	84	39 min	β⁺; K; 98% γ 0.793; 78% γ 0.974; 57% γ 1.040	−73.69		
	Y	84	4.6 s	β⁺; K; 35% γ 0.793			
	Y	85 (g)	2.7 h	β⁺; 45% K; (100% m); 85% γ (m) 0.232; 64% γ 0.504; 7% γ 0.914	−77.86		
	Y	85 (m)	4.7 h	β⁺ 2.2; 30% K; (4% m); 23% γ 0.232; 5% γ 2.124; 5% 0.768 (double)	−77.84		

(Continued)

Table IX.39 (Continued)

Z	Nuclide	A†	half-life and/or abundance	decay scheme⊕	mass excess/ MeV	cross section*/barn	J^P
	Y	86	14.74 h	66% K; β⁺ 3.2; 83% γ 1.077; 33% γ 0.628; 31% γ 1.153	−79.24		4⁻
	Y	86 m	48 min	99% IT; 1% K; 94% γ 0.208	−79.02		8⁺
	Y	87	80.3 h	100% K; (100% m); 92% γ 0.485; 82% γ (m) 0.388	−83.007		$\frac{1}{2}^-$
	Y	87 m	13.2 h	98% IT; 1% β⁺ 1.2; 1% K; 78% γ 0.381	−82.626		$\frac{9}{2}^+$
	□Y	88	106.61 d	100% K; 99% γ 1.836; 91% γ 0.898; 1% γ 2.734	−84.298		4⁻
	Y	89	100%		−87.695	1.2 (γ)	$\frac{1}{2}^-$
	Y	89 m	16.06 s	100% IT; 99% γ 0.909	−86.786		$\frac{9}{2}^+$
	Y	90	64.1 h	100% β⁻ 2.28	−86.481		2⁻
	Y	90 m	3.19 h	100% IT; 96% γ 0.203; 91% γ 0.480	−85.799		7⁺
	Y	91	58.5 d	100% β⁻ 1.54	−86.350	1.4 (γ) (r)	$\frac{1}{2}^-$
	Y	91 m	49.71 min	100% IT; 95% γ 0.556	−85.794		$\frac{9}{2}^+$
	Y	92	3.54 h	86% β⁻ 3.63; 14% γ 0.934; 5% γ 1.405; 2% γ 0.449, 0.561	−84.82		2⁻
	Y	93	10.25 h	90% β⁻ 2.89; 7% γ 0.267; 2% γ 0.947; 1% γ 1.918	−84.23		$\frac{1}{2}^-$
	Y	93 m	0.82 s	100% IT; γ	−83.47		$\frac{9}{2}^+$
	Y	94	18.7 min	41% β⁻ 4.88; 40% β⁻ 3.96; 56% γ 0.919; 6% γ 1.139; 5% γ 0.551	−82.38		2⁻
	Y	95	10.3 min	58% β⁻ 4.43; 19% γ 0.954; 8% γ 2.176; 8% γ 3.577	−81.23		
	Y	96	9.8 s	9% β⁻ 3.3; 91% β⁻ 2.7; 89% γ 1.751; 60% γ 0.915; 55% γ 0.617			
	Y	96	6.0 s	75% β⁻ 7.0; 25% β⁻ 5.4; 25% γ 1.594	−78.4		
	Y	97 (g)	3.7 s	40% β⁻ 6.7; 27% β⁻ 3.4; 18% γ 3.288; 14% γ 3.401; 7% γ 1.997	−76.3		
	Y	97 (m)	1.21 s	37% β⁻ 6.1; 45% β⁻ 5.1; 92% γ 1.103; 71% γ 0.161; 40% γ 0.970	−75.6		
	Y	98	0.6 s	51% β⁻ 8.1; 26% β⁻ 7.2; 11% γ 1.223; 5% γ 2.941; 4% γ 1.591	−73.2		
	Y	98	2.0 s	12% β⁻ 6.3; 46% β⁻ 3.8; 96% γ 1.223; 74% γ 0.621; 55% γ 0.647	−73.2		
	Y	99	1.5 s	β⁻; 1% n; γ	−71.5		
	Y	100	0.8 s	β⁻; γ	−68.0		
70	Yb	154	0.39 s	α 5.32	−50.1		0⁺
	Yb	155	1.7 s	α 5.19	−50.5		
	Yb	156	24 s	α 4.80	−53.1		0⁺
	Yb	157	34 s	α 4.50	−53.3		
	Yb	158	1.1 min	β⁺; K; γ	−55.5		0⁺
	Yb	160	4.8 min	β⁺; K; γ	−57.6		0⁺
	Yb	161	4.2 min	β⁺; K; γ	−57.4		
	Yb	162	18.9 min	>98% K; β⁺; 40% γ 0.163; 28% γ 0.119; 2% γ 0.045	−59.3		0⁺
	Yb	163	11.0 min	β⁺; K; γ	−59.6		
	Yb	164	76 min	100% K	−60.9		0⁺

(Continued)

Table IX.39 (Continued)

Z	Nuclide	A†	half-life and/or abundance	decay scheme	mass excess/ MeV	cross section*/barn	J^P
	Yb	165	10 min	K; β^+ 1.58; 37% γ 0.080; 6% γ 0.069; 3% γ 1.090	−60.16		
	Yb	166	56.7 h	100% K; 15% γ 0.082	−61.58		0^+
	Yb	167	17.5 min	100% K; 54% γ 0.113; 22% γ 0.106; 20% γ 0.176	−60.58		$\frac{5-}{2}$
	Yb	168	0.135%		−61.57	3500 (γ)	0^+
	Yb	169	32.0 d	100% K; 44% γ 0.063; 36% γ 0.198; 22% γ 0.177	−60.36		$\frac{7+}{2}$
	Yb	169 m	46 s	100% IT	−60.34		$\frac{1-}{2}$
	Yb	170	3.1%		−60.76	10 (γ)	0^+
	Yb	171	14.4%		−59.30	53 (γ)	$\frac{1-}{2}$
	Yb	172	21.9%		−59.25	1.3 (γ)	0^+
	Yb	173	16.2%		−57.55	17 (γ)	$\frac{5-}{2}$
	Yb	174	31.6%		−56.94	65 (γ)	0^+
	Yb	175	4.19 d	87% β^- 0.47; 6% γ 0.396; 3% γ 0.283; 2% γ 0.114	−54.69		$\frac{7+}{2}$
	Yb	176	12.6%		−53.49	2.4 (γ)	0^+
	Yb	176 m	11.7 s	100% IT; 93% γ 0.293; 91% γ 0.390; 81% γ 0.190	−52.44		
	Yb	177	1.9 h	54% β^- 1.40; 20% γ 0.151; 6% γ 1.080; 3% γ 0.123	−50.99		$\frac{9+}{2}$
	Yb	177 m	6.40 s	100% IT; 77% γ 0.104	−50.66		$\frac{1-}{2}$
	Yb	178	74 min	β^-; γ	−49.66		0^+
30	Zn	57	0.04 s	β^+; p	−32.6		
	Zn	60	2.42 min	β^+ 3.1; 3% K; 68% γ 0.670; 23% γ 0.062; 10% γ 0.273	−54.18		0^+
	Zn	61	89.1 s	68% β^+ 4.4; 16% γ 0.475; 7% γ 1.661; 2% γ 0.970	−56.2		$\frac{3-}{2}$
	Zn	62	9.13 h	93% K; 7% β^+ 0.61; 25% γ 0.041; 24% γ 0.597; 14% γ 0.548	−61.17		0^+
	Zn	63	38.0 min	β^+ 2.3; 7% K; 8% γ 0.670; 6% γ 0.962; 1% γ 1.412	−62.211		$\frac{3-}{2}$
	Zn	64	48.6%		−66.001	0.78 (γ)	0^+
	Zn	65	244.0 d	99% K; 1% β^+ 0.33; 50% γ 1.116	−65.910		$\frac{5-}{2}$
	Zn	66	27.9%		−68.898	1 (γ) (r)	0^+
	Zn	67	4.10%		−67.880	7 (γ) (r)	$\frac{5-}{2}$
	Zn	68	18.8%		−70.006	0.81 (γ); 0.07 (γm)	0^+
	Zn	69	56 min	100% β^- 0.90	−68.417		$\frac{1-}{2}$
	Zn	69 m	13.76 h	100% IT; 95% γ 0.439	−67.978		$\frac{9+}{2}$
	Zn	70	0.62%		−69.560	0.09 (γ); 0.008 (γm)	0^+
	Zn	71	2.5 min	57% β^- 2.82; 30% β^- 2.31; 30% γ 0.512; 7% γ 0.910; 3% γ 0.390	−67.32		$\frac{1-}{2}$
	Zn	71 m	3.92 h	β^-; 92% γ 0.386; 62% γ 0.487; 56% γ 0.620	−67.17		
	Zn	72	46.5 h	86% β^- 0.30; 83% γ 0.145; 9% γ 0.192; 8% γ 0.016	−68.13		0^+
	Zn	73	24 s	β^-; γ	−65.0		
	Zn	74	95 s	39% β^- 2.3; 57% β^- 2.1; (90% m); 80% γ 0.057; 34% γ 0.140; 23% γ 0.190	−65.7		0^+
	Zn	75	10.2 s	β^-	−62.5		
	Zn	76	5.7 s	β^-	−62.6		0^+

(Continued)

Table IX.39 (Continued)

Z	Nuclide □ $A^\dagger$	half-life and/or abundance	decay scheme ⊕	mass excess/ MeV	cross section*/barn	J^P
40	Zr 83	0.7 min	β^+; 10% γ 0.106; 9% γ 0.475; 9% γ 1.525	−65		
	Zr 84	5 min	K; β^+	−71.4		0^+
	Zr 85	7.85 min	β^+; K; (98% m); 41% γ 0.454; 24% γ 0.416; 4% γ 1.198	−73.2		
	Zr 85 m	10.9 s	IT; β^+; K; γ	−72.9		
	Zr 86	16.5 h	100% K; 96% γ 0.243; 20% γ 0.028; 5% γ 0.612	−77.9		0^+
	Zr 87	1.6 h	β^+ 2.26; K; (99% m); 78% γ (m) 0.381; 4% γ 1.228; 1% γ 1.210	−79.4		
	Zr 87 m	14.0 s	100% IT; 97% γ 0.201; 26% γ 0.135	−79.1		
	Zr 88	83.4 d	100% K; 97% γ 0.394	−83.62		0^+
	Zr 89	78.4 h	78% K; 22% β^+ 0.9; (99% m); 98% γ (m) 0.909; 1% γ 1.713	−84.860		$\frac{9^+}{2}$
	Zr 89 m	4.18 min	94% IT; 5% K; 1% β^+ 0.9; 90% γ 0.588; 6% γ 1.507	−84.272		$\frac{1^-}{2}$
	Zr 90	51.5%		−88.765	0.03 (γ) (r)	0^+
	Zr 90m	0.809 s	100% IT; 84% γ 2.319; 16% γ 0.426; 4% γ 0.133	−86.446		5^-
	Zr 91	11.2%		−87.893	1.1 (γ) (r)	$\frac{5^+}{2}$
	Zr 92	17.1%		−88.456	0.2 (γ) (r)	0^+
	Zr 93	1.5×10^6 yr	β^- 0.06; (> 95% m)	−87.117	1 (γ) (r)	$\frac{5^+}{2}$
	Zr 94	17.4%		−87.264	0.06 (γ)	0^+
	Zr 95	64.0 d	1% β^- 0.88; 44% β^- 0.40; 55% β^- 0.36; (1% m); 55% γ 0.757; 44% γ 0.724	−85.663		$\frac{5^+}{2}$
	Zr 96	2.80%		−85.445	0.02 (γ)	0^+
	Zr 97	16.9 h	86% β^- 1.92; (95% m); 93% γ (m) 0.743; 5% γ 0.508; 3% γ 1.148	−82.954		$\frac{1^+}{2}$
	Zr 98	30.7 s	100% β^- 2.2	−81.29		0^+
	Zr 99	2.1 s	39% β^- 3.5; 58% β^- 3.4; (36% m); 56% γ 0.469; 45% γ 0.546; 27% γ 0.594	−77.9		
	Zr 100	7.1 s	β^-; γ	−76.6		0^+
	Zr 101	2.0 s	β^-	−73.1		
	Zr 102	2.9 s	β^-	−72.4		0^+

□ Indicates a popular radioactive source.
† m following indicates a conventional isomer; f, a fission isomer
⊕ The numerical term before (or after) a particle symbol gives respectively the percentage yield (or energy in MeV) for that particle type; some weaker emissions are omitted; a missing numerical term means that its value is unknown; (dist) signifies a spread of energies; (double) signifies two particles of very close energy.
  IT denotes an electromagnetic transition from an isomeric state to the ground state of the same nuclide
  (m) denotes decay resulting in a daughter isomeric state; a number before the m is the % age of all decay events
  SF denotes spontaneous fission
  K denotes a β^- capture process to which capture from the K shell contributes; if a K process is energetically impossible, the β^- capture process is denoted by L
  First listing for α, p or n particles indicates they are direct particle-emission modes; when such particles arise from decay of a daughter state, they are listed after the emission giving rise to the daughter
* Unless otherwise stated, this is the cross section for neutrons of speed 2200 m s^{-1}: (r) indicates the use of neutrons with a thermal reactor spectrum; (s) indicates the use of subcadmium neutrons, upper limit of energy spectrum 0.5 eV
  The processes nγ, nγ leading to an isomeric state, neutron-induced fission, and n α are denoted by (γ), (γm), (f) and (α) respectively.
  (t) signifies the sum of all absorption processes.
Reference: Lederer and Shirley, *Table of Isotopes*, 7th edn., 1978.

Table IX.40 Average values of the components of the total energy released by slow-neutron induced fission in some fission materials

Material	Energy/MeV released instantaneously:			Energy/MeV from decay products:			Total energy Mev
	kinetic energy of fission fragments	prompt neutrons	energy from prompt γs	β⁻	anti neutrino	delayed γs	
$^{239}_{94}$Pu	175.8	5.9	7.8	5.3	7.1	5.2	207.1
$^{233}_{92}$U	168.2	4.9	7.7	5.2	6.9	5	197.9
$^{235}_{92}$U	169.1	4.8	7	6.5	8.8	6.3	202.5

Reference: Sher, *BNL–NCS 51363* vol.2, 835, 1981

Table IX.41 Half-life of principal activity, principal resonance energy, average activation cross section and resonance integral for some neutron foil resonance detectors

Nuclide and reaction	half-life	resonance energy/eV	average activation cross section/barn over spectrum:		resonance integral/barn
			thermal	fission	
$^{107}_{47}$Ag (n, γ) $^{108}_{47}$Ag	2.3 min	16.5	27	0.096	74
$^{197}_{79}$Au (n, γ) $^{198}_{79}$Au	2.7 d	4.9	87.9	0.097	1560
$^{59}_{27}$Co (n, γ) $^{60}_{27}$Co	5.27 yr	130	33.2	0.0031	71
$^{63}_{29}$Cu (n, γ) $^{64}_{29}$Cu	12.8 h	580	4	0.012	3.2
$^{164}_{66}$Dy (n, γ) $^{165}_{66}$Dy	2.35 h	−1.89	2100	0.098	300
$^{115}_{49}$In (n, γ) $^{116}_{49}$In†	54 min	1.46	142	0.12	2600
$^{176}_{71}$Lu (n, γ) $^{177}_{71}$Lu	6.74 d	0.142	3120		900
$^{55}_{25}$Mn (n, γ) $^{56}_{25}$Mn	2.58 h	337	11.7	0.0038	14.1
$^{23}_{11}$Na (n, γ) $^{24}_{11}$Na	15.4 h	2800	0.47	0.00026	0.31
$^{103}_{45}$Rh (n, γ) $^{104}_{45}$Rh†	4.4 min	1.26	9.8	0.015	83
$^{152}_{62}$Sm (n, γ) $^{153}_{62}$Sm	47 h	8	190		3000
$^{186}_{74}$W (n, γ) $^{187}_{74}$W	23.7 h	18.8	37	0.041	500

† Conventional isomer
Reference: *UK Nuclear Data Library* (Ref. AEEW-M824), HMSO

Table IX.42 Half-life of principal activity, mean response energy in fission spectrum, average activation cross section and resonance integral for some neutron foil threshold detectors

Nuclide and reaction	half-life	mean response energy/MeV	average activation cross section/ millibarn over fission spectrum	resonance integral/ millibarn
$^{27}_{13}$Al (n, α) $^{24}_{11}$Na	15 h	8.5	0.62	0
$^{27}_{13}$Al (n, p) $^{27}_{12}$Mg	9.4 min	5.8	3.8	0
$^{63}_{29}$Cu (n, 2n) $^{62}_{29}$Cu	9.7 min	13.7	0.05	0
$^{56}_{26}$Fe (n, p) $^{56}_{25}$Mn	2.58 h	7.4	0.81	0
$^{115}_{49}$In (n, n') $^{115}_{49}$In†	4.5 h	2.8	173	70
$^{58}_{28}$Ni (n, p) $^{58}_{27}$Co	71.3 d	4.3	109	9.7
$^{31}_{15}$P (n, p) $^{31}_{16}$S	2.63 h	3.7	34	0
$^{32}_{16}$S (n, p) $^{32}_{15}$P	14.2 d	4.2	65	0.087

† Conventional isomer
Reference: as for Table IX.41.

Table IX.43 Average values, for neutron thermal spectrum, of cross section for elastic scattering and for radioactive capture by cladding and structural elements in nuclear reactors

Element	Al	Co	Cr	Cu	Fe	Mn	Mo	Nb	Ni	Pb	Si	Ta	W	Zr
scattering cross section/barn	1.4		4.1	7	11.3	1.7	4.5	5.7	17.4	8.6	2.1	5.1	5	6.2
capture cross section/barn	0.2	33.2	2.7	3.3	2.26	11.7	2.4	1	3.42	0.15	0.16	18	16	0.16

Reference: UK *Nuclear Data Library*, Ref. AEEW-M824, HMSO

Table IX.44 Values, for neutron slowing-down region, of cross section for elastic scattering by, and of capture resonance integral for, cladding and structural elements in nuclear reactors

Element	Al	Co	Cr	Cu	Fe	Mn	Mo	Nb	Ni	Pb	Si	Ta	W	Zr
scattering cross section/barn	2.3		5.1	8.4	10.3	41	7.9	6.9	16	10	2.5	15	6.6	7.8
integral/barn	0.18	71	1.4	4.1	1.28	14.1	38	11	2.03	0.2	0.01	720	373	1

Reference: as for Table IX.43

Table IX.45 Average values, for neutron fission spectrum, of cross section for elastic scattering, for absorption, and for inelastic scattering by cladding and structural elements in nuclear reactors

Element	Al	Co	Cr	Cu	Fe	Mn	Mo	Nb	Ni	Pb	Si	Ta	W	Zr
elastic scattering cross section/barn	2.9		2.8	2.9	2.5	2.8	4.6	4.5	3.1	5.6	3	4.1	4.6	5.3
absorption cross section/millibarn	4.8	3	3.1	28	7.9	5.8	30	30	8.5	2.9	9.3	100	51	6.3
inelastic scattering cross section/barn	0.29		0.5	0.74	0.62	0.92	1.3	1.3	0.42	0.7	0.15	2.5	2	0.56

Reference: as for Table IX.43

Table IX.46 Average values, for neutron thermal spectrum, of cross section for elastic scattering and for absorption for coolant components in nuclear reactors

Component	C	D in D_2O	H in H_2O	He	K	N	Na	O
scattering cross section/barn	4.69	5.1	45.3	0.73	2.2	11	3.3	3.35
absorption cross section/millibarn	3	0.46	294	6.1	1800	1690	470	0.28

Reference: as for Table IX.43

Table IX.47 Values, for neutron slowing-down region, of cross section for elastic scattering by, and of absorption resonance integral for, coolant components in nuclear reactors

Component	C	D in D_2O	H in H_2O	He	K	N	Na	O
scattering cross section/barn	4.42	3.27	17.2	1.12	2.2	7.7	8.8	3.74
integral/millibarn	1.5	0.24	141.6	3	120000	900	310	0.13

Reference: as for Table IX.43

Table IX.48 Average values, for neutron fission spectrum, of cross section for elastic and inelastic scattering and for radiative capture by coolant components in nuclear reactors

Component	C	D in D_2O	H in H_2O	He	K	N	Na	O
elastic scattering cross section/barn	2.35	2.56	4	3.66	2.6	1.9	2.72	2.8
inelastic scattering cross section/barn	0.011	0	0	0	0.1	0.0058	0.48	0.0031
capture cross section/millibarn	0.62	0.007	0.039	0.001	100	130	2.1	7.1

Reference: as for Table IX.43

Table IX.49 Average values, for neutron thermal spectrum, of cross section for elastic scattering and for absorption for nuclide or element poisons and absorbers in nuclear reactors

Nuclide or element	B	$^{10}_{5}B$	Cd	Hf	$^{6}_{3}Li$	Li	$^{135}_{54}Xe$
scattering cross section/barn	4.4	2.13	10.1	8	1.04	0.72	3.8×10^5
absorption cross section/barn	672	3398	2900	84	63	833	2.7×10^6

Reference: as for Table IX.43

Table IX.50 Values, for neutron slowing-down region, of cross section for elastic scattering by, and of absorption resonance integral for, nuclide or element poisons and absorbers in nuclear reactors

Nuclide or element	B	$^{10}_{5}B$	Cd	Hf	$^{6}_{3}Li$	Li	$^{135}_{54}Xe$
scattering cross section/barn	4.2	2.3	6.8	16	1.2	0.99	560
integral/barn	325	1640	61	2000	30	405	5890

Reference: as for Table IX.43

Table IX.51 Average values, for neutron thermal spectrum, of number of neutrons produced per fission, of cross section for elastic scattering, for radiative capture and for fission by fissile and fertile nuclides in nuclear reactors

Nuclide	fissile				fertile[†]					
	$^{239}_{94}Pu$	$^{241}_{94}Pu$	$^{233}_{92}U$	$^{235}_{92}U$	$^{237}_{93}Np$	$^{233}_{91}Pa$	$^{238}_{94}Pu$	$^{240}_{94}Pu$	$^{232}_{90}Th$	$^{238}_{92}U$
neutrons per fission	2.88	2.934	2.487	2.432						
scattering cross section/barn	8.6	12	11.6	17		11	13.8	2.12	12.1	7.94
capture cross section/barn	271.9	328.7	42.2	86.3	170	36.8	323	256	6.63	2.42
fission cross section/barn	693.2	936.1	472.6	502.2		15				

[†] No available values for $^{234}_{92}U$ and $^{236}_{92}U$
Reference: as for Table IX.43

Table IX.52 Values, for neutron slowing-down region, of cross section for elastic scattering by, and of capture and fission resonance integrals for, fissile and fertile nuclides in nuclear reactors, together with average number of neutrons per fission

Nuclide	fissile				fertile†							
	$^{239}_{94}Pu$	$^{241}_{94}Pu$	$^{233}_{92}U$	$^{235}_{92}U$	$^{237}_{93}Np$	$^{233}_{91}Pa$	$^{238}_{94}Pu$	$^{240}_{94}Pu$	$^{232}_{90}Th$	$^{234}_{92}U$	$^{236}_{92}U$	$^{238}_{92}U$
neutrons per fission	2.882	2.934	2.487	2.424		2.62	2.88	3.055	2.28	2.46	2.6	2.62
scattering cross section/barn	11.5	10.8	10.3	10.7		11.1	11.1	57.1	10.6	11.4	11.3	19.4
capture resonance integral/barn	171	165	144	139	720	918	150	8011	88	665	320	269
fission resonance integral/barn	280	555	740	264	2.1	1.3	72	2.41	0.035	1.78	0.51	0.17

Reference: as for Table IX.43

Table IX.53 Average values, for neutron fission spectrum, of number of neutrons produced per fission, of cross section for elastic and inelastic scattering, for radiative capture and for fission by fissile and fertile nuclides in nuclear reactors

Nuclide	fissile				fertile†							
	$^{239}_{94}Pu$	$^{241}_{94}Pu$	$^{233}_{92}U$	$^{235}_{92}U$	$^{237}_{93}Np$	$^{233}_{91}Pa$	$^{238}_{94}Pu$	$^{240}_{94}Pu$	$^{232}_{90}Th$	$^{234}_{92}U$	$^{236}_{92}U$	$^{238}_{92}U$
neutrons per fission	3.169	3.24	2.673	2.658		2.78	3.11	3.3	2.28	2.62	2.77	2.81
elastic scattering cross section/barn	4.41	4.76	4.44	4.73		5	4.7	5.09	4.75	5.14	5.09	5.06
inelastic scattering cross section/barn	1.79	1.33	1.08	1.46		2.17	0.63	1.4	2.4	1.46	1.94	2.38
capture cross section/barn	0.049	0.04	0.043	0.081		0.12	0.079	0.0907	0.12	0.028	0.17	0.087
fission cross section/barn	1.78	1.66	1.89	1.24	1.35	0.96	2.19	1.24	0.074	1.15	0.57	0.3

Reference: as for Table IX.43

Table IX.54 Average values, for core and blanket, of number of neutrons produced per fission and of neutron fission cross section for nuclides in a typical fast nuclear reactor fuelled by plutonium-uranium oxide and with liquid-metal coolant

Nuclide	Core		Blanket	
	neutrons per fission	fission cross section/barn	neutrons per fission	fission cross section/barn
$^{239}_{94}Pu$	2.93	1.7	2.9	2
$^{240}_{94}Pu$	3.06	0.37	3.02	0.25
$^{241}_{94}Pu$	2.98	2.3	2.95	3
$^{242}_{94}Pu$	3.02	0.29	2.99	0.18
$^{232}_{90}Th$	2.3	0.0089	2.28	0.0045
$^{233}_{92}U$	2.52	2.6	2.51	3.2
$^{235}_{92}U$	2.45	1.8	2.43	2.3
$^{238}_{92}U$	2.73	0.041	2.71	0.021

Reference: Kaye and Laby, *Tables of Physical and Chemical Constants*, 15 ed., Longmans, 1986

Table IX.55 Thermal diffusion length and average diffusion coefficient over thermal neutron spectrum for neutrons in some moderators and shielding materials at temperature 293 K, together with mean thermal-neutron lifetimes

Material	Al	Be	BeO	Bi	concrete	D_2O	Fe	graphite	H_2O	paraffin wax	Pb	U
diffusion length/mm	200	208	290	320	77	940	12.7	520	27	21	121.8	13.7
diffusion coefficient/mm	55	5	5	11.2	6	8.4	3.4	8.5	1.4	1.1	9.2	7
lifetime/ μs	8800	3460	7000	3600	400	c.10	19	13000	205	160	640	11

Reference: Kaye and Laby, *Tables of Physical and Chemical Constants*, 15 ed., Longmans, 1986

Table IX.56 Some fusion reactions of possible significance for electrical energy generation
Figures in brackets are particle energies in MeV.

$$^2_1H + ^2_1H \overset{50\%}{\underset{50\%}{\rightleftarrows}} \begin{array}{l} ^3_2He\ (0.82) + n\ (2.449) \\ ^3_1H\ (1.011) + p\ (3.002) \end{array}$$

$$^3_1H + ^3_2He \rightarrow \begin{array}{l} ^4_2He + n + p + 12.096\ MeV \\ ^4_2He\ (4.8) + ^2_1H\ (9.52) \end{array}$$

$$^2_1H + ^3_1H \rightarrow ^4_2He\ (3.56) + n\ (14.029)$$
$$^2_1H + ^3_2He \rightarrow ^4_2He\ (3.713) + p\ (14.64)$$
$$^3_1H + ^3_1H \rightarrow ^4_2He + 2n + 11.332\ MeV$$

$$^3_1H + p \rightarrow ^3_2He + n - 0.764\ MeV$$
$$^3_2He + ^3_2He \rightarrow ^4_2He + 2p + 12.86\ MeV$$

Reference: Kaye and Laby, *Tables of Physical and Chemical Constants*, 15 ed., Longmans, 1986
Note: Current views on cold fusion are summarized by Morrison, *Physics World*, Vol. 3, 2/35, IOP, 1990
JET fusion prospects at Culham have recently (11.91) been greatly enhanced (Rodgers, *Physics World,* Vol. 4, 12/5, IOP, 1991)

Table IX.57 Half-life of principal activity, threshold response energy and activation cross section for 14 MeV fusion neutrons for some neutron foil detectors

Nuclide and reaction	halflife	energy/MeV	cross section/barn	Nuclide and reaction	halflife	energy/MeV	cross section/barn
$^{27}_{13}Al(n,\alpha)^{24}_{11}Na$	15 h	5.2	0.12	$^{58}_{28}Ni(n,p)^{58}_{27}Co$	70.8 d	1	0.38
$^{27}_{13}Al(n,p)^{27}_{12}Mg$	9.462 min	2.5	0,08	$^{60}_{28}Ni(n,p)^{60}_{27}Co$	5.27 yr	4.3	0.13
$^{59}_{27}Co(n,\alpha)^{56}_{25}Mn$	2.578 h	5	0.03	$^{31}_{15}P(n,p)^{31}_{14}Si$	157.3 min	1.5	0.09
$^{59}_{27}Co(n,2n)^{58}_{27}Co$	70.8 d	10.5	0.5	$^{103}_{45}Rh(n,n')^{103m}_{45m}Rh^\dagger$	56.1 min	0.1	0.25
$^{50}_{24}Cr(n,2n)^{49}_{24}Cr$	41.93 min	13.6	0.01	$^{32}_{16}S(n,p)^{32}_{15}P$	14.29 d	1.2	0.22
$^{63}_{29}Cu(n,2n)^{62}_{29}Cu$	9.74 min	10.9	0.53	$^{232}_{90}Th(n,f)F.P^{\dagger\dagger}$		1.2	0.36
$^{65}_{29}Cu(n,2n)^{64}_{29}Cu$	12.7 h	10.1	0.93	$^{232}_{90}Th(n,2n)^{231}_{90}Th$	25.52 h	6.5	1.3
$^{63}_{29}Cu(n,\alpha)^{60}_{27}Co$	5.27 yr	5.5	0.036	$^{46}_{22}Ti(n,p)^{46}_{21}Sc$	83.83 d	3	0.26
$^{54}_{26}Fe(n,2n)^{53}_{26}Fe$	8.51 min	13.9	0.005	$^{47}_{22}Ti(n,np)^{46}_{21}Sc$	83.83 d	13.5	0.012
$^{54}_{26}Fe(n,p)^{54}_{25}Mn$	312.5 d	1.5	0.36	$^{47}_{22}Ti(n,p)^{47}_{21}Sc$	3.35 d	1	0.115
$^{56}_{26}Fe(n,p)^{56}_{25}Mn$	2.578 h	4.5	0.11	$^{48}_{22}Ti(n,p)^{48}_{21}Sc$	43.7 h	4.5	0.063
$^{127}_{53}I(n,2n)^{126}_{53}I$	13.02 d	9.3	1.6	$^{238}_{92}U(n,f)F.P^{\dagger\dagger}$		1	1.14
$^{115}_{49}In(n,n')^{115}_{49}In^\dagger$	4.486 h	0.4	0.068	$^{64}_{30}Zn(n,2n)^{63}_{30}Zn$	38.1 min	12.2	0.16
$^{24}_{12}Mg(n,p)^{24}_{11}Na$	15 h	5.5	0.18	$^{64}_{30}Zn(n,p)^{64}_{29}Cu$	12.7 h	1.8	0.16
$^{55}_{25}Mn(n,2n)^{54}_{25}Mn$	312.5 d	10.4	0.8	$^{90}_{40}Zr(n,2n)^{89}_{40}Zr$	78.4h	12.1	0.65
$^{93}_{41}Nb(n,n')^{93}_{41}Nb^\dagger$	16.4 yr	0.1	0.036	$^{90}_{40}Zr(n,p)^{90}_{39}Y$	64 h	3.5	0.09
$^{58}_{28}Ni(n,2n)^{57}_{28}Ni$	36.08 h	12.5	0.026				

†Conventional isomer
††F.P indicates fission products
Reference: *ENDF/B–IV Dosimetry File BNL-NCS-50446,446*, 1975

Table IX.58 The more important radionuclides produced in each chemical element likely to occur in the structural materials of a fusion-reactor vacuum vessel and breeding blanket
For half-lives of radionuclides see Table IX.39.

Element	Radionuclides					Element	Radionuclides						
Al	$^{26}_{13}Al$	$^{27}_{12}Mg$	$^{24}_{11}Na$			Nb	$^{92}_{41}Nb^\dagger$	$^{93}_{41}Nb^\dagger$		$^{94}_{41}Nb^\dagger$	$^{93}_{40}Zr$		
C	$^{10}_{4}Be$	$^{14}_{6}C$				Ni	$^{57}_{27}Co$	$^{58}_{27}Co$	$^{60}_{27}Co$	$^{60}_{27}Co^\dagger$	$^{55}_{26}Fe$	$^{59}_{28}Ni$	$^{63}_{28}Ni$
Co	$^{58}_{27}Co$	$^{58}_{27}Co^\dagger$	$^{60}_{27}Co$	$^{59}_{26}Fe$	$^{60}_{26}Fe$	Si	$^{26}_{13}Al$	$^{28}_{13}Al$	$^{27}_{12}Mg$	$^{22}_{11}Na$	$^{24}_{11}Na$	$^{31}_{14}Si$	
Cr	$^{51}_{24}Cr$	$^{54}_{25}Mn$	$^{49}_{23}V$	$^{52}_{23}V$		Ti	$^{42}_{18}Ar$	$^{45}_{20}Ca$	$^{46}_{21}Sc$	$^{47}_{21}Sc$	$^{48}_{21}Sc$		
Cu	$^{60}_{27}Co$	$^{62}_{29}Cu$	$^{64}_{29}Cu$	$^{60}_{26}Fe$	$^{63}_{28}Ni$	V	$^{46}_{21}Sc$	$^{48}_{21}Sc$	$^{51}_{22}Ti$	$^{48}_{23}V$	$^{49}_{23}V$	$^{52}_{23}V$	
Fe	$^{55}_{26}Fe$	$^{53}_{25}Mn$	$^{54}_{25}Mn$	$^{56}_{25}Mn$		W	$^{182}_{72}Hf$	$^{181}_{74}W$	$^{185}_{74}W$	$^{187}_{74}W$			
Mg	$^{22}_{11}Na$	$^{24}_{11}Na$				Zn	$^{60}_{27}Co$	$^{64}_{29}Cu$	$^{60}_{26}Fe$	$^{63}_{28}Ni$	$^{63}_{30}Zn$	$^{65}_{30}Zn$	
Mn	$^{53}_{25}Mn$	$^{54}_{25}Mn$	$^{56}_{25}Mn$			Zr	$^{94}_{41}Nb$	$^{95}_{41}Nb$	$^{90}_{38}Sr$	$^{88}_{39}Y$	$^{90}_{39}Y$	$^{89}_{40}Zr$	$^{93}_{40}Zr$	$^{95}_{40}Zr$
Mo	$^{93}_{42}Mo$	$^{99}_{42}Mo$	$^{92}_{41}Nb^\dagger$	$^{93}_{41}Nb^\dagger$	$^{94}_{41}Nb$								
	$^{95}_{41}Nb$	$^{98}_{43}Tc$	$^{99}_{43}Tc$	$^{99}_{43}Tc^\dagger$									

†Conventional isomer
Reference: Kaye and Laby, *Tables of Physical and Chemical Constants*, 15 ed., Longmans, 1986

Table IX.59 Values of average reaction cross section over energy spectrum of fusion neutrons emitted from a deuterium-tritium plasma at 20 KeV ion temperature, for various target elements and reaction types
Average neutron energy is 14 MeV.

Element	reaction type				
	(n, α)	$(n, 2n)$	$(n, n'\alpha)$	$(n, n'p)$	(n, p)
			cross section/barn		
Al	0.123	0.035		0.34	0.077
C	0.026		0.262		0.002
Co	0.029	0.642	0.001	0.078	0.081
Cr	0.038	0.298	0.001	0.109	0.082
Cu	0.03	0.647	0.008	0.149	0.087
Fe	0.045	0.427	0.001	0.072	0.123
Mg	0.087	0.034		0.004	0.149
Mn	0.033	0.774	0.001	0.152	0.047
Mo	0.023	0.961	0.001	0.04	0.028
Nb	0.009	0.487	0.003	0.011	0.04
Ni	0.095	0.152	0.016	0.402	0.292
Pb	0.002	1.979	0.001	0.001	0.002
Si	0.109	0.01		0.002	0.295
Ti	0.034	0.284	0.001	0.053	0.072
V	0.017	0.63		0.063	0.036
W	0.001	2.14	0.001	0.001	0.001
Zn	0.018	0.374	0.001	0.21	0.101
Zr	0.009	0.88	0.001	0.038	0.031

Reference: Gardner and Howerton, UCRL–50400, Lawrence Radiation Laboratory, California, 1978

Table IX.60 Charge, mass, mean life and spin for the various leptons together with their principal decay modes.
For value of charge e, see Table I.7.

Lepton	symbol[†]	charge	mass/MeV	mean life/s	spin	decay mode[†]
electron	e	$-e$	0.5110034 ± 0.0000014	stable	$\frac{1}{2}$	stable
electron-neutrino	υ_e	0	< 0.000 046	stable	$\frac{1}{2}$	stable
muon	μ	$-e$	105.65932 ± 0.00029	$2.19709 \times 10^{-6} \pm 0.00005 \times 10^{-6}$	$\frac{1}{2}$	$e\upsilon\bar{\upsilon}$
muon-neutrino	υ_μ	0	0.52	stable	$\frac{1}{2}$	stable
tauon	τ	$-e$	1784.2 ± 3.2	$(3.4 \pm 0.5) \times 10^{-13}$	$\frac{1}{2}$	$\pi^-\pi^0\upsilon, \mu\upsilon\bar{\upsilon}, e\upsilon\bar{\upsilon}$
tauon neutrino	υ_τ	0	< 74	stable	$\frac{1}{2}$	stable

[†]A bar above the symbol of a particle denotes the corresponding antiparticle; see also Table IX.62
Reference: Kaye and Laby, *Tables of Physical and Chemical Constants*, 15 ed., Longmans, 1986

Table IX.61 Baryon number, charge, mass and Z-component of isospin for the various quarks together with name and value of other quantum numbers

Quark	symbol[†]	baryon number	charge	mass–GeV	Z-component	other quantum numbers	
						name	value
down quark	d	$\frac{1}{3}$	$-\frac{1}{3}e$	0.35	$-\frac{1}{2}$		
up quark	u	$\frac{1}{3}$	$2e/3$	0.35	$\frac{1}{2}$		
strange quark	s	$\frac{1}{3}$	$-\frac{1}{3}e$	0.5		strangeness	-1
charmed quark	c	$\frac{1}{3}$	$2e/3$	1.5		charm	1
bottom quark	b	$\frac{1}{3}$	$-\frac{1}{3}e$	4.7		bottomness	1
top quark	t	$\frac{1}{3}$	$2e/3$	30–50		topness	1

[†]A bar above the symbol of a quark denotes the corresponding antiquark
Reference: Close, *An Introduction to Quarks and Partons*, Academic Press, 1979

Table IX.62 Baryon number, mass, mean life, quark content and spin of nonstrange hadrons together with their principal decay modes
Nonstrange hadrons have zero charm and strangeness.

Hadron	symbol[⊗,†]	baryon number	mass/MeV	mean life/s	quark content[†]	spin	decay mode[⊗]
pion	π^+ (π^-)	0	139.5673 ± 0.0007	(2.603 ± 0.0023) × 10^{-8}	$u\bar{d}$ ($d\bar{u}$)	0	($\mu^\pm \nu$)
	π^0	0	134.963 ± 0.0038	(0.83 ± 0.06) × 10^{-16}	$u\bar{u}$ & $d\bar{d}$	0	$\gamma\gamma$
eta	η^0	0	548.8 ± 0.6	(7.93 ± 1.1) × 10^{-19}	$u\bar{u}$, $d\bar{d}$ & $s\bar{s}$	0	$\gamma\gamma$, $\pi^0\pi^0\pi^0$ $\pi^+\pi^-\pi^0$
proton	p	1	938.2796 ± 0.0027	stable	uud	$\frac{1}{2}$	stable
neutron	n	1	939.5731 ± 0.0027	898 ± 16	ddu	$\frac{1}{2}$	$pe\bar{\nu}$

†A bar above the symbol of a particle denotes the corresponding antiparticle
⊗ The superscript denotes charge state: $^+$ ($^-$) indicates charge equal to that on proton (electron); 0 indicates no charge
Reference: Particle Data Group, *Rev. Mod. Phys.*, 1984 (updated biennially)

Table IX.63 Baryon number, strangeness, mass, mean life, quark content and spin of strange hadrons together with their principal decay modes
Strange hadrons have zero charm.

Hadron	symbol[⊗]	baryon number	strangeness	mass/MeV	mean life/s	quark content[†]	spin	decay mode[⊗]
K-meson	K^+ (K^-)	0	1 (−1)	493.667 ± 0.015	(1.2371 ± 0.0026) × 10^{-8}	$u\bar{s}$ ($s\bar{u}$)	0	$\mu^\pm\nu$, $\pi^\pm\pi^0$
	K^0_S	0	1	497.67 ± 0.13	(0.8923 ± 0.0022) × 10^{-10}	(c 50% $s\bar{d}$)	0	$\pi^+\pi^-$, $\pi^0\pi^0$
	K^0_L	0	−1		(5.183 ± 0.04) × 10^{-8}	(c 50% $d\bar{s}$)	0	$\pi^0\pi^0\pi^0$, $\pi^+\pi^-\pi^0$, $\pi^\pm e\nu$, $\pi^\pm\mu\nu$
hyperons:	Λ	1	−1	115.6 ± 0.05	(2.632 ± 0.02) × 10^{-10}	uds	$\frac{1}{2}$	$p\pi^-$, $n\pi^0$
	Σ^+	1	−1	1189.36 ± 0.06	(0.8 ± 0.004) × 10^{-10}	uus	$\frac{1}{2}$	$p\pi^0$, $n\pi^+$
	Σ^-	1	−1	1197.34 ± 0.05	(1.482 ± 0.011) × 10^{-10}	dds	$\frac{1}{2}$	$n\pi^-$
	Σ^0	1	−1	1192.46 ± 0.08	(5.8 ± 1.3) 10^{-20}	uds	$\frac{1}{2}$	$\Lambda\gamma$
	Ξ^-	1	−2	1321.32 ± 0.13	(1.641 ± 0.016) × 10^{-10}	dss	$\frac{1}{2}$	$\Lambda\pi^-$
	Ξ^0	1	−2	1314.9 ± 0.6	(2.9 ± 0.1) × 10^{-10}	uss	$\frac{1}{2}$	$\Lambda\pi^0$
	Ω^-	1	−3	1672.45 ± 0.32	(0.819 ± 0.027) × 10^{-10}	sss	$\frac{3}{2}$	ΛK^-, $\Xi^0\pi^-$, $\Xi^-\pi^0$

†⊗, See footnotes † and ⊗, Table IX.62
Reference: as for Table IX.62

Table IX.64 Baryon number, charm, strangeness, mass, mean life, quark content and spin of charmed hadrons together with their principal decay modes.

Symbol[†⊗]	baryon number	charm	strangeness	mass/MeV	mean life/s	quark content[†]	spin	decay mode[†⊗]
D^+ (D^-)	0	1 (−1)	0	1869.4 ± 0.6	(8 to 10.7) × 10^{-13}	$c\bar{d}$ ($d\bar{c}$)	0	$\bar{K}^0\pi^\pm\pi^+\pi^-$ $\bar{K}^0\pi^\pm\pi^0$
D^0 ($\bar{D}^0$)	0	1 (−1)	0	1864.7 ± 0.6	(3.8 to 5.2) × 10^{-13}	$c\bar{u}$ ($u\bar{c}$)	0	$K^-\pi^+\pi^0$
F^+ (F^-)	0	1 (−1)	1 (−1)	2021 ± 15	(1.2 to 3.2) × 10^{-13}	$c\bar{s}$ ($\bar{c}s$)	0	$\eta\pi\pi\pi$ $\eta\pi$
Λ_c^+	1	1	0	2282.2 ± 3.1	(0.7 to 2) × 10^{-13}	cud	$\frac{1}{2}$	$pK^-\pi^+$

†, ⊗ See footnotes † and ⊗, Table IX.62
Reference: as for Table IX.62

Table IX.65 Mass, spin and width of beautyonium states together with their $b\bar{b}$ configuration and principal modes of decay

State	mass†/MeV	spin	width/MeV	configuration	decay mode ⊗
Y (9460)	9456	1	0.042 – 0.015	1^3S_1	e^+e^-, μ^+, μ^-, hadrons
χ_b (9873)	9873	0		1^3P_0	γ Y(9460)
χ_b (9894)	9894	1		1^3P_1	γ Y(9460)
χ_b (9914)	9914	2		1^3P_2	γ Y(9460)
Y (10020)	10 016	1	0.103 – 0.01	2^3S_1	Y(9460)
χ'_b (10231)	10 231	0		2^3P_0	
χ'_b (10249)	10 249	1		2^3P_1	
χ_b' (10264)	10 264	2		2^3P_2	
Y (10350)	10 347	1		3^3S_1	
Y (10570)	10 569	1	14 – 54	4^3S_1	

† To ± 10%
⊗ See footnote ⊗ Table IX.62
Reference: Duff, *Fundamental Particles*, Taylor & Francis, 1986

Table IX.66 Mass, spin and width of charmonium states together with their $c\bar{c}$ configuration and principal modes of decay

State	mass/MeV	spin	width/MeV	configuration	decay mode⊗,†
η_c(2980)	2981 ± 6	0	20	$^1S_0$	
J/ψ (3100)	3096.9 ± 0.1	1	0.063 ± 0.009	$^3S_1$	$e^+e^-, ^+\mu^-\gamma$ hadrons
χ(3415)	3415 ± 1	0		$^3P_0$	$2(\mu^+\pi^-)$
P_c or χ (3510)	3510 ± 0.6	1		$^3P_1$	$\gamma J/\psi$(3100)
χ(3555)	3555.8 ± 0.6	2		$^3P_2$	$\gamma J/\psi$(3100)
η_c(3590)	3594 ± 5	0		$^1S_0$	
ψ(3685)	3686 ± 0.1	1	0.215 ± 0.04	$^3S_1$	$J/\psi\pi^+\pi^-, J/\psi\pi^0\pi^0, \gamma\chi$(3415), $\gamma\chi$(3510), $\gamma\chi$(3555)
ψ(3770)	3770 ± 3	1	25 ± 3	$^3D_1$	$D\bar{D}$

⊗, † See footnotes ⊗, †, Table IX.62
Reference: as for Table IX.65

Table IX.67 Charge, mass and width of gauge bosons of spin 1, together with the force transmitted and principal decay modes

Gauge boson	symbol	charge	mass/GeV	width/GeV	force transmitted	decay mode⊗
boson (weak)	$W^\pm$	±e	80.8 ± 2.7	7	charged weak (radio activity)	$e^\pm \upsilon, \mu^\pm \upsilon, \tau^\pm \upsilon$ hadrons
	Z^0	0	92.9 ± 1.6	8.5	neutral weak	$e^+e^-\mu^+\mu^-, \tau^+\tau^-$ hadrons
gluon	g	0	0		interquark colour and strong	stable
photon	γ	0	0		electromagnetic	stable

⊗ See footnote ⊗, Table IX.62
Reference: as for Table IX.65

Section X
Acoustics

		Page
X.1	Speed of sound, at about standard pressure, in some gases at stated temperature	273
X.2	Variation of the attenuation of sound in air with relative humidity at temperature 293 K and standard pressure, for various frequencies	273
X.3	Quotient of sound absorption coefficient and square of sound frequency for some gases at temperature 293 K and standard pressure	274
X.4	Speed of sound in distilled and sea water at various temperatures and standard pressure	274
X.5	Quotient of sound absorption coefficient and square of sound frequency for distilled water at various temperatures and standard pressure	274
X.6	Variation of the attenuation of sound with temperature and frequency at standard pressure in sea water of 3.5% salinity	274
X.7	Speed of sound and its rate of change with temperature for various liquids at stated temperature and standard pressure	275
X.8	Quotient of sound absorption coefficient and square of sound frequency for various liquids at stated temperature and standard pressure	275
X.9	Typical values of the speed of longitudinal bulk sound waves, irrotational rod sound waves, shear sound waves and Rayleigh sound waves in various solids, at temperature of about 293 K (unless otherwise stated)	276
X.10	Typical values of absorption coefficient for longitudinal sound waves in various solids at a temperature of about 293 K	277
X.11	Characteristics of transverse vibrations of a rod free at each end	277
X.12	Characteristics of transverse vibrations of a rod free at one end and clamped at the other	278
X.13	Preferred octave-band, 1/2-octave band and 1/3-octave band centre frequencies for acoustic measurements	278
X.14	Typical octave-band analysis of sound in various types of environment	278
X.15	Variation of minimum audible field with sound frequency	278
X.16	Classification of hearing impairment by average permanent threshold shift for best ear for sound frequencies 500 Hz, 1000 Hz and 2000 Hz	278
X.17	Recommended maximum noise levels at various frequencies	279
X.18	Variation of loudness level of pure tones with sound pressure level and frequency	279
X.19	Approximate loudness sensation equivalents for various sound pressure levels	279
X.20	Noise loudness in some common types of environment	279
X.21	Typical variation of reverberation absorption coefficients with sound frequency for various materials	280
X.22	Variation of typical reverberation times of a room for various 1/3-octave bandwidths	280
X.23	Variation of sound reduction index with sound frequency for some types of partition	281
X.24	Variation of sound reduction index with sound frequency for some types of window	281
X.25	Variation of typical ambient noise level, at which speech interference occurs for a normal male voice, with distance between speaker and hearer	281
X.26	Characteristics of some concert halls	282
X.27	Consonant musical frequency intervals	282
X.28	Frequency ratios of musical intervals in the diatonic major scale	282
X.29	Frequencies of notes in equal temperament	282

Section X
Acoustics

Table X.1 Speed of sound, at about standard pressure, in some gases at stated temperature

Gas	temperature/K	speed/m s^{-1}	Gas	temperature/K	speed/m s^{-1}
air (dry)	273	331.45	cyclo C_6H_{12}	303	191
Ar	273	307.85	D_2	273	929
Br_2	331	149	D_2O	373	450
CCl_2F_2	290	146	F_2	375	336
CCl_2FCClF_2	326	132	HCl	273	296
CCl_3F	291	149	HBr	273	200
CCl_4	295	140	HI	273	157
$CHClF_2$	308	186	H_2O	373	477.5
$CHCl_3$	295	160	H_2S	297	309
CH_3CH_3	304	326	He	273	972.5
CH_4	314	466	Kr	303	224
CO_2	324	287	N_2	302	354.4
CS_2	308	206	NH_3	303	440
C_2H_2	273	329	NO	289	334
C_2H_4	293	327	N_2O	298	275
CH_3CHO	273	278	Ne	303	461
$(C_2H_5)_2O$	313	187	O_2	303	332.2
C_6H_6	363	200	SF_6	284	140

Reference: Kaye and Laby, *Tables of Physical and Chemical Constants*, 15ed., Longmans, 1986
Smith, *J. Acoust. Soc. Am.*, vol. 25, 81, 1953

Table X.2. Variation of the attenuation of sound in air with relative humidity at temperature 293 K and standard pressure, for various frequencies

frequency/ kHz	\multicolumn{9}{c}{relative humidity/%}
	10	20	30	40	50	60	70	80	90
	\multicolumn{9}{c}{attenuation/dB km^{-1}}
1	12	6	5	4	4	5	5	5	6
1.25	18	9	7	6	5	6	6	6	6
1.6	28	14	9	8	7	7	7	7	7
2	41	21	14	11	9	9	9	9	9
2.5	59	31	20	16	13	12	11	11	10
3.15	82	47	31	23	19	17	15	14	14
4	110	72	48	36	29	25	22	20	19
5	139	106	72	54	43	36	32	28	26
6.3	169	153	110	83	66	55	48	42	39
8	197	216	167	128	103	86	74	65	59
10	220	286	241	192	156	130	112	99	89
12.5	243	362	337	281	233	197	170	150	134
16	269	448	469	419	360	310	270	239	214
20	298	526	606	580	520	459	405	362	326
25	338	605	752	775	731	666	601	543	493
31.5	400	697	913	1005	1002	951	883	813	749
40	499	818	1094	1268	1336	1327	1276	1206	1131
50	643	977	1297	1544	1693	1751	1743	1694	1625
69	878	1221	1575	1887	2121	2266	2332	2336	2298
80	1266	1616	1994	2356	2668	2908	3070	3160	3189
100	1841	2194	2587	2983	3352	3672	3927	4112	4231

Reference: Bazley, NPL Report, Ac. 74, 1976

Table X.3 Quotient of sound absorption coefficient and square of sound frequency for some gases at temperature 293 K and standard pressure

Gas quotient/ 10^{-7} dB km^{-1} Hz^{-2}	Ar	He	O_2	N_2
	1.62	0.45	1.67	1.42

Reference: Kinsler et al., *Fundamentals of Acoustics*, 3rd edn. Wiley 1982

Table X.4 Speed† of sound in distilled and sea water at various temperatures and standard pressure

Distilled H_2O		3.5% salinity sea H_2O	
temperature/K	speed/m s^{-1}	temperature/K	speed/m s^{-1}
273	1402.39	273	1449
283	1447.27	278	1470.6
293	1482.34	283	1489.8
303	1509.13	288	1506.7
313	1528.86	293	1521.5
323	1542.55	298	1534.4
333	1550.99	303	1545.5
343	1554.8		
353	1554.49		
363	1550.48		

Reference: Kaye and Laby, *Tables of Physical and Chemical Constants*, 15ed., Longmans, 1986
†For elastic wave speeds in the Earth's interior see Table XIV.8.

Table X.5 Quotient of sound absorption coefficient and square of sound frequency for distilled water at various temperatures and standard pressure

temperature/K	quotient/10^{-11} dB km^{-1} Hz^{-2}	temperature/K	quotient/10^{-11} dB km^{-1} Hz^{-2}
273	46.9	323	10.4
283	32.1	333	8.6
293	22.6	343	7.5
303	16.5	353	6.7
313	13	363	6.3

Reference: Pinkerton, *Proc. Phys. Soc.*, vol. 62, 129, 1949

Table X.6 Variation of the attenuation of sound with temperature and frequency at standard pressure in sea water of 3.5% salinity

temperature/K	frequency/kHz					
	0.5	1	2	5	10	20
	attenuation/dB km^{-1}					
273	0.03	0.07	0.14	0.41	1.3	4.6
283	0.02	0.07	0.14	0.33	0.92	3.2
293	0.02	0.06	0.13	0.3	0.7	2.2
303	0.01	0.05	0.13	0.29	0.58	1.6

Reference: Schulkin and Marsh, *J. Acoust. Soc. Am.* vols. 63, 43, 1978; 35, 739, 1963; 34, 864, 1962

Table X.7 Speed of sound and its rate of change with temperature for various liquids at stated temperature and standard pressure

Liquid	temperature/ K	speed/ m s^{-1}	rate of change/ m s^{-1} K^{-1}	Liquid	temperature/ K	speed/ m s^{-1}	rate of change/ m s^{-1} K^{-1}
Ar	30	840	−6.5	Cs	313	980	−0.31
Bi	553	1651	−0.13	D$_2$O	293	1383.6	3.2
CCl$_4$	298	921	−3	Freon (C51-12)	298	524	−3.4
CHCl$_3$	298	984	−3.5	glycerol	298	1920	−1.9
n-CH$_3$CH$_2$CH$_2$OH	298	1207	−3.3	H$_2$	18	1246	−26
CH$_3$(CH$_2$)$_4$CH$_2$OH	298	1303	−3.4	Hg	298	1449	−0.46
CH$_3$COCH$_3$	298	1170	−4.5	In	433	2313	−0.29
CH$_3$COOH	292.6	1173		K	353	1869	−0.49
CH$_3$OH	298	1103	−3.3	N$_2$	71	912	−9.8
CH$_4$	103	1420	−9.7	Na	383	2520	−0.52
CS$_2$	298	1141	−3.2	Ne	30	540	−16.9
C$_2$H$_5$OH	298	1145	−3.3	O$_2$	71	1056	−7.8
n-C$_4$H$_9$OH	298	1242	−3.4	oil (castor)	298	1490	−3.1
C$_5$H$_5$N	298	1417	−4.2	oil (lubricating)	298	1461	−3.4
n-C$_5$H$_{11}$OH	298	1277	−3.3	oil (sperm)	305	1411	
C$_6$H$_5$CH$_3$	298	1306	−4.3	Pb	613	1766	−0.28
C$_6$H$_5$Cl	298	1270	−3.9	Rb	323	1247	−0.38
C$_6$H$_5$NH$_2$	298	1640	−3.6	Sn	513	2471	−0.25
C$_6$H$_6$	298	1300	−4.7	Zn	723	2780	−4.3
cyclo C$_6$H$_{11}$OH	298	1465	−3.7				
cyclo C$_6$H$_{12}$	293	1280	−5.4				

Reference: Kaye and Laby, *Tables of Physical and Chemical Constants*, 15ed., Longmans, 1986

Table X.8 Quotient of sound absorption coefficient and square of sound frequency for various liquids at stated temperature and standard pressure

Liquid	temperature/ K	quotient/ 10^{-11} dB km^{-1} Hz^{-2}	measurement frequency/ MHz	Liquid	temperature/ K	quotient/ 10^{-11} dB km^{-1} Hz^{-2}	measurement frequency/ MHz
Bi	553	c 7	20	cyclo C$_6$H$_{11}$OH	298	c 434	< 45
CCl$_4$	298	465		cyclo C$_6$H$_{12}$	293	c 156	
CHCl$_3$	298	321		Cs	313	97	30
n-CH$_3$CH$_2$CH$_2$OH	298	58		D$_2$O	293	28	50
CH$_3$COCH$_3$	298	c 30		glycerol	298	c 2606	4–12
CH$_3$OH	298	28		Hg	298	4.9	100
CS$_2$	298	c 4864	3	K	353	30	30
C$_2$H$_5$OH	298	44		Na	383	10	30
n-C$_4$H$_9$OH	298	74	25	oil (castor)	298	c 4604	3
C$_6$H$_5$CH$_3$	298	73	15–200	Pb	613	c 8	20
C$_6$H$_5$Cl	298	122	< 200	Rb	323	68	30
C$_6$H$_5$NH$_2$	298	c 43		Sn	513	c 5	20
C$_6$H$_6$	298	756	< 70	Zn	723	c 3	20

Reference: Kaye and Laby, *Tables of Physical and Chemical Constants*, 15ed., Longmans, 1986

Table X.9 Typical values of the speed of longitudinal bulk sound waves, irrotational rod sound waves, shear sound waves and Rayleigh sound waves in various solids, at temperature of about 293 K (unless otherwise stated)

Solid	speed of sound/m s^{-1}			
	longitudinal	irrotational	shear	Rayleigh
ADP crystal, X cut	6250			
ADP crystal, Y cut	6250			
ADP crystal, Z cut	4300			
Ag	3704	2806	1698	1592
Al	6374	5102	3111	2906
Au (hard drawn)	3240	2030	1200	
BaTiO$_3$ ceramic	4000			
Be	12 890		8880	
bone (human tibia)	4000		1970	
brass	4372	3451	2100	1964
brick		3650		
†butyl rubber/C (100/40)	1600			
Cd	2780	2400		
cellulose acetate butyrate	2080			
concrete	4250–5250			
Constantan	5177	4276	2625	2445
cork		500		
Cr	6608	6229	4005	3655
Cu	4759	3813	2325	2171
duralumin	6398	5120	3122	2917
ebonite	2500			
Fe (cast)	4994	4477	2809	2590
Fe (soft)	5957	5189	3224	2986
glass (crown)	5660	5342	3420	3127
glass (heavy flint)	5260	4717	2960	2731
glass (Pyrex)	5640	5170	3280	
granite	5400			
ice at 253 K	3840			
†Invar (0.2 C, 63.8 Fe, 36 Ni)	4657	4216	2658	2447
marble		3810		
Mg	5823	5082	3163	2930
Mn	4600	3830		
Mo	6475	5636	3505	3248
Monel metal	5350	4400	2720	
Nb	5068	3497	2092	1970
neoprene	1510			
†neoprene/C (100/60)	1690			
Ni (unmagnetized hard)	5814	4974	3078	2857
Ni (unmagnetized soft)	5608	4787	2929	2722
Ni-Span-C		4831	2799	
nylon	2680			
Pb	2160	1188	700	
Perspex	2700	2177	1330	1242
polythene	2000			
polystyrene	2350	1840	1120	1047
polyvinyl chloride	2300			
polyvinyl chloride acetate	2250			
polyvinyl formal	2680			
polyvinylidene chloride	2400			
Pt	3260	2800	1730	
quartz (crystal, X cut)	5720	5440		
quartz (fused)	5970	5759	3765	3410
†rubber/C (100/40)	1680			
rubber (natural)	1600			
rubber (RTV silicone)	900–1050			
sandstone	2920	2820	1840	
SiO$_2$ (fused)	5968	5760	3764	
slate		4510		
Sn	3380	2626	1594	1491

(Continued)

Table X.9 (Continued)

Solid	speed of sound/m s^{-1}			
	longitudinal	irrotational	shear	Rayleigh
steel (mild)	5960	5196	3235	2996
steel (stainless)	5980	5282	3297	3049
steel (tool) hardened	5874	5116	3179	2945
Ta	4159	3337	2036	1902
Teflon	1400			
Ti	6130	5164	3182	2958
U	3370		1940	
V	6023	4584	2774	2600
tourmaline (crystal, Z cut)	7250	7170		
W (annealed)	5221	4619	2887	2668
W (drawn)	5410	4320	2640	
WC	6655	6223	3984	3643
wood (ash, with grain)		4670		
wood (ash, across grain)		1390		
wood (oak, with grain)		4100		
wood (pine, with grain)		3600		
Zn (rolled)	4187	3826	2421	2225
Zr	4650	2250		

†Figures in brackets are percentage mass composition
Reference: Bradfield, *NPL Notes on Applied Science No. 30*, HMSO 1964
       Synge, *Proc. Roy. Irish Acad., A* vol. 58, 13, 1956

Table X.10 Typical values of absorption coefficient for longitudinal sound waves in various solids at a temperature of about 293 K

Solid	coefficient/ dB m^{-1}	measurement frequency/MHz	Solid	coefficient/ dB m^{-1}	measurement frequency/MHz
ADP crystal, X cut	93.8	10	Perspex	495.1	2.5
ADP crystal, Z cut	84.2	10	polythene	469	1
Al	3.47	10	polystyrene	199.8	2.5
bone (human tibia)	3995.6	2.9	polyvinyl chloride	30.4	0.35
†butyl rubber/C (100/40)	1155.2	0.35	polyvinyl chloride acetate	11 031	10
cellulose acetate butyrate	894.7	2.5	polyvinyl formal	998.8	2.5
duralumin	10.7	10	polyvinylidene chloride	1798	2.5
glass (crown)	17.4	10	quartz (crystal, X cut)	0.11	10
granite	521.2	0.6	†rubber/C (100/40)	317.9	0.35
neoprene	1997.8	2.5	rubber (natural)	130.3	0.35
nylon	99.9	1	steel (tool) hardened	42.9	10

†Figures in brackets are percentage mass composition
Reference: Kaye and Laby, *Tables of Physical and Chemical Constants*, 15ed., Longmans, 1986

Table X.11 Characteristics of transverse vibrations of a rod free at each end

vibration frequency/ fundamental frequency	number of nodes	fraction of length from an end at which node occurs
1	2	0.224, 0.776
2.76	3	0.132, 0.5, 0.868
5.4	4	0.094, 0.357, 0.643, 0.906
8.93	5	0.073, 0.277, 0.5, 0.723, 0.927

Reference: Poynting and Thomson, *Sound*, Griffin, 1899

Table X.12 Characteristics of transverse vibrations of a rod free at one end and clamped at the other

vibration frequency/ fundamental frequency	number of nodes	fraction of length from clamped end at which node occurs
1	1	0
6.267	2	0, 0.783
17.55	3	0, 0.504, 0.868
34.39	4	0, 0.358, 0.644, 0.906

Reference: Kinsler *et al.*, *Fundamentals of Acoustics*, 3rd edn., Wiley, 1982

Table X.13 Preferred octave-band, $\frac{1}{2}$-octave-band and $\frac{1}{3}$-octave-band centre frequencies for acoustic measurements

	frequency/Hz
octave	16, 31.5, 63, 125, 250, 500, 1000, 2000, 4000, 8000
$\frac{1}{2}$ octave	16, 22.4, 31.5, 45, 63, 90, 125, 180, 250, 355, 500, 710, 1000, 1400, 2000, 2800 4000, 5600, 8000
$\frac{1}{3}$ octave	10, 12.5, 16, 20, 25, 31.5, 40, 50, 63, 80, 100, 125, 160, 200, 250, 315, 400, 500, 630, 800, 1000, 1250, 1600, 2000, 2500, 3150, 4000, 5000, 6300, 8000

Reference: Smith, *Acoustics*, Longman, 1979

Table X.14 Typical octave-band analysis of sound in various types of environment

Environment	centre frequency of octave band/Hz							
	63	125	250	500	1000	2000	4000	8000
				noise/dB				
kerbside (heavy traffic)	96	93	90	88	89	84	78	76
kerbside (light traffic)	81	81	75	70	72	71	63	60
machine shop	68	72	90	87	86	88	90	84
students' refectory	68	70	75	75	68	64	56	49

Reference: as for Table X.13

Table X.15 Variation of the minimum audible field with sound frequency

frequency/Hz	field/dB	frequency/Hz	field/dB
20	74.3	500	6
30	58.1	700	4.7
40	48.4	1000	4.2
60	36.8	2000	1
80	29.8	3000	−2.9
100	25.1	4000	−3.9
140	18.9	6000	4.6
200	13.8	8000	15.3
250	11.2	10 000	16.4
300	9.4	12 000	12
400	7.2	15 000	24.1

Reference: Robinson and Dadson, *Br. J. Appl. Phys.*, vol. 7, 166, 1956

Table X.16 Classification of hearing impairment by average permanent threshold shift for best ear for sound frequencies 500 Hz, 1000 Hz and 2000 Hz

shift/dB	Classification	shift/dB	Classification
under 25	within normal limits	56–70	moderately severe
26–40	mild or slight	71–90	severe
41–55	moderate	above 91	profound

Reference: as for Table X.12

Table X.17 Recommended maximum noise levels at various frequencies

frequency/Hz	37.5–150	150–300	300–600	600–1200	1200–2400	2400–4800
noise level/dB	100	90	85	85	80	80

Reference: as for Table X.13

Table X.18 Variation of loudness level of pure tones with sound pressure level and frequency.

sound pressure level/dB	\multicolumn{12}{c}{frequency/Hz}

sound pressure level/dB	25	50	100	200	500	1000	2000	3000	4000	6000	8000	10 000	15 000
						\multicolumn{12}{c}{loudness level/phon}					
0							3	7	8				
10					9	10	13	16	18	10			
20				12	20	20	22	26	27	20	10	9	
30			11	25	32	30	32	36	37	30	21	21	14
40		1	25	38	43	40	42	46	48	40	32	33	30
50		18	38	49	53	50	52	57	58	50	42	44	45
60		34	51	61	64	60	62	67	69	61	53	55	58
70	15	49	63	72	74	70	73	78	79	71	63	65	69
80	34	63	75	82	84	80	84	90	90	81	74	75	79
90	53	76	86	92	93	90	95	101	102	92	84	85	88
100	69	88	97	102	102	100	106	113	113	102	94	94	94
110	84	100	107	110	111	110	117	124	125	112	103	103	100
120	98	111	116	119	120	120	129	137	136	123	113	112	103
130	109	121	125	127	128	130							

Reference: Kaye and Laby, *Tables of Physical and Chemical Constants*, 15ed., Longmans, 1986

Table X.19 Approximate loudness sensation equivalents for various sound pressure levels

sound pressure level/dB	sensation	sound pressure level/dB	sensation	sound pressure level/dB	sensation
0	threshold of hearing	50	quiet conversation	100	near loud motor horn
10	virtual silence	60	quiet motor at 1 m	110	pneumatic drill
20	quiet room	70	loud conversation	120	near plane engine
30	watch ticking at 1 m	80	door slamming	130	threshold of pain
40	quiet street	90	busy typing room		

Reference: Tennent, *Science Data Book*, Oliver and Boyd, 1986

Table X.20 Noise loudness in some common types of environment

Environment	loudness/phon	Environment	loudness/phon
80 m below large jet plane	134	machine shop	106
inside noisy motor car	94	normal speech (male) at 1 m	75
kerbside (heavy traffic)	103	students' refectory	84
kerbside (light traffic)	80		

Reference: as for Table X.13

Table X.21 Typical variation of reverberation absorption coefficients with sound frequency for various materials

Material	thickness/mm	frequency/Hz					
		125	250	500	1000	2000	4000
		absorption coefficient					
acoustic plaster	13	0.15	0.20	0.35	0.60	0.60	0.50
acoustic tiles (perforated fibreboard)	18	0.10	0.35	0.70	0.75	0.65	0.50
asbestos (sprayed)	25	0.10	0.30	0.65	0.85	0.85	0.80
brickwork		0.02	0.02	0.03	0.04	0.05	0.07
carpet (Axminster)	8		0.05	0.15	0.30	0.45	0.55
carpet on underlay	14		0.05	0.20	0.40	0.60	0.65
curtain (velour, draped)		0.14	0.35	0.55	0.72	0.70	0.65
glass fibre (resin-bonded)	25	0.10	0.25	0.55	0.70	0.80	0.85
glass wool (uncompressed)	25	0.10	0.25	0.45	0.60	0.70	0.70
mineral wool	25	0.10	0.25	0.50	0.70	0.85	0.85
polystyrene, expanded (rigid backing)	13	0.05	0.05	0.10	0.15	0.15	0.20
polystyrene, expanded (on 50 mm battens)	13	0.05	0.15	0.40	0.35	0.20	0.20
polyurethane foam (flexible)	50	0.25	0.50	0.85	0.95	0.90	0.90
snow	25	0.15	0.40	0.65	0.75	0.80	0.85
wood panelling (oak, on 25 mm battens)	13	0.20	0.10	0.05	0.05	0.05	0.05

Reference: Evans and Bazley, *Sound Absorbing Materials*, NPL, 1978

Table X.22 Variation of typical reverberation times of a room for various $\frac{1}{3}$-octave bandwidths

bandwidth centre frequency/Hz	reverberation time/s	bandwidth centre frequency/Hz	reverberation time/s
100	1.55	800	1
125	1.6	1000	0.9
160	1.45	1250	1.05
200	1.3	1600	1.05
250	1.2	2000	1.05
315	1.05	2500	1
400	1.05	3150	0.95
500	1	4000	0.95
630	1.1		

Reference: as for Table X.13

Table X.23 Variation of sound reduction index with sound frequency for some types of partition

Partition type[†]	frequency/Hz						
	100	200	400	800	1600	3150	
	sound reduction index/dB						mean/dB
brick, unplastered (115)	34	37	40	41	51	57	43
brick (225) plastered both faces	48	40	45	50	61	67	52
cavity wall, brick (50 × 115) airspace (50) plastered both faces, butterfly ties	30	36	45	58	65	82	53
clinker concrete wall (75) unplastered	16	15	20	23	29	38	23
clinker concrete wall (75) plastered both faces	22	33	36	47	55	56	42
concrete floor (100) reinforced	36	40	38	47	55	65	47
wood joist floor, boarding (22) nailed to joists (225 × 50) plasterboard ceiling (9.5) skim coat plaster	16	26	34	35	42	41	32
fibreboard (12)	11	15	18	22	27	30	21
plasterboard (9.5)	13	19	22	28	34	35	25
plasterboard (9.5) each side of (100 × 50) studs (airspace 100)	12	23	32	36	44	46	32
plywood (6)	8	10	15	20	25	28	18
steel, 16G	14	20	25	31	37	43	28

[†]Figures in brackets give thickness in mm
Reference: Bazley, *The Airborne Sound Insulation of Partitions*, NPL, 1978

Table X.24 Variation of sound reduction index with sound frequency for some types of window

Window type[†]	frequency/Hz						
	100	200	400	800	1600	3150	
	sound reduction index/dB						mean/dB
A: glass (2.5) in metal frames (500 × 300) closed openable sections	23	12	18	22	29	27	21
as A, except sealed	18	12	20	27	30	28	23
as A, but wood frames	14	17	22	26	29	30	23
as A, but glass (6) in wood frames	16	20	24	29	29	36	26
glass (25) in wood frames (650 × 540)	28	25	30	33	41	47	34
plate glass (6)	18	23	24	30	30	35	27
B: as A, but double, airspace (50)	19	19	27	37	47	52	33
as B, but airspace (100)	22	20	35	43	51	53	37
as B, but airspace (180)	28	25	36	43	50	53	39
as B, but airspace (180) & acoustic tiles on reveals	25	34	41	45	53	57	42

[†]Figures in brackets give dimensions in mm
Reference: as for Table X.13

Table X.25 Variation of typical ambient noise level, at which speech interference occurs for a normal male voice, with distance between speaker and hearer

distance/m	0.1	0.2	0.3	0.4	0.5	0.6	0.8	1	1.5	2	3	4
noise level/dB	73	69	65	63	61	59	56	54	51	48	45	42

For a female voice, each level should be reduced by 5 dB
For a raised voice, a very loud voice, and shouting, each level should be raised by 6 dB, 12 dB and 18 dB respectively
Reference: as for Table X.13

Table X.26 Characteristics of some concert halls

Name	hall volume/m³	number of seats	full hall mid-frequency reverberation time/s
Academy of Music, Philadelphia	15 090	2836	1.35
Beethovenhalle, Bonn	15 700	1407	1.7
Carnegie Hall, New York	24 250	2760	1.7
F. R. Mann Concert Hall, Tel Aviv	21 200	2715	1.55
Liederhalle, Grosser Saal, Stuttgart	16 000	2000	1.62
Metropolitan Opera House, New York	19 520	3639	1.2
Philharmonic Hall, Berlin	26 030	2200	2
Philharmonic Hall, New York	24 430	2644	2
Royal Festival Hall, London	22 000	3000	1.47
Royal Opera House, London	12 240	2180	1.1
St Andrew's Hall, Glasgow	16 100	2133	1.9
Symphony Hall, Boston	18 740	2631	1.8
Tanglewood Music Shed, Lennox, Mass.	42 450	6000	2.05
Teatro alla Scala, Milan	11 245	2289	1.2
Theatre National de L'Opera, Paris	9960	2131	1.1

Reference: as for Table X.13

Table X.27 Consonant musical frequency intervals.

Interval name	octave	fifth	fourth	major third	major sixth	minor third	minor sixth
Frequency ratio	1:2	2:3	3:4	4:5	3:5	5:6	5:8

Reference: as for Table X.19

Table X.28 Frequency ratios of musical intervals in the diatonic major scale.

	major tone	minor tone	diatonic semitone	major tone	minor tone	major tone	diatonic semitone	
ratio for adjacent steps	9/8	10/9	16/15	9/8	10/9	9/8	16/15	
ratio to keynote	1	1.125	1.25	1.33	1.5	1.66	1.875	2
	keynote	major 2nd	major 3rd	perfect 4th	perfect 5th	major 6th	major 7th	octave

Reference: Kaye and Laby, *Tables of Physical and Chemical Constants*, 15ed., Longmans, 1986

Table X.29 Frequencies of notes in equal temperament

Note	frequency/Hz	Note	frequency/Hz	Note	frequency/Hz
C'	261.6	F	349.2	A	440.0
C#	277.2	F#	370.0	A#	466.2
D	293.7	G	392.0	B	493.9
D#	311.1	G#	415.3	C"	523.2
E	329.6				

Reference: as for Table X.19

Section XI
Physical Properties of Inorganic and Organic Compounds

		Page
XI.1	Physical constants of selected inorganic compounds	285
XI.2	Physical constants of selected organic compounds	307
XI.3	Physical constants of selected organo-metallic compounds	360
XI.4	Physical constants of amino acids and derivatives	366
XI.5	Physical constants of carbohydrates and related compounds	376
	1 Trioses, tetroses and related alcohols	376
	2 Pentoses and related alcohols	376
	3 Hexoses and related alcohols	380
	4 Di- and tri-saccharides	384
	5 Amino sugars	386
	6 Uronic acids	388
XI.6	Properties of selected steroids (bile acids, corticosteroids, sex hormones, sterols)	392
XI.7	Properties of constituents of nucleic acids and related compounds (nucleosides, nucleotides, purines, pyrimidines)	395
XI.8	Physical constants of polymers and plastics	398
XI.9	Physical constants of fluorocarbon refrigerants	400
XI.10	Physical constants of some common refrigerants	402
XI.11	Properties of gases available in cylinders (bottled gases)	402
XI.12	Critical constants	408
	1 Elements and inorganic compounds	408
	2 Organic compounds	409
XI.13	van der Waals constants and Boyle temperatures for some gases	410
XI.14	Second virial coefficients of selected gases	411
XI.15	Solubility of gases in water	412
XI.16	Solubility of some gases in sea water as a function of temperature and salinity	413
XI.17	Vapour pressure of water at different temperatures	414
	1 Vapour pressure of ice, –90 to 0 °C	414
	2 Vapour pressure of water, –15 to 100 °C	414
	3 Vapour pressure of water above 100 °C	414
XI.18	Vapour pressure of mercury from –38 to 400 °C	415
XI.19	Boiling point of water at different pressures	415
XI.20	Ionization potentials of selected molecules	416
XI.21	Hardness of materials	417
	1 Mohs hardness scale	417
	2 Hardness of materials (on Mohs original sclae)	417
XI.22	Critical concentrations for micelle formation	418
	1 Ionic surfactants	418
	2 Non-ionic surfactants	418
XI.23	Physical constants of proteins in water at 293 K	419
XI.24	Solubility of some compounds in water	419
	1 Inorganic compounds	419
	2 Organic compounds	419
XI.25	Density of aqueous ethanol solutions at 20 °C	420
XI.26	Density of some aqueous solutions at 20 °C	420
	1 Acids	420
	2 Bases	420
	3 Alcohols, dioxane and other hydroxylic compounds	420
	4 Some electrolyte solutions	421
	5 Some sugar solutions	421
XI.27	Vapour pressures of selected elements and inorganic compounds	422
XI.28	Vapour pressures of selected organic compounds	424
XI.29	Data on the properties of commercial laboratory reagents	427
XI.30	Preparation of volumetric solutions	428

Section XI
Physical Properties of Inorganic and Organic Compounds

Table XI.1 Physical Constants of Selected Inorganic Compounds
For more extensive data see: Dictionary of Inorganic Compounds (1992). Chapman and Hall, London.
Column 2. Atomic mass of elements quoted to 3 places of decimals (see Table III.2 for more accurate data)
Columns 3 (m.p.) and 4 (b.p.). d: decomposes; s: sublimes.
Columns 5 (density) and 6 (refractive index). Values quoted at 25 °C unless otherwise stated. (g): gas; (l): liquid; (s): solid
Column 7 (solubility in water). g of solid, cm^3 of gas in 100 cm^3 of water, at 25 °C unless otherwise stated. d: decomposes, hydrolysed; i: insoluble; s:soluble; ss: slightly soluble; vs: very soluble; vss: very slightly soluble.
Column 8 (solubility in other solvents). Abbreviations as for column 7. In addition, a: mineral acids; ac: acetone; alc: ethanol; alk: alkalis; bz: benzene; chl: chloroform; ct: carbon tetrachloride; et: ether; pyr: pyridine; aq: aqua regia (HCl/HNO$_3$, 3:1); h: hot
Column 9. General hazards.
C: corrosive; CSA: cancer suspect agent; D: deliquescent; E: explosive; Ef: efflorescent; FG: flammable gas; FL: flammable liquid; FS: flammable solid; FP: flammable when in powder form; I: irritant; L: lachrymator; LS: light-sensitive; MS: moisture-sensitive, e.g. hydrolysed by moisture, deliquescent; O: (powerful) oxidizing agent; P: poisonous; R: radioactive; T: toxic or very toxic.
For further details of other hazards, including toxicity, LD$_{50}$ and approved occupational exposure limits see references in Section XVI and Tables XVI.1 to 4.
Colours blk: black; bl: blue; br: brown; col: colourless (gas or liquid); gn: green; gy: grey; o: orange; p: pink; r: red; viol: violet; w: white or colourless (solid); y: yellow.
Crystal structure amorph: amorphous; cr: crystal or crystalline; cub: cubic; hex: hexagonal; leaf: leaflets; monocl: monoclinic; need: needles; oct: octahedral; pl: plates; powd: powder; pr: prisms; rhbdr: rhombohedral; rhomb: rhombic; tetr: tetragonal; triclin: triclinic; trig: trigonal.

Compound/formula	M_r	m. p. /°C	b. p. /°C	ρ /kg m^{-3}	n_D	Solubility Water	Solubility Other solvents	Hazards and general description
ALUMINIUM								
Al	26.982	660.46	2 467	2 698.9		i	s alk, HCl, H$_2$SO$_4$	silver-white, ductile metal
Al$_2$Br$_6$	266.71	97.5	263.6^{747} (fused)	2 640^{10}		d	s alc, ac CS$_2$	C, MS, w rhomb pl
AlBr$_3$. 6H$_2$O	374.80	93	d 135	2 540		s	s alc ss CS$_2$	C, MS, w – y need
Al$_4$C$_3$	143.96	d > 1 400		2 360		d to CH$_4$		FS, MS y – gy
Al$_2$Cl$_6$	133.34	190 (2.5 atm)	s 180	2 440 (fused)		69.9 with violence	s alc, chl, et, ct	C, MS, w hex
AlCl$_3$. 6H$_2$O	241.43	d 100		2 398	1.6	45	s alc, et	C, MS w rhomb
AlF$_3$	83.98	s 1 291		2 882		0.6		w tricl
AlF$_3$. 3.5H$_2$O	147.03	d 100		1 914		i		T, I, w powder
AlH$_3$	30.0	d 100				d		polymer
Al(OH)$_3$	78.00	– H$_2$O 300		2 420		i	s a, alk	w monocl
Al$_2$I$_6$	407.69	191	360	3 980		d	s alc, et, CS$_2$	C, M, br powd
AlI$_3$. 6H$_2$O	515.79	d 185	d	2 630		vs	s alc, CS$_2$	y – w cr
Al(NO$_3$)$_3$. 9H$_2$O	375.13	73.5	d 150	2 600	1.54	63.7	s alc, alk, a, ac	T, O, w rhomb
AlN	40.99	> 2 000 in N$_2$	s 2 000	3 260		d	d alc	I, MS w cr hex
Al$_2$O$_3$	101.96	2 045	2 980	3 965	1.768	i	ss a, alk	w hex
AlPO$_4$	121.95	> 1 500		2 566	1.546 1.556 1.578	i	s a, alk	w rhomb pl
Al$_2$(SO$_4$)$_3$	342.15	d 770		2 710	1.47	31.3	s a	w powder
Al$_2$S$_3$	150.16	1 100	s 1500	2 020^{13}		d	s a	MS (smells of H$_2$S), y hex
AMMONIA								
NH$_3$	17.03	– 77.7	– 33.35	771	1.000 376^0	89.9	s alc, et	col gas
CH$_3$COONH$_4$	77.08	114	d	1 170^{20}		148	s alc ss ac	MS w cr

(Continued)

Table XI.1 (Continued)

Compound/formula	M_r	m.p. /°C	b.p. /°C	ρ / kg m^{-3}	n_D	Solubility Water	Solubility Other solvents	Hazards and general description
NH$_4$Al(SO$_4$)$_2$ · 12H$_2$O	453.33	93.5	–H$_2$O at 120	1 640	459	15	s a	w cub
NH$_4$Br	97.95	s 452		2 429	1.712	97	s alc, ac, et	MS, w cub
(NH$_4$)$_2$CO$_3$ · H$_2$O	114.10	d 58				100		w cub
NH$_4$HCO$_3$	79.06	d 36		1 580	1.423 1.536 1.555	11.9		w powd
NH$_4$Cl	53.49	s 340	520	1 527	1.642	29.7	s alc, liq NH$_3$	I, MS w cub
(NH$_4$)$_2$CrO$_4$	152.08	d 180		1 910^{12}		40.5	ss ac	CSA, O, y monocl
(NH$_4$)$_2$Cr$_2$O$_7$	252.06	d 170		2 150		30.8	s alc	CSA, O, or
NH$_4$F	37.04	s		1 000		100	s alc	MS w hex
NH$_4$OH	35.05		soln only at 25 °C					C, T
NH$_4$I	144.94	s 551	220 (vac)	2 514	1.7031	154.2^0	vs alc, ac, NH$_3$	MS, I, w cub
(NH$_4$)$_2$Fe(SO$_4$)$_2$ · 6H$_2$O	392.14	d 100		1 864^{20}	1.487 1.492 1.499	26.9^{20}		I, LS, gn monocl
NH$_4$Fe(SO$_4$)$_2$ · 12H$_2$O	482.19	39–41	–H$_2$O, 230	1 710	1.485 4	124.0^{25}		I, viol cub oct
(NH$_4$)$_2$MoO$_4$	196.01	d		2 276		d		w monocl pr
NH$_4$NO$_3$	80.04	169.6	d	1 725		118.3^0	s alc, ac	I, O, w rhomb
(NH$_4$)$_2$C$_2$O$_4$ · H$_2$O	142.11	d		1 500	1.439 1.546 1.594	2.54^0		I, w rhomb
(NH$_4$)$_2$HPO$_4$	132.05	d 155		1 619	1.52	57.5^{10}		w monocl
NH$_4$H$_2$PO$_4$	115.03	190		1 803^{19}	1.525 1.479	22.7^0		w tetr
(NH$_4$)$_2$SO$_4$	132.14	d 235		1 769^{50}	1.521 1.523 1.533	70.6^0		w rhomb
(NH$_4$)$_2$S$_2$O$_8$	228.18	d 120		1 982	1.498 1.502 1.587	58.2^0		O, MS, w monocl
(NH$_4$)$_2$S	68.14	d				vs	s alc, vs NH$_3$	MS, y cr
(NH$_4$)$_2$S$_5$	196.40	d 115				vs	s alc	y, pr
NH$_4$SCN	76.12	149.6	d 170	1 305		128^0	s alc, ac NH$_3$	MS, w monocl

ANTIMONY

Compound/formula	M_r	m.p. /°C	b.p. /°C	ρ / kg m^{-3}	n_D	Solubility Water	Solubility Other solvents	Hazards and general description
Sb	121.757	630.755	1 635	6 691		i	s hot conc H$_2$SO$_4$, aq reg	silv metal, hex
SbBr$_3$	361.48	96.6	280	4 148	1.74	d	s HCl, HBr, alc, ac	MS, w rhomb
SbCl$_3$	228.11	73.4	283	3 140		d	s alc, HCl, chl	C, MS, w rhomb
SbCl$_5$	299.02	2.8	79$^{22\,mm}$	2 336^{20} (l)	1.601^{14}	d	s HCl, chl	C, MS, w liq or monocl
SbF$_3$	178.75	292	s 319	4 379^{21}		384.7^0		C, T, w rhomb
SbF$_5$	216.74	7	149.5	2 990^{23} (l)		s	s KF	C, MS, w oily liq
SbH$_3$	124.77	–88	–17.1	4 360^{15} (g)		0.41^0		FG
SbI$_3$	502.46	170	401	4 917^{17}	2.78	d	s CS$_2$, bz,	
SbI$_5$	756.27	79	400.6					br
Sb$_2$O$_3$	291.5	656	s 1 550	5 200	2.087	vss	s KOH, HCl	I, w cub
Sb$_2$O$_4$ (Sb$_2$O$_3$, Sb$_2$O$_5$)	305.5	d 930		5 820	2.00	vss	vss KOH, HCl	I, w powd
Sb$_2$O$_5$	323.50	d 380		3 780		i	vss KOH, HCl	I, O, y powd
Sb$_2$S$_3$	339.69	550	c. 1 150	4 640	3.194 4.064 4.303	i	s NH$_4$HS	T, blk, rhomb

ARGON

Compound/formula	M_r	m.p. /°C	b.p. /°C	ρ / kg m^{-3}	n_D	Solubility Water	Solubility Other solvents	Hazards and general description
Ar	39.948	–189.2	–185.856	1.783 7^0	1.40^{-186} 1.000 281^0	5 cm^3 ^{0}		col inert gas

(Continued)

Table XI.1 (Continued)

Compound/formula	M_r	m. p. /°C	b. p. /°C	ρ /kg m^{-3}	n_D	Solubility Water	Solubility Other solvents	Hazards and general description
ARSENIC								
As	74.922	817 (28 atm)	s 613	5 727		i	s HNO$_3$	T, gy metal
H$_3$AsO$_4$. 0.5H$_2$O	150.95	35.5	–H$_2$O, 230	2 000–2 500		17	s alc, alk	T, CSA, w cr
AsBr$_3$	314.65	32.8	221	3 540		d	s HBr, HCl	T, MS, w–y pr
AsCl$_3$	181.28	–8.5	130.2	2 163^{20}		d	s HCl, HBr, alc, et	T, MS, oily liq
AsF$_3$	131.92	–8.5	63	2 666		d	s alc, et, bz	T, MS, oily liq
AsF$_5$	169.91	–80	–53	7.71 (g)		d	s alc, et, bz	T, MS col gas
AsH$_3$	77.95	–116.3	–55	2.695 (g)		0.07	s chl, bz	T, col gas
As$_2$O$_3$	197.84	315	459	3 738		3.7^{20}	s alk, HCl	CSA, w amorph
As$_2$O$_5$	229.84	d 315		4 320		150^{16}	s a, alk, alc	T, CSA, w amorph
As$_2$S$_3$	246.04	300	707	3 430	2.4, 2.81	5 × 10^{-5}	s a, alk	T, CSA, y monocl
BARIUM								
Ba	127.327	729	1 637	3 510^{20}		d	s alc	FS, MS, w metal
BaBr$_2$	297.16	850	d	4 781^{24}	1.75	104^{20}	vs alc, MeOH	T, w cr
BaBr$_2$. 2H$_2$O	333.19		–2H$_2$O 120	3 580^{24}	1.713 1.727 1.744	151^{20}	vs MeOH, s alc	w cr
BaCO$_3$	197.35	d 1 350		4 430		0.002^{20}	a a	T, I, w hex
BaCl$_2$	208.25	963	1 560	3 856^{24}	1.730 3 1.736 7 1.742 0	37.5	ss HCl	T, w monocl
BaCl$_2$. 2H$_2$O	244.28	–	–2H$_2$O 113	3 097^{24}	1.629 1.642 1.658	37.5	ss HCl	T, w monocl
BaCrO$_4$	253.33			4 498^{15}		0.000 34^{16}	s a	T, y rhomb
BaF$_2$	175.34	1 355	2 137	4 890	1.474 1	0.12	s a	T, w cub
Ba(OH)$_2$. 8H$_2$O	315.48	78	–8H$_2$O 780	2 180	1.471 1.502 1.50	5.6^{15}	ss alc	T, C, w monocl
BaI$_2$. 2H$_2$O	427.18	–H$_2$O, 98	–2H$_2$O, 539 d 740	5 150		200^{15}	s alc, ac	MS, LS, w rhomb
Ba(NO$_3$)$_2$	261.35	d 592		3 240^{22}	1.572	8.7^{20}		O, T, w cub
BaO	153.34	1 923	c. 2 000	5 720	1.98	3.48^{20}	s a, alc	C, MS, w–y powd
BaO$_2$	169.34	450	–O$_2$ 800	4 960		vss	s a	T, O, w–gy powd
BaSO$_4$	233.40	1580		4 500^{15}	1.637 1.638 1.649	0.000 22^{18}		w rhomb
BERYLLIUM								
Be	9.012	1 278	2 970	1 850^{20}		i	s a, alk	T, CSA, gy metal
BeBr$_2$	168.83	s 490		3 465		s	s alc, et	T, MS, w need
BeCl$_2$	79.92	405	520	1 899		72	vs alc, et, pyr	T, MS, w need
BeF$_2$	47.01	s 800		1 986		∞		T, w, amorph
Be(OH)$_2$	43.0	d 138				i		T
BeI$_2$	262.82	510	590	4 325		d	s alc, et	T, MS
Be(NO$_3$)$_2$3H$_2$O	187.07	60	142	1 557		vs	vs alc	T, MS, w–y cr
BeO	25.01	2 530±30	c. 3 900	3 010	1.719 1.733	i	s conc. H$_2$SO$_4$	T, CSA, w hex
BeSO$_4$. 4H$_2$O	177.14	–2H$_2$O, 100	–4H$_2$O, 400	1 713^{10}	1.472 1.440	42.5		T, CSA, w tetr
BISMUTH								
Bi	208.980	271.442	1 551	9 747		i	s h H$_2$SO$_4$	FP, w–r metal rhomb

(Continued)

Table XI.1 (Continued)

Compound/formula	M_r	m. p. / °C	b. p. / °C	ρ / kg m^{-3}	n_D	Solubility Water	Solubility Other solvents	Hazards and general description
BiBr$_3$	448.71	218	453	5 720		d to BiOBr	s HCl, HBr, et	C, MS, y powd
BiCl$_3$	315.34	230–232	427	4 750		d to BiOCl	s a, alc, et	C, MS, w cr
Bi(OH)$_3$	260.0	d 100		4 360		0.000 14	s a	w amorph powd
BiI$_3$	589.69	408	d 500	5 778[15]		i	s alc	r hex
Bi(NO$_3$)$_3$. 5H$_2$O	485.07	d 30		2 830		d	vs HNO$_3$ s HCl	O, I, w triclin
Bi$_2$O$_3$	465.96	820	1 900	8 900		i	s a	y rhomb
BiOCl	260.43	red heat		7 720[15]	2.15	i	s a	I, w powd
Bi$_2$S$_3$	514.15	d 685		7 390	1.340 1.456 1.459	i	s HNO$_3$	br–blk powd
BORON								
B	10.811	2 027	3 802 s	2 340		i		FP, I, y monocl, or br amorph powd
H$_3$BO$_3$	61.83	d 169 to HBO$_2$		1 435[15]	1.337 1.461 1.462	6.35[20]	s alc, et	MS, w tricl in
BBr$_3$	250.54	−46	91.3	2 643.1[18]	1.531 2	d	s alc, ct	C, T, MS, fuming liq
B$_4$C	55.26	2350	> 3 500	2 520		i	s fused alk	blk rhbdr
BCl$_3$	117.17	−107.3	12.5	1 349[11](l)	1.419 5[6]	d		C, T, MS, fuming liq
BF$_3$	67.81	−126.7	−99.9	2.99 (g)		0.24		T, col gas
B$_2$H$_6$	27.67	−165.5	−92.5	447^{-11}(l)		d to H$_3$BO$_3$ + H$_2$		FG, col gas
B$_4$H$_{10}$	53.32	−120.8	16	560^{-35}(l)		d slow	s bz	T, col gas
B$_5$H$_9$	63.13	−46.8	58.4	660[0]		d		FG, col liq
B$_6$H$_{10}$	74.95	−65	0 (72 mm)	690		d		col liq
B$_{10}$H$_{14}$	122.22	99.5	213	940		ss	s alc, et, bz	w cr
BI$_3$	391.52	49.9	210	3 350[50]		d	vs bz, ct	MS, w pl
BN	24.82	s c.3 000		2 250		i		w hex
B$_2$O$_3$	69.62	460	c. 1 860	2 460	1.64 1.61	ss		MS, rhomb cr
B$_2$S$_3$	117.81	310		1 550		d		w cr
BROMINE								
Br$_2$	159.808	−7.2	58.78	3 119[20]	1.661	3.58[20]	vs alc, et, chl, CS$_2$	T, O, C, dark r liq
BrCl	115.36	−66	d 10			d	s et, CS$_2$	C, T, MS, r liq or gas
BrF	98.91	d −33						C, T, r–br gas
BrF$_3$	136.90	8.8	135	2 490[135]		d violent		C, T, MS, y liq
BrF$_5$	174.90	−61.3	40.5	2 466		d		C, T, MS, col liq
BrO$_2$	111.91	d 0						y solid unstable from −40°C
CADMIUM								
Cd	112.411	321.108	765	8 650		i	s a, h, H$_2$SO$_4$	FS, CSA, w malleable metal
CdBr$_2$	272.22	567	863	5 192		57[10]	s alc, et, HCl	T, CSA, y cry
CdBr$_2$. 4H$_2$O	344.28	d 36				121[10]	s alc, ac	T, CSA, w need
CdCO$_3$	172.41	d < 500		4 258[4]		i	s a, KCN	T, CSA, w trig
CdCl$_2$	183.32	568	960	4 047		140[20]	s alc, MeOH	T, CSA, w hex
CdCl$_2$. 2 . 5H$_2$O	228.35	d 34		3 327	1.651 3	168[20]	s MeOH	T, CSA, w monocl
CdF$_2$	150.40	1 100	1 758	6 640	1.56	4.35	s a, HF	T, CSA, w cub
Cd(OH)$_2$	146.41	d 300		4 790[15]		0.000 26	s a	T, CSA, w trig
CdI$_2$	366.21	387	796	5 670[20]		86.2	s alc, ac	T, CSA, gn–y powd
Cd(NO$_3$)$_2$. 4H$_2$O	308.47	59.4	132	2 455[17]		215	s alc	T, CSA, O, w powd

(Continued)

Table XI.1 (Continued)

Compound/formula	M_r	m. p. /°C	b. p. /°C	ρ /kg m^{-3}	n_D	Solubility Water	Solubility Other solvents	Hazards and general description
CdO	128.40	d 900		8 150	2.49	i	s a	T, CSA, br cub
CdS	144.46	1 750 (100 atm)	s 980 (in N$_2$)	4 820	2.506 2.529	0.000 13[18]	s a	T, CSA, y–or hex
CdSO$_4$	208.46	1000		4 691[20]		75.5		T, CSA, w rhomb
CdSO$_4$.H$_2$O	226.48	d 108		3 790[20]		s		T, CSA, MS, monocl
CdSO$_4$.7H$_2$O	334.57	d 4		2 480		s		T, CSA, w monocl
3CdSO$_4$.8H$_2$O	769.51	d 41		3 090	1.565	113[0]		T, CSA, w monocl
CAESIUM								
Cs	132.905	28.4	666	1.8785[15]		d	s liq NH$_3$	FS, MS, w metal hex
CsBr	212.81	636	1 300	4 400	1.698 4	124.3	s a	MS, w cub
Cs$_2$CO$_3$	325.82	d 610				260.5[15]	s a, alc, et	MS, w cr
CsCl	168.36	645	1 290	3 988	1.641 8	162.2[1]	s alc, MeOH	MS, w cub
CsF	151.90	682	1 251	4 115	1.478	367[18]	s MeOH	MS, I, w cub
CsH	133.91	d		3 410		d	d a	MS, w cr
CsOH	149.91	272.3		3 675		395.5[18]	s alc	C, MS, w to y
CsI	259.81	626	1 280	4 510	1.787 6	44[0]	s alc	MS, w rhomb
CsNO$_3$	194.91	414	d	3 685	1.55 1.56	9.16[0]	s ac	MS, w hex or cub
Cs$_2$O	281.81	d 400		4 250		d	s a	MS, or need
Cs$_2$SO$_4$	361.87	1 010		4 243	1.560 1.564 1.566	167[0]		MS, w rhomb
CALCIUM								
Ca	40.078	839	1 492	1 540		d to H$_2$	s a, liq NH$_3$	FS, MS, w soft metal
Ca(OOCCH$_3$)$_2$	158.17	d			1.55 1.56 1.57	37.4[0]	ss alc	MS, w cr
CaBr$_2$	199.90	sl d 730	806 – 812	3 353		142[20]	s alc, ac, a	MS, w rhomb
CaC$_2$	64.10	2 300		2 220	1.75	d		FS, C, MS, w tetr
CaCO$_3$	100.00	d 825		2 930	1.530 1.681 1.685	0.001 5	s a	w rhomb
CaCl$_2$	110.99	782	>1 600	2 150	1.52	74.5[20]	s alc, ac, a	I, MS, w cub
CaCl$_2$.6H$_2$O	219.08	– 4H$_2$O, 30	– 6H$_2$O, 200	1 710	1.417 1.393	279[0]	s alc	I, w trig
Ca(ClO)$_2$. CaCl$_2$. xCa(OH)$_2$	Composition varies	d				d	d a	I, O, MS, w powd smells of Chlorine
CaCrO$_4$.2H$_2$O	192.09	– 2H$_2$O, 200				16.3[20]	s a, alc	y monocl pr
CaCN$_2$	80.10	s >1 150				d to NH$_3$		MS, T, y hex
CaF$_2$	78.08	1 423	c. 2 500	3 180	1.434	0.001 6[18]		I, MS, w cub
CaH$_2$	42.10	d 600		1 900		d to H$_2$	d a	I, MS, C, w rhomb cr
Ca(OH)$_2$	74.10	–H$_2$O, 580		2 240	1.574 1.545	0.185[0]	s a	I, w hex
CaI$_2$	293.89	784	c. 1 100	3 956		209[20]	s MeOH, alc, ac, a	MS, y–w hex
CaI$_2$.6H$_2$O	401.98	d 42		2 550		757[0]	s a, alc	y hex need
Ca(NO$_3$)$_2$	164.09	561		2 504[18]		121[18]	s a, alc, MeOH	O, I, MS, w cub
Ca(NO$_3$)$_2$.3H$_2$O	218.14	51.1						O, I, w tricl
Ca(NO$_3$)$_2$.4H$_2$O	236.15	42.7	d 132	1896	1.465 1.498 1.504	266[0]	s alc	O, I, MS, w monocl
CaC$_2$O$_4$	128.10	d		2 200[4]		0.000 67[13]	s a	T, w cub
CaO	56.08	2 580	2 850	3 300	1.838	0.131[10] d	s a	MS, I, w cub
Ca$_3$(PO$_4$)$_2$	310.18	1 670		3 140	1.629 1.626	0.002	s a	w amorph powd

(Continued)

Table XI.1 (Continued)

Compound/formula	M_r	m.p. / °C	b.p. / °C	ρ / kg m^{-3}	n_D	Solubility Water	Solubility Other solvents	Hazards and general description
CaSO$_4$	136.14	1 450		2 960	1.575 1.569 1.613	0.209^{30}	s a	MS, w rhomb or monocl
CaSO$_4$2H$_2$O	172.17	−1.5 H$_2$O, 128	−2 H$_2$O, 163	2 320	1.521 1.523 1.530	0.241	s a	MS, w cub
CaS	72.14	d		2 500	2.137	0.021^{15} d	d a	MS, T, w cub
CARBON								
C diamond	12.011			3 510	2.417 3	i		w cub
graphite	12.011	3 827	4 200	2 250		i		blk hex
CBr$_4$	331.65	90.1	189.5	3 420		0.024^{20}	s alc, et, chl	I, w monocl
CCl$_4$	153.81	−23	76.8	1 586.7	1.460 1	0.08	s alc, bz, et, chl	CSA, T, col liq
CF$_4$	87.99	−184	−128	1 960^{-184}		0.001 8		T, col gas
CI$_4$	519.63	d 171		4 340^{20}		i	s alc, CS$_2$ et, bz, MeOH	dk r cub
CO$_2$	44.01	−56.6$^{5.2\,at}$	s−78.51	1.977^0	1.000 45^0	171 cm^{30}	s alc, ac	col gas
CO	28.01	−199	−191.47	1.250^0	1.000 34^0	3.5 cm^{30}	s alc, bz, ac	T, P col gas
C$_3$O$_2$	68.03	−111.3	7	1 140^0	1.438	d		T, col gas or liq
COS	60.07	−138.2	−50.2	1.073^0		54 cm^{30}	s alc, CS$_2$	T, FG, P
CS$_2$	76.14	−110.8	46.3	1 261^{22}	1.629 5^{18}	0.22^{22}	s alc, et	T, FL, w liq
COBr$_2$	187.83		64.5			d		T, MS, w liq
COCl$_2$	98.92	−104	8.3	1 392		d	s bz,	T, MS, col gas
CERIUM								
Ce	140.115	799	3 257	6 773		sl d	s dilute a	FP, MS, gy metal, cub or hex I
CeCl$_3$	246.48	848	1 727	3 920^0		100	s alc, ac	MS, w cr
Ce(NO$_3$)$_3$.6H$_2$O	434.23	−3H$_2$O 150	d 200			vs	s alc, ac	I, O, r cr
Ce$_2$O$_3$	328.24	1690		6 860		i	s H$_2$SO$_4$	gy–gn trig
CeO$_2$	172.12	c. 260 0		7 132^{23}		i	s H$_2$SO$_4$	br–w cub
Ce(SO$_4$)$_2$	332.24	d 195		3 910^{15}		d to basic salts		y cry
CHLORINE								
Cl$_2$	70.905	−100.98	−34.6	3.214^0	1.000 768 (g) 1.367 (l)	310 cm^{310}	s alk	T, O, I, gn–y gas
HClO$_3$.7H$_2$O	210.57	< −20	d 40	1 282^{14}		vs		known only as col soln
HClO$_4$	100.46	−112	39 (56 mm)	1 764^{22}		∞		C, O, col unstable liq
HClO$_4$.H$_2$O	118.47	50	explodes 110	1 880		vs	s alc	C, O, fairly stable, need
HClO$_4$.2H$_2$O	136.49	−17.8	200	1 650		vs	s alc	C, O, stable liq
ClF	54.45	−154	−100.8	1 620^{-100}		d		T, MS, I, col gas
ClF$_3$	92.45	−83	11	1.77^{13}		d		T, MS, I, col gas
Cl$_2$O	86.91	−20	explodes 3.8	3.89^0		200 cm^3	s alk	T, MS, I, y–r gas, r–br liq
ClO$_2$	67.45	−59.5	explodes 10	3.09^{11}		2000 cm^{34}	s alk	T, MS, I, E, y–r gas
Cl$_2$O$_7$	182.9	−91.5	82			d	s bz	T, O, I, col oil
HSO$_3$Cl	116.52	−80	158	1 766^{18}	1.433 1	d to H$_2$SO$_4$ + HCl		T, C, MS, col fum liq
CHROMIUM								
Cr	51.996	1 875	2 482	7 200^{28}		i	s dil H$_2$SO$_4$, HCl	I, gy cub hard metal
Cr(CO)$_6$	220.06	d 110	explodes 210	1 770		i		E, T, w orthorhomb
CrCl$_2$	122.90	824		2 878		vs		MS, I, w need
CrCl$_3$	158.35	c. 1 150	s 1 300	2 760^{15}		i		I, MS, T, viol trig

(Continued)

Table XI.1 (Continued)

Compound/formula	M_r	m. p. /°C	b. p. /°C	ρ /kg m^{-3}	n_D	Solubility Water	Solubility Other solvents	Hazards and general description
CrCl$_3$.6H$_2$O	266.45	83		1 760		58.5	s alc	I, MS, T, viol monocl
Cr(NO$_3$)$_3$.9H$_2$O	400.15	60	d 100			s	s alc, alk, ac	CSA, O, purple monocl
CrO	68.00					i		blk powd
Cr$_2$O$_3$	151.99	2 435	4 000	5 210	2.551	i		CSA, MS, gn hex
CrO$_2$	84.00	d 300				i	s HNO$_3$	CSA, br–blk powd
CrO$_3$	99.99	195	d	2 700		61.7^0	s H$_2$SO$_4$	CSA, O, MS, r rhomb
CrO$_2$Cl$_2$	154.9	−96.5	117	1 911		d	d alc s et, ac	O, C, MS, dk r liq
Cr$_2$(SO$_4$)$_3$.15H$_2$O	662.41	−10 H$_2$O, 100		1 867^{17}		s		viol amorph
Cr$_2$S$_3$	200.18	−S, 1 350		3 770^{19}		i	s HNO$_3$	CSA, br–blk powd
COBALT								
Co	58.933	1 495	2 956	8 900		i	s a	CSA, FP, gy metal, cub
Co(OOCCH$_3$)$_2$.4H$_2$O	249.08	−4H$_2$O 140		1 705^{19}	1.542	s	s a, alc	I, MS, r–viol monocl
CoBr$_2$	218.75	678 in N$_2$		4 909		66.7^{59}	s alc, MeOH, et, ac	MS, gn hex
CoBr$_2$.6H$_2$O	326.84	47–48	−6H$_2$O 130	2 460		119	s alc, et	MS, r–viol pr
CoCO$_3$	118.94	d		4 130	1.855 1.60	i	s a	I, r trig
CoCl$_2$	129.84	724 in HCl gas	1 049	3 356		45^7	s alc, ac MeOH	T, MS, bl hex
CoCl$_2$.6H$_2$O	237.93	86	−6H$_2$O 110	1 924		76.7^0	s alc, ac, et	T, MS, r monocl
CoF$_2$	96.93	c. 1 200	1 400	4 460		1.5		I, MS, p monocl
CoF$_3$	115.93			3 880		d		O, MS, br hex
Co(OH)$_2$	92.95	d		3 597^{15}		0.000 32		r rhomb
CoI$_2$	312.74	515 vac	570 vac	5 860		159^0	s alc, ac	MS, blk hex
CoI$_2$.6H$_2$O	420.83	d 27		2 900		s	s alc, et, chl	MS, r hex
Co(NO$_3$)$_2$.6H$_2$O	291.04	55–56	−3H$_2$O 56	1 870	1.52	133.8^0	s alc, ac	O, I, r monocl
CoO	74.93	1 935		6 450		i	s a	blk cub
Co$_2$O$_3$	165.86	d 895		5 180		i	s a	MS, blk hex or rhomb
CoSO$_4$	155.00	d 735		3 710		36^{20}	s MeOH	I, MS, bl cub
CoSO$_4$.6H$_2$O	263.09	−2H$_2$O 95		2 019^{15}	1.531 1.549 1.552	s		I, MS, r monocl
CoS	91.00	>1 116		5 450^{18}		0.000 38^{18} ss a		I, r oct
Co(SCN)$_2$.3H$_2$O	229.14	−3H$_2$O, 105				s	s alc, MeOH	T, MS, viol rhomb
COPPER								
Cu	63.546	1 084.88	2 582	8 960		i	s HNO$_3$, h H$_2$SO$_4$	FP, I, r metal, cub
Cu(OOCCH$_3$)$_2$.H$_2$O	199.65	115	d 240	1 882	1.545 1.550	7.2	s alc, et	I, T, gn powd
CuBr	143.45	492	1 345	4 980	2.116	i	s HBr, HCl	LS, w cub
CuBr$_2$	223.31	498		4 770		126	s alc, ac, pyr, NH$_3$	I, MS, blk monocl
CuCl	98.99	430	1 490	4 140	1.93	0.006 2	s HCl, et	T, MS, w cub
CuCl$_2$	134.45	620	d 993	3 386		70^0	s alc, MeOH	T, I, MS, br–y powd
CuCl$_2$.2H$_2$O	170.47	−2H$_2$O 100	d	2 540	1.644 1.683 1.731	110^4	s alc	T, I, MS, bl–gn rhomb
CuCN	89.56	473 in N$_2$	d	2 920		i	s HCl, KCN, NH$_4$OH	T, w monocl powd
CuF	82.54	908	s 1100			i	s HCl, HF	MS, r cry
CuF$_2$	101.54	d 950		4 230		4.7^{20}	s a	MS, w tricl
Cu(OH)$_2$	97.56	d		3 368		i	s a, KCN, NH$_4$OH	bl powd
CuI	190.44	605	1 290	5 620	2.346	0.000 8^{18}	s HCl, KCN, KI	LS, T, w cub

(Continued)

Table XI.1 (Continued)

Compound/formula	M_r	m. p. /°C	b. p. /°C	ρ / kg m^{-3}	n_D	Solubility Water	Solubility Other solvents	Hazards and general description
Cu(NO$_3$)$_2$·6H$_2$O	295.64	−3H$_2$O 26		2 074		243.7^0	s alc	O, I, T, MS, bl cr
Cu$_2$O	143.08	1 235	−O$_2$ 1 800	6 000	2.705	i	s HCl, NH$_4$OH	MS, T, r oct cub
CuO	79.54	1 326		6 315	2.63	i	s a	I, blk monocl
CuSO$_4$	159.60	sl d > 200	d 650 to CuO	3 603	1.733	14.3^0	ss MeOH	T, I, MS, w powd
CuSO$_4$·5H$_2$O	249.68	−4H$_2$O, 110	−5H$_2$O, 150	2 284	1.514 1.537 1.543	31.6^0	s MeOH	T, I, bl tricl
Cu$_2$S	159.14	1 100		5 600		10^{-14}	s HNO$_3$	T, I, blk rhomb
CuS	95.6	d 220		4 600	1.45	0.000 033^{18}	s HCl, KCN	T, blk monocl
CuSCN	121.62	1 084		2 843		0.000 5^{18}	s NH$_4$OH	T, w
DEUTERIUM								
D$_2$	4.032	−254.6	−249.7	169$^{-250.9}$ (l)		ss		col gas
DCl	37.47	−114.8	−81.6			11.9 cm^{3,25}		T, C, gas
D$_2$O	20.031	3.82	101.42	1 105^{20}	1.338 44^{20}			col liq
FLUORINE								
F$_2$	37.997	−219.6	−188.1	1.69^{15}	1.000 195	d		MS, T, C, g–y gas
F$_2$O	54.00	−223.8	−144.8	1650^{-190}		d		T, C, MS, col gas
GALLIUM								
Ga	69.723	29.8	2 237	5 904$^{29.6}$ (s) 6 095$^{29.8}$ (l)		i	s a	MS, gy–blk metal
GaBr$_3$	309.45	121.5	278.8	3 690		s		w cr
GaCl$_2$	140.63	164	535			d	s bz	MS, w cr
GaCl$_3$	176.03	77.9	201.3	2 470		vs	s bz, CS$_2$	MS, w cr
GaF$_3$	126.72	s 800	c. 1000	4 470		0.002	s HF	w powd
Ga(OH)$_2$	120.74	d 440				i	s a	w powd
GaI$_3$	450.43	212	s 345	4 150		d		MS, y–w cr
Ga(NO$_3$)$_3$·xH$_2$O	255.74	d 100				vs	s alc	O, I, MS, w cr
GaN	83.73	s 800		6 100		i		g powd
Ga$_2$O$_3$	187.44	1900		6 440	1.92 1.85	i	s alk	w hex rhomb
Ga$_2$O	155.44	> 600	s >500	4 770		i	s a, alk	br powd
Ga$_2$(SO$_4$)$_3$	427.63	d 690				vs	s alc	MS, w powd
Ga$_2$S$_3$	235.63	1245–1265		3 650		d	s a, alk	MS, y cr
GERMANIUM								
Ge	72.61	937	2830	5 350^{20}		i	s h H$_2$SO$_4$	gy metal, cub
GeCl$_2$	143.50	d 450				d		MS, w powd
GeCl$_4$	214.41	−49.5	84	1 844^{30}	1.464	d	s alc, et, HCl	MS, C, L, col liq
GeH$_4$	76.72	−165	−88.5	1 523^{-142}		i	s liq NH$_3$	T, col gas
Ge$_2$H$_6$	151.23	−109	29	1 980^{-100}		d	s liq NH$_3$	T, MS, col gas
GeO	88.59	s 710			1.607	i	s Cl$_2$ water	blk powd
GeO$_2$	104.59	1 086		6 239		i		w tetr
GeS	104.65	530	s 430	4 010		0.24	s HCl, alk	y–r amorph
GeS$_2$	136.72	c. 800	s>600	2 940^{14}		0.45	s alk	MS, T, w powd
GOLD								
Au	196.967	1064.43	2967	19 320		i	s aq reg	y ductile metal, cub
AuCl	232.42	d 170		7 400		ss	s HCl, HBr	y cr
AuCl$_3$	303.33	d 254		3 900		68	s alc, et	T, r cr
HAuCl$_4$·xH$_2$O	339.79			3 900		s		C, MS
AuCN	222.98	d		7 120		ss	s KCN	T, y powd
Au$_2$O$_3$	441.93	−O$_2$ 160	−3O$_2$ 250			i	s HCl, HNO$_3$, NaCN	
HELIUM								
He	4.003	−269.65	−268.93	0.178 5^0 (g) 147.0^{-270} (l)	1.000 036^0	0.94 cm^3		col inert gas

(Continued)

Table XI.1 (Continued)

Compound/formula	M_r	m. p. / °C	b. p. / °C	ρ / kg m^{-3}	n_D	Solubility Water	Solubility Other solvents	Hazards and general description
HYDROGEN								
H_2	2.016	−258.14	−252.87	0.089 9 (g) 70.0 (1)	1.000 132^0	1.91 cm^3		FG, T, col gas
HBr	80.92	−88.5	−67	3.5^0(g) 2.770^{-67}(1)	1.325 (1)	221^0	s alc	T, C, col gas, y liq
HBr.H_2O	98.93		−3.6	1 780		s		T, C, col liq
HCl	36.46	−114.8	−84.9	1.000 45 (g) 1 187^{-85}(1)		82.3^0	s alc, et, bz	T, P, C, col gas or liq
HCN	27.03	−14	26	0.901 (g) 699^{22} (1)	1.267 5 ^{10}	vs	s alc, et	T, P, col gas or liq
HF	20.01	−83.1	19.54	0.991 (g)		vs		C, T, col fuming liq or gas.
HI	127.91	−50.8	−35.38	5.66^0 (g) 2 850^{-47}(1)		42.5	s alc	C, T, col gas or y liq
H_2O	18.015	0.00	100.00	1 000^0	1.333		s alc	col liq or hex cr
H_2O_2	34.01	−0.41	150.2	1 442.2	1.414	∞	s alc, et	C, O, col liq
H_3P	34.00	−133.5	−87.4	1. 529 (g) 746.0^{-90}(1)	1.317 (1)	26 cm^3^{17}	s alc, et	FG, P, T, col gas
H_2S	34.08	−85.5	−60.7	1.539^0 (g)	1.000 644^0 1.374 (1)	437 cm^{30}	s alc, CS_2	P, FG, T, col gas
H_2Te	129.62	− 49	− 2	5.81 (g) 2 570^{-20} (1)		2	s alc, alk	P, T, col gas
INDIUM								
In	114.818	156.634	2 006	7 300^{20}		i	s a	FP, I, soft w metal
InCl	150.27	225	608	4 180		d	s a	T, I, MS, r
$InCl_3$	221.18	s 300		3 460		vs	ss alc, et	T, I, MS, w pl
$In(NO_3)_3.3H_2O$	354.88	−2H_2O 100	d			vs	s alc	O, I, MS, w, pl
In_2O_3	277.64		volatizes 850	7 179		i	s a	I, y amorph
IODINE								
I_2	253.809	113.5	184.35	4 953	3.34	0.034	s alc, et, bz, KI, ct, CS_2	C, L, viol–blk met lustre
HIO_3	175.91	d 110		4 629^0		310^{16}	HNO_3	C, O, I, w–y rhomb
$HIO_4 . 2H_2O$	227.94	122	d 140			113	s alc, et	O, C, MS, w monocl
IBr	206.81	42–50	d 116	4 416^0		d	s alc, et, chl, CS_2	C, MS, gy cr
ICl	162.36	27.2	97.4	3 182^0		d	s alc, et, CS_2, HCl	C, MS, r need, r–br oil
ICl_3	233.26	101 (16 at)	d 77	3 117^{15}		d	s alc, et, ct, bz	C, MS, y–br rhomb, r liq
IF_5	221.90	9.6	98	3 750		d	d alc	C, MS, col liq
IF_7	259.89	5.5	s 4.5	2 800^6		d		C, MS, col, cr or liq
I_2O_5	333.81	d 300		4 799		187.4^{15}	s alc, et, chl, CS_2	O, I, w cr
IRON								
Fe	55.847	1 535	2 887	7 874		i	s a	FP, MS, silv metal, cub
$FeBr_2$	215.67	d 684		4 636		109^{10}	s alc	g hex
$FeBr_3$	295.57	d				s	s alc, et	MS, r
Fe_3C	179.55	1 837		7 694		i	s a	gy cub
$FeCO_3$	115.85	d		3 800	1.875 1.633	0.006 7	s a	gy trig
$Fe_2(CO)_9$	363.79	d 80		2 085^{18}		i	ss alc	T, y hex cr
$Fe(CO)_5$	195.90	− 21	102.8^{749mm}	1 457^{21}		i	s alc, et, bz, alk	FL, T, visc y liq
$Fe(CO)_4$	167.89	d 140		1 996^{18}		i	s org solv	T, gn cr
$FeCl_2$	126.75	670–674	s	3 160	1.567	64.4^{10}	s alc, ac	I, MS, gn–y hex
$FeCl_2.4H_2O$	198.81			1 930		160^{10}	s alc	I, MS, bl–gn monocl
$FeCl_3$	162.21	306	d 315	2 898		74.4^0	s alc, et, MeOH, ac	T, I, MS, br hex

(Continued)

Table XI.1 (Continued)

Compound/formula	M_r	m. p. /°C	b. p. /°C	ρ / kg m^{-3}	n_D	Solubility Water	Solubility Other solvents	Hazards and general description
FeCl$_3$.6H$_2$O	270.3	37	280–5			92^{20}	s alc, et	T, I, MS, br–y cr
Fe(OH)$_2$	89.86	d		3 400		0.000 15^{18}	s a	gn hex
Fe(OH)$_3$	106.90	d 500		3 900		i	s a	br powd
FeI$_2$	309.66	red heat		5 315		s	s alc, ac	I, MS, gy hex
Fe(NO$_3$)$_3$.9H$_2$O	404.02	47.2	d 125	1 684		s	s alc, ac	O, I, MS, viol monocl
Fe(–CO$_2$)$_2$.2H$_2$O	179.90	d 190		2 280		0.022	s a	T, I, y rhomb
FeO	71.85	1 420		5 700	2.32	i	s a	blk cub
Fe$_2$O$_3$	159.69	1 565		5 240	3.01	i	s HCl, H$_2$SO$_4$	r–blk powd
Fe$_3$O$_4$	231.54	d 1 538		5 180	2.42	i	s conc a	r–blk powd
FeSO$_4$.7H$_2$O	278.05	–6H$_2$O, 64	–7H$_2$O, 300	1 898	1.471 1.478 1.486	15.65	ss alc	bl–gn monocl
Fe$_2$(SO$_4$)$_3$	399.87	d 480		3 097^{18}	1.814	d		I, MS, y romb
FeS	87.91	1 195	d	4 740		0.000 6	s a	MS, T, blk hex
FeS$_2$ (Pyrite)	119.98	1 171		5 000		0.000 5	s a	I, y cub
KRYPTON								
Kr	83.80	–156.6	–152.9	3.736 (g) 2 155^{-153}(1)		6 cm^3		inert col gas
LEAD								
Pb	207.2 (1)	327.50	1 749	11 343.7^{16}	2.01	i	s HNO$_3$	FP, gy soft metal cub
Pb(OOCCH$_3$)$_2$	325.28	280	d	3 250^{20}		44.3^{20}		CSA, w cr
PbBr$_2$	367.01	373	916	6 660		0.84^{20}	s a, KBr	T, w rhomb
PbCO$_3$	267.20	d 315		6 600	1.804 2.076 2.078	0.000 1^{20}	s a, alk	T, w rhomb
PbCl$_2$	278.10	501	950	5 850	2.199 2.217 2.260	1.0^{20}	ss HCl	T, w rhomb
PbCl$_4$	349.00	–15	exp 105	3 180^0		d	s conc HCl	MS, T, y oil
PbCrO$_4$	323.18	844	d	6 120^{15}	2.31 2.37 2.66	i	s a, alk	T, y monocl
PbF$_2$	245.19	855	1 290	8 240		0.064^{20}	s HNO$_3$	T, P, C, w rhomb
Pb(OH)$_2$	241.20	d 145				0.016^{20}	s a, alk	T, w powd
PbI$_2$	461.00	402	954	6 160		0.063^{20}	s alk, KI	T, LS, P, y hex powd
Pb(NO$_3$)$_2$	331.20	d 470		4 530^{20}	1.782	60	s alc, alk	O, I, T, w cub
PbO	223.19	888		9 530		0.001 7^{20}	s HNO$_3$, alk	T, y tetr
PbO$_2$	239.19	d 290		9 375		i	s HCl	O, I, T, br tetr
Pb$_3$O$_4$	685.57	d 500		9 100		i	s HCl	O, I, T, r cr
PbSO$_4$	303.25	1 170		6 200	1.877 1.822 1.894	0.0045		T, C, w monocl or rhomb
PbS	239.25	1 114		7 500	3.921	i	s a	T, bl metallic cub
Pb(C$_2$H$_5$)$_4$	323.44	–136.8	d 200	1 659^{11}	1.519 5^{30}	i	s alc, bz, et	T, col liq
Pb(CH$_3$)$_4$	267.33	–27.5	110	1 995	1.512 0^{20}	i	s alc, bz	T, col liq
LITHIUM								
Li	6.941	180.5	1 317	534		d		FS, MS, T, C, w soft metal
Li(OOCCH$_3$).2H$_2$O	102.01	70	d		1.40	300^{15}	s alc	w rhomb
LiAlH$_4$	37.95	d 125		917		d	s et	FS, MS, T, w cr powd
LiBH$_4$	21.78	d 279		660		d	s et	FL, C, T,
LiBr	86.85	550	1 265	3 464	1.784	145^4	s alc, MeOH et,	MS, w cr
Li$_2$CO$_3$	73.89	618	d 1 310	2 110	1.428 1.567 1.572	1.54^0		C, w monocl
LiCl	42.39	605	1 347	2 068^{15}	1.662	63.7^0	s alc, MeOH, ac,	MS, T, w cub

(Continued)

Table XI.1 (Continued)

Compound/formula	M_r	m. p. /°C	b. p. /°C	ρ / kg m^{-3}	n_D	Solubility Water	Solubility Other solvents	Hazards and general description
LiF	25.94	845	1 676	2 635^{20}	1.391 5	0.27^{10}	s HF	T, MS, w cub
LiH	7.95	680	820			d		T, C, MS, FS, w cr
LiD	8.96	680	820			d		T, C, MS, FS
LiOH.H$_2$O	41.96			1 510	1.460 1.524	22.3^{10}	s a	C, MS, T, w monocl
LiI	133.84	449	1 180	3 494	1.955	165^{20}	s alc, MeOH, ac, NH$_4$OH	MS, w cub
LiI.3H$_2$O	187.89	−H$_2$O, 73	−3H$_2$O, 300	3 480		151^0	s alc, ac	LS, MS, w hex
LiNO$_3$	68.94	264	d 600	2 380	1.735	90^{27}	s alc, NH$_4$OH	O, I, w trig
Li$_3$N	34.82	d 840				d		MS, T, r amorph
Li$_2$O	29.88	>1 700		2 013	1.644	d		MS, C, T, w cub
Li$_2$SO$_4$	109.94	845		2 221	1.465	26.1^0		MS, w monocl
Li$_2$S	45.94	900–975		1 660		vs	vs alc	FS, MS, w cub
MAGNESIUM								
Mg	24.305	648.8	1 100	1 740^5		i	s min a	FS, MS, silver metal, hex
MgBr$_2$	184.13	711		3 720		101.5	s alc, MeOH	MS, w hex cr
MgBr$_2$.6H$_2$O	292.22	d 172		2 000		316^0	s alc, ac	MS, w hex pr
MgCO$_3$	84.32	d 350		2 958	1.717 1.515	0.01	s a	w trig
MgCl$_2$	95.22	714	1 412	2 320	1.675 1.59	54^{20}	s alc	MS, w hex cr
MgCl$_2$.6H$_2$O	203.31	d 118		1 569	1.495 1.507 1.528	167	s alc	MS, w monocl
MgF$_2$	62.31	1 261	2 239	3 000	1.378 1.390	0.007 6^{18}	s HNO$_3$	T, w tetr
Mg(OH)$_2$	58.33	−H$_2$O 350		2 360	1.559 1.580	0.000 9^{18}	s a	w hex pl
MgI$_2$	278.12	d <637		4 430		148^{18}	s alc, et	MS, w hex
Mg(NO$_3$)$_2$.6H$_2$O	256.41	89	d 330	1 636		125	s alc, liq NH$_3$	O, I, MS, w monocl
MgO	40.31	2 800	3 600	3 580	1.736	0.000 62	s a, NH$_4$ salts	MS, w cub
MgSO$_4$	120.37	d 112 4		2 660	1.56	20^0	s alc, et	MS, w rhomb cr
MgSO$_4$.7H$_2$O	246.48	−6H$_2$O, 150	−7H$_2$O, 200	1 670	1.433 1.455 1.461	71^{20}	ss alc	w rhomb or monocl
MgS	56.38	d >2 000		2 840	2.271	d	s a	r–br cub
MANGANESE								
Mn	54.938	1 244	2 100	7 473		d	s a	FP, MS, T, gy– metal cub
Mn(OOCCH$_3$)$_2$.4H$_2$O	245.08			1 589		s	s alc	p monocl
MnBr$_2$	214.76	d		4 385		127.3^0		MS, T, p cr
MnCO$_3$	114.95	d		3 125		0.006 5	s a	MS, p rhomb
MnCl$_2$	125.84	650	1 190	2 977		72.3	s alc	MS, I, p cub cr
MnCl$_2$.4H$_2$O	197.91	58	−4H$_2$O, 106	2 010		151^0	s alc	MS, I, p monocl
MnF$_2$	92.93	856		3 980		0.66^{40}	s a	T, I, r tetr or powd
MnF$_3$	111.93	d		3 540		d	s a	T, MS, r cr
Mn(OH)$_2$	88.95	d		3 258^{13}	1.723 1.681	0.000 2^{18}	s a, NH$_4$ salts	p trig
Mn(NO$_3$)$_2$.4H$_2$O	251.01	25.8	129.4	1 820		426.4^0	vs alc	O, I, p monocl
MnO	70.94	1 650	d	5 430	2.16	i	s a	gn powd
Mn$_3$O$_4$	228.81	1 705		4 856	2.46	i	s HCl	T, blk, tetr
Mn$_2$O$_3$	157.87	d 1080		4 500		i	s a	T, blk cub
MnO$_2$	86.94	d 535		5 026		i	s a	O, I, blk rhomb
Mn$_2$O$_7$	221.87	5.9	d 55	2 396^{20}		vs		MS, O, I, E, r oil
MnSO$_4$	151.00	700	d 850	3 250		52^5	s alc	MS, I, p
MnSO$_4$.7H$_2$O	277.11	−7H$_2$O, 280		2 090		172		MS, I, p monocl
MnS	87.00	d		3 990	2.70	0.000 47^{18}	s a	T, I, p amorph

(Continued)

Table XI.1 (Continued)

Compound/formula	M_r	m. p. / °C	b. p. / °C	ρ / kg m^{-3}	n_D	Solubility Water	Solubility Other solvents	Hazards and general description
MERCURY								
Hg	200.59 (2)	−38.836	356.66	13 593.9^{20}		i	s HNO$_3$	T, gy metal, liq
Hg(OOCCH$_3$)$_2$	318.76	d		3 270		25^{10}	s alc, ac, a	T, LS, MS, w powd
Hg$_2$Br$_2$	561.00	s 345		7 307		0.000 004	s a	T, C, w tetr
HgBr$_2$	360.41	236	322	6 109		0.61	s alc, MeOH	T, I, w rhomb
Hg$_2$Cl$_2$	472.09	s 400		7 150	1.973 2.656	0.002	s aq. req	T, I, w tetr
HgCl$_2$	271.5	276	302	5 440	1.859	6.9^{20}	s alc, et	T, C w powd
Hg(CN)$_2$	252.63	d		3 996	1.645	9.3^{14}	s alc, MeOH	T, MS, P, w powd
HgF$_2$	238.59	d 645		8 950^{15}		d	s HF	T, MS, w cub
HgI$_2$	454.40	259	354	6 094^{127}		ss	s et, KI	T, LS, y rhomb
Hg$_2$(NO$_3$)$_2$.2H$_2$O	561.22	70	d	4 790		d	s HNO$_3$	T, O, MS, w monocl
Hg(NO$_3$)$_2$.H$_2$O	342.61			4 300		s	s HNO$_3$	T, O, MS, w powd
Hg$_2$O	417.18	d 100		9 800		i	s HNO$_3$	T, blk powd
HgO	216.59	d 500		11 100^4	2.37 2.5 2.65	0.005	s a	T, LS, y–r rhomb
Hg$_2$SO$_4$	479.24	d		7 560		0.06	s HNO$_3$	T, w–y powd
HgSO$_4$	296.65	d 850		6 470		d	s a	T, LS, w powd
HgS α	232.65	s 583.5		8 100	2.854 3.201	0.000 001^{18}	s aq. reg, Na$_2$S	T, LS, r hex
HgS β	232.65	583.5		7 730		i	s aq. reg, Na$_2$S	T, LS, blk cub
MOLYBDENUM								
Mo	95.94 (1)	2 623	5 560	10 222		i	s h conc HNO$_3$, h. conc H$_2$SO$_4$, aq reg	FP, gy metal, cub
Mo(CO)$_6$	264.00	d 150		1 960		i	s bz	T, w rhomb
MoCl$_5$	273.21	194	268	2 928		d	a, a, chl, ct, liq NH$_3$	MS, T, I, gn–blk trig
MoF$_6$	209.93	17.5^{406mm}	35	2 550$^{17.5}$(l)		d	s alk	MS T, w cr
MoO$_2$	127.94			6 470		i	ss h conc H$_2$SO$_4$	T, I, gy tetr
MoO$_3$	143.94	795	s 1155	4 692^{21}		0.11^{18}	s a, alk	T, I, w rhomb
MoO$_3$.2H$_2$O[H$_2$MoO$_4$].H$_2$O	179.97	−H$_2$O, 70	d	3 124^{15}		0.133^{18}	s alk	T, I, y monocl
Mo$_2$S$_3$	288.07	d 1 100		5 910^{15}		i		T, I, gy need
MoS$_2$	160.07	1 185		5 060		i	s h conc HNO$_3$, aq reg	T, I, blk hex
NEON								
Ne	20.180	−248.67	−246.048	0.9002^0		1.47 cm$^3{}^{20}$		inert col gas
NICKEL								
Ni	58.693	1 455	2 782	8 902		i	s HNO$_3$	FP, CSA, gy metal, cub
Ni(OOCCH$_3$)$_2$.4H$_2$O	248.86	d		1 744		16	s a	T, CSA, gn powd
NiBr$_2$	218.53	963		5 098^{27}		112.8^0	s alc, et	I, CSA, MS, y–br
NiBr$_2$.3H$_2$O	272.57	−3H$_2$O, 300				199^0	s alc, et	CSA, I, MS, y–gn need
NiCO$_3$	118.72	d				0.009	s a	CSA, gn rhomb
Ni(CO)$_4$	170.75	−25	43	1 320^{17}		0.018^{10}	s alc, et, bz	CSA, FL, col liq
NiCl$_2$	129.62	1 001	s 973	3 550		64.2^{20}	s alc, NH$_4$OH	CSA, MS, I, y
NiCl$_2$.6H$_2$O	237.71				1.57	254^{20}	vs alc	CSA, I, gn monocl
NiF$_2$	96.71	s 1 000		4 630		4	s alk, a, et	CSA, T, gn tetr
Ni(OH)$_2$	92.72	d. 230		4 150		0.013	s a, NH$_4$OH	CSA, T, g cr
NiI$_2$	312.52	797		5 834		124.2^0	s alc	CSA, MS, blk cr
Ni(NO$_3$)$_2$.6H$_2$O	290.81	56.7	136.7	2 050		238.5^0	s alc, NH$_4$OH	CSA, O, MS, gn monocl
NiO	74.71	1 990		6 670	2.181 8	i	s a, NH$_4$OH	CSA, I, gn–blk cub
NiSO$_4$	154.78	d 848		3 680		29.3^0		CSA, y cub

(Continued)

Table XI.1 (Continued)

Compound/formula	M_r	m. p. / °C	b. p. / °C	ρ / kg m^{-3}	n_D	Solubility Water	Solubility Other solvents	Hazards and general description
NiSO$_4$.7H$_2$O	280.88	–H$_2$O, 99	–6H$_2$O, 103	1 948	1.467 1.489 1.492	75.6^{15}	s alc	CSA, gn rhomb
NiS	90.77	797		5 300– 5 650		0.000 36^{18}	s aq reg, HNO$_3$,KHS	CSA, blk amorph
NITROGEN								
N$_2$	28.013	–209.86	–195.806	1.2506 (g) 808.1$^{-195.8}$ (l)		2.33 cm^{30}	ss alc	col g, liq, w cub cr
HN$_3$	43.03	–80	37	1 090		vs	s alc, alk,et	T, col liq
HNO$_3$	64.02	–42	83	1 502.7	1.397^{16}	vs	d alc	T, O, col liq
HNO$_2$	47.01	only known in solution						
NCl$_3$	120.37	<–40	exp 95	1 653		i	s chl, bz ct, CS$_2$	T, y oil
NF$_3$	71.00	–206.6	–128.8	1 537^{-129}(l)		i		T, col gas
N$_2$H$_4$	32.05	1.4	113.5	1 011^{15}	1.470^{22}	vs	s alc	T, CSA, col liq, or w cr
N$_2$H$_4$.H$_2$O	50.07	–40	118.5	1 030^{21}	1.428 4	vs	s alc	T, CSA col fum liq
N$_2$H$_4$.HCl	68.51	89	d 240			vs	vs liq NH$_3$	T, CSA, w need
N$_2$H$_4$.2HCl	104.97	–HCl 198	d 200	1 420		27.2^{22}		T, CSA, w cub
N$_2$H$_4$.H$_2$SO$_4$	130.13	254	d	1 370		3.4		T, CSA, w rhomb
NH$_2$OH	33.03	33.05	56.5	1 204		s	s a,alc,MeOH	C, T, MS, w need
NH$_2$OH.HCl	69.49	151	d	1 670		83^{17}	s alc, MeOH	C, T, MS, w monocl
N$_2$O	44.01	–90.8	–88.5			130^0 cm^3	s alc,et	col gas
NO	30.01	–163.6	–151.8	1.3402	1.330-^{90}(l)	7.34 cm^{30}	s alc	col gas, bl liq
N$_2$O$_3$	76.01	–102	d 3.5	1 447 (l)^0		s	s et, a, alk	T, r–br gas, bl liq
N$_2$O$_4$	92.02	–11.2	21.2	1 449 (l)^{20}	1.40^{20}	s+d	s alk,chl	T, y liq, br gas
N$_2$O$_5$	108.01	30	d 47	1 642^{18}		s	s chl	T, MS, w rhomb
NOCl	65.46	–64.5	–5.5	2.99 (g)		d		T, MS, y gas
N$_2$S$_4$	184.28	d 178		2 240^{18}	2.046		s alc, bz,CS$_2$	T, y cry
OSMIUM								
Os	190.23(3)	3 045	5 027	22 480^{20}		i	ss aq.reg, HNO$_3$	FP, I, gy metal hex
Os$_3$(CO)$_{12}$	906.73	224						T, I
OsCl$_3$	296.56	d >500				vs	s a, alc, alk	T, MS, br cub
OsO$_4$	254.10	39–41	130	4 900		5.7^{10}	s alc, et, ct NH$_4$OH	T, I, w monocl or y powd
OXYGEN								
O$_2$	31.999	–218.4	–182.962	1.429^0 (g) 1149-^{183} (l)		3.16 cm^{30}	ss alc	col gas, hex cry
OF$_2$	54.00	–223.8	–144.8	1 900-^{223} (l)		ss, d	ss a, alc	T, col unstable gas
O$_3$	47.998	–192.7	–111.9	2.144^0 (g)	1.222 6(l)	49 cm^{30}	s alk	T, O, col gas, bl liq
PALLADIUM								
Pd	106.42(1)	1 554	2 927	12 020^{20}		i	s aq.reg, h HNO$_3$	FP, gy – w metal, cub
PdCl$_2$	177.31	d 500		4 000^{18}		s	s HBr	T, MS, r cub
PdO	122.40	870		8 700^{20}		i	i aq.reg	blk powd
PHOSPHORUS								
P (red)	30.974	590 (43 atm)	417 s	2 340		i		FS, r amorph
P (white)	30.974	44.1	280.4	1 820	2.144	0.000 3^{15}	s alc, CS$_2$, bz, alk, et	FS, y–w cub
H$_3$PO$_4$	98.00	42.35	–H$_2$O, 213	1 834^{18}		548	s alc	C, MS, col liq
H$_3$PO$_3$	82.00	73.6	d 200	1 651^{21}		309^0	s alc	C, MS, col liq
H$_3$PO$_2$	66.00	26.5	d 130	1 493^{19}		s	vs alc, et	C, MS, col oil

(Continued)

Table XI.1 (Continued)

Compound/formula	M_r	m. p. / °C	b. p. / °C	ρ / kg m^{-3}	n_D	Solubility Water	Solubility Other solvents	Hazards and general description
PBr$_3$	270.70	−40	172.9	2 852[15]	1.697[27]	d	s ct, chl, et, CS$_2$	C, T, MS, col fum liq
PBr$_5$	430.52	d 106				d	s bz, ct, CS$_2$	C, T, MS, y rhomb
PCl$_3$	137.33	−112	75.5	1574	1.514 8	d	s bz, et, CS$_2$, chl, ct	C, T, MS, col fum liq
PCl$_5$	208.24	s 162	d 167	1 600		d	s CS$_2$, ct	C, T, MS, y−w tetr, fum
POCl$_3$	153.33	1.25	105.3	1 675	1.460	d		C, T, MS, col fum liq
PF$_3$	87.97	−151.5	−101.5	3.907		d	s alc	C, T, MS, col gas
PF$_5$	125.97	−83	−75	5.805		d		C, T, MS, col gas
PH$_4$Br	114.91	s 30		2.464 (g)		d		T, MS, w cub
PH$_4$Cl	70.46	s 28				d		T, MS, w cub
PH$_4$I	161.91	18.5	80	2 860		d		T, MS, w tetr
PI$_3$	411.68	61	d	4 180		d	vs CS$_2$	C, T, MS, r hex
P$_4$O$_6$	219.89	23.8	175.4	2 135[21]		d	s chl, bz, et, CS$_2$	C, T, MS, w powd
P$_4$O$_{10}$	283.88	580	s 300	2 390		d		C, T, MS, w powd
P$_2$S$_5$	222.27	288	514	2 030		i	s CS$_2$, alk	C, T, MS, y powd

PLATINUM

Compound/formula	M_r	m. p. / °C	b. p. / °C	ρ / kg m^{-3}	n_D	Water	Other solvents	Hazards and general description
Pt	195.08(3)	1769	3 827	21 450[20]		i	s aq reg	FP, gy metal cub
H$_2$PtCl$_6$ · 6H$_2$O	517.92	60		2 431		vs	s alc, et	T, I, MS, r−br
PtCl$_2$	266.00	d 581		6 050		ss	s HCl	T, I, g hex
PtCl$_4$	336.90	d 370		4 303		58.7	s ac	T, I, br−r cry
PtO$_2$	227.09	450		10 200		i	i aq reg	blk powd

POTASSIUM

Compound/formula	M_r	m. p. / °C	b. p. / °C	ρ / kg m^{-3}	n_D	Water	Other solvents	Hazards and general description
K	39.098	63.65	754	862		d	s a, Hg, NH$_3$	FS, MS, gy metal cub
KOOCCH$_3$	98.15	292		1 570		253	s alc, MeOH	MS, w powd
KAl(SO$_4$)$_2$ · 12H$_2$O	474.39	−9H$_2$O 92.5	−12H$_2$O 200	1 757[20]	1.454 1.456 4	11.4[20]		w cub
KSbO(C$_4$H$_4$O$_6$) · 0.5H$_2$O	333.93	−H$_2$O, 100		2 607	1.620 1.636 1.638	5.26[7]		T, w rhomb
K$_3$AsO$_4$	256.23	1 310				19	s alc	T, MS, w need
K$_2$HAsO$_4$	218.13	d 300				19[6]		T, w monocl
KH$_2$AsO$_4$	180.04	288		2 867	1.567 1.518	19[6]	s a, NH$_3$	T, w tetr
KBH$_4$	53.95	d 500		1 178	1.494	19.3[20]	s alc, MeOH	FS, C, w cub
KBrO$_3$	167.01	d 370		3 270[17]		13.3[40]	ss alc	O, I, w trig
KBr	119.01	734	1 435	2 750	1.559	53.5[0]	s alc, ss et	MS, w cub
K$_2$CO$_3$	138.21	891	d	2 428[19]	1.531	112[20]		I, MS, w monocl
KHCO$_3$	100.12	d 100		2 170	1.482	22.4		MS, w monocl
KClO$_3$	122.55	356	d 400	2 320	1.409 1.517 1.524	7[20]	s alk	O, I, w monocl
KClO$_4$	138.55	d 400		2 520[10]	1.471 7 1.472 4 1.476	2		O, I, w rhomb
KCl	74.56	770	s 1 500	1 984	1.490	35[20]	s et, alk ss alc	MS, w cub
K$_2$CrO$_4$	194.20	968.3		2 732[18]	1.74	63[20]		CSA, O, I, y rhomb
K$_2$Cr$_2$O$_7$	294.19	398	d 500	2 676	1.738	5[0]		CSA, O, I, r monocl
KCr(SO$_4$)$_2$ · 12H$_2$O	499.41	89	−12H$_2$O, 400	1 826	1.481 4	24.4		I, viol−r cub
KCN	65.12	634.5		1 520[16]	1.410	50	s alc, MeOH	T, I, MS, w cub
K$_3$Fe(CN)$_6$	329.26	d		1 850	1.566 1.569 1.583	33[4]	s ac	LS, r monocl
K$_4$Fe(CN)$_6$ · 3H$_2$O	422.41	70	d	1 850	1.577	28[12]	s ac	y monocl
KF	58.10	858	1 505	2 480	1.363	92	s HF, NH$_3$	C, T, MS, w cub

(Continued)

Table XI.1 (Continued)

Compound/formula	M_r	m. p. /°C	b. p. /°C	ρ / kg m^{-3}	n_D	Solubility Water	Solubility Other solvents	Hazards and general description
KBF$_4$	125.91	d 350		2 498[20]	1.324 1.325 1.325	0.4	ss alc, et	I, T, w rhomb
K$_2$SiF$_6$	220.25	d		3 080	1.399 1	0.12[17]	s HCl	T, w hex
KH	40.11	d		1 470	1.453	d		FS, C, MS, w need
KOH	56.11	361	1 322	2 044		107[15]	vs alc	C, T, MS, w rhomb
KIO$_3$	214.00	560	d>1 000	3 930[22]		4.7[0]	s KI	O, I, w monocl
KI	166.01	681	1 330	3 130	1.677	127[0]	s alc ac	MS, LS, w cub
K$_2$MnO$_4$	197.14	d 190				d	s KOH	O, I, gn rhomb
KMnO$_4$	158.04	d <240		2 703	1.59	6.4[20]	vs MeOH, ac	O, I, viol rhomb
KNO$_3$	101.11	334	d 400	2 109[16]	1.335 1.505 6 1.506 4	13[0]	s liq NH$_3$	O, I, w rhomb
KNO$_2$	85.11	d 350		1 915		320	s h alc	O, T, MS, w pr
K$_2$C$_2$O$_4$.H$_2$O	184.24	–H$_2$O, 100		2 127[4]	1.440 1.485 1.550	38		T, MS, w monocl
K$_2$O	94.20	d 350		2 320[0]		d	s alc, et	C, T, MS, w, cub
K$_2$O$_2$	110.20	490	d			d		C, T, MS, w amorph
K$_3$PO$_4$	212.28	1 340		2 564[17]		90[20]		MS, w rhomb
KH$_2$PO$_4$	136.09	252.6	d 256	2 338	1.510 1.486 4	33		MS, w tetr
K$_2$HPO$_4$	174.18	d				167[20]	vs alc	MS, w amorph
K$_4$P$_2$O$_7$.3H$_2$O	384.40	–2H$_2$O, 180	–3H$_2$O, 300	2 330		s		MS, w rhomb
K$_2$SeO$_4$	221.16			3 066	1.535 1.539 1.545	110		T, MS, w rhomb
KNaC$_4$H$_4$O$_6$.4H$_2$O	282.23	70–80	–4H$_2$O, 215	1 790	1.492 1.493 1.496	26[0]	ss alc	w rhomb
K$_2$SO$_4$	174.27	1 069	1 689	2 662	1.494 1.495 1.497	12		w rhomb
KHSO$_4$	136.17	214	d	2 322	1.480	52		I, MS, w rhomb
K$_2$S$_2$O$_8$	270.33	d<100		2 477	1.461 1.467 1.566	6.2		O, I, w tricl
K$_2$S$_2$O$_7$	254.33	>300	d	2 512		s		C, w need
K$_2$S	110.27	840		1 805[14]		s	s alc	T, MS, y–br cub
K$_2$SO$_3$.2H$_2$O	194.30	d				100	ss alc	T, MS, w–y hex
K$_2$TeO$_3$	253.80	d 460				vs	s KOH	I, MS, w cr
KSCN	97.18	173.2	d 500	1 886[14]		239	s alc ac	MS, I, w rhomb pr
RADIUM								
Ra	(226.025)	700	1 630	(5 000)		d		R, T, CSA, w metal
RaBr$_2$	385.82	728	s 900	5 790		s	s alc	R, T, CSA, w–y monocl
RaCl$_2$	296.91	1 000		4 910		s	s alc	R, T, CSA, w–y monocl
RADON								
Rn	(222.00)	–71	–61.8	9.73		22.4 cm^3		R, T, CSA, col gas
RUBIDIUM								
Rb	85.468	38.89	688	1 532		d	s a	FS, MS, soft w metal
RbBr	165.38	693	1 340	3 350	1.553 0	116		MS, w cub
Rb$_2$CO$_3$	230.95	837	d 740			450[20]	ss alc	I, MS, w cr
RbCl	120.92	718	1 390	2 800	1.493	94	ss alc, MeOH	MS, w cub
RbF	104.47	795	1 410	3 557	1.398	131[20]	s HF	T, I, w cub
RbOH	102.48	301		3 203[11]		180	s alc	C, T, MS, w
RbI	212.37	674	1 300	3 550	1.647 4	163	s liq NH$_3$	MS, LS, w cub

(Continued)

Table XI.1 (Continued)

Compound/formula	M_r	m.p. /°C	b.p. /°C	ρ / kg m^{-3}	n_D	Solubility Water	Solubility Other solvents	Hazards and general description
RbNO$_3$	147.47	310	d	3 110	1.51 1.52 1.524	65	vs HNO$_3$	O, I, w hex cub
Rb$_2$O	186.94	d 400		3 720		d	s liq NH$_3$	C, I, w–y cub
Rb$_2$SO$_4$	267.00	1 060	d 1 700	3 613^{20}	1.513 1.513 1.514	51		I, w rhomb
Rb$_2$S	203.00	d 530		2 912		vs		T, w–y
RUTHENIUM								
Ru	101.07(2)	2 310	4 200	12 450		i	s fused alk	FP, gy brittle metal
RuCl$_3$	207.43	d>500		3 110		i	s HCl	MS, br cr
RuO$_2$	133.07	d		6 970		i	s fused alk	MS, bl–blk tetr
SCANDIUM								
Sc	44.956	1 541	2 832	2 989		d		MS, w metal cub
ScCl$_3$	151.32	939	s 850	2 390		vs		MS, w cr
Sc(OH)$_3$	95.98	d				i	s a	w amorph
Sc$_2$O$_3$	137.91			3 864		i	s a	w powd
SELENIUM								
Se	78.96(3)	217	684.9	4 810		i	s H$_2$SO$_4$	T, gy metal hex
H$_2$SeO$_4$	144.97	58	d 260	3 004^{15}		1 060	s H$_2$SO$_4$	T, C, MS, w hex pr
H$_2$SeO$_3$	128.97	d 70		3 004^{15}		167^{20}	vs alc	T, C, MS w hex
SeCl$_4$	220.77	s 196		c. 3 800	1.807	d		T, MS, w–y cub
SeF$_6$	192.95	–39	–34.5	3.25 (g)$^{-28}$		d		C, MS, col gas
H$_2$Se	80.98	–60.4	–41.5	3.664 (g)		0.68	s CS$_2$	T, col gas
SeO$_2$	110.96	s 315		3 950^{15}		38^{14}	s alc, ac, bz	T, C, w monocl
SILICON								
Si	28.086	1 410	2 480	2 330		i	s HF + HNO$_3$	FP, gy cub
SiBr$_4$	347.72	5.4	154	2 771.5		d		T, C, MS, col fum liq
SiCl$_4$	169.90	–70	57.6	1 483 (l)^{20}		d		T, C, MS
SiF$_4$	104.08	–90.2	–86	4.69 (g)		d	s HF	T, C, MS, col gas
SiH$_4$	32.12	–185	–112	1.44 (g)		i		FG, E, col gas
Si$_2$H$_6$	62.22	–132.5	–14.5	2.865 (g)		d	s alc, bz, CS$_2$	FG, E, col gas
Si$_3$H$_8$	92.32	–117.4	52.9	743 (l)^0		d		FG, MS, col liq
Si$_4$H$_{10}$	122.42	–108	84.3	790 (l)^0		d		FG, MS, col liq
SiI$_4$	535.70	120.5	287.5	4 198		d	s CS$_2$	MS, T, w cub
Si$_3$N$_4$	140.28	1 900 (press)		3 440		i	s HF	gy amorph powd
SiO$_2$	60.09	1 713	2 230	2 190–2 600	1.41–1.55	i	s HF	w amorph powd
SiS$_2$	92.21	s 1 090		2 020		d	s alk	T, MS, w rhomb
SILVER								
Ag	107.868	961.93	2 195	10 500		i	s HNO$_3$	FP, w metal cub
AgOOCCH$_3$	166.92	d				1.11		w pl
AgBrO$_3$	235.78	d		5 206	1.874 1.920	0.196	s NH$_4$OH	O, w tetr
AgBr	187.78	432	d>1 300	6 473	2.253	8.4 × 10^{-6}	s KCN ss NH$_4$OH	LS, y powd
Ag$_2$CO$_3$	275.75	d 218		6 077		0.003	s NH$_4$OH	LS, y powd
AgCl	143.32	455	1 550	5 560	2.071	8.9×10^{-4}	s NH$_4$OH	LS, w cub
Ag$_2$CrO$_4$	331.73			5 625		1.3×10^{-3}	s NH$_4$OH, KCN	CSA, MS, r monocl
AgCN	133.89	d 320		3 950		2.3×10^{-5}	s HNO$_3$, KCN, NH$_4$OH	T, I, w hex
AgF	126.87	435	c. 1 159	5 852^{16}		180		C, T, MS, y cub
AgI	234.77	558	1 506	5 683^{20}	2.21 2.22	1.22×10^{-7}	s KCN, KI	LS, y hex

(Continued)

Table XI.1 (Continued)

Compound/formula	M_r	m. p. /°C	b. p. /°C	ρ / kg m^{-3}	n_D	Solubility Water	Solubility Other solvents	Hazards and general description
AgNO$_3$	169.87	212	d 444	4 352[19]	1.729 1.744 1.788	245	s et	T, O, w rhomb
AgNO$_2$	153.88	d 140		4 453		0.41	s NH$_4$OH	I, O, w rhomb
Ag$_2$C$_2$O$_4$	303.76	exp 140		5 029[4]		0.004	s KCN, NH$_4$OH, a	I, T, w cr
Ag$_2$O	231.74	d 300		7 143[17]		0.0022	s a, KCN	LS, br–blk cub
AgClO$_4$	207.32	d 486		2 806		557	s alc, bz	O, I, D, w cr
Ag$_3$PO$_4$	418.58	849		6 370		0.0006	s a, KCN	y cub
Ag$_2$SO$_4$	311.80	652	d 1 085	5 450[29]	1.7583 1.7748 1.7852	0.83	s a, NH$_4$OH	LS, w rhomb
Ag$_2$S	247.80	825	d	7 317		3 × 10^{-17}	s KCN, a	T, LS, blk cub
AgSCN	165.95	d				2 × 10^{-5}	s NH$_4$OH	T, w cr

SODIUM

Compound/formula	M_r	m. p. /°C	b. p. /°C	ρ / kg m^{-3}	n_D	Solubility Water	Solubility Other solvents	Hazards and general description
Na	22.990	97.81	883	971	4.22	d		MS, FS, gy metal cub
NaOOCCH$_3$	82.03	324		1 528	1.464	324	ss alc	MS, w monocl
NaOOCCH$_3$. 3H$_2$O	136.08	58	−3H$_2$O, 123	1 450	1.464	100		w monocl pr
NaNH$_2$	39.01	210	400			d		FS, MS, w powd
Na$_2$HAsO$_4$. 7H$_2$O	312.01	−5H$_2$O, 50	d 180	1 880	1.462 1.466 1.478	41.4	ss alc	T, CSA, w monocl
NaH$_2$AsO$_4$. H$_2$O	181.94	130	d >200	2 530	1.583 1.553 1.507	s		T, CSA, w rhomb
Na$_3$AsO$_4$. 12H$_2$O	423.93	86.3		1 752– 1 804	1.457 1.466	39[15]	s alc	T, CSA, w trig
NaN$_3$	65.01	d		1 846		41.7[17]	s alc, liq NH$_3$	T, w hex
NaBO$_2$	65.80	966	1 434	2 464		26[20]		w hex pr
Na$_2$B$_4$O$_7$. 10H$_2$O	381.47	−8H$_2$O, 60	−10H$_2$O, 320	1 730	1.447 1.469 1.472	3.2		Ef, w monocl
NaBH$_4$	37.83	d 400		1 074	1.542	55	s alc, MeOH	FS, C, w cub
NaBrO$_3$	150.90	381		3 339[17]	1.594	39		O, I, w cub
NaBr	102.90	747	1 390	3 203	1.641 2	116[50]	ss alc	MS, w cub
Na$_2$CO$_3$	105.99	851	d	2 532	1.535	7.1[0]	ss alc	MS, I, w powd
Na$_2$CO$_3$. 10H$_2$O	286.14	−H$_2$O, 33.5		1 440[15]	1.405 1.425 1.440	29.4		I, w monocl
NaHCO$_3$	84.01	d 270		2 159	1.500	10.3	ss alc	MS, w monocl pr
NaClO$_3$	106.44	248–261	d	2 490[15]	1.513	100	s alc, liq NH$_3$	O, C, w cub
NaClO$_4$	122.44	d 482			1.460 6 1.461 7 1.473 1	s	s alc	O, I, D, w rhomb
NaClO$_4$. H$_2$O	140.46	130	d 482	2 020		210	s alc	O, I, D, w rhbdr
NaCl	58.44	801	1 413	2 165	1.544 2	36		I, MS, w cub
NaClO$_2$	90.44	d 180–200				39[17]		O, I, MS, w cr
NaOCl	74.44	in solution only						O, I
NaOCl. 2. 5H$_2$O	119.48	57.5	d			96		O, I, MS, w
Na$_2$Cr$_2$O$_4$	161.97			2 720		85	s MeOH	CSA, O, y rhomb
Na$_2$Cr$_2$O$_7$. 2H$_2$O	298.00	−2H$_2$O, 100	d 400	2 348	1.661 1.699 1.751	185		CSA, O, D, r monocl
NaOCN	65.01	>300	d	1 937		s		T, I, w need
NaCN	49.01	563.7	1 496		1.452	63		T, I, D, w cub
CH$_3$(CH$_2$)$_{11}$OSO$_3$Na	288.38	204–207						I, w powd
NaOC$_2$H$_5$	68.05	>300		868	1.385 0[20]			FS, C, MS
NaF	41.99	993	1 695	2 558[41]	1.336	4.1	s HF	T, I, w cub

(Continued)

Table XI.1 (Continued)

Compound/formula	M_r	m. p. / °C	b. p. / °C	ρ / kg m^{-3}	n_D	Solubility Water	Solubility Other solvents	Hazards and general description
NaH	24.00	d 800		920	1.470	d		FS, MS, gy need
NaOH	40.00	318.4	1 390	2 130	1.357 6	114	vs alc	C, T, D, w
NaIO$_3$	197.89	d		4 277^{17}		9		O, I, w rhomb
NaIO$_4$	213.89	d 300		4 174		14	s H$_2$SO$_4$, HNO$_3$	O, I, w tetr
NaI	149.89	661	1 304	3 667	1.774 5	184	s alc, ac	I, MS, w cub
NaOCH$_3$	54.02	>300		945			d	FS, MS, w powd
NaNO$_3$	84.99	306.8	d 380	2 261	1.587 1.336	92	s alc, MeOH	O, I, w trig
NaNO$_2$	69.00	271	d 320	2 168^0		85	ss alc, MeOH et	T, O, MS, w rhomb pr
Na$_2$C$_2$O$_4$	134.00	d 250		2 340		3.6		T, MS, w powd
Na$_2$O	61.98	s 127 5		2 270		d	d alc	T, C, MS, w powd
Na$_2$O$_2$	77.98	d 460		2 805		d	d alc	T, C, O, MS, y powd
Na$_3$PO$_4$. 12H$_2$O	380.12	d 73		1 620^{20}	1.446 1.452	15		w trig
Na$_2$HPO$_4$. 12H$_2$O	358.14	−5H$_2$O, 35	−12H$_2$O, 100	1 520	1.432 1.436 1.437	11.8		I, Ef, w rhomb
NaH$_2$PO$_4$. H$_2$O	137.99	−H$_2$O, 100	d 204	2 040	1.456 1.458 1.487	60^0		MS, w rhomb
Na$_4$P$_2$O$_7$. 10H$_2$O	446.06	−H$_2$O 94 mp 880		1 820	1.450 1.453 1.460	7.1		w monocl
Na$_3$P	99.94	d				d		T, MS, r
Na$_2$HPO$_3$. 5H$_2$O	216.04	53	d 200		1.443	460		MS, w rhomb
NaH$_2$PO$_2$. H$_2$O	105.99	d (violent)				100	vs alc	MS, w monocl
Na$_2$SeO$_4$. 10H$_2$O	369.09	d 32		1 610		58		T, w monocl
Na$_2$SeO$_3$. 5H$_2$O	263.01					90		T, MS, w cr
Na$_2$SiO$_3$. 9H$_2$O	284.20	44	−6H$_2$O 100			vs	s NaOH	Ef, w rhomb
Na$_2$SO$_4$	142.04	884		2 680	1.484 1.477 1.471	4.8^0		I, MS, w orthorhomb
Na$_2$SO$_4$. 10H$_2$O	322.19	32.38	−10H$_2$O 100	1 464	1.394 1.396 1.398	28		I, Ef, w monocl
NaHSO$_4$. H$_2$O	138.07	58.5	d	2 103^{14}	c. 1.46	c. 67, d		C, MS, w monocl
Na$_2$S . 9H$_2$O	240.18	d 920		1 427^{16}		47		FS, C, MS, w tetr
Na$_2$SO$_3$	126.04	d red heat		2 633^{16}	1.565 1.515	12.05^0	ss alc	I, MS, w powd
NaHSO$_3$	104.06	d		1 480	1.526	vs	ss alc	I, T, w monocl
NaSCN	81.07	287			1.625	143	vs alc, ac	T, LS, MS, w rhomb
Na$_2$S$_2$O$_4$	174.11	d 300		2 189		s		FS, MS, w powd
Na$_2$S$_2$O$_3$. 5H$_2$O	248.18	d 48		1 729^{17}	1.489 1.508 1.536	76		I, MS, w monocl
STRONTIUM								
Sr	87.62(1)	769	1 384	2 583		d	s a, alc, liq NH$_3$	FS, MS, w metal
SrBr$_2$	247.44	643	d	4 216	1.575	100	s alc	MS, w hex need
SrCO$_3$	147.63	1 497 (69 at)	d 1 340	3 700	1.516 1.664 1.666	0.001	s a	w powd
SrCl$_2$. 6H$_2$O	266.62	−4H$_2$O, 60	−6H$_2$O, 100	1 930	1.536 1.487	106^0	ss alc	MS, w trig
SrCrO$_4$	203.61			3 895^{15}		0.09	s HCl, HNO$_3$	O, T, y monocl
SrF$_2$	125.62	1 473	2 489	4 240	1.442	0.012	s h HCl	w powd

(Continued)

Table XI.1 (Continued)

Compound/formula	M_r	m. p. /°C	b. p. /°C	ρ /kg m^{-3}	n_D	Solubility Water	Solubility Other solvents	Hazards and general description
Sr(OH)$_2$	121.63	373	d 710	3 625		0.4^0	s a, NH$_4$Cl	C, MS, w
Sr(OH)$_2$. 8H$_2$O	265.76	$-$8H$_2$O, 100		1 900	1.499 1.476	1.0	s a, NH$_4$Cl	C, MS, w tetr
SrI$_2$	341.43	515	d	4 549		180	s alc, MeOH, NH$_4$OH	w pl
Sr(NO$_3$)$_2$	211.63	570		2 986		71^{18}		O, I, w cub
SrO	103.62	2 430	c. 3 000	4 700	1.810	0.8	ss alc	C, MS, w cub
SrSO$_4$	183.68	1 605		3 960	1.622 1.624 1.631	0.01		w rhomb
SULPHUR								
S$_8$	256.533	112.8	444.674	2 070	1.957	i	s CS$_2$	I, y rhomb
H$_2$SO$_4$	98.08	10.36	338	1 841		vs	d alc	T, C, O, MS, col liq
H$_2$S$_2$O$_7$	178.14	35	d	1 900^{20}		d	d alc	T, C, O, MS, w cr
H$_2$SO$_3$	82.08	in solution only		c. 1 030			s alc et	T, C
S$_2$Br$_2$	223.95	$-$40	d 200	2 630	1.730	d	s CS$_2$	C, L, MS, r liq
SOBr$_2$	207.88	$-$52	138	2 683	1.675	d	s bz, chl, CS$_2$	C, L, MS, o fuming liq
SCl$_2$	102.97	$-$78	d 59	1 621	1.557^{11}	d	s bz, ct	C, L, MS, r fuming liq
S$_2$Cl$_2$	135.03	$-$80	138	1 678	1.666^{14}	d	s bz, et, CS$_2$	C, L, MS, o fuming liq
SO$_2$Cl$_2$	134.97	$-$54.1	69	1 680	1.444	d	s bz	C, L, MS, col fuming liq
SOCl$_2$	118.97	$-$105	79	1 655	1.514	d	s bz chl	C, L, MS, col fuming liq
SOClF	102.51	$-$139.5	12.2			d		C, MS, col gas
SF$_6$	146.05	$-$50.5	s $-$63	6.602(g)		ss	s alc, KOH	col gas
SO$_2$F$_2$	102.06	$-$136.7	$-$55.4	3.72 (g)		10^0	s alc, ct	col gas
SOF$_2$	86.06	$-$110.5	$-$48.3	2.93 (g)		d	s et, bz, chl	C, L, MS, col gas
SO$_2$	64.06	$-$72.7	$-$10	2.927 (g)		9.4	s alc, H$_2$SO$_4$	C, T, col gas or liq
SO$_3$	80.06	16.83	44.8	1 970		d	s H$_2$SO$_4$	C, T, O, MS, w need
TELLURIUM								
Te	127.60 (3)	449.5	989.8	6 240	1.0025	i	s H$_2$SO$_4$, HNO$_3$, aq reg, KCN KOH	T, FP, w metal
H$_6$TeO$_6$. Te(OH)$_6$	229.64	136		3 158		53	ss HNO$_3$	T, C, w monocl pr
TeCl$_2$	198.50	209	327	7 050		d	s a	T, MS, blk cr, unstable
TeCl$_4$	269.41	224	380	3 260^{18}		d	s HCl, bz, alc, chl, ct	T, I, MS, w cr
TeF$_6$	241.59	$-$36	35.5	2 560^{-25} (l)		d		T, MS, col gas
H$_2$Te	129.62	$-$48.9	$-$2.2	4.49 (g)		vs		T, poisonous col gas
TeO$_2$	159.60	733	1 245	5 670	2.00 2.18 2.35	i	s HCl, alk	T, w tetr
TeO$_3$	175.60	d 395		5 075^{105}		i		T, y amorph
THALLIUM								
Tl	204.383	303.5	1 457	11 850		i	s HNO$_3$, H$_2$SO$_4$	FP, T, w metal, tetr
TlBr	284.28	480	815	7 557^{17}	2.4–2.8	0.05	s alc	T, y cub
Tl$_2$CO$_3$	468.75	272		7 110		4.03^{15}		T, w monocl
TlCl	239.82	430	720	7 000	2.247	0.39		T, LS, w cr
TlF	223.37	327	655	8 230^4		78.5^{15}	ss alc	T, w cub
TlOOCH	249.39	101		4 967^{104}		500^{10}	vs MeOH	T, MS, w need
TlOH	221.38	d 139				60	s alc	T, w–y need
TlI	331.27	440	823	7 098^{15}		0.0006^{20}	s liq NH$_3$	T, LS, y rhomb
TlNO$_3$	226.38	206	430	5 556^{21}		11.9	s ac	T, MS, w powd
Tl$_2$O	424.74	300	d 1 865	9 520^{16}		vs to TlOH	s a, alc	T, MS, blk powd
Tl$_2$O$_3$	456.74	717	d 875	1 019^{22}		i	s a	T, w amorp pr
Tl$_2$SO$_4$	504.80	632	d	6 770	1.860 1.867 1.885	5.5		T, w rhomb
Tl$_2$S	440.80	448.5	d	8 460		0.02	s a	T, blk tetr

(Continued)

Table XI.1 (Continued)

Compound/formula	M_r	m. p. / °C	b. p. / °C	ρ / kg m^{-3}	n_D	Solubility Water	Solubility Other solvents	Hazards and general description
THORIUM								
Th	232.038	1 750	4 227	11 720		i	s HCl, H$_2$SO$_4$, aq reg	T, R, gy metal cub
ThCl$_4$	373.85	771	d 928	4 590		124	s alc, KCl	T, R, MS, w rhomb
Th(NO$_3$)$_4$. 12H$_2$O	696.24	d				194		T, R, MS, w leaf
ThO$_2$	264.04	3 050	4 400	9 860		i	s h H$_2$SO$_4$	T, R, w cub
Th(SO$_4$)$_2$. 9H$_2$O	568.29	−9H$_2$O, 400		2 770		1.6		T, R, w monocl
TIN								
Sn	118.710	231.968 1	2 270	7 280 (w) 5 770 (gy)		i	s HCl, H$_2$SO$_4$, aq reg, alk	FP, MS, gy metal
SnBr$_2$	278.51	215.5	620	5 117^{17}		85.2^0	s alc, ac, et	C, MS, y−w rhomb
SnBr$_4$	438.33	31	202	3 340		d	s ac, PCl$_3$	C, MS, w rhomb
SnCl$_2$	189.60	246	652	3 950		83.9^0	s alc, et ac, pyr	C, MS, w rhomb
SnCl$_2$. 2H$_2$O	225.63	37.7	d	2 710		d	s alc, et, ac	C, MS, w monocl
SnCl$_4$	260.50	−33	114.1	2 226	1.512	d	s et	C, MS, col liq
SnCl$_4$. 5H$_2$O	350.58	d 56				s		C, MS, w monocl
SnH$_4$	122.72	−150	−52				s conc H$_2$SO$_4$, conc alk	T, gas
SnI$_2$	372.50	320	717	5 285		0.98	vs NH$_4$OH, HI soln	T, o monocl need
SnI$_4$	626.31	144.5	364.5	4 473^0	2.106	d	s CS$_2$, bz, ct	T, MS, o−r cub
SnO	134.69	d 1 080		6 446^0		i	s a, alk	blk cub
SnO$_2$	150.69	1 127	s 1 800	6 950	1.997 2.093	i		w tetr
SnS	150.75	882	1 230	5 220		2 × 10^{-6}	d HCl	T, gy−blk cub
SnS$_2$	182.82	d 600		4 500		2 × 10^{-4}		T, y hex
SnSO$_4$	214.75	>360				33	s H$_2$SO$_4$	MS, w hex pr
Sn(C$_2$H$_5$)$_4$	234.94	−112	181	1 187	1.472 5	i	s org solv	T, col liq
Sn (CH$_3$)$_4$	178.83	−54.8	78	1 291	1.438 6	i	s org solv	T, FL, col liq
TITANIUM								
Ti	47.88 (3)	1 660	3 313	4 508		i	s dil a	FP, MS, gy metal
TiBr$_4$	367.54	39	230	2 600		d	s alc, et	T, MS, o−l
TiC	59.91	c. 3 140	4 820	4 930		i	s aq reg, HNO$_3$	gy cub
TiCl$_2$	118.81	d 475		3 130		d	s alc	C, MS, br−blk hex
TiCl$_3$	154.26	d 440	660 (108 mm)	2 640		s	vs alc	MS, C, FS, viol
TiCl$_4$	189.71	−24	136.4	1 730	1.61^{10}	s	s HCl, alc	C, y liq
Ti(OC$_2$H$_5$)$_4$	228.15		150 −152 (10 mm)	1 106.6	1.504 3			FL, MS, oily liq
TiN	61.91	2 930		5 220			ss hot aq reg + HF	y cub
TiO$_2$	79.90	1 825		4 170	2.583 2.586 2.741	i	s H$_2$SO$_4$, alk	w rhomb
TUNGSTEN								
W	183.84 (1)	3 422	5 727	19 254		i	s HNO$_3$+ HF, fus NaOH + NaNO$_3$	FP, gy cub metal
WBr$_6$	663.30	232		6 900		i	s alc, et	bl−blk need
WC	195.86	2 870 ± 50	6 000	15 630		i	s HNO$_3$ + HF, aq reg	blk hex
W(CO)$_6$	351.91	d 150		2 650		i	s fuming HNO$_3$	T, w rhomb cr
WCl$_4$	325.66	d		4 624		d		C, MS, gy
WCl$_6$	396.57	275	346.7	3 520		d	s alc, bz, et	C, MS, bl cub
WF$_6$	297.84	2.5	17.5	12.9 (g)		d	s alk	T, C, col gas

(Continued)

Table XI.1 (Continued)

Compound/formula	M_r	m. p. /°C	b. p. /°C	ρ / kg m^{-3}	n_D	Solubility Water	Solubility Other solvents	Hazards and general description
WO$_2$	215.85	1 500–1 600 in N$_2$		12 110		i	s a, alk	I, br cub
WO$_3$	231.85	1 473		7 160		i	s hot alk	I, y rhomb
WOCl$_4$	341.66	211	227.5			d	s bz, CS$_2$	I, MS, r need
H$_2$WO$_4$	249.86	–H$_2$O, 100	1 473	5 500	2.24	i	s alk, HF	I, y powd
URANIUM								
U	238.029	1 132.3	3 677	18 950		i	s a	R, gy metal cub
UO$_2$(OOC.CH$_3$)$_2$.2H$_2$O	422.19	–2H$_2$O, 110	d 275	2 893^{15}		8.3^{17}	vs alc	R, T, y rhomb
UO$_2$Cl$_2$	340.93	578	d			320^{15}	s alc, et	R, T, MS, y
UF$_6$	352.02	64.6	56.2	4 680^{21}		d	s ct, chl	R, T, MS, w monocl
UO$_2$(NO$_3$)$_2$.6H$_2$O	502.13	60.2	118	2 807^{13}	1.496 7	127	vs alc, et, ac, MeOH	R, T, MS, y rhomb
UO$_2$	270.03	2 500		10 960		i	s HNO$_3$, conc H$_2$SO$_4$	R, T, br rhomb
UO$_3$	286.03	d		7 290		i	s HNO$_3$, HCl	R, T, y–r powd
U$_3$O$_8$ (pitchblende)	842.09	d 1 300		8 300		i	s HNO$_3$	R, T, gn
UO$_2$SO$_4$.3H$_2$O	420.14	d 100		3 280^{17}		20.5^{16}	s H$_2$SO$_4$, HCl	R, T, y–gn cry
VANADIUM								
V	50.942	1 890	3 000	6 110	3.03	i	s aq reg, HF, H$_2$SO$_4$, HNO$_3$	I, gy metal cub
VCl$_2$	121.85	s 1 000		3 230^{18}		d	s alc, et	T, MS, gn hex
VCl$_3$	157.30	d		3 000		s	s alc, et	T, MS, p cry
VCl$_4$	192.75	–28	148.5	1 816^{30}		d	s alc, et, chl	T, r liq
VO	66.94	ignites		5 758^{14}		i	s a	I, gy cry
V$_2$O$_3$	149.88	1 970		4 870^{18}		i	s HNO$_3$, HF, alk	T, I, blk cry
VO$_2$	82.94	1 967		4 339		i	s a, alk	T, I, blk cry
V$_2$O$_5$	181.88	690	d 1 750	3 357	1.46 1.52 1.76	0.07	s a, alk	T, I, y–r rhomb
VOCl$_3$	173.30	–77	126.7	1 840		d	s alc, et	C, MS, y liq
VSO$_4$.7H$_2$O	273.11	d						viol monocl
XENON								
Xe	131.29 (2)	–111.9	–108.1	5.887 (g)		11.9 cm^3		col inert gas
XeF$_2$	169.3	129		4 300		d	vs HF	MS, w cry
XeF$_4$	207.3	117		4 100		d		MS, stable, w cr
XeF$_6$	245.3	49.6				d		MS, stable, w cr
XeO$_3$	179.3			4 600				MS, E, stable in soln
ZINC								
Zn	65.39 (2)	419.58	907	7 135		i	s a, alk	MS, FP, gy metal
Zn(OOC.CH$_3$)$_2$.2H$_2$O	219.49	–2H$_2$O, 100		1 735	1.494	40	s alc	w monocl
ZnBr$_2$	225.19	394	650	4 201	1.545 2^{15}	470	s NH$_4$OH	MS, I, T, w rhomb
ZnCO$_3$	125.39	d 300		4 398	1.818 1.618	0.021	s a, alk	MS, w trig
ZnCl$_2$	136.28	283	732	2 907	1.681 1.713	432	s alc, et	MS, T, I, w hex
ZnCrO$_4$	181.36			3 400		i	s a, liq NH$_3$	T, y pr
Zn(CN)$_2$	117.41	d 800		1 852		0.000 5^{20}	s alk, KCN	T, w rhomb
ZnF$_2$	103.37	872	c. 1 500	4 950		1.62^{20}	s hot a	T, w monocl
Zn(OH)$_2$	99.38	d 125		3 053		0.001	s a, alk	T, w rhomb
ZnI$_2$	319.18	446	d 624	4 736		430	s a, alc, et	T, MS, LS, w hex
Zn(NO$_3$)$_2$.6H$_2$O	297.47	36.4	–6H$_2$O, 105	2 065		127	vs alc	O, C, w tetr
Zn$_3$N$_2$	224.12			6 220		d	s HCl	MS, T, gy

(Continued)

Table XI.1 (Continued)

Compound/formula	M_r	m.p. /°C	b.p. /°C	ρ / kg m^{-3}	n_D	Solubility Water	Solubility Other solvents	Hazards and general description
ZnO	81.37	1 975		5 606	2.008 2.029	0.000 16	s a, alk,	w hex
ZnSO$_4$. 7H$_2$O	287.54	100	$-$7H$_2$O, 280	1 957	1.457 1.480 1.484	96.5^{20}	ss alc	MS, I, w rhomb
ZnS	97.43	1 700 (50 atm)		4 100	2.368	6.5 × 10^{-5}	vs a	MS, T w cub
ZIRCONIUM								
Zr	91.224	1 852	4 377	6 506		i	s HF, aq reg	FP, gy metal
ZrCl$_4$	233.03	437 (25 atm)	s 331	2 803		s	s alc, et, conc HCl	C, MS, w cr
ZrH$_2$	93.24						s dil HF, conc a	FS, MS, gy–blk powd
ZrO$_2$	133.22	2 715	c. 5 000	5 890		i	s H$_2$SO$_4$, HF	w monocl
ZrOCl$_2$. 8H$_2$O	322.25	$-$6H$_2$O, 150	$-$8H$_2$O, 210		1.552 1.563	s	s alc, et	C, MS, w need

Table XI.2 Physical Constants of Selected Organic Compounds

For more extensive data see: *Dictionary of Organic Compounds* (5th edn. 1982) and later Supplements, Chapman & Hall, London; *CRC Handbook of Data on Organic Compounds* (4th edn (HODOC II), CRC Press, c/o Wolfe Publishing, London; *Beilsteins Handbuch der Organischen Chemie* and Supplements. (Springer-Verlag, Berlin).

Column 1. Commonly used name of compound, together with synonyms; IUPAC recommended names denoted by*.

Columns 4 (m.p.) and 5 (b.p.). d: decomposes; s: sublimes.

Columns 6 (density reference to water at 4°C) and 7 (refractive index for the sodium D line). Values quoted at 20°C unless otherwise stated.

Column 8. Flash point determined in closed cup.

Column 9. Solubility in water (w), ethanol (alc), ether (et), acetone (ac) and benzene (bz). Solubility in other solvents: alkali (alk), chloroform (chl), dioxane (diox), pyridine (py) etc. at room temperature or on heating (^{h}).

d: decomposes, hydrolyses; i: insoluble; δ: very slightly soluble; s: soluble; ss: slightly soluble; vs: very soluble; ∞: soluble in all proportions.

Column 10. General hazards.

AS: air-sensitive; C: corrosive; CSA: cancer suspect agent; E: explosive; FG: flammable gas; FL: flammable liquid; FS: flammable solid; H: harmful; I: irritant; L: lachrymator; LS: light-sensitive; MS: moisture-sensitive, e.g. hydrolysed by moisture, deliquescent, O: (powerful) oxidizing agent; P: poisonous;

For further details of other hazards, including toxicity, LD_{50} and approved occupational exposure limits see references in Section XVI and Tables XVI 1 to 4.

T: toxic; vT: very toxic. Other hazards, e.g. mutagen, are spelt out.

Column 11. References:

Beilsteins Handbuch der Organischen Chemie: B. volume (supplement). page.

The Aldrich Library of FT-IR Spectra, 1st edn: (1985) Aldrich Chem. Co., Gillingham, Dorset: IR. 1 (volume). page.

The Aldrich Library of NMR Spectra, 2nd edn: (1983) Aldrich Chem. Co., Gillingham, Dorset: NMR. 2 (volume). page.

(Table begins overleaf)

Table XI.2 (Continued)

Compound/synonym(s)	Formula	M_r	m.p. /°C	b.p. /°C	Density / kg m^3	Refractive index
acenaphthene						
1:8-ethylenenaphthalene	$C_{10}H_6CH_2CH_2$	154.21	96.2	279	1 225^0	1.604 8^{95}
acetaldehyde						
ethanal*	CH_3CHO	44.05	−121	20.8	783.4^{18}	1.331 6
acetaminde						
ethanamide*	CH_3CONH_2	59.07	82.3	221.2	1 159.0	1.427 8^{78}
acetanilide	$C_6H_5NHCOCH_3$	135.17	115–116	304	1 219.0^{15}	
acetic acid						
ethanoic acid*	CH_3COOH	60.5	16.60	117.9	1 049.2	1.371 6
acetic anhydride						
ethanoic anhydride*	$(CH_3CO)_2O$	102.9	−73.1	139.55	1 082.0	1.390 0
acetone						
propanone*	CH_3COCH_3	58.08	−95.35	56.2	789.9	1.358 8
acetone dicarboxylic acid						
3-oxo-pentanedioic acid*	$(HOOCCH_2)_2CO$	146.1	d 135	d		
acetonitrile, methyl						
cyanide, ethanenitrile*	CH_3CN	41.05	−45.72	81.6	785.7	1.344 2
acetophenone						
methylphenyl ketone	$C_6H_5COCH_3$	120.16	20.5	202	1 028.1	1.537 18
acetylacetone						
2,4-pentanedione	$(CH_3CO)_2CH_2$	100.13	−23	140.4	972.1^{25}	1.449 4
acetyl chloride						
ethanoyl chloride*	CH_3COCl	78.5	−112	50.9	1 105.1	1.389 76
acetylene						
ethyne*	C_2H_2	26.04	−80.8	−84.5	618.1^{-32}	1.000 5^0
acetylene dicarboxylic acid						
2-butynedioic acid	$HOOCC\!:\!CCOOH$	114.06	d 179			
N-acetyl-N-phenylhydrazine						
1-acetyl-2-phenylhydrazine	$C_6H_5NHNHCOCH_3$	150.18	128–130			
2-acetylsalicylic acid,						
2-acetoxybenzoic acid	$2\text{-}(CH_3COO)C_6H_4COOH$	180.12	138–140			
acridine	$C_6H_4CHC_6H_4N$	179.22	107–110	346	1 005	
acrolein						
propenal*	$CH_2\!:\!CHCHO$	56.07	−86.95	53	841.0	1.401 7
acrylamide						
propenamide*	$CH_2\!:\!CHCONH_2$	71.08	84–85	125/25 mm		
acrylic acid						
propenoic acid*	$H_2C\!:\!CHCOOH$	72.06	13	141.6	1 051	1.422 4
acrylonitrile						
propenenitrile*	$H_2C\!:\!CHCN$	53.06	−83.5	77.5	806	1.391 1
adenine						
6-aminopurine	$C_5H_5N_5$	135.13	>360			
adipic acid						
hexanedioic acid*	$HOOC(CH_2)_4COOH$	146.14	153	265/100 mm	1 360^{25}	
adipoyl chloride						
hexanedioic acid dichloride*	$ClOC(CH_2)_4COCl$	183.03		126/12 mm	1 259	1.470 6

Flash point/°C	Solubility						Hazards	References	
	w	alc	et	ac	bz	Other			
	i	ss			vs		H, I	B.5(2).495	IR.1(1).960D
−27	∞	∞	∞		∞		FL, I, CSA	B.1(3).2617	IR.1(1).465B NMR.2(1).357A
173	s	vs	i		ss	chl	I, CSA	B.2(2).177	NMR.2(1).629B
		vs	s	vs	s	chl	H, I	B.12(2).137	IR.1(2).356B
	∞	∞	∞	∞	∞	CS_2	C	B.2(2).91 B.5(2).761	IR.1(1).481B NMR.2(1).419B
	vs	ss	∞		s	chl	C, FL, L	B.2(2).170	IR.1(1).711A NMR.2(1).601A
−20	∞	∞	∞		∞	chl	FL, I	B.1(3).2696	
	s	s	ss	i			MS	B.3(3).482	IR.1(1).532C
2	∞	∞	∞	∞	∞	CCl_4	C, FL, L	B.2(2)181	
82	i	s	s	s	s	chl	H, I	B.7(2).208	IR.1(2).8B NMR.2(2).7D
34	vs	∞	∞	∞	∞	chl	H, FL, I	B.1.777	IR.1(1).425A NMR.2(1).388B
4	d	d	∞	∞	∞	chl	C, FL, MS	B.2(2).142,175	
	δ	δ		s	s	chl	FG	B.1(3).887	
	vs	vs	vs				T, I	B.2(2).670	IR.1(2).943D
	vs[h]	vs	ss		s	chl	T	B.15(2).92	IR.1(2).356A NMR.2(2).334B
	s[h]	vs	s		ss	chl	H, T, I	B.10(2).41	
	δ[h]	vs	vs		vs	CS_2	H, I, FS	B.20(2).300	IR.1(2).874A NMR.2(2).749C
−19	vs	s	s	s			vT, FL	B.1(3).2953	IR.1(1).470D
	vs	s	s			chl	T, CSA	B.2(2).388	IR.1(1).749B NMR.2(1).630C
52	∞	∞	∞	s	s		C, T, FL	B.2(2).1215	IR.1(1).498B NMR.2(1).434B
0	s	∞	∞	s	s		vT, FL, CSA	B.2(3).1234	IR.1(1).842D NMR.2(1).60B
	ss	ss	i				H	B.26(2).252	IR.1(2).713C NMR.2(2).589C
196	ss	vs	s				I	B.2(2).572	IR.1(1).495B
>110	d	d					C, MS, L	B.2(2).575	IR.1(1).741A NMR.2(1).624B

(Continued)

Table XI.2 (Continued)

Compound/synonym(s)	Formula	M_r	m. p. /°C	b. p. /°C	Density / kg m³	Refractive index
DL-alanine						
(α)-aminopropionic acid	$CH_3CH(NH_2)COOH$	89.09	d 289			
β-alanine						
β-aminopropionic acid	$H_2NCH_2CHCOOH$	89.09	d 205		1 437[19]	
allene						
propadiene*	$H_2C{:}C{:}CH_2$	40.07	-136	-34	1 787	1.416 8
allyl alcohol						
2-propen-1-ol*	$H_2C{:}CHCH_2OH$	58.08	-129	97	854	1.413 5
4-aminobenzoic acid	$4\text{-}(NH_2)C_6H_4COOH$	137.14	188–189		1 374[25]	
1-aminonaphthalene*						
α-naphthylamine	$C_{10}H_7NH_2$	143.19	50	301	1 114	1.670 3[51]
2-aminonaphthalene*						
β-naphthylamine	$C_{10}H_7NH_2$	143.19	113	306.1	1 061[98]	1.649 3[98]
2-aminophenol*						
2-hydroxyaniline	$2\text{-}(H_2N)C_6H_4NH_2$	109.13	174	s 153/11 mm	1 328	
3-aminophenol*						
3-hydroxyaniline	$3\text{-}(H_2N)C_6H_4NH_2$	109.13	123	164/11 mm		
4-aminophenol*						
4-hydroxyaniline	$4\text{-}(H_2N)C_6H_4NH_2$	109.13	186.7			
2-aminopyridine	$N{:}CHCH{:}CHCH{:}CNH_2$	94.12	59–60	204		
3-aminopyridine	$N{:}CHCH{:}CHC(NH_2){:}CH$	94.12	59–63	248		
4-aminopyridine	$N{:}CHCH{:}C(NH_2)CH{:}CH$	94.12	160–162	273		
t-amyl acetate						
2-methyl-2-butyl ethanoate	$CH_3COO(CH_2)_4CH_3$	130.19	-100	142	876	1.401 9
n-amyl alcohol (see 1-pentanol)						
t-amyl alcohol						
2-methyl-2-butanol	$CH_3CH_2C(CH_3)_2OH$	88.15	-12	102	805	1.405 2
aniline						
aminobenzene	$C_6H_5NH_2$	93.13	-6	184	1 022	1.586 3
aniline hydrochloride	$C_6H_5NH_2{\cdot}HCl$	129.59	198	245	1 221.5[4]	
anisole						
methoxybenzene	$C_6H_5OCH_3$	108.14	-37.5	154	995	1.516 0
anthracene	$C_6H_4CHC_6H_4CH$	178.23	216.2	340	1 283[25]	
anthranilic acid						
2-aminobenzoic acid	$2\text{-}(H_2N)C_6H_4COOH$	137.14	146–147	s	1 412	
9,10-anthraquinone*	$C_6H_4COC_6H_4CO$	208.22	283–285	379.8	1 438[4]	
anthrone						
9,10-dihydroanthracene*	$C_6H_4COC_6H_4CH_2$	194.23	152–154			
D-($-$)-arabinose	$CH_2(CHOH)_4O$	150.13	156–160		1 585	
L-arginine						
L-2-amino-5- guanidopentanoic acid	$H_2NC({:}NH)NH(CH_2)_3\text{-}CH(NH_2)COOH$	174.20	d 223			
L-(+)-ascorbic acid vitamin C	$OCOC(OH){:}C(OH)CH\text{-}CH(OH)CH_2OH$	176.12	d 192			
DL-asparagine monohydrate DL-2-aminosuccinamic acid	$H_2NCOCH_2CH\text{-}(NH_2)COOH{\cdot}H_2O$	150.14	220 d		1 543[15]	

Flash point/°C	\multicolumn{6}{c	}{Solubility}	Hazards	References	
	w	alc	et	ac	bz	Other			
	s	ss	i					B.4(2).814	IR.1(1).571A NMR.2(1).485D
	s	ss	i	i				B.4(2).827	
	i			s			FG	B.1(3).922	
22	∞	∞	∞				vT, FL, I	B.1(3).1874	
	s^h	s	s					B.14(2).246	IR.1(2).198C NMR.2(2).195B
>110	ss	vs	vs				CSA, LS	B.12(2).675	NMR.2(1).1058B
	s	s	s				CSA, LS	B.12(2).710	
	s	vs	s		ss		H, I	B.13(2).164	IR.1(1).1195D NMR.2(1).997B
	s^h	vs	vs		ss		H, I	B.13(2).209	IR.1(1).1200A NMR.2(1).1004A
	ss	vs		i			H, I	B.13(2).220	IR.1(1).1210B NMR.2(1).1013A
	s	s	s	s			vT, I	B.22.428	
	s	s	s				T, I	B.22.431	IR.1(2).767C NMR.2(2).646D
	s	vs	s		s		vT, I	B.22.433	
23	ss	s	s	s			FL, I	B.2(2).143	IR.1(1).602B NMR.2(1).509C
21	s	∞	∞	vs	s	chl	T, FL	B.1(2).422	IR.1(1).125D
70	s	∞	∞	∞	∞	CCl$_4$	vT, CSA	B.12(2).44	IR.1(1).1189A
193	vs	vs	i				vT, H, I, MS, CSA	B.12(2).65	IR.1(1).1189B
51	i	s	s	vs	vs		FL, MS	B.6(2).139	
	i	ss	ss	ss	ss		H, I	B.5(2).569	IR.1(1).961D NMR.2(1).763D
	s	s	s		ss		I	B.14(2).205	IR.1(2).189A NMR.2(2).186B
185	i	ss	i		ss		I	B.7(2).709	NMR.2(2)87D
				s	s^h			B.7(2).414	
	vs	ss	i	i				B.31.34	IR.1(1).189A NMR.2(2).905D
	s	ss	i					B.4(2).845	IR.1(1).786A NMR.2(1).659B
	vs	s	i		i			B.18(3).3038	
	ss	i	i		i			B.4(2).900	IR.1(1).782A NMR.2(1).656B

(Continued)

Table XI.2 (Continued)

Compound/synonym (s)	Formula	M_r	m.p. /°C	b.p. /°C	Density / kg m^3	Refractive index
D-aspartic acid (R)-(−)-aminosuccinic acid	HOOCCH$_2$CH(NH$_2$)COOH	133.10	251		1 661[15]	
azobenzene (trans)	C$_6$H$_5$N:NC$_6$H$_5$	182.23	68.5	293	1 203	1.626 6[70]
barbitone (veronal) 5, 5-diethylbarbituric acid	(C$_2$H$_5$)$_2$CCONHCONHCO	184.19	190		1 220	
barbituric acid	CH$_2$CONHCONHCO	128.099	248	d 260		
benzaldehyde benzenecarbinol*	C$_6$H$_5$CHO	106.12	−26	178	1 044	1.546 3
benzamide	C$_6$H$_5$CONH$_2$	121.14	128–129	290	1 341[4]	
benzanilide N-phenylbenzoic acid	C$_6$H$_5$CONHC$_6$H$_5$	197.24	163	117/10 mm	1 315	
benzene* phene*	C$_6$H$_6$	78.11	5.5	80.1	878.65	1.5011
benzene sulphonamide	C$_6$H$_5$SO$_2$NH$_2$	157.19	151–152			
benzene sulphonic acid*	C$_6$H$_5$SO$_3$H	158.18	65–66			
benzene sulphonyl chloride	C$_6$H$_5$SO$_2$Cl	176.62	d 14.5	d 251 117/10 mm	1 384	1.551 8
benzil diphenylglyoxal*	C$_6$H$_5$COCOC$_6$H$_5$	210.23	94–95	188/12 mm	1 084[102]	
benzoic acid benzene carboxylic acid	C$_6$H$_5$COOH	122.13	122.4	249	1 265.9[15]	1.504[132]
benzoic anhydride	(C$_6$H$_5$CO)$_2$O	226.23	39–40	360	1 199	1.576 7[15]
benzoin 2-hydroxy-2-phenylacetphenone	C$_6$H$_5$CH(OH)COC$_6$H$_5$	212.25	133–135	194/12 mm		
benzonitrile phenyl cyanide	C$_6$H$_5$CN	103.13	−13	190.7	1 010	1.528 9
benzophenone diphenyl ketone	(C$_6$H$_5$)$_2$CO	182.22	49–51	305	1 146	1.607 7
1, 4-benzoquinone* p-quinone	O:C$_6$H$_4$:O	108.10	115	s	1 318	
benzoyl chloride	C$_6$H$_5$COCl	140.57	−1	197.2	1 212	1.553 7
benzoyl peroxide dibenzoyl peroxide	(C$_6$H$_5$CO)$_2$O$_2$	242.23	106	exp		1.545
benzyl acetate	CH$_3$COOCH$_2$C$_6$H$_5$	150.18	−51	206	1 055	1.500 6
benzyl alcohol α-hydroxytoluene	C$_6$H$_5$CH$_2$OH	108.15	−15.3	205.4	1 045	1.540 3
benzylamine α-aminotoluene	C$_6$H$_5$CH$_2$NH$_2$	107.16	10	185	981	1.542 4
benzyl benzoate	C$_6$H$_5$COOCH$_2$C$_6$H$_5$	212.25	18–20	323	1 112	1.568 0
benzyl bromide α-bromotoluene	C$_6$H$_5$CH$_2$Br	171.04	−3 to −1	198–199	1 438	1.575 2

Flash point/°C	w	alc	et	ac	bz	Other	Hazards	References	
	s[h]	i	i		i	HCl		B.4(2).900	IR.1(1).589A NMR.2(1).496D
	ss	s	s		s	Py	H, CSA	B.16(2).4	IR.1(2).965A NMR.2(2).978C
	ss	s[h]	s	s		chl	H	B.24(2).279	
	s[h]	ss	s				I	B.24(2).267	IR.1(1).811C
62	ss	∞	∞	vs	vs		H, T, I	B.7(2).145	IR.1(2).104A NMR.2(2).103C
	ss	vs	ss	ss		CCl$_4$ CS$_2$		B.9(2).163	IR.1(2).365C NMR.2(2).343D
	i	ss	ss					B.12(2).152	IR.1(2).375C NMR.2(2).344D
−11	i	∞	∞	∞		chl	T, FL, CSA	B.5(2).119 B.11(2).24	IR.1(2).258B NMR.2(2).847B
	ss	s[h]	s						
	vs	vs	i	ss			C	B.11(2).18	IR.1(2).491D NMR.2(2).818B
>110	d	vs	s				C, MS	B.11(2).23	IR.1(2).516A NMR.2(2).835C
	i	vs	vs	s	vs		I	B.7(2).764	IR.1(2).58B
121	s[h]	vs	vs	s	vs[h]	chl	I	B.9(2).72	IR.1(2).186A
>110	i	s	s				MS, I	B.9(2).147	IR.1(2).331A NMR.2(2).315A
		vs[h]		vs		py		B.8(2).193	IR.1(2).375B NMR.2(2)45D
71	ss[h]	∞	∞	vs	vs		T, I	B.9(2).196	
>110	i	vs	vs	vs	s		I	B.7(2).349	
	s[h]	s	s				vT, I	B.7.609	NMR.2(1).414C
68	d	d	∞		s	CS$_2$	C, L, MS	B.9(2).159	IR.1(2).340B
	ss	s	s	s	s	CS$_2$	O, I, E	B.9(2).157	IR.1(2).331B NMR.2(2).315B
102	ss	∞	s	s			T, I, CSA	B.6(2).415	IR.1(2).277A NMR.2(2).269A
100	s	s	s	s	s	chl	H, I, MS	B.6(3).1445	
60	∞	∞	∞	vs	s		C, L, I, AS	B.12(2).540	IR.1(1).1266B
147	i	s	s	s	s	chl	H, T, I	B.9(2)100	IR.1(2).292D NMR.2(2).281D
86	i	∞	∞				T, C, I, L	B.5(3).709	IR.1(1).973A NMR.2(1).774D

(Continued)

Table XI.2 (Continued)

Compound/synonym (s)	Formula	M_r	m. p. /°C	b. p. /°C	Density / kg m^3	Refractive index
benzyl chloride						
α-chlorotoluene	$C_6H_5CH_2Cl$	126.59	−43	179.3	1 100	1.538 0
bibenzyl, dibenzyl,						
1, 2-diphenylethane	$C_6H_5CH_2CH_2C_6H_5$	182.27	52.2	284	1 014	1.547 6^{60}
biphenyl, diphenyl,						
phenylbenzene	$C_6H_5C_6H_5$	154.21	71	255.9	866	1.475
2, 2'-bipyridyl, 2, 2'-						
dipyridyl 2, 2'-bipyridine	$(C_5H_4N)_2$	156.19	70–73	273		
[(1S)-endo]-(−)borneol	$C_{10}H_{18}O$	154.26	206–278	s 212	1 011	
bromoacetic acid						
bromoethanoic acid*	$BrCH_2COOH$	138.95	49–51	208	1 934^{50}	1.480 4^{50}
bromobenzene*						
phenyl bromide	C_6H_5Br	157.02	−31	156	1 491	1.559 7
1-bromobutane*						
n-butyl bromide	$CH_3CH_2CH_2CH_2Br$	137.03	−112	101.6	1 276	1.439 0
2-bromobutane*						
sec-butyl bromide	$CH_3CHBrCH_2CH_3$	137.03	−111.9	91	1 255	1.436 9
bromoethane*						
ethyl bromide	C_2H_5Br	108.97	−118.6	38.4	1 460	1.423 5
bromoform						
tribromomethane	$CHBr_3$	252.75	8.3	149.5	2 890	1.597 6
1-bromohexane*						
hexyl bromide	$CH_3(CH_2)_4CH_2Br$	165.08	−84.7	154–158	1 176	1.447 8
bromomethane*						
methyl bromide	CH_3Br	94.94	−93.6	4	1 675	1.421 8
1-bromo-2-methylpropane*						
isobutyl bromide	$(CH_3)_2CHCH_2Br$	137.03		90–92	1 260	1.435 0
2-bromo-2-methylpropane*						
t-butyl bromide	$(CH_3)_3CBr$	137.03	−20	73	1 221	1.427 8
1-bromonaphthalene*	$C_{10}H_7Br$	207.08	−1	281	1 489	1.657 6
1-bromopentane*						
amyl bromide	$CH_3(CH_2)_4Br$	151.05	−95	130	1 218	1.443 6
2-bromopentane*						
sec-amyl bromide	$CH_3CH_2CH_2CH(Br)CH_3$	151.05	−95	117	1 208	1.441 3
1-bromopropane*						
n-propyl bromide	$CH_3CH_2CH_2Br$	123.00	−110	71	1 354	1.434 3
2-bromopropane*						
isopropylbbromide	$(CH_3)_2CHBr$	123.00	−89	59	1 314	1.425 1
N-bromosuccinimide	$OCCH_2CH_2CO$ $\underset{Br}{\overset{\;}{\vert\,N\,\vert}}$	177.99	d 173		2 098	
2-bromotoluene*						
o-tolyl bromide	$BrC_6H_5CH_3$	171.04	−27	59/10 mm	1 422	1.557
3-bromotoluene*						
m-tolyl bromide	$BrC_6H_5CH_3$	171.04	−40	183.7	1 410	1.551 0

Flash point/°C	Solubility						Hazards	References	
	w	alc	et	ac	bz	Other			
73	i	∞	∞			chl	vT, I, CSA	B.5(3).685	IR.1(1).972D
>110	i	s	s			CS$_2$		B.5(2).506	IR.1(1).936A NMR.2(1).739B
	i	s	s		vs	CCl$_4$ CS$_2$	H, I	B.5(2).479	IR.1(1).948A NMR.2(1).755A
	ss	vs	vs		vs	chl	T, I	B.23(2).211	IR.1(2).731C NMR.2(2).612D
65	ss	s	s		vs		T, FS	B.6(2).82	IR.1(1).169B
>110	∞	∞	∞	s	s		T, C, L	B.2(2).201	IR.1(1).508D
51	i	vs	vs		vs	CCl$_4$	FL, I	B.5(2).158	
23	i	∞	∞	∞		chl	FL, I	B.1(2).290	
21	i		∞	∞		chl	FL, I	B.1(3).293	IR.1(1).66C NMR.2(1).68C
−23	ss	∞	∞			chl	FL, I	B.1(3).171	
	ss	∞	∞		s	chl	T, L	B.1(3).88	
57	i	∞	∞	s		chl	FL, I	B.1(3).391	IR.1(1).58C NMR.2(1).62A
	ss	∞	∞			chl	vT, FG	B.1(3).79	
18	i						FL, CSA	B.1.126	IR.1(1).68D NMR.2(1).70A
18	i						FL	B.1(3).322	IR.1(1).67C NMR.2(1).69B
>110	s^h	∞	∞	s	∞	chl	H	B.5(3).1580	IR.1(1).1027C NMR.2(1).826C
31	i	s	∞		s	chl	FL, I	B.1(3).344	IR.1(1).57D NMR.2(1).61B
20	i	s	s		s	chl	FL, I	B.1(3).345	IR.1(1).68A NMR.2(1).71B
25	ss	s	s	s	s	chl	FL, T, I	B.1(3).239	IR.1(1).56A NMR.2(1).59D
19	ss ss	∞	∞ ss	s vs	s	chl	FL, T, I I, MS	B.1(3).242 B.21(2).306	IR.1(1).798B
78	i	vs	vs		vs	CCl$_4$	H, I	B.5(3).704	IR.1(1).979A NMR.2(1).780D
60	i	s	∞	s		chl	H, T	B.5(3).706	IR.1(1).985A NMR.2(1)787D

(Continued)

Table XI.2 (Continued)

Compound/synonym (s)	Formula	M_r	m. p. /°C	b. p. /°C	Density / kg m³	Refractive index
4-bromotoluene*						
p-tolyl bromide	BrC$_6$H$_5$CH$_3$	171.04	26–29	184.35	1 399	1.547 7
1,3-butadiene*						
bivinyl	H$_2$C:CHCH:CH$_2$	54.09	−109	−4.5	621	1.429 2^{-25}
butane*	CH$_3$CH$_2$CH$_2$CH$_3$	58.13	−138	−0.5	601^0	1.354 3^{-13}
(±)-1, 2-butanediol*						
1, 2-dihydroxybutane	CH$_3$CH$_2$CH(OH)CH$_2$OH	90.12		190	1 002	1.437 8
(±) -1, 3-butanediol*						
1, 3-butylene glycol	CH$_3$CH(OH)CH$_2$CH$_2$OH	90.12		203–204	1 005	1.441 0
1, 4-butanediol*						
tetramethylene glycol	HO(CH$_2$)$_4$OH	90.12	16	230	1 017	1.445 0
DL-2, 3-butanediol	CH$_3$CH(OH)CH(OH)CH$_3$	90.12	7.6	182.5	1 003	1.431 0
1-butanol*						
butyl alcohol	CH$_3$(CH$_2$)$_3$OH	74.12	−90	117.7	809.8	1.399 3
(±)-2-butanol*						
sec-butyl alcohol	CH$_3$CH$_2$CH(OH)CH$_3$	74.12	−114.7	99.5	808	1.397 8
2-butanone*						
ethyl methyl ketone	CH$_3$CH$_2$COCH$_3$	72.11	−87	79.6	805.4	1.378 8
1-butene*						
α-butylene	CH$_3$CH$_2$CH:CH$_2$	56.11	−185.3	−6.3	595	1.396 2
2-butene (cis)*						
cis-β-butylene	CH$_3$CH:CHCH$_3$	56.11	−139	3.7	621	1.393 1^{-25}
2-butene (trans)*						
trans-β-butylene	CH$_3$CH:CHCH$_3$	56.11	−105	1	604	1.384 8^{-25}
butyl acetate	CH$_3$COO(CH$_2$)$_3$CH$_3$	116.16	−78	124–126	882.5	1.394 0
n-butylamine						
1-aminobutane*	CH$_3$(CH$_2$)$_3$NH$_2$	73.14	−49	78	741.4	1.403 1
(±)-sec-butylamine						
DL-2-aminobutane	CH$_3$CH$_2$CH(NH$_2$)CH$_3$	73.14	−72	63	724.6	1.392 8
tert-butylamine						
2-amino-2-methylpropane*	(CH$_3$)$_3$CNH$_2$	73.14	−67	44.4	696	1.378 4
butyraldehyde						
butanal*	CH$_3$CH$_2$CH$_2$CHO	72.11	−99	75.7	817	1.384 3
butyric acid						
butanoic acid*	CH$_3$CH$_2$CH$_2$COOH	88.12	−4	163.5	957.7	1.398 0
butyric anhydride						
butanoic acid anhydride*	(CH$_3$CH$_2$CH$_2$CO)$_2$O	158.20	−75	199–201	966.8	1.4070
(+)-camphene	C$_{10}$H$_{16}$	136.24	44–48	159–160	850	1.457 0^{25}
(±)-camphor	C$_{10}$H$_{16}$O	152.24	178.8	s		
carbon disulphide	CS$_2$	76.14	−112	46.2	1 263	1.631 9
carbon tetrabromide						
tetrabromomethane*	CBr$_4$	331.65	88–90	190	2 961^{100}	1.594 2^{100}
carbon tetrachloride						
tetrachloromethane*	CCl$_4$	153.82	−22.99	76.54	1 594	1.459 5
carbon tetrafluoride						
tetrafluoromethane*	CF$_4$ (Freon-14)	88.01	−150	−129		
carbon tetraiodide						
tetraiodomethane*	CI$_4$	519.63	d 170		4 230	
carbonyl sulphide						
carbon oxysulphide	COS	60.08	−138	−50	21	

Flash point/°C	Solubility						Hazards	References	
	w	alc	et	ac	bz	Other			
85	i	s	s	s	s	chl	H, I	B.5(3).707	IR.1(1).991A NMR.2(1).795C
	i i	s vs	s vs	vs	s	chl	FG, CSA FG, I	B.1(3).929 B.1(2).261	
93	s	s		s				B.1(2).544	IR.1(1).130B NMR.2(1).120D
121	s	s	i				I, MS	B.1(2).545	IR.1(1).131B NMR.2(1).123C
>110 85	∞ ∞	s ∞	ss s	s			I, MS MS	B.1(2).545 B.1(2).546	IR.1(1).131C NMR.2(1).124A
35	s	∞	∞	vs	s		FL, I	B.1(2).387	
26	vs	∞	∞	vs	s		FL, I	B.1(2).404	
−3	vs	∞	∞	∞	∞		FL, I	B.1(2).726	
	i	vs	vs		s		FG	B.1(3).715	
	i	vs	vs		s		FG	B.1(3).728	
22	ss	∞	∞	s	s		FG FL, I	B.1(3).730 B.2(2).140	
−14	∞	s	s				FL, C, I	B.4(2).631	IR.1(1).281A
−19	s	∞	∞	vs		chl	FL, C, I	B.4(2).636	IR.1(1).284D NMR.2(1).243A
−8	∞	∞	∞				vT, FL	B.4(3).323	IR.1(1).285C NMR.2(1).244A
−11	s	∞	∞	vs	vs		FL, C	B.1(2).721	IR.1(1).466A NMR.2(1).357D
76	∞	∞	∞				T, C	B.2(3).576	IR.1(1).482A NMR.2(1).419D
87	d	d	s				C, MS	B.2(2).251	IR.1(1).711C NMR.2(1).601D
36	i	ss	s				FL,	B.5(3).380	IR.1(1).45A NMR.2(1).48C
64	i	vs	vs	s	s	chl	FS, I	B.7(2).104	IR.1(1).441D NMR.2(1).404D
−33	s	∞	∞			chl	FL, T	B.3(2).139	
none	i	s	s			chl	H, I	B.1(3).92	IR.1(1).86B
none	i	s	∞	s	∞	chl	T, CSA	B.1(3).65	
none	ss				s	chl	H,	B.1(3).35	
	i	i				chl	H, I	B.1(3).104	IR.1(1).86C
	ss	s				KOH	vT, FG	B.3(3).104	

(Continued)

Table XI.2 (Continued)

Compound/synonym(s)	Formula	M_r	m.p. /°C	b.p. /°C	Density / kg m³	Refractive index
catechol 1,2-dihydroxybenzene*	C_6H_4-$(OH)_2$	110.11	105	245	1 149	1.604
cetylpyridinium bromide 1-hexadecylpyridinium bromide*	$CH_3(CH_2)_{15}\overset{+}{N}C_5H_5Br^- \cdot H_2O$	402.47	66–68			
cetyltrimethylammonium bromide, 1-hexadecyltrimethyl ammonium bromide*	$CH_3(CH_2)_{15}N^+(CH_3)_3Br^-$	364.46	d >230			
chloral trichloroacetaldehyde*	CCl_3CHO	147.39	−57.5	97.75	1 512	1.455 7
chloramine-T hydrate N-chloro-4-toluene sulphonamide sodium salt	$CH_3C_6H_4SO_2N(Cl)Na \cdot 3H_2O$	281.69	d 161–170			
chloroacetic acid	$ClCH_2COOH$	94.50	63	187.9	1 404[40]	1.435 1[55]
chlorobenzene* phenyl chloride	C_6H_5Cl	112.56	−45.6	132	1 106	1.524 1
2-chloro-1,3-butadiene* chloroprene	$CH_2:CClCH:CH_2$	88.54		59–60	958	1.458 3
1-chlorobutane* n-butyl chloride	$CH_3(CH_2)_3Cl$	92.57	−123	77–78	886	1.402 1
2-chlorobutane* sec-butyl chloride	$CH_3CHClCH_2CH_3$	92.57	−131	68.25	873	1.397 1
chlorodifluoromethane*	$ClCHF_2$ (Freon-22)	86.47	−146	−40.8	1 491[−69]	
1-chloro-2,3-epoxypropane* epichlorohydrin	$\overset{\frown}{OCH_2CHCH_2Cl}$	92.53	−48	116.5	1 183	1.438 0
chloroethane* ethyl chloride	C_2H_5Cl	64.52	−136.4	12.3	891	1.367 6
chloroform trichloromethane*	$CHCl_3$	119.38	−63.5	61.7	1 483	1.445 9
chloromethane* methyl chloride	CH_3Cl	50.49	−97.7	−24.2	916	1.338 9
1-chloro-2-methylpropane* isobutyl chloride	$(CH_3)_2CHCH_2Cl$	92.57	−131	68–69	883	1.397 5
2-chloro-2-methylpropane* tert-butyl chloride	$(CH_3)_3CCl$	92.57	−25	52	842	1.385 7
1-chloronaphthalene*	$C_{10}H_7Cl$	162.62	−20	111–113/5mm	1 194	1.632 6
1-chloropentane* n-amyl chloride	$CH_3(CH_2)_4Cl$	106.60	−99	107.8	882	1.412 7
2-chlorophenol*	ClC_6H_4OH	128.56	9	175	1 263	1.552 4
3-chlorophenol*	ClC_6H_4OH	128.56	33–35	214	1 268[25]	1.563 2
4-chlorophenol*	ClC_6H_4OH	128.56	43–45	220	1 306	1.557 9[40]
1-chloropropane* n-propyl chloride	$CH_3CH_2CH_2Cl$	78.54	−123	46.6	891	1.388 0
2-chloropropane* isopropyl chloride	$(CH_3)_2CHCl$	78.54	−117.2	34–36	862	1.377 7
cholesterol 5-cholesten-3β-ol	$C_{27}H_{44}O$	386.67	148.5	360	1 067	
(+)-cinchonine	$C_{19}H_{22}N_2C$	294.40	258–260	s		
1,8-cineole eucalyptol	$C_{10}H_{18}O$	154.26	1.5	176.4	927	1.458 6

Flash point/°C	Solubility						Hazards	References	
	w	alc	et	ac	bz	Other			
137	s	s	s	vs	s	KOH chl	C, T, H, I	B.6(3).4187	IR.1(1).1101B NMR.2(1).870B
	s						I, MS	B.20(3).2317	
	s						T, I		IR.1(1).394D NMR.2(1).344A
	vs[h]	s[h]	s[h]				T, I	B.1(2).677	
	s		i		i		I		IR.1(2).530D NMR.2(2).849C
	vs	s	s		s	chl	vT, C	B.2(2).187	
23	i	∞	∞	vs		chl	FL, H, I	B.5(2).148	
	ss		∞	∞	∞		FL, H	B.1(3).949	
−6	i	∞	∞				FL	B.1(3).275	
−15	i	∞	∞	vs		chl	FL, H	B.1(3).278	IR.1(1).66B NMR.2(1).68B
none	vs		s	s		chl		B.1(3).41	
33	ss	∞	∞		s		vT, FL, CSA	B.17(2).13	
	ss	vs	∞				FG, I	B.1(3).133	
none	ss	∞	∞	s	∞		vT, CSA	B.1(2)14	
	s	s	∞	∞	∞	chl	FG, MS, H	B.1(3).36	
21	i	s	s				FL	B.1.124	IR.1(1).68C NMR.2(1).69C
18	ss	∞	∞		s	chl	FL	B.1(3).316	IR.1(1).67B NMR.2(1).69A
121	i	s	s		s	CS$_2$	H, I	B.5(3).1570	IR.1(1).1027B
11	i	∞	∞	s		chl	FL, I	B.1(3).339	IR.1(1).57C NMR.2(1).61A
63	ss	s	s	vs			C, H, T	B.6(3).671	IR.1(1).1072A
110	ss s[h]	s		vs			C, H, T	B.6(3).681	IR.1(1).1075A NMR.2(1).872C
115	i	vs	vs	vs			C, H, T	B.6(3).684	IR.1(1).1078B
18	ss	∞	∞	s		chl	FL, H, I	B.1(3).219	IR.1(1).55D NMR.2(1).59C
−35	ss	∞	∞	s		chl	FL, H, I	B.1(3).221	
	i	ss[h]	vs	s		chl		B.6(3).2507	IR.1(2).1048B NMR.2(2).919B
	i	ss	ss	i	i	chl	T, H, LS	B.23(2).369	IR.1(2).864C NMR.2(2).741B
50	i	s	s			chl		B.17(2).32	IR.1(1).245C

(Continued)

Table XI.2 (Continued)

Compound/synonym (s)	Formula	M_r	m. p. /°C	b. p. /°C	Density / kg m³	Refractive index
trans cinnamic acid						
trans-3-phenylpropenoic acid*	$C_6H_5CH:CHCOOH$	148.16	135–136	300	1 248	
citric acid						
2-hydroxy-1,2,3-propanetri-carboxylic acid*	$HOOCCH_2C(OH)(COOH)-CH_2COOH$	192.13	153	d	1 542^{18}	
coumarin						
1,2-benzopyrone	$C_6H_4O.CO.CH:CH$	146.15	71	302	935	
creatine						
(α-methylguanido) acetic acid	$NH:C(NH_2)N(CH_3)CH_2-COOH$	131.14	303		1330	
o-cresol, 2-hydroxytoluene						
2-methylphenol*	$CH_3C_6H_4OH$	108.15	32	191	1027	1.536 1
m-cresol, 3-hydroxytoluene,						
3-methylphenol*	$CH_3C_6H_4OH$	108.15	11.5	203	1 034	1.543 8
p-cresol, 4-hydroxytoluene,						
4-methylphenol*	$CH_3C_6H_4OH$	108.15	32–34	202	1 018	1.531 2
crotonaldehyde						
2-butenal* (trans)	$CH_3CH:CHCHO$	70.09	−74	104	849	1.436 5
crotonic acid						
2-butenoic acid (trans)	$CH_3CH:CHCOOH$	86.09	72–74	185	1 027	1.424 9^{77}
cumene						
isopropylbenzene	$C_6H_5CH(CH_3)_2$	120.20	−96	152–154	862	1.491 5
1, 3-cyclohexadiene*						
1, 2-dihydrobenzene	$CH:CHCH:CHCH_2CH_2$	80.13	−89	80.5	841	1.474 1
1, 4-cyclohexadiene*						
1, 4-dihydrobenzene	$CH:CHCH_2CH:CHCH_2$	80.13	−49.2	85.6	847	1.472 0
cyclohexane*						
hexahydrobenzene	$CH_2(CH_2)_4CH_2$	84.16	6.55	80.74	779	1.426 6
cyclohexanol*						
hexahydrophenol	$CH_2(CH_2)_4CHOH$	100.16	25.1	160–161	963	1.464 1
cyclohexanone*	$CH_2(CH_2)_4CO$	98.15	−47	155	948	1.450 7
cyclohexene*						
3, 4, 5, 6-tetrahydrobenzene	$CH:CH(CH_2)_3CH_2$	82.15	−104	83	811	1.446 5
cyclopentane*	$CH_2(CH_2)_3CH_2$	70.14	−93.9	49.26	746	1.406 5
cyclopropane*	$CH_2CH_2CH_2$	42.08	−128	−32.7	720^{-79}	1.379 9^{-42}
DL-cysteine						
DL-2-amino-3-mercaptopropionic acid	$HSCH_2CH(NH_2)COOH$	121.16	d 225			
DL-cystine	$[-SCH_2CH(NH_2)COOH]_2$	240.30	260			
dansyl chloride, 5-dimethylamino-1-naphthalene sulphonyl chloride	$(CH_3)_2NC_{10}H_6SO_2Cl$	269.75	69–71			
decalin (cis), decahydro-naphthalene, bicyclo [4, 4, 0] decane	$C_{10}H_{18}$	138.25	−43	192–194	897	1.481 0
decane*	$CH_3(CH_2)_8CH_3$	142.29	−30	174	730	1.410 2

Flash point/°C	Solubility					Hazards	References	
	w	alc	et	ac	bz	Other		
	i	vs	s	s	s	chl	I	B.9(2).377	NMR.2(2).174A
	vs	vs	s	i			I	B.3(2).359	IR.1(1).524B
	s^h	s	vs			chl	T, H, CSA	B.17(1).22	IR.1(2).322B NMR.2(2).309A
	s	ss	i				I	B.4(2).796	
81	s	vs	vs	∞	∞	CCl$_4$	vT, C	B.6(3).1233	IR.1(1).1069B NMR.2(1).867B
86	ss	∞	∞	∞	∞	CCl$_4$	vT, C	B.6(3).1286	IR.1(1).1073B NMR.2(1).871A
89	ss	∞	∞	∞	∞	CCl$_4$	vT, C	B.6(3).1341	IR.1(1).1075D NMR.2(1).873A
8	s	vs	vs	vs	∞		vT, FL, L	B.1(2).787	
87	vs	vs	s	s			C, T	B.2(2).390	
46	i	∞	∞	∞	∞	CCl$_4$	I	B.5(2).306	IR.1(1).932A
26	i	s	vs		s	chl	FL, I	B.5(2).79	IR.1(1).46C NMR.2(1).49D
−6	i	∞	∞		s	chl	FL, I	B.5(2).80	IR.1(1).46D NMR.2(1).50A
−18	i	∞	∞	∞	∞	CCl$_4$	FL, I	B.5.21	
67	s	s	s	s	∞	CS$_2$	I, MS	B.6(2).5	IR.1(1).157C NMR.2(1).146C
46	s	s	s	s	s	chl	C, T, H	B.7(2).5	IR.1(1).432C NMR.2(1).394C
−12	i	∞	∞	∞	∞		FL, I	B.5(2).37	
−37	i	∞	∞	∞	∞	CCl$_4$	FL, I	B.5(2).4	
	s	vs	vs	s			FG	B.5(2).3	
	vs	vs	i	i	i			B.4(2).920	IR.1(1).593B
	s^h	i				alk		B.4(2).936	
	d						C, MS, I		IR.1(2).527C NMR.2(2).846C
58	i	∞	vs	vs	∞	chl	FL, I	B.5(2).5657	IR.1(1).38D NMR.2(1).43A
46	i	∞	s				FL, I	B.1(3).519	IR.1(1).2C NMR.2(1).10C

(Continued)

Table XI.2 (Continued)

Compound/synonym (s)	Formula	M_r	m. p. /°C	b. p. /°C	Density / kg m³	Refractive index
decanoic acid*						
capric acid	$CH_3(CH_2)_8COOH$	172.27	30–32	268–270	886	1.428 8
1-decanol*						
decyl alcohol	$CH_3(CH_2)_9OH$	158.29	7	229–231	829	1.437 2
n-decylamine						
1-aminodecane*	$CH_3(CH_2)_9NH_2$	157.30	12–14	220.5	794	1.436 9
diacetone alcohol						
4-hydroxy-4-methyl-2-pentanone*	$(CH_3)_2C(OH)CH_2COCH_3$	116.16	−44	166	939	1.421 3
1,2-diaminoethane*						
ethylenediamine	$H_2NCHCH_2NH_2$	60.11	8.5	117	899	1.456 5
1,6-diaminohexane*						
hexamethylenediamine	$H_2N(CH_2)_6NH_2$	116.21	41–42	204–205		
1,2-diaminopropane*						
1,2-propanediamine*	$CH_3CH(NH_2)CH_2NH_2$	74.13		120	870	1.446 0
1,3-diaminopropane*						
trimethylenediamine	$H_2N(CH_2)_3NH_2$	74.13	−12	140	888	1.460 0
1,2-dibromoethane*						
ethylene dibromide	$BrCH_2CH_2Br$	187.87	9.8	131–132	2 179	1.538 7
dibromomethane*						
methylene bromide	CH_2Br_2	173.85	−52	97	2 497	1.541 0
1,3-dibromopropane*						
trimethylene bromide	$Br(CH_2)_3Br$	201.90	−34	167	1 982	1.522 5
dibutyl phthalate	$C_6H_4[COO(CH_2)_3CH_3]_2$	278.35	−35	340	1 047	1.491 1
dichloroacetic acid	$Cl_2CHCOOH$	128.94	9–11	194	1 563	1.465 8
1,2-dichlorobenzene*	$C_6H_4Cl_2$	147.00	−15	179–180	1 305	1.551 5
1,3-dichlorobenzene*	$C_6H_4Cl_2$	147.00	−24.7	173	1 288	1.545 9
1,4-dichlorobenzene*	$C_6H_4Cl_2$	147.00	54–56	174	1 248	1.528 5
dichlorodifluoromethane*	CCl_2F_2 (Freon-12)	120.91	−158	−29.8	1 750⁻¹¹⁵	
1,2-dichloroethane*						
ethylene dichloride	$ClCH_2CH_2Cl$	98.96	−35	83	1 253	1.444 8
1,1-dichloroethylene						
1,1-dichloroethene*	$H_2C{:}CCl_2$	96.94	−122	30–32	1 218	1.424 9
1,2-dichloroethylene (cis)						
1,2-dichloroethene*	$ClCH{:}CHCl$	96.94	−80	60	1 284	1.449 0
dichlorofluoromethane*	$FCHCl_2$ (Freon-21)	102.92	−135	8.9	1 405	1.372 4
dichloromethane*						
methylene chloride	CH_2Cl_2	84.93	−97	40	1 327	1.424 2
2,4-dichlorophenol*	$C_6H_3Cl_2OH$	163.00	42–43	210		
1,2-dichloropropane*						
propylene chloride	$CH_3CHClCH_2Cl$	112.99	−100	95–96	1 156	1.438 4
1,3-dichloropropane*						
trimethylene chloride	$ClCH_2CH_2CH_2Cl$	112.99	−99	120.4	1 188	1.448 7
2,2-dichloropropane*						
acetone dichloride	$CH_3CCl_2CH_3$	112.99	−34	69	1 112	1.414 5
α,α – dichlorotoluene*						
benzal chloride	$C_6H_5CHCl_2$	161.03	−16.4	205	1 256	1.550 2
diethanolamine						
2,2′-iminodiethanol	$(HOCH_2CH_2)_2NH$	105.14	28	217/150 mm	1 097	1.477 6

Flash point/°C	Solubility						Hazards	References	
	w	alc	et	ac	bz	Other			
>110	ss	∞ʰ	s	vs	vs	chl	I	B.2(2).309	IR.1(1).483C NMR.2(1).421B
82	i	∞	∞	∞	∞	chl	I	B.1(2).459	
85	ss	∞	∞	∞	∞	chl	C, T	B.4.199	IR.1(1).282C NMR.2(1).240D
61	∞	∞	∞				I	B.1(3).3234	IR.1(1).421C NMR.2(1).382D
100	vs	∞	i		i		FL, C	B.4(2).676	IR.1(1).289C
81	vs	s			s		C, MS	B.4(2).710	IR.1(1).292C NMR.2(1).251A
33	vs		i			chl	FL, C,	B.4(2).697	
48	s	∞	∞				vT, C, FL	B.4(3).552	
none	ss	vs	∞	∞	s		T, CSA	B.1(3).182	
none	ss	∞	∞	∞			H	B.1(3).85	
54		s	s			chl	H, I	B.1(3).248	IR.1(1).73B NMR.2(1).74C
171	i	∞	∞		∞		I	B.9(2).586	
>110	∞	∞	∞	s			T, C	B.2(2).194	IR.1(1).508B NMR.2(1).444B
65	i	s	s	∞	∞	CCl₄	T, I	B.5(2).153	
63	i	s	s	∞	s	CCl₄	T, I	B.5(2).154	IR.1(1).982D NMR.2(1).785D
65	i s	∞ s	s s	∞	s	chl	T, I, CSA	B.5(2).154 B.1.61	IR.1(1).988B
15	ss	vs	∞	s	s	chl	H, FL, CSA	B.1(3).141	
−22	i	s	vs	s	s	chl	H, FL, L	B.1(3).647	IR.1(1).98B NMR.2(1).90B
6	ss i	∞ s	∞ s	∞	vs	chl chl	H, FL, MS	B.1(3).651 B.1(3).47	IR.1(1).98A
none	ss	∞	∞				I, T	B.1(2).13	
113	ss	s	s	s		chl	H, I	B.6(3).699	IR.1(1).1089C NMR.2(1).885C
4	ss	s	s	s		chl	FL, I, H	B.1(3).225	IR.1(1).80C NMR.2(1).80C
32	ss	vs	vs	s		chl	FL, I, H	B.1(3).227	IR.1(1).73A NMR.2(1).74B
−5	i	s	∞	s		chl	FL, I, H	B.1(3).228	IR.1(1).81B
92	i	s	s				C, L	B.5(3).696	IR.1(1).973B
137	vs	vs	ss	ss			C, I, MS	B.4(2).729	IR.1(1).340D NMR.2(1).297D

(Continued)

Table XI.2 (Continued)

Compound/synonym (s)	Formula	M_r	m. p. /°C	b. p. /°C	Density / kg m^3	Refractive index
diethylamine	$(C_2H_5)_2NH$	73.14	−48	54–55	707	1.386 1
N,N-diethylaniline diethylphenylamine	$C_6H_5N(C_2H_5)_2$	149.24	−38	217	935	1.540 9
diethyl ether, ether ethoxyethane*	$(C_2H_5)_2O$	74.12	−116	34.51	714	1.352 6
dimethylamine	$(CH_3)_2NH$	45.09	−93	7	680 ^{0}	1.350 ^{17}
N,N-dimethylaniline	$C_6H_5N(CH_3)_2$	121.18	2.5	194	956	1.558 2
dimethyl carbonate carbonic acid dimethyl ester*	$(CH_3O)_2CO$	90.08	2–4	90	106 9	1.368 7
dimethyl ether, methyl ether, methoxymethane*	$(CH_3)_2O$	46.07	−140	−23		
N,N-dimethylformamide	$HCON(CH_3)_2$	73.10	−61	153	949	1.430 5
dimethylglyoxime 2,3-butanone dioxime*	$CH_3C(:NOH)C-(:NOH)CH_3$	116.13	241–242	s		
1,2-dimethyl phthalate	$C_6H_4-(COOCH_3)_2$	194.19	2	284	1 190	1.513 8
dimethyl sulphate, sulphuric acid, dimethyl ester*	$(CH_3O)_2SO_2$	126.13	−32	188	1 329	1.387 4
dimethyl sulphide	$(CH_3)_2S$	62.13	−98	38	848	1.435 5
dimethyl sulphoxide (DMSO) methyl sulphoxide	$(CH_3)_2SO$	78.13	18.4	189	1 101	1.4787
1,2-dinitrobenzene	$C_6H_4(NO_2)_2$	168.11	118.5	194/30 mm	1 312^{120}	
1,3-dinitrobenzene*	$C_6H_4(NO_2)_2$	168.11	90	297	1 368	
1,4-dinitrobenzene*	$C_6H_4(NO_2)_2$	168.11	174	183/34 mm	1 625	
2,4-dinitro-1-fluorobenzene*	$(O_2N)_2C_6H_3F$	186.10	28–30	178/25 mm	1 482	1.569 0
2,4-dinitrophenol*	$(O_2N)_2C_6H_3OH$	184.11	106–108	s	1 683 ^{24}	
2,4-dinitrophenylhydrazine*	$(O_2N)_2C_6H_3NHNH_2$	198.14	194	d		
1,4-dioxane* diethylene dioxide	$CH_2CH_2OCH_2CH_2O$	88.11	11.8	101	1 034	1.422 5
diphenylamine N-phenylaniline	$(C_6H_5)_2NH$	169.23	53–55	302	1 160^{22}	
1-dodecanol lauryl alcohol, dodecyl alcohol	$CH_3(CH_2)_{11}OH$	186.34	26	260–262	820	
dodecyl sulphate, Na salt sodium lauryl sulphate	$CH_3(CH_2)_{11}OSO_3Na$	288.38	204–207			
(1R,2S)-(−)-ephedrine, (1R,2S)-(−)-α-(1-methylaminoethyl)-benzyl alcohol	$C_6H_5CH[CH(NHCH_3)CH_3]$ $-OH$	165.24	37–39	218–220		
erythritol (meso)	$HOCH_2[CH(OH)]_2CH_2OH$	122.12	121.5	329–231	1 451	
ethane	CH_3CH_3	30.07	−183.3	−88.6	572$^{−108}$	1.037 7 ^{0}
ethanesulphonic acid* ethylsulphonic acid	$C_2H_5SO_3H$	110.13	−17	123/1 mm	1 350	1.433 5
ethanethiol* ethyl mercaptan	C_2H_5SH	62.13	−144.4	35	839	1.431 0

Flash point/°C	Solubility						Hazards	References	
	w	alc	et	ac	bz	Other			
−28	vs	∞	s				FL, C, I	B. 4(2). 590	IR. 1(1). 295B NMR. 2(1). 253D
97	ss	s	vs	s		chl	vT, I	B. 12(2). 92	IR. 1(1). 1191D
−40	ss	∞		vs	∞	chl	FL, I	B. 1(2). 311	
	vs	s	s				FG, C, I	B. 4(2). 550	
62	ss	s	s	s	s	chl	vT, I	B. 12(2). 82	IR. 1(1). 1191B NMR. 2(1). 990D
18	i	s	s				FL, MS, H	B. 3(2). 3	IR. 1(1). 610D NMR. 2(1). 516C
	s	s	s	s	ss	chl	FG	B. 1(3). 1188	
57	∞	∞	∞	∞	∞	chl	H, I	B. 4. 58	
	i	vs	vs		ss		H, I	B. 1(2). 826	IR. 1(2). 922D NMR. 2(2). 968C
146	i	∞	∞	s			I	B. 9(2). 584	
83	s	∞	s	s			vT, CSA	B. 1. 283	IR. 1(1). 903D NMR. 2(2). 804C
−36	i	s	s			MeOH	FL, stench	B. 1. 288	
95	s	s	s	s			I, MS	B. 1. 289	
	i	s			s	chl	vT	B. 5(2). 193	
	ss[h]	vs[h]	s	vs	s[h]	chl	vT	B. 5(2). 193	IR.1(1). 1338A NMR. 2(1). 1139D
	i	ss		s	s		vT	B. 5(2). 195	IR. 1(1). 1344B NMR. 2(1). 1146A
>110		s[h]					vT, I, CSA	B. 5(3). 634	IR. 1(1). 1369D NMR. 2(1). 1173C
	ss	s	s	s	s	chl	vT, FS	B. 6(3). 854	IR. 1(1). 1370C NMR. 2(1). 1174C
	i	s[h]	ss		ss		vT, FS, I	B. 15. 489	IR. 1(1). 1371D NMR. 2(1). 1175C
12	∞	∞	∞	∞	∞		FL, H, CSA	B. 19(2). 4	
152	i	vs	s	vs	vs	py CCl$_4$	T, I	B. 12(2). 101	IR. 1(1). 1190B
>110	i	s	s				I	B. 1(2). 463	IR. 1(1). 112A NMR. 2(1). 103D
	s						I	B. 1(3). 1786	
85	s	s	s	s		chl	T, I,	B. 13. 636	IR. 1(1). 1277C NMR. 2(1). 1078C
	vs	ss	i			py	I	B. 1(3). 2356	IR. 1(1). 184C NMR. 2(2). 901A
	i	ss		ss	s		FG	B. 1(3). 120	
>110	s	s			i	alk	C, T	B. 4(2). 525	IR. 1(1). 885C NMR. 2(2). 788C
−17	ss	s	s	s		alk	FL, stench	B. 1(2). 341	IR. 1(1). 255A NMR. 2(1). 216D

(Continued)

Table XI.2 (Continued)

Compound/synonym(s)	Formula	M_r	m. p. /°C	b. p. /°C	Density / kg m^3	Refractive index
ethanol,* ethyl alcohol, alcohol	C_2H_5OH	46.07	−117.3	78.3	789	1.361 1
ethanolamine 2-aminoethanol*	$H_2NCH_2CH_2OH$	61.08	10.5	170	1 018	1.454 0
2-ethoxyethanol,* cellosolve, glycol monoethyl ether	$C_2H_5OCH_2CH_2OH$	90.12	−90	135	930	1.408 0
ethyl acetate	$CH_3COOC_2H_5$	88.12	−84	77.06	902	1.372 3
ethyl acetoacetate ethyl-3-oxobutanoate*	$CH_3COCH_2COOC_2H_5$	130.14	−43	181	1 028	1.419 4
ethylamine aminoethane*	$C_2H_5NH_2$	45.09	−81	16.6	683	1.366 3
ethyl benzene*	$C_6H_5C_2H_5$	106.17	−95	136.2	867	1.495 9
ethyl benzoate	$C_6H_5COOC_2H_5$	150.18	−34.6	212	1 047	1.500 7
ethylene ethene*	$H_2C{:}CH_2$	28.05	−169	−104	1.26^0	
ethylenediaminetetraacetic acid,	$(HOOCCH_2)_2NCH_2CH_2N(CH_2COOH)_2$	292.24	d 252			
ethylene glycol 1, 2-ethanediol*	$HOCH_2CH_2OH$	62.07	−11.5	198	1 109	1.431 8
ethylene oxide 1, 2-epoxyethane*	CH_2CH_2O	44.05	−111	13.5/ 746 mm	882^{10}	1.359 7^7
ethyl formate ethyl methanoate*	$HCOOC_2H_5$	74.08	−80	54.5	917	1.359 8
ethyl nitrite nitroethane*	CH_3CH_2ONO	75.07	−90	115	1 045	1.391 7
ethyl salicylate ethyl-2-hydroxybenzoate*	$2\text{-}(OH)C_6H_4COOC_2H_5$	166.18	2–3	234	1 131	1.521 9
ethyl sulphone diethyl sulphone*	$(C_2H_5)_2SO_2$	122.19	73–74	248	1 357	
eugenol 4-allyl-2-methoxyphenol	$(H_2C{:}CHCH_2)C_6H_3\text{-}(OCH_3)OH$	164.21	−8	253–254	1 065	1.540 5
fluorescein 3', 4'-dehydroxyfluoran	$C_{20}H_{12}O_5$	332.32	d 320			
fluorobenzene	C_6H_5F	96.11	−42	85	1 023	1.465 3
formaldehyde methanal*	HCHO	30.03	−92	−21	815	
formamide	$HCONH_2$	45.04	2.5	111/20 mm	1 133	1.447 2
formic acid methanoic acid*	HCOOH	46.03	8.4	100.7	1 220	1.371 4
D-(−)-fructose levulose	$C_6H_{12}O_6$	180.16	d 103		1 600	
fumaric acid trans-butenedioic acid*	HOOCCH:CHCOOH	116.07	299–300 sub		1 635	
2-furaldehyde furfural	OCH:CHCH:C.CHO	96.09	−38	162	1 159	1.526 1
furan 1, 4-epoxy-1, 3-butadiene*	CH:CHCH:CHO	68.08	−85.7	31.4	951	1.421 4

Flash point/°C	Solubility						Hazards	References	
	w	alc	et	ac	bz	Other			
8	∞		∞	∞	s	chl	FL, T	B. 1(3). 1223	
93	∞	∞	ss		ss	chl	C, T, I, MS	B. 4(2). 717	
44	∞	∞	∞	vs			T, I, teratogen	B. 1(2). 518	IR. 1(1). 223A
−3	s	∞	∞	vs	vs	chl	FL, I	B. 2(2). 129	
84	vs	∞	∞		s	chl	I	B. 3(2). 1204	
	∞	∞	∞				FG, C, I	B. 4(2). 586	NMR. 2(1). 237D
22	i	∞	∞				FL, I	B. 5(2). 274	
84	i	s	∞	s	s	chl	I	B. 9(2). 88	IR. 1(2). 291C
									NMR. 2(2). 281B
	i	ss	s	ss	ss		FG	B. 1(1). 75	
							I	B. 4(3). 1187	
110	∞	∞	s	∞	ss	chl	H, I, MS	B. 1(3). 2053	
−19	s	s	s	s	s		FL, T	B. 17(2). 9	
−19	s	∞	∞	s			FL, MS	B. 2(3). 31	IR. 1(1). 599B
									NMR. 2(1). 507B
15	ss	∞	∞	s		alk	FL, O	B. 1(3). 199	
107	i	∞	vs				I	B. 10 (2). 47	IR. 1(2). 295D
									NMR. 2(2). 283D
	s		s[h]		vs			B. 1(2). 345	IR. 1(1). 880D
									NMR. 2(2). 782D
>110	i	∞	∞			chl	I	B. 6(2). 921	IR. 1(1). 1091A
									NMR. 2(1). 896D
	ss[h]	ss	ss	vs	ss	MeOH py		B. 19(2). 249	IR. 1(2). 1010A
−12	i	∞	∞	∞	∞	CCl_4	FL	B. 5(2). 147	IR. 1(1). 971A
									NMR. 2(1). 773A
	s	s	∞	∞	∞	chl	CSA, T	B. 1(2). 619	
								B. 6(2). 1244	
154	∞	∞	δ	s	i		H, I, T, teratogen	B. 2. 26	
68	∞	∞	∞	vs	s		C	B. 2(2). 3	
	vs	s		ss		MeOH py		B. 31. 321	
	ss	s	ss	ss			I	B. 2(2). 631	IR. 1(1). 502D
54	s	vs	∞	vs	s	chl	C, T, AS	B. 17(2). 305	
−35	i	vs	vs	s	s		FL, H, I	B. 17(2). 34	IR. 1(2). 579D

(Continued)

Table XI.2 (Continued)

Compound/synonym(s)	Formula	M_r	m. p. /°C	b. p. /°C	Density / kg m^3	Refractive index
furfuryl alcohol (2-furyl)-methanol*	OCH:CHCH:CCH$_2$OH	98.10	−29	170	1 135	1.486 8
D-(+)-galactose	O.(CH.OH)$_4$CH.CH$_2$OH	180.16	159–161			
D-(+)-glucosamine D, β -2-amino-glucose	C$_6$H$_{13}$O$_5$N	179.17	d 110			
D-(+)-glucosamine hydrochloride	C$_6$H$_{13}$NO$_5$.HCl	215.64	d 190			
α-D-glucose dextrose	O.(CH.OH)$_4$CH.CH$_2$OH	180.16	156–158		1 562^{18}	
D-glucuronic acid	O.(CH.OH)$_4$CH.COOH	194.14	159–161			
L-glutamic acid, (S)-(+)-glutamic acid, 2-aminopentanedioic acid*	HOOCCH$_2$CH$_2$ –CH(NH$_2$)COOH	147.13	d 205		1 538	
L-glutamine, (S)-(+)-glutamine	H$_2$NOCCH$_2$CH$_2$CH –(NH$_2$)COOH	146.15	d 185			
glutaraldehyde 1, 5-pentanedial*	OCH(CH$_2$)$_3$CHO	100.13	−6	d 187–188		
glutaric acid pentanedioic acid*	HOOC(CH$_2$)$_3$COOH	132.13	95–98	200/20 mm	1 424^{25}	
D-glyceraldehyde [(R)-(+)]- D-2, 3-dihydroxypropanal*	HOCH$_2$CH(OH)CHO	90.08	145	140/0.8 mm	1 455^{18}	1.494 0
glycerol 1, 2, 3-propanetriol*	HOCH$_2$CH(OH)CH$_2$OH	92.09	20	182/20 mm	1 261	1.471 6
glycerol triacetate triacetin	CH$_3$COOCH(CH$_2$OOCCH$_3$)$_2$	218.21	−4	258	1 159	1.430 1
glycerol trinitrate nitroglycerin	O$_2$NOCH$_2$CH –(ONO$_2$)CH$_2$ONO$_2$	227.09	13	256 explodes	1 593	1.478 6^{12}
glycerol trioleate, glyceroltri (cis-9-octadecenoate)*	C$_{17}$H$_{33}$COOCH –(CH$_2$OOCC$_{17}$H$_{33}$)$_2$	885.47	−32	235–240/ 18mm	899	1.462 1^{40}
glycerol tripalmitate glycerol trihexadecanoate*	C$_{15}$H$_{31}$COOCH –(CH$_2$OOCC$_{15}$H$_{31}$)$_2$	807.35	66–67	310–320	875	1.438 1^{80}
glycerol tristearate glycerol trioctadecanoate*	C$_{17}$H$_{35}$COOCH –(CH$_2$OOCC$_{17}$H$_{35}$)$_2$	891.51	54		856	1.439 9^{80}
glycine aminoacetic acid	H$_2$NCH$_2$COOH	75.07	d 245		828^{17}	
glycogen	(C$_6$H$_{10}$O$_5$)$_n$	(162.14)$_n$	d 255			
glycolic acid hydroxyacetic acid	HOCH$_2$COOH	76.05	80	d		
glycylglycine diglycine	H$_2$NCH$_2$CONHCH$_2$COOH	132.12	d 215			
N-(N-glycylglycyl) glycine, triglycine	H$_2$NCH$_2$CONHCH$_2$ –CONHCH$_2$COOH	189.17	d 246			
glyoxylic acid oxo-acetic acid*	OHCCOOH	74.04	70–75			
guaiacol 2-methoxyphenol*	2-(CH$_3$O)C$_6$H$_4$OH	124.15	27–29	205	1 129	1.542 9

Flash point/°C	w	alc	et	ac	bz	Other	Hazards	References	
65	∞	vs	vs				T, I	B. 17 (2). 113	IR. 1(2). 581C
	vs[h]	ss	i		i	py		B. 1. 909	IR. 1(1). 192A
								B. 31. 295	
	vs	ss	i		i			B. 31. 167	
	vs	s[h]						B. 31. 169	IR. 1(1). 350A
									NMR. 2(1). 306A
	vs	ss		i				B. 1. 879	
	s	s						B. 31. 261	
	ss	i	ss	i	i			B. 4(2). 910	IR. 1(1). 590A
									NMR. 2(1). 497B
	s	i	i	i				B. 4(3). 1540	IR. 1(1). 782D
	∞	∞		s			C, T	B. 1(2). 831	IR. 1(1). 480C
	vs	vs	vs		i	chl	I	B. 2(3). 1685	IR. 1(1). 493C
									NMR. 2(1). 430A
>110	s	ss	ss	i				B. 1(1). 427	IR. 1(1). 479C
160	∞	∞	ss	i		chl	I, MS	B. 1(3). 2297	IR. 1(1). 183A
148	ss	∞	∞	vs	∞	chl		B. 2(2). 160	IR. 1(1). 620C
	ss	s	∞	vs	s	MeOH	E	B. 1(3). 2328	
i	i	ss	vs			chl		B. 2(2). 440	
	i	ss	vs		s	chl		B. 2(2). 340	IR. 1(1). 621A
									NMR. 2(1). 525C
	i	i		s	ss	chl		B. 2(3). 1035	
	vs	i	i	ss				B. 4(1). 462	
	vs	i	i						
	s	s	s				H, I, MS	B. 3(2). 167	IR. 517A
									NMR. 2(1) 451A
	vs	ss	i					B. 4. 371	IR. 1(1). 778B
									NMR. 2(1). 655C
	s[h]	i	i						IR. 1(1). 778C
>110	vs	ss	ss	ss			C, MS	B. 3(2). 385	IR. 1(1). 530B
82	ss	s	s			chl	T, H, I	B. 6(2). 776	IR. 1(1). 1072C
									NMR. 2(1). 870C

(Continued)

Table XI.2 (Continued)

Compound/synonym(s)	Formula	M_r	m.p. /°C	b.p. /°C	Density / kg m^3	Refractive index
heptane*	$CH_3(CH_2)_5CH_3$	100.21	−91	98.4	684	1.3878
heptanoic acid*	$CH_3(CH_2)_5COOH$	130.19	−7.5	223	918	1.4170
1-heptanol* n-heptyl alcohol	$CH_3(CH_2)_5CH_2OH$	116.21	−34	176	822	1.4249
hexachlorobenzene*	C_6Cl_6	284.79	227–229	332	1 569	
hexachloro-1,3-butadiene*	$Cl_2C{:}CClCCl{:}CCl_2$	260.76	−21	215	1 682	1.5550
hexachloroethane* perchloroethane	Cl_3CCCl_3	236.74	186–187		2 091	
hexafluoroethane* perfloroethane	F_3CCF_3 (Freon 116)	138.01	−100	−78	1 590^{-78}	
hexamethylenetetramine hexamine	$(CH_2)_6N_4$	140.19	s 280		1 331^{-5}	
hexane*	$CH_3(CH_2)_4CH_3$	86.18	−95	68.95	660	1.3750
hexanoic acid* n-caproic acid	$CH_3(CH_2)_4COOH$	116.16	−3	205	927	1.4163
1-hexanol* n-hexyl alcohol	$CH_3(CH_2)_4CH_2OH$	102.18	−46.7	156.5	814	1.4178
2-hexanone* n-butyl methyl ketone	$CH_3(CH_2)_3COCH_3$	100.16	−57	127	812	1.4007
1-hexene*	$CH_3CH_2CH_2CH_2CH{:}CH_2$	84.16	−140	64	673	1.3837
hippuric acid N-benzoylglycine	$C_6H_5CONHCH_2COOH$	179.18	189–193		1 371	
histamine 4-imidoazolethylamine	$H_2NCH_2CH_2C{:}CHNHCH{:}N$	111.15	86	167/0.8 mm		
L-histidine [(S)-(+)-] L-2-amino-3-(4-imidazoyl)- propionic acid	$N{:}CHNHCH{:}CCH_2$ $-CH(NH_2)COOH$	155.16	d 282			
homophthalic acid (2-carboxyphenyl) acetic acid*	$HOOCCH_2C_6H_4COOH$	180.17	185–87			
hydroquinone 1,4-dihydroxybenzene*	$C_6H_4(OH)_2$	110.11	172–175	285	1 328^{15}	
cis-4-hydroxy-L-proline, (2S-4S)-(−)-4-hydroxy-2- pyrrolidinecarboxylic acid	$NH.CH_2CH(OH)CH_2CH$ $-COOH$	131.13	d 257			
8-hydroxyquinoline, oxine	$N{:}CHCH{:}CHC_6H_3(OH)$	145.16	75–76	266/752 mm	1 034^{209}	
imidazole 1,3-diazole; glyoxaline	$CH{:}CHN{:}CHNH$	68.08	89–91	257	1 030^{101}	1.4801^{101}
indene	$C_6H_4CH_2CH{:}CH$	116.16	−2	181.6	996	1.5768
indole 1-benzo[b]pyrrole	$C_6H_4NHCH{:}CH$	117.15	52–54	254	1 220	
iodobenzene* phenyl iodide	C_6H_5I	204.01	−31	188	1 823	1.6195
iodoethane* ethyl iodide	C_2H_5I	155.97	−108	69–73	1 936	1.5133

Flash point/°C	Solubility					Hazards	References	
	w	alc	et	ac	bz	Other		
−1	i	vs	∞	∞	∞	chl	FL,	B. 1(3). 415	
>110	ss	s	s	s			C	B. 2(2). 294	
73	ss	∞	∞				T	B. 1(3). 1679	IR. 1(1). 110D NMR. 2(1). 102C
	i	ss	s		vs^h	chl	T, I, CSA	B. 5(2). 157	IR. 1(1). 1016B
none	i	s	s				C, CSA	B. 1(3). 955	IR. 1(1). 102A
none	i	vs	vs		s		I, CSA	B. 1(3). 168	
none	i	ss	ss					B. 1(3). 132	
250	vs	s	ss	s	ss	chl	FS, I	B. 1(2). 745	IR. 1(1). 378B
−23	i	vs	s			chl	FL, I	B. 1(3). 374	IR. 1(1). 1B NMR. 2(1). 9B
104	i	s	s				C, T	B. 2(2). 281	NMR. 2(1). 420B
60	ss	s	∞	s	∞	chl	I	B. 1(3). 1650	IR. 1(1). 110C NMR. 2(1). 102B
35	ss	∞	∞	s			T, I, FL	B. 1(2). 745	IR. 1(1). 408A NMR. 2(1). 371C
−26	i	s	s		s	chl	I, FL	B. 1(3). 800	
	s	s	ss	s				B. 9(2). 174	IR. 1(2). 376D NMR. 2(2). 353C
	s	s	ss			chl	H, I, MS	B. 25(2). 302	
	s^h	s^h	i					B. 25(2). 404	IR. 1(2). 621D NMR. 2(2). 493B
	s^h	s	ss		i			B. 9(2). 617	IR. 1(2). 153B NMR. 2(2). 149A
	s	vs	s	vs	i	CCl$_4$	T, I	B. 6(2). 832	IR. 1(1). 1107D
	vs	ss						B. 22(1). 546	IR. 1(1). 584B NMR. 2(1). 504D
	i	vs	i	s	vs^h	chl alk	H, I, LS	B. 21. 91	IR. 1(2). 859D
145	vs	vs	s	s	ss	py chl	C, H, I	B.23(2). 34	IR. 1(2). 611D NMR. 2(2). 484D
58	i	∞	∞	s	s	py CS$_2$	H, LS	B. 5(2). 410	IR. 1(1). 955B NMR. 2(1). 759C
>110	s^h	vs	vs	s			T, stench	B. 20(2). 196	IR. 1(2). 653A NMR. 2(2). 519A
74	i	s	∞	∞	∞	CCl$_4$	H, I, LS	B. 5(2). 165	IR. 1(1).971D NMR. 2(2). 773D
none	ss	∞	s				C, I, MS	B. 1(3). 193	IR. 1(1). 55C NMR. 2(1). 59B

(Continued)

Table XI.2 (Continued)

Compound/synonym (s)	Formula	M_r	m. p. /°C	b. p. /°C	Density / kg m^3	Refractive index
iodoform triiodomethane*	CHI$_3$	393.73	123	c. 218	4 008	
iodomethane* methyl iodide	CH$_3$I	141.94	−66	41–43	2 279	1.538 0

L-isoleucine;(2S,3S)-(+)-
  2-amino-3-methylpentanoic acid
	C$_2$H$_5$CH(CH$_3$)CH(NH$_2$)
	  –COOH
		131.18	d 288			

isoniazid; 4-pyridine- carboxylic acid hydrazide	C$_6$H$_7$N$_3$O	137.14	171–173			
isophthalic acid 1, 3-benzenedicarboxylic acid*	C$_6$H$_4$(COOH)$_2$	166.13	341–343	s		
isoprene 2-methyl-1, 3-butadiene*	H$_2$C:CHC(CH$_3$):CH$_2$	68.12	−146	34	681	1.416 6
isoquinoline 2-benzazine*	C$_6$H$_4$CH:NCH:CH	129.16	26–28	242	1 099	1.614 8
2-ketoglutaric acid 2-oxopentanedioic acid*	HOOCCH$_2$CH$_2$COCOOH	146.10	113–116			
L-lactic acid; (S)-(+)- 2-hydroxypropanoic acid*	CH$_3$CH(OH)COOH	90.08	52.8	19/12 mm	1 206	1.427 0
α-D-lactose monohydrate	C$_{12}$H$_{22}$O$_{11}$.H$_2$O	360.12	d 219			
lauric acid dodecanoic acid*	CH$_3$(CH$_2$)$_{10}$COOH	200.33	44–46	225/100 mm	883	1.430 4^{50}

L-leucine, (S)-(+)-leucine
  L-α-aminoisocaproic acid
	(CH$_3$)$_2$CHCH$_2$CH
	  –(NH$_2$)COOH
		131.18	d 293			

2, 4-lutidine 2, 4-dimethylpyridine	C$_5$H$_3$N(CH$_3$)$_2$	107.16	−60	159	931	1.499 1
2, 6-lutidine 2, 6-dimethylpyridine	CH$_3$C:CHCH:CHC(CH$_3$):N	107.16	−6	145.7	923	1.497 6
L-lysine, (S)-(+)-lysine, L-2, 6-diaminohexanoic acid	H$_2$N(CH$_2$)$_4$CH(NH$_2$)COOH	146.19	d 212			
maleic acid cis-butenedioic acid*	HOOCCH:CHCOOH	116.07	134–136		1590	
maleic anhydride	COCH:CHCO.O	98.06	54–56	200	1314^{60}	
maleimide	CH:CHCONHCO	97.07	93–5	s		
L-malic acid, (S)-(−)-malic acid, (S)-(−)-hydroxysuccinic acid	HOOCCH$_2$CH(OH)COOH	134.09	128–129	150 d	1601	
malonic acid propanedioic acid*	CH$_2$(COOH)$_2$	104.06	d 135		1619^{16}	
malononitrile dicyanomethane	CH$_2$(CN)$_2$	66.06	32	218–220	1191	1.4146
D-maltose monohydrate, glucose-4-α-glucoside	C$_{12}$H$_{22}$O$_{11}$.H$_2$O	360.32	102–103		1540	

Flash point/°C	Solubility						Hazards	References	
	w	alc	et	ac	bz	Other			
none	i	s^h	s	s	i	chl CS$_2$	H, T, I	B. 1(3). 102	IR. 1(1). 84A NMR. 2(1). 82C
none	ss	∞	∞	s	s		vT, CSA	B. 1(2). 35	
	s	ss	i					B. 4. 454	IR. 1(1). 575D NMR. 2(1). 489C
							T, H, CSA	B. 22(2). 37	IR. 1(2). 798B NMR. 2(2). 676D
	ss	s	i		i		I	B. 9(2). 608	IR. 1(2). 201A NMR. 2(2). 198B
–53	i	∞	∞	∞	∞		I, FL	B. 1(3). 966	IR. 1(1). 30D NMR. 2(1). 35A
107	i	vs	∞	vs	∞	chl	T, H, I	B. 20(2). 236	IR. 1(2). 877D NMR. 2(2). 752B
	vs	vs	vs	s			I	B. 3(2). 481	IR. 1(1). 532B NMR. 2(1). 469A
>110	∞	∞	ss				C	B. 3(3). 442	IR. 1(1). 517C
	vs	ss	i					B. 31. 408	IR. 1(1). 194C NMR. 2(2). 910B
>110	i	vs	vs	s	vs		C, I	B. 2(3). 868	IR. 1(1). 484A NMR. 2(1). 421D
	ss	i	i					B. 4(2). 859	IR. 1(1). 575B NMR. 2(1). 489B
37	vs	vs	vs	s			FL, T	B. 20(2). 160	IR. 1(2). 741D NMR. 2(2). 621D
33	∞	ss	s	s		chl	FL, I	B. 20(2). 160	IR. 1(2). 740B NMR.2(2). 621B
	vs	i	i	i	i			B. 4(2). 857	IR. 1(1). 587C NMR. 2(1). 495A
	vs	vs	s	vs	i		C, I	B. 2(2). 641	IR. 1(1). 503A NMR. 2(1). 440B
103	d^h		s	s		chl	C, I, MS	B. 17(2). 445	IR. 1(1). 718C NMR. 2(1). 607A
	s	s	s				C, I, L	B. 21(2). 311	IR. 1(1). 797B NMR. 2(1). 668C
	vs	vs	vs					B. 3. 419	IR. 1(1). 522C NMR. 2(1). 456C
	vs	s	s		i	py	H, I	B. 2(2). 516	IR. 1(1). 490C NMR. 2(1). 427A
112	s	vs	vs	s	s	chl	vT, I	B. 2(2). 535	IR. 1(1). 840C
	vs	ss	i					B. 31. 386	IR. 1(1). 195A NMR. 2(2). 910C

(Continued)

Table XI.2 (Continued)

Compound/synonym(s)	Formula	M_r	m.p. /°C	b.p. /°C	Density / kg m^3	Refractive index
D-+-mandelic acid, (S)-(+)- (+) hydroxyphenyl acetic acid	C$_6$H$_5$CH(OH)COOH	152.16	131–134			
D-mannitol	HOCH$_2$(CH.OH)$_4$CH$_2$OH	182.17	168	295/3.5 mm	1489	1.330
D-(+)-mannose	O(CH.OH)$_4$CH.CH$_2$OH	180.16	d 132		1539	
melamine, 2, 4, 6-triamino-1, 3, 5-triazine*	H$_2$NC:NC(NH$_2$):NC(NH$_2$):N	126.12	d 354		1573^{14}	1.872
mellitic acid benzenehexacarboxylic acid*	C$_6$(COOH)$_6$	342.18	d 286			
menthol, (1S, 2R, 5S)-(+)-menthol	C$_{10}$H$_{20}$O	156.27	43–44	103/9 mm		
mercaptoacetic acid* thioglycolic acid	HSCH$_2$COOH	92.12	–16	96/5 mm	1325	1.5030
2-mercaptoethanol* thioglycol	HSCH$_2$CH$_2$OH	78.13		157	1114	1.4996
mesaconic acid methylfumaric acid*	HOOCCH:C(CH$_3$)COOH	130.10	200–202	s	1466	
mesitylene 1,3,5-trimethylbenzene*	C$_6$H$_3$(CH$_3$)$_3$	120.20	–45	162–164	865	1.497 8
mesityl oxide 4-methyl-3-penten-2-one*	(CH$_3$)$_2$C:CHCOCH$_3$	98.15	–53	129–130	858	1.444 0
metaldehyde	(CH$_3$CHO)$_n$ n = 4–6	(44.05)$_n$	190–196 s			
methacrylic acid 2-methylpropenoic acid*	H$_2$C:C(CH$_3$)COOH	86.09	16	163	1 015	1.431 4
methane	CH$_4$	16.04	–183	–164	555^0	
methanesulphonic acid* methylsulphonic acid	CH$_3$SO$_3$H	96.11	20	167/10 mm	1 481	1.430 3
methanesulphonyl chloride*	CH$_3$SO$_2$Cl	114.55		60/21 mm	1 480	1.457 3
methanethiol* methylmercaptan	CH$_3$SH	48.11	–123	6	867	
methanol* methyl alcohol	CH$_3$OH	32.04	–97.8	64.96	791	1.328 8
L-methionine, (S)-(+)-methionine, L-2-amino-4-(methylthio)-butanoic acid	CH$_3$S(CH$_2$)$_2$CH(NH$_2$)COOH	149.21	d 284			
2-methoxyethanol* methyl cellosolve	CH$_3$OCH$_2$CH$_2$OH	76.11	–85	125	965	1.402 4
methyl acetate methyl ethanoate*	CH$_3$COOCH$_3$	74.08	–98	57.5	933	1.359 3
methyl acrylate methylpropenoate*	H$_2$C:CHCOOCH$_3$	86.09	–75	80	954	1.404 0
methylamine aminomethane*	CH$_3$NH$_2$	31.06	–93	–6.3	699^{-11}	
methyl benzoate	C$_6$H$_5$COOCH$_3$	136.15	–12	199	1 089	1.516 4
2-methylbutane* iso-pentane	(CH$_3$)$_2$CHCH$_2$CH$_3$	72.15	–160	27.9	620	1.353 7
3-methyl-1-butanol* iso-amyl alcohol	(CH$_3$)$_2$CHCH$_2$CH$_2$OH	88.15	–117	130	809	1.405 4

Flash point/°C	Solubility						Hazards	References	
	w	alc	et	ac	bz	Other			
	s	s	vs			chl	LS, I	B. 10(2). 114	IR. 1(2). 145C
	vs	ss[h]	i					B. 1(3). 2392	
	vs	ss	i		i			B.31. 284	IR. 1(1). 188B
									NMR. 2(2). 904C
	ss vs[h]	ss	i				CSA, I	B. 26(2). 132	
	vs	s					I	B.9(2). 741	IR. 1(2). 202C
91		s	s	s	s			B. 6(2). 49	IR. 1(1). 163D
									NMR. 2(1). 153D
>110	∞	∞	∞				C, T, stench	B. 3(2). 175	IR. 1(1). 527B
									NMR. 2(1). 462B
73	s	s	s		s		vT, stench	B. 1(2). 523	IR. 1(1). 262B
									NMR. 2(1). 228B
	ss vs[h]	vs	s		ss		H	B. 2(2). 651	IR. 1(1). 503B
									NMR. 2(1). 440D
44	i	∞	∞	∞	∞	CCl$_4$	FL, I	B. 5(2). 313	IR. 1(1). 942C
30	s	∞	∞	s			FL, L, H, I	B. 1(3). 2995	IR. 1(1). 414D
									NMR. 2(1). 378B
55	i	ss[h]	ss[h]	i	ss		FS, I	B. 1(2). 671	
76	s	∞	∞				C, T	B. 2(3). 1278	IR. 1(1). 499C
									NMR. 2(1). 434C
	s	s	s		s		FG	B. 1(3). 1	
>110	vs	s	vs				C, T, MS	B. 4(2). 524	NMR. 2(2). 788A
110	i	s	s				vT, C, MS	B. 4(2). 525	IR. 1(1). 900B
									NMR. 2(2). 802B
	ss[h]	vs	vs				FG, MS	B. 1(3). 1212	
11	∞	∞	∞	∞	vs	chl	FL, T	B. 1(3). 1147	
	s	i	i	i	i			B. 4(2). 938	IR. 1(1). 594C
									NMR. 2(1). 501B
46	∞	vs	∞	s	∞		T, H, teratogen	B. 1(3). 2069	
−9	vs	∞	∞	vs	s	chl	FL, I	B. 2(2). 125	
6	ss	s	s	s	s		FL, L, I	B. 2(3). 1218	IR. 1(1). 638B
									NMR. 2(1). 537A
	vs	s	∞	s	s		FG, C, I	B. 4(2). 546	
82	i	s	∞			MeOH	H, I	B. 9(2). 87	IR. 1(2). 291B
									NMR. 2(2). 281A
−56	i	∞	∞				FL, I	B. 1(3). 352	
45	ss	s	s				I	B. 1. 392	

(Continued)

Table XI.2 (Continued)

Compound/synonym(s)	Formula	M_r	m. p. /°C	b. p. /°C	Density / kg m^3	Refractive index
3-methyl-2-butanone* isopropyl methyl ketone	$(CH_3)_2CHCOCH_3$	86.14	−92	94–95	805	1.388 0
methyl chloroacetate	$ClCH_2COOCH_3$	108.53	−32	129–130	1 234	1.421 8
methyl crotonate methyl 2-butenoate* (trans)	$CH_3CH{:}CHCOOCH_3$	100.13	−42	118–120	944	1.424 2
methylcyclohexane* hexahydrotoluene	$CH_3CH(CH_2)_4CH_2$	98.19	−126	101	769	1.4231
2-methylcyclohexanol (±)-cis-	$CH_3CHCH(OH)(CH_2)_3CH_2$	114.19	6–8	165	936	1.4640
D-2-methylcyclohexanone*	$CH_3CH(CH_2)_4CO$	112.17		162–163	926	1.447 8
N-methylformamide	$HCONHCH_3$	59.07	−40	180–185	1 011	1.431 9
methyl formate	$HCOOCH_3$	60.05	−99	32–33	974	1.343 3
methyl isocyanate methylcarbylamine	CH_3NCO	57.05	−17	43–45	967	1.369 5
methyl isothiocyanate methyl mustard oil	CH_3NCS	73.12	30–34	117–118	1 069	1.525 8
methyl methacrylate methyl 2-methylpropenoate*	$H_2C{:}C(CH_3)COOCH_3$	100.12	−48	100–101	944	1.414 2
1-methylnaphthalene*	$C_{10}H_7CH_3$	142.20	−22	240–244	1 020	1.615 9
2-methylnaphthalene*	$C_{10}H_7CH_3$	142.20	34–36	241–242	1 006	1.601 9^{40}
2-methylpentane* isohexane	$CH_3CH_2CH_2CH(CH_3)_2$	86.18	−154	60–61	653	1.371 6
2-methylpropane* isobutane	$HC(CH_3)_3$	58.12	−160	−12	720^{-79}	1.316 9
2-methyl-1-propanol* isobutyl alcohol	$(CH_3)_2CHCH_2OH$	74.12	−108	107.9	803	1.396 0
2-methyl-2-propanol* tert-butyl alcohol	$(CH_3)_3COH$	74.12	25	82.2	789	1.386 0
N-methylpyrrole 1-methylpyrrole	$CH{:}CHCH{:}CHN(CH_3)$	81.12	−57	112–113	914	1.487 5
methyl salicylate methyl 2-hydroxybenzoate*	$2\text{-}(HO)C_6H_4COOCH_3$	152.16	−8 to −7	222	1 174	1.536 9
methyl sulphone	$(CH_3)_2SO_2$	94.13	110	238	1 170^{110}	1.422 6
morpholine tetrahydro-1,4-isoxazine	$NHCH_2CH_2OCH_2CH_2$	87.12	−5	129	1 000	1.454 8
myristic acid tetradecanoic acid*	$CH_3(CH_2)_{12}COOH$	228.38	53–55	250/100 mm	844	1.433 5^{70}
1-naphthaldehyde 1-naphthalenecarbonal*	$C_{10}H_7CHO$	156.18	1–2	160/15 mm	1 150	1.650 7
2-naphthaldehyde 2-naphthalenecarbonal*	$C_{10}H_7CHO$	156.18	59–62	160/19 mm	1 078^{99}	1.621 1^{99}
naphthalene*	$C_{10}H_8$	128.17	80–82	217.7	1 025	1.400 3

Flash point/°C	\multicolumn{6}{c}{Solubility}	Hazards	References	
	w	alc	et	ac	bz	Other			
6	ss	∞	∞	vs			FL, T	B. 1(2). 741	IR. 1(1). 406B NMR. 2(1). 369D
51	ss	∞	∞	∞	∞		C, T, FL, L	B. 2(2). 191	
4	i	vs	vs				FL, I, L	B. 2(2). 392	IR. 1(1). 641C NMR. 2(1). 540D
−3	i	s	s	∞	∞	CCl₄	FL, I	B. 5. 29	
58	ss	∞	s				H	B. 6(2). 20	IR. 1(1). 159B NMR. 2(1). 146D
46	i	s	s				H	B. 7(2). 15	IR. 1(1). 433A NMR. 2(1). 394D
98	s	s	i	∞			H, I, teratogen	B. 4(2). 563	IR. 1(1). 753D NMR. 2(1). 634D
−26	vs	∞	s			MeOH	FL, I	B.2(2). 25	
−6	s						vT, FL	B. 4(2). 578	IR. 1(1). 869C NMR. 2(1). 723C
32	ss	∞	vs				vT, FL, L,	B. 4(2). 579	IR. 1(1). 873D NMR. 2(1). 727A
10	ss	∞	∞	∞			FL, C, L, I	B. 2(3). 1280	IR. 1(1). 640C NMR. 2(1). 539A
82	i	vs	vs		s		I	B. 5(3). 1620	IR. 1(1). 956B NMR. 2(1). 760A
97	i	vs	vs		s		I	B. 5(3). 1627	IR. 1(1). 956C NMR. 2(1). 760B
−23	i	s	s	∞	∞	chl	FL, H, I	B. 1(3). 396	IR. 1(1). 8C NMR. 2(1). 14B
	i	vs	vs			chl	FG, I	B. 1. 124	
37	ss	∞	vs				FL, I	B. 1. 373	
4	∞	∞	∞				FL, I	B. 1(3). 1568	
15	i	∞	∞				FL, H	B. 20. 163	IR. 1(2). 565D
>110	ss	vs	vs				T, I	B. 10 (2). 44	
143	s	s			s			B. 1. 289	IR. 1(1). 880C NMR. 2(2). 782C
35	∞	s	s	s	s		FL, C	B. 27(2). 3	NMR. 2(1). 335C
>110	i	s	s				I	B. 2(3). 911	IR. 1(1). 484C NMR. 2(1). 422B
>110	i	s	s	s	s		L, I, AS	B. 7. 400	IR. 1(2). 133C NMR. 2(2). 131B
	ssʰ	vs	vs	s			L, I, AS	B. 7. 401	IR. 1(2).133D NMR. 2(2). 131C
78	i	s	vsʰ	vs	vs	CS₂	FS, I	B. 5(3). 1549	IR. 1(1). 956A

(Continued)

Table XI.2 (Continued)

Compound/synonym(s)	Formula	M_r	m.p. /°C	b.p. /°C	Density / kg m³	Refractive index
2-naphthalenesulphonic acid*	$C_{10}H_7SO_3H$	208.24	124–125	d	1 441^{75}	
1-naphthoic acid						
1-naphthalenecarboxylic acid*	$C_{10}H_7COOH$	172.20	161	300	1 398	
2-naphthoic acid						
2-naphthalenecarboxylic acid*	$C_{10}H_7COOH$	172.20	185–187	>300	1 077^{100}	
1-naphthol						
1-hydroxynaphthalene*	$C_{10}H_7OH$	144.19	96	278–280	1 099^{99}	1.622 4^{99}
2-naphthol						
2-hydroxynaphthalene*	$C_{10}H_7OH$	144.19	122–124	295	1 280	
1, 2-naphthoquinone*						
1, 2-dihydro-1, 2-diketonaphthalene	$C_6H_4CO.CO.CH:CH$	158.16	135d		1 450	
1, 4-naphthoquinone*						
1, 4-dihydro-1, 4-diketonaphthalene*	$C_6H_4CO.CH:CHCO$	158.16	123–126	s		
(1-naphthyl)-acetic acid*	$C_{10}H_7CH_2COOH$	186.21	130–133	d		
nicotinamide, 3-						
pyridinecarboxylic acid amide*	$CH:NCH:CHCH:CCONH_2$	122.13	130–132		1 400	1.466
(S)-(–)-nicotine; α-N-methyl-						
(L)-pyridyl-pyrrolidine	$C_{10}H_{14}N_2$	162.24	–79	124/18 mm	1 010	1.5289
nicotinic acid						
3-pyridinecarboxylic acid*	$CH:NCH:CHCHC.COOH$	123.11	236–239	s	1 473	
ninhydrin; 1, 2, 3-triketohydrindene hydrate	$C_6H_4CO.CO.CO.H_2O$	178.15	d 250			
nitrilotriacetic acid	$N(CH_2COOH)_3$	191.14	d 246			
2-nitroaniline	$O_2NC_6H_4NH_2$	138.13	71–73	284	1 442^{15}	
3-nitroaniline	$O_2NC_6H_4NH_2$	138.13	112–114	100/0.16 mm	1 175^{160}	
4-nitroaniline	$O_2NC_6H_4NH_2$	138.13	149	260/100 mm	1 424	
nitrobenzene*	$C_6H_5NO_2$	123.11	5.7	210.8	1 204	1.556 2
1-nitrobutane	$CH_3(CH_2)_3NO_2$	103.12	–81	152–153	973	1.410 3
nitroethane*	$CH_3CH_2NO_2$	75.07	–90	114–115	1 045	1.391 6
2-nitroethanol*	$O_2NCH_2CH_2OH$	91.07	–80	194/765 mm	1 270	1.444 5
nitromethane*	CH_3NO_2	61.04	–29	100.8	1 131	1.381 7
1-nitronaphthalene*	$C_{10}H_7NO_2$	173.17	59–61	304	1 332	
2-nitronaphthalene*	$C_{10}H_7NO_2$	173.17	79	312/734 mm		
2-nitrophenol*	$O_2NC_6H_4OH$	139.11	44–46	216	1 485^{14}	1.5723^{50}
3-nitrophenol*	$O_2NC_6H_4OH$	139.11	96–98	194/70 mm	1 280^{100}	
4-nitrophenol*	$O_2NC_6H_4OH$	139.11	114–116	d 279	1 479	
1-nitropropane*	$CH_3CH_2CH_2NO_2$	89.09	–108	131–132	998	1.401 6
2-nitropropane*	$CH_3CH(NO_2)CH_3$	89.09	–93	120	988	1.394 4
4-nitrosophenol*	ONC_6H_4OH	123.11	d 132			

Flash point/°C	Solubility						Hazards	References	
	w	alc	et	ac	bz	Other			
	vs	s	s			ss[h]	C, T, H	B. 11(2). 96	IR. 1(2). 497C
	ss	s	s			chl	I	B. 9. 647	IR. 1(2). 241C NMR. 2(2). 238C
	ss[h]	s	s			chl	I	B. 9. 656	IR. 1(2). 241D NMR. 2(2). 238D
	i	vs	vs	s	s	chl	T, I	B.6.596	IR.1(1).1109A NMR.2(1).908D
	i	vs	vs		s	chl	H, I, LS	B. 6. 627	IR. 1(1). 1109B
	s	s	s				I	B. 7(2). 645	IR. 1(2). 79A NMR. 2(2). 85A
	ss	vs[h]	s		s	chl	T, I, H	B. 7(2). 651	IR. 1(2). 79B NMR. 2(2). 84B
	ss	ss	vs	vs	s	chl	C, H	B. 9(2). 456	IR. 1(2). 173A NMR. 2(2). 170A
	vs	vs						B. 22(2). 34	IR. 1(2). 796B
101	∞	vs	vs			chl	vT, MS	B. 23. 117	IR. 1(2). 768B NMR. 2(2). 647B
	ss	ss	ss					B. 22(2). 32	IR. 1(2). 785B NMR. 2(2). 664D
	vs[h]	s	ss			alk	I, LS, H	B. 7(2). 831	IR. 1(2). 55C
	ss	s[h]					H, I, CSA	B. 4(2). 801	IR. 1(1). 566A NMR. 2(1). 483D
	ss	s	vs	vs	vs	chl	vT, I	B. 12(2). 367	IR. 1(1). 1333A NMR. 2(1). 1134D
	ss	s	s	s	ss	MeOH chl	vT, I	B. 12(2). 374	IR. 1(1). 1337A NMR. 2(1). 1139A
165	i	s	s	s	ss	MeOH, chl	vT, I	B. 12(2). 383	IR. 1(1). 1343C
87	ss	vs	vs	vs	vs		vT	B. 5(2). 171	IR. 1(1). 1328D
47	ss	∞	∞			alk		B. 1(3). 303	IR. 1(1). 399D
30	ss	∞	∞	s		alk	FL, H, I	B.1(3).199	IR.1(1).399B
>110	∞	∞	∞		i		I	B.1(3).1364	IR.1(1).401D NMR.2(1).353B
35	s	s	s	s		alk	FL, H	B.1(2).40	
	i	vs	vs		vs	chl, py	FS, T	B.5(3).1593	IR.1(1).1388D NMR.2(1).1192C
	i	vs	vs				FS, CSA	B.5(3).1596	IR.1(1).1389A
	ss	vs[h]	vs	vs	vs	chl, py, alk	T, I	B.6(3).794	IR.1(1).1331C NMR.2(1).1134B
	ss[h] vs[h]	vs	vs	vs	vs[h]	chl, alk	T, I	B.6(3).805	IR.1(1).1336B NMR.2(1).1138D
	ss[h] vs[h]	vs	vs	vs		chl, py	T, I	B.6(3).811	
33	ss	∞	∞			chl	FL, T, I	B.1(3).256	IR.1(1).399C
37	ss					chl	FL, H, CSA	B.1(3).258	IR.1(1).400C NMR.2(1).352D
	ss	s	s	s	s[h]	alk	FS, I	B.6(2).205 B.7(2).574	IR.1(1).1327A NMR.2(1).1128A

(Continued)

Table XI.2 (Continued)

Compound/synonym (s)	Formula	M_r	m. p. /°C	b. p. /°C	Density / kg m^3	Refractive index
4-nitrotoluene*	CH$_3$C$_6$H$_4$NO$_2$	137.14	52–54	238	1 392	
nonane*	CH$_3$(CH$_2$)$_7$CH$_3$	128.26	−53	151	718	1.405 4
L-norleucine						
(S)-(+)-2-aminohexanoic acid	CH$_3$(CH$_2$)$_3$CH(NH$_2$)COOH	131.18	d 301			
L-norvaline						
(S)-(+)-2-aminopentanoic acid	CH$_3$CH$_2$CH$_2$CH(NH$_2$)COOH	117.15	d 291			
octadecane*	CH$_3$(CH$_2$)$_{16}$CH$_3$	254.51	28–30	317	777	1.439 0
1-octadecanol*						
stearyl alcohol	CH$_3$(CH$_2$)$_{17}$OH	270.50	59–60	170/2 mm	812^{59}	
octane*	CH$_3$(CH$_2$)$_6$CH$_3$	114.23	−57	125.7	703	1.397 4
octanoic acid*						
caprylic acid	CH$_3$(CH$_2$)$_6$COOH	144.21	16.5	237–239	909	1.428 5
1-octanol*						
octyl alcohol	CH$_3$(CH$_2$)$_7$OH	130.23	−15	196	827	1.429 5
2-octanol*						
(R)-(−)-2-octanol	CH$_3$(CH$_2$)$_5$CH(OH)CH$_3$	130.23		86/20 mm	820	1.426 4
oleic acid						
9-octadenenoic acid (cis-)*	CH$_3$(CH$_2$)$_7$CH:CH(CH$_2$)$_7$–COOH	282.47	13.4	228–229/ 15 mm	893	1.458 2
orcinol						
3, 5-dihydroxytoluene*	CH$_3$C$_6$H$_3$–(OH)$_2$	124.15	107–108	290	1 290^4	
L-ornithine,						
(S)-(+)-2, 5-diaminopentanoic acid*	H$_2$N(CH$_2$)$_3$CH(NH$_2$)COOH	132.16	140			
oxalic acid						
ethanedioic acid*	HOOCCOOH	90.04	d 190		1 900	
oxamide						
oxalic acid diamide	H$_2$NCOCONH$_2$	88.07	d 419			
oxazole	OCH:NCH:CH	69.06	−87 to −84	69–70	1 050	1.425 6
palmitic acid						
hexadecanoic acid*	CH$_3$(CH$_2$)$_{14}$COOH	256.43	63	267/100 mm	853^{62}	1.433 5^{60}
paracetamol						
N(4-hydroxyphenyl) acetamide	CH$_3$CONHC$_6$H$_4$OH	151.17	169–171		1 293	
paraformaldehyde	(HCHO)$_n$		d 164			
paraldehyde						
2, 4, 6-trimethy-1, 3, 5-trioxane	C$_6$H$_{12}$O$_3$	132.16	12.6	128	994	1.404 9
pentaerythritol						
tetrakis(hydroxymethyl) methane*	C(CH$_2$OH)$_4$	136.15	258–260			1.548
pentane*	CH$_3$(CH$_2$)$_3$CH$_3$	72.15	−130	36.07	626	1.357 5
pentanedioic acid*						
glutaric acid	HOOC(CH$_2$)$_3$COOH	132.13	99	200/20 mm	1 424^{24}	1.418 8^{106}
1-pentanol*						
n-amyl alcohol	CH$_3$(CH$_2$)$_4$OH	88.15	−79	136–138	814	1.410 1
2-pentanol*						
methyl n-propyl carbinol	CH$_3$CH$_2$CH$_2$CH(OH)CH$_3$	88.15		118–119	810	1.405 5
2-pentanone*						
methyl propyl ketone	CH$_3$CH$_2$CH$_2$COCH$_3$	86.14	−78	102	809	1.389 7

Flash point/°C	Solubility						Hazards	References	
	w	alc	et	ac	bz	Other			
106	i	s	vs	vs	vs	chl, py	vT, I	B.5(3).736	IR.1(1).1338A
									NMR.2(1).1140A
31	i	vs	vs	∞	∞	chl	FL, I	B.1(3).502	
	s	i						B.4.432	IR.1(1).576D
	vs[h]	i	i			chl		B.4(3).1532	IR.1(1).574C
165	i	ss	s	s				B.1(3).565	IR.1(1).4C
									NMR.2(1).12A
	i	s	s	ss	ss	chl	I	B.1(3).1833	
15	i	∞	s	∞	∞	chl	FL, I	B.1(3).457	
110	ss[h]	∞				chl	C	B.2(2).301	IR.1(1).483A
									NMR.2(1).420D
81	i	∞	∞				I	B.1(3).1703	
71	ss	s	s	s			I	B.1(2).451	IR.1(1).120C
									NMR.2(1).112C
>110	i	∞	∞	∞	∞	chl	I	B.2(3).1387	
	s	s	s		s		I, LS	B.6(3).4531	IR.1(1).1106A
									NMR.2(1).899C
	s	s	ss					B.4(2).844	IR.1(1).587B
									NMR.2(1).494D
	s	vs	ss		i		C, T, I	B.2(3).1534	
	ss	ss	i				T, I, H	B.2(3).1586	IR.1(1).766B
18							FL, I	B.27(2).9	IR.1(2).637C
	i	s	∞	s	s	chl	I	B.2.370	
	i	vs						B.13(2).243	IR.1(2).358B
									NMR.2(2).337C
71	ss						MS, I	B.1.566	IR.1(1).205D
44	ss	∞	∞			chl	FL, I	B.19(2).394	
	s		i		i		MS	B.1(2).601	IR.1(1). 184B
−49	ss	∞	∞	∞	∞	chl	FL, I	B.1(3).238	
	vs	vs	vs		i	chl	I	B.2(3).1685	IR.1(1).493C
									NMR.2(1).430A
48	i	∞	∞	s			FL, T, I	B.1(3)1598	IR.1(1).110B
									NMR.2(1).102A
33	vs	s	s				FL, H, I	B.1(3).1609	IR.1(1).118D
									NMR.2(1).110C
7	ss	∞	∞				FL	B.1(3)2800	IR.1(1).406A
									NMR.2(1).369C

(Continued)

Table XI.2 (Continued)

Compound/synonym(s)	Formula	M_r	m. p. /°C	b. p. /°C	Density / kg m^3	Refractive index
3-pentanone* diethyl ketone	$CH_3CH_2COCH_2CH_3$	86.14	−40	102	814	1.392 4
1-pentene* propylethylene	$CH_3CH_2CH_2CH:CH_2$	70.14	−138	30	640	1.371 5
2-pentene* (cis) cis-β-n-amylene	$CH_3CH_2CH:CHCH_3$	70.14	−151	37–38	656	1.379 5
1-pentyne* n-propyl acetylene	$CH_3CH_2CH_2C:CH$	68.13	−105 to −106	40	691	1.385 2
2-pentyne* ethyl-methyl acetylene	$CH_3CH_2C:CCH_3$	68.13	−109	55–56	710	1.403 9
phenacetin N-(4-ethoxyphenyl)-acetamide	$CH_3CONHC_6H_4OC_2H_5$	179.22	134–136	d		1.571
phenanthrene*	$C_6H_4CH:CHC_6H_4$	178.24	99–101	340	980^4	1.594 3
9,10-phenanthrenequinone* 9,10-phenanthroquinone	$C_6H_4CO.COC_6H_4$	208.23	209–211	>360 s	1 405	
1,10-phenanthroline 4,5-diazaphenanthrene*	$C_{12}H_{10}N_2$	180.22	114–117	> 300		
phenazine azophenylene	$C_{12}H_8N_2$	180.21	174–117	>360 s		
phenazone 2,3 dimethyl-1-phenyl-5-pyrazolone	$C_6H_5NCOCH:C(CH_3)NCH_3$	188.23	114	211–212/ 10 mm	1 075^{130}	1.569 7
phenetole ethoxybenzene*	$C_6H_5OC_2H_5$	122.17	−30	169–170	966	1.507 6
phenol* hydroxybenzene*	C_6H_5OH	94.11	40.9	182	1 071	1.550 9^{21}
phenolphthalein 2,2-bis(4-hydroxyphenyl)-phthalide	$O.COC_6H_4C(C_6H_4OH)_2$	318.33	261–263		1 277^{22}	
phenothiazine thiodiphenylamine	$C_6H_4SC_6H_4NH$	199.28	182–185	371		
phenoxyacetic acid phenoxyethanoic acid*	$C_6H_5OCH_2COOH$	152.16	98–100	d 285		
2-phenoxyethanol* phenoxytol	$C_6H_5OCH_2CH_2OH$	138.17	11–13	237	1 102	1.534 0
phenyl acetate phenyl ethanoate*	$CH_3COOC_6H_5$	136.15		196	1 078	1.503 3
phenylacetic acid phenylethanoic acid*	$C_6H_5CH_2COOH$	136.15	77	265	1 081	
L-phenylalanine (S)-(−)-phenylalanine	$C_6H_5CH_2CH(NH_2)COOH$	165.19	d 270			
phenyl benzoate	$C_6H_5COOC_6H_5$	198.22	68–70	298–299	1 235	
1,2-phenylenediamine 1,2-diaminobenzene*	$C_6H_4(NH_2)_2$	108.14	102–105	256–258		

Flash point/°C	Solubility						Hazards	References	
	w	alc	et	ac	bz	Other			
12	vs	∞	∞				FL	B.1(3).2806	
−28	i	∞	∞		s		FL, H, I	B.1(3).770	IR.1(1).13B NMR.2(1).20A
−45	i	∞	∞		s		FL, I	B.1(3).776	IR.1(1).17B
−34	i	vs	∞		s	chl	FL, I	B.1(3).957	IR.1(2).931B
−30	i	vs	∞		s	chl	FL, I	B.1(3).958	
	ss	s	ss	s	ss	py, chl	H, CSA	B.13(2).244	IR.1(2).359A NMR.2(2).338B
	i	s	s	s	s	CS$_2$, chl	H, I	B.5.667	
	i	ss	s		ss		I	B.7.796	IR.1(2).82B
	vs^h	s		s	s		T, MS	B.23.227	IR.1(2).889D NMR.2(2).763D
	ss	s^h	ss		s		H, I, mutagen	B.23.223	IR.1(2).884D NMR.2(2).760A
	vs	vs	ss		s	chl	H	B.24(2).11	IR.1(2).396B NMR.2(2).375B
57	i	s	s					B.6.140	IR.1(1).1035B NMR.2(1).833B
79	s	s	vs	∞	∞^{h}	CCl$_4$, chl	vT, C, I	B.6(2).116	
	i	vs	s	vs	i	py, chl		B.18.143	
	ss	ss	s	vs	vs		I, LS	B.27.63	IR.1(1).1316A NMR.2(1).1116B
	s	vs	vs			CS$_2$	I	B.6(2).157	IR.1(2).142C NMR.2(2).138D
110	i	s	s			alk	H, I	B.6(3).567	IR.1(1). 1157A NMR.2(1).956C
76	ss	∞	∞			chl	I	B.6(2).153	IR.1(2).312B NMR.2(2).301C
	ss	vs	vs	s		CS$_2$	H	B.9(2).294	IR.1(2).137A NMR.2(2).135A
	s	i	i					B.14(2).296	IR.1(2).251A NMR.2(2).248C
	i	s^h	s^h					B.9(2).96	IR.1(2).292C NMR.2(2).303B
	s^h	vs	s		s	chl	T, I, CSA	B.13.6	

(Continued)

Table XI.2 (Continued)

Compound/synonym(s)	Formula	M_r	m.p. /°C	b.p. /°C	Density / kg m^3	Refractive index
1,3-phenylenediamine						
1,3-diaminobenzene*	$C_6H_4(NH_2)_2$	108.14	63–66	282–284	1 070[58]	1.633 9[58]
1,4-phenylenediamine						
1,4-diaminobenzene*	$C_6H_4(NH_2)_2$	108.14	139–141	267		
phenyl ether						
diphenyl ether	$(C_6H_5)_2O$	170.21	27–28	259	1 075	1.579 5
phenylhydrazine*	$C_6H_5NHNH_2$	108.14	19.8	238—241	1 099	1.608 4
phenylisocyanate						
phenylcarbylamine	C_6H_5NCO	119.12	–30	162–163	1 096	1.535 0
phenyl isothiocyanate						
phenyl mustard oil	C_6H_5NCS	135.19	–21	221	1 130	1.651 5
phenyl sulphide						
diphenyl sulphide	$(C_6H_5)_2S$	186.28	–26	296	1 114	1.633 4
phenyl sulphone						
diphenyl sulphone	$(C_6H_5)_2SO_2$	218.28	126–129	379	1 252	
phenyl sulphoxide						
diphenyl sulphoxide	$(C_6H_5)_2SO$	202.28	69–71	210/15 mm		
phloroglucinol						
1,3,5-trihydroxybenzene*	$C_6H_3(OH)_3$	126.11	218–219	s	1 460	
phosgene						
carbonyl chloride	$COCl_2$	98.92	–118	7.6	1 381	
phthaldehyde						
1,2-benzenedicarbinol*	$C_6H_4(CHO)_2$	134.14	55–58			
phthalamide						
phthalic diamide	$C_6H_4(CONH_2)_2$	164.18				
phthalic acid						
1,2-benzenedicarboxylic acid*	$C_6H_4(COOH)_2$	166.14	d 210		1 593	
phthalic anhydride	$CO.C_6H_4CO.O$	148.12	131–133	284	1 527[4]	
phthalimide	$CO.C_6H_4CO.NH$	147.14	232–236			
2-picoline						
2-methylpyridine	$N:C(CH_3)CH:CHCH:CH$	93.13	–70	128–129	944	1.495 7
3-picoline						
3-methylpyridine	$N:CHC(CH_3):CHCH:CH$	93.13	–18	144	957	1.504 0
4-picoline						
4-methylpyridine	$N:CHCH:C(CH_3)CH:CH$	93.13	2–4	145	955	1.503 7
picolinc acid						
2-pyridinecarboxylic acid	$N:CHCH:CHCH:CCOOH$	123.11	136–138	s		
picric acid						
2,4,6-trinitrophenol*	$(O_2N)_3C_6H_2OH$	229.11	122–123	exp		1.763
pimelic acid						
heptanedioic acid*	$HOOC(CH_2)_5COOH$	160.17	104–106	212/10 mm	1 329[15]	
pinacol						
2,3-dimethyl-2,3-butanediol*	$HOC(CH_3)_2C(CH_3)_2OH$	118.18	40–43	174		
pinacolone						
3,3-dimethyl-2-butanone*	$(CH_3)_2CCOCH_3$	100.16	–50	106	801	1.395 2

Flash point/°C	Solubility						Hazards	References	
	w	alc	et	ac	bz	Other			
	vs	s	s		s		T, I	B.13(1).10	IR.1(1).1239A NMR.2(1).1048A
	s^h	s^h	s		s^h	chl	T, I	B.13.61	
>110	i	s	s		s			B.6.146	IR.1(1).1056C
88	s^h	∞	∞	vs	∞	chl	vT, CSA, I	B.15(2).44	IR.1(1).1256D NMR.2(1).987B
55	d	d	vs				vT, L, MS	B.12(2).244	IR.1(2).470A
87	i	s	s				T, C, L, MS	B.12(2).247	
>110	i	s^h	∞		∞	CS$_2$	T, stench	B.6.299	IR.1(1).1183A NMR.2(1).973C
	i	s^h	s		s			B.6(3).992	IR.1(2).488C NMR.2(2).813A
		vs	vs		vs			B.6(2).290	IR.1(2).483B NMR.2(2).806A
	ss	vs	vs		vs	py	I, LS	B.6.1075	IR.1(1).1104B NMR.2(1).906B
	d	d			s	CCl$_4$, chl	vT, L, I,	B.3(3).31	
>110	s	s	s				AS	B.7(2).605	IR.1(2).106D NMR.2(2).105D
	ss	ss	i				I	B.9(2).601	
	ss	s	ss				I	B.9(2).580	IR.1(2).200D
151	ss	s	ss		s^h		C, MS, I	B.17(2).463	
								B.21(2).466	
	ss				i	alk	I	B.21(2).348	
26	vs	∞	∞	vs			FL, T	B.20(2).155	NMR.2(2).613B
26	∞	∞	∞	vs			FL, C, H	B.20(2).157	
56	∞	∞	∞	s			FL, T, I	B.20(2).158	
	ss	s	i		ss^h		I	B.22(2).30	IR.1(2).784C
	ss	s	s	vs	s	py	FL, T, E	B.6(3).873	
	s	s	s		i		I	B.2(2).586	IR.1(1).496A NMR.2(1).432A
77	ss	vs	vs				I	B.1(2).553	IR.1(1).132A
23	ss	s	s	s			FL	B.1(3).354	IR.1(1).405D NMR.2(1).370A

(Continued)

Table XI.2 (Continued)

Compound/synonym(s)	Formula	M_r	m. p. /°C	b. p. /°C	Density / kg m^3	Refractive index
(1R)-(+)-α-pinene	$C_{10}H_{16}$	136.24	−62	155−156	858	1.465 2
piperazine	$NHCH_2CH_2NHCH_2CH_2$	86.14	108−110	145−146		1.446[113]
piperidine hexahydropyridine	$NH(CH_2)_4CH_2$	85.15	−9	106	861	1.453 0
procaine; 2-(diethylamino)-ethyl-4-aminobenzoate	$H_2NC_6H_4COOCH_2CH_2$ $-N(C_2H_5)_2$	236.32	61−62			
L-proline, (S)-(−)-proline	$NHCH_2CH_2CH_2CHCOOH$	115.13	d 228			
propane*	$CH_3CH_2CH_3$	44.11	−189.7	−42.1	500	1.289 8
1, 2-propanediol* propylene glycol	$CH_3CH(OH)CH_2OH$	76.11	−60	187	1 036	1.432 4
1, 3-propanediol* trimethylene glycol	$HOCH_2CH_2CH_2OH$	76.11	−27	214	1 059	1.439 8
1-propanol* n-propyl alcohol	$CH_3CH_2CH_2OH$	60.11	−127	97.4	804	1.385 0
2-propanol* isopropyl alcohol	$CH_3CH(OH)CH_3$	60.11	−89.5	82.4	785	1.377 6
propene* propylene	$CH_3CH:CH_2$	42.08	−185	−47.7	519	1.356 7[-70]
propiolic acid propynoic acid*	$HC:CCOOH$	70.05	18	102/200 mm	1 138	1.432 0
propionaldehyde propanal*	CH_3CH_2CHO	58.08	−81	48.8	805	1.363 6
propionic acid propanoic acid*	CH_3CH_2COOH	74.08	−23	141	993	1.386 0
propionitrile propanonitrile*	CH_3CH_2CN	55.08	−93	97	782	1.365 5
propylamine 1-aminopropane*	$CH_3CH_2CH_2NH_2$	59.11	−83	48	718	1.388 5
iso-propylamine 2-aminopropane*	$(CH_3)_2CHNH_2$	59.11	−95	32.4	889	1.374 2
propylene oxide (DL) 1, 2-epoxypropane*	CH_3CHCH_2O	58.08	−112	34.3	830	1.367 0
propyne* methylacetylene	$CH_3C:CH$	40.07	−101	−23	706[-50]	1.386 3[-40]
purine; 7-imidazo-[4, 5-d]-pyrimidine	$N:CHN:CHC:CNHCH:N$	120.11	216−217	s		
pyrazine 1, 4-diazine*	$N:CHCH:NCH:CH$	80.09	54−56	115−116	1 031	1.495 3[61]
pyrazole 1, 2-diazole	$CH:CHCH:NNH$	68.08	67−70	186−188		1.420 3
pyridine azine*	$N:CHCH:CHCH:CH$	79.10	−42	115.5	982	1.510 2
2, 6-pyridinedicarboxylic acid dipicolinic acid	$C_7H_5NO_4$	167.12	d 248−250			
pyrimidine 1, 3-diazine*	$N:CHCH:CHN:CH$	80.09	20−22	123−124	1 016	1.499 8

Flash point/°C	Solubility					Hazards	References	
	w	alc	et	ac	bz	Other		
32	i	∞	∞	∞		chl	FL, I	B.5(3).366	
109	vs	s	i				C, MS	B.23(2).3	IR.1(1).371A NMR.2(1).326A
4	∞	∞	s	s	s	chl	vT, FL, I	B.20(2).6	IR.1(1).358C NMR.2(1).314A
	ss	s	s		s	chl	T, I	B.14.424	IR.1(2).303C NMR.2(2).288B
	vs	ss	i	ss	ss			B.22(2).3	IR.1(1).583C
	s	s	vs	ss	vs	chl	FG	B.1(3).204	
107	∞	∞	s	s				B.1(3).2142	
79	∞	∞	vs	ss				B.1(2).540	IR.1(1).128A
15	∞	∞	∞	s	vs		FL, I	B.1(3).1397	
22	∞	∞	∞	s	vs		FL, I	B.1(3).1439	
	vs	vs					FG	B.1(3).677	
58	∞	∞	∞	vs		chl	C, L	B.2(3).1447	IR.1(2).943A NMR.2(2).959C
−26	s	∞	∞				FL, C, I,	B.1(3).2682	
51	∞	∞	s				C, T	B.2(3).502	
6	vs	s	s	s	s		vT, FL, I	B.2(3).547	
−37	s	vs	vs	vs	s	chl	FL, C	B.4(3).250	
−27	∞	∞	∞	vs	s	chl	FL, C, I	B. 4(3).271	IR.1(1).284B NMR.2(1).242C
−37	∞	∞	∞				FL, H, CSA	B.17.6	
	ss	vs			s	chl	FG	B.1(3).919	
	vs	vs[h]	ss	s				B.26.354	IR.1(2).707C NMR.2(2).582A
55	s	s	s	s				B.23.91	IR.1(2).839D NMR.2(2).719B
	s	s	s	s			H, I, MS	B.23.39	IR.1(2).606C NMR.2(2).480C
20	∞	∞	∞	∞	∞	chl	FL, H, I	B.20.181	
	ss	ss[h]					I	B.22(2).106	IR.1(2).791D NMR.2(2).670A
31	∞	s					FL, MS	B.23.89	IR.1(2).806C NMR.2(2).686A

(Continued)

Table XI.2 (Continued)

Compound/synonym (s)	Formula	M_r	m. p. /°C	b. p. /°C	Density / kg m^3	Refractive index
pyrogallol						
1, 2, 3-trihydroxybenzene*	$C_6H_3(OH)_3$	126.11	133–134	309	1 453^4	1.561^{134}
pyrrole						
azole*	NHCH:CHCH:CH	67.09	−23	131	967	1.508 5
2-pyrrolecarboxylic acid	NHCH:CHCH:CCOOH	111.10	d 201			
pyrrolidine						
tetrahydropyrrole	$NHCH_2CH_2CH_2CH_2$	71.12		87–89	852	1.443 1
2-pyrrolidone	$CH_2CH_2CH_2CONH$	85.11	23–25	245	1 120	1.480 6
pyruvic acid						
2-oxopropanoic acid*	$CH_3COCOOH$	88.06	11–13	165	1 267	1.428 0
quinaldic acid; quinaldinic acid; 2-quiniline carboxylic acid*	C_6H_4N:C(COOH)CH:CH	173.17	157–159			
quinhydrone						
benzoquinhydrone	$O:C_6H_4:O.OHC_6H_4OH$	218.21	170–172	s	1 401	
quinine	$C_{20}H_{24}N_2O_2.3H_2O$	378.47	173–175	s		1.625
quinoline						
1-benzazine*	C_6H_4CH:CHCH:N	129.16	−15 to −16	236–238	1 093	1.626 8
iso-quinoline						
2-benzazine*	C_6H_4CH:NCH:CH	129.16	26–28	242	1 099	1.614 8
resorcinol						
1, 3-dihydroxybenzene*	$C_6H_4(OH)_2$	110.11	110–113	178/16 mm	1 272	
L-(+)-rhamnose hydrate						
6-deoxy-L-mannose	$CH_3CH(CHOH)_4O.H_2O$	182.18	92–95		1 471	
D-(−)-ribose	$HOCH_2CH(CHOH)_3O$	150.13	87			
saccharin						
2-sulphobenzoic acid imide	$CO.C_6H_4SO_2NH$	183.19	228–230		828	
salicylaldehyde						
2-hydroxybenzaldehyde*	$C_6H_4(OH)CHO$	122.13	1–2	197	1 167	1.571 9
salicylic acid						
2-hydroxybenzoic acid*	$C_6H_4(OH)COOH$	138.12	159	s	1 443	1.565
sarcosine						
N-methylglycine	CH_3NHCH_2COOH	89.09	d 208			
sebacic acid						
decanedioic acid*	$HOOC(CH_2)_8COOH$	202.25	134–136	273/50 mm	1 271	
semicarbazide*						
amino urea	$H_2NCONHNH_2$	75.07	96			
seimcarbazide hydrochloride*	$H_2NCONHNH_2.HCl$	111.52	d 175			
L-serine, (S)-(+)-serine						
L-2-amino-3-hydroxy-propanoic acid*	$HOCH_2CH(NH_2)COOH$	105.10	d 222			
skatole						
3-methylindole*	C_9H_9N	131.18	95–97	265–6/ 755 mm		
D-sorbitol						
D-glucitol	$HOCH_2(CHOH)_4CH_2OH$	182.17	93–97		1 489	1.333 0

Flash point/°C	Solubility						Hazards	References	
	w	alc	et	ac	bz	Other			
	vs	vs	vs		i	NH$_3$	T, I	B.6(2).1059	
33	ss	s	s	s	s		FL, MS, LS	B.20.159	IR.1(2).565A NMR.2(2).449A
	s	s	s				I	B.22(2).15	IR.1(2).568D NMR.2(2).452A
2	∞	s	s			chl	FL, C, I	B.20(2).3	IR.1(1).353C NMR.2(1).309C
110	vs	vs	vs		vs	CS$_2$, chl	I, MS	B.21(2).213	
83	∞	∞	∞	s			C, I, LS	B.3(3).1146	NMR.2(1).465D
	vs^h			vs^h				B.22(2).55	IR.1(2).869B NMR.2(2).745B
	s^h	vs	vs				T, I	B.7(2).572	
	ss	vs	s	ss	ss	py, MeOH	LS, H	B.23(1).166	IR.1(2).864D NMR.2(2).741C
101	s^h	∞	∞	∞	∞	CS$_2$	T, H, I	B.20(2).222	
107	i	vs	∞	vs	∞	chl	T, I	B.20(2).236	IR.1(2).877D NMR.2(2).752B
	s	s	s		ss	CCl$_4$	T, I	B.6(2).802	
	s	s				MeOH		B.1(2).901 B.31.66	IR.1(1).191A NMR.2(2).908B
	s	ss						B.1(1).434	IR.1(1).189C NMR.2(2).906C
	ss	s	ss	s	ss	alk	CSA	B.27.168	
76	ss	∞	∞	vs	vs		T, H, I, AS	B.8.31	IR.1(2).106A NMR.2(2).105B
	ss	vs	vs	vs	ss	MeOH	T, H, I, MS	B.10(2).25	IR.1(2).188C NMR.2(2).185A
	s	ss	i				MS	B.4.345	IR.1(1).570A
	ss	s	s	i				B.2(2).608	
	vs	s	i	i			CSA, mutagen	B.3(2).80	
	vs	i	i				T, CSA, mutagen	B.3.100	IR.1(1).804D
	vs	i	i	i				B.4(2).919	IR.1(1).580B
	s	s	s	s	s	chl	H, stench	B.20(2).203	IR.1(2).653D NMR.2(2).519D
	vs	ss^h	i	s		py		B.1(3).2385	IR.1(1).186A

(Continued)

Table XI.2 (Continued)

Compound/synonym(s)	Formula	M_r	m.p. /°C	b.p. /°C	Density / kg m^3	Refractive index
L-(-)-sorbose	OCH$_2$(CHOH)$_3$C(OH)CH$_2$OH	180.16	165		1 612^{17}	
stearic acid						
octadecanoic acid*	CH$_3$(CH$_2$)$_{16}$COOH	284.48	67–69	183–4/1 mm	941	1.429 9^{80}
stilbene (trans)						
trans-1, 2-diphenylethene*	C$_6$H$_5$CH:CHC$_6$H$_5$	180.25	122–124	305/720 mm	970	1.626 4^{17}
styrene; vinylbenzene						
ethenylbenzene*	C$_6$H$_5$CH:CH$_2$	104.15	−31	145–146	906	1.546 8
suberic acid						
octanedioic acid*	HOOC(CH$_2$)$_6$COOH	174.20	142–144	230/15 mm		
succinamide						
succinic acid diamide	H$_2$NCOCH$_2$CH$_2$CONH$_2$	116.12	d 260–265			
succinic acid						
butanedioic acid*	HOOCCH$_2$CH$_2$COOH	118.09	186–188	d 235	1 572^2	1.450
succinic anhydride	(CH$_2$CO)$_2$O	100.07	119–220	261	1 234	
succinimide	HNCOCH$_2$CH$_2$CO	99.09	123–125	287–288 d	1 418	
succinonitrile						
succinodinitrile	NCCH$_2$CH$_2$CN	80.09	54–57	265–287	985	1.417 3^{60}
sucrose	C$_{12}$H$_{22}$O$_{11}$	342.30	185–187		1 581^{17}	1.537 6
sulphadiazine						
2-sulphanilamidopyrimidine	C$_{10}$H$_{10}$N$_4$O$_2$S	250.28	d 253			
sulphanilamide						
4-aminobenzenesulphonamide*	H$_2$NC$_6$H$_4$SO$_2$NH$_2$	173.19	165–167		1 080	
sulpholane						
tetramethylene sulphone	CH$_2$CH$_2$CH$_2$CH$_2$SO$_2$	120.17	27	285	1 261	1.484 0
tannic acid, gallotannic acid	C$_{76}$H$_{52}$O$_{46}$	1701.23	218			
L-tartaric acid; (2R, 3R)-(+)-						
2, 3-dihydroxybutanedioic acid*	HOOCCH(OH)CH(OH)–COOH	150.09	172–174		1 760	1.495 5
taurine						
2-aminoethanesulphonic acid*	H$_2$NCH$_2$CH$_2$SO$_3$H	125.15	d 317			
terephthalic acid						
1, 4-benzenedicarboxylic acid*	C$_6$H$_4$(COOH)$_2$	166.14	>300	s		
1, 1, 1, 2-tetrabromoethane*	BrCH$_2$CBr$_3$	345.67	0	106–107/ 10 mm	2 875	1.633 3
1, 1, 2, 2-tetrabromoethane*						
acetylene tetrabromide	Br$_2$CHCHBr$_2$	345.67	0	119/15 mm	2 967	1.637 0
1, 1, 1, 2-tetrachloroethane*	ClCH$_2$CCl$_3$	167.85	−70	130	1 588	1.482 1
1, 1, 2, 2-tetrachloroethane*						
acetylene tetrachloride	Cl$_2$CHCHCl$_2$	167.85	−43	146	1 595	1.493 5
tetrachloroethene*						
tetrachloroethylene	Cl$_2$C:CCl$_2$	165.83	−22	121	1 623	1.505 6
tetraethylammonium bromide	(CH$_3$CH$_2$)$_4$NBr	210.16	d 285		1 397	
tetraethylene glycol	HOCH$_2$(CH$_2$OCH$_2$)$_3$CH$_2$OH	194.23	−6	327–328	1 128	1.457 7
tetrahydrofuran						
1, 4-epoxybutane*	CH$_2$CH$_2$CH$_2$CH$_2$O	72.11	−108	67	886	1.405 0
tetrahydrofurfuryl alcohol						
(2-tetrahydrofuryl)-methanol	OCH$_2$CH$_2$CH$_2$CHCH$_2$OH	102.13	−80	178	1 054	1.451 2

Flash point/°C	Solubility						Hazards	References	
	w	alc	et	ac	bz	Other			
	s	ss[h]	ss					B.31.346	IR.1(1).193C NMR.2(2).909C
	i	s[h]	vs	s	ss	CCl$_4$, chl	I	B.2(2).346	
	i	ss	vs		vs		H	B.5(2).537	IR.1(1).952B NMR.2(1).752B
31	i	s	s	s	∞	MeOH, CS$_2$	FL, T, I	B.5(2).362	IR.1(1).945B
	ss	s	ss			MeOH	I	B.2(2).595	IR.1(1).496B NMR.2(1).432B
	ss	i	i					B.2(2).554	IR.1(1).767D
	ss	s	s	s	i	MeOH	I	B.2(2).540	
	i	s	ss			chl	MS, I	B.17(2).429	
	s	ss	ss				I	B.21.369	IR.1(1).795C NMR.2(1).667A
>110	vs	s	ss	s	s	chl	T, H, I	B.2.615	IR.1(1).840D NMR.2(1).699A
	s	ss	i			py		B.31.424	IR.1(1).196B
	ss	ss		ss			I		
	s[h]	s	s	s		MeOH	H	8.14.698	
165	ss			ss			H, I	B.17(1).5	IR.1(1).882C NMR.2(2).784C
	s	vs	i	vs	i		CSA, I	B.31.133	IR.1(1).544C
	vs	vs	ss	s	i			B.3(3).994	IR.1(1).523D
	s	i	i				I	B.4.528	IR.1(1).890A NMR.2(2).792A
	i	i	i				I	B.9.841	IR.1(2).201C
>110		s	s	s	s	chl	I	B.1(3).191	IR.1(1).87B NMR.2(1).84A
none	i	∞	∞	s	s	chl	T, mutagen	B.1(3).192	IR.1(1).88A
none	ss	∞	∞	s	s	chl	vT, CSA	B.1(3).158	IR.1(1).87A NMR.2(1).83D
none	ss	∞	∞	s	s	chl	vT, CSA	B.1(3).159	
none	i	∞	∞		∞		H, I, CSA	B.1(3).664	
	vs	vs				chl, MeOH	I, MS	B.4(2).597	
176	vs	s	s			diox	I	B.1(3).2106	IR.1(1).226C NMR.2(1).192D
−17	s	vs	vs	vs	vs		FL, I	B.17(2).15	
83			s	s			I, MS	B.17(2).106	IR.1(1).240B

(Continued)

Table XI.2 (Continued)

Compound/synonym (s)	Formula	M_r	m. p. /°C	b. p. /°C	Density / kg m^3	Refractive index
1, 2, 3, 4-tetrahydronaphthalene* tetraline	$C_{10}H_{12}$	132.21	−36	207	970	1.541 3
tetrahydropyran pentamethylene oxide	$CH_2CH_2CH_2CH_2CH_2O$	86.14	−45	88	881	1.420 0
tetrahydrothiophene tetramethylene sulphide	$CH_2CH_2CH_2CH_2S$	88.18	−96	119	1 000	1.504 8
tetramethylammonium-borohydride	$(CH_3)_4NBH_4$	88.99	> 300		813^{25}	
tetramethylammonium bromide	$(CH_3)_4NBr$	154.06	d 230		1 560	
tetramethylene sulphoxide tetrahydrothiophene oxide	$CH_2CH_2CH_2CH_2SO$	104.17			1 158	1.520 0
N, N, N', N'-tetramethyl-1, 2-diaminoethane	$(CH_3)_2NCH_2CH_2N(CH_3)_2$	116.21	−55	120–122	770	1.417 9
tetramethylsilane	$Si(CH_3)_4$	88.23	−99	26.5	651	1.358 5
tetrapropylammonium bromide	$(CH_3CH_2CH_2)_4NBr$	266.27	d 270			
thiazole	CH:CHN:CHS	85.13		117–118	1 200	1.536 9
thioacetamide	CH_3CSNH_2	75.13	112–114			
thioacetic acid thioloethanoic acid*	CH_3COSH	76.12	<−17	87–90	1 064	1.463 0
1-thioglycerol 3-mercapto-1, 2-propanediol	$HSCH_2CH(OH)CH_2OH$	108.17		118/5 mm	1 295	1.526 8
thiophene	CH:CHCH:CHS	84.14	−38	84	1 065	1.527 8
2-thiophenecarboxylic acid	S.CH:CHCH:C(COOH)	128.15	129–130	d 260		
thiophenol, benzenethiol; mercaptobenzene*	C_6H_5SH	110.18	−15	169	1 077	1.588 0
thiourea* 2-thiourea*	H_2NCSNH_2	76.12	175–178		1 405	
L-threonine; (2S, 3R)-(−)- 2-amino-3-hydroxybutyric acid	$CH_3CH(OH)CH(NH_2)$ −COOH	119.12	253			
L-allo-threonine (2S, 3S)-(+)- 2-amino-3-hydroxybutyric acid	$CH_3CH(OH)CH(NH_2)$ −COOH	119.12	d 272			
thymol; 2-hydroxy-1-iso-propyl-4-methylbenzene*	$[(CH_3)_2CH]C_6H_3(CH_3)OH$	150.22	49–51	233	965	1.522 7
o-tolualdehyde 2-methylbenzaldehyde	$CH_3C_6H_4CHO$	120.15		199–200	1 039	1.547 2
m-tolualdehyde 3-methylbenzaldehyde	$CH_3C_6H_4CHO$	120.15		199	1 019	1.541 3
p-tolualdehyde 4-methylbenzaldehyde	$CH_3C_6H_4CHO$	120.15		204–205	1 019	1.545 4

Flash point/°C	w	alc	et	ac	bz	Other	Hazards	References
77	i	vs	vs				H, I	B.5(3).1219; IR.1(1).937D; NMR.2(1).738D
−15		s	s		s		FL, I	B.17(2).18
12	i	∞	∞	∞	∞		FL, stench	B.17(2).15; IR.1(1).271A; NMR.2(1).232D
	s	s					T, I	B.4(4).148
	vs	ss	i		i		T, I	B.4(2).558; IR.1(1).386B; NMR.2(1).340B
>110								IR.1(1).879C; NMR.2(2).781C
10							FL, C	B.4.250; IR.1(1).305A; NMR.2(1).262D
−27			s				FL, MS	B.4.625; NMR.2(2).983A
	s					chl	MS	B.4(1).364; IR.1(1).389B; NMR.2(1).343A
22	ss	s	s	s			FL	B.27(2).9; IR.1(2).642D; NMR.2(2).510C
	vs	vs	ss	ss			CSA, I, stench	B.2(2).210; IR.1(1).827B; NMR.2(1).684C
11	s	vs	∞	vs			FL, C, stench	B.2(2).208; IR.1(1).528A; NMR.2(1).463A
>110	ss	∞	ss	vs	ss		H, I, stench	B.1(3).2339; IR.1(1).262C; NMR.2(1).229A
−1		∞	∞	∞	∞	CCl$_4$, py, diox	FL, stench	B.17(2).35; IR.1(2).591B; NMR.2(2).466B
	vs[h]	vs	vs			chl		B.18.289; IR.1(2).600D; NMR.2(2).474D
50	i	s	s		s		vT, stench	B.6(2).284; IR.1(1).1173A
	s	s	i				T, CSA	B.3(2).128
	s	i	i					B.4(3).1623; IR.1(1).582A; NMR.2(1).492D
	s	i	i					B.4(3).1629; IR.1(1).581C
102	i	vs	vs			chl	H, I	B.6(2).494; IR.1(1).1087D; NMR.2(1).882A
67	ss[h]	s	s	vs	s	chl	I, AS	B.7.295; IR.1(2).104B; NMR.2(2).103D
78	ss	∞	∞	vs	s	chl	I, AS	B.7.296; IR.1(2).107(A); NMR.2(2).106A
80	ss	∞	∞	∞		chl	I, AS	B.7.297; IR.1(2).109B; NMR.2(2).108A

(Continued)

354 Physical Properties of Inorganic and Organic Compounds

Table XI.2 (Continued)

Compound/synonym(s)	Formula	M_r	m. p. /°C	b. p. /°C	Density / kg m^3	Refractive index
toluene, methylbenzene* phenylmethane*	$C_6H_5CH_3$	92.15	−93	110.6	867	1.496 1
4-toluenesulphonamide	$CH_3C_6H_4SO_2NH_2$	171.22	138–139			
4-toluenesulphonic acid	$CH_3C_6H_4SO_3H$	172.21	104–105	140/20 mm		
4-toluenesulphonyl chloride tosyl chloride	$CH_3C_6H_4SO_2Cl$	190.65	67–69	145–146/ 15 mm		
o-toluic acid 2-methylbenzoic acid	$CH_3C_6H_4COOH$	136.16	103–105	258–259	1 062	1.512[115]
m-toluic acid 3-methylbenzoic acid	$CH_3C_6H_4COOH$	136.16	108–110	263	1 054	1.509
p-toluic acid 4-methylbenzoic acid	$CH_3C_6H_4COOH$	136.16	180–182	274–275		
o-toluidine 2-aminotoluene	$CH_3C_6H_4NH_2$	107.16	−28	199–200	998	1.570 9
m-toluidine 3-aminotoluene	$CH_3C_6H_4NH_2$	107.16	−30	203–204	999	1.566 9
p-toluidine 4-aminotoluene	$CH_3C_6H_4NH_2$	107.16	45–47	200	962	1.563 6
1, 3, 5-triazine*	N:CHN:CHN:CH	81.08	81–83	114	1 380	
1, 2, 4-triazole	HNCH:NCH:N	69.07	119–121	260	1 132[153]	1.485 4[25]
trichloroacetic acid trichloroethanoic acid*	Cl_3CCOOH	163.39	54–58	197	1 620[25]	1.460 3[61]
1, 2, 3-trichlorobenzene*	$C_6H_3Cl_3$	181.45	53 55	218–219		
1, 2, 4-trichlorobenzene*	$C_6H_3Cl_3$	181.45	16–17	213–214	1 454	1.571 7
1, 3, 5-trichlorobenzene*	$C_6H_3Cl_3$	181.45	63–64	208		
1, 1, 1-trichloroethane* methylchloroform	CH_3CCl_3	133.41	−35	74	1 339	1.437 9
1, 1, 2-trichloroethane*	$ClCH_2CHCl_2$	133.41	−37	113.8	1 440	1.471 4
trichloroethylene trichloroethene*	$ClCH:CCl_2$	131.39	−84.8	87	1 464	1.477 3
2, 4, 6-trichlorophenol	$Cl_3C_6H_2OH$	197.45	64–66	246		
1, 1, 1-trichloro-2, 2, 2-trifluoroethane*	CF_3CCl_3	187.38	13–14	46	1 579	1.359 9
1, 1, 2-trichloro-1, 2, 2-trifluoroethane*	$Cl_2CFCClF_2$ (Freon 113)	187.38	−35	47–48	1 575	1.357 8
triethanolamine 2, 2', 2''-nitrilotriethanol	$(HOCH_2CH_2)_3N$	149.19	21	190–193/ 5 mm	1 124	1.485 2
triethylamine	$(CH_3CH_2)_3N$	101.19	−114.7	88.8	727	1.400 0
triethylenetetramine	$H_2NCH_2(CH_2NHCH_2)_2$ $-CH_2NH_2$	146.24	12	266–267	982	1.497 1
trifluoroacetic acid trifluoroethanoic acid*	F_3CCOOH	114.02	−15.4	72.4	1 480	
trifluoromethane	CHF_3 (Freon-23)	70.01	−160	−82	1 520[−100]	
trifluoromethanesulphonic acid	F_3CSO_3H	150.08		162	1 696	1.327 0
triglycine N-(N-glyclyglycyl)glycine	$H_2NCH_2CONHCH_2$ $-CONHCH_2COOH$	189.17	d 246			

Flash point/°C	Solubility						Hazards	References	
	w	alc	et	ac	bz	Other			
4	i	∞	∞	s	∞	CS$_2$	FL, H, I	B.5(3).651	
	ss	s	ss				H, I	B.11(2).55	IR.1(2).530C
	vs	s	s				C, T, I, MS	B.11(2).43	IR.1(2).493A
									NMR.2(2).818C
	i	s	s		vs		C, MS	B.11(2).54	IR.1(2).520B
	s^h	vs	vs			chl	T, H, I	B.9.462	IR.1(2).186C
									NMR.2(2).183B
	ss	vs	vs				H, I	B.9.475	IR.1(2).190B
									NMR.2(2).187C
	i	vs	vs			MeOH	H, I	B.9.483	IR.1(2).193B
									NMR.2(2).190C
85	ss	∞	∞			CCl$_4$	vT, H, I, CSA	B.12(2).429	IR.1(1).1192D
85	ss	∞	∞	∞	∞	CCl$_4$	vT, H, I	B.12(2).463	IR.1(1).1197C
									NMR.2(1).998C
88	ss	vs	s	s		py	vT, CSA	B.12(2).482	IR.1(1).1202B
		s	s				MS		NMR.2(1).727B
	vs	vs	ss		ss			B.26(2).7	IR.1(2).626B
									NMR.2(2).496D
>110	vs	s	s				C, MS, H	B.2(2).196	IR.1(1).508C
126	i	ss	vs		vs	CS$_2$	I	B.5(2).156	IR.1(1).1000B
									NMR.2(1).800C
>110	i	ss	vs				T, I	B.5(2).156	
126	i	ss	vs		vs	CS$_2$	H, I	B.5(2).156	IR.1(1).1008C
									NMR.2(1).808A
none	i	∞	∞			chl	I, MS	B.1(2).55	NMR.2(1).82D
none	ss	s	∞			chl	CSA, I, H	B.1(3).154	IR.1(1).85A
									NMR.2(1).83A
none	ss	∞	∞	s		chl	CSA, I	B.1(3).656	
none	ss^h	vs	vs		vs		T, H, CSA	B.6(3).716	IR.1(1).1093A
									NMR.2(1).888D
none	i	s	s			chl		B.1(3).157	IR.1(1).90A
none	i	s	∞		∞		I	B.1(3).157	
185	∞	∞	ss		ss	chl	I, MS	B.4(2).729	IR.1(1).344A
									NMR.2(1).301D
−6	s	s	s	vs	vs	chl	FL, C, I	B.4(2).593	
143	s^h	s					C, T	B.4(3).542	IR.1(1).311B
none	s	s	s	s			C, T	B.2(2).186	
	s	vs		s	s			B.1(3).34	
none	d	d	∞				C, MS	B.3(4).34	
	s^h	i	i					B.4(3).1198	

(Continued)

Table XI.2 (Continued)

Compound/synonym (s)	Formula	M_r	m. p. /°C	b. p. /°C	Density / kg m³	Refractive index
trimethylacetic acid						
2, 2-dimethylpropanoic acid	$(CH_3)_3CCOOH$	102.13	33–35	163–164	905^{50}	
trimethylamine	$(CH_3)_3N$	59.11	−117	2.9	656	1.344 3
2, 2, 4-trimethylpentane*						
isooctane	$(CH_3)_2CHCH_2C(CH_3)_3$	114.23	−107	98–99	692	1.391 5
1, 3, 5-trinitrobenzene*	$C_6H_3(NO_2)_3$	213.11	121–122	315		
2, 4, 6-trinitrotoluene (TNT)	$CH_3C_6H_2(NO_2)_3$	227.13	82	240 exp	1 654	
1, 3, 5-trioxane						
metaformaldehyde	$OCH_2OCH_2OCH_2$	90.08	61–62	114	1 170^{65}	
triphenylamine	$(C_6H_5)_3N$	245.33	125–127	365	774^0	1.353^{16}
1, 3, 5-triphenylbenzene*	$(C_6H_5)_3C_6H_3$	306.41	172–174	460	1 199^{30}	
triphenylmethane*	$(C_6H_5)_3CH$	244.34	92–94	358–359/ 754 mm	1 014	1.583 9^{99}
triphenylmethanol*						
triphenylcarbinol	$(C_6H_5)_3COH$	260.34	161–163	360	119 9^0	
tris (hydroxymethyl) aminomethane; 2-amino-2-(hydroxymethyl)-1, 3-propanediol*	$H_2NC(CH_2OH)_3$	121.14	171.2	219–220/ 10 mm		
tryptamine						
3-(2-aminoethyl) indole	$C_{10}H_{12}N_2$	160.22	114–119	137/0.15 mm		
L-tryptophan, (S)-(−)-						
L-2-amino-3-(3-indole)-proprionic acid	$C_{11}H_{12}N_2O_2$	204.23	d 280–285			
L-tryosine, (S)-(−)-,						
2-amino-3(4-hydroxyphenyl) propanoic acid*	$HOC_6H_4CH_2CH(NH_2)COOH$	181.19	d 295			
undecane						
hendecane*	$CH_3(CH_2)_9CH_3$	156.32	−26	196	740	1.417 2
uracil						
2, 4-dihydroxypyrimidine	$NHCONHCOCH:CH$	112.10	338			
urea*	H_2NCONH_2	60.06	133–135	d	1 335	
urethane						
ethyl carbamate*	$H_2NCOOC_2H_5$	89.10	48.5–50	182–184	986	1.414 4^{52}
uric acid						
2, 6, 8-trihydroxypurine	$CONHCONHC:CNHCONH$	168.12	> 300	d	1 890	
valeraldehyde						
pentanal*	$CH_3(CH_2)_3CHO$	86.13	−92	103	810	1.394 2
valeric acid						
pentanoic acid*	$CH_3(CH_2)_3COOH$	102.13	−34.5	185	939	1.408 0
iso-valeric acid						
3-methylbutanoic acid*	$(CH_3)_2CHCH_2COOH$	102.13	−29	175–177	929	1.403 3
L-valine; (S)-(+)- 2-amino-3-methylbutanoic acid	$(CH_3)_2CHCH(NH_2)COOH$	117.15	d 315			
L-norvaline; (S)-(+)- 2-aminopentanoic acid*	$CH_3CH_2CH_2CH(NH_2)COOH$	117.15	d 291	s		

Flash point/°C	Solubility						Hazards	References	
	w	alc	et	ac	bz	Other			
63	ss	vs	vs				C	B.2(3).708	IR.1(1).487D NMR.2(1).425A
−6	vs	s	s		s	chl	FG, C	B.4(2).553	
−7	i	∞	s	∞	∞	chl	FL, I	B.1(3).492	
	ss[h]	ss	ss	vs	s	py, chl	T, O, E	B.5(2).203	
	i	ss	s	vs	vs	py	T, O, E	B.5(3).767	
45	vs	s	s		s	CCl$_4$ chl	FS, H	B.19(2).392	IR.1(1).250C NMR.2(1).211C
	i	ss	s	s				B.12(2).106	IR.1(1).1192A NMR.2(1).991D
	i	s	s		vs	CS$_2$		B.5(2).670	IR.1(1).949A NMR.2(1).755D
	i	ss	vs		s	chl, py		B.5(2).613	IR.1(1).936C NMR.2(1).739D
	i	vs	vs	s	s			B.6(2).686	IR.1(1).1125D NMR.2(1).929A
	vs					MeOH	I, MS	B.4.303	IR.1(1).347C NMR.2(1).304A
	i	s	i	s	i		H, I	B.22(2).346	IR.1(2).660C NMR.2(2).527B
	ss	s[h]	i					B.22(2).466	IR.1(2).671D
	ss[h]	i	i					B.14(2).366	IR.1(2).255D NMR.2(2).254B
60	i	∞	∞				FL, I	B.1(3).534	IR.1(1).2D NMR.2(1).10D
	ss[h]/vs	vs	vs					B.24(2).168	IR.1(2).811B NMR.2(2).689D
	vs	vs	i		i		I	B.3(2).35	IR.1(1).799C NMR.2(1).670B
92	vs	vs	vs		vs	chl, py	CSA	B.3(2).19	IR.1(1).771D NMR.2(1).647C
	i	i	i			alk		B.26(2).293	IR.1(2).710C
12	ss	s	s				FL, I	B.1(3).2797	IR.1(1).467B NMR.2(1).358D
88	s	s	s				C, I, stench	B.2(3).664	
70	s	∞	∞			chl	C, T, I	B.2(2).271	IR.1(1).488A NMR.2(1).425B
	s	i	i		i			B.4(2).852	IR.1(1).573D
	vs[Sh]	i	i			chl		B.4(3).1549	IR.1(1).574C

(Continued)

Table XI.2 (Continued)

Compound/synonym(s)	Formula	M_r	m. p. /°C	b. p. /°C	Density / kg m³	Refractive index
vanillin 4-hydroxy-3-methoxybenzaldehyde	$(OH)C_6H_3$-$(OCH_3)CHO$	152.16	81–83	170/15 mm	1 056	
veratrole 1, 2-dimethoxybenzene	$C_6H_4(OCH_3)_2$	138.17	22–23	206–207	1 084	1.533 7
vinyl acetate ethenyl acetate	$CH_3COOCH:CH_2$	86.09	−93	72–73	932	1.395 4
vinylacetic acid 3-butenoic acid*	$CH_2:CHCH_2COOH$	86.09	−39	163	1 009	1.424 9
vinyl bromide bromoethene*	$H_2C:CHBr$	106.96	−139	15.8	1 517	1.435 0
vinyl chloride chloroethene*	$CH_2:CHCl$	62.50	−153.8	−13.37	911	1.370 0
xanthine 2, 6-purinedione	CONHCONHC:CN:CHNH	152.11	d >300			
o-xylene 1, 2-dimethylbenzene*	$C_6H_4(CH_3)_2$	106.17	−25.18	144.4	880	1.505 5
m-xylene 1, 3-dimethylbenzene*	$C_6H_4(CH_3)_2$	106.17	−48	139.1	864	1.497 2
p-xylene 1, 4-dimethylbenzene*	$C_6H_4(CH_3)_2$	106.17	13.3	138	861	1.495 8
2, 3-xylenol; 1, 2-dimethyl-3-hydroxybenzene*	$(CH_3)_2C_6H_3OH$	122.17	75	217		1.542 0
2, 4-xylenol; 2, 4-dimethyl-1-hydroxybenzene*	$(CH_3)_2C_6H_3OH$	122.17	22–23	210	965	1.542 0
2, 5-xylenol; 1, 3-dimethyl-5-hydroxybenzene*	$(CH_3)_2C_6H_3OH$	122.17	71–73	212	971	
2, 6-xylenol; 1, 3-dimethyl-2-hydroxybenzene*	$(CH_3)_2C_6H_3OH$	122.17	45–46	203		
xylitol	$HOCH_2(CHOH)_3CH_2OH$	152.15	95–97			
L-xylose	$OCH_2(CHOH)_3CHOH$	150.13	150–152		1 525	

Flash point/°C	Solubility						Hazards	References	
	w	alc	et	ac	bz	Other			
	ss^h vs	vs	vs	vs	s^h	chl, CS$_2$	MS, LS, AS	B.8(2).278	IR.1(2).120D
87	ss	s	s					B.6(2).779	IR.1(1).1037C NMR.2(1).836B
−6	s^h	∞	s	s	s	chl, CCl$_4$	FL, I	B.2(2).147	IR.1(1).632D NMR.2(1).533C
65	s	∞	∞				I	B.2(2).389	IR.1(1).498D NMR.2(1).435B
none	i	s	s	s	s	chl	T, CSA	B.1(2).162	IR.1(1).93A NMR.2(1).86C
	ss	s	vs				CSA, I	B.1(2).157	
	ss	ss	i			alk	T	B.26(2).260	IR.1(2).709C NMR.2(2).585A
32	i	∞	∞	∞	∞	CCl$_4$	FL, H, I	B.5(2).281	
25	i	∞	∞	∞	∞		FL, H, I	B.5(2).287	
27	i	∞	∞	∞	∞		FL, H, I	B.5(2).296	
	ss	s	s				vT, C	B.6(2).453	IR.1(1).1080D NMR.2(1).876A
>110	ss	∞	∞				vT, C	B.6(2).455	IR.1(1).1087B NMR.2(1).881C
	s	s					vT, C	B.6(2).462	IR.1(1).1087C NMR.2(1).881D
73	s	s	s				vT, C	B.6(2).466	IR.1(1).1081D NMR.2(1).876C
	vs	s				py		B.31(2).604	IR.1(1).185B NMR.2(2).901C
	vs	s^h	ss					B.31.55	IR.1(1).187D NMR.2(2).904B

Table XI.3 Physical Constants of Selected Organo-metallic Compounds

For more extensive data see: *Dictionary of Organometallic Compounds* (1984) and Supplements, Chapman & Hall, London.
Column 1. Commonly used name of compound, together with synonyms; IUPAC recommended names denoted by*.
Columns 4 (m.p.) and 5 (b.p.). d : decomposes.
Columns 6 (density reference to water at 4 °C) and 7 (refractive index for the sodium D line). Values quoted at 20 °C unless otherwise stated.
Column 8. Flash point determined in closed cup.
Columns 9. Solubility in water (w), ethanol (alc), ether (et), acetone (ac) and benzene (bz). Solubility in other solvents: alkali (alk), chloroform (chl), pyridine (py) etc. at room temperature or on heating (^{h}).
d: decomposes, hydrolysed; i: insoluble; s: soluble; ss: slightly soluble; vs: very soluble; ∞: soluble in all proportions.
Column 10. General hazards. C: corrosive; CSA: cancer suspect agent; FL: flammable liquid; FS: flammable solid; I: irritant; LS: light-sensitive; MS: moisture sensitive, e.g. hydrolysed by moisture, deliquescent; P: pyrophoric; T: toxic; vT: very toxic. Other hazards, e.g. mutagen, are spelt out.
For further details of other hazards, including toxicity, LD_{50} and approved occupational exposure limits see references in Section XVI and Tables XVI 1 to 4.

Compound/Synonym(s)	Formula	M_r	m. p. /°C	b. p. /°C	ρ /kg m^{-3}	n_D
aluminium isopropoxide	[(CH$_3$)$_2$CHO]$_3$Al	204.25	118			
aluminium tri-sec-butoxide	[C$_2$H$_5$CH(CH$_3$)O]$_3$Al	246.33		200–206 /30 mm	967	1.4381
dimethylaluminium chloride	(CH$_3$)$_2$AlCl	92.51	−21	126–127	996	
triethoxyaluminium	Al(OC$_2$H$_5$)$_3$	162.17	150–160	d	1142	
triethylaluminium	Al(C$_2$H$_5$)$_3$	114.17	−50	128–130 /50 mm	835	
trimethylaluminium	Al(CH$_3$)$_3$	72.09	15	125–126	752	1.432
triethylantimony	Sb(C$_2$H$_5$)$_3$	208.94	−98	159–160	1234[16]	
trimethylantimony	Sb(CH$_3$)$_3$	166.86		80.6	1523[15]	
triphenylantimony	Sb(C$_6$H$_5$)$_3$	353.07	52–54	377	1434	
cacodyl oxide	[(CH$_3$)$_2$As]$_2$O	225.98	−25	149–151	1486[15]	
dimethylarsine	(CH$_3$)$_2$AsH	106.00		35.6 /747 mm	1213[29]	
dimethylarsinic acid, cacodylic acid	(CH$_3$)$_2$As(O)OH	138.00	195–196			
methylarsine	CH$_3$AsH$_2$	91.97		2		
phenylarsonic acid, benzenearsonic acid	C$_6$H$_5$AsO$_3$H$_2$	202.04	d 160		1760	
trimethylarsine	(CH$_3$)$_3$As	120.03		70	1124	
triphenylarsine	(C$_6$H$_5$)$_3$As	306.24	60–62	>360	1223[48]	1.6139[48]
dimethylboric acid	(CH$_3$)$_2$BOH	57.89		0/36 mm		
dimethylboron bromide, bromodimethylborane	(CH$_3$)$_2$BBr	120.79	−129	31–32	1238	
methylboric acid	CH$_3$B(OH)$_2$	59.86	d			
tetramethylammonium borohydride	(CH$_3$)$_4$NBH$_4$	88.99			813	
triethylboron, triethylborane	(C$_2$H$_5$)$_3$B	98.00	−93	95	677	1.397 1
trimethoxyboroxine, trimethoxyborozole	OB(OCH$_3$)OB(OCH$_3$)OB(OCH$_3$)	173.53	10		1195	1.399 6
trimethylborate, trimethoxyborine	(CH$_3$O)$_3$B	103.92	−34	68–69	915	1.356 8
trimethylboron, trimethylborine	(CH$_3$)$_3$B	55.92	−161	−20	1.91	
triphenylboron	(C$_6$H$_5$)$_3$B	242.13	145	203/15 mm		
diethylcadmium	(C$_2$H$_5$)$_2$Cd	170.52	−21	64/19 mm	1656	
dimethylcadmium	(CH$_3$)$_2$Cd	142.47	−4.5	105–106	1985	
benzenechromium tricarbonyl	C$_6$H$_5$Cr(CO)$_3$	214.14	163–166			
trimethylgallium	(CH$_3$)$_3$Ga	114.83	−19	53–57		
trimethylgallium monammine	(CH$_3$)$_3$GaNH$_3$	131.86	31	s		
tetraethylgermanium	(C$_2$H$_5$)$_4$Ge	188.84	−90	162–163	1198[0]	1.554[0]
triethylgermanium chloride	(C$_2$H$_5$)$_3$GeCl	195.23	<−50	176		
trimethylgermanium bromide	(CH$_3$)$_3$GeBr	197.60	−25	114	1540	1.470 5
triphenylgermanium chloride, chlorotriphenylgermane	(C$_6$H$_5$)$_3$GeCl	339.36	114–117	285/12 mm		

Flash point/°C	Solubility						Hazards
	w	alc	et	ac	bz	other	
27							FL, C
							MS, P
		i	ss		ss		MS, P
	d						MS, P
	d	s	s				MS, P
	i	s	s				
	ss	i	s				
>110	i	ss			s		T, I
	ss	s	s				
		s	s		s	chl, CS_2	P
	vs	s	i				vT, MS
	ss	s	s				vT, I
	s	ss					vT, I
	ss		s				vT, I
	i	ss	s		s		vT, I
	vs						
−37	d						FL, C
	ss	s	s				
	vs	s					T, I
	i	s	s				T, P
10					vs	CCl_4	FL, MS
−8	d	s	s				FL, MS
	ss	vs	vs				FG
	d	d			s		FS
	d		vs				T, MS
	d						MS, T
	i				s		T
	d		s			NH_3	MS, T
	d		s			NH_3	MS, T
	d		s		s	HCl	MS, T
	d		s		s	chl, CCl_4	MS, T
37	d		s		s		MS, FL, C
	i		s		s		I

(Continued)

Table XI.3 (Continued)

Compound/Synonym(s)	Formula	M_r	m. p. /°C	b. p. /°C	ρ /kg m^{-3}	n_D
ferrocene, dicyclopentadienyliron	C$_5$H$_5$FeC$_5$H$_5$	186.07	174–176	249		
tetraethyllead	(C$_2$H$_5$)$_4$Pb	323.44	–137	91/19 mm	1659[11]	1.519 5
tetramethyllead	(CH$_3$)$_4$Pb	267.33	–27 to –28	110	1995	1.512 0
tetraphenyllead	(C$_6$H$_5$)$_4$Pb	515.61	227–228	126/13 mm	1530	
triphenyllead bromide	(C$_6$H$_5$)$_3$PbBr	518.42	166			
n-butyllithium	CH$_3$(CH$_2$)$_3$Li	64.06	s 80	d 150	765	
ethyllithium	C$_2$H$_5$Li	36.00	95 (in N$_2$)	s		
diphenylmagnesium	(C$_6$H$_5$)$_2$Mg	178.53	d 280			
diethylmercury	(C$_2$H$_5$)$_2$Hg	258.71		159	2 444	1.539 9
dimethylmercury	(CH$_3$)$_2$Hg	230.66		96	3 069	1.532 7
diphenylmercury	(C$_6$H$_5$)$_2$Hg	354.81	128–129	204/10.5 mm	2 318	
ethylmercuric chloride	C$_2$H$_5$HgCl	265.10	193		3 482	
ethylmercurithiosalicylic acid-sodium salt; thimerosal	2-(C$_2$H$_5$HgS)C$_6$H$_4$COONa	404.81	d 232			
methylmercuric chloride	CH$_3$HgCl	251.09	170		4 063	
phenylmercuric acetate	C$_6$H$_5$HgOOCCH$_3$	336.74	150–152			
phenylmercuric chloride	C$_6$H$_5$HgCl	313.15	d 248			
dicyclopentadienylnickel, nickelocene	C$_5$H$_5$NiC$_5$H$_5$	188.90	171–173			
tetrakis (triphenylphosphine) palladium (O)	[(C$_6$H$_5$)$_3$P]$_4$Pd	1155.58				
diethylphosphate	(C$_2$H$_5$O)$_2$POOH	154.10		d 203	1 186[25]	1.417 0
diethylphosphine	(C$_2$H$_5$)$_2$PH	90.11		85		
diethylphosphinic acid	(C$_2$H$_5$)$_2$POOH	122.11	18.5	320		
diethylphosphite	(C$_2$H$_5$O)$_2$POH	138.10		50/2 mm	1 072	1 407 6
dimethylphosphate	(CH$_3$O)$_2$POOH	126.05		d 172	1 335[25]	1.408[25]
dimethylphosphine	(CH$_3$)$_2$PH	62.05		25		
dimethylphosphinic acid	(CH$_3$)$_2$POOH	94.05	92	377		
dimethylphosphite	(CH$_3$O)$_2$POH	110.05		170–171	1 200	1.403 5
ethylphosphine	CH$_3$CH$_2$PH$_2$	62.05		25		
ethylphosphonic acid ethanephosphonic acid*	C$_2$H$_5$P(O)(OH)$_2$	110.05	61–63			
methyldichlorophosphate	Cl$_2$P(O)OCH$_3$	148.91		62–64 /15 mm	1 488	1.435 9
methyldichlorophosphite	Cl$_2$POCH$_3$	132.91	–91	93	1 406	1.473 6
methylphosphine	CH$_3$PH$_2$	48.02		–14		
methylphosphonic acid	CH$_3$P(O)(OH)$_2$	96.02	105–107	d		
methylphosphonic dichloride	CH$_3$P(O)Cl$_2$	132.91	35–37	163		
phenylphosphine	C$_6$H$_5$PH$_2$	110.10		160–161	1 001[15]	1.579 6
triethylphosphate*	(C$_2$H$_5$O)$_3$PO	182.16	–56	215	1 070	1.405 3
triethylphosphine	(C$_2$H$_5$)$_3$P	118.16	–88	127–128	800	1.456 3
triethylphosphite	(C$_2$H$_5$O)$_3$P	166.16		156–158	969	1.413 3
trimethylphosphate	(CH$_3$O)$_3$PO	140.08	–46	197	1 214	1.396 0
trimethylphosphine	(CH$_3$)$_3$P	76.08	–86	38–39	735	
trimethylphosphite	(CH$_3$O)$_3$P	124.08		111–112	1 052	1.409 5
triphenylphosphine	(C$_6$H$_5$)$_3$P	262.29	79–81	377	1 075[80]	1.635 8[80]
tetramethylplatinum	(CH$_3$)$_4$Pt	255.23	d			
trimethylplatinum iodide	(CH$_3$)$_3$PtI	367.09	d 250			
potassium phthalimide	COC$_6$H$_4$CONK	185.23	>300			
dichloromethylsilane	CH$_3$SiHCl$_2$	115.04	–93	41	1 105	
diethyldichlorosilane	(C$_2$H$_5$)$_2$SiCl$_2$	157.12		128–129	1 050	1.430 1
diethylsilanediol	(C$_2$H$_5$)$_2$Si(OH)$_2$	120.23	96	d 140		
dimethylsilicane	SiH$_2$(CH$_3$)$_2$	60.17	–150	–20	680^{-80}	
ethyltrichlorosilane, trichloroethylsilane	C$_2$H$_5$SiCl$_3$	163.51	–106	99	1 238	1.425 2
methylsilicane	SiH$_3$CH$_3$	46.14	–156	31	620^{-67}	
tetraethylsilane, tetraethylsilicane	Si(C$_2$H$_5$)$_4$	144.33	–82.5	153–154	766	1.426 5

Flash point/°C	Solubility						Hazards
	w	alc	et	ac	bz	other	
	i	s	s		s	MeOH	
	i	s	s		s		T
	i	s	s		s		T
	i				s		vT
	i						vT
	d		s		s		MS
	d		d		s		MS, T
			i		i		
	i	ss	vs				C, T
		s	s				C, T
	i	s^h	ss		s	chl, CS$_2$	C, T
	i	vs^h	ss				T
							vT, I
							vT, I
	ss	s		s	s		vT, I
	ss^h	s^h			s^h		vT, I
		s					T, CSA
							LS
	i		s				
	vs	vs	vs				
90	d^h	s	s				MS, I
	s	s	i	s			
	i	s	s				
	vs	vs	vs		vs^h		I
29		s				py	FL, CSA
	d						
	vs	vs	vs		i		C
>110							C, MS
40							C, MS
	i	ss	vs				
	vs	vs	vs		i		C
>110							vT, C
115	s	vs	s		s		
−17	i	∞	∞				P
54	i	vs	vs				MS, I
none	vs	ss	s				CSA, mutagen
	i		s				FL, stench
	d^h	vs	vs				FL, C
181	i	s	vs		s	chl	I
	i		vs	vs	vs		
	i	ss	ss	ss	ss		
							MS
−32							FL, C
28	d	s					FL, C
	s						
−10							vT, FL
26							FL, MS

(Continued)

Table XI.3 (Continued)

Compound/Synonym(s)	Formula	M_r	m. p. /°C	b. p. /°C	ρ /kg m^{-3}	n_D
tetramethoxysilane, tetramethyl orthosilicate	Si(OCH$_3$)$_4$	152.22	−4	121−122	1 023	1.368 0
tetramethylsilane, tetramethylsilicane	Si(CH$_3$)$_4$	88.23	−99	26−28	648	1.358 5
triethylchorosilane, chlorotriethylsilane	(C$_2$H$_5$)$_3$SiCl	150.72		142−144	898	1.430 1
triethysilane	(C$_2$H$_5$)$_3$SiH	116.28		107−108	728	1.413 0
trimethoxysilane	(CH$_3$O)$_3$SiH	122.20	−115	81	960	1.357 9
trimethylsilylacetonitrile	(CH$_3$)$_3$SiCH$_2$CN	113.24		65−70 /20 mm	827	1.420 7
sodium acetamide	NaNHCOCH$_3$	81.05	d 300−350			
sodium acetylide	NaC : CH	48.02	d > 210			
sodium ethoxide	C$_2$H$_5$ONa	68.05	>300	d		
ethyl sodium	NaC$_2$H$_5$	52.05	d			
sodium methoxide	CH$_3$ONa	54.02	>300			
sodium phenoxide trihydrate	C$_6$H$_5$ONa . 3H$_2$O	170.14	61−64			
tetraphenylboron sodium	(C$_6$H$_5$)$_4$BNa	342.23	>300			
triphenylmethylsodium	(C$_6$H$_5$)$_3$CNa	266.32				
butyltin trichloride, butyltrichlorotin	CH$_3$(CH$_2$)$_3$SnCl$_3$	282.17	−63	93/10 mm	1 693	1.522 9
dibutyltin diacetate, diacetoxydibutyltin	[CH$_3$(CH$_2$)$_3$]$_2$Sn(OOCCH$_3$)$_2$	351.01	10	142−145 /10 mm	1 320	1.477 3
dibutyltin dibromide, dibromodibutyltin	[CH$_3$(CH$_2$)$_3$]$_2$SnBr$_2$	392.74	20		1 739	1.544 2
dibutyltin oxide dibutyloxotin	[CH$_3$(CH$_2$)$_3$]$_2$SnO	248.92	>300		1 580	
diethyl tin	(C$_2$H$_5$)$_2$Sn	176.81	<−12	d 150	1 654	
diethyltin dibromide, dibromodiethyltin	(C$_2$H$_5$)$_2$SnBr$_2$	336.63	63	232−233	2 068[14]	
diethyltin oxide	(C$_2$H$_5$)$_2$SnO	192.81	infus			
dimethyltin dichloride, dichlorodimethyltin	(CH$_3$)$_2$SnCl$_2$	219.67	103−105	188−190		
dimethyltin oxide	(CH$_3$)$_2$SnO	164.76	infus		1 269	
diphenyltin dichloride, dichlorodiphenyltin	(C$_6$H$_5$)$_2$SnCl$_2$	343.81	42	333−337		
ethyltin tribromide	C$_2$H$_5$SnBr$_3$	387.48	310			
hexabutylditin, bis (tributyltin)	([CH$_2$(CH$_2$)$_3$]$_3$Sn-)$_2$	580.08	198		1 148	
methyltin trichloride, trichloromethylstannane	CH$_3$SnCl$_3$	240.08	48−51	171		
phenyltin trichloride, trichlorophenylstannane	C$_6$H$_5$SnCl$_3$	302.16		142−143 /25 mm	1 839	1.585 1
tetraethyltin	(C$_2$H$_5$)$_4$Sn	234.94	−112	181	1 187	1.472 5
tetramethyltin	(CH$_3$)$_4$Sn	178.83	−54	74−75	1 291	1.441 0
tetraphenyltin	(C$_6$H$_5$)$_4$Sn	427.11	224−227	>420	1 490[0]	
tributlytin bromide, bromotributyltin	[CH$_3$(CH$_2$)$_3$]$_3$SnBr	369.95		120−122 /2 mm	1 338	1.507 0
tributyltin hydride	[CH$_3$(CH$_2$)$_3$]$_3$SnH	291.05		80/0.4 mm	1 082	1.473 1
triethyltin bromide, bromotriethylstannane	(C$_2$H$_5$)$_3$SnBr	285.79	−13.5	222−224	1 630	1.525 6
trimethyltin chloride, chlorotrimethylstannane	(CH$_3$)$_3$SnCl	199.24	37−39	154		
triphenyltin chloride	(C$_6$H$_5$)$_3$SnCl	385.46	106	240/13.5 mm		
tetraethoxytitanium, titanium (IV) ethoxide	Ti(OC$_2$H$_5$)$_4$	228.15		150−152 /10 mm	1 088	1.504 3
titanium (IV) isopropoxide	Ti[OCH(CH$_3$)$_2$]$_4$	284.25	18−120	218 /10 mm	955	1.465 4
diethylzinc	Zn(C$_2$H$_5$)$_2$	123.49	−28	117−118	1 205	1.498 3
dimethylzinc	Zn(CH$_3$)$_2$	95.44	−42	46	1 386[10]	

Flash point/°C	Solubility						Hazards
	w	alc	et	ac	bz	other	
20	d						FL, C
−27			s				FL, MS
29	d		s				FL, C
−3							FL
−4							FL, MS
50							MS, T
	d	d				NH$_3$	MS
	d	d				NH$_3$	MS, T
	d	vs					FS, MS, C
	d	d	d				MS
	d	s					FS, C, MS
							C, MS
	d		s		s	NH$_3$	MS
81		s		s			C, MS
>110	i	s		s			vT, MS
110	i						C, MS
	i	i	i				vT, I
	i		s		s	chl, CCl$_4$	T, I
	s		s		s		T
	i	i	i	i	i	HCl, conc alk	
	s	s	s		s		C, MS
	i	i	i	i	i	HCl, NaOH	
>110	ss	s	s				MS, I
	s	s					I
>110	i	s	s				MS, T
40	s	s	s				C, MS
>110	i				s	MeOH	C, MS
53	i	s	s		s		vT
−12	i	s	s		s		FL, T
110	i	ss			s^h	py, chl, CCl$_4$	I, LS
>110							C
40							MS, I
99	ss	s	s		s		C, MS
97	s	s	s				vT, C
	i	s	s		s		vT, C
28							FL, MS
22							FL, MS
	d						P, C
	d	d	s				MS, C

Table XI.4 Physical Constants of Amino Acids and Derivatives

For more extensive data see: *Dictionary of Organic Compounds* (5 edn, 1982, and later supplements), Chapman & Hall, London.
Column 1. Common name of acid, full chemical name and configurational symbols.
Column 3. Density at 25 °C, reference to water at 4 °C.
Column 4. (m.p.). d: decomposes.
Column 5. pK values in water and 72% aqueous ethanol at 25 °C; (pI): isoelectric point.
Column 6. Optical rotatory power (customarily called specific rotation) measured at 589.3 nm in cell of path length 1 dm quoted in the units deg dm^{-1} g^{-1} cm^3. Concentration, c, in units of g per 100 cm^3 of specified solvent.
Column 7. Solubility (g per 100 g of solvent) in water and ethanol at the specified temperature. Solubility in other solvents: acetone (ac), benzene (bz), chloroform (chl). d: decomposes, hydrolysed; i: insoluble; s: soluble; ss: slightly soluble; vs: very soluble.
Column 8. Physical properties of derivatives.
DNP: dinitrophenyl-
Column 9. References for amino acids:
Beilsteins Handbuch der Organischen Chemie: (Spinger-Verlag, Berlin) B. volume (supplement). page.
The Aldrich Library of FT-IR Spectra, 1st edn. (1985) Aldrich Chem. Co. Gillingham, Dorset. IR.1 (volume). page.
The Aldrich Library of NMR Spectra, 2nd edn. (1983) Aldrich Chem. Co. Gillingham, Dorset. NMR.2 (volume). page.

Amino acid	M_r	ρ /kg m^{-3}	m.p. /°C	pK_1, pK_2,..., (pI) water	72% EtOH	Specific rotation [α]	t/°C	c	Solvent
DL-alanine (Ala) α-aminopropionic acid	89.09	1 424	d 289	2.348 9.866 (6.10)	3.55 10.02				
D-alanine (R)-(−)-alanine	89.09		d 295			−14.6 −1.8	25 25	2 2	5 M HCl H$_2$O
L-alanine (S)-(+)-alanine	89.09	1 401	d 314			+14.6 +1.8 +3.0	25 25 20	2 2 1.8	5 M HCl H$_2$O 3 M NaOH
DL-arginine hydrochloride monohydrate (Arg) 2-amino-5-guanidopentanoic acid*	228.68		128–130						
D-arginine (R)-(−)-arginine	174.20		d 226			−23.0 −12.5	23 25	1.6 2	6 M HCl H$_2$O
L-arginine (S)-(+)-arginine	174.20	1 100	d 223	2.01 9.04 12.48 (10.76)	3.34 9.40 14.1	+26.1 +12.5 +11.8	24 25 20	1.6 2 0.87	6 M HCl H$_2$O 0.5 M NaOH
DL-asparagine monohydrate (Asn) 2, 4-diamino-4-oxobutanoic acid	150.14		d 220						
D-asparagine monohydrate (R)-(−)-asparagine	150.14		d 275			−27 +5.6	20 25	1 2	1 M HCl H$_2$O
L-asparagine monohydrate (S)-(+)-asparagine	150.14		d 233	2.02 8.80 (5.41)		+27.8 −5.6	20 25	1 2	1 M HCl H$_2$O
DL-aspartic acid (Asp) aminosuccinic acid	133.10		270–280						
D-aspartic acid (R)-(−)-aspartic acid	133.10		251			−24 −5.05	20 25	2.3 2	6 M HCl H$_2$O
L-aspartic acid (S)-(+)-aspartic acid	133.10	1 660	269–271	2.10 3.86 9.82 (2.98)	2.85 5.20 10.51	+25 +5.05 −1.7	20 25 18	2 2 1.33	5 M HCl H$_2$O 3 M NaOH
DL-cysteine (Cys) 2-amino-3-mercaptopropanoic acid	121.16		d 225						
D-cysteine hydrochloride monohydrate (S)-(−)-cysteine	157.61					+16.5 −5	25 20	2 2	H$_2$O 5 M HCl

Solubility			Derivatives					References
water	EtOH	Other Solvents		m.p./°C	[α]	t/°C	solvent	
16.72[25]	0.0087[25]	i et	N-acetyl-	137–138				B.4(2).814
31.89[75]			N-benzoyl-	165–166				IR.1(1).570D
			DNP-	172–173				
15.78[20]	0.2 in 80% EtOH	i et	N-acetyl-	125	+66.5		water	B.4(2).812
			N-benzoyl-	151	−37.3	20	KOH	IR.1(1).571A
								NMR.2(1).485D
16.65[25]	0.16[20]	i et, ac,	N-acetyl-	125	−66.2		water	B4(2).809
28.51[75]			N-benzoyl	152–154	+37.1	20	KOH	IR.1(1).571B
			DNP-	178				NMR.2(1).486A
			N-acetyl- picrate (di-)	266 d 200				B.4(2).850
			N,N' dibenzoyl- picrate	176 205				B.4.424
			N,N'-dibenzoyl methyl ester	176	−17	25	water	IR.1(1).785C
15.0[21]	i	i et	picrate N,N'-dibenzoyl DNP-	d 217 d 235 260				B.4(2).845 IR.1(1).786A NMR.2(1).659B
2.16[25]	i	i et	DNP- phenylthio- hydantoin	191–192 234				B.4(2).900 IR.1(1).782A NMR.2(1).656B
3.0[25]	0.02[20]							B.4(1).531 IR.1(1).782B NMR.2(1).656C
3.0[25]	0.000 3[25]	s NH$_4$OH i et	N-acetyl- N-benzoyl DNP-	203–204 189 180–182	+18.9	20	water	B.4(2).896 IR.1(1).782C NMR.2(1).495C
0.78[25] 4.46[75]	0.032[25] in 75% EtOH	i et	N-acetyl- N-benzoyl- DNP-	150 165–166 196				B.4(2).900 IR.1(1).588D NMR.2(1).496C
s	i	i et	N-benzoyl- N-2, 4-dichloro phenoxyacetyl	184–185 201–202	−37.4 −20	18 20	KOH NaOH	B.4(1).531 IR.1(1).589A NMR.2(1).496D
0.5[25] 2.875[75]	0.000 196[25]		N-acetyl- N-benzoyl- DNP-	139–141 171–173 186–187	+37	20	NaOH	B.4(2).892 B.14(2).653 IR.1(1).589B NMR.2(1).497A
s		s NH$_4$OH	N-benzoyl- S-benzyl- N,S-diacetyl	137 215 112–112				B.4.513
vs	vs	s NH$_4$OH	S-benzyl	212–215	−25	25		B.4(3).1618

(Continued)

Table XI.4 (Continued)

Amino acid	M_r	ρ /kg m^{-3}	m.p. /°C	pK_1, pK_2,..., (pI)		Specific rotation			
				water	72% EtOH	[α]	t/°C	c	Solvent
L-cysteine hydrochloride monohydrate (R)-(+)-cysteine	157.61		d 175–178	1.71 8.33 10.78 (5.02)		+5.2 –16.5	23 25	2 2	5 M HCl H$_2$O
DL-cystine (Cys-Cys) (RS-RS)- 3,3'-dithiobis (2-aminopropanoic acid)	240.30		260						
D-cystine (S,S)-(+)-cystine	240.30		247–249			+212	20	1	1 M HCl
L-cystine (R,R)-(–)-cystine	240.30		258–260 d	1.65 2.26 7.85 9.85 (5.06)		–217.8 –70	20 18.5	1 0.4	1 M HCl 0.2 M NaOH
meso-diaminopimelic acid (Dap) (2RS, 6SR)-2, 6-diaminoheptanedioic acid	190.2		313–315	1.8 2.2 8.8 9.9					
DD-diaminopimelic acid hydrate (2R, 6R)-diaminopimelic acid	208.2		309–310			–45.5 –8.4	26 24	1 5	1 M HCl H$_2$O
LL-diaminopimelic acid hydrate (2S, 6S)-diaminopimelic acid	208.2		310–312	1.8 2.2 8.8 9.9		+45.0 +8.1	26 25	1 1	1 M HCl H$_2$O
DL-glutamic acid (Glu) 2-aminopentanedioic acid*	147.13	1 460	d 225						
D-glutamic acid (R)-(–)-glutamic acid	147.13		s 200			–29 –12	23 25	1 2	6 M HCl H$_2$O
L-glutamic acid (S)-(+)-glutamic acid	147.13	1538	d 205	2.19 4.25 9.67 (3.22)	3.16 5.63 10.75	+29 +12 +11.5	23 25 18	1 2 1.47	6 M HCl H$_2$O 1 M NaOH
glycine (Gly) aminoethanoic acid*	75.07	1 601	d 245	2.35 9.60 (5.97)	3.46 9.82				
DL-histidine (His) 2-amino-3-(4-imidazolyl) propanoic acid	155.16		d 273						
D-histidine (R)-(–)-histidine	155.16		d 287			–11.8 +38.5	25 25	2 2	5 M HCl H$_2$O
L-histidine (S)-(+)-histidine	155.16		d 282	1.77 6.10 9.18 (7.64)	3.00 5.85 9.45	+11.8 –38.5 –10.9	25 25 20	2 2 0.775	5 M HCl H$_2$O 0.5 M NaOH
4-hydroxy-DL-proline (Hyp) 4-hydroxy-2-pyrrolidine carboxylic acid (2RS, 4RS)	131.13		d 247						
cis-4-hydroxy-D-proline (2R, 4R)-(+)-hydroxyproline (D-allo-form)	131.13		d 251			+58	20	2	H$_2$O
cis-4-hydroxy-L-proline (2S, 4S)-(–)-hydroxyproline (L-allo-form)	131.13		d 257	1.92 9.73 (5.82)		–18.8 –59	25 20	2 2	5 M HCl H$_2$O
trans-4-hydroxy-L-proline (2S, 4R)-(–)-hydroxyproline (L-trans-form)	131.13		d 273			–47.3 –75.3 –70.6	20 25 20	1.31 1 0.655	1 M HCl H$_2$O 0.5 M NaOH
(2RS, 4SR)-trans-hydroxy-L-proline	131.13		247						

Solubility			Derivatives					References
water	EtOH	Other Solvents		m.p./°C	[α]	t/°C	solvent	
vs	vs	s NH₄OH	N-acetyl S-benzyl- bis-DNP S-carboxymethyl	109–110 d 214 d 155 204–207	+5 +28 +0.5	25 24	water NaOH HCl	B.4(2).920
0.003²⁰ 0.01⁵⁰			diacetyl- N,N'-dibenzoyl	121 165–166				B.4(2).936
0.011²⁵	i	s min acids, alk	N,N'-dimethyl-	216–218	−75	26	HCl	B.4(2).919
0.011²⁵ 0.052⁷⁵		s min acids, alk. i et, bz	N,N'-dimethyl- N,N'-bis-DNP N,N'-dibenzoyl-	216–218 109 190–192	+75.7 −202	22 20	HCl EtOH	B.4(2).925 IR.1(1).596B NMR.2(1).503A
0.92²¹	i	s min acids, alk	N,N'-dibenzoyl-	195				B.4.497
vs	i		N,N'-dibenzoyl- dianilide	d 307				
vs	i		N,N'-dibenzoyl- dianilide	d 307				
2.054²⁵ 11.86⁷⁵	ss	ss et	N-acetyl- DNP-	185 148–149				B.4(2).911 IR.1(1).589C
1.5²⁰	0.07¹⁵	i et	N-acetyl- N-carbamoyl-	201 160				B.4(2).910 IR.1(1).589D
0.864²⁵ 5.53⁷⁵	0.000347²⁵ 0.0056⁴⁵	s MeOH	N-acetyl- DNP- N-carbamoyl-	199 134–136 160				B.4(2).902 IR.1(1).590A NMR.2(1).497B
24.99²⁵ 54.47⁷⁵	0.0037²⁵	0.61²⁰ py 0.032²⁵ M MeOH	N-acetyl- N-benzoyl- DNP-	206 187 203				B.4(1).462 IR.1(1).563B NMR.2(1).481C
s	i	i et, ac chl	N-acetyl- methyl ester	148 191–193				B.25(2).409 IR.1(2).621B NMR.2(2).492D
s	i	i et, ac chl	methyl ester	203	−14.2	20	MeOH	B.25(2).404 IR.1(2).621C NMR.2(1).493A
4.19²⁵	ss	i et	N-acetyl- N,N'-bis DNP- methyl ester	157–159 d 232–234 200–201				B.25(2).409 IR.1(2).621D NMR.2(2).493B
s	ss	i et	N-acetyl-	143–144				B.22(2).144
s	ss	i et						B.22(1).546 IR.1(1).584A NMR.2(1).504C
36.11²⁵ 51.67²⁵	ss	i et	N-acetyl-	144–145	−91.5		water	B.22(1).546 IR.1(1).584B NMR.2(1).504D
s	ss	i et	N-acetyl- N-3,5-dinitro- benzoyl N-3,5-dinitro- benzoyl-	135 205–207	−118.5 −147.2	25	water	B.22(2).143 IR.1(1).583D NMR.2(1).504B

(Continued)

Table XI.4 (Continued)

Amino acid	M_r	ρ /kg m^{-3}	m.p. /°C	pK$_1$, pK$_2$,..., (pI) water	72% EtOH	Specific rotation [α]	t/°C	c	Solvent
DL-isoleucine (Ile) (2RS, 3SR)-2-amino-3-methylpentanoic acid	131.18		d 292	2.318 9.758 (6.038)	3.69 9.81				
D-isoleucine (2R, 3S)-(−)-isoleucine	131.18		d 283			−39.5 −12.4	25 25	1 1	5 M HCl H$_2$O
L-isoleucine (2S, 3R)-(+)-isoleucine	131.18		d 288			+39.5 +12.4 +11.09	25 25 20	1 1 3.34	5 M HCl H$_2$O 0.33 M NaOH
DL-leucine (Leu) 2-amino-4-methylpentanoic acid	131.18	1 191	s 293–296	2.328 9.744 (6.036)					
D-leucine (R)-(−)-leucine	131.18		d 293			−15.6 +11.0	20 25	4 2	6 M HCl H$_2$O
L-leucine (S)-(+)-leucine	131.18	1 165	d 293			+16.0 −11.0 +7.6	25 25 20	2 2 1.31	5 M HCl H$_2$O 3 M NaOH
DL-lysine (Lys) 2,6-diaminohexanoic acid	146.19		d 170						
D-lysine (R)-(−)-lysine	146.19		d 218			−25.9 −13.5	25 25	2 2	5 M HCl H$_2$O
L-lysine (S)-(+)-lysine	146.19		d 212	2.18 8.95 10.53 (9.47)	3.56 8.95 10.49 (84%EtOH)	+25.9 +13.5	25 25	2 2	5 M HCl H$_2$O
DL-methionine (Met) 2-amino-4-methylmercapto-butanoic acid	149.21	1 340	d 271	2.28 9.21 (5.74)					
D-methionine (R)-(−)-methionine	149.21		d 271			−23.2 +10.0	25 25	2 2	5 M HCl H$_2$O
L-methionine (S)-(+)-methionine	149.21		d 284			+23.2 −10.0	25 25	2 2	5 M HCl H$_2$O
DL-norleucine 2-aminohexanoic acid	131.18		d 295	2.34 9.83 (6.08)					
D-norleucine (R)-(−)-norleucine	131.18		d 301			−21.2 −4.7	22 25	4.7 2	6 M HCl H$_2$O
L-norleucine (S)-(+)-norleucine	131.18		d 301			+23.3 +4.7	24 25	4.2 2	5 M HCl H$_2$O
DL-norvaline 2-aminopentanoic acid	117.15		303 (sealed tube)	2.31 9.81 (6.06)					
D-norvaline (R)-(−)-norvaline	117.15		>300			−23.0	25	10	2.4 M HCl
L-norvaline (S)-(+)-norvaline	117.15		d 291			+24.0 +7.0	20 25	10 2	6 M HCl H$_2$O
DL-ornithine 2,5-diaminopentanoic acid	132.16		195	1.94 8.65 10.76 (9.7)					
D-ornithine (R)-(−)-ornithine	132.16					−28.4 −12.1	25 25	2 2	5 M HCl H$_2$O

Solubility			Derivatives					References
water	EtOH	Other Solvents		m.p./°C	[α]	t/°C	solvent	
2.23[25]	s[h]	i et	N-acetyl-	117–123				B.4.456
4.61[75]			DNP-	168–172				
			N-4-toluenesulphonyl	139–140				
s		i et	N-acetyl	150	−15	25	EtOH	B.4.456
					+82.6	25	water	
4.117[25]	0.09[20]	i et	N-acetyl-	150	+15	25	EtOH	B.4.454
6.076[75]	0.13[80]		DNP-	113–114	+84		NaOH	IR.1(1).575D
								NMR.2(1).489C
0.991[25]	0.155[25]		N-acetyl-	159				B.4(2).870
2.276[75]	in 75% EtOH		N-benzoyl-	79				IR.1(1).575C
			DNP-	203				
2.20[25]			N-acetyl-	112–114	+21.3	25	EtOH	B.4.446
2.66[50]			N-benzoyl-	105–107	−6.4	20	KOH	IR.1(1).575A
								NMR.2(1).489A
2.19[25]	0.217[25]	i et	N-acetyl-	189–190	−21.3	25	EtOH	B.4(2).859
3.82[75]			N-benzoyl	105–107(anhyd)				IR.1(1).575B
			DNP-	94–95				NMR.2(1).489B
s	ss		diacetyl-	141				B.4.436
			α-N-benzoyl-	235				
			ε-N-benzoyl-	254				
			N,N'-dibenzoyl-	149–150	+3.06	20	NaOH	
s	ss	ss et, chl						B.4(3).1404
v	ss	i et	α-N-acetyl-	d 256–258				B.4(2).857
			ε-N-acetyl-	d 250				IR.1(1).587C
			ε-N-benzoyl-	235	20.1	19	HCl	NMR.2(1).495A
			α,ε-bis-DNP-	170–174				
3.38[25]			N-acetyl-	117–119				B.4(2).938
10.52[75]			N-benzoyl-	143–145				IR.1(1).593A
			DNP-	117–118				NMR.2(1).500D
			N-acetyl-	105	+20.5		water	IR.1(1).594B
			methyl ester	149–150	−26.3	21	water	NMR.2(1).501A
s	i	i et	N-acetyl	104–107	−20.1		water	B.4(2).938
			methyl ester	150	+26.8	19	water	IR.1(1).594C
			N-formyl-	98–99	−10	25	water	NMR.2(1).501B
1.07[20]	0.014[25]	s MeOH	N-acetyl-	104–105				B.4.433
5.23[100]	75% EtOH	ss ac	N-benzoyl-	132–136				IR.1(1).576A
			N-dimethyl-	161–162				NMR.2(1).489D
1.6[19]			N-acetyl-	114–115	+5.0	25	water	B.4.433
			N-benzoyl-	53	+19.6	25	MeOH	IR.1(1).576A
					+21.4	20	NaOH	NMR.2(1).493C
1.5[25]	i		N-acetyl-		−19.4	25	MeOH	B.4.432
			N-benzoyl-	53				IR.1(1).576D
10.7[15]	ss	i et, chl, ac	N-acetyl-	115				B.4(2).843
			N-benzoyl-	152				IR.1(1).574A
								NMR.2(1).488C
			N-benzoyl-	95	+14	20	NaOH	B.4(3).1532
			N-formyl-	136	−2.1	18	EtOH	IR.1(1).574B
								NMR.2(1).493B
10.75[15]	ss	i et, chl, ac	N-acetyl-	100	+35	25	water	B.4(3).1549
			N-benzoyl-	97	−15	19	NaOH	IR.1(1).574C
			N-formyl-	137	+2.05	28	HCl	
mono-HCl	mono-HCl	mono-HCl	dichloroacetyl-	105				
s	i	i et, MeOH	dibenzoyl-	184–188				
			α-N-DNP	d 227				
s			dichloroacetyl-	112	−15.1		EtOH	B.4(3).1357
			dihydrochloride-	194–195	−16.4	27	water	IR.1(1).587A
								NMR.2(1).494C

(Continued)

Table XI.4 (Continued)

Amino acid	M_r	ρ /kg m^{-3}	m.p. /°C	pK$_1$, pK$_2$,..., (pI)		Specific rotation			
				water	72% EtOH	[α]	t/°C	c	Solvent
L-ornithine	132.16		140			+28.4	25	2	5 M HCl
(S)-(+)-ornithine						+12.1	25	2	H$_2$O
DL-phenylalanine (Phe)	165.19		d 266–267	2.58					
2-amino-3-phenylpropanoic acid				9.24					
				(5.91)					
D-phenylalanine	165.19		273–276			+4.47	25	2	5 M HCl
(R)-(+)-phenylalanine						+34.5	25	2	H$_2$O
L-phenylalanine	165.19		d 270–273			−4.47	25	2	5 M HCl
(S)-(−)-phenylalanine						−34.5	25	2	H$_2$O
DL-proline (Pro)	115.13		d 205						
pyrrolidine-2-carboxylic acid									
D-proline	115.13		d 223			+60.4	25	2	5 M HCl
(R)-(+)-proline						+86.2	25	2	H$_2$O
L-proline	115.13		d 228	1.99	3.04	−60.4	25	2	5 M HCl
(S)-(−)-proline				10.6	10.55	−86.2	25	2	H$_2$O
				(6.3)		−93.0	20	2.4	0.6 M KOH
DL-serine (Ser)	105.09	1 537	d 240	2.21					
2-amino-3-hydroxypropanoic acid				9.15					
				(5.68)					
D-serine	105.09		d 220			−15.1	25	2	5 M HCl
(R)-(−)-serine						+7.5	25	2	H$_2$O
L-serine	105.09		d 222			+15.1	25	2	5 M HCl
(S)-(+)-serine						−7.5	25	2	H$_2$O
DL-threonine (2RS, 3SR) (Thr)	119.12		d 234	2.09					
2-amino-3-hydroxybutanoic acid				9.10					
				(5.60)					
D-threonine	119.12		251–252			+15.0	25	2	5 M HCl
(2R, 3S)-(+)-threonine						+28.5	25	2	H$_2$O
L-threonine	119.12		251–253			−15.0	25	2	5 M HCl
(2S, 3R)-(−)-threonine						−28.5	25	2	H$_2$O
DL-allo-threonine	119.12		244–245	2.11					
(2RS, 3RS)-allo-threonine				9.10					
				(5.60)					
D-allo-threonine	119.12		d 276			−8.8	23	2	H$_2$O
(2R, 3R)-(−)-allo-threonine									
L-allo-threonine	119.12		d 272			+9.0	23	2	H$_2$O
(2S, 3S)-(+)-allo-threonine									
DL-tryptophan (Trp)	204.23		d 289						
2-amino-3-(3-indole) propanoic acid									
D-tryptophan	204.23		d 282			−2.8	25	2	M HCl
(R)-(+)-tryptophan						+33.7	25	2	H$_2$O
L-tryptophan	204.23		d 280	2.38		+2.8	25	2	M HCl
(S)-(−)-tryptophan				9.39		−33.7	25	2	H$_2$O
				(5.88)					
DL-tyrosine (Tyr)	181.19		d 325						
3-(4-hydroxyphenyl) alanine									

Solubility			Derivatives					References
water	EtOH	Other Solvents		m.p./°C	[α]	t/°C	solvent	
v	v	ss et	α-N-acetyl-	226–227	−7.5		HCl	B.4(2).844
			α-N-benzoyl-	225–226				IR.1(1).587B
			δ-N-DNP-	d 223				NMR.2(1).494D
1.41²⁵	ss	ss et	N-acetyl-	152–153				B.14(2).299
3.71⁷⁵			N-benzoyl-	187–188				
			DNP-	d 204–206				
3.0²⁵	ss^h	i et, ac	N-acetyl-	170–171	−49.3		water	B.14(2).297
			N-benzoyl-	146–148	−17.1	20	KOH	IR.1(2).250D
			N-formyl-	167	−72.7	26		
2.97²⁵	i	i et	N-acetyl-	170–171	+51.4	26	water	B.14(2).296
6.62⁷⁵			N-benzoyl-	142–143	+19.8		KOH	IR.1(2).251A
			DNP-	189				NMR.2(2).248C
s	s	ss chl, ac, bz	N-acetyl-	106				B.22(2).5
		i et	N-3-nitro-benzoyl-	90–92				IR.1(1).583A
								NMR.2(1).504A
s	s		N-3-nitro-benzoyl-	137–140	+120	20	NaOH	B.22.1
			methyl ester		+34	25		IR.1(1).583B
162.3²⁵			N-acetyl-	118	−115	23	water	B.22(2).3
239⁶⁵	1.5¹⁹	i et	DNP-	138				IR.1(1).583C
5.02²⁵	0.074²⁵		N-acetyl-	132				B.4(2).934
19.21⁷⁵	in 75% EtOH	i et	DNP-	200–202				IR.1(1).579D
								NMR.2(1).491C
			N-benzyl-oxy-carbonyl-	115–119	−5.1	26	acetic acid	B.4(2).919
25²⁰	i	i et						IR.1(1).580A
								NMR.2(1).491D
25²⁰			DNP-	173				B.4(2).919
			N-glycyl-	201–202	−9.2	27	conc. H₂SO₄	IR.1(1).580B
20²⁵	0.07	i chl	N-benzoyl-	143–144				B.4.514
c. 55⁸⁰			N-chloroacetyl-	124				
			DNP-	177–178				
		i et, chl	N-benzoyl-	147–148	−25.5	26		B.4(3).1625
			N-formyl-	164–165	−11.9			
s	i	i et, chl	N-benzoyl-	147–148	+25.1	26		B.4(3).1623
			DNP-	145	+106		4%NaHCO₃	IR.1(1).582A
			N-formyl	163–164	+11.8	26		NMR.2(1).492D
13.9²⁵	0.3 in 95% EtOH		N-benzoyl-	175–176				
								B.4(3).1629
								IR.1(1).581B
s	ss		N-acetyl-	153–154				B.4(3).1629
			DNP-	152				IR.1(1).581C
0.25³⁰	ss		N-acetyl-	d 205				B.22(2).470
			N-benzoyl-	188–191				IR.1(2).671B
								NMR.2(2).539C
								IR.1(2).671C
								NMR.2(2).539D
1.14²⁵	ss	i et	N-benzoyl-	104–105				B.22(2).466
2.80⁷⁵		chl	DNP-	175				IR.1(2).671D
			N-acetyl-	189–190	+25	15	95% EtOH	
0.035²⁵	ss	i et	N-acetyl-	92–95				B.14(2).379
0.084⁵⁰			O,N-bis-DNP	84				IR.1(2).255B
			N-benzoyl-	122–123				

(Continued)

Table XI.4 (Continued)

Amino acid	M_r	ρ /kg m^{-3}	m.p. /°C	pK_1, pK_2, ..., (pI)		Specific rotation			
				water	72% EtOH	[α]	t/°C	c	Solvent
D-tyrosine (R)-(+)-tyrosine	181.19		d 310			+10.9	23	4	1 M HCl
L-tyrosine (S)-(−)-tyrosine	181.19		d 295	2.20 9.11 10.07 (5.63)		−7.6 −13.2	25 18	2 0.9	5 M HCl 3 M NaOH
DL-valine (Val) 2-amino-3-methylbutanoic acid	117.15	1 316	d 295	2.286 9.719 (6.002)	3.60 9.73				
D-valine (R)-(−)-valine	117.15		d 293			−28.3 −5.63	25 25	2 2	5 M HCl H$_2$O
L-valine (S)-(+)-valine	117.15	1 230	d 315			+28.3 +5.63	25 25	2 2	5 M HCl H$_2$O

Solubility			Derivatives					References
water	EtOH	Other Solvents		m.p./°C	[α]	t/°C	solvent	
0.045[25]			N-acetyl-	153–154	−48.3	25	MeOH	B.14(2).365
0.105[50]			N-benzoyl-	165–166	−19.6	20		IR.1(2).255C
0.045[25]	0.01[17]		N-acetyl-	153–154	+47.3	20		B.14(2).366
0.244[75]	in 95% EtOH	i et, ac	N-benzoyl-	165–166	+17	29	MeOH	IR.1(2).255D
			O,N-bis-DNP-	178				NMR.2(2).254B
			methyl ester	136–137	+25.75	20	MeOH	
7.09[25]	0.019[25]	i et	N-acetyl-	148				B.4(2).854
12.61[75]			N-benzoyl-	132–133				IR.1(1).573C
			DNP-	183				NMR.2(1).488D
			N-formyl-	140–145				
5.3[20]	0.45[25]		N-acetyl-	164–165	−3.4	25	EtOH	B.4(2).853
	in 80% EtOH		N-formyl-	156				IR.1(1).573B
			N-benzoxycarbonyl	60–62	+5.0		MeOH	NMR.2(1).488B
8.85[25]	ss	i et	N-acetyl-	164–165	+4.0	20	water	B.4(2).852
10.24[65]			N-benzoyl-	127	+17.8		water	IR.1(1).573D
			DNP-	132				
			N-formyl-	156				
			N-benzoxycarbonyl	59–62	−4.3		acetic acid	

376 Physical Properties of Inorganic and Organic Compounds

Table XI.5 Physical Constants of Carbohydrates and Related Compounds

For more extensive data see: *Dictionary of Organic Compounds*, (5th edn, 1982 and later supplements), Chapman & Hall, London, Carbohydrates, Ed. P. M. Collins (1987), Chapman and Hall, London.
Column 1. Common name of acid, full chemical name, configurational symbols and structural formula.
Column 4. Optical rotatory power (customarily called specific rotation) measured at 589.3 nm in cell of path length 1 dm quoted in the units deg dm^{-1} g^{-1} cm^3. Concentration, c, in units of g per 100 cm^3 of specified solvent.
Column 5. Solubility at room temperature. Solvents: acetone (ac), benzene (bz), chloroform (chl), ether (et), pyridine (py).
  i: insoluble; s: soluble; ss: slightly soluble; vs: very soluble.
Column 6. Physical properties of derivatives.
Column 7. References for carbohydrates:
Beilsteins Handbuch der Organischen Chemie: (Springer-Verlag, Berlin). B. volume (supplement). page
The Aldrich Library of FT-IR Spectra, 1st edn: (1985) Aldrich Chem Co. Gillingham, Dorset IR.1(volume). page.
The Aldrich Library of NMR Spectra, 2nd edn: (1983), Aldrich Chem Co. Gillingham, Dorset NMR. 2(volume). page.

Name, formula	M_r	m.p./°C	$[\alpha]_D$	Solubility
1. Trioses, Tetroses and Related Alcohols				
D-glyceraldehyde, D-2, 3-dihydroxypropanal CHO | HCOH | CH$_2$OH	90.08	syrup	+9.4[23] (c = 2, water)	s water
glycerol, 1, 2, 3-propanetriol CH$_2$OH | CHOH | CH$_2$OH	92.094	17.8	inactive	vs water, EtOH ss et i chl, bz
erythritol, meso (R, S) 1, 2, 3, 4-butanetetrol CH$_2$OH | HCOH | HCOH | CH$_2$OH	122.12	120–123	inactive	vs water ss EtOH, py i et
D-erythrose, (2R, 3R)-(−)-erythrose CHO | HCOH | HCOH | CH$_2$OH	120.12	syrup	+1→ −14.8[20] (c = 1, water)	s water, MeOH, EtOH
D-threose, (R, S)-(+)-threose 2, 3, 4-trihydroxybutanal CHO | HOCH | HCOH | CH$_2$OH	120.105	126–132	+29.1→ +19.6 (water)	s water, MeOH, EtOH i et
2. Pentoses and Related Alcohols				
D-arabinose (β-form structure)	150.131	159–160	−175 → −105[20] (water)	s water ss EtOH i et, bz

	Derivatives			References
Name	m. p./°C		$[\alpha]_D$	
2,4-dinitrophenylhydrazone	155–156			B.1(1).427
methylphenylhydrazone	108–110		-20.5^{19} (py)	IR.1(1).479B
dibenzoate	80		-35.6	
tribenzoyl-	72			B.1(3).2297
tris-4-nitro-benzoyl-	197.5			IR.1(1).183A
tetra-O-acetyl-	89			B.1(2).2356
tetra-O-benzoyl-	188–189			IR.1(1).184C
1,3:2,4-di-O-benzylidene-	201–202			NMR.2(2).901A
phenylosazone	164			B.31.12
benzylphenylhydrazone	105		-32 (EtOH)	IR.1(1).187B
				NMR.2(2).904A
phenylosazone	164–165			B. 1.855
benzylphenylhydrazone	194–195			B. 31.13
2,4-O-benzylidene	160–161			
di-O-acetyl-	140–142		$+83.5 \rightarrow +34.5$ (chl)	
tetra-O-ocetyl	95–97		-43 (chl)	B.31.32
benzylphenylhydrazone	177–178		$+14.4$	IR.1(1). 189A
2,3:4,5-di-O-isopropylidene–			-17.1 (chl)	NMR.2(2).905D
phenylosazone	163			
tetra-O-benzoyl-	160–161		-323 (chl)	

(Continued)

Table XI.5 (Continued)

Name, formula	M_r	m.p./°C	$[\alpha]_D$	Solubility
L-arabinose (β-form)	150.131	159–160	+190.6 → +104.5 (c = 3, water)	s water ss EtOH i et
D-arabitol 1, 2, 3, 4, 5-pentanepentol	152.147	103–104	+7.7^{20} (c = 9.26, sat borax soln)	vs water ss EtOH i et
L-arabitol	152.147	102	−5^{20} (sat borax soln)	vs water ss EtOH i et
2-deoxy-D-ribose	134.132	96–98	−59^{20} (c =1, water, 72 h)	s water
ribitol, adonitol	152.147	102–104	inactive	s water, hot EtOH i et
β–D–ribose	150.131	95	−21.5 → −19.5 (c = 4, water)	s water ss EtOH
xylitol (meso)	152.147	93–94.5	inactive	vs water s EtOH, py
α-D-xylose	150.131	144–145	+92 → +19^{20} (c = 4, water)	vs water s EtOH ss et

	Derivatives		References
Name	m. p./°C	[α]$_D$	
4-nitrophenylhydrazone	187–188	+14.9 (EtOH)	B.31.34
2, 5-dichlorophenylhydrazone	170		IR.1(1).188D
tetra-O-acetyl-	86	+147.2 (chl)	NMR.2(2).906A
tetra-O-benzoyl	157–158	+298 (chl)	
penta-O-acetyl	74–75	+37.2 (chl)	B.1.531
2, 3, 5-tribenzyl-	50–52	+1 (chl)	IR.1(1).185D
2, 3:4, 5-di-O-benzylidene-	101–106	+2.5 (py)	
penta-O-acetyl-	76		B.1(2).604
			IR.1(1).185C
4-nitrophenylhydrazone	160	–11.1 (EtOH)	
1, 3, 4-tri-O-acetyl-	98	–171.8 (chl)	IR.1(1).189B
1, 3, 4-tribenzoyl–	127	–65 (chl)	NMR.2(2).906.B
2, 3 : 4, 5-di-O-benzylidene	174–175	–35.7 (chl)	B.1(2).604
penta-O-acetyl-	51		IR.1(1).185A
			NMR.2(2).901B
2-methylphenylhydrazone	164–165	+10 (EtOH)	B.1(1).434
3, 4-O-isopropylidene	115–117	–85 → –82 (water)	IR.1(1). 189C
tetra-O-acetyl-	109–110	–57^{20}(water)	NMR.2(2).906C
benzylphenylhydrazone	127–128		
1, 2:3, 4-di-O-isopropylidene	36		B. 1. 531
penta-O-acetyl-	62–63		NMR.2(2).901C
penta-O-benzoyl-	106–107		
4-bromophenylhydrazone	128	–21	B.31.47
phenylosazone	165–167	–26 → +46 (EtOH)	IR.1(1).188C
tetra-O-acetyl	58.5	+88.9 (chl)	NMR.2(2).905A
tetra-O-benzoyl	119–121	+149 (chl)	

(Continued)

Table XI.5 (Continued)

Name, formula	M_r	m.p./°C	$[\alpha]_D$	Solubility

L-xylulose
CH$_2$OH
  C=O
H—C—OH
HO—C—H
  CH$_2$OH
	150.131	syrup	+33.1 (c = 1, water)	s water

3. Hexoses and Related Alcohols

β-D-allose	180.157	148–150	+0.58 → +14.4 (c = 5, water)	s water i EtOH
α-6-deoxy-D-glucose	164.158	146	+75 → +30[28] (c = 8, water)	s water, EtOH
β-2-deoxy-D-glucose	164.158	146–147	+38.5 → +45.9[17] (c = 0.5, water)	s water EtOH
α-2-deoxy-D-ribose	134.132	91	−56.2[22]	s water ss iso–PrOH
β-D-fructose	180.157	102–104.d	133.5 → −92.4[20] −133.5 water	vs water s EtOH, MeOH, py ss ac.
α-L-fucose, 6-deoxy-L-galactose	164.158	140–141	−124 → −76[17] water	vs water ss EtOH i et

Name	Derivatives		References
	m. p./°C	$[\alpha]_D$	
4-bromophenylhydrazone	131	−25.8 → +31.5 (py)	B.31.58
phenylosazone	161–163		
penta-O-acetyl-	95–96	-15^{25} (chl)	B.1(1).443
phenylosazone	189–191	−77 (EtOH/py)	
1, 2, 3, 4-tetra-O-acetyl-	118	$+122^{25}$(chl)	
methylphenylhydrazone	157		
benzylphenylhydrazone	158		IR.1(1)192D
			NMR.2(2).909B
4-nitrophenylhydrazone	160	-11.1^{14} (EtOH)	
1, 3, 4-tribenzoyl-	127	-65^{23} (chl)	IR.1(1). 189B
1, 3, 4-triacetyl-	98	-171.8^{23} (chl)	NMR.2(2).906B
4-nitrophenylhydrazone	180–181		B.31.321
phenylosazone	204–205		
1, 3, 4, 5, 6-pentaacetyl-(keto)	68–69	+34.8 (chl)	
1, 3, 4, 5, 6-pentabenzoyl-(keto)	125	+40.9 (chl)	
methylphenylhydrazone	180–182	$+6^{23}$ (py)	B.31.78
benzylphenylhydrazone	178	+14.9 (MeOH)	IR.1(1).190A
tribenzyl-	87–89	-29.7^{24} (chl)	NMR.2(2).907B

(Continued)

Table XI.5 (Continued)

Name, formula	M_r	m.p./°C	$[\alpha]_D$	Solubility
galacticol, dulcitol CH$_2$OH H–C–OH HO–C–H HO–C–H H–C–OH CH$_2$OH	182.173	188–189	inactive	s water ss EtOH i et
α-D-galactose	180.157	170 (anhyd) 118–120 (hyd)	+80.2[20] (c = 5, water)	s water, py ss EtOH MeOH i et
α-D-glucose	180.157	146 (anhyd) 83 (hyd)	+111.2 → +52.5[20] (c = 3.9, water)	vs water ss EtOH i et, ac
β-D-glucose	180.157	148–150	+17.5 → +52.5[20] (c = 3.9, water)	vs water ss EtOH i et, ac
D-mannitol CH$_2$OH HO–C–H HO–C–H H–C–OH H–C–OH CH$_2$OH	182.173	166	−0.49[25] (water) +28.6[25] (c = 10, borate)	vs water ss hot EtOH, py i et
α-D-mannose	180.157	133	+29.3 → +14.5[25] (c = 4, water)	vs water ss EtOH i et
α-L-rhamnose, 6-deoxy-L-mannose	164.158	93–95	+8.9[20] (c = 2, water)	s water, EtOH, MeOH i et

	Derivatives		References
Name	m. p./°C	$[\alpha]_D$	
hexa-O-acetyl-
1, 3:4, 6-di-O-benzylidene	168–169
215–220		B.1(2).612
IR.1(1).186D
NMR.2(2).902D
penta-O-acetyl-
4-nitrophenylhydrazone
phenylhydrazone
2, 4, 6-tri-O-benzoyl-	96
192
159–160
123–124	+106.6 (chl)
+70 (EtOH/py)
+21 → +9 (py)
+40.4 → +37.6 (chl)	B.1.909
B.31.295
IR.1(1).192A
NMR.2(2).908C
phenylhydrazone
1, 2, 3, 4, 6-penta-O-acetyl-
phenylosazone	159–160
112–114
207	$-87 \to -50^{20}$ (water)
$+102^{20}$ (chl)
−35 (py)	B.31.83
phenylhydrazone
1, 2, 3, 4, 6-penta-O-acetyl-	140–142
132–135	$-5 \to -50^{20}$ (water)
$+4^{20}$ (chl)	
1, 2, 3, 4, 5, 6-hexa-O-acetyl-
1, 2, 3, 4, 5, 6-hexa-O-benzoyl
2, 3:4, 5-di-O-benzylidene-	126
149
203–205	+25 (chl)
+50.7 (chl)
+76.7 (py)	B.1(3).2392
phenylhydrazone
1, 2, 3, 4, 6-penta-O-acetyl-
4-nitrophenylhydrazone	198–200
64
202–203	+27 → +28 (EtOH)
+55 (chl)
+56 (EtOH/py)	B.31.284
IR.1(1).188B
NMR.2(2).904C
phenylhydrazone
phenylosazone
2, 4-dinitrophenylhydrazone	159
190
164–165	+54.2 → +27 (water)
$+94^{20}$ (py)	B.1(2).901
B.31.66
IR.1(1).191A
NMR.2(2).908B

(Continued)

Table XI.5 (Continued)

Name, formula	M_r	m.p./°C	$[\alpha]_D$	Solubility
D-sorbitol, D-glucitol CH$_2$OH H–C–OH HO–C–H H–C–OH H–C–OH CH$_2$OH	182.173	110–112	2.0^{20} (water) +7 (borax)	vs water s ac ss EtOH i et
α-L-sorbose	180.157	165	−43.2^{20} ($c = 12$, water)	s water ss hot EtOH, MeOH, et
4. Di- and Tri-saccharides				
cellobiose, 4-O-β- glucopyranosyl-D- glucopyranose	342.299	225 d	+14.2 → +34.6^{20} ($c = 8$, water)	s water, i EtOH, et, ac
β-D-gentiobose, 6-O-β-D- glucopyranosyl-D-glucose	342.299	190–195	−0.8 → +10^{22} (water)	s water, hot EtOH, hot MeOH
isomaltose, 6-O-α-D- glucopyranosyl-D-glucose	342.299	120	+120^{25} (water)	s water, MeOH
lactose, 4-O-β-D- galactopyranosyl-D- glucopyranose	342.299	α 202 (hyd) α 252 (anhyd) β 252	α + 85 → + 52.6^{20} ($c = 4.5$, water) β + 34 → + 52.6^{20} ($c = 4$, water)	vs water ss EtOH i et
maltose, 4-O-α-D- glucopyranosyl-D-glucose	342.299	102–103 (hyd) 160–165 (anhyd)	+112 → + 130^{20} ($c = 4$, water)	s water i EtOH, et

Name	Derivatives		References
	m. p./°C	$[\alpha]_D$	
1, 2, 3, 4, 5, 6-hexa-O-acetyl-	99	+6.8[18] (ac)	B.1(3).2385
1, 2, 3, 4, 5, 6-hexa-O-benzoyl-	128–130	+28[18] (chl)	IR.1(1).186A
1, 2, 4, 6-tetra-O-benzoyl-	164–166	+16.8[18] (DMF)	
phenylosazone	164	−13[15] (EtOH/py)	B.31.346
1, 3, 4, 6-penta-O-acetyl-(keto)	69–70		IR.1(1).193C
2, 5-dichlorophenylhydrazone	116–117	−36.2[22] (EtOH)	NMR.2(2).909C
octa-O-acetyl- (β-form)	202	−14.5[20] (chl)	B.31.380
octa-O-acetyl- (α-form)	229	+43.6[20] (chl)	IR.1(1).195B
phenylosazone	208–210	−6.5[18] (EtOH/py)	NMR.2(2).911A
octa-O-acetyl-	196	−5.3[20] (chl)	
phenylosazone	184–186	−14.8[20] (py)	
octa-O-acetyl-	143–144	+97 (chl)	
octa-O-acetyl-(α-form)	152	+53.6[20] (chl)	B.31.408
octa-O-acetyl-(β-form)	90	−4.7[20] (chl)	IR.1(1).194C, D
octa-O-benzoyl-(β-form)	140–142	+38.1[20] (chl)	NMR.2(2).910B
phenylosazone	200–212		
octa-O-acetyl-(α-form)	125	+123[20] (chl)	B.31.386
octa-O-acetyl-(β-form)	159–160	+62.6[20] (chl)	IR.1(1).195A
phenylosazone	206	+82.6[20] (EtOH/py)	NMR.2(2).910C

(Continued)

Table XI.5 (Continued)

Name, formula	M_r	m.p./°C	$[\alpha]_D$	Solubility
maltotriose, O-α-D-gluco-pyranosyl-(1 → 4)-O-α-D-glucopyranosyl-(1 → 4)-D-glucose 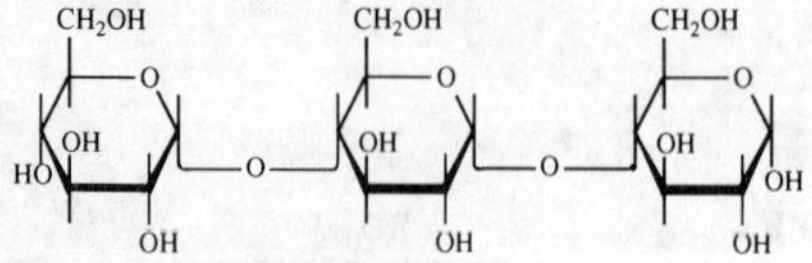	504.441	132–135	+160[22] ($c = 2$, water)	s water
melibiose, 6-O-α-D-galacto-pyranosyl-D-glucose 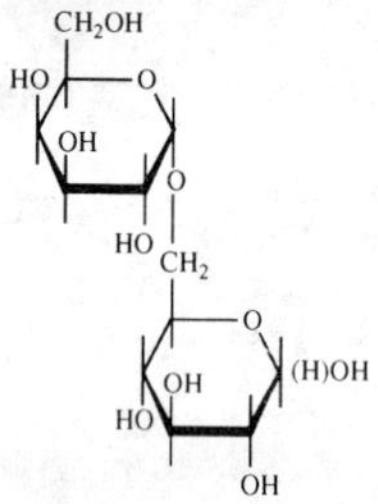	342.299	85 d	+110.5 → +126.5[20] ($c = 4$, water)	vs water ss EtOH i et, ac
raffinose, β-D-fructo-furanosyl-O-α-D-galacto-pyranosyl-(1 → 6)-α-D-glucopyranoside 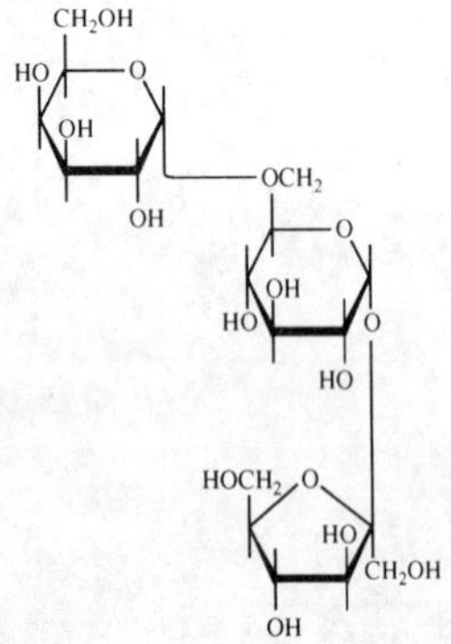	494.362	80 (hyd) 118 (anhyd)	+104[20] (water)	vs, water, ss. MeOH, EtOH, i et
sucrose, β-D-fructo-furanosyl-α-D-gluco-pyranoside 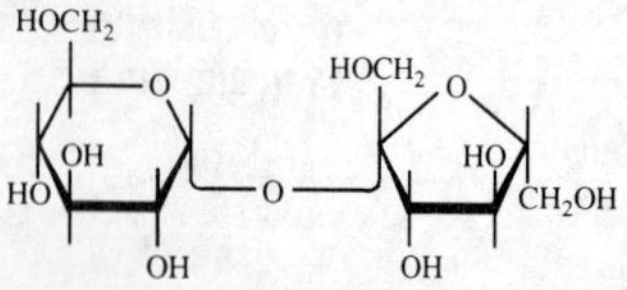	342.299	160–180 (chars)	+66.47[25] ($c = 26$, water)	s water, py, ss EtOH, i et

5. Amino Sugars

Name, formula	M_r	m.p./°C	$[\alpha]_D$	Solubility
N-acetyl-D-galactosamine, 2-acetamido-2-deoxy-D-galactose 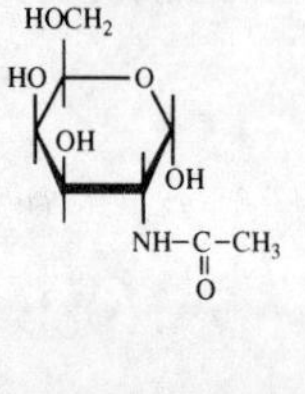	221.210	160–161	+85[25] ($c = 1$, water)	s water

Name	Derivatives		References
	m. p./°C	$[\alpha]_D$	
β-undecaacetyl-	134–136	+86[25] (chl)	IR.1(1).196D NMR.2(2).912D
octa-O-acetyl- phenylosazone	177 176–178	+102.5[20] (chl) +43.2[21] (py)	B.31.421 IR.1(1).196C NMR.2(2).912B
undecsacetyl-	101	+100 (water)	B.31.462 IR.1(1).197B NMR.2(2).913C
octa-O-acetyl octa-O-benzoyl-	89 60–63	+58.5[25] (EtOH) +32.6 (chl)	B.31.424 IR.1(1).196B
			IR.1(1).757C

(Continued)

Table XI.5 (Continued)

Name, formula	M_r	m.p./°C	$[\alpha]_D$	Solubility
N-acetyl-α-D-glucosamine, 2-acetamido-2-deoxy-D-glucose 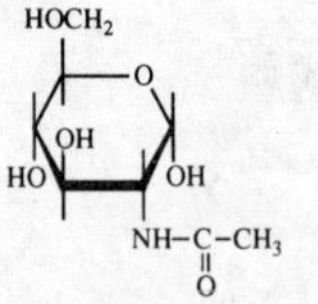	221.210	205	+82 → +40.2[25] (c = 1, water)	s water
α-D-galactosamine, 2-amino-2-deoxy-D-galactose 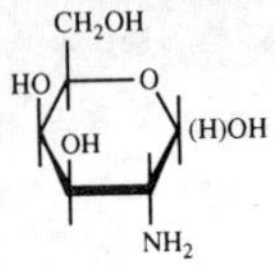	179.172	185	+59.3[25] (water) hydrochloride: +95[20] (water)	s water, acids
α-D-glucosamine, 2-amino-2-deoxy-D-glucose 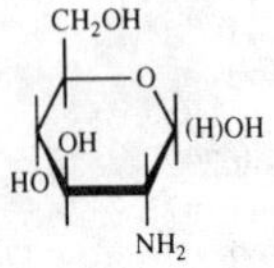	179.172	88	+100 → +47.5[20] (c = 1, water)	vs water, acids ss EtOH, i et, chl
β-D-mannosamine, 2-amino-2-deoxy-D-mannose 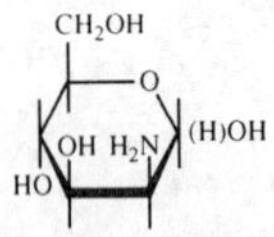	179.172	179.2	−4.6[20] (c = 10, aq HCl)	s water, MeOH, acids

6. Uronic Acids

Name, formula	M_r	m.p./°C	$[\alpha]_D$	Solubility
N-acetylneuraminic acid 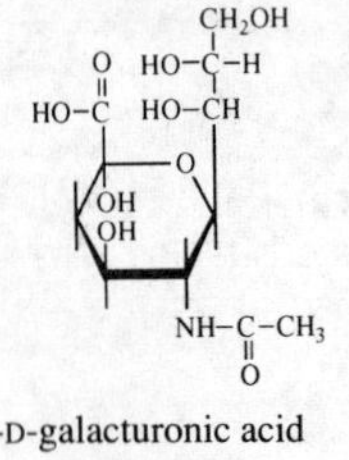	309.272	d 185–187	−32[22] (c = 1, water)	s water, MeOH ss EtOH i ac, et, chl
α-D-galacturonic acid 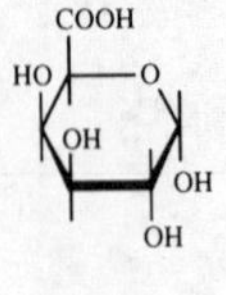	194.141	d 156–159	+98 → +50.9[20] (c = 10, water)	s water ss EtOH i et
β-D-glucuronic acid 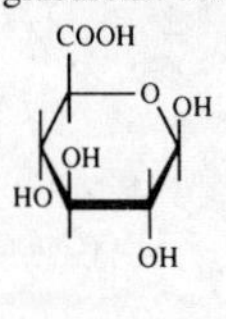	194.141	165	+11.7 → +36.3 (c = 6, water)	s water EtOH

Name	Derivatives m. p./°C	$[\alpha]_D$	References
1, 3, 4, 6-tetra-O-acetyl- 1-benzoyl–	139 192–193	+92 (chl) +190^{20} (MeOH)	B.31.170 IR.1(1).757D NMR.2(1).638C
N-acetyl- penta-O-acetyl- DNP	160–161 178 184–186	+85^{25} (water) +102^{20} (chl) +84 (aq. EtOH)	IR.1(1).349D NMR.2(1).305D
N-acetyl- N-benzoyl- penta-O-acetyl- DNP semicarbazone	205 204–206 139 202–204 165	+82 → +40.2 (water) +31^{16}(water) +92^{20}(chl) +65 (aq EtOH)	B.31.167 IR.1(1).350A NMR.2(1).306A
N-acetyl- N, 1, 3, 4, 6-penta-O-acetyl-	128–129 162–163	−9.4 → +9.7 (water) −17^{24} (chl)	B.4(4).2022
4-O-acetyl- 4, 7, 8, 9-tetra-O-acetyl, methyl ester methyl glycoside	199–200 174–175 200	−62 (water) −5.2 (chl) −55	IR.1(1).758C
4-bromophenylhydrazone 1, 2:3, 4-di-O-benzylidene- methyl ester	145–146 199–201 147	+9 (MeOH) +94 → +34 (MeOH)	B.3(1).306 IR.1(1).525C
4-nitrophenylhydrazone phenylhydrazone 2, 3, 4,-tribenzyl- γ-lactone	224–225 182 91–92 177–178	 +9.9 +19.8 (water)	B.31.261

(Continued)

Table XI.5 (Continued)

Name, formula	M_r	m.p./°C	$[\alpha]_D$	Solubility
α-D-mannuronic acid	194.141	d 120–130	+16 → –6^{25} (water)	s water i et
(+)-muramic acid, (R)-2-amino-3-O-(1-carboxyethyl)-2-deoxy-D-glucose	251.236	d 160–162	+146 → +116^{22} (c = 2, water)	s water

Name	Derivatives		References
	m. p./°C	$[\alpha]_D$	
methyl glycoside	108	+65.6 (water)	
lactone (β-form)	143–144	+94 (water)	
N-acetyl-	119–120	+60 → +40^{20} (water)	
N-acetyl-benzyl-α-pyranoside	162–163	+168^{25} (MeOH)	

Table XI.6 Properties of Selected Steroids (bile acids, corticosteroids, sex hormones, sterols)

For more extensive data see: *Dictionary of Organic Compounds* (5th edn, 1982, and later supplements). Chapman & Hall, London.
The numbering of the cyclopentanophenanthrene nucleus is given in the structure of cholesterol.
Column 4. Optical rotatory power [α] (customarily called specific rotation) measured at 589.3 nm in cell of path length 1 dm at the specified temperature, quoted in units deg dm^{-1} g^{-1} cm^3. Concentration, c, in units of g per 100 cm^3 of specified solvent.
Column 5. Solubility at 25°C. i: insoluble; s:soluble; ss: slightly soluble; vs: very soluble.
Solvents; ac: acetone; alk: alkalies; bz: benzene; chl: chloroform; diox: dioxane; et: ether; py: pyridine.
Column 6. Comments. Other properties, toxicity etc. CSA: cancer suspect agent, LS: light-sensitive. Beilstein reference: B.volume (supplement) page.

Name, synonym, formula	M_r	m.p. /°C	[α]^t	Solubility	Reference and comments
1. Bile Acids					
cholic acid, 3α,7α,12α-trihydroxy-5(β)-cholanoic acid	408.58	200–201	+36^{23} $c = 0.6$ 95% EtOH	s H$_2$O s EtOH, ac, chl alk	pK_a = 6.4 B.10(3)2162
deoxycholic acid, 3,12-dihydroxycholanic acid	392.58	174–176	+54.5^{20} $c = 1.0$ EtOH	ss H$_2$O vs EtOH s alk, ss et, chl	pK_a = c.6
chendeoxycholic acid, 3α,7α-dihydroxy-5β-cholanic acid	392.58	143	+11.2^{20} $c = 2.0$ EtOH	ss H$_2$O s MeOH, EtOH	
lithocolic acid, 3α-hydroxycholanic acid	376.58	184–186	+33.7^{20} $c = 1.5$ EtOH	i H$_2$O, et s EtOH, chl, NaOH	
2. Corticosteroids					
cortisone, 17-hydroxy-11-dehydrocorticosterone	360.45	d 224	+212^{20} $c = 1.2$ EtOH	i H$_2$O, s MeOH, EtOH, ac ss et.	B.8(3)4057
9-α-fluorohydrocortisone,	380.40	d 261	+139^{23} $c = 0.55$ 95% EtOH	ss H$_2$O s EtOH, MeOH, chl	
hydrocortisone, 17-hydroxycorticosterone	362.47	220	+167^{22} $c = 1$ EtOH	ss H$_2$O s EtOH, MeOH, chl, diox	
deoxycorticosterone Δ^4-pregnene-3,20-dion-21-ol	330.47	141–142	+178^{20} $c = 1$ EtOH	ss H$_2$O, s EtOH, ac, et	
corticosterone, 11,21-dihydroxyprogesterone	346.40	180–182	+222^{15} $c = 1.1$ EtOH	i H$_2$O, s EtOH, ac, et	
11-dehydrocorticosterone,	344.46	183–184	+258^{25} $c = 1$ EtOH	i H$_2$O s EtOH, ac, bz	

Table XI.6 (Continued)

Name, synonym, formula	M_r	m.p. /°C	$[\alpha]^t$	Solubility	Reference and comments
aldosterone, 18-oxocorticosterone	360.45	164 (anhyd)	$+150^{25}$ $c = 0.1$ chl	s ac, chl	

3. Sex Hormones

Name, synonym, formula	M_r	m.p. /°C	$[\alpha]^t$	Solubility	Reference and comments
androsterone, androstane-3α-ol-17-one	290.45	181–184	$+96^{20}$ $c = 0.7$ EtOH	i H_2O, s EtOH, ac, et, bz	
β-estradiol, 1, 3, 5-estratriene-3, 17β-diol	272.39	178–179	$+80.4^{20}$ $c = 1$ diox	i H_2O s EtOH, ac, diox, alk	B.6(3) 5332
estriol, 1,3,5-estratriene-3, 16α, 17β-triol	288.39	282	$+62^{20}$ $c = 1$ 95% EtOH	ss H_2O s EtOH, chl, et, py, alk	CSA B.6(3) 6520
estrone, 1, 3, 5-estratriene-3-ol-17-one	270.37	258–260	$+161^{25}$ $c = 1$ diox	i H_2O s EtOH, ac, chl, alk	CSA
17α-ethynylestradiol, 17α-ethynyl-1, 3, 5-estratriene-3, 17β-diol	296.41	182–183	$+1- +10^{20}$ $c = 1$ diox	i H_2O s EtOH, MeOH, ac, chl, et, diox	CSA
ethynylestradiol 3-methyl ether, menstranol,	310.40	153–155	$+2- +8^{25}$ $c = 2$ diox	i H_2O, s EtOH, MeOH, diox	CSA
progesterone	314.47	129–130	$+182^{23}$ $c = 2$ diox	i H_2O s EtOH, et, diox, conc. H_2SO_4.	CSA

(Continued)

Table XI.6 (Continued)

Name, synonym, formula	M_r	m.p. /°C	$[\alpha]^t$	Solubility	Reference and comments
testosterone, 17β-hydroxy-4-androsten-3-one	288.43	153–155	+109[24] $c=4$ EtOH	i H_2O, s EtOH, et, ac, veg oils	CSA
4. Sterols cholesterol, 5 cholesten-3β-ol	386.66	145.8 (anhyd)	−40[24] $c=2$ chl −31.5[20] $c=2$ et	i H_2O, ss EtOH, ac, s et, bz, diox, py	
ergosterol, 24-methyl-5, 7, 22-cholestatrien-3β-ol	396.67	156–158	−127.5[23] $c=1.2$ chl	i H_2O s[h] EtOH, et s bz, chl	LS
lanosterol, isocholesterol	386.67	140–141	+62[20] $c=1$ chl	i H_2O s[h] EtOH s chl, et	
β-sitosterol, 24-ethylcholesterol	414.72	139–142	−37[25] $c=2$ chl	s EtOH, et, chl	
stigmasterol, 24-ethyl 5, 22-cholestadien-3β-ol	412.71	160–164	−51[22] $c=2$ chl	i H_2O s[h] EtOH s ac, bz, chl, et	

Table XI.7. Properties of Constituents of Nucleic acids and Related Compounds (nucleosides, nucleotides, purines and pyrimidines)
Reference: *The Merck Index*, 11th ed. (1989).
Column 5. Solubility at 25°C.
i:insoluble; s: soluble; ss: slightly soluble; vs: very soluble; h: hot
Solvents: a: mineral acids; ac: acetone; alk: alkalies; bz: benzene; chl: chloroform; et: ether; py: pyridine.
Column 6. Optical rotatory power [α] (customarily called specific rotation), measured at 589.3 nm in cell of path length 1 dm at the specified temperature, quoted in units deg dm^{-1} g^{-1} cm^3. Concentration, c, in units of g per 100 cm^3 of specified solvent.
FL: flammable liquid; MS: moisture-sensitive.
Beilstein reference: B. volume (supplement). page.

Name, synonym, formula	M_r	m.p. /°C	pK_a	Solubility	References and comments
adenine, 6-aminopurine C(NH$_2$):NCH:NC:CNHCH:N	135.13	d 360	1.0 4.1 9.8	vs^h H$_2$O vs. alk ss EtOH i et, chl	B.26(1).252
adenosine, adenine-9β- D-ribofuranoside	267.25	235–236	3.6 12.5	s H$_2$O i EtOH	B.31.27 [α]^{20} = –60 c = 1, H$_2$O
5'-adenylic acid, adenosine-5'-phosphate	347.23	d 195	<1 3.8 6.2	vs^h H$_2$O s EtOH HCl	B.31.27 [α]^{20} = –47.5 in aq NaOH
caffeine, 1,3,7- trimethylxanthine C(=O)N(CH$_3$)C(=O)N(CH$_3$)C:CN(CH$_3$)CH:N	194.19	234–236		s^h H$_2$O, chl, s py ss ac, EtOH, bz i et	B. 26 (2). 266
cytosine, 2-oxo-4-aminopyrimidine CH:NC(=O)NHC(NH$_2$):CH	111.10	d 320	4.4 12.2	s^h H$_2$O ss EtOH s a i et	B.24.314
cytidine, cytosine-3β- D-ribofuranoside	243.22	d 215	4.1 c. 12.3	vs H$_2$O ss EtOH	B.31.24 [α]^{20} = +35.3 c = 1, H$_2$O
3'-cytidylic acid, cytidine-3'-phosphate	323.20	d 232	0.8 4.28 6.0	s^{h}H$_2$O, EtOH	B.31.25 [α]^{20} = +49.4 c = 1, H$_2$O
guanine, 6-oxy-2-aminopurine C(=O)NHC(NH$_2$):NC:CNHCH:N	151.13	d 360 9.92	3.2 ss EtOH, et 12.3	i H$_2$O s a,alk	B.26(2).262
guanosine, guanine-9β- D-ribofuranoside	283.24	d 239	1.6 9.2 12.3	s^h H$_2$O s a, alk i EtOH, chl, et	B.31.28 [α]^{20} = –60.5 in aq NaOH
5'-guanylic acid, guanosine-5'-phosphate	363.23	d 208	<1 2.4 6.1 9.4	s H$_2$O	B.31.29
hypoxanthine, 6-hydroxypurine C(=O)NHCH:NC:CNHCH:N	136.11	d 150	2.0 8.9 12.1	s^h H$_2$O s a, alk	B.26(2).252
inosine, hypoxanthine-9β-D-ribofuranoside	268.23	d 218	1.2 8.9 12.5	s^{h}H$_2$O s alc, alk	B.31.25
5'-inosinic acid, inosine-5'-phosphate	348.22	syrup	1.2 1.5 6.0 8.9	s H$_2$O ss EtOH, et	[α]^{20}= –18.5 in HCl

(Continued)

Table XI.7 (Continued)

Name, synonym, formula	M_r	m.p. /°C	pK_a	Solubility	References and comments
purine, 7-imidazo-[4,5-d]pyrimidine CH:NCH:NC:CNHCH:N	120.11	216–217	2.4 8.9	vs H_2O vs^h EtOH s ac ss et, chl	B.26(2).252
pyrimidine, 1,3-diazin N:CHN:CHCH:CH	80.09	22 b.p.123–124		vs H_2O s EtOH, et	B.23.89 FL, MS
theobromine, 3,7-dimethylxanthine C(=O)NHC(=O)N(CH₃)C:CN(CH₃)CH:N	180.18	351	10.0 13.9	ss H_2O EtOH, i et, bz, chl	B.26(2).264
theophylline, 1,3,-dimethylxanthine C(=O)N(CH₃)C(=O)N(CH₃)C:CNHCH:N	180.17	274–275		vs$^h H_2O$ ss EtOH, et, chl	B.26(2).263
thymine, 5-methyl uracil NHC(=O)NHC(=O)C(CH₃):CH	126.12	320–322	c. 0 9.9 >13	ss H_2O, EtOH i et	B.24.353
thymidine, thymine-2-deoxy-β-D-ribo-furanoside	242.23	186–187	9.8 >13	s H_2O s^h EtOH, et s MeOH, py ss chl	B.24(4).1297 $[\alpha]^{25} = +30.6$ $c = 1$, H_2O
5'-thymidylic acid, thymidine-5'-phosphate	322.20	175	c.1.6 6.5 10.0		
uracil, 2, 6-dioxytetrahydropyrimidine NHC(=O)NHC(=O)CH:CH	112.09	d 338	c. 0.5 9.9 >13.0	vs^h H_2O s alk ss EtOH i et	B.24(2).168
uridine, 3β-D-ribofuranosyluracil	244.20	165	9.2 12.3	s H_2O, py ss EtOH	B.31.23 $[\alpha]^{20} = +4$ in H_2O
5'-uridylic acid, uridine-5'-phosphate	324.19	198–199	c.1 6.4 9.5	vs H_2O	
uric acid, 2, 6, 8-trihydroxypurine C(=O)NHC(=O)NHC:CNHC(=O)NH	168.11	d 250	5.4 11.3	i H_2O, alc, et vs alk	B.26(2).293
xanthine, -2,6-purinedione C(=O)NHC(=O)NHC:CNHCH:N	152.11	d	9.9 13.2	ss H_2O, EtOH s a, alk i et, chl	B.26(2).260

(Table XI.8 begins on p.398)

Table XI.8 Physical Constants of Polymers and Plastics

The physical and chemical properties of polymers vary from sample to sample and are very dependent on such factors as the composition and pretreatment of the sample and the amount and kind of filler and plasticizer. For these reasons a range of values for a particular property is given. For further information see: J. Brandup and E. H. Immergut (eds), *Polymer Handbook* (3rd edn, 1989), Wiley, Chichester.
Type of polymer: E, elastomer; TP, thermoplastic; TS, thermosetting resin.
Chemical resistance: A, attacked; D, decomposed; E, excellent; F, fair; G, good; P, poor; S, soluble. Two letters with bar, indicates variability between samples F/G means fair to good resistance depending on sample.

Name of polymer or Plastic	Type	Density ρ/kg m^{-3}	Tensile strength /MN m^{-2}	Relative Permittivity ε_r	Resistivity ρ/ohm m	Heat capacity C/kJ kg^{-1} K^{-1}	Thermal conductivity /W m^{-1} K^{-1}
1. Polyolefins							
acrylonitrilebutadiene	E	1060–1080	48–55	2.4–3.8		1.25–1.67	
butyl rubber							
polyisoprene							
natural rubber	E	906–913			10^4	1.88	0.13
hard rubber	E	1130–1180	39		10^{14}	1.38	0.16
polychloroprene	E	1230–1250			10^8–10^{11}	2.18	0.19
poly-2-chloro-1,3-butadiene (neoprene)							
polyethene, polyethylene							
low density	TP	910–925	7–10	2.3–2.35	10^{13}–10^{16}	2.3	0.33
medium density	TP	926–941	8–24	2.3	10^{13}–10^{16}	2.3	0.33–0.42
high density	TP	941–965	21–38	2.3–2.35	10^{13}–10^{16}	2.3	0.45–0.52
polypropene, polypropylene	TP	902–906	29–40	2.2–2.6	$>10^{13}$	2.0	0.12
polystyrene	TP	1040–1080	34–70	2.5–2.65	$>10^{14}$	1.3–1.5	0.04–0.14
polystyrene-acronitrile	TP	1050–1100	62–83	2.5–3.1	10^{11}–10^{15}	1.3–1.5	
polytetrafluoroethene, PTFE	TP	2100–2300	13–31	2	10^{16}	1.05	0.25
polychlorotrifluoroethene, PCTFE	TP	2100–2300	31–41	2.3–2.5	10^{16}	0.92	0.2–0.22
poly(tetrafluoroethene)-co-[per-fluoro-(methyl vinyl ether)] PFA	TP	2120–2170	28–31	2.06	10^{12}–10^{14}		
fluorinated ethenepropene, FEP	TP	2120–2170	14–21	2.1	$>10^{16}$	1.17	0.20
ethene-co-tetrafluoroethene, ETFE	TP	1900	38–48	2.5	10^{16}	1.9–2	0.238
2. Polyvinyl Resins							
polyacrylonitrile	TP	1160–1180	200 (fibres)	6.5	10^{12}		
polymethylmethacrylate	TP	1170–1200	48–76	3.3–4.5	$>10^{12}$	1.46	0.17–0.25
polyvinylchloride, PVC	TP	1300–1400	34–62	3.0–4.0	10^{14}	0.84–1.17	0.12–0.17
poly(vinylidenefluoride), PVDF	TP	1750–1780	36–56	2.9	2×10^{12}		0.1
polyvinyl fluoride, PVF	TP	1400					
3. Polyester Resins							
alkyd resins (fibre filled)	TS	1240–2600	31–45	3.6–4.7			
polycarbonate	TP	1200	55–65	2.96–3.17	10^{14}	1.25	0.19
polyester, styrene-alkyd (glass fibre mat reinforced)	TS	1500–2100	207–350	4.0–5.5	10^9	0.8–1.6	
4. Nitrogen Polymers							
nylon-6, poly-ϵ-caprolactam	TP	1120–1170	45–90	3.7–5.5	10^{10}–10^{13}	1.6	0.25
nylon-66, polyhexamethyleneadipamide	TP	1130–1150	62–83	4.0–4.6	10^{12}–10^{13}	1.7	0.25
polyurethane	TS, E	1110–1250	29–55	5.4–7.6	10^9–10^{11}	1.8	0.07
urea-formaldehyde, urea-methanal (cellulose filled)	TS	1470–1520	38–89	6.7–8.0	0.005–0.05	2.5	
melamine-formaldehyde (cellulose filled)	TS	1470–1520	48–90	7.2–8.4	10^{10}–10^{12}	1.7	
5. Polyethers							
epoxy resins	TS	1110–1400	27–90	3.5–5	10^{10}–10^{15}	1.05–1.7	0.17–0.21
polyacetals	TP	1430	69–83	3.7	10^{15}	1.46	0.23

$10^5 \times$ Coeff. of linear expansion/°C	Max. continuous service temperature/°C	Refractive index n_D	Chemical resistance to:						
			Strong mineral acids	Oxidizing acids (conc)	Strong alkalies	Alcohols	Aromatic hydrocarbons	Chlorinated hydrocarbons	Oils
6.0–6.5	88–110		G	P	G	G	F		G
		1.52	G	G	G	F	F	F	
		1.6	G	G	G	F	F	F	
		1.56	G	G	G	P	P	P	F
10–20	60–77	1.50–1.54	G	G/P	G	E/P	F	F	G
14–16	71–93	1.52–1.54	E	G/P	E	E/P	G	G	E
11–13	92–200	1.54	E	G/P	E	E/P	F	F	G
6–10	softens 160	1.49	E	G/P	E/G	E/G	G/F	F	G
6–8	66–62	1.59	E	P	E	E	S	S	F/P
3.6–3.8	77–88	1.57	G/E	P	G/E	G/E	F/G	F	G/E
10	260	1.3–1.4	E	E	E	E	E	E	E
4.5–7	200	1.43	E	E	E	E	E	F/G	E
17	260	1.350	G	G	G	G	G	G	G
	200	1.344	G	G	G	G	G	G	G
9	150	1.403	G	F	G	G	F		
7		1.52	G	S	F				
5–9	60–93	1.48–1.50	F/P	A	P		softens	S	G
5–18	70–74	1.54	G/E	F/G	G	E	P	P	E
8–14	150	1.42	G	G	G	G	G	G	G
	150		G	G	G	G	G	G	
4.0–5.5	150–220		F		F	F/G	F/G	F	
6.6	140	1.59	F		P	P	P/S	S	P
1.8–3.0	120–204		P	P	P	G	P/F	F	G
28	100		P	P	G	G	F/G	F	G
8	80–150	1.53	P	P	E	G	F/G	F	G
10–20			P	P	P	G	F/G	F	
2.2–3.6	77	1.55	P	P	P	G	G	G	
2.0–5.7	99		P	P	P	G	G	G	G
4.5–9	80	1.55–1.61	F/G	P	E	E	E	G	E
8.1	84	1.48	P	P	P	G	G	G	G

(Continued)

Table XI.8 (Continued)

Name of polymer or Plastic	Type	Density ρ/kg m^{-3}	Tensile strength /MN m^{-2}	Relative Permittivity ϵ_r	Resistivity ρ/ohm m	Heat capacity C/kJ kg^{-1} K^{-1}	Thermal conductivity /W m^{-1} K^{-1}
6. Other Polymers							
cellulose (cotton, wood pulp)		1480–1530	80–240	6.7–7.5	10^5–10^{12}	1.3–1.5	0.23
cellulose acetate							
soft	TP	1270–1340	13–32	3.5–7.5	10^8–10^{12}	1.25–1.8	0.17–0.33
hard	TP	1270–1340	31–59	3.5–7.5	10^8–10^{11}	1.25–1.8	0.17–0.33
cellulose nitrate	TP	1350–1400	50	7.0–7.5	10^8	1.3–1.7	0.12–0.21
phenol-formaldehyde phenol-methanal	TS	1360–1430	41–62	5.0–6.5	10^9–10^{10}	1.59–1.76	0.13–0.25
silicones (mineral filled)	TS, E	1800–2800	21–28	3.4–3.6	>10^{12}	0.8–1.3	0.15–0.32

Table XI.9 Physical Properties of Fluorocarbon Refrigerants
Reference: *Handbook of Chemistry and Physics* (1991–92), CRC Press.
The density, ρ; refractive index, n_D; viscosity of the liquid, η; surface tension, γ; relative permittivity of the liquid, ϵ_r; molar heat capacity, C_P, of liquid and vapour; molar heat of vaporization, $\Delta_{vap}H$; and thermal conductivity of the liquid, λ, are at 25°C unless otherwise indicated. [a] Estimated value.

Compound	Formula	M_r	m.p. /°C	b.p. /°C	ρ /kgm^{-3}	n_D	η /mPa s
fluorotrichloromethane (Freon 11)	CFCl$_3$	137.37	–111	23.82	1467	1.374	0.42
dichlorodifluoromethane (Freon 12)	CCl$_2$F$_2$	120.91	–158	–29.8	1311	1.287	0.26
chlorotrifluoromethane (Freon 13)	CClF$_3$	104.46	–181	–81.4	1298^{-30}	1.199^{-75}	0.016
bromotrifluoromethane (Freon 13B1)	CBrF$_3$	148.92	–168	–57.75	1538	1.238	0.15
carbon tetrafluoride (Freon 14)	CF$_4$	88.01	–184	–130	1317^{-80}	1.151^{-73}	0.02
dichlorofluoromethane (Freon 21)	CHCl$_2$F	102.92	–135	8.9	1366	1.354	0.34
chlorodifluoromethane (Freon 22)	CHClF$_2$	86.47	–160	–40.75	1194	1.256	0.23
trifluoromethane (Freon 23)	CHF$_3$	70.01	–160	–82	670	1.251^{-73}	0.016[a]
1,1,2,2-tetrachlorodifluoroethane (Freon 112)	Cl$_2$FCCFCl$_2$	203.83	26	92.8	1634^{30}	1.413	1.21
1,1,2-trichlorotrifluoroethane (Freon 113)	ClF$_2$CCFCl$_2$	187.38	–35	47.6	1565	1.354	0.68
1,2-dichlorotetrafluoroethane (Freon 114)	ClF$_2$CCF$_2$Cl	170.91	–94	3.77	1456	1.288	0.38
1,2-dibromotetrafluoroethane (Freon 114B2)	BrF$_2$CCF$_2$Br	259.83	–110	47.26	2163	1.367	0.72
chloropentafluoroethane (Freon 115)	ClF$_2$CCF$_3$	154.47	–106	–39	1291	1.214	0.26
hexafluoroethane (Freon 116)	F$_3$CCF$_3$	138.01	–100	–78	1587^{-73}	1.206^{-73}	
1-chloro-1,1-difluoroethane (Freon 142B)	CH$_3$CClF$_2$	100.50	–131	–9.2			
1,1-difluoroethane (Freon 152A)	CH$_3$CHF$_2$	66.05	–117	–25			
octafluoropropane (Freon 218)	CF$_3$CF$_2$CF$_3$	188.02	–183	–37			
octafluorocyclobutane (Freon C318)	CF$_2$CF$_2$CF$_2$CF$_2$	200.03	–41	–6.0			
chlorotrifluoroethylene (Freon 1113)	CClF:CF$_2$	116.47	–158	–28			
tetrafluoroethylene (Freon 1114)	CF$_2$:CF$_2$	100.02	–143	–76			
fluoromethane (Freon 41)	CH$_3$F	34.03	–142	–78			

N.B. Where b.p. of liquid is below 25 °C, the data for the liquid state refer to the saturated liquid (i.e. equilibrium pressure equal to vapour pressure at 25 °C).

$10^5 \times$ Coeff. of linear expansion/°C	Max. continuous service temperature/°C	Refractive index n_D	Chemical resistance to:						
			Strong mineral acids	Oxidizing acids (conc)	Strong alkalies	Alcohols	Aromatic hydrocarbons	Chlorinated hydrocarbons	Oils
		1.53–1.56	P	P	P	F	G		G
8–16		1.46–1.50	F/G	vP	vP	P	P/F	S	F/G
8–16		1.46–1.50	F/G	vP	vP	P	P/F	S	F/G
8–12		1.49–1.51	D	D	D	S	S		
2.5–6	121	1.5–1.7	P	P	P	G	F/G	F	G
2–4	288	1.4	P/G		P	P	P	P	G

γ /mN^{-1}	ϵ_r	Critical constants			C_p/J K^{-1}mol^{-1}		$\Delta_{vap}H$ /kJ mol^{-1}	λ/W m^{-1}K^{-1}
		P_c/atmos	T_c/°C	V_c/cm^{3}mol^{-1}	liq	vap		
18	2.28[29]	43.5	198	247	119.5	81.6[38]	24.8	0.087
9	2.13[29]	40.6	112	217	117.4	73.4	19.97	0.071
14[-73]		38.2	28.9	181	107.9[-30]	69.1	15.50	0.035
4		39.1	67	200	129.6	69.79	17.68	0.043
14[-73]		36.96	–45.67	141	108.2[-80]	62.2	11.96	0.069[-100]
18	5.34[28]	51	178.5	197	110.2	60.29	24.92	0.11
8	6.11[24]	49.12	96	165	108.5	56.80	20.19	0.090
15[-73]		47.7	25.9	133	454.9	51.55	16.76	0.014
23[30]	2.52	34	278	370			31.55[a]	0.069
19	2.41	33.7	214.1	325	170.9	126.22	27.49	0.066
12	2.26	32.2	145.7	293	173.8	121.6	23.24	0.059
18	2.34	34	214.5	329	180.5		27.2[a]	0.047
5		30.8	80	259	184.2	106.0	19.46	0.045
16[-73]		29.4	19.7	225	134.0[-73]	105.1	16.15	0.078[-73]
		49.1	96					
		44.4	114					
		26.6	72					
		27.6	115					
		40.3	106					
		39.5	33					
		62.2	45					

Table XI.10 Physical Properties of Some Common Refrigerants
Reference: as for Table XI.9.

Refrigerant	b.p. /°C	Flash point /°C	Ignition temp. /°C	Explosive limits lower/upper	†Vapour density (air = 1)	*TWA /ppm	Water soluble
ammonia	−33.35		651	16/25	0.59	25	yes
butane	−0.5	−60	430	1.8/8.4	2.04	600	no
carbon dioxide	−78.5		non-flammable		1.53	5 000	yes
carbon tetrachloride	76.8		non-flammable		5.32	10	no
ethane	−88.6	<−7	510	3.0/12.5	1.04		no
ethylene	−104	<−7	450	3.1/32	0.97		yes
isobutane	−10	<−7	540	1.8/8.4	2.01		no
chloromethane	−24.2	330	630	10.7/11.4	1.78	50	yes
propane	−42.1	<−7	465	2.2/9.5	1.56		no
propylene	−47.7	<−7	500	2.4/10.3	1.49		yes
sulphur dioxide	−10		non-flammable		2.2	2	yes
trichloroethylene	87		non-flammable at normal temperatures		4.53	100	no

*: Approved long-term (8-h) occupational exposure limit. See also Tables XVI.1 & XVI.2
†: At 25 °C and 1 atmosphere pressure.

Table XI.11 Properties of Gases Available in Cylinders (bottled gases)
Data from *Properties of Gases Chart*, 4th ed. (1990), BDH Chemicals Ltd, Poole, Dorset.
All gases, except oxygen, must be treated as asphyxiants in addition to their other properties.
Column 1. * denotes IUPAC recommended name.
Column 9. Flammability limits. The figures given are the lower and upper concentrations (percentage by volume) of flammable gas in mixtures with air within which combustion will occur at atmospheric pressure and temperature. A flammable gas mixed with air will not inflame if there is either too little gas or too little air present.
Column 10. TWA. Approved long term (8-h) occupational exposure limit (see also Tables XVI.1 and XVI.2 for explanation of terms). The first figure is the parts of vapour or gas per million parts of contaminated air by volume at 25 °C and 1 atmosphere pressure; the figure in parentheses is the approximate concentration expressed as milligrams of contaminant per cubic metre of air.
Column 11. Materials attacked by the gas at 20 °C.
Column 12. Hazards. C: corrosive; CSA: cancer suspect agent; E: explosive; FG: flammable gas; H: harmful; I: irritant; MS: moisture sensitive, e.g. hydrolysed by moisture; O: (powerful) oxidizing agent; T: toxic; vT: very toxic.

Name	Formula	M_r	m.p. /°C	b.p. /°C	Specific gravity (air = 1)	T_c /°C
acetylene: ethyne*	CH:CH	26.04	−80.8	−84 (s)	0.91	36
allene; propadiene*	$H_2C:C:CH_2$	40.07	−136	−34	1.4	120
ammonia	NH_3	17.03	−77.7	−33.35	0.60	132.4
argon	Ar	39.95	−189.2	−185.7	1.38	−122.3
arsine	AsH_3	77.95	−116.3	−55	2.70	100
boron trichloride	BCl_3	117.17	−107.3	12.5		179
boron trifluoride	BF_3	67.81	−126.7	−99.9	2.37	−12.3
bromomethane	CH_3Br	94.94	−93.6	4	3.35	194
bromotrifluoroethylene	$CBrF:CF_2$	160.92		−3	5.5	185
bromotrifluoromethane (Freon 13B1)	$CBrF_3$	148.92	−168	−57.8	5.3	67
1, 3-butadiene	$H_2C:CHCH:CH_2$	54.09	−109	−4.5	1.9	152
butane	$CH_3CH_2CH_2CH_3$	58.12	−138	−0.5	2.1	152
1-butene	$CH_3CH_2CH:CH_2$	56.11	−185	−6.3	2.0	146
cis-2-butene	$CH_3CH:CHCH_3$	56.11	−139	3.7	2.0	162
trans-2-butene	$CH_3CH:CHCH_3$	56.11	−105	1.0	2.0	155
1-butyne	$CH_3CH_2C:CH$	54.09	−126	8.1	1.93	191
carbon dioxide	CO_2	44.01	−56.6 (5.2 atm)	−78.5 (s)	1.53	31.1
carbon monoxide	CO	28.01	−199	−191.5	0.97	−140.2
carbon tetrafluoride (Freon 14)	CF_4	87.99	−184	−128	3.0	−46
carbonyl chloride, phosgene	$COCl_2$	98.92	−104	8.3	3.5	182
carbonyl fluoride	COF_2	66.01	−114	−85	2.3	23

P_c /atm	Flammability limits vol %	TWA /ppm (/mg m^{-3})	Materials attacked	Hazards
61.2	2.5–81		Cu, Ag, Hg	FG, C
51.3			Cu, Ag	FG, C
111.5	15–28	25 (18)	Cu, Sn, Zn	T, C
48.4				
		0.05 (0.2)		FG, T
38.5			most metals when moist	T, C
49.3		1 (3)	rubber, nylon, PVC cellulose	vT, C
51.3		5 (20)	Al	FG, vT, C
44.4	pyrophoric		Mg	FG, T, C
39.1		1000 (6000)		T, I
42.4	2–11.5	10 (22)		FG, CSA
37.5	1.9–8.5	600 (1430)		FG, I
39.5	1.6–9.3			FG
41.4	1.6–9.7			FG
40.5	1.6–9.7			FG
48.4			Cu, Ag	FG, C
73.0		5000 (9000)		
34.5	12.5–74	50 (55)	Fe, Ni etc. at high pressures	FG, T
37.0				H
56.3		0.1 (0.4)	most metals when moist	T, C, MS
57.2		5 (15)	most metals when moist	T, C, MS

(Continued)

Table XI.11 (Continued)

Name	Formula	M_r	m.p. /°C	b.p. /°C	Specific gravity (air = 1)	T_c /°C
carbonyl sulphide	COS	60.07	−138.2	−50.2	2.1	102
chlorine	Cl_2	70.91	−100.98	−34.6	2.47	144
chlorine trifluoride	ClF_3	92.45	−83	11.1	3.2	174
1-chloro-1, 1-difluoroethane (Freon 142B)	CH_3CClF_2	100.50	−131	−9.2	3.7	137
chlorodifluoromethane (Freon 22)	$CHClF_2$	86.47	−160	−41	3.1	96
chloroethane	CH_3CH_2Cl	64.52	−136.4	12.3	2.22	187
chloromethane	CH_3Cl	50.49	−97.7	−24.2	1.78	143
chloropentafluoroethane (Freon 115)	$CClF_2CF_3$	154.47	−106	−39	5.5	80
chlorotrifluoroethylene (Freon 1113)	$CClF:CF_2$	116.47	−158	−28	4.1	106
chlorotrifluoromethane (Freon 13)	$CClF_3$	104.46	−181	−81	3.6	28.9
cyanogen	$(CN)_2$	52.04	−27.9	−21.2	1.8	127
cyclopropane	$CH_2CH_2CH_2$	42.08	−128	−32.7	1.45	124
deuterium	D_2	4.032	−254.6	−249.6	0.14	−234.8
diborane	BH_3BH_3	27.67	−165.5	−92.5	0.96	16.7
dichlorodifluoromethane (Freon 12)	CCl_2F_2	120.91	−158	−29.8	4.2	112
dichlorofluoromethane (Freon 21)	$CHCl_2F$	102.92	−135	8.9	3.6	178.5
dichlorosilane	SiH_2Cl_2	101.01	−122	8	3.5	176
1, 2-dichlorotetrafluoroethane (Freon 114)	$CClF_2CClF_2$	170.92	−94	3.77	5.9	145.7
1, 1-difluoroethane (Freon 152A)	CHF_2CH_3	66.05	−117	−25	2.3	114
1, 1-difluoroethylene (Freon 1132A)	$CF_2:CH_2$	64.03	−144	−86	2.2	30
dimethylamine	$(CH_3)_2NH$	45.09	−93	7	1.55	165
dimethyl ether	CH_3OCH_3	46.07	−140	−23	1.6	127
2, 2-dimethylpropane	$CH_3C(CH_3)_3$	72.15	−16.6	9.5	2.5	161
dinitrogen oxide	N_2O	44.01	−90.8	−88.5	1.53	36.5
ethane	CH_3CH_3	30.07	−183.3	−88.6	1.05	32.2
ethylamine	$CH_3CH_2NH_2$	45.09	−81	16.6	1.6	183
ethylene, ethene*	$CH_2:CH_2$	28.05	−169	−104	0.98	10
ethylene oxide	CH_2CH_2O	44.05	−111	10	1.5	196
fluorine	F_2	38.00	−219.6	−188.1	1.3	−128.5
fluoroethane	CH_3CH_2F	48.06	−143	−38	1.7	102
fluoromethane	CH_3F	34.03	−142	−78	1.20	45
germane	GeH_4	76.72	−165	−88.5	2.6	35
helium	He	4.003	−272.2	−268.9	0.137	−267.9
hexafluoroethane	CF_3CF_3	138.01	−100	−78	4.8	19.7
hexafluoropropene	$CF_3CF:CF_2$	150.02	−156	−29	5.2	94
hydrogen	H_2	2.016	−259.1	−252.9	0.069	−239.95
hydrogen bromide	HBr	80.92	−88.5	−67	2.8	89.75
hydrogen chloride	HCl	36.46	−114.8	−84.9	1.27	51.45
hydrogen cyanide	HCN	27.03	−14	26	0.95	184
hydrogen fluoride	HF	20.01	−83.1	19.5	1.86	187.9
hydrogen iodide	HI	127.91	−50.8	−35.4	4.4	150.75
hydrogen selenide	H_2Se	80.98	−66	−42	2.8	138
hydrogen sulphide	H_2S	34.08	−85.5	−60.7	1.19	100.5
krypton	Kr	83.80	−156.6	−152.3	2.9	−63.8
methane	CH_4	16.04	−183	−162	0.56	−82
methylamine	CH_3NH_2	31.06	−93	−6.3	1.07	157
2-methylpropane	$CH_3CH(CH_3)_2$	58.12	−160	−12	2.1	135
neon	Ne	20.18	−248.7	−246.05	0.7	−228.8

P_c/atm	Flammability limits vol %	TWA /ppm (/mg m^{-3})	Materials attacked	Hazards
58.2	11.9–29			FG, vT
76.0		1 (3)	most metals when moist	vT, C
57.2		0.1 (0.4)	most metals form passive coatings	FG, T, C, MS
40.5	9.0–14.8			FG
49.1		1000 (3500)		
52.3	3.8–15.4	1000 (2600)		FG, I
66.1	10.7–17.4	50 (105)	Al, Mg, Zn	FG, C, H, I, MS
30.8		1000 (6320)		
40.5	8.4–38.7			FG, T
38.2				H
58.2	6–32	10 (20)		FG, vT
54.3	2.4–10.4			FG
16.43				FG
39.5	0.8–98	0.1 (0.1)	rubber	FG, T
40.6		1000 (4950)	natural rubber	
51		10 (40)		H
46.4	4.1–98.8		most metals when moist	FG, vT, MS
32.2		1000 (7000)	natural rubber	
44.4	5.1–17.1			FG
44.4	5.5–21.3			FG
52.3	2.8–14.4	10 (18)	Cu, Sn, Zn	FG, C, H
52.3	3.4–18			FG
31.4	1.4–7.5			FG
71.6				O
48.4	3.0–12.5			FG
55.3	3.5–14.0	10 (18)	Cu, Sn, Zn	FG, C, I
50.3	3.1–32			FG
70.1	3.0–100	5 (10)		FG, T
51.5		0.1 (0.2)	most metals form passive coatings	T
49.3				FG, I
62.2				FG, T
55.3		0.2 (0.6)		FG, T
2.26				
29.6				H
28.6				T
12.8	4.0–75			FG
84.0		3 (10)	most metals when moist	T, C
81.5		5 (7)	most metals when moist	vT, C
53.3	6.0–41	10 (10)		FG, T
64.0		3 (2.5)	most metals when moist	T, C
81.0			most metals when moist	T, C
87.8		0.05 (0.2)		FG, T
88.9	4.3–46	10 (14)		FG, vT
54.3				
45.4	5.0–15.4			FG
74.0	4.9–20.7	10 (12)	Cu, Sn, Zn	FG, C
35.5	1.8–8.4			FG, I
26.8				

(Continued)

Table XI.11 (Continued)

Name	Formula	M_r	m.p. /°C	b.p. /°C	Specific gravity (air = 1)	T_c /°C
nitrogen	N_2	28.01	−209.86	−195.8	0.97	146.8
nitrogen dioxide	NO_2	46.01	−11.2	21.2	2.6	158.3
nitrogen monoxide	NO	30.01	−163.6	−151.8	1.04	−92.9
nitrogen trifluoride	NF_3	71.00	−206.6	−128.8	2.5	−39
nitrosyl chloride	NOCl	65.46	−64.5	−5.5	2.31	168
octafluorocyclobutane (Freon C318)	$CF_2CF_2CF_2CF_2$	200.03	−41	−6.0	7.3	115
octafluoropropane (Freon 218)	$CF_3CF_2CF_3$	188.02	−183	−37	6.6	72
oxygen	O_2	32.00	218.4	−182.96	1.1	−118.35
ozone	O_3	48.00	−192.7	−111	1.66	−121.5
phosphine	PH_3	34.00	−133.5	−87.4	1.2	51
phosphorus pentafluoride	PF_5	125.97	−83	−75	4.5	19
propane	$CH_3CH_2CH_3$	44.11	−189.7	−42.1	1.55	97
propene	$CH_3CH:CH_2$	42.08	−185	−47.7	1.48	92
propyne	$CH_3C:CH$	40.07	−101	−23	1.41	129
silane	SiH_4	32.11	−185	−112	1.1	−3
silicon tetrafluoride	SiF_4	104.08	−90.2	−86	3.6	−14
sulphur dioxide	SO_2	64.06	−72.6	−10	2.26	157.6
sulphur hexafluoride	SF_6	146.05	−50.5	−63.8.s	5.1	45.6
sulphur tetrafluoride	SF_4	108.06	−121	−40	3.8	91
sulphuryl fluoride	SO_2F_2	102.06	−136.7	−55.4	3.7	92
tetrafluoroethylene (Freon 1114)	$CF_2:CF_2$	100.02	−143	−76	3.5	33
tetrafluorohydrazine	NF_2NF_2	104.01	−168	−73	3.7	36
trifluoromethane (Freon 23)	CHF_3	70.01	−160	−82	2.4	25.9
trimethylamine	$(CH_3)_3N$	59.11	−117	2.9	2.1	160
vinyl bromide bromoethene *	$CH_2:CHBr$	106.96	−139	15.8	3.8	190
vinyl chloride chloroethene *	$CH_2:CHCl$	62.50	−153.8	−13.37	2.2	152
vinyl fluoride fluoroethene *	$CH_2:CHF$	46.04	−161	−72	1.6	55
xenon	Xe	131.30	−111.9	−107.1	4.6	16.6

P_c /atm	Flammability limits vol %	TWA /ppm (/mg m^{-3})	Materials attacked	Hazards
33.6				
100.0		3 (5)	most metals when moist	vT, C
64.6		25 (30)		vT
44.4	explosive	10 (30)	most metals when moist	T, C, E
89.5			most metals when moist	T, C
27.6				
26.6				
50.14				O
54.6		0.1 (0.2)	Cu, rubber	O
64.1		0.3 (0.4)		FG, T
33.6			most metals when moist	T, C, MS
42.4	2.2–9.5	1000 (1800)		FG
42.4	2.0–11.1			FG
55.3	1.7–11.7	1000 (1650)	Cu, Ag, Mg	FG
47.4	pyrophoric	0.5 (0.7)		FG
36.5			most metals when moist	T, C, MS
77.9		2 (5)	several metals when moist	T, C
37.1		1000 (6000)		H
		0.1 (0.4)	most metals and glass when moist	T, C, MS
50.3		5 (20)		T, C, MS
39.5	11–60			FG
77.0	explosive			T, E
47.7		1000 (5600)		H
40.5	2.0–11.6		Cu, Sn, Zn when moist	FG, C, I
68.1	9–14	5 (20)	Cu	FG, T, CSA
57.2	4–22	1	Cu	FG, T, CSA
51.3	2.6–22		Cu	FG, T, CSA
57.6				

Table XI.12 Critical Constants

For more extensive data see K. A. Kobe and R. E. Lynn, *Chem. Rev.* (1953), **52**, 117; *Handbook of Chemistry and Physics* (1991–92), CRC Press.

1. Elements and Inorganic Compounds

Compound	P_c /MN m^{-2}	ρ_c /kg m^{-3}	V_c /cm^3 mol^{-1}	T_c /K
air	3.77	313		132.6
ammonia	11.28	235	72	405.6
argon	4.86	530.7	75	150.8
bromine	10.34	1180	135	584
carbon dioxide	7.38	468	94	304.2
carbon disulphide	7.90	440	173	552.2
carbon monoxide	3.50	301	93	132.9
chlorine	7.71	573	124	417.1
deuterium	1.665		60	38.34
deuterium oxide	21.86	363	55	644.1
dinitrogen oxide	7.25	452.5	97	309.6
fluorine	5.22		66	144.3
helium	0.229	69.3	58	5.2
hydrogen	1.297	31.0	65.5	33.2
hydrogen bromide	8.51			362.9
hydrogen chloride	8.26	420	86.8	324.6
hydrogen deuteride	1.48	48.1	63	35.9
hydrogen fluoride	6.48	290	69	461
hydrogen iodide	8.21			423.9
hydrogen sulphide	9.01	349	98	373.6
krypton	5.50	908.5	92	209.4
mercury	18.24		43	1173
neon	2.72	483.5	42	44.4
nitrogen	3.40	311	90	126.3
nitrogen dioxide	10.13	557	83	431.4
nitrogen monoxide	6.55	520	58	180.3
oxygen	5.08	419	76	154.8
ozone	5.53	436	110	261.0
sulphur	11.75			1313
sulphur dioxide	7.89	524	122	430.7
sulphur hexafluoride	3.76	752	194	318.7
sulphur trioxide	8.48	633	127	491.4
tin(IV) chloride	3.75	742	351	591.9
water	22.12	325	55.4	647.4
xenon	5.84	1105	119	289.7

(Continued)

Table XI.12 (Continued)
2. Organic Compounds

Compound	P_c /MN m^{-2}	ρ_c /kg m^{-3}	V_c /cm^3 mol^{-1}	T_c /K
acetic acid	5.79	351	171	594.5
acetone	4.78	278	209	509.7
acetylene	6.1	231	113	308.3
aniline	5.31	340	274	699.2
benzene	4.90	302	259	562.1
1,3-butadiene	4.33	245	221	425
butane	3.82	228	255	425.2
iso-butane	3.65	228	255	408.2
1-butanol	4.41	270	275	562.9
carbon tetrachloride	4.56	558	276	556.3
chlorobenzene	4.52	365	308	632.4
chloroform	5.47	500	239	536.6
chloromethane	6.68	353	143	416.3
cyclohexane	4.07	273	308	553.5
cyclopentane	4.51	270	260	511.7
cyclopropane	5.58	258	163	398.3
diethyl ether	3.64	265	280	466.7
dimethyl ether	5.37	242	190	400.1
1,4-dioxane	5.21	370	238	587
ethane	4.88	203	148	305.4
ethanol	6.38	276	167	516.3
ethyl acetate	3.85	308	286	523.3
ethylamine	5.62	248	182	456
ethylene	5.03	218	129	283.1
ethylene oxide	7.19	314	140	469
ethyl methyl ether	4.40	272	221	437.8
ethyl methyl ketone	4.00	260	277	533
n-heptane	2.74	231	432	540.3
n-hexane	3.01	233	370	507.5
methane	4.60	162	99	190.55
methanol	8.10	272	118	512.6
methyl acetate	4.69	325	228	506.8
methylamine	7.46			430.0
naphthalene	4.05	310	413	748.4
n-pentane	3.37	237	304	469.7
propane	4.25	217	203	369.8
1-propanol	5.17	275	219	536.7
propylene	4.62	233	180	364.9
propyne	5.63	245	163	402.4
pyridine	5.63	312	254	619.9
toluene	4.11	292	316	591.7
m-xylene	3.54	282	376	619.1
o-xylene	3.62	288	369	632.1
p-xylene	3.43	280	379	618.1

Table XI.13 Van der Waals Constants and Boyle Temperature for Some Gases
For more extensive data see: *Handbook of Chemistry and Physics* (1991–92), CRC Press.

Gas	a /N m^4 mol^{-1}	$10^5 b$ /m^3 mol^{-1}	T_B/K	Gas	a /N m^4 mol^{-1}	$10^5 b$ /m^3 mol^{-1}	T_B/K
acetone	14.09	9.94		hydrogen	0.247 6	2.661	110.04
acetylene	4.448	5.136		hydrogen bromide	4.510	4.431	
ammonia	4.225	3.707		hydrogen chloride	3.716	4.081	
argon	1.363	3.219	411.52	hydrogen sulphide	4.490	4.287	
benzene	18.24	11.54		krypton	2.349	3.978	577
n-butane	14.66	12.26		mercury	8.200	1.696	
iso-butane	13.04	11.42		methane	2.283	4.278	509.66
carbon dioxide	3.640	4.267	714.81	methanol	9.649	6.702	
carbon monoxide	1.505	3.985		methylamine	7.224	5.992	
carbon tetrachloride	20.66	13.83		methyl ether	8.180	7.246	
chlorine	6.579	5.622		neon	0.213 5	1.709	122.11
chloroethane	11.05	8.651		nitrogen	1.408	3.913	327.22
chloromethane	7.570	6.483		nitrogen dioxide	5.354	4.424	
chloroform	15.37	10.22		nitrogen monoxide	1.358	2.789	
n-decane	49.19	29.05		n-octane	37.81	23.68	
1,2-dichloroethane	17.13	10.86		oxygen	1.378	3.183	405.88
dinitrogen oxide	3.832	4.415		n-pentane	19.26	14.60	
ethane	5.562	6.380		phosphine	4.692	5.156	
ethanol	12.18	8.407		propane	8.779	8.445	
ethylamine	10.74	8.409		propylene	8.490	8.272	
ethyl ether	17.61	13.44		sulphur dioxide	6.803	5.636	
ethylene	4.530	5.714		xenon	4.250	5.105	768.03
helium	0.034 57	2.370	22.64	water	5.536	3.049	
n-hexane	24.71	17.35					

Table XI.14 Second Virial Coefficients of Selected Gases

The values of B quoted relate to the expansion:
$$PV_m/RT = 1 + B/V_m + C/V_m^2. \quad (1)$$
Within the temperature range quoted the value of B can, within the limits of experimental error, be fitted to an equation of the form:
$$B/(cm^3\ mol^{-1}) = a - b\ exp[c\ (K/T)]. \quad (2)$$
For more extensive data see J. H. Dymond and E. B. Smith, *The Virial Coefficients of Pure Gases and Mixtures* (1980), Clarendon, Press Oxford.

	B/cm³ mol⁻¹ at temperature/K				Coefficients of equation (2)			Temperature range/K
	100	273	373	600	a	b	c	
Ar	−187.0	−21.7	−4.2	11.9	154.2	119.3	105.1	80–1000
He	11.4	12.0	11.3	10.4	114.1	98.7	3.245	7–150
Kr		−62.9	−28.7	1.7	189.6	148.0	145.3	110–700
Ne	−6.0	10.4	12.3	13.8	81.0	63.6	30.7	44–973
Xe		−153.7	−81.7	−19.6	247.0	192.9	199.8	160–650
H_2	−2.0	13.7	15.6		315.0	289.7	9.47	14–400
D_2	−5.0	13.1	15.7		155.1	132.5	18.9	18–423
N_2	−160.0	−10.5	6.2	21.7	185.4	141.8	88.7	75–700
O_2	−197.5	−22.0	−3.7	12.9	152.8	117.0	108.8	90–400
F_2	−202	−16.4			71.4	48.0	165.0	80–300
H_2O			−451.0	−99.8	33.0	15.2	1300.7	293–1250
D_2O				−100.6	−16.8	3.37	1928.2	423–773
HCl		−174.8	−85.2		57.7	37.8	495.9	190–480
H_2S			−117.6		47.7	30.3	632.9	278–493
CO_2		−142	−72.2	−12.4	137.6	87.7	325.7	220–1100
CS_2			−497.3		211.0	167.1	538.7	280–430
CO	−192.9	−15.1	4.1		202.6	154.2	94.2	90–573
NH_3		−345	−142		44.3	23.6	766.6	273–573
N_2O		−170.7	−79.6		180.7	114.8	305.4	200–423
NO		−23.0			15.9	11.0	372.3	120–310
SO_2		−534.2	−234.2		134.4	72.5	606.5	265–473
SF_6		−344.0	−163.5		422.1	281.3	273.5	200–520
CH_4		−53.6	−21.2	8.1	206.4	159.5	133.0	110–600
CH_3OH			−680.9		123.9	17.5	1428	298–573
CH_3NH_2			−273.8		85.3	57.1	685.9	293–550
CH_3Cl			−243.6	−58.1	396.7	259.2	337.3	280–600
CCl_4			−985.8		607.6	463.0	461.0	315–420
CF_4		−112.0	−44.3	14.1	302.9	213.4	181.5	200–770
C_2H_4		−166.4	−84		278.9	193.5	231.2	180–450
C_2H_6		−219.5	−113.0	−25.4	311.7	230.6	227.8	200–600
C_2H_5OH			−806.8		864.6	45.4	1345	313–400
CH_3OCH_3		−613.9			371.0	267.4	351.3	273–330
$C_2H_5NH_2$			−438.1		533.6	289.0	452.3	293–400
C_2H_5Cl			−423.4	−110.1	182.4	88.4	717.9	273–600
CH_3CHO			−530.0		−71.8	12.3	1351	288–470
C_3H_8		−476.1	−243.1		433.6	301.7	301.3	240–550
n-C_4H_{10}		−924.1	−444.9		557.0	344.7	398.0	250–550
$C_2H_5OC_2H_5$			−555.4		968.3	346.6	552.3	280–400
n-C_5H_{12}			−714.3		694.5	407.5	462.7	300–550
n-C_6H_{14}			−1036.2		839.6	404.1	572.6	300–450
n-C_7H_{16}			−1568.0	−454.7	928.0	523.9	582.3	300–700
n-C_8H_{18}			−2122	−506.4	1138	536.1	672.5	300–700
C_6H_6			−858.2	−290	571.4	348.4	526.6	300–600
$C_6H_5CH_3$			−1532		719.1	373.3	646.4	340–580
$CH_2(CH_2)_4CH_2$			−957.0		677.3	393.2	531.4	300–560
C_5H_5N			−1073.5		796.1	398.8	576.3	350–440

Table XI.15 Solubility of Gases in Water

For more extensive data see: *Encyclopédie des Gaz* (1976), Elsevier Scientific Publishing Co and L'Air Liquide Amsterdam.
Bunsen coefficient, α/m^3 of gas at STP per m^3 water, partial pressure = 1 atm

Temperature/°C Gas	0	10	20	25	30	40	60	80	100
acetylene	1.720	1.332	1.047	0.934			0.712 (50)	0.640 (75)	
air	0.028 6	0.022 4	0.018 3	0.016 7	0.015 4	0.013 1	0.009 8	0.006	
ammonia (mass of gas/g in 1g water)	0.897 5	0.683 5	0.529			0.316	0.168	0.065	0.0298 (90)
argon	0.053 7	0.041 6	0.034 0	0.031 2	0.028 8	0.025 1	0.020 6		
bromomethane		6.34	3.75	3.16	2.80				
n-butane			0.032 5		0.023 1				
carbon dioxide	1.716 2	1.188 7	0.870 4		0.667 8	0.502 0	0.274 7		
carbon monoxide	0.035 16	0.027 82	0.022 66	0.020 76		0.016 47	0.011 97	0.007 62	
carbon tetrafluoride				0.0038					
chlorine	4.610	3.095	2.260		1.769	1.414	1.006	0.672	0.380 (90)
chloroethane			1.991						
chloromethane			3.17 (15)						
dinitrogen oxide	1.140 (5)	0.948	0.665		0.472	0.358			
ethane	0.096 0 (2)	0.066 6	0.049 2		0.037 1	0.030 4			
ethylene	0.226	0.162	0.122	0.108	0.098				
fluorine oxide	0.068		0.043						
helium	0.009 8		0.008 6		0.008 4				
hydrogen	0.021 4	0.019 3	0.017 8		0.016 3	0.015 3	0.012 5	0.008 5	
hydrogen bromide	611.6			532.1			468.6 (50)		344.6
hydrogen chloride	506	478	448		422	395	341	293	245
hydrogen cyanide	482	328	224		156	109	57		
hydrogen selenide			1.86		1.58				
hydrogen sulphide	4.67	3.399	2.582		2.037	1.661	1.190	0.917	0.840 (90)
krypton	0.099		0.059		0.049				
methane	0.054 (2)	0.043	0.035		0.029	0.025			
methylpropane			0.032 5		0.023 1				
neon	0.014		0.010		0.009 9				
nitrogen	0.023 48	0.018 75	0.015 57		0.013 43	0.011 83	0.010 27	0.009 57	0.009 47
nitrogen monoxide	0.074	0.057	0.047			0.035	0.029	0.027	0.026
oxygen	0.048 9	0.038 0	0.031 0		0.0261	0.0231	0.0195	0.0176	0.0170
ozone	0.45	0.37	0.24		0.11				
phosphine				1.183 (23)					
propane			0.039		0.029				
propylene	0.434	0.282	0.230						
sulphur dioxide	79.79	56.65	39.37		27.16	18.79			
sulphur hexafluoride		0.0076	0.0056		0.0046				
trifluoromethane				0.319					
xenon	0.203		0.108		0.085				

Table XI.16 Solubility of Some Gases in Sea Water as a Function of Temperature and Salinity
Reference: as for Table XI.15.

Salinity, %w/v	$t/°C$	Bunsen coefficient, α/ m³ of gas at STP per m³ water				
		0	4	12	16	20
argon	0	0.053 7	0.050 60	0.046 03	0.043 81	0.041 61
	10	0.041 6	0.039 50	0.036 17	0.034 58	0.033 06
	20	0.034 0	0.032 20	0.029 72	0.028 55	0.027 32
	30	0.028 8	0.027 13	0.025 19	0.024 28	0.023 40
carbon dioxide	0	1.716 2			1.478 1	1.424 7
	10	1.188 7			1.037 4	1.004 0
	20	0.870 4			0.770 2	0.748 0
	30	0.667 8			0.598 8	0.583 2
nitrogen	0	0.023 48	0.022 48	0.020 27	0.019 18	0.018 10
	10	0.018 75	0.017 83	0.016 21	0.015 42	0.014 64
	20	0.015 57	0.014 82	0.013 56	0.012 95	0.012 35
	30	0.013 43	0.012 83	0.011 79	0.011 29	0.010 79
oxygen	0	0.048 9	0.045 8	0.041 4	0.039 4	0.037 6
	10	0.038 0	0.036 4	0.033 2	0.031 8	0.030 4
	20	0.031 0	0.029 8	0.027 3	0.026 2	0.025 2
	30	0.026 1	0.025 3	0.023 3	0.022 4	0.021 6

Table XI.17 Vapour Pressure of Water at Different Temperatures. Reference: as for Table XI.13.

1. Vapour Pressure of Ice, −90 to 0 °C

Temperature /°C	Vapour pressure /N m^{-2}	Temperature /°C	Vapour pressure /N m^{-2}
−90	0.009 33	−30	38.1
−80	0.053 3	−25	63.5
−70	0.259	−20	103.5
−60	1.077	−15	165.5
−50	3.940	−10	260.0
−40	12.88	−5	401.7
		0	610.5

2. Vapour Pressure of Water, −15 to 100 °C

Temperature /°C	Vapour pressure /kN m^{-2}	Temperature /°C	Vapour pressure /kN m^{-2}
−15	0.1915	35	5.623
−10	0.2865	40	7.376
−5	0.4217	45	9.583
0	0.6105	50	12.333
5	0.8723	55	15.737
10	1.228	60	19.916
11	1.312	65	25.003
12	1.402	70	31.157
13	1.497	75	38.543
14	1.598	80	47.343
15	1.705	85	57.808
16	1.818	90	70.095
17	1.937	91	72.800
18	2.063	92	75.592
19	2.197	93	78.473
20	2.338	94	81.446
21	2.486	95	84.512
22	2.643	96	87.675
23	2.809	98	94.295
24	2.983	99	97.757
25	3.167	100	101.325
30	4.243		

3. Vapour Pressure of Water above 100°C

Temperature /°C	Vapour pressure /kN m^{-2}	Temperature /°C	Vapour pressure /kN m^{-2}
100	101.325	200	155.44
101	104.99	210	190.72
102	108.77	220	231.92
103	112.67	230	279.69
104	116.66	240	334.65
105	120.80	250	397.54
106	125.05	260	469.13
107	129.40	270	550.10
108	133.91	280	631.33
109	138.51	290	743.93
110	143.26	300	859.03
120	198.54	310	986.90
130	270.13	320	1129.06
140	361.43	330	1286.72
150	476.02	340	1461.10
160	618.08	350	1653.21
170	792.06	360	1865.08
180	1002.61	370	2102.39
190	1255.00		

Table XI.18 Vapour Pressure of Mercury from −30 to 400 °C
Reference: as for Table XI.13.

Temperature /°C	Vapour pressure /mN m⁻²	Temperature /°C	Vapour pressure /N m⁻²	Temperature /°C	Vapour pressure /kN m⁻²
−30	0.64	60	3.365	210	3.163
−20	2.41	70	6.433	220	4.284
−10	8.08	80	11.84	230	5.731
0	24.66	90	21.09	240	7.580
10	65.32	100	36.38	250	9.916
20	160.12	110	60.96	260	12.84
30	370.24	120	99.42	270	16.46
40	810.5	130	158.1	280	20.91
50	1689.2	140	246.0	290	26.34
		150	374.2	300	32.90
		160	558.5	310	40.78
		170	817.0	320	50.17
		180	1172.7	330	61.29
		190	1656.3	340	74.38
		200	2304.7	350	89.68
				360	107.49
				370	128.08
				380	151.77
				390	178.90
				400	209.86

Table XI.19 Boiling Point of Water at Different Pressures
Reference: as for Table XI.13.
For other pressures see Table XI.17 (Vapour Pressure of Water at Different Temperatures)

Pressure /mm Hg	Pressure /kN m⁻²	Boiling point/ °C	Pressure /mm Hg	Pressure /kN m⁻²	Boiling point/ °C
700	93.32	97.714	755	100.66	99.815
705	93.99	97.910	760	101.325	100.000
710	94.66	98.106	765	101.99	100.184
715	95.33	98.300	770	102.66	100.366
720	95.99	98.493	775	103.32	100.548
725	96.66	98.686	780	103.99	100.728
730	97.33	98.877	785	104.66	100.908
735	97.99	99.067	790	105.32	101.087
740	98.66	99.255	795	105.99	101.264
745	99.32	99.443	800	106.66	101.441
750	99.99	99.630			

Table XI.20 Ionization Potentials of Selected Molecules

Reference: as for Table XI.13.
$\Delta_f H$ is the enthalpy of formation of the positive ion.
Uncertainties (+/–) in parentheses.

Molecule	Ionization potential/eV	$\Delta_f H$ /kJ mol^{-1}	Molecule	Ionization potential/eV	$\Delta_f H$ /kJ mol^{-1}
H_2	15.427 (0.002)	1490	C_2H_5OH	10.49	774
D_2	15.46 (0.01)	1490	CH_3COCH_3	9.69	715
B_2H_6	12.0	1197	C_4H_4O (furan)	8.89	824
CH_4	12.6	1146	$C_2H_5OC_2H_5$	9.6	674
CD_4	12.888	1172	C_6H_5OH	8.51	724
C_2H_2	11.4	1326	C_6H_5CHO	9.52	874
C_2H_4	10.5	1059	HCOOH	11.05 (0.01)	686
C_2H_6	11.5	1025	CH_3COOH	10.36	565
cyclo-C_3H_6	10.09 (0.02)	1025	$HCOOCH_3$	10.815 (0.005)	695
C_3H_8	11.1	967	$C_4H_8O_2$ (1, 4-dioxane)	9.13 (0.03)	527
n-C_4H_{10}	10.63	900	F_2	15.7	1515
C_6H_6	9.24	975	HF	15.77 (0.02)	1251
cyclo-C_6H_{12}	9.8	824	BF_3	15.5	364
$C_6H_5CH_3$	8.82	900	PH_3	9.98	967
$C_6H_5CH=CH_2$	8.47 (0.02)	971	H_2S	10.4	983
1,2-$C_6H_4(CH_3)_2$	8.56	845	CS_2	10.080	1092
1,3-$C_6H_4(CH_3)_2$	8.58	845	SO_2	12.34 (0.02)	895
1,4-$C_6H_4(CH_3)_2$	8.44	833	CH_3SH	9.440 (0.005)	887
$C_{10}H_8$ (naphthalene)	8.12	920	C_4H_4S (thiophene)	8.860 (0.005)	958
$C_{12}H_{10}$ (biphenyl)	8.27 (0.01)	962	Cl_2	11.48 (0.01)	1109
$C_{14}H_{10}$ (anthracene)	7.55	954	HCl	12.74	1138
$C_{14}H_{10}$ (phenanthrene)	8.1	975	CCl_4	11.47 (0.01)	1004
N_2	15.576	1502	CH_3Cl	11.3	1000
NH_3	10.2	933	CH_2Cl_2	11.35 (0.02)	1004
N_2H_4	8.74 (0.06)	937	$CHCl_3$	11.42 (0.03)	1000
CH_3NH_2	8.97	841	$Cr(CO)_6$	8.03 (0.03)	−230
$(CH_3)_2NH$	8.24 (0.02)	778	$Fe(CO)_5$	7.95 (0.03)	33
$(CH_3)_3N$	7.82 (0.02)	732	$Mo(CO)_6$	8.12 (0.03)	−130
C_4H_5N (pyrrole)	8.20 (0.01)	900	$Ni(CO)_4$	8.28 (0.03)	197
C_5H_5N (pyridine)	9.3	1033	$W(CO)_6$	8.18 (0.03)	−84
$C_6H_5NH_2$	7.7	845	Br_2	10.54 (0.03)	1046
O_2	12.063 (0.001)	1163	HBr	11.62 (0.03)	1084
O_3	12.3 (0.1)	1331	CH_3Br	10.53	979
H_2O	12.6	975	CH_2Br_2	10.49 (0.02)	1008
CO	14.013 (0.004)	1243	$CHBr_3$	10.51 (0.02)	1029
CO_2	13.769 (0.03)	933	I_2	9.28 (0.02)	958
NO	9.25	983	HI	10.39	1029
N_2O	12.894	1326	CH_3I	9.54	933
NO_2	9.79	975	$Hg(CH_3)_2$	9.0	975
CH_3OH	10.84	845	$Pb(CH_3)_4$	8.0 (0.4)	908
CH_3CHO	10.2	820			

Table XI.21 Hardness of Materials
Reference: as for Table XI.13.
1. Mohs Hardness Scale

Hardness number	Original scale	Modified scale
1	talc	talc
2	gypsum	gypsum
3	calcite	calcite
4	fluorite	fluorite
5	apatite	apatite
6	orthoclase	orthoclase
7	quartz	vitreous silica
8	topaz	quartz or stellite

Hardness number	Original scale	Modified scale
9	corundum	topaz
10	diamond	garnet
11		fused zirconia
12		fused aluminia
13		silicon carbide
14		boron carbide
15		diamond

2. Hardness of Materials (on Mohs original scale)

Material	Hardness	Material	Hardness
agate	6–7	gypsum	2
alabaster	1.7	haematite	6
alum	2–2.5	indium	1.2
aluminium	2–2.9	iridium	6–6.5
amber	2–2.5	iron	4–5
anthracite	2.2	kaolinite	2–2.5
antimony	3–3.3	lead	1.5
apatite	5	lithium	0.6
arsenic	3.5	magnesium	2
asbestos	5	magnetite	6
asphalt	1–2	manganese	5
barite	3.3	marble	3–4
Bell metal	4	mica	2.8
beryl	7.8	opal	4–6
bismuth	2.5	orthoclase	6
boric acid	3	osmium	7
boron	9.5	palladium	4.8
brass	3–4	phosphorus	0.5
cadmium	2	phosphorbronze	4
caesium	0.2	platinum	4.3
calamine	5	potassium	0.5
calcite	3	pumice	6
calcium	1.5	pyrite	6.3
carbon	10	quartz	7
carborundum	9–10	rock salt	2
chromium	9	rubidium	0.3
copper	2.5–3	ruthenium	6.5
corundum	9	selenium	2
diamond	10	silicon	7
diatomaceous earth	1–1.5	silver	2.5–4
dolomite	3.5–4	silver chloride	1.3
emery	7–9	sodium	0.4
feldspar	6	steel	5–8.5
flint	7	strontium	1.8
fluorite	4	sulphur	1.5–2.5
galena	2.5	talc	1
gallium	1.5	tellurium	2.3
garnet	6–7	tin	1.5–1.8
glass	4.5–6.5	topaz	8
gold	2.5–3	Wood's metal	3
graphite	0.5–1	zinc	2.5

Table XI.22 Critical Concentrations (c.m.c.) for Micelle Formation

For more extensive data see: K. Shinoda, T. Nakagawa, B.-I. Tamamushi and T. Isemura, "Colloidal Surfactants" (1963), Academic Press, New York.

1. Ionic Surfactants

Surfactant	$t/°C$	c.m.c /mmol dm^{-3}
Anionic agents		
$C_{12}H_{25}OSO_3^-$ Na$^+$, K$^+$, Ag$^+$	25	6.2
	40	8.65
	60	9.5
Ca^{2+}, Zn^{2+}, Cu^{2+}	60	2.6
$C_{14}H_{29}OSO_3^-$ Na$^+$, K$^+$, Ag$^+$	40	2.4
	60	2.7
$C_{16}H_{33}OSO_3^-$ Na$^+$, K$^+$, Ag$^+$	25	0.40
	40	0.58
	60	0.77
$C_{18}H_{37}OSO_3^-$ Na$^+$, K$^+$, Ag$^+$	40	0.165
	60	0.42
$C_{16}H_{33}(CH_2CH_2O)_2OSO_3^-$ Na$^+$	25	0.21
$C_{16}H_{33}(CH_2CH_2O)_3OSO_3^-$ Na$^+$	25	0.1
$C_{16}H_{33}(CH_2CH_2O)_4OSO_3^-$ Na$^+$	25	0.08
$C_{18}H_{37}(CH_2CH_2O)OSO_3^-$ Na$^+$	25	0.19
$C_{18}H_{37}(CH_2CH_2O)_2OSO_3^-$ Na$^+$	25	0.08
$C_{18}H_{37}(CH_2CH_2O)_3OSO_3^-$ Na$^+$	25	0.05
$C_{18}H_{37}(CH_2CH_2O)_4OSO_3^-$ Na$^+$	25	0.04
$C_{10}H_{21}SO_3^-$ Na$^+$	60	43
		32.6
$C_{12}H_{25}SO_3^-$ H$^+$	25	8.5
Na$^+$	20	10
	60	12
K$^+$	25	9
$C_{14}H_{29}SO_3^-$ Na$^+$	20	1.6
	60	3.3
H$^+$	40	2.4
4-$C_8H_{17}C_6H_4SO_3^-$ Na$^+$	25	11
	35	14.7
	60	15
2-$C_8H_{17}C_6H_4SO_3^-$ Na$^+$	55	19
4-$C_{10}H_{21}C_6H_4SO_3^-$ Na$^+$	50	3.8
4-$C_{12}H_{25}C_6H_4SO_3^-$ Na$^+$	30	1.19
	60	1.2
4-$C_{14}H_{29}C_6H_4SO_3^-$ Na$^+$	75	0.6
C_6H_{13}⟩CHSO$_3^-$ Na$^+$ C_7H_{15}	20	16
$C_7H_{15}COO^-$ Na$^+$		351
$C_{11}H_{23}COO^-$ K$^+$	26	23
$C_{12}H_{25}COO^-$ K$^+$	25	12.5
$C_{13}H_{27}COO^-$ Na$^+$	58	7
$C_{16}H_{35}COO^-$ K$^+$	50	2.2
$C_{17}H_{35}COO^-$ K$^+$	55	0.45
$C_8H_{17}CH:CHC_7H_{14}COO^-$ K$^+$	25	1.5
4-$C_{10}F_{19}OC_6H_4SO_3^-$ Na$^+$	25	2.6
$H(CF_2)_6COO^-$ NH$_4^+$	25	250
$H(CF_2)_8COO^-$ NH$_4^+$	25	38
$H(CF_2)_{10}COO^-$ NH$_4^+$	25	9
$C_5F_{11}COO^-$ K$^+$	25	500
$C_7F_{15}COO^-$ K$^+$	25	27
$C_9F_{19}COO^-$ K$^+$	25	0.9
sodium cholate		14
sodium deoxycholate		5
sodium taurocholate		10–15
sodium taurodeoxycholate		2–6
Cationic agents		
$C_{12}H_{25}NH_3^+$ Cl$^-$	26	13
$C_{10}H_{21}(CH_3)_3N^+$ Br$^-$	25	62
$C_{12}H_{25}(CH_3)_3N^+$ Br$^-$	25	14
$C_{14}H_{29}(CH_3)_3N^+$ Br$^-$		0.28
$C_{16}H_{33}(CH_3)_3N^+$ SO$_4^{2-}$	25–35	0.55
Br$^-$	25–35	0.85
Cl$^-$	30	1.3
$C_{18}H_{37}(CH_3)_3N^+$ Cl$^-$	30	0.3
$C_{16}H_{33}(CH_3)_2(C_2H_4OH)N^+$ Cl$^-$	30	1.2
$C_{16}H_{33}(CH_3)(C_2H_4OH)_2N^+$ Cl$^-$	30	1.0
$C_{16}H_{33}(C_2H_4OH)_3N^+$ Cl$^-$	30	0.8
$C_{16}H_{33}(C_5H_5N)^+$ Br$^-$	35	0.77
Aerosols		
(OT) ($C_8H_{17}OOC)_2C_2H_3SO_3^-$ Na$^+$	25	4
(MA) ($C_6H_{13}OOC)_2C_2H_3SO_3^-$ Na$^+$	25	50
(AY) ($C_5H_{11}OOC)_2C_2H_3SO_3^-$ Na$^+$	25	100

(Continued)

2. Non-ionic Surfactants

Surfactant	$t/°C$	c.m.c. /mmol dm^{-3}
n-decyl-(β)-D-glucopyranoside		2.2
n-dodecyl-(β)-D-glucopyranoside		0.19
n-octyl-(β)-D-glucopyranoside		25
4-$C_9H_{19}C_6H_4O(CH_2CH_2O)_{9.5}H$	25	0.078–0.092
4-$C_9H_{19}C_6H_4O(CH_2CH_2O)_{10.5}H$	25	0.075–0.09
4-$C_9H_{19}C_6H_4O(CH_2CH_2O)_{15}H$	25	0.11–0.13
4-$C_9H_{19}C_6H_4O(CH_2CH_2O)_{20}H$	25	0.13–0.17
4-$C_9H_{19}C_6H_4O(CH_2CH_2O)_{30}H$	25	0.25–0.3
4-$C_9H_{19}C_6H_4O(CH_2CH_2O)_{100}H$	25	1
$C_{10}F_{19}O(CH_2CH_2O)_{23}C_{10}F_{19}$	25	0.32
Nonidet 40 (octylphenol-ethylene oxide condensate)		0.29
Triton X-100 (iso-octylphenoxypolyethoxyethanol)		0.24

Table XI.23 Physical Constants of Proteins in Water at 293 K
Reference: *Handbook of Biochemistry* (1970), CRC Press.
Sv: sedimentation coefficient (the unit 10^{-13} s is known as a svedberg); D: diffusion coefficient; $\bar{v}$: specific volume; η: intrinsic viscosity.

	$10^{13}Sv$ /s	$10^{11}D$ /m² s⁻¹	$\bar{v}$ /cm³ g⁻¹	$10^3\eta$ /m³ kg⁻¹	M_r
beef insulin	1.7	15	0.72		12 000
cytochrome c₁	1.71	11.4	0.728		13 370
lysozyme	1.91	11.20	0.703	2.7	13 930
lactalbumin	1.9	10.6	0.75		17 400
ribonuclease	2.00	13.10	0.707	3.4	12 640
myoglobin	2.06	11.3	0.749	3.1	16 890
chymotrypsinogen	2.58	9.48	0.741	2.5	23 650
ovalbumin	3.6	7.8	0.75		44 000
human serum albumin	4.3	6.15	0.735	3.7	68 460
hemoglobin	4.6	6.9	0.749	3.6	64 400
alcohol dehydrogenase	4.88	6.50	0.751		73 050
bovine serum albumin	5.04				
serum gobulin	7.1	4.0	0.75		167 000
urease	18.6	3.4	0.73		490 000
collagen		0.69	0.695	1150	345 000
myosin	6.43	1.10	0.73	217	524 800
cytochrome a		3.58	0.72		529 800
tobacco mosaic virus	185	0.53	0.72	37	40 590 000
sucrose		45.86			342

Table XI.24 Solubility of Some Compounds in Water.
For more extensive data see: W. F. Linke (ed.), *Solubilities*, (4th edn; vol. I, 1958; vol. II, 1965), American Chemical Society, Washington, DC.
Solubility is expressed as g of anhydrous compound which when dissolved in 100 g of water produces a saturated solution at the stated temperature.

1. Inorganic Compounds

Compound	Temperature/°C						
	0	10	20	40	60	80	100
NH_4Cl	29.4	33.3	37.2	45.8	55.2	65.6	77.3
$BaCl_2 \cdot 8H_2O$	31.6	33.3	35.7	40.7	46.4	52.4	58.8
$Ba(OH)_2$	1.67	2.48	3.89	8.22	20.94	101.4	
$Ca(OH)_2$	0.137	0.132	0.118	0.104	0.087	0.063	0.051
$CuSO_4 \cdot 5H_2O$	14.3	17.4	20.7	28.5	40.0	55.0	75.4
$HgCl_2$	3.6	4.8	6.5	10.3	16.3	30.0	61.3
KCl	28.07	31.23	34.2	40.0	45.8	51.3	56.3
KBr	53.5	59.5	65.2	75.5	85.5	95.0	104
KI	127.5	136	140	160	176	192	208
KOH^a	95.7	103	119	138		161	183
KNO_3	13.3	20.9	31.6	63.9	110	169	246
$AgNO_3$	117	162	218	312	441	588	733
$Na_2CO_3 \cdot 10H_2O^b$	7.0	12.5	21.5	48.5	46.4	45.8	45.5
$NaCl$	35.7	35.8	36.0	36.6	37.3	38.4	39.8
$Na_2SO_4 \cdot 10H_2O^c$	5.0	9.0	19.4	48.8	45.3	43.7	42.5

Note change of solid phase at transition points: ^{a}: $KOH \cdot 2H_2O$ at 32.5 °C; ^{b}: $Na_2CO_3 \cdot H_2O$ at 35 °C; ^{c}: Na_2SO_4 at 32.38 °C.

2. Organic Compounds

Compound	Temperature/°C						
	0	10	20	40	60	80	100
benzoic acid	0.170	0.210	0.290	0.555	1.155	2.715	5.875
fumaric acid*	0.231	0.330	0.499	1.077	2.328	5.028	
oxalic acid*	3.416	5.731	8.69	17.71	30.71	45.80	
malic acid*	28.2	34.6	41.1	54.1	64.0	73.9	
succinic acid	2.80	4.51	6.9	16.2	35.8	70.8	125
sucrose	179.2	190.5	203.9	238.1	287.3	362.1	487.2

*: g anhydrous compound per 100 g of saturated solution.

Table XI.25 Density of Aqueous Ethanol Solutions at 20 °C. Reference: Kaye and Laby, *Tables of Physical and Chemical Constants*, 15th ed. (1986), Longman, England. Densities, ρ^{20}/kg m^{-3}, are listed for increasing mass of alcohol in 100 g of solution (mass%); e.g. a 55% solution has a density of 902.6 kg m^{-3}.

	Density, ρ/kg m^{-3}									
%	0	1	2	3	4	5	6	7	8	9
0	998.2	996.3	994.5	992.7	991.0	989.3	987.8	986.2	984.7	983.3
10	981.9	980.5	979.2	977.8	976.5	975.2	973.9	972.6	971.3	970.0
20	968.7	967.3	966.0	964.3	963.2	961.7	960.2	958.7	957.1	955.5
30	953.9	952.1	950.4	948.6	946.8	944.9	943.1	941.1	939.2	937.2
40	935.2	933.1	931.1	929.0	926.9	924.7	922.7	920.4	918.3	916.0
50	913.9	911.6	909.5	907.1	904.9	902.6	900.4	898.0	895.8	893.4
60	891.1	888.8	886.5	884.2	881.8	879.5	877.1	874.8	872.4	870.0
70	867.6	865.3	862.9	860.5	858.1	855.6	853.3	850.8	848.5	845.9
80	843.6	841.0	838.7	836.0	833.5	831.0	828.4	825.8	823.2	820.6
90	818.0	815.3	812.5	809.8	807.0	804.2	801.3	798.5	795.4	792.4
100	789.3									

Table XI.26 Densities of Some Aqueous Solutions at 20 °C
Reference: as for Table XI.13. Densities, ρ^{20}/kg m^{-3}, are listed for increasing mass of anhydrous compound in 100 g of solution (wt%). The density of 0.00 wt% is the same for all solutions: 998.2 kg m^{-3}.

1. Acids

	Density, ρ/kg m^{-3}									
Acid (wt%)	10	20	30	40	50	60	70	80	90	100
acetic	1012.1	1025.0	1036.9	1047.4	1056.2	1062.9	1067.3	1068.0	1064.4	1047.7
citric	1040.2	1085.8	1134.6							
formic	1022.4	1046.7	1070.0	1093.5	1114.0	1136.4	1158.6			
hydrochloric	1047.6	1098.0	1149.2	1197.7						
lactic	1019.9	1043.9	1068.0	1091.5	1115.3	1139.2	1162.4	1184.8		
nitric	1054.3	1115.0	1180.1	1246.6						
phosphoric	1053.1	1113.5	1180.4	1253.6						
sulphuric	1066.1	1139.8	1219.1	1302.8	1395.2	1498.7	1610.5	1727.2	1814.4	1830.5

2. Bases

	Density, ρ/kg m^{-3}								
Base (wt%)	2	4	8	12	16	20	26	32	40
ammonia (NH$_3$)	989.5	981.1	965.1	950.2	936.1	922.8	904.0		
potassium hydroxide	1015.5	1033.0	1069.0	1105.9	1143.5	1181.8	1240.8	1302.0	1388.1
sodium hydroxide	1020.7	1042.8	1086.9	1130.9	1175.1	1219.2	1284.8	1348.8	1429.9

3. Alcohols, Dioxane and Other Hydroxylic Compounds

	Density, ρ/kg m^{-3}									
Alcohol (wt%)	10	20	30	40	50	60	70	80	90	100
1,4-dioxane		1016.7		1035.3 (45%)			1042.6	1038.7 (82%)		1033.7
1,2-dihydroxy-ethane	1010.8	1024.1	1037.8	1051.4	1064.4	1076.5				
glycerol	1021.5	1045.9	1071.7	1098.4	1125.4	1153.0	1181.0	1208.5	1235.2	1261.1
methanol	981.6	966.6	951.4	934.7	915.6	894.4	871.5	846.8	820.4	791.7
1-propanol	982.9	968.6	949.2	928.8	908.5	887.5	866.5	847.0	826.2	803.4
2-propanol	981.6	969.6	951.6	930.2	906.6	882.4	858.3	834.1	809.5	784.8

(Continued)

Table XI.26 (Continued)
4. Some Electrolyte Solutions

Salt (wt%)	Density, ρ/kg m⁻³								
	2	4	8	12	16	20	26	32	40
ammonium sulphate	1010.1	1022.0	1045.6	1069.1	1092.4	1115.4	1149.6	1183.3	1227.7
caesium chloride	1013.5	1029.3	1062.5	1097.8	1135.5	1175.6	1241.1	1313.5	1422.6
magnesium sulphate	1018.6	1039.2	1081.6	1125.7	1171.7	1219.8	1296.1		
potassium bromide	1012.7	1027.5	1058.1	1090.3	1124.2	1160.1	1217.9	1281.0	1374.6
potassium chloride	1011.0	1023.9	1050.0	1076.8	1104.3	1132.8			
potassium iodide	1013.1	1028.2	1059.8	1093.1	1128.4	1165.9	1227.0	1294.4	1395.9
sodium bromide	1013.9	1029.8	1063.0	1098.1	1135.2	1174.5	1238.2	1308.3	1413.8
sodium chloride	1012.5	1026.8	1055.9	1085.7	1116.2	1147.8	1197.2		

5. Some Sugar Solutions

Sugar (wt%)	Density, ρ/kg m⁻³								
	5	10	20	28	36	44	52	60	70
D-fructose	1018.1	1038.5	1081.6	1118.2	1156.8	1197.5	1240.4	1285.4	1344.3
D-glucose	1017.5	1037.5	1079.7	1115.4	1152.9	1192.4	1234.2	1279.3	
lactose	1018.4	1039.0							
maltose	1018.4	1038.5	1080.1	1117.1	1156.1	1197.9	1241.6	1285.5	
sucrose	1017.8	1038.1	1081.0	1117.5	1156.2	1197.2	1240.6	1286.4	1347.2

Table XI.27 Vapour Pressures of Selected Elements and Inorganic Compounds

The table gives the temperatures (T/K) at which the vapour of the compound at the top of the column has the vapour pressure listed at the left. s indicates the solid state.

For more extensive data see: *Handbook of Chemistry and Physics* (1991–92), CRC Press, Cleveland; and D. R. Stull, *Ind. and Eng. Chem.* (1947), **39**, 517.

Pressure p/kPa	ammonia	argon	bromine	carbon dioxide	carbon disulphide	carbon monoxide	carbonyl chloride
				T/K			
0.2	166 s	55.0 s	227 s	140 s	204	52 s	184
1	178.6 s	60.7 s	244.1 s	151.2 s	224.2	57 s	201
2	184.7 s	63.6 s	252.1 s	156.4 s	234.4	59.3 s	209
5	193.5 s	67.8 s	263.6 s	163.8 s	249.5	62.8 s	222
10	201.8 s	71.4 s	275.7	170.0 s	262.4	66.0 s	233
20	211.8	75.4 s	290	176.7 s	276.9	69.7	245
50	226.6	81.4 s	312	186.4 s	299.2	76.0	264
101.325	239.8	87.3	332	194.7 s	319.6	81.7	281
200	254.3	94.3	354	203.3 s	342	88.0	299
500	277.3	105.8	389	216.0 s	377	98.8	330
1000	298.0	116.6	421	233.0	409	108.9	359
2000	322.5	129.8	459	253.6	448	121.2	393
5000	362.0		523	287.4	512		447

Pressure p/kPa	chlorine	fluorine	helium	hydrogen	hydrogen bromide	hydrogen chloride	hydrogen fluoride
				T/K			
0.2	158 s	53	1.3	10.0 s	137 s	123 s	183 s
1	170 s	58	1.66	11.4 s	149.3 s	134.8 s	202.0
2	176.3	61	1.85	12.2 s	155.3 s	140.4 s	212.1
5	187.4	65.3	2.17	13.4 s	164.1 s	148.6 s	226.9
10	197.0	68.9	2.48	14.5	171.6 s	155.5 s	239.4
20	207.8	73.0	2.87	16.0	179.9 s	163.6	253.3
50	224.3	79.3	3.54	18.2	193.3	176.5	274.2
101.325	239.2	85.0	4.22	20.3	206.4	188.1	292.7
200	256			22.8	221	201	312
500	283			27.1	244	222	344
1000	308			31.2	265	241	371
2000	338				291	264	402
5000	387				334	301	448

(Continued)

Table XI.27 (Continued)

Pressure p/kPa	hydrogen iodide	hydrogen sulphide	krypton	neon	nitrogen	nitrogen dioxide	nitrogen monoxide
				T/K			
0.2	159 s	142 s	77 s	16.3 s	48.1 s	218 s	87 s
1	172.4 s	154.3 s	84.3 s	18.1 s	53.0 s	232.1 s	93.8 s
2	179.2 s	160.4 s	88.1 s	19.0 s	55.4 s	238.9 s	96.9 s
5	189.3 s	169.5 s	93.8 s	20.4 s	59.0 s	248.6 s	100.1
10	197.9 s	177.2 s	98.6 s	21.6 s	62.1 s	256.6 s	104.3
20	207.6 s	185.7 s	103.9 s	22.9 s	65.8	266.6	108.9
50	222.2 s	199.8	112.0 s	24.9	71.8	285.4	115.6
101.325	237.6	212.8	119.7	27.1	77.4	302.2	121.4
200	255	227	129.2	29.6	83.6	310	128
500	282	251	144.7	33.7	94.0	332	138
1000	307	273	159.2	37.5	103.8	352	148
2000	338	299	176.8	42.2	115.6	373	159
5000	390	341	206.0			404	175

Pressure p/kPa	oxygen	ozone	sulphur dioxide	sulphur hexafluoride	sulphur trioxide (α)	xenon
			T/K			
0.2	55.4	105	180 s	144 s	237 s	107 s
1	61.3	114.8	193.1 s	156.6 s	252.8 s	117.3 s
2	64.3	119.7	199.3 s	162.6 s	260.3 s	122.5 s
5	68.8	127.1	211.2	171.4 s	271.0 s	130.1 s
10	72.7	133.5	221.0	178.8 s	279.8 s	136.6 s
20	77.1	140.7	231.9	186.9 s	289.2 s	143.8 s
50	83.9	151.8	248.4	198.8 s	304.6	154.7 s
101.325	90.2	161.8	263.2	209.2 s	317.9	165.0
200	97.2		279	219.6	333	177.8
500	108.8		306	243.4	358	198.9
1000	119.6		330	265.2	381	218.6
2000	132.7		358	291.1	411	242.5
5000	154.4		404		460	282.2

Table XI.28 Vapour Pressures of Selected Organic Compounds

The table gives the temperatures (T/K) at which the vapour of the compound at the top of the column has the vapour pressure listed at the left. s indicates the solid state.

For more extensive data see: *Handbook of Chemistry and Physics* (1991–92), CRC Press; and D. R. Stull, *Ind. and Eng. Chem.* (1947), 39, 517.

Pressure p/kPa	acetaldehyde	acetic acid	acetone	aniline	benzene	bromobenzene	bromoethane	butane
					T/K			
0.2	193		217	313	240 s	282	203	174
1	210.3		237.3	337.3	258.1 s	308.1	222.2	191.8
2	219.2	297.5	247.2	349.7	266.8 s	321.1	231.7	200.4
5	232.4	314.4	261.9	368.7	279.5	340.2	245.9	213.2
10	243.7	328.9	274.4	385.3	293.1	356.6	258.0	224.0
20	256.5	345.0	288.5	404.2	308.4	375.0	271.6	236.4
50	276.1	369.4	309.9	433.0	331.9	403.3	292.6	255.4
101.325	293.5	391.2	329.6	457.6	353.2	429.4	311.5	272.7
200		416.6	351.7	485.9	377.0	459.3	333.3	292.0
500		453.4	386.1	527.9	415.6	505.6	368.1	323.1
1000		487.1	417.6	565.8	452.0	547.6	400.0	352.6
2000		525	454.1	615	494.6	600.1	437.4	389.1
5000		582	497	694			490	

Pressure p/kPa	1-butanol	2-butanol	butanoic acid	carbon tetrabromide	carbon tetrachloride	carbon tetrafluoride	chlorobenzene	chloroethane
					T/K			
0.2	279	267	307			93	265	185
1	299.8	286.7	330.9			101.6	289.8	202.9
2	309.9	296.2	342.7		259.4	106.2	302.1	211.7
5	324.8	310.2	356.0	368.0	275.3	113.0	320.4	224.8
10	337.3	322.0	374.6	385.1	289.0	118.9	336.0	236.0
20	351.3	335.1	390.8	404.3	304.5	125.6	353.6	248.6
50	372.2	354.9	415.2	434.2	328.2	135.8	380.6	267.9
101.325	390.9	372.7	436.9	462.7	349.8	145.1	405.3	285.4
200	413.0	391.4			375.1		433.3	305.6
500	445.7	420.7			414.8		478.1	337.1
1000	476.2	445.1			451.1		518.4	365.8
2000	510	477.1			495.1		566	400.5
5000								457

(Continued)

Table XI.28 (Continued)

Pressure p/kPa	chloroform	chloromethane	cyclohexane	cyclohexanol	dichloromethane	diethylamine	diethyl ether	dimethylamine
					T/K			
0.2	219		232 s	299	206		202	190
1	239.4	176.9	253.2 s	324.0	225.3	237.6	220.3	206.1
2	249.5	184.6	263.6 s	336.4	234.8	247.3	229.6	214.1
5	264.5	196.0	287.7 s	354.7	248.7	261.8	243.4	225.9
10	277.6	205.2	292.4	370.2	260.7	274.2	255.3	236.0
20	292.0	216.9	308.0	387.2	274.0	288.1	268.6	247.2
50	314.2	233.7	332.1	412.5	294.5	309.4	289.1	264.5
101.325	334.4	249.1	353.8	434.2	313.0	328.6	307.8	280.0
200	357.0	266.7	379.2			351.0	329.2	298.2
500	393.1	295.1	419.6			386.2	363.1	327.1
1000	425.5	320.4	457.2			418.5	395.0	353.1
2000	465.0	350.4	501.6			457.7	432	385.0
5000	531	398						

Pressure p/kPa	dimethyl aniline	dimethyl ether	1,4-dioxane	ethane	ethanol	ethyl acetate	ethylamine	ethylene
					T/K			
0.2	309	161		116	247	235	196	107
1	337.4	176.3		127.8	266.8	255.2	213.0	117.6
2	351.7	184.1		133.7	276.4	265.4	221.3	123.1
5	372.9	195.7	297.1	142.6	290.5	280.5	233.5	131.2
10	390.9	205.6	311.7	150.3	302.3	293.5	244.0	138.2
20	410.8	216.6	328.2	159.0	315.3	308.0	255.7	146.1
50	440.5	233.3	352.9	172.4	334.6	330.2	273.6	158.3
101.325	466.3	249.5	374.2	184.6	351.5	350.3	289.8	169.4
200		266.7		198.2	370.6	373.8	308.9	182.3
500		294.0		220.4	399.1	409.8	338.4	202.0
1000		318.6		241.1	425.0	442.8	365.0	220.3
2000		349.0		266.8	456.1	482.6	397.1	244.0
5000					502			282

(Continued)

Table XI.28 (Continued)

Pressure p/kPa	formaldehyde	formic acid	hexane	methane	methanol	methyl acetate	methylamine	naphthalene
				T/K				
0.2			223	69 s	233	221		331 s
1	182.6		243.9	76.1 s	252.8	240.0	196.3	353.4
2	190.2		254.2	79.7 s	262.5	249.8	204.0	368.2
5	201.6	295.7	269.7	85.0 s	276.6	264.2	215.4	390.1
10	211.4	310.5	283.0	89.6 s	288.4	276.4	225.1	408.8
20	222.4	326.9	297.9	95.1	301.4	290.2	235.8	429.9
50	239.1	351.7	321.0	103.7	320.7	311.2	252.2	462.1
101.325	254.1	373.8	341.9	111.6	337.7	331.8	266.8	491.2
200			366.1	120.9	357.1	352.6	283.3	
500			404.9	135.0	385.6	386.2	309.2	
1000			439.7	148.4	411.1	417.3	332.6	
2000			482.5	164.6	441.0	454.1	361.0	
5000								

Pressure p/kPa	pentane	phenol	propane	1-propanol	2-propanol	pyridine	toluene	o-toluidine
				T/K				
0.2	200	315	147	263	254	258	252	323
1	218.9	340.3	161.6	283.2	272.8	280.7	274.7	349.2
2	228.5	352.8	168.9	293.2	282.2	292.2	286.3	362.4
5	242.8	371.2	179.8	307.5	295.8	309.4	303.6	381.9
10	255.0	386.8	189.2	319.6	307.4	324.1	318.4	398.7
20	268.8	404.4	199.8	332.9	319.9	340.5	335.1	417.6
50	290.0	431.0	216.2	352.8	338.8	365.7	360.6	446.6
101.325	309.2	455.0	231.0	370.9	355.6	388.4	383.8	472.9
200	331.1	481.1	247.6	390.2	374.5		409.6	
500	365.6	521.4	274.6	422.2	403.4		451.1	
1000	398.0	557.0	300.0	450.1	428.8		488.9	
2000	437.5	601.9	331.2	484.0	493.3		535.6	
5000		674						

(Continued)

Table XI.28 (Continued)

Pressure p/kPa	o-xylene	m-xylene	p-xylene
		T/K	
0.2	275	272	
1	300.1	296.3	295.4
2	312.6	308.8	307.7
5	331.3	327.1	326.2
10	347.3	342.9	342.0
20	365.2	360.6	359.7
50	392.7	387.7	386.9
101.325	417.6	412.3	411.5
200			
500			
1000			
2000			
5000			

Table XI.29 Data on the Properties of Commercial Laboratory Reagents
Reference: as for Table XI.13.

Reagent	Density ρ/kg m^{-3}	Approx. percentage by mass	Concentration c/mol dm^{-3}	Volume to make 1 dm^3 of approx. 1 mol dm^{-3} soln /cm^3
ammonia solution	880	35	16	71
acetic acid, glacial	1049	99.8	17	58
hydrobromic acid	1490	48	9	111
hydrochloric acid	1180	35	11	90
hydrofluoric acid	1150	46	26	38
hydriodic acid	1700	55	7	143
nitric acid	1420	70	16	63
nitric acid, fuming	1500	95	24	
orthophosphoric acid	1700	85	15	67
perchloric acid	1700	70	12	83
sulphuric acid	1835	98	18	55
sulphuric acid fuming (oleum)	1920 (c. 30% SO_3)			

Table XI.30 Preparation of Volumetric Solutions
Reference: as for Table VI.26.
*: These solutions can be prepared as standard solutions. (All the remainder must be standardized.)

Reagent	Relative molecular mass	Conc. /mol dm^{-3}	Weight of reagent (g) to make 1 dm^3 of solution	Remarks
ammonium cerium(IV) sulphate $(NH_4)_4 Ce(SO_4)_2 \cdot 2H_2O$	632.6	0.1	63.26	Use 1 mol dm^{-3} H_2SO_4
ammonium iron (II) sulphate $(NH_4)_2 Fe(SO_4)_2 \cdot 6H_2O^*$	392.14	0.1	39.214	Use 1 mol dm^{-3} H_2SO_4
arsenic trioxide $As_2O_3^*$	197.84	0.025	4.946	Dissolve in 200 cm^3 20% NaOH and make up to 1 dm^3 with water
barium hydroxide $Ba(OH)_2 \cdot 8H_2O$	315.48	0.05	15.77	
cerium (IV) sulphate $Ce(SO_4)_2$	332.24	0.1	33.224	Use 1 mol dm^{-3} H_2SO_4
copper sulphate $CuSO_4 \cdot 5H_2O^*$	249.68	0.1	24.968	
iodine I_2	253.81	0.05	12.7	Use 2% KI solution
oxalic acid $H_2C_2O_4 \cdot 2H_2O^*$	126.7	0.05	6.303	
potassium dichromate $K_2Cr_2O_7$	294.19	0.02	5.884	
potassium hydrogen phthalate $KHC_8H_4O_4^*$	204.23	0.1	20.423	
potassium iodate KIO_3^*	214.00	0.02	4.28	
potassium permanganate $KMnO_4$	158.04	0.02	3.16	
potassium thiocyanate KCNS	97.18	0.1	9.718	
silver nitrate $AgNO_3^*$	169.87	0.1	16.987	
sodium carbonate $Na_2CO_3^*$	105.99	0.05	5.2995	
sodium chloride $NaCl^*$	58.44	0.1	5.844	
sodium salt of ethylenediamine tetra-acetic acid $[CH_2N(CH_2COONa) CH_2COOH]_2 \cdot 2H_2O$	372.1	0.01	3.721	Use deionized water
sodium hydroxide NaOH	40.00	0.1	4.000	
sodium oxalate $Na_2C_2O_4^*$	134.00	0.05	6.700	
sodium thiosulphate $Na_2S_2O_3 \cdot 5H_2O$	248.18	0.1	24.82	
zinc sulphate $ZnSO_4 \cdot 6H_2O^*$	269.43	0.01	2.6943	Use deionized water

Section XII
Heat and Temperature

		Page
XII.1	Volume and pressure coefficients of cubic expansion for some gases over temperature interval 273–373 K (unless otherwise stated) and various initial pressures	431
XII.2	Coefficient of cubic expansion for various liquids at temperature 293 K	431
XII.3	Variation with temperature of the coefficient of linear expansion of some solid elements	432
XII.4	Variation with temperature of the coefficient of linear expansion of some solid alloys and compounds	432
XII.5	Guide values for the coefficient of linear expansion of some miscellaneous solid substances	432
XII.6	Variation of thermal conductivity with temperature for various gases at standard pressure	433
XII.7	Thermal conductivity of some vapours and liquids at equilibrium saturation pressure	433
XII.8	Thermal conductivity for various liquids at stated temperatures	433
XII.9	Thermal conductivity of solid elements at various temperatures	434
XII.10	Thermal conductivity along various axes of some non-cubic metal crystals at standard temperature, unless otherwise stated	435
XII.11	Typical values of thermal conductivity of some alloys at various temperatures	436
XII.12	Thermal conductivity of some refractory materials at various temperatures	438
XII.13	Typical values of thermal conductivity of some everyday materials at standard temperature, unless otherwise stated	438
XII.14	Ratio of specific heat capacity at constant pressure to that at constant volume for various gases at stated temperatures and, unless otherwise stated, standard pressure	439
XII.15	Variation with temperature of the specific heat capacity at constant pressure of water	439
XII.16	Specific heat capacity at constant pressure of some metallic elements at various temperatures	440
XII.17	Typical values of the specific heat capacity at constant pressure of some alloys at various temperatures	441
XII.18	Typical values of the specific heat capacity at constant pressure of some miscellaneous substances at various temperatures	442
XII.19	Addition correction to observed surface temperature as recorded by spectral pyrometer (effective wavelength 0.65 μm) to give true surface temperature as a function of spectral emissivity	443
XII.20	Typical values of the normal emissivity of a smooth polished surface at several wavelengths for various substances at stated temperature	443
XII.21	Variation with temperature of the thermal electromotive force for various elements relative to platinum, one junction being at a constant temperature of 273 K	444
XII.22	Variation with temperature of the thermal electromotive force for some alloys relative to platinum, one junction being at a constant temperature of 273 K	445
XII.23	Variation with temperature of the absolute negative thermo-electric power of platinum	445
XII.24	Fixed points of the International Temperature Scale of 1990	445
XII.25	Definition of the International Practical Temperature Scale of 1990	446

Section XII
Heat and Temperature

Table XII.1 Volume and pressure coefficients of cubic expansion for some gases over temperature interval 273–373 K (unless otherwise stated) and various initial pressures

Gas	initial pressure/MN m^{-2}	coefficients/10^{-3} K^{-1} volume	coefficients/10^{-3} K^{-1} pressure	Gas	initial pressure/MN m^{-2}	coefficients/10^{-3} K^{-1} volume	coefficients/10^{-3} K^{-1} pressure
air	0.031		3.6643 over 273–1340 K	N$_2$	†0.1333	3.6735	3.6744
	†0.1333	3.6728	3.6744		20.3	4.34	
	2.67		3.887		101	2.18	
Ar	0.069		3.668	NH$_3$	0.1013	3.854 over 273–323 K	
CO	0.031		3.6648 over 273–1340 K	N$_2$O	0.1013	3.719	3.676
	0.1013	3.669	3.667	Ne	†0.1333	3.66	3.6617
CO$_2$	0.032		3.6756 over 273–1340 K	O$_2$	0.024–0.031		3.6652 over 273–1340 K
	0.069	3.7073	3.6981		0.088		3.6738
	0.133	3.741	3.7262		10.1	4.86	
H$_2$	†0.1333	3.6588	3.662	SO$_2$	0.1013	3.903	3.845
He	†0.1333	3.658	3.6605				

† indicates a thermometric gas
Reference: Kaye and Laby, *Tables of Physical and Chemical Constants*, 15 ed., Longmans, 1986

Table XII.2 Coefficient of cubic expansion for various liquids at temperature 293 K

Liquid	coefficient/10^{-5} K^{-1}	liquid	coefficient/10^{-5} K^{-1}	liquid	coefficient/10^{-5} K^{-1}
Br$_2$	112	CS$_2$	119	C$_6$H$_6$	121
CCl$_4$	122	C$_2$H$_5$Br	141	glycerol	49
CHCl$_3$	127	C$_2$H$_5$OC$_2$H$_5$	163	ethylene glycol	57
CH$_3$(CH$_2$)$_2$CH$_3$	158	C$_2$H$_5$OH	109	H$_2$O	21
CH$_3$COCH$_3$	143	C$_6$H$_4$(CH$_3$)$_2$	99	H$_2$SO$_4$ (100%)	56
CH$_3$COOH	107	C$_6$H$_5$CH$_3$	107	Hg	18.2
CH$_3$I	120	C$_6$H$_5$NH$_2$	85	turpentine	96
CH$_3$OH	118				

Reference: Kaye and Laby, *Tables of Physical and Chemical Constants*, 15 ed., Longmans, 1986

Table XII.3 Variation with temperature of the coefficient of linear expansion of some solid elements

Element	temperature/K 100	293	500	800	Element	temperature/K 100	293	500	800	Element	temperature/K 100	293	500	800
	coefficient/10^{-6} K^{-1}					coefficient/10^{-6} K^{-1}					coefficient/10^{-6} K^{-1}			
Ag	14.2	18.9	20.6	23.7	Cu	10.3	16.5	18.3	20.3	Rh	5	8.2	9.3	10.8
Al	12.2	23.1	26.4	34	Fe	5.6	11.8	14.4	16.2	Sb	9.1	11	11.7	11.7
Au	11.8	14.2	15.4	17	Ge	2.4	5.7	6.5	7.2	Si	−0.4	2.6	3.5	4.1
B		4.7	5.4	6.2	In	25.4	32.1			Sn	16.5	22	27.2	
Be	1.3	11.3	15.1	19.1	Ir	4.4	6.4	7.2	8.1	Ta	4.8	6.3	6.8	7.2
Bi	12.3	13.4	12.7		Mg	14.6	24.8	29.1	35.4	Ti	4.5	8.6	9.9	11.1
C (diamond)	0.05	1	2.3	3.7	Nb	5.2	7.3	7.8	8.2	Tl	25.2	29.9	34.7	
C (glassy)		3.1	3.3	3.6	Ni	6.6	13.4	15.3	16.8	U	10	13.9	16.9	24.3
Cd	26.9	30.8	36		Pb	25.6	28.9	33.3		V	5.1	8.4	9.9	10.9
Co	6.8	13	15	15.2	Pd	8	11.8	13.2	14.5	W	2.6	4.5	4.6	5
Cr	2.3	4.9	8.8	11.8	Pt	6.6	8.8	9.6	10.3	Zn	24.5	30.2	32.8	

Reference: Touloukian and Ho (eds) *Thermophysical Properties of Matter*, vols. 12 and 13, IFI/Plenum, 1974

Table XII.4 Variation with temperature of the coefficient of linear expansion of some solid alloys and compounds

† Material	temperature/K 100	293	500	800	† Material	Temperature/K 100	293	500	800
	coefficient/10^{-6} K^{-1}					coefficient/10^{-6} K^{-1}			
aluminium bronze (5 Al, 90.5 Cu, 4.5 Ni)	12–14	15.9	18.1	20.3	MgF$_2$ (poly)	2.2	11.1	13.3	16.8
brass (67 Cu, 33 Zn)		17.5	20	22.5	nickel steel (50 Fe, 50 Ni)		9.4	9.6	12.5
bronze (85 Cu, 15 Sn)		17.3	19.3	21.9	phosphor bronze (90 Cu, trace P, 10 Sn)		17	20	
cast iron (3 C, 95 Fe, 2 Si)		11.9	13.1	14.5	quartz, ∥ axis	4	6.8	11.4	31.4
Constantan (65 Cu, 35 Ni)	11.2	15	17.4	19.2	quartz, ⊥ axis	9.1	12.2	19.5	37.6
cupro-nickel (31 Cu, 1.5 Fe, 1 Mn, 66.5 Ni)	9.8	12.7	15.4	18.2	quartz, polycrystalline	7.3	10.3	16.8	32.2
duralumin (95–96 Al, 4–5 Cu)	13.1	21.6	27.5	30.1	SiC	0.14	3.3	4.2	4.9
GaAs	1.9	5.7	6.5	7.1	stainless steel (18 Cr, 74 Fe, 8 Ni)	11.4	14.7	17.5	20.2
GaP		4.7	5.5	6	stainless steel (13–17 Cr, 83–87 Fe)	6	9.5	12.1	13.8
Inconel	8.7	11.6	14.4	17.6	steel, carbon (0.7–1.4 C, 98.6–99.3 Fe)	6.7	10.7	13.7	16.2
InSb	2.8	5	6.1		stellite (65 Co, 25 Cr, 10 W)	6.9	11.2	14.6	17.2
Invar (64 Fe, 36 Ni)	1.4	0.13	5.1	17.1	WC (poly)		3.7	4.3	4.8
MgF$_2$, ∥ axis	3.9	14.5	17	19.2					
MgF$_2$, ⊥ axis	1.4	9.5	11.5	15.8					

† Mass percentage composition given in brackets where appropriate
Reference: as for Table XII.3

Table XII.5 Guide values for the coefficient of linear expansion of some miscellaneous solid substances

Substance	coefficient/10^{-6} K^{-1}	Substance	coefficient/10^{-6} K^{-1}
Alundum	5.5	porcelain	2–6
borosilicate crown glass	7–8	Portland stone	c 3
brick	3–10	Pyrex	2.8
concrete	7–14	sandstone, slate	5–12
dense flint glass	8–9	Si, fused	0.4–0.55
granite, limestone	4–10	woods across grain	35–60
marble	3–15	woods along grain	3–6
plastics	see Table XI.8	Zerodur	0.1
polytetrafluoroethylene	525		

Reference: as for Table XII.3

Table XII.6 Variation of thermal conductivity with temperature for various gases at standard pressure

Gas	temperature/K conductivity/10^{-2} W m^{-1} K^{-1}					Gas	temperature/K conductivity/10^{-2} W m^{-1} K^{-1}				
	73	173	273	373	1273		73	173	273	373	1273
Air		1.58	2.41	3.17	7.6	He	5.95	10.45	14.22	17.77	41.9
Ar		1.09	1.63	2.12	5	Kr		0.57	0.87	1.15	2.9
Br			0.4	0.6		N_2		1.59	2.4	3.09	7.4
CH_3CH_3			1.8			NH_3			2.18	3.38	
CH_4		1.88	3.02			NO		1.54	2.38		
C_3H_4			1.64			N_2O			1.51		
Cl_2			0.79	1.15		Ne	1.74	3.37	4.65	5.66	12.8
CO		1.51	2.32	3.04		O_2		1.59	2.45	3.23	8.6
CO_2			1.45	2.23	7.9	$R_{12}(CF_2Cl_2)$			0.85	1.35	
F_2		1.56	2.54	3.47		Rn			0.33	0.45	
H_2	5.09	11.24	16.82	21.18		SO_2			0.77		
H_2O			1.58	2.35		Xe		0.34	0.52	0.7	1.9
H_2S			1.2								

Reference: Touloukian and Ho (eds), *Thermophysical Properties of Matter*, vol. 3, IFI/Plenum Data. Corp. New York, 1970

Table XII.7 Thermal conductivity of some vapours and liquids at equilibrium saturation pressure

Substance	Ar	C_6H_6	H	H_2O	He	KNO_3
temperature/K	90	298	20	373	4	683
conductivity/10^{-2} W m^{-1} K^{-1}						
for vapour	0.57	0.7	1.45	2.17	0.0125	
for liquid	11.98	14.63	11.78	68.19	2.75	42.5

Reference: as for Table XII.6

Table XII.8 Thermal conductivity for various liquids at stated temperatures

Liquid	† conductivity/10^{-2} W m^{-1} K^{-1}	Liquid	† conductivity/10^{-2} W m^{-1} K^{-1}
Al	9200 (973)	glycerine	28.6 (273), 29.2 (333)
Ar	12.6 (84.2), 12.16 (87.3)	H_2O	56.1 (273), 67.3 (353), 68.6 (378–433), 59.8 (542)
Bi	1300 (573), 1700 (973)	Hg	780 (273), 940 (373), 1170 (573)
CCl_4	11.5 (253), 10.2 (333)	I	9 (573)
CCl_2F_2	9 (253), 7.3 (293)	In	4200 (573)
CH_3COCH_3	19.8 (193), 14.6 (333)	K	5300 (373), 4500 (573), 3200 (973)
CH_3OH	22.3 (233), 18.6 (333)	Li	4700 (573), 5900 (973)
CH_4	21.53 (93), 19.64 (108)	N	15.11 (69), 14.8 (71.4)
C_2H_5OH	18.9 (233), 15 (353)	Na	8800 (373), 7800 (573), 6000 (973)
n-C_3H_7OH	16.8 (233), 14.8 (353)	oil, cylinder	15.2 (293), 14.2 (473)
n-C_4H_9OH	16.7 (213), 10.6 (353)	oil, transformer	13.6 (273), 12.7 (373)
$C_6H_5CH_3$	15.9 (193), 11.9 (353)	P (yellow)	18 (373), 16 (573)
$C_6H_5C_2H_5$	15.2 (193), 11.7 (353)	paraffin, medicinal	12.7 (273), 12.5 (423)
$C_6H_5NH_2$	17.2 (293)	Pb	2100 (973)
C_6H_6	14.7 (293), 13.7 (323)	Rb	3200 (373), 2900 (573), 2200 (973)
CO	15.89 (72), 14.21 (80.8)	S	17 (573)
Cd	4500 (973)	Sb	2700 (973)
Cs	2000 (373), 2060 (573), 1770 (973)	Sn	3200 (573), 4000 (973)
ethylene glycol	25.2 (273), 26.4 (373)	Te	630 (973)
Ga	3300 (373), 4500 (573)	Xe	7 (173), 5 (223)
		Zn	6600 (973)

† Figures in brackets are temperatures in kelvin.
Reference: Kaye and Laby, *Tables of Physical and Chemical Constants*, 15 ed., Longmans, 1986

Table XII.9 Thermal conductivity of solid elements at various temperatures

Element	\\	temperature/K \\	\\	\\	\\
	173	273	373	573	973
		conductivity/ W m^{-1} K^{-1}			
Ag	432	428	422	407	377
Al	241	236	240	233	
As		53.9	42.7		
Au	324	319	313	299	272
B	72	32	19	11	10
Be	367	218	168	129	93
Bi	11	8.2	7.2		
C (amorphous)	1.1	1.5	1.8	2.2	2.5
C (diamond)	1700–4900	1000–2600	700–1700		
C (graphite)	70–220	80–230	75–195	50–130	35–70
C (pyrolytic graphite, ∥ planes)	3870	2130	1510	936	549
C (pyrolytic graphite, ⊥ planes)	10.8	6.4	4.4	2.8	1.6
Ca		206	192		
Cd	100	97	95	89	
Ce	8	11	13	16	
Co	130	105	89	69	53
Cr	120	96.5	92	82	66
Cs	37	36			
Cu	420	403	395	381	354
Dy	9	10.5			
Er	14	15			
Eu		14			
Fe	99	83.5	72	56	34
Ga	43	41			
Gd	12	10			
Ge	113	67	46.5	29	17.5
Hf	25	23	22	21	21
Hg	29.5				
Ho	14	16	17		
I		0.5	0.4		
In	92	84	76		
Ir	156	147	145	139	
K	105	104			
La	12	13	14.5		
Li	94	86	82		
Lu	18	17			
Mg	160	157	154	150	
Mn	7	8			
Mo	145	139	135	127	113
Na	141	142			
Nb	53	53	55	58	64
Nd		16.5	16.7		
Ni	113	94	83	67	71
Np		6.3 at 300 K			
Os	93	88	87	87	
P (black)	20	13			
P (white or yellow)		0.25			
Pb	37	36	34	32	
Pd	72	72	73	79	93
Pm			18.4		
Pr	9.9	12	13.4		
Pt	73	72	72	73	78
Pu	4	6	8		
Ra		18.6 at 293K			
Rb	59	58			
Re	52	49	47	44	45
Rh	156	151	147	137	

(Continued)

Table XII.9 (Continued)

Element	temperature/K				
	173	273	373	573	973
	conductivity/ W m^{-1} K^{-1}				
Ru	123	117	115	108	98
S (amorphous)	0.18	0.2			
S (polycrystalline)	0.39	0.29	0.15		
Sb	33	25.5	22	19	
Sc	15	16			
Se (amorphous)	0.23	0.43			
Se (crystalline, ∥ c axis)	6.8	4.8	4.8		
Se (crystalline, ⊥ c axis)	2	1.4	1.4		
Si	330	168	108	65	32
Sm	10	13	13	14	
Sn	76	68	63		
Sr		36.4	32.5		
Ta	58	57	58	58.5	60
Tb	11	10.5			
Tc		51	50	50	
Te (∥ c axis)	5.1	3.6	2.9	2.4	
Te (⊥ c axis)	2.9	2.1	1.7	1.5	
Th	55	54	54	56	58
Ti	26	22	21	19	21
Tl	51	47	44		
Tm	16	17			
U	24	27	29	33	43
V	32	31	31	33	38
W	188	177	163	139	119
Y	16.5	17			
Yb		35.4	34.3		
Zn	117	117	112	104	
Zr	26	23	22	21	23

Reference: Touloukian and Ho (eds), *Thermophysical Properties of Matter* (vols. 1 and 2), IFI/Plenum Data. Corp. New York, 1970

Table XII.10 Thermal conductivity along various axes of some non-cubic metal crystals at standard temperature, unless otherwise stated.

Metal conductivity/W m^{-1} K^{-1}	Bi	Cd	Dy	Er	Ga	Gd	Hg at 228 K	Ho	Lu	Sn	Tb	Tm
a axis	9.3	104	10.25	12.6	40.8	10.3	25.9	13.6	13.8	74	9.45	14.1
b axis					88.3							
c axis	5.4	83.05	11.65	18.4	16	10.7	33	22.1	23.3	51.8	14.5	24.2

Reference: Kaye and Laby, *Tables of Physical and Chemical Constants*, 15 ed., Longmans, 1986

Table XII.11 Typical values of the thermal conductivity of some alloys at various temperatures

Alloy †	\multicolumn{11}{c}{temperature/K}
	4	20	50	100	173	273	373	573	773	973	1273
	\multicolumn{11}{c}{conductivity/W m^{-1} K^{-1}}
Al ALLOYS:											
Al	1075	4100	870	295		232	240	233			
Alpax gamma						188	188	184			
1 C	45	225	315	228		205					
Lo-Ex						172	175	173			
N 3	11	58	117	143	158	170					
N 8	3.3	17	39	66	92	110					
RR 57						161	171	178			
RR 59						168	176	186			
T 61	14.7	30	64	96		150					
T 81		26	46	68		118					
Y						180	188	194			
Cu ALLOYS:											
Cu	1450	3700	1230	480		401	395	381		354	
brass A	4	21	45	70	92	106	128	146			
brass B				59		113					
bronze						53	60	80			
Constantan						22	24	27			
Cu-Be	2	10	24								
cupro-nickel	1	14	31								
German silver					20	23	29	45			
manganin						21	26				
Ni ALLOYS:											
Ni	138	865	320	154	115	94	83	67		71	
Alumel						30	32	35			
Chromel P						19	23				
Hastelloy R-235							11.7	14.8	17.6	20	25.5
Hastelloy X	0.5	3	5	6.5	8	9.9					
Incoloy 800						11.3	12.8	16.4	19.4	22.8	31
Incoloy 802						11.3	13.1	16.2	19.2	22.1	26
Inconel 600						14.6	15.8	19.1	22.1	25.7	
Inconel 718	0.4	2.6	4.8	7	8.7	10					
Inconel X					8.7	10	11.4				
Inconel X-750						11.3	13	16.5	20.1	23.6	27.9
K Monel					12	14	17				
Nimonic 75						13.9	17.5	21	24.3		
Nimonic 80							12.1	15.5	18.4	23.5	
Nimonic 90							13	16.5	20	23.7	
Nimonic 105							11.6	14.7	17.4	21.2	27.6
STEELS:											
CARBON											
A						59	58	49	40	32	28
B						52	51	46	39	32	27
C						52	51	46	38	30	27
D						51	49	42	36	31	27
E						45	45	40	35	28	26
F						46	46	43	37	31	27
HIGH ALLOY											
A						13	15	18	21	23	26
B						13	15	18	21	23	28
C						24	26	28	28	27	28
Kovar						14.1	14.7	15.6	17.5	19.3	

(Continued)

Table XII.11 (Continued)

Alloy †	4	20	50	100	173	273	373	573	773	973	1273
						conductivity/W m^{-1} K^{-1}					
LOW ALLOY											
Armco iron	13	65	109	95	83.6	76					
A						49	46	42	36	29	28
B						43	43	41	36	31	28
C						37	38	37	34	29	
D						36	38	38	34	28	28
E						34	36	37	35	29	28
F						33	36	37.5	35	28	
G						33	34	36	34	28	28
H						25	28	31	31	28	26
NICKEL											
A						33	38				
B					21	29	34				
C			10.5	19	26	31					
D	1	3.5	8.5	16	23	28					
Invar		1.6	4.3	7.7	11	13.5					
STAINLESS											
A	0.4	2.4	6	9.3	11.5	13.6	15.7	18.9	21.5	23.8	26.8
B	1	2.2	5	7.6		11.2					
Era-ATV						11	12.5	15.5		21.5	
304, 321, 347	0.3	2	5.5	9	11.5	14.5	16.5	20	22.5	25.5	29.5
310		1.7	4.3	6.9	8.3	11.5	13.5	17.5	20.5	23	
316	0.3	2	5.5	9	11.5	14	15	18.5	21.5	24	28.5
403, 405, 409							25	26	27		
410, 420								23.6	24.6	26.3	28
430, 434							22.2	22.9	23.7	24.4	
Ti ALLOYS:											
Ti	14	28	40	31	26	22					
A		2	3.6	4.8	6.1	7.4	8.3	10.5			
B					9.3	10.5	10.7				
C		1.3	2.5	4	5.4	7					
D		0.9	1.9	3.4	5	7.4					
UNCLASSIFIED ALLOYS:											
A						31					
B						38					
C						46	51	58	69		
D							8.7	12			
E							11.3	13			

† For mass percentage composition, see Table XVII.3
Reference: as for Table XII.9 (vol.1)

Table XII.12 Thermal conductivity of some refractory materials at various temperatures.

Material †	temperature/K					Material †	temperature/K				
	298	373	773	1273	1773		298	373	773	1273	1773
	conductivity/W m^{-1} K^{-1}						conductivity/W m^{-1} K^{-1}				
AlN	36	33	23			Si	150	110	45	26	
Al_2O_3	38	35	11	7	6	SiC	110	90	65	45	40
95% Al_2O_3	23	13	9	6	5	Si_3N_4 + 1% MgO	30	28	21	14.5	13
90% Al_2O_3	17	12	7	5	4	SiO_2 (vitreous*)	1.6	1.7	2.1	5	
85% Al_2O_3	15	12	7	4	3.5	SiO_2 (quartz)					
$Al_6Si_2O_{13}$		6	4.5	4		single crystal,					
B_4C	30	25	21	17	15	$\parallel c$ axis	11	8.3	5		
BeO	300	220	70	18	14	$\perp c$ axis	6.5	5	3.6		
C (graphite)	108	107	76			ThO_2	14	12	6	2	2
CaO		15	8.7	7.8		ThO_2 (sintered)	8–10	6–8	3–5	2–3	
$MgAl_2O_4$	16	15	9	6		TiB_2		70	64		
MgO	40	35	16	7	6.5	TiC*	30	32	36	40	45
30% porous MgO	10–14	5–8				TiN		25	27		
						TiO_2		9.2	4.5	3.3	
PORCELAINS & CLAY-BASED:						TiO_2 (sintered)	2.5–4				
cordierite	1.5–2.5	1.5–2.5				UO_2	12	8	4.5	3.2	
mullite	2–6	2–6				UO_2 (sintered)	8–10	6–8	4–5	2–3	2
refractory	2.4	2.3	2.1	2		WC*	40			45	50
siliceous	1.7–2.1	1.7–2	1.8–2	1.9–2		ZrB_2		73	67		
steatite	5.5–6		2.8–3.7			ZrC			31	35	
zircon	7	6	4	3.5		ZrO_2 (stabilized)	1.8	1.8	2	2.2	2.4

† Dense polycrystalline single-phase compounds, unless otherwise stated
* Direct transmission of radiation influences values at high temperatures
Reference: Morrell, *Handbook of Properties of Technical and Engineering Ceramics, Part I*, HMSO, 1985.

Table XII.13 Typical values of thermal conductivity of some everyday materials at standard temperature, unless otherwise stated

Material	conductivity/ W m^{-1} K^{-1}	Material	conductivity/ W m^{-1} K^{-1}	Material	conductivity/ W m^{-1} K^{-1}
asbestos board	0.11	glass	0.55–1.1	plasticine	0.65–0.8
beeswax	0.25	glass fibre	0.03–0.07	plastics	see Table XI.8
bitumen	0.17	hardboard	0.125	plywood	0.125
brick	0.8–1.2	ice at 173 K	3.9	porcelain	1.5
cardboard	0.21	ice at 268 K	2.3	rubber	0.15
charcoal & coal	0.2	insulating fibreboard	0.055	rubber (foam)	0.045
concrete (low density)	0.1	kapok	0.035	silver sand	0.3–0.4
concrete (high density)	1.8	mica	0.6–0.7	soil (clay)	1.1
cotton wool	0.03	mineral wool	0.04	wood	0.055–0.17
cork	0.04	paper	0.06	vermiculite granules	0.065
ebonite (solid)	0.17	paraffin wax	0.25	wool	0.05
felt	0.04	plaster board	0.16		

Reference: Kaye and Laby, *Tables of Physical and Chemical Constants*, 15 ed., Longmans, 1986

Table XII.14 Ratio of the specific heat capacity at constant pressure to that at constant volume for various gases at stated temperatures and, unless otherwise stated, standard pressure

Gas	temperature/K	ratio	Gas	temperature/K	ratio	Gas	temperature/K	ratio
air (dry)	193.7	1.405	CO_2	277–284	1.3	H_2O (vapour)	373	1.334
	273–290	1.401/2		573	1.22	H_2S		1.34
	773	1.357		773	1.2	He	273	1.63
	1173	1.32	CS_2		1.239	Hg (vapour)	583	1.666
at 20 MPa	193.7	2.333	C_2H_2		1.26	Kr	292	1.689
	273	1.828	C_2H_4		1.264	NH_3		1.336
Ar	273	1.667	C_2H_5Br		1.188	NO		1.394
CCl_4		1.13	C_2H_5Cl	295.7	1.187	NO_2	293	1.172
$CHCl_3$	297–315	1.11	$C_2H_5OC_2H_5$	285–293	1.024		423	1.31
	372.8	1.15		372.7	1.112	N_2	293	1.401
CH_3Br		1.274	C_2H_5OH	326	1.133	N_2O		1.324
CH_3COOH	409.5	1.147		372.8	1.134	Ne	292	1.642
CH_3Cl	292–303	1.279	C_2H_6		1.22	O_2	278–287	1.4
CH_3I		1.286	C_3H_8		1.13	O_3		1.29
CH_3OH	372.7	1.256	C_6H_6	293	1.4	SO_2	289–307	1.26
CH_4		1.313		372.7	1.105		773	1.2
CO	2073	1.297	H_2	277–290	1.407/8	Xe	292	1.666

Reference: Kaye and Laby, *Tables of Physical and Chemical Constants*, 15 ed., Longmans, 1986

Table XII.15 Variation with temperature of the specific heat capacity at constant pressure of water

Temperature/K	specific heat capacity/ J kg^{-1} K^{-1}	Temperature/K	specific heat capacity/ J kg^{-1} K^{-1}	Temperature/K	specific heat capacity/ J kg^{-1} K^{-1}
273	4217.4	308	4177.9	343	4189.3
278	4201.9	313	4178.3	348	4192.5
283	4191.9	318	4179.2	353	4196.1
288	4185.5	323	4180.4	358	4200.2
293	4181.6	328	4182.1	363	4204.8
298	4179.3	333	4184.1	368	4210
303	4178.2	338	4186.5	372	4214.5

Reference: Kaye and Laby, *Tables of Physical and Chemical Constants*, 15 ed., Longmans, 1986

Table XII.16 Specific heat capacity at constant pressure of some metallic elements at various temperatures

Metal	temperature/K						Metal	temperature/K					
	77	173	273	373	573	773		77	173	273	373	573	773
	specific heat capacity/J kg^{-1} K^{-1}							specific heat capacity/J kg^{-1} K^{-1}					
Ag	162	219	235	239	249	257	Pd			130	132	136	
Al	336	743	880	937	1000	1130	Pt	85	123	132	135	141	146
Au	97	121	128	131	135	140	Pu			146	c 170	c 200	c 180
Be	80	852	1750	2010	2440	c 2880	Rb			346			
Bi	103	119	123				Re	78	125	138	142	145	146
Cd	179	218	229	238	255		Rh	88	209	238	252	269	
Ce			190	199	219		Ru	94		231	237	251	
Co			411	443	490	540	Sb	150					c 280
Cr	117	351	438	481	527	550	Sm		174	218	263	296	
Cu	195	341	379	397	419	c 430	Sn	170	211	221	243		
Er	192	140	166				Ta	95	130	139	143	150	152
Fe (α)	144	336	435	480	560	670	Tb	178	255	186	180		
Gd	166		300	350	380	390	Th	91		113	120	130	140
Hf	101		142	148	157	164	Ti	218	438	511	536	c 580	c 620
Hg	115	132					Tl	115	126	131	138		
Ho			164				Tm	146	156	159	161		
In	191	222	231				U	85	106	114	119	144	167
K			732	812	770		V	188	409	489	510	531	573
La	156			194	210		W	68	120	133	135	140	143
Li	1320		3480				Y	200	273	294	304	320	
Lu		602	640				Zn	251	357	385	402	437	
Mg	488	887	1000	1060	1140	1270	Zr	170	257	273	287	296	c 310
Mn (α)	201	394	467	502	565	630							
Mo	96	209	242	256	269	280							
Na	900	1110	1210										
Nb	167	247	265	272	280	290							
Ni	163	357	429	465	569	527							
Os			130	132	136								
Pb	114	124	129										

Reference: Kaye and Laby, *Tables of Physical and Chemical Constants*, 15 ed., Longmans, 1986

Table XII.17 Typical values of the specific heat capacity at constant pressure of some alloys at various temperatures

Alloy†	\multicolumn{6}{c}{temperature/K}
	77	173	273	373	573	773
	\multicolumn{6}{c}{specific heat capacity/J kg^{-1} K^{-1}}
Al alloys c.90 Al	480	690	820	910	1030	
Alumel (2 Al, 25 Mn, 72 Ni)			481	496		
brass	200	340	387	390	448	
Chromel (10 Cr, 90 Ni)				473	521	
Constantan (60 Cu, 40 Ni)	175		c 400	420	450	
Inconel (15 Cr, 10 Fe, 75 Ni)			440	465	506	
Monel (29 Cu, 4 Fe, 67 Ni)		350	406	448	489	514
Nichrome			432	464	509	548
solder (50 Pb, 50 Sn)	142	167	177			
STEELS:						
high alloy (12 Cr)				480	570	700
low alloy				480–510	550–590	670–700
mild				490	560	680
nickel:						
3.5 Ni				496	569	694
5 Ni	155	365	445			
9 Ni	154	364	454		557	
20 Ni					536	654
30 Ni					546	570
50 Ni					561	582
stainless 18 Cr, 74 Fe, 8 Ni				519	555	611
18 Cr, 70 Fe, 12 Ni	197	401	470			
24 Cr, 56 Fe, 20 Ni	195	393	463			
tool (1.2 C, 85 Fe, 13 Mn, traces)				500	570	650
Ti alloy (6 Al, 90 Ti, 4 V)	210	450	527	565	612	

†Numbers refer to approximate mass percentage composition; more details in Table XVII.3.
Reference: Kaye and Laby, *Tables of Physical and Chemical Constants*, 15 ed., Longmans, 1986

Table XII.18 Typical values of the specific heat capacity at constant pressure of some miscellaneous substances at various temperatures

Substance	temperature/K					
	77	173	273	373	573	773
	specific heat capacity/J kg^{-1} K^{-1}					
Al_2O_3	61	400	730	910		
asbestos			mean value c 840 for 293–373 K			
B	64	480	c 960	1370	1840	2220
basalt			mean value c 840–1000 for 293–373 K			
				1000		
BN	150	430	720	990	1380	1600
C (diamond)	8	140	420	770	1300	1590
C (graphite)	9	340	640	940	1380	1600
CaF_2			830	930	990	1070
$CaSiO_3$			680	840	970	1030
†cermets: BN + C				1040	1410	1620
99 Be, 1 BeO				2420	2560	2700
10 Be, 90 BeO				1300	1650	1830
20 C, 80 SiC				870	1100	1220
77 C, 23 SiC				960	1270	1460
5 Co, 95 WC				240	270	290
ebonite			mean value 1380 for 293–373 K			
glass: crown			mean value c 670 for 283–323 K			
flint			mean value c 500 for 283–323 K			
Pyrex			700	850	1100	
ice	690	1380	2100			
KCl	440	630	680			
marble (white)			c 900			
MgO	100	570	880	1030	1120	1220
NaCl	480	770	840			
paraffin wax			c 2900			
porcelain			mean value c 750 for 288–473 K and			
			mean value c 1070 for 288–1273 K			
plastics	see section XI, Table 8					
pyroceram			710	900	1030	1120
rubber: india			mean value 1100–2000 for 293–373 K			
natural	520	1000	1800			
Si	180	490	680	770	850	880
SiC	50			820	1010	1120
SiO_2: fused				c 840		1200
quartz	190	470	730	860	1060	1110
quartz glass			700	830	1020	
sand			c 800			
Teflon	300	650	1020			
WC					c 240	

†Numbers refer to mass percentage composition
Reference: Kaye and Laby, *Tables of Physical and Chemical Constants*, 15 ed., Longmans, 1986

Table XII.19 Addition correction to observed surface temperature as recorded by spectral pyrometer (effective wavelength 0.65 μm) as a function of spectral emissivity

observed temperature/K	emissivity								
	0.1	0.2	0.3	0.4	0.5	0.6	0.7	0.8	0.9
	addition/K								
873	87	59	44	33	25	18	12	8	4
1073	135	91	67	50	37	27	19	12	6
1273	194	130	95	71	53	39	27	17	8
1473	267	177	128	96	71	52	36	22	10
1673	353	232	168	125	93	67	46	29	13
1873	453	295	212	157	117	85	58	36	17
2073	570	368	263	195	144	104	72	44	21
2273	704	450	321	236	174	126	86	53	25
2773	1124	700	493	360	264	190	130	80	37
3273	1690	1022	709	513	374	267	182	112	52

Reference: Kaye and Laby, *Tables of Physical and Chemical Constants*, 15 ed., Longmans, 1986

Table XII.20 Typical values of the normal emissivity of a smooth polished surface at several wavelengths for various substances at stated temperature

Substance	temperature/K	wavelength/μm		
		0.65	1	5
		emissivity		
Ag	1000	0.05	0.05	0.05
Al	500	0.1 at 2 μm		0.05
Al$_2$O$_3$	1200	0.1	0.06	0.39
	1400	0.15	0.07	0.43
	1600	0.25	0.1	0.46
	1800	0.4	0.2	0.6
Au	1000	0.15	0.05	0.03
*BeO	1100	0.5	0.35	0.8
Co	1300	0.35	0.25	
Cr	1550	0.35		
†20 Cr, 80 Ni	1200	0.4		
oxidized	1200	0.9	0.85	0.8
Fe	1400	0.35	0.3	0.15
Hf	1200	0.45		
Ir	1500	0.3	0.23	0.1
kanthal A (oxidized)	1300	0.85	0.85	0.75
*MgO	1400	0.25	0.2	0.37
Mo	2000	0.4	0.3	0.15
Nb	2000	0.4	0.32	0.2
Ni	1100	0.45	0.35	0.15
Ni (oxidized)	1300	0.88	0.84	0.75
Os	1500	0.43		
Pd	1500	0.35		
Pt	1200	0.35	0.25	0.08
† 87 Pt, 13 Rh	1100	0.28		
Re	1400	0.4	0.35	0.2
Rh	1400	0.2		
Ru	1500	0.34		
Si	1500	0.4	0.25	0.25
stainless steel	1200	0.33	0.3	0.2
oxidized	1200	0.8	0.8	0.7
Ta	2200	0.41	0.3	0.18
*ThO$_2$	1400	0.4	0.35	0.37
Ti	1000	0.6	0.3	0.18
U	1100	0.3		
V	800		0.65	0.28
W	2400	0.43	0.38	0.12
Zr	1600	0.5	0.45	0.3
*ZrO$_2$	1400	0.4	0.2	0.45

†Numbers signify mass percentage composition
*Displays a temperature-emissivity variation similar to that of Al$_2$O$_3$
Reference: Touloukian and De Witt (eds) *Thermophysical Properties of Matter*, vols. 7, 8, 9, Plenum Press, 1972

Table XII.21 Variation with temperature of the thermal electromotive force for various elements relative to platinum, one junction being at a constant temperature of 273 K
Minus sign denotes current flow from element to Pt at hot end.

Element	\multicolumn{10}{c}{temperature/K}
	73	173	373	473	673	873	1073	1273	1473	1673
	\multicolumn{10}{c}{electromotive force/µV}
Ag	−210	−390	740	1770	4570	8410	13 330			
Al	−450	−60	420	1060	2840	5150				
Au	−210	−390	780	1840	4630	8120	12 260	17 050		
Bi	12 390	7540	−7340	−13 570						
C			700	1540	3720	6790	10 980	16 460		
Ca			−510	−1130						
Cd	−40	−310	900	2350						
Ce			1140	2460						
Co			−1330	−3080	−7240	−11 280	−13 990	−14 210	−10 700	
Cs	220	−130								
°Cu	−190	−370	760	1830	4680	8340	12 810	18 160		
°Fe	−2920	−1940	1980	3540	5880	7800	10 840	14 280		
Ge	−44 000	−26 620	33 900	72 400	82 300	43 900				
Hg			−600	−1330						
In			690							
Ir	−250	−350	650	1490	3550	6100	9100	12 570	16 450	20 470
K	1610	780								
Li	−1120	−1000	1820							
Mg	370	−90	440	1100						
Mo			1450	3190	7570	13 130	19 830	27 740	36 860	
Na	1000	−290								
Ni	2280	1220	−1480	−3100	−5450	−7040	−9330	−12 110		
Pb	240	−130	440	1090						
Pd	810	480	−570	−1230	−2820	−5030	−7960	−11 610	−15 860	−20 400
Rb	1090	460								
Rh	−200	−340	700	1610	3910	6770	10 140	14 020	18 390	22 990
Sb			4890	10 140	20 530	28 870				
Si	63 130	37 170	−41 560	−80 570						
Sn	260	−120	420	1070						
Ta	210	−100	330	930	2910	5950	10 020	15 150	21 370	
Th			−130	−260	−500	−450	220	1720	4030	
Tl			580	1300						
W	430	−150	1120	2620	6700	12 260	19 250	27 730	37 720	
Zn	−70	−330	760	1890	5290					

° Thermocouple material: see note below
Reference: Anderson (ed.) *Physics Vade Mecum*, 2nd edn., AIP, 1989
Note: thermoelectric thermometry is now largely automated, so thermocouple tables are mainly of academic interest; extensive data is given in BS 4937 (1973) and IEC 584–1 (1977)

Table XII.22 Variation with temperature of the thermal electromotive force for some alloys relative to platinum, one junction being at a constant temperature of 273 K

Minus sign denotes current flow from alloy to Pt at hot end.

Alloy†	temperature/K									
	73	173	373	473	673	873	1073	1273	1473	1673
	electromotive force/µV									
3.5 Ag, 96.5 Sn			450							
90 Ag, 10 Cu			800	1900	4810	8640				
°Alumel	2390	1290	−1290	−2170	−3640	−5280	−7070	−8780	−10330	−11770
Au-Cr			−170	−320	−550	−660				
Be-Cu			670	1620	4190					
brass, yellow			600	1490	3850	6960				
°Chromel P	−3360	−2200	2810	5960	12750	19610	26200	32470	38480	44040
°Constantan	5350	2980	−3510	−7450	−16190	−25460	−34810	−43850		
16 Cr, 24 Fe, 60 Ni			850	2010	5000	8680	13030	18060		
20 Cr, 80 Ni			1140	2620	6250	10530	15410	20870		
75 Cu, 25 Ni			−2760	−6010	−13780	−22590				
95 Cu, 4 Sn, 1 Zn			600	1480	3910	7140				
manganin			610	1550	4250	7840				
50 Pb, 50 Sn			460							
phosphor bronze			550	1340	3500	6300				
steel, spring			1320	2630	4840	6860				
steel, stainless			440	1040	2600	4670	7350			

° See Table XII.21 footnote and note
† Numbers indicate mass percentage composition
Reference: as for Table XII. 21

Table XII. 23 Variation with temperature of the absolute negative thermoelectric power of platinum

temperature/K	300	400	500	600	700	800	900	1000	1100	1200
power/µV K^{-1}	5.05	7.66	9.69	11.33	12.87	14.38	15.97	17.58	19.03	20.56

Reference: Kaye and Laby, *Tables of Physical and Chemical Constants*, 15 ed., Longmans, 1986

Table XII.24 Fixed points of the International Temperature Scale of 1990

Introduced January 1990 to replace previous scales; the purpose is to improve worldwide harmonization of standards. The changes are small (zero at 273.15 K) and at no temperature exceed 0.4 K. For most *practical* purposes it is usually adequate to assume 273 K ≡ 0° C.

Defining equilibrium state	temperature/K	Defining equilibrium state	temperature/K
BOILING POINTS:		MELTING POINT:	
e − H$_2^\ddagger$ at 101 292 Pa pressure	20.27	Ga	302.9146
e − H$_2^\ddagger$ at 33 321.3 Pa pressure	17.035		
		TRIPLE POINTS:	
FREEZING POINTS:		Ar	83.8058
Ag	1234.93	e − H$_2^\ddagger$	13.8033
Al	933.473	H$_2$O	273.16 (≡0.01° C)
Au	1337.33	Hg	234.3156
Cu	1357.77	Ne	24.5561
In	429.7485	O$_2$	54.3584
Sn	505.078		
Zn	692.677		

†e − H$_2$ is H$_2$ with equilibrium concentration of ortho and para H$_2$ at that temperature.
Reference: Rusby, *The International Temperature Scale of 1990*, ITS 90, NPL Special Report QU S45, 1989 and leaflets

Table XII.25 Definition of the International Practical Temperature Scale of 1990
See note, Table XII.24.
Only the calibration, not the use, of secondary thermometric devices is affected by these definitions.

Temperature range/K	Method
0.65–5	By measurement of vapour pressure of ^{3}He and ^{4}He and using specified equations.
3–24.5561	By use of an interpolating constant-volume gas thermometer, calibrated at a temperature between 3 and 5 K using a He vapour pressure as in range 0.65–5 K, and at H and Ne triple points.
13.8033–273.16	By use of a suitable platinum-resistance thermometer, the comparison being made for the 273.16 value. Calibration is by measurements at suitable (specified) fixed points, the choice depending on the actual range. Interpolation is done using a specified reference function, together with a deviation equation.
273.15–1234.93	As for range 13.8033–273.16 K, but different fixed points are specified.
234.3156–302.9146	By use of a suitable platinum-resistance thermometer. The specified fixed points are the melting point of gallium and the triple points of water and mercury. For measurements above 273.15 K, the Hg measurement may be omitted.
> 1234.93	Radiation pyrometry based on Planck's law is the prescribed method. The reference source is a black-body radiator at any of the silver, gold or copper freezing points.

Reference: as for Table XII.24

Section XIII
Thermodynamic Properties of Elements and Compounds

		Page
XIII.1	Thermodynamic data for the elements	449
XIII.2	Thermodynamic data for selected inorganic compounds	451
XIII.3	CODATA recommended values for thermodynamic functions of various chemical species at 101 325 Pa	466
XIII.4	Thermodynamic data for selected organic compounds at 298.15 K	470
XIII.5	Thermodynamic data for selected biological molecules at 298.15 K and 1 atmosphere pressure	474
XIII.6	Enthalpies of combustion for selected organic compounds	477
XIII.7	Enthalpies of fusion and vaporization at the transition temperature of some inorganic compounds	478
XIII.8	Enthalpies of fusion and vaporization at the transition temperature for some organic compounds	479
XIII.9	Enthalpies of formation of gaseous atoms from elements in their standard states at 298.15 K	481
XIII.10	Enthalpies of formation of free radicals at 298.15 K	482
XIII.11	Enthalpy of solution of selected electrolytes at 298.15 K	482
XIII.12	Enthalpy of neutralization at 298.15 K	483
XIII.13	Integral enthalpies of solution at 298.15 K	483
XIII.14	Mean bond dissociation enthalpies (single bond enthalpies)	484
XIII.15	Bond dissociation enthalpies of diatomic molecules at 298.15 K	484
XIII.16	Bond dissociation enthalpies in some polyatomic molecules at 298.15 K	485
XIII.17	Enthalpy of hydration of some ions at 298.15 K	486
XIII.18	Free energy function $[(G^{\circ}(T) - H^{\circ}(0)]/T$ for selected gaseous carbon compounds based on $H^{\circ}(0)$	486
XIII.19	Molar thermodynamic functions of a harmonic oscillator	487
XIII.20	Lattice enthalpies of some metal halides	488
XIII.21	Thermodynamic data for some ion association reactions at 298.15 K	488
XIII.22	Molar heat capacity of elements	489
XIII.23	Molar heat capacity of some solid inorganic compounds	490
XIII.24	Molar heat capacity of some gaseous organic compounds	491
XIII.25	Temperature dependence of heat capacities	491
XIII.26	Debye characteristic temperatures	493
XIII.27	Standard Gibbs (free) energy function of hydrolysis of some phosphate and other biologically important compounds at pH 7 and 298.15 K	493
XIII.28	Thermodynamic constants for phase changes of water and deuterium oxide	493
XIII.29	Molal cryoscopic and ebullioscopic constants for some common solvents	494
XIII.30	Henry's law constants for gases in solution	495
	1 Henry's law constant for inorganic gases at 25 °C	495
	2 Henry's law constant for nitrogen in water at different temperatures	495
	3 Henry's law constant for some organic compounds in aqueous solution at 25 °C	495
XIII.31	Joule-Thompson coefficients for some gases at 273 K and 1 atmosphere pressure	496
XIII.32	Standard molar Gibbs energies of transfer of selected ions from water into aqueous alcohols at 298.15 K	496
XIII.33	Calorific values of solid, liquid and gaseous fuels	497
XIII.34	Properties of some common binary azeotopes at 1 atmosphere pressure	498
XIII.35	Critical solution temperatures for some partially miscible binary liquid mixtures	499
XIII.36	Systems with simple eutectic points	500
XIII.37	Cryohydrates	501
XIII.38	Equilibrium constants for gaseous reactions at various temperatures	501

Section XIII
Thermodynamic Properties of Elements and Compounds

Table XIII.1 Thermodynamic Data for the Elements
Enthalpy of fusion, $\Delta_{fus}H$, and enthalpy of vaporization, $\Delta_{vap}H$, at the melting and boiling points respectively; enthalpy of atomization, $\Delta_{at}H$, standard entropy and molar heat capacity at 298.15 K.
For more extensive data see Thermodynamic Tables, Thermodynamic Research Centre, Texas, A & M; DIPPR Data Compilation, Hemisphere.
*: CODATA recommended values (for more extensive data at different temperatures see Table XIII.19).
(s): element sublimes.
Reference: J.D. Cox, D.D. Wagman and V.A. Medvedev, *Codata Key Values for Thermodynamics* (1989), Hemisphere Pub. Corp., New York.

Element	m.p. /K	$\Delta_{fus}H$ /kJ mol^{-1}	b.p. /K	$\Delta_{vap}H$ /kJ mol^{-1}	$\Delta_{at}H$ /kJ mol^{-1}	$S^{\ominus}$ /J K^{-1} mol^{-1}	$C_p^{\ominus}$ /J K^{-1} mol^{-1}
H$_2$	14.0	0.117	20.3	0.9163	216.0	130.571*	28.836*
He	0.95	0.021	4.2	0.084	0	126.044*	20.786*
Li	453.7	3.0	1620.2	147.1	159.4	29.10	24.77
Be	1557.2	11.71	3243.2	297	324.3	9.50	16.44
B	2300	22.6	4075 (s)	507.8	562.7	5.86	11.09
C (graphite)	4000		5100	715 ($\Delta_s H$)	716.7	5.740*	8.517*
N$_2$	63.3	0.719	77.4	5.586	472.7	191.500*	29.124*
O$_2$	54.8	0.444	90.2	6.820	249.2	205.043*	29.378*
F$_2$	53.5	0.52	85.0	6.54	79.0	202.682*	31.304*
Ne	24.6	0.34	27.1	1.77	0	146.219*	20.786*
Na	371.0	2.601	1156.1	98.01	107.3	51.21	28.24
Mg	922.0	8.95	1363	127.6	147.7	32.68	24.89
Al	933.6	10.7	2740	290.8	326.4	28.32	24.33
Si	1683	50.2	2628	359.0	455.6	18.83	20.00
P (white)	317.3	0.63	553.2	12.4	314.6	41.09	23.8
S (β)	392.2	1.73	717.8	10	278.8	32.054*	22.750*
Cl$_2$	172.2	6.41	238.6	20.41	121.3	222.972*	33.949*
Ar	83.9	1.188	87.5	6.506		154.74	20.786*
K	336.8	2.33	1047.2	76.9	89.24	64.18	29.58
Ca	1118.2	8.53	1757.2	154.67	178.2	41.42	25.31
Sc	1814.2	16.11	3104.2	304.80	377.8	34.64	25.52
Ti	1933	18.6	3560	425.2	469.9	30.63	25.02
V	2163	22.8	3653	446.7	514.2	28.91	24.89
Cr	2130	20	2945	339.5	396.6	23.77	23.35
Mn	1517	14.64	2235	219.74	280.8	32.00	26.32
Fe	1808	13.8	3023	349.5	416.3	27.28	25.10
Co	1768	16.2	3143	373.3	424.7	30.04	24.80
Ni	1726	17.2	3005	377.5	429.7	29.87	26.07
Cu	1356.6	13.14	2840	300.5	338.3	33.15	24.43
Zn	692.7	7.38	1180	115.31	130.7	41.63	25.40
Ga	302.9	5.59	2676	256.06	277.0	40.88	25.86
Ge	1210.6	31.80	3103	334.30	376.6	31.09	23.35
As	1090 (28 at)	27.7	886 (s)	32.4 ($\Delta_s H$)	302.5	35.15	24.98
Se	490	5.44	958	26.32	227.07	42.44	25.36
Br$_2$	266.0	10.57	331.9	29.45	111.88	152.23	75.69
Kr	116.6	1.64	120.9	9.03	0	163.97	20.786*
Rb	312.0	2.34	961	69.20	80.88	76.77	31.06
Sr	1042	8.2	1657	136.9	164.4	52.3	26.4
Y	1795	17.15	3611	393.30	421.3	44.43	26.53
Zr	2130	21	4650	590.5	608.8	38.99	25.36
Nb	2741	26.9	5015	690.1	725.9	36.40	24.60
Mo	2890	36	4885	590.4	658.1	28.66	24.06

(Continued)

Table XIII.1 (Continued)

Element	m.p. /K	$\Delta_{fus}H$ /kJ mol^{-1}	b.p. /K	$\Delta_{vap}H$ /kJ mol^{-1}	$\Delta_{at}H$ /kJ mol^{-1}	$S^\ominus$ /J K^{-1} mol^{-1}	$C_p^\ominus$ /J K^{-1} mol^{-1}
Tc	2445	23	5150	502	677		
Ru	2583	25.52	4173	567.77	642.7	28.53	24.06
Rh	2239	21.76	4000	495.39	556.9	31.51	24.98
Pd	1825	16.74	3413	393.30	378.2	37.57	25.98
Ag	1235	11.30	2485	250.63	284.55	42.55	25.35
Cd	594	6.07	1038	99.87	112.0	51.75	25.98
In	429.8	3.26	2353	226.35	243.3	57.82	26.74
Sn	505.1	7.20	2543	290.37	302.1	51.55	26.99
Sb	903.9	19.83	2023	67.97	262.3	45.69	25.23
Te	722.7	17.49	1262.9	50.63	196.73	49.71	25.73
I$_2$	386.7	15.52	457.5	41.80	106.84	116.14	54.44
Xe	161.3	2.30	166.1	12.64	0	169.57	20.786*
Cs	301.6	2.09	951.6	67.77	76.06	85.23	31.17
Ba	998.2	8.01	1913	140.2	180	66.94	26.36
La	1194	11.30	3730	399.57	431	56.9	27.11
Ce	1072	9.20	3700	313.80	422.6	71.96	26.94
Pr	1204	10.04	3785	332.63	355.6	73.2	27.20
Nd	1294	10.88	3341	283.68	327.6	71.5	27.45
Pm	c 1353		2730?				
Sm	1350	11.09	2064	191.63	206.7	69.6	29.54
Eu	1095	10.46	1870	175.73	175.3	77.78	27.66
Gd	1586	15.48	3539	311.71	397.5	68.1	37.03
Tb	1629		3396		388.7	73.22	28.91
Dy	1685		2835		290.4	74.77	28.16
Ho	1747	17.15	2968	251.04	300.8	75.3	27.15
Er	1802	17.15	3136	292.88	317.1	73.18	28.12
Tm	1818		2220		232.2	74.01	27.03
Yb	1092		1467		152.3	59.87	26.74
Lu	1936		3668		427.6	50.96	26.86
Hf	2500	21.76	4875	661.07	619.2	43.56	25.73
Ta	3269	36	5698	737.0	782.0	41.51	25.36
W	3683	35.4	5933	422.58	849.4	32.64	24.27
Re	3453	33.05	5900	707.10	769.9	36.86	25.48
Os	3318	29.29	5300	627.6	791	32.6	24.7
Ir	2683	26.36	4403	563.58	665.3	35.48	25.10
Pt	2045	19.66	4100	510.45	565.3	41.63	25.86
Au	1337.6	12.36	3080	324.43	366.1	47.40	25.42
Hg	234.3	2.292	629.7	59.30	61.32	76.02	27.98
Tl	577	4.27	1730	162.09	182.21	64.64	26.32
Pb	600.65	4.77	2013	177.9	195	64.81	26.44
Bi	544.5	11.0	1833	179	207.1	56.73	25.52
Po	527	13	1235	120			
At	575	(12)	610	30			
Rn	202	2.9	211	16.4	0	176.1	20.786*
Fr	300	(2.1)	950	(64)	(64)	95.4	
Ra	973	8.37	1413	136.82	159	71	27.0
Ac	1320		3470		406	56.5	27.2
Th	2023	15.65	c 5063	543.92	598.3	53.39	27.32
Pa	< 1870				607	51.9	
U	1405	15.48	4091	422.58	535.6	50.21	27.66

Table XIII.2 Thermodynamic Data for Selected Inorganic Compounds

For a pure solid (cr) or liquid (l) the standard state is the substance in the condensed phase at 298.15 K and a pressure of 1 atmosphere. For a gas (g) the standard state is the hypothetical state at unit fugacity, in which state the enthalpy is that of the real gas at the same temperature and zero pressure. For dissolved substances (aq) the standard state is taken as the hypothetical state of unit molality at 298.15 K.

The compounds listed represent only a small fraction of those for which data are available in the *National Bureau of Standards Technical Notes* 270–3, 270–4, 270–5, 270–6, 270–7 and 270–8. For more extensive data see: reference of Table XIII.1 and D.D. Wagman, W.H. Evans, V.B. Parker, R.H. Schuman, I Halow, S.M. Bailey, K.L. Churney and R.L. Nuttall, *J. Phys. Chem. Ref. Data* (1982), **11**, Supp. 2; (1989), **18**, 1807.

*: CODATA recommended key values (1989).

Substance	State	$\Delta_f H^\ominus$ /kJ mol^{-1}	$\Delta_f G^\ominus$ /kJ mol^{-1}	$S^\ominus$ /J K^{-1} mol^{-1}	$C_p^\ominus$ /J K^{-1} mol^{-1}	$H^\ominus(298) - H^\ominus(0)$ /kJ mol^{-1}
ALUMINIUM						
Al	cr	0	0	28.300*	24.200*	4.540*
	g	330.0*		164.445*	21.391*	6.919*
Al^{3+}	g	5481				
	aq	−538.4*	−485.3	−325*		
AlBr$_3$	cr	−527			101.67	
AlCl$_3$	cr	−704.2	−628.9	110.7	89	
AlF$_3$	cr	−1510.4*		66.500*	75.122*	11.620*
Al$_2$O$_3$	cr	−1675.7*	−1576.41	50.920*	79.033*	10.016*
Al$_2$(SO$_4$)$_3$	cr	−3440.84	−3100.1	239.32	259.41	
NH$_4$Al(SO$_4$)$_2 \cdot$ 12H$_2$O	cr	−5942.37	−4938.00	697.05	683.25	
AMMONIA AND AMMONIUM						
NH$_3$	g	−45.94*	−16.48	192.660*	45.630*	10.046*
	aq	−80.8	−26.61	110.0		
NH$_4^+$	aq	−133.26*	−79.37	111.17*	79.9	
NH$_4$Br	cr	−270.83	−175.31	113.0	96.23	
NH$_4$Cl	cr	−314.42	−202.97	94.56	84.10	
	aq	−299.66	−210.6	169.9	−56.5	
NH$_4$OH	aq	−366.1	−263.76	181.17	−68.62	
NH$_4$NO$_3$	cr	−365.56	−184.01	151.08	139.33	
(NH$_4$)$_2$SO$_4$	cr	−1180.85	−901.90	220.08	187.48	
ANTIMONY						
Sb	cr	0	0	45.69	25.23	
	g	262.3	222.2	180.2	20.79	
Sb^{3+}	g	5130.0				
SbBr$_3$	cr	−259.4	−239.3	207.1		
	g	−194.6	−223.8	372.75	80.21	
SbCl$_3$	cr	−382.17	−323.71	184.1	107.9	
	g	−313.8	−301.2	337.70	76.69	
SbCl$_5$	l	−440.2	−350.2	301.2		
SbH$_3$	g	145.1	147.7	232.67	41.0	
Sb$_2$O$_5$	cr	−972.9	−829.3	125.1		
Sb$_2$S$_3$	cr	−174.9	−173.6	182.0	119.87	
ARGON						
Ar	g	0	0	154.732*	20.786*	6.197*
ARSENIC						
As	cr (α)	0	0	35.15	24.98	
	(γ)	14.6				
	(β)	4.2				
As^{3+}	g	5950				
As$_2$	g	222.2	171.9	239.3	35.00	
AsBr$_3$	g	−129.7	−159.0	363.75	79.16	
AsCl$_3$	l	−305.0	−259.4	207.5		
	g	−261.5	−248.9	327.06	75.73	
AsI$_3$	g	66.14	68.91	222.67	38.07	
As$_2$O$_5$	cr	−924.87	−782.41	105.43	116.52	
H$_3$AsO$_4$	cr	−906.25				
AsO$_4^{3-}$	aq	−888.1	−648.5	−162.8		
As$_2$S$_3$	cr	−169.0	−168.6	163.6	116.3	

(Continued)

Table XIII.2 (Continued)

Substance	State	$\Delta_f H^\ominus$ /kJ mol^{-1}	$\Delta_f G^\ominus$ /kJ mol^{-1}	$S^\ominus$ /J K^{-1} mol^{-1}	$C_p^\ominus$ /J K^{-1} mol^{-1}	$H^\ominus(298) - H^\ominus(0)$ /kJ mol^{-1}
BARIUM						
Ba	cr	0	0	66.94	26.36	
Ba^{2+}	g	1660				
	aq	−537.64	−560.74	9.62		
BaBr$_2$	cr	−757.3	−736.8	146.44		
BaCO$_3$	cr	−1216.28	−1137.6	112.13	85.35	
BaCl$_2$	cr	−858.6	−810.4	123.68	75.14	
BaCl$_2 \cdot$ 2H$_2$O	cr	−1460.1	−1296.45	202.9	161.96	
BaF$_2$	cr	−1207.1	−1156.9	96.35	71.21	
Ba(NO$_3$)$_2$	cr	−922.07	−796.71	213.8	151.38	
BaO	cr	−558.15	−528.44	70.29	47.45	
BaSO$_4$	cr	−1473.19	−1362.3	132.21	101.75	
BERYLLIUM						
Be	cr	0	0	9.500*	16.443*	1.950*
	g	324*	286.6	136.166*	20.786*	6.197*
Be^{2+}	g	2993.2				
BeO	cr	−609.4*	−580.3	13.770*	25.565*	2.837*
BeSO$_4$	cr	−1205.2	−1093.86	77.91	86.69	
BISMUTH						
Bi	cr	0	0	56.73	25.52	
	g	207.1	168.2	186.81	20.79	
Bi^{3+}	g	5005				
	aq		82.84			
BiCl$_3$	cr	−379.1	−315.05	176.98	104.6	
Bi$_2$O$_3$	cr	−573.88	−493.71	151.46	113.51	
BiOCl	cr	−366.94	−322.2	120.5		
Bi$_2$S$_3$	cr	−143.1	−140.6	200.4	122.2	
BORON						
B	cr	0	0	5.900*	11.087*	1.222*
	g	565*	518.8	153.327*	20.786*	6.316*
BBr$_3$	l	−239.7	−238.5	229.7		
	g	−205.64	−232.46	324.13	67.78	
B$_4$C	cr	−71	−71	27.11	52.8	
BCl$_3$	l	−423.4	−387.4	206.3	106.7	
	g	−403.76	−388.73	290.0	62.72	
BF$_3$	g	−1136.0*	−1120.3	254.314*	50.463*	11.650*
BF$_4^-$	aq	−1574.9	−1487.0	179.9		
B$_2$H$_6$	g	35.6	86.6	232.0	56.90	
BI$_3$	g	71.13	20.75	349.07	70.79	
BN	cr	−254.4	−228.4	14.81	19.71	
B$_2$O$_3$	cr	−1273.5*	−1193.7	53.970*	62.761*	9.301*
	amorph	−1254.5	−1182.4	77.82	61.1	
H$_3$BO$_3$	cr	−1094.8*	−969.0	89.950*	86.060*	13.520*
BROMINE						
Br$_2$	l	0	0	152.210*	75.69	24.52*
	g	30.91*	3.14	245.359*	36.057*	9.725*
	aq	−2.58	3.93	130.54		
Br	g	111.87*	82.43	174.909*	20.786*	6.197*
Br$^-$	g	−234				
	aq	−121.41*	−103.97	82.55*	−141.8	
BrCl	g	14.64	−0.96	240.0	34.98	
BrF$_3$	l	−300.8	−240.6	178.2	124.6	
	g	−255.60	−29.45	292.42	66.61	
HBr	g	−36.29*	−53.43	198.591*	29.141*	8.648*
BrO$^-$	aq	−94.1	−33.5	41.8		
BrO$_3^-$	aq	−83.7	1.7	161.7		

(Continued)

Table XIII.2 (Continued)

Substance	State	$\Delta_f H^\ominus$ /kJ mol^{-1}	$\Delta_f G^\ominus$ /kJ mol^{-1}	$S^\ominus$ /J K^{-1} mol^{-1}	$C_p^\ominus$ /J K^{-1} mol^{-1}	$H^\ominus(298) - H^\ominus(0)$ /kJ mol^{-1}
CADMIUM						
Cd	cr	0	0	51.800*	26.020*	6.247*
	g	111.80*	77.45	167.640*	20.786*	6.197*
Cd^{2+}	g	2623.5				
	aq	−75.92*	−77.58	−72.8*		
CdBr$_2$	cr	−316.18	−296.31	137.24	76.65	
CdBr$_2 \cdot$ 4H$_2$O	cr	−1492.6	−1248.0	316.3		
CdCO$_3$	cr	−750.6	−669.4	92.47		
CdCl$_2$	cr	−391.50	−343.97	115.27	74.68	
CdCl$_2 \cdot$ 5H$_2$O	cr	−1131.9	−944.09	227.2		
CdF$_2$	cr	−700.4	−647.7	77.4		
CdI$_2$	cr	−203.3	−201.37	161.1	79.96	
CdO	cr	−258.35*	−228.4	54.8*	43.43	8.41*
Cd(OH)$_2$	cr	−560.7	−473.6	96.2		
CdS	cr	−161.9	−156.5	64.9		
CdSO$_4$	cr	−933.28	−822.78	123.04	99.58	
CdSO$_4 \cdot$ 8/3H$_2$O	cr	−1729.30*		229.65*		35.56*
[Cd(NH$_3$)$_2$]$^{2+}$	aq	−266.1	−159.0	144.8		
[Cd(NH$_3$)$_4$]$^{2+}$	aq	−450.2	−226.4	336.4		
CAESIUM						
Cs	cr	0	0	85.230*	31.210*	7.711*
	g	76.5*	49.15	175.491*	20.786*	6.197*
Cs$_2$	g			284.571*	38.255*	11.029*
Cs$^+$	g	452.3				
	aq	−258.00*	−292.0	132.1*	−10.46	
CsCl	cr	−443.04	−414.6	101.17	52.48	
	g	−240.20	−257.82	255.85	36.94	
CsF	cr	−553.5	−525.5	92.80	51.09	
	g	−359.0	−375.7	243.13	35.86	
CsI	cr	−346.60	−340.58	123.05	52.80	
	g	−153.1	−191.2	275.18	37.45	
Cs$_2$O	cr	−345.77	−308.15	146.86	75.98	
CsNO$_3$	cr	−494.2	−406.60	155.23		
CsOH	cr	−417.23				
	g	−246.9	−247.3	254.72	49.71	
Cs$_2$SO$_4$	cr	−1443.0	−1323.65	211.92	134.89	
CALCIUM						
Ca	cr	0	0	41.588*	25.929*	5.736*
	g	177.8*	144.3	154.778*	20.786*	6.197*
Ca^{2+}	g	1925.9				
	aq	−543.00*	−553.54	−56.2*		
CaBr$_2$	cr	−682.8	−663.6	129.7		
CaC$_2$	cr	−59.8	−64.9	69.96	62.72	
CaCO$_3$	cr	−1206.9	−1128.8	92.88	81.88	
CaCl$_2$	cr	−795.8	−748.1	113.8	72.59	
CaCl$_2 \cdot$ 6H$_2$O	cr	−2607.9				
CaF$_2$	cr	−1219.6	−1167.3	66.87	67.03	
CaI$_2$	cr	−533.5	−528.9	142.3		
Ca(NO$_3$)$_2$	cr	−938.39	−743.20	193.3	149.37	
Ca(NO$_3$)$_2 \cdot$ 4H$_2$O	cr	−2132.3	−1700.8	338.9		
CaO	cr	−634.92*	−604.04	38.100*	42.049*	6.750*
Ca(OH)$_2$	cr	−986.09	−898.56	76.2	87.49	
Ca$_3$(PO$_4$)$_2$	cr	−4120.8	−3884.8	235.98	227.82	
CaS	cr	−482.4	−477.4	56.48	47.40	
CaSO$_4$	cr	−1434.1	−1321.9	106.69	99.66	
CaSO$_4 \cdot$ 2H$_2$O	cr	−2022.6	−1797.4	194.1	186.02	

(Continued)

Table XIII.2 (Continued)

Substance	State	$\Delta_f H^\ominus$ /kJ mol^{-1}	$\Delta_f G^\ominus$ /kJ mol^{-1}	$S^\ominus$ /J K^{-1} mol^{-1}	$C_p^\ominus$ /J K^{-1} mol^{-1}	$H^\ominus(298) - H^\ominus(0)$ /kJ mol^{-1}
CARBON						
C (graphite)	cr	0	0	5.740*	8.517*	1.050*
(diamond)	cr	1.89	2.899	2.377	6.116	
	g	716.68*	671.29	157.991*	20.839*	6.536*
CO	g	−110.53*	−137.15	197.551*	29.141*	8.671*
CO_2	g	−393.51*	−394.36	213.676*	37.135*	9.365*
H_2CO_3	aq	−699.64	−623.16	187.4		
HCO_3^-	aq	−689.93*	−586.85	98.4*		
CO_3^{2-}	aq	−675.23*	−527.90	−50.0*		

(For other carbon compounds see Table XIII.4.)

Substance	State	$\Delta_f H^\ominus$ /kJ mol^{-1}	$\Delta_f G^\ominus$ /kJ mol^{-1}	$S^\ominus$ /J K^{-1} mol^{-1}	$C_p^\ominus$ /J K^{-1} mol^{-1}	$H^\ominus(298) - H^\ominus(0)$ /kJ mol^{-1}
CERIUM						
Ce	cr	0	0	71.96	27.11	
	g	422.6	384.9	191.66	23.07	
Ce^{3+}	g	3963.9				
	aq	−696.2	−671.9	−205.0		
$CeCl_3$	cr	−1053.5	−977.8	150.6	87.45	
Ce_2O_3	cr	−1796.2	−1706.2	150.6	114.6	
CeO_2	cr	−1082.8	−1024.7	62.3	61.63	
CHLORINE						
Cl_2	g	0	0	222.972*	33.949*	9.181*
	aq	−23.4	6.90	121.33		
Cl	g	121.301*	105.70	165.081*	21.838*	6.271*
Cl^-	g	−246.0				
	aq	−167.080*	−131.26	56.60*	−136.40	
ClF	g	−54.48	−55.94	217.78	32.05	
HCl	g	−92.31*	−95.30	186.793*	29.136*	8.640*
Cl_2O	g	80.3	97.9	266.10	45.40	
ClO^-	aq	−107.1	−36.8	41.8		
ClO_2	g	102.5	120.5	256.73	41.97	
ClO_3^-	aq	−99.2	−3.3	162.3		
ClO_4^-	aq	−128.10*	−8.62	184.0*		
$HClO_4$	l	−40.6				
CHROMIUM						
Cr	cr	0	0	23.77	23.35	
	g	396.6	351.9	174.39	20.79	
Cr^{2+}	g	2653				
	aq	−143.5				
$CrCl_2$	cr	−395.4	−356.1	115.31	71.17	
$CrCl_3$	cr	−556.5	−486.2	123.0	91.80	
Cr_2O_3	cr	−1139.7	−1058.1	81.17	118.74	
CrO_2Cl_2	l	−579.5	−510.9	221.8		
CrO_4^{2-}	aq	−881.1	−727.8	50.21		
$Cr_2O_7^{2-}$	aq	−1490.3	−1301.2	261.9		
COBALT						
Co	cr	0	0	30.04	24.80	
	g	424.7	378.7	179.40	23.02	
Co^{2+}	g	2841.6				
	aq	−58.2	−54.4	−113		
Co^{3+}	g	6080.1				
	aq	92	134	−305		
$CoCl_2$	cr	−312.7	−269.9	109.16	78.49	
$CoCl_2 \cdot 6H_2O$	cr	−2115.4	−1725.5	343		
CoO	cr	−237.94	−214.22	52.97	55.22	
Co_3O_4	cr	−891	−774	102.5	123.4	
$CoSO_4$	cr	−888.3	−782.4	118.0		
$CoSO_4 \cdot 7H_2O$	cr	−2979.93	−2473.15	406.06	390.49	

(Continued)

Table XIII.2 (Continued)

Substance	State	$\Delta_f H^\ominus$ /kJ mol^{-1}	$\Delta_f G^\ominus$ /kJ mol^{-1}	$S^\ominus$ /J K^{-1} mol^{-1}	$C_p^\ominus$ /J K^{-1} mol^{-1}	$H^\ominus(298) - H^\ominus(0)$ /kJ mol^{-1}
COPPER						
Cu	cr	0	0	33.150*	24.440*	5.004*
	g	337.4*	298.61	166.289*	20.786*	6.197*
Cu$^+$	g	1090.0				
	aq	71.67	50.0	40.58		
Cu^{2+}	g	3054				
	aq	63.545*	65.52	−98*		
CuBr	cr	−104.6	−100.83	96.11	54.72	
CuBr$_2$	cr	−141.8				
CuCl	cr	−137.2	−119.87	86.19	48.53	
CuCl$_2$	cr	−220.1	−175.7	108.07	71.88	
CuCl$_2 \cdot$ 2H$_2$O	cr	−821.3	−656.0	167		
CuI	cr	−67.8	−69.5	96.7	54.06	
Cu$_2$O	cr	−168.6	−146.0	93.13	63.64	
CuO	cr	−157.3	−129.7	42.63	42.30	
Cu$_2$S	cr	−79.5	−86.2	120.9	76.32	
CuS	cr	−53.1	−53.6	66.5	47.82	
CuSO$_4$	cr	−771.4*	−661.9	109.2*	100.0	16.86*
CuSO$_4 \cdot$ 5H$_2$O	cr	−2279.65	−1880.06	300.4	280.33	
[Cu(NH$_3$)$_4$]$^{2+}$	aq	−348.5	−111.3	273.6		
FLUORINE						
F$_2$	g	0	0	202.682*	31.304*	8.825*
F	g	79.38*	61.92	158.642*	22.746*	6.518*
F$^-$	g	−270.7				
	aq	−335.35*	−278.82	−13.8*	−106.7	
F$_2$O	g	−21.76	−4.6	247.32	43.30	
HF	l	−299.78		75.4 + x^+	51.67	
	g	−273.30*	−273.2	173.670*	29.137*	8.590*
	(x^+ = undetermined residual entropy)					
GALLIUM						
Ga	cr	0	0	40.88	25.86	
	l	5.56				
	g	276.98	238.9	169.95	25.35	
Ga^{3+}	g	5816				
	aq	−211.7	−159.0	−331		
GaBr$_3$	cr	−386.6	−359.8	180		
GaCl$_3$	cr	−524.7	−454.8	142		
Ga$_2$O$_3$	cr	−1089.1	−998.3	84.97	92.05	
Ga(OH)$_3$	cr	−964.4	−831.4	100.4		
GERMANIUM						
Ge	cr	0	0	31.090*	23.222*	4.636*
	g	372*	336.0	167.795*	30.733*	7.398*
Ge^{2+}	g	2687.2				
GeBr$_4$	l	−347.7	−331.4	280.7		
	g	−300.0	−318.0	396.06	101.84	
GeCl$_4$	l	−531.8	−462.8	245.6		
GeF$_4$	g	−1190.20*		301.819*	81.602*	17.293*
GeH$_4$	g	90.8	113.4	217.02	45.02	
GeO	cr	−261.9	−237.2	50.2		
GeO$_2$	cr	−580.0*		39.71*	50.166*	7.230*
GeS	cr	−69.0	−71.5	71		
GOLD						
Au	cr	0	0	47.40	25.42	
	g	366.1	326.4	180.39	20.79	
AuCl$_3$	cr	−117.6				
AuCl$_4^-$	aq	−322.2	−235.2	266.9		

(Continued)

Table XIII.2 (Continued)

Substance	State	$\Delta_f H^\ominus$ /kJ mol^{-1}	$\Delta_f G^\ominus$ /kJ mol^{-1}	$S^\ominus$ /J K^{-1} mol^{-1}	$C_p^\ominus$ /J K^{-1} mol^{-1}	$H^\ominus(298) - H^\ominus(0)$ /kJ mol^{-1}
HELIUM						
He	g	0	0	126.044*	20.786*	6.197*
HYDROGEN						
$^1H_2$	g	0	0	130.571*	28.836*	8.469*
$^2H_2$ (D$_2$)	g	0	0	144.85	29.196	
$^1H^2H$ (HD)	g	0.318	−1.46	143.69	29.196	
H	g	217.998*	203.26	114.608*	20.786*	6.197*
H$^+$	g	1536.2				
	aq	0	0	0	0	
H$^-$	g	139.7				
OH	g	38.95	34.22	183.64	29.89	
OH$^-$	g	140.88				
	aq	−230.015*	−157.29	−10.90*	−148.53	
H$_2$O	l	−285.830*	−237.18	69.950*	75.291	13.293*
	g	−241.826*	−228.59	188.726*	33.609*	9.905*
^2H_2O (D$_2$O)	l	−294.60	−243.49	75.94	84.35	
	g	−249.20	−234.55	198.22	34.27	
$^1H^2HO$ (DOH)	l	−289.89	−241.91	79.29		
	g	−245.30	−233.13	199.40	33.81	
H$_2$O$_2$	l	−187.78	−120.41		89.12	
	g	−136.31	−105.60	232.6	43.1	
INDIUM						
In	cr	0	0	57.82	26.74	
	g	243.30	208.74	173.68	20.84	
In$^+$	g	807.8				
In^{3+}	g	5327.1				
	aq	−105	−98	−151		
InBr	g	−56.9	−94.3	259.37	36.65	
In$_2$O$_3$	cr	−925.79	−830.73	104.2	92	
In$_2$S$_3$	cr	−427	−412.5	163.6	118	
In$_2$(SO$_4$)$_3$	cr	−2787	−2439	272	280	
IODINE						
I$_2$	cr	0	0	116.139*	54.44	13.196*
	g	62.421*	19.36	260.578*	36.888*	10.116*
	aq	22.59	16.40	137.2		
I	g	106.76*	70.28	180.678*	20.786*	6.197*
I$^-$	g	−196.6				
	aq	−56.78*	−51.59	106.45*	−142.3	
I$_3^-$	aq	−51.46	−51.46	239.32		
IBr	g	40.84	3.72	258.66	36.44	
ICl	l	−23.89	−13.60	135.1		
	g	17.78	−5.44	247.44	35.56	
ICl$_3$	s	−89.5	−22.34	167.4		
IF	g	−95.64	−118.49	236.06	33.43	
HI	g	26.50*	1.71	206.481*	29.157*	8.657*
IO$^-$	aq	−107.5	−38.5	−5.4		
IO$_3^-$	aq	−221.3	−128.0	118.4		
IO$_4^-$	aq	−151.5	−58.6	−221.8		
IRON						
Fe	cr	0	0	27.28	25.10	
	g	416.3	370.7	180.38	25.68	
Fe^{2+}	g	2752.2				
	aq	−89.12	−78.87	−137.7		
Fe^{3+}	g	5714.9				
	aq	−48.5	−4.6	−315.9		

(Continued)

Table XIII.2 (Continued)

Substance	State	$\Delta_f H^\ominus$ /kJ mol^{-1}	$\Delta_f G^\ominus$ /kJ mol^{-1}	$S^\ominus$ /J K^{-1} mol^{-1}	$C_p^\ominus$ /J K^{-1} mol^{-1}	$H^\ominus(298) - H^\ominus(0)$ /kJ mol^{-1}
Fe$_3$C (cementite)	cr	25.1	20.1	104.6	105.9	
FeCO$_3$	cr	−740.57	−666.72	92.88	82.13	
FeCl$_2$	cr	−341.79	−302.33	117.95	76.65	
FeCl$_3$	cr	−399.49	−334.05	142.3	96.65	
Fe$_{0.95}$O (wustite)	cr	−266.27	−246.40	57.49	48.11	
Fe$_2$O$_3$ (haematite)	cr	−824.2	−742.2	87.40	103.85	
Fe$_3$O$_4$ (magnetite)	cr	−1118.4	−1015.5	146.4	143.4	
Fe(OH)$_2$ (p.p.t.)	cr	−569.0	−486.6	88		
Fe(OH)$_3$ (p.p.t.)	cr	−823.0	−696.6	106.7		
FeS	cr	−100.0	−100.4	60.29	50.54	
FeS$_2$ (pyrite)	cr	−178.2	−166.9	52.93	62.17	
FeSO$_4$	cr	−928.4	−820.9	107.5	100.58	
FeSO$_4$. 7H$_2$O	cr	−3014.6	−2510.3	409.2	394.47	
Fe(CO)$_5$	l	−774.0	−705.4	338.1	240.6	
	g	−733.9	−697.3	445.2		
KRYPTON						
Kr	g	0	0	163.976*	20.786*	6.197*
LEAD						
Pb	cr	0	0	64.800*	26.650*	6.870*
	g	195.2*	161.9	175.266*	20.786*	6.197*
Pb^{2+}	g	2373				
	aq	0.92*	−24.39	18.5*		
PbBr$_2$	cr	−278.7	−261.9	161.5	80.12	
PbCO$_3$	cr	−699.1	−625.5	131.0	87.40	
PbCl$_2$	cr	−359.41	−317.90	136.0		
PbF$_2$	cr	−664.0	−617.1	110.5		
PbI$_2$	cr	−175.48	−173.64	174.85	77.36	
PbO (yellow)	cr	−215.33	−187.90	68.70	45.77	
(red)	cr	−218.99	−188.95	66.5	45.81	
PbO$_2$	cr	−277.4	−217.36	68.6	64.64	
Pb$_3$O$_4$	cr	−718.4	−601.2	211.3	146.9	
PbS	cr	−100.4	−98.7	91.2	49.5	
PbSO$_4$	cr	−919.97*	−813.20	148.50*	103.21	20.050*
LITHIUM						
Li	cr	0	0	29.120*	24.860*	4.632*
	g	159.3*	126.7	138.673*	20.786*	6.197*
Li$_2$	g			196.890*	36.103*	9.675*
Li$^+$	g	687.16				
	aq	−278.47*	−293.30	12.24*	68.6	
LiAlH$_4$	cr	−117	−49	88	86	
LiBH$_4$	cr	−190.8	−125.1	75.86	82.55	
	aq	−230.33	−179.1	124.3		
LiBr	cr	−351.21	−342.00	74.27		
Li$_2$CO$_3$	cr	−1216.0	−1132.11	90.37	99.12	
LiCl	cr	−408.61	−384.38	59.33	47.99	
LiF	cr	−615.97	−587.73	35.65	41.59	
LiH	cr	−90.62	−68.37	20.01	27.87	
Li^{2}H (LiD)	cr	−90.92	−67.70	23.60	34.31	
LiI	cr	−270.41	−270.29	86.78	51.04	
LiNO$_3$	cr	−482.3	−381.1	90.0		
Li$_2$O	cr	−597.94	−561.20	37.57	54.10	
LiOH	cr	−484.93	−438.99	42.80	49.66	
Li$_2$SO$_4$	cr	−1436.49	−1321.77	115.1	117.57	

(Continued)

Table XIII.2 (Continued)

Substance	State	$\Delta_f H^\ominus$ /kJ mol^{-1}	$\Delta_f G^\ominus$ /kJ mol^{-1}	$S^\ominus$ /J K^{-1} mol^{-1}	$C_p^\ominus$ /J K^{-1} mol^{-1}	$H^\ominus(298) - H^\ominus(0)$ /kJ mol^{-1}
MAGNESIUM						
Mg	cr	0	0	32.71*	24.869*	4.998*
	g	147.1*	113.14	148.539*	20.786*	6.197*
Mg^{2+}	g	2348.5				
	aq	−467.0*	−454.8	−137*		
MgBr$_2$	cr	−524.3	−503.8	17.2		
MgCO$_3$	cr	−1095.8	−1012.1	65.7	75.5	
MgCl$_2$	cr	−641.32	−591.83	89.62	71.38	
MgCl$_2$. 6H$_2$O	cr	−2499.00	−2114.97	366.1	315.06	
MgF$_2$	cr	−1124.2*	−1070.3	57.238*	61.512*	9.911*
MgI$_2$	cr	−364.0	−358.2	129.7		
Mg(NO$_3$)$_2$	cr	−790.65	−589.5	164.0	149.92	
MgO	cr	−601.5*	−569.44	26.950*	37.237*	5.160*
Mg(OH)$_2$	cr	−924.54	−833.58	63.18	77.03	
MgS	cr	−346.0	−341.8	50.33	45.56	
MgSO$_4$	cr	−1284.9	−1170.7	91.6	96.48	
MgSO$_4$. 7H$_2$O	cr	−3388.71	−2874.1	372		
MANGANESE						
Mn	cr	0	0	32.00	26.32	
	g	280.75	238.5	173.59	20.79	
Mn^{2+}	g	2519.2				
	aq	−220.75	−228.0	−73.64	50	
MnCO$_3$	cr	−894.1	−816.7	85.8	81.50	
MnCl$_2$	cr	−481.29	−440.53	118.24	72.93	
MnO	cr	−385.22	−362.92	59.71	45.44	
MnO$_2$	cr	−520.03	−465.18	53.05	54.14	
Mn$_2$O$_3$	cr	−959.0	−881.2	110.5	107.65	
Mn$_3$O$_4$	cr	−1387.8	−1283.2	155.6	139.66	
Mn(OH)$_2$	cr	−695.4	−615.0	99.1		
MnO$_4^-$	aq	−541.1	−447.3	191.2		
MnS	cr	−214.2	−218.4	78.24	49.96	
MnSO$_4$	cr	−1065.25	−957.42	112.1	100.50	
MERCURY						
Hg	l	0	0	75.90*	27.98	9.342*
	g	61.38*	31.85	174.862*	20.786*	6.197*
Hg$^+$	g	1074.5				
Hg^{2+}	g	2890.4				
	aq	170.21*	164.4	−36.19*		
Hg$_2^{2+}$	aq	166.87*	153.55	65.74*		
Hg$_2$CO$_3$	cr	−553.5	−468.2	180		
Hg$_2$Br$_2$	cr	−206.90	−181.08	218		
HgBr$_2$	cr	−170.7	−153.1	171		
Hg$_2$Cl$_2$	cr	−265.37*	−210.78	191.6*		23.25*
HgCl$_2$	cr	−224.3	−178.7	146.0		
HgI$_2$ (red)	cr	−105.4	−101.7	180		
HgO (red)	cr	−90.79*	−58.56	70.25*	44.06	9.117*
(yellow)	cr	−90.46	−58.43	71.1		
HgS (red)	cr	−58.2	−50.6	82.4	48.41	
(black)	cr	−53.6	−47.7	88.3		
Hg$_2$SO$_4$	cr	−743.09*	−625.88	200.70*	131.96	26.070*
MOLYBDENUM						
Mo	cr	0	0	28.66	24.06	
	g	658.1	612.5	181.84	20.79	
Mo(CO)$_6$	cr	−982.8	−877.8	325.9	242.3	
MoF$_6$	l	−1585.5	−1473.1	259.66	169.79	
	g	−1557.7	−1472.3	350.41	120.58	
MoO$_2$	cr	−588.94	−533.04	46.28	55.98	

(Continued)

Table XIII.2 (Continued)

Substance	State	$\Delta_f H^\ominus$ /kJ mol^{-1}	$\Delta_f G^\ominus$ /kJ mol^{-1}	$S^\ominus$ /J K^{-1} mol^{-1}	$C_p^\ominus$ /J K^{-1} mol^{-1}	$H^\ominus(298) - H^\ominus(0)$ /kJ mol^{-1}
MoO_3	cr	−745.09	−668.02	77.74	74.98	
MoS_2	cr	−235.1	−225.9	62.59	63.55	
NEON						
Ne	g	0	0	146.219*	20.786*	6.197*
NICKEL						
Ni	cr	0	0	29.87	26.07	
	g	429.7	384.5	182.08	23.36	
Ni^{2+}	g	2930				
	aq	−54.0	−45.6	−128.9		
$Ni(CO)_4$	l	−633.0	−588.3	313.4	204.6	
	g	−602.91	−587.27	410.5	145.18	
$NiCl_2$	cr	−305.33	−259.06	97.65	71.67	
$NiCl_2 \cdot 6H_2O$	cr	−2103.17	−1713.52	344.3		
NiF_2	cr	−651.4	−604.2	73.60	64.06	
NiO	cr	−239.7	−211.7	37.99	44.31	
$Ni(OH)_2$	cr	−529.7	−447.3	88		
NiS	cr	−82.0	−79.5	52.97	54.64	
$NiSO_4$	cr	−872.9	−759.8	97.1	138.3	
NITROGEN						
N_2	g	0	0	191.500*	29.124*	8.670*
N	g	472.68*	455.58	153.192*	20.786*	6.197*
NF_3	g	−124.7	−83.3	260.62	53.1	
NH_3	g	−45.94*	−16.48	192.660*	35.630*	10.043*
NH_4^+	aq	−133.26*	−79.37	111.17*	79.9	
N_2H_4	l	50.63	149.24	121.21	98.87	
	g	95.40	159.28	238.36	49.58	
HN_3	l	264.0	327.2	140.6		
	g	294.1	328.0	238.86	43.68	
NO	g	90.25	86.57	210.65	29.84	
NOCl	g	51.71	66.07	261.58	44.69	
NO_2	g	33.18	51.30	239.95	37.20	
NO_2^-	aq	−104.6	−32.2	140.2	−97.5	
NO_3^-	aq	−206.85*	−111.34	146.70*	−86.6	
N_2O	g	82.05	104.18	219.74	38.45	
N_2O_3	g	83.72	139.41	312.17	65.61	
N_2O_4	l	−19.50	97.45	209.2	142.7	
	g	9.16	97.82	304.18	77.28	
N_2O_5	cr	−43.1	113.8	178.2	143.1	
	g	11.3	115.1	355.6	84.5	
HNO_2 (cis)	g	−78.00	−42.97	248.66	44.77	
(trans)	g	−80.12	−45.27	249.11	46.07	
HNO_3	l	−174.10	−80.79	155.60	109.87	
	g	−135.06	−74.77	266.27	53.35	
OXYGEN						
O_2	g	0	0	205.043*	29.378*	8.680*
O	g	249.18*	231.75	160.950*	21.912*	6.725*
O_3	g	142.7	163.2	238.82	39.20	
PHOSPHORUS						
P (white)	cr	0	0	41.090*	23.824*	5.360*
(red)	cr	−17.57	−12.1	22.80	21.21	
	g	316.5*	278.28	163.085*	20.786*	6.197*
P_2	g	144.0*		218.014*	32.032*	8.904*
P_4	g	58.9*	24.48	279.902*	67.081*	14.100*
PBr_3	l	−184.5	−175.7	240.2		
	g	−139.3	−162.8	347.98	75.98	

(Continued)

Table XIII.2 (Continued)

Substance	State	$\Delta_f H^\ominus$ /kJ mol^{-1}	$\Delta_f G^\ominus$ /kJ mol^{-1}	$S^\ominus$ /J K^{-1} mol^{-1}	$C_p^\ominus$ /J K^{-1} mol^{-1}	$H^\ominus(298) - H^\ominus(0)$ /kJ mol^{-1}
PBr$_3$	gr	−269.9				
PCl$_3$	l	−319.7	−272.4	217.1		
	g	−287.0	−267.8	311.67	71.84	
PCl$_5$	cr	−443.5				
	g	−374.9	−305.0	364.5	112.8	
PF$_3$	g	−918.8	−897.5	273.13	58.70	
PF$_5$	g	−1595.8				
PH$_3$	g	5.44	13.4	210.12	37.11	
PH$_4$I	cr	−69.9	0.8	123.0	109.6	
PN	g	109.87	87.74	211.08	29.71	
P$_4$O$_6$	cr	−1640				
POCl$_3$	l	−597.1	−520.9	222.46	138.78	
	g	−542.2	−502.5	325.35	84.94	
P$_4$O$_{10}$	cr	−2984.0	−2697.8	228.86	211.71	
H$_3$PO$_4$	cr	−1266.9	−1112.5	110.50	106.06	
	l	−1254.4	−1111.7	150.6	145.06	
PO$_4^{3-}$	aq	−1277.4	−1018.8	−221.8		
HPO$_4^{2-}$	aq	−1299.0*	−1089.2	−33.5		
H$_2$PO$_4^-$	aq	−1302.6*	−1130.4	92.5*		

PLATINUM

Substance	State	$\Delta_f H^\ominus$	$\Delta_f G^\ominus$	$S^\ominus$	$C_p^\ominus$	$H^\ominus(298) - H^\ominus(0)$
Pt	cr	0	0	41.63	25.86	
	g	565.3	520.5	192.30	25.53	
PtO$_2$	g	171.5	167.8			
PtCl$_6^{2-}$	aq	−673.6	−489.5	220.1		

POTASSIUM

Substance	State	$\Delta_f H^\ominus$	$\Delta_f G^\ominus$	$S^\ominus$	$C_p^\ominus$	$H^\ominus(298) - H^\ominus(0)$
K	cr	0	0	64.680*	29.600*	7.008*
	g	89.0*	60.63	160.232*	20.786*	6.197*
K$_2$	g			249.650*	37.982*	10.741*
K$^+$	g	514.19				
	aq	−252.14*	−283.26	101.20*	21.8	
KBr	cr	−393.80	−380.66	95.90	52.30	
	aq	−373.92	−387.23	184.9	−120.1	
KBrO$_3$	cr	−332.2	−243.5	149.16	120.16	
K$_2$CO$_3$	cr	−1151.0	−1063.6	155.52	114.43	
KHCO$_3$	cr	−963.2	−863.6	115.5		
KCN	cr	−113.0	−101.88	128.49	66.27	
KCNS	cr	−200.16	−178.32	124.26	88.53	
	aq	−175.94	−190.58	246.9	−18.4	
KCl	cr	−436.75	−409.15	82.59	51.30	
	aq	−419.53	−414.51	159.0	−114.6	
KClO$_3$	cr	−397.73	−296.31	143.1	100.25	
KClO$_4$	cr	−432.75	−303.17	151.0	112.38	
K$_2$CrO$_4$	cr	−1403.7	−1295.8	200.12	145.98	
K$_2$Cr$_2$O$_7$	cr	−2061.5	−1882.0	289	219.2	
	aq	−1994.9	−1867.7	466.9		
KF	cr	−567.27	−537.77	66.57	49.04	
K$_3$Fe(CN)$_6$	cr	−249.8	−129.7	426.06		
K$_4$Fe(CN)$_6$	cr	−594.1	−453.1	418.8	332.21	
KI	cr	−327.90	−324.89	106.32	52.93	
KI$_3$	aq	−303.8	−334.7	341.8		
KIO$_3$	cr	−501.37	−418.4	151.46	106.48	
	aq	−473.6	−411.3	220.9		
KMnO$_4$	cr	−837.2	−737.6	171.71	117.57	
KNO$_2$	cr	−369.82	−306.60	152.09	107.40	
KNO$_3$	cr	−494.63	−394.93	133.05	96.40	
	aq	−459.74	−394.59	248.9	−64.9	
K$_2$O	cr	−363.2	−322.2	94.1	92.05	

(Continued)

Table XIII.2 (Continued)

Substance	State	$\Delta_f H^\ominus$ /kJ mol^{-1}	$\Delta_f G^\ominus$ /kJ mol^{-1}	$S^\ominus$ /J K^{-1} mol^{-1}	$C_p^\ominus$ /J K^{-1} mol^{-1}	$H^\ominus(298) - H^\ominus(0)$ /kJ mol^{-1}
K_2O_2	cr	−495.8	−425.1	102.1		
KOH	cr	−424.76	−379.11	78.87	64.89	
	aq	−482.37	−440.53	91.63	−126.8	
K_2S	cr	−380.7	−364.0	104.6		
K_2SO_4	cr	−1437.79	−1321.43	175.56	131.46	
$KHSO_4$	cr	−1160.6	−1031.4	138.1		
RADIUM						
Ra	cr	0	0	71	27.0	
Ra^{2+}	aq	−527.6	−561.5	54		
$RaCl_2 \cdot 2H_2O$	cr	−1464	−1302.9	213		
$RaSO_4$	cr	−1471.1	−1365.7	138		
RADON						
Rn	g	0	0	176.1	20.79	
RHENIUM						
Re	cr	0	0	36.86	25.48	
$ReCl_3$	cr	−264	−188	123.8	92.4	
$HReO_4$	cr	−762.3	−656.5	158.2		
RUBIDIUM						
Rb	cr	0	0	76.780*	31.060*	7.489*
	g	80.9*	53.09	169.985*	20.786*	6.197*
Rb_2	g			270.887*	38.027*	10.922*
Rb^+	g	490.10				
	aq	−251.12*	−283.97	121.75*		
RbBr	cr	−394.59	−381.79	109.96	52.84	
Rb_2CO_3	cr	−1136.0	−1051.0	181.33	117.61	
RbCl	cr	−435.35	−407.81	95.90	52.38	
RbI	cr	−333.80	−328.86	118.41	53.18	
$RbNO_3$	cr	−489.7	−395.85	147.3	102.1	
Rb_2O	cr	−339				
RbOH	cr	−418.2				
Rb_2SO_4	cr	−1435.61	−1316.96	197.44	134.06	
SCANDIUM						
Sc	cr	0	0	34.64	25.52	
	g	377.8	336.06	174.68	22.09	
Sc^{3+}	aq	−614.2	−586.6	−255		
ScO	g	−57.24	−83.26	224.47	30.88	
Sc_2O_3	cr	−1908.8	−1819.4	77.0	94.22	
SELENIUM						
Se (black)	cr	0	0	42.44	25.36	
	g	227.07	187.07	176.61	20.83	
Se_2	g	146.0	96.2	251.9	35.40	
$SeCl_4$	cr	−183.3				
SeF_6	g	−1117	−1017	313.76	110.5	
SeO_2	cr	−225.35				
SILICON						
Si	cr	0	0	18.810*	19.789*	3.217*
	g	450*	411.3	167.872*	22.251*	7.550*
$SiBr_4$	l	−457.3	−443.9	277.8		
	g	−415.5	−431.8	377.77	97.11	
SiC	cr	−65.3	−62.8	16.61	26.86	
$SiCl_4$	l	−687.0	−619.90	239.7	145.31	
	g	−657.01	−617.01	330.62	90.25	
SiF_4	g	−1615.0*	−1572.68	282.652*	73.622*	15.356*

(Continued)

Table XIII.2 (Continued)

Substance	State	$\Delta_f H^\ominus$ /kJ mol^{-1}	$\Delta_f G^\ominus$ /kJ mol^{-1}	$S^\ominus$ /J K^{-1} mol^{-1}	$C_p^\ominus$ /J K^{-1} mol^{-1}	$H^\ominus(298) - H^\ominus(0)$ /kJ mol^{-1}
SiF_6^{2-}	aq	−2389.1	−2199.5	122.2		
SiH_4	g	34.3	56.9	204.51	42.84	
Si_2H_6	g	80.3	127.2	272.55	80.79	
Si_3N_4	cr	−743.5	−642.7	101.3		
SiO_2 (quartz)	cr	−910.7*	−856.67	41.460*	44.602*	6.916*
(cristobalite)	cr	−909.48	−855.46	42.68	44.18	
H_2SiO_3	cr	−1188.7	−1092.4	134		
SILVER						
Ag	cr	0	0	42.550*	25.350*	5.745*
	g	284.9*	245.68	172.888*	20.786*	6.197*
Ag^+	g	1019.2				
	aq	105.79*	77.12	73.45*	21.8	
AgBr	cr	−100.37	−96.90	107.1	52.38	
AgCN	cr	146.0	156.9	107.19	66.73	
AgCl	cr	−127.070*	−109.80	96.25*	50.79	12.033*
AgI	cr	−61.84	−66.19	115.5	56.82	
$AgNO_2$	cr	−45.06	19.08	129.20	80.21	
$AgNO_3$	cr	−124.39	−33.47	140.91	93.05	
	aq	−101.80	−34.23	219.2	−64.9	
Ag_2O	cr	−31.05	−11.21	121.3	65.86	
Ag_2S	cr	−32.59	−40.67	144.01	76.53	
AgSCN	cr	87.9	101.38	131.0	63	
Ag_2SO_4	cr	−715.88	−618.48	200.4	131.38	
$[Ag(NH_3)_2]^{2+}$	aq	−111.29	−17.24	245.2		
SODIUM						
Na	cr	0	0	51.300*	28.230*	6.460*
	g	107.5*	76.79	153.609*	20.786*	6.197*
Na_2	g			230.136*	37.575*	10.403*
Na^+	g	609.84				
	aq	−240.30*	−261.89	58.45*	46.4	
$Na_2B_4O_7$	cr	−3291.1	−3096.2	189.54	186.77	
$Na_2B_4O_7 \cdot 10H_2O$	cr	−6288.6	−5516.6	586	615	
NaBr	cr	−361.06	−348.98	86.82	51.38	
Na_2CO_3	cr	−1130.68	−1044.49	134.98	112.30	
$Na_2CO_3 \cdot 10H_2O$	cr	−4081.22	−3428.20	564.00	550.32	
$NaHCO_3$	cr	−950.81	−851.0	101.67	87.61	
NaCN	cr	−87.49	−76.44	115.60	70.37	
NaCNO	cr	−405.39	−358.15	96.7	86.6	
NaCl	cr	−411.15	−384.15	72.13	50.50	
	g	−176.65	−196.65	229.70	35.77	
	aq	−407.27	−393.15	115.5	−90.0	
NaClO	aq	−347.3	−298.7	100.4		
$NaClO_3$	cr	−365.77	−262.32	123.4		
$NaClO_4$	cr	−383.30	−254.93	142.3		
Na_2CrO_4	cr	−1342.2	−1235.0	176.61	142.13	
NaF	cr	−573.65	−543.51	51.46	46.86	
NaI	cr	−287.78	−286.06	98.53	52.09	
NaN_3	cr	21.25	93.76	96.86	76.61	
$NaNH_2$	cr	−123.84	−64.02	76.90	66.15	
$NaNO_2$	cr	−358.65	−284.60	103.76		
$NaNO_3$	cr	−467.85	−367.06	116.52	92.88	
Na_2O	cr	−414.22	−375.47	75.06	69.12	
Na_2O_2	cr	−531.2	−447.69	94.98	89.24	
NaOH	cr	−425.61	−379.53	64.45	59.54	
Na_3PO_4	cr	−1917.4	−1788.9	173.80	153.47	
Na_2HPO_4	cr	−1748.1	−1608.3	150.50	135.31	
NaH_2PO_4	cr	−1536.8	−1386.2	127.49	116.86	
Na_2S	cr	−373.2	−359.8	97.9	79.45	

(Continued)

Table XIII.2 (Continued)

Substance	State	$\Delta_f H^\ominus$ /kJ mol^{-1}	$\Delta_f G^\ominus$ /kJ mol^{-1}	$S^\ominus$ /J K^{-1} mol^{-1}	$C_p^\ominus$ /J K^{-1} mol^{-1}	$H^\ominus(298) - H^\ominus(0)$ /kJ mol^{-1}
Na_2SO_3	cr	−1100.8	−1012.5	145.94	120.25	
Na_2SO_4	cr	−1387.08	−1270.22	149.58	128.20	
$Na_2SO_4 \cdot 10H_2O$	cr	−4327.3	−3647.4	592.0		
$NaHSO_4$	cr	−1125.5	−992.9	113.0		
$Na_2S_2O_3$	cr	−1123.0	−1028.0	155		
$Na_2S_2O_3 \cdot 5H_2O$	cr	−2607.9	−2230.0	372.4		
STRONTIUM						
Sr	cr	0	0	52.3	26.4	
	g	164.4	131.0	164.51	20.79	
Sr^{2+}	g	1790.6				
	aq	−545.80	−559.44	−32.64		
$SrBr_2$	cr	−717.6	−697.1	135.10	75.35	
$SrCO_3$	cr	−1220.0	−1140.1	97.1	81.42	
$SrCl_2$	cr	−828.9	−781.2	114.85	75.60	
SrF_2	cr	−1216.3	−1164.8	82.13	70.00	
$Sr(NO_3)_2$	cr	−978.22	−780.15	194.56	149.91	
SrO	cr	−592.0	−561.9	54.4	45.02	
SrS	cr	−453.1	−448.5	68.2	48.70	
$SrSO_4$	cr	−1453.1	−1341.0	117		
SULPHUR						
S (rhombic)	cr	0	0	32.054*	22.750*	4.412*
(monoclinic)	cr	0.3	0.1			
	g	277.17*	238.28	167.720*	23.674*	6.657*
S_2	g	128.60*		228.058*	32.505*	9.131*
S^{2-}	aq	33.1	85.8	−14.6		
SF_6	g	−1209	−1105.4	291.71	97.28	
H_2S	g	−20.6*	−33.56	205.696*	34.248*	9.957*
HS^-	aq	−16.3*	12.05	67*		
SO_2	g	−296.81*	−300.19	248.114*	39.842*	10.549*
SO_3	cr	−454.51	−368.99	52.3		
	l	−441.04	−368.36	96.60		
	g	−395.72	−371.08	256.65	50.67	
SO_3^{2-}	aq	−635.5	−486.6	−29.3		
HSO_3^-	aq	−626.22	−527.81	139.7		
H_2SO_4	l	−813.99	−690.10	156.90	138.91	
SO_4^{2-}	aq	−909.34*	−744.63	18.50*	−293	
HSO_4^-	aq	−886.9*	−756.00	131.7*	−84	
$S_2O_4^{2-}$	aq	−753.5	−600.4	92		
$S_2O_5^{2-}$	aq	−1338.9	−1110.4	248.1		
$SOCl_2$	g	−212.5	−198.3	309.66	66.5	
SO_2Cl_2	g	−364.0	−320.1	311.83	76.99	
SCN^-	aq	76.44	92.68	144.3	−40.2	
TELLURIUM						
Te	cr	0	0	49.71	25.73	
	g	196.73	157.11	182.63	20.79	
TeO	cr	65.3	38.5	241.4	30.08	
TeO_2	cr	−322.6	−270.3	79.5		
THALLIUM						
Tl	cr	0	0	64.64	26.32	
	g	182.21	147.44	180.85	20.79	
Tl^+	g	777.73				
	aq	5.36	−32.38	125.5		
Tl^{3+}	g	5639.2				
	aq	196.6	214.6	−192		
TlBr	cr	−173.2	−167.36	120.5		
Tl_2CO_3	cr	−700.0	−614.6	155.2		

(Continued)

Table XIII.2 (Continued)

Substance	State	$\Delta_f H^\ominus$ /kJ mol^{-1}	$\Delta_f G^\ominus$ /kJ mol^{-1}	$S^\ominus$ /J K^{-1} mol^{-1}	$C_p^\ominus$ /J K^{-1} mol^{-1}	$H^\ominus(298) - H^\ominus(0)$ /kJ mol^{-1}
TlCl	cr	−204.14	−184.93	111.25	50.92	
TlI	cr	−123.8	−125.39	127.6		
Tl$_2$O	cr	−178.7	−147.3	126		
TlOH	cr	−238.9	−195.8	88		
Tl$_2$S	cr	−97.1	−93.7	151		
Tl$_2$SO$_4$	cr	−931.8	−830.48	230.5		
THORIUM						
Th	cr	0	0	51.830*	26.230*	6.350*
	g	602*	557.56	190.060*	20.789*	6.197*
Th^{4+}	aq	−769.0	−705.0	−422.6		
ThBr$_4$	cr	−965.2	−927.2	230		
	g	−762.7	−784.5	431	105.0	
ThCl$_4$	cr	−1186.6	−1094.5	190.4		
ThI$_4$	cr	−664.8	−655.2	255		
	g	−460.7	−515.1	469	106.27	
ThO$_2$	cr	−1226.4*	−1168.80	65.23*	61.76	10.560*
TIN						
Sn (white)	cr	0	0	51.180*	27.112*	6.323*
(grey)	cr	−2.1	0.13	44.14	25.77	
	g	301.2*	267.4	168.383*	21.259*	6.215*
Sn^{2+}	g	2434.9				
	aq	−8.9*	−27.2	−16.7*		
SnBr$_2$	cr	−243.5				
SnBr$_4$	cr	−377.4	−350.2	264.4		
SnCl$_2$	cr	−325.1				
	aq	−329.7	−299.6	172		
SnCl$_4$	l	−511.3	−440.2	258.6	165.3	
SnO	cr	−280.71*	−256.9	57.170*	47.783*	8.736*
SnO$_2$	cr	−577.63*	−519.7	49.04*	53.219*	8.384*
Sn(OH)$_2$	cr	−561.1	−491.6	155		
SnS	cr	−100	−98.3	77.0	49.25	
TITANIUM						
Ti	cr	0	0	30.63	25.02	
	g	473*	425.1	180.189*	24.430*	7.539*
TiBr$_4$	cr	−616.7	−589.5	243.5	131.5	
	g	−549.4	−568.2	398.3	100.8	
TiC	cr	−184.5	−180.7	24.23	33.64	
TiCl$_4$	l	−804.2	−737.2	252.34	145.18	
	g	−763.2*		353.131*	95.408*	21.487*
TiN	cr	−338.1	−309.6	30.25	37.07	
TiO$_2$ (rutile)	cr	−944.7	−889.5	50.620*	55.080*	8.680*
TUNGSTEN						
W	cr	0	0	32.64	24.27	
	g	849.4	807.1	173.84	21.31	
WC	cr	−40.54		32.38	35.35	
WO$_2$	cr	−589.69	−533.92	50.54	56.11	
WO$_3$	cr	−842.87	−764.08	75.90	73.76	
URANIUM						
U	cr	0	0	50.200*	27.665*	6.364*
	g	533*	491.2	199.680*	23.694*	6.499*
UCl$_6$	cr	−1092	−962	285.8	175.7	
UF$_6$	cr	−2197.0	−2068.6	227.6	166.77	
	g	−2112.9	−2029.2	379.7	167.49	
UO$_2$	cr	−1085.0*	−1031.8	77.030*	63.600*	11.280*
	g	−465.7	−471.5	274.5	51.38	

(Continued)

Table XIII.2 (Continued)

Substance	State	$\Delta_f H^\ominus$ /kJ mol^{-1}	$\Delta_f G^\ominus$ /kJ mol^{-1}	$S^\ominus$ /J K^{-1} mol^{-1}	$C_p^\ominus$ /J K^{-1} mol^{-1}	$H^\ominus(298) - H^\ominus(0)$ /kJ mol^{-1}
UO$_3$ (γ)	cr	−1223.8*	−1146.0	96.110*	81.670*	14.585*
U$_3$O$_8$	cr	−3574.8*	−3369.8	282.550*	238.000*	42.740*
UO$_2$Cl$_2$	cr	−1243.9	−1146.4	150.54	107.86	
UO$_2$(NO$_3$)$_2$	cr	−1349.3	−1105.0	243		
UO$_2^{2+}$	aq	−1019.0*	−953.5	−98.2*		
VANADIUM						
V	cr	0	0	28.91	24.89	
	g	514.21	453.26	182.19	26.01	
VCl$_2$	cr	−452	−406	97.1	72.22	
VCl$_3$	cr	−580.7	−511.3	131.0	93.18	
VCl$_4$	l	−569.4	−503.8	255		
VF$_5$	l	−1480.3	−1373.2	175.7		
VO	cr	−431.8	−404.2	38.9	45.44	
VO^{2+}	aq	−486.6	−446.4	−133.9		
V$_2$O$_3$	cr	−1218.8	−1139.3	98.3	103.22	
V$_2$O$_4$	cr	−1427.2	−1318.4	102.5	116.98	
V$_2$O$_5$	cr	−1550.6	−1419.6	131.0	127.65	
XENON						
Xe	g	0	0	169.576*	20.786*	6.917*
XeF$_2$	g	−84	−48	254	54.1	
XeF$_4$	cr	−261.5	−123	146	118	
	g	−215.5	−209.2	327.8	118.5	
XeF$_6$	cr	−360				
	g	−297				
ZINC						
Zn	cr	0	0	41.630*	25.390*	5.657*
	g	130.40*	95.18	160.881*	20.786*	6.197*
Zn^{2+}	g	2782.7				
	aq	−153.39*	−147.03	−109.8*	46	
ZnBr$_2$	cr	−328.65	−312.13	138.5		
ZnCO$_3$	cr	−812.78	−731.57	82.4	79.7	
ZnCl$_2$	cr	−415.28	−369.43	111.46	71.34	
ZnF$_2$	cr	−764.4	−713.4	73.68	65.65	
ZnI$_2$	cr	−208.28	−208.95	161.1		
Zn(NO$_3$)$_2$	cr	−483.7				
Zn(NO$_3$)$_2 \cdot$ 6H$_2$O	cr	−2306.64	−1773.14	456.9	323.0	
ZnO	cr	−350.46*	−318.32	43.65*	40.25	6.933*
Zn(OH)$_2$	cr	−643.25	−555.13	81.6	72.4	
ZnS (wurtzite)	cr	−192.63				
(zinc blende)	cr	−205.98	−201.29	57.7	46.0	
ZnSO$_4$	cr	−982.8	−874.5	119.7		
ZnSO$_4 \cdot$ 7H$_2$O	cr	−3077.75	−2563.08	388.7	383.4	
ZIRCONIUM						
Zr	cr	0	0	38.99	25.36	
	g	608.8	566.5	181.25	26.65	
ZrCl$_4$	cr	−980.52	−889.9	181.6	119.79	
ZrF$_4$	cr	−1911.3	−1810.0	104.6	103.72	
ZrO$_2$	cr	−1100.56	−1042.82	50.38	56.19	

Table XIII.3 CODATA Recommended Values for Thermodynamic Functions of Chemical Species at 101 325 Pa
For more extensive data see: J. D. Cox, D. D. Wagman and V. A. Medvedev, *Codata Key Values for Thermodynamics* (1989), Hemisphere Publishing Corporation, New York.
g = gaseous state, l = liquid state, s = solid state

Species	Temperature /K	$C_p^\ominus$ /J K^{-1} mol^{-1}	$-[G^\ominus - H^\ominus(0)]/T$ /J K^{-1} mol^{-1}	$S^\ominus$ /J K^{-1} mol^{-1}	$H^\ominus - H^\ominus(0)$ /kJ mol^{-1}
O (g)	100	23.703	113.769	135.839	2.207
	298.15	21.912	138.393	160.950	6.725
	600	21.124	153.974	175.952	13.187
	1000	20.915	165.096	186.682	21.585
	2000	20.827	179.920	201.140	42.439
	3000	20.937	188.497	209.597	63.300
	4000	21.302	194.564	215.665	84.401
O$_2$ (g)	100	29.112	144.188	173.199	2.901
	298.15	29.378	175.928	205.043	8.680
	600	32.093	196.467	226.342	17.925
	1000	34.880	212.086	243.473	31.387
	2000	37.783	234.716	268.657	67.883
	3000	39.986	248.806	284.405	106.798
	4000	41.695	259.235	296.156	147.681
H (g)	100	20.786	71.114	91.900	2.079
	298.15	20.786	93.822	114.608	6.197
	600	20.786	108.358	129.144	12.472
	1000	20.786	118.976	139.762	20.786
	2000	20.786	133.384	154.170	41.572
	3000	20.786	141.812	162.598	62.359
	4000	20.786	147.792	168.578	83.145
H$_2$ (g)	100	28.155	70.625	100.617	2.999
	298.15	28.836	102.170	130.571	8.469
	600	29.327	122.170	150.967	17.278
	1000	30.204	136.958	166.105	29.147
	2000	34.277	157.598	188.308	61.419
	3000	37.076	170.377	202.778	97.201
	4000	39.085	179.901	213.729	135.313
H$_2$O (g)	100	33.310	119.380	152.272	3.289
	298.15	33.609	155.504	188.726	9.905
	600	36.289	178.930	212.926	20.398
	1000	41.544	196.729	232.699	35.970
	2000	51.600	223.388	264.873	82.971
	3000	57.274	241.086	286.985	137.698
	4000	60.428	254.762	303.933	196.682
He (g)	100	20.786	82.580	103.336	2.079
	298.15	20.786	105.257	126.044	6.197
	600	20.786	116.004	136.790	10.393
	1000	20.786	130.412	151.198	20.786
	2000	20.786	144.820	165.606	41.572
	3000	20.786	153.248	174.034	62.359
	4000	20.786	159.228	180.014	83.145
Ne (g)	100	20.786	102.726	123.512	2.079
	298.15	20.786	125.433	146.219	6.197
	600	20.786	139.969	160.756	12.472
	1000	20.786	150.588	171.374	20.786
	2000	20.786	164.995	185.782	41.572
	3000	20.786	173.423	194.210	62.359
	4000	20.786	179.403	200.190	83.145
F (g)	100	21.204	113.512	134.371	2.086
	298.15	22.746	136.779	158.642	6.518

(Continued)

Table XIII.3 (Continued)

Species	Temperature /K	$C_p^\ominus$ /J K^{-1} mol^{-1}	$-[G^\ominus - H^\ominus(0)]/T$ /J K^{-1} mol^{-1}	$S^\ominus$ /J K^{-1} mol^{-1}	$H^\ominus - H^\ominus(0)$ /kJ mol^{-1}
	600	21.833	152.188	174.259	13.243
	1000	21.266	163.412	185.254	21.842
	2000	20.925	178.411	199.855	42.888
	3000	20.851	187.066	208.322	63.769
	4000	20.823	193.165	214.316	84.604
F_2 (g)	100	29.114	141.192	170.263	2.907
	298.15	31.304	173.683	202.682	8.825
	600	35.172	194.392	225.963	18.942
	1000	37.065	211.00	244.450	33.451
	2000	38.935	235.018	270.818	71.601
	3000	38.006	249.760	286.548	110.365
	4000	34.728	260.353	297.056	146.813
HF (g)	100	29.133	113.577	141.845	2.827
	298.15	29.137	144.826	173.670	8.590
	600	29.230	165.056	194.062	17.403
	1000	30.169	179.919	209.162	29.243
	2000	34.070	200.600	231.321	61.443
	3000	36.394	213.358	245.622	96.793
	4000	37.805	222.814	256.297	133.933
Cl (g)	100	20.788	121.281	142.067	2.070
	298.15	21.838	144.047	165.081	6.271
	600	22.782	159.011	180.797	13.072
	1000	22.234	170.235	192.322	22.086
	2000	21.341	185.505	207.397	43.784
	3000	21.063	194.334	215.989	64.963
	4000	20.950	200.539	222.030	85.962
Cl_2 (g)	100	29.299	159.781	188.900	2.912
	298.15	33.949	192.179	222.972	9.181
	600	36.547	214.547	247.742	19.917
	1000	37.442	231.915	266.661	34.747
	2000	38.468	256.581	292.935	72.707
	3000	40.921	271.514	308.913	112.195
	4000	44.094	282.449	321.159	154.844
HCl (g)	100	29.117	126.293	154.979	2.869
	298.15	29.136	157.816	186.793	8.640
	600	29.576	178.122	207.246	17.474
	1000	31.640	193.111	222.799	26.689
	2000	35.690	214.349	246.169	63.640
	3000	37.506	227.578	261.022	100.331
	4000	38.689	237.367	271.981	138.455
Br (g)	100	20.786	131.415	152.201	2.079
	289.15	20.786	154.123	174.909	6.197
	600	20.833	168.660	189.451	12.475
	1000	21.365	179.298	200.194	20.896
	2000	22.710	193.973	215.511	43.075
	3000	22.669	202.794	224.737	65.827
	4000	22.298	209.130	231.208	88.314
Br_2 (g)	100	30.899	179.028	208.513	2.948
	298.15	36.057	212.740	245.359	9.725
	600	37.305	236.333	271.080	20.848
	1000	37.793	254.386	290.265	35.879
	2000	39.089	279.680	316.783	74.205
	3000	41.619	294.925	333.126	114.601
	4000	41.375	306.053	345.160	156.429

(Continued)

Table XIII.3 (Continued)

Species	Temperature /K	$C_p^\ominus$ /J K^{-1} mol^{-1}	$-[G^\ominus-H^\ominus(0)]/T$ /J K^{-1} mol^{-1}	$S^\ominus$ /JK^{-1} mol^{-1}	$H^\ominus-H^\ominus(0)$ /kJ mol^{-1}
HBr (g)	100	29.115	138.004	166.775	2.877
	298.15	29.141	169.583	198.591	8.649
	600	29.873	189.914	219.108	17.516
	1000	32.335	204.981	234.928	29.946
	2000	36.230	226.483	258.758	64.550
	3000	37.936	239.898	273.801	101.710
	4000	39.107	249.818	284.884	140.261
I (g)	100	20.786	137.185	157.971	2.079
	298.15	20.786	159.892	180.678	6.197
	600	20.786	174.429	195.215	12.472
	1000	20.795	185.047	205.834	20.787
	2000	21.308	199.472	220.354	41.764
	3000	22.191	207.990	229.167	63.530
	4000	22.679	214.128	235.629	86.004
I$_2$ (g)	100	33.135	191.544	221.896	3.035
	298.15	36.888	226.650	260.578	10.116
	600	37.613	251.015	286.655	21.384
	1000	38.070	269.456	305.979	36.522
	2000	40.474	295.168	332.968	75.597
	3000	41.630	310.739	349.742	117.010
	4000	39.323	322.038	361.452	157.657
HI (g)	100	29.114	145.816	174.665	2.885
	298.15	29.157	177.447	206.481	8.657
	600	30.351	197.813	227.126	17.587
	1000	33.157	212.999	243.302	30.302
	2000	36.787	234.825	267.627	65.604
	3000	38.390	248.453	282.873	103.259
	4000	39.244	258.518	294.054	142.144
S (s)	100	12.770	5.640	12.489	0.685
	298.15	22.750	17.256	32.054	4.412
(l)	600	34.379	32.478	60.103	16.575
	1000	32.000	47.160	76.686	29.526
S (g)	100	21.356	121.898	142.783	2.088
	298.15	23.674	145.391	167.720	6.657
	600	22.338	161.195	183.860	13.600
	1000	21.490	172.699	195.034	22.335
	2000	21.276	187.980	209.773	43.585
	3000	21.999	196.793	218.525	65.196
	4000	22.671	203.064	224.953	87.556
SO$_2$ (g)	100	33.543	175.642	208.934	3.329
	298.15	39.842	212.731	248.114	10.549
	600	48.938	238.946	279.062	24.070
	1000	54.290	260.646	305.541	44.894
	2000	57.891	293.877	344.598	101.443
	3000	59.152	314.992	368.331	160.018
	4000	60.144	330.568	385.484	219.666
H$_2$S (g)	100	33.293	136.035	169.116	3.308
	298.15	34.248	172.298	205.696	9.957
	600	39.052	196.070	231.032	20.977
	1000	45.968	214.627	252.650	38.023
	2000	55.084	243.196	287.888	89.385
	3000	58.553	262.160	310.966	146.416
	4000	60.551	276.594	328.102	206.030

(Continued)

Table XIII.3 (Continued)

Species	Temperature /K	$C_p^\ominus$ /J K^{-1} mol^{-1}	$-[G^\ominus - H^\ominus(0)]/T$ /J K^{-1} mol^{-1}	$S^\ominus$ /J K^{-1} mol^{-1}	$H^\ominus - H^\ominus(0)$ /kJ mol^{-1}
N (g)	100	20.786	109.698	130.485	2.079
	298.15	20.786	132.406	153.192	6.197
	600	20.786	146.942	167.728	12.472
	1000	20.786	157.560	178.347	20.786
	2000	20.790	171.968	192.755	41.573
	3000	20.963	180.398	201.204	62.416
	4000	21.809	186.397	207.330	83.730
N$_2$ (g)	100	29.104	130.679	159.701	2.902
	298.15	29.124	162.419	191.500	8.670
	600	30.109	182.794	212.067	17.564
	1000	32.696	197.929	228.061	30.132
	2000	35.969	219.562	251.964	64.805
	3000	37.027	232.986	266.780	101.383
	4000	37.547	242.837	277.510	138.693
NH$_3$ (g)	100	33.284	122.653	155.731	3.308
	298.15	35.630	158.975	192.660	10.043
	600	45.229	183.442	220.471	22.218
	1000	56.244	203.665	246.277	42.612
	2000	72.067	236.958	290.937	107.959
	3000	78.758	260.329	321.590	183.781
	4000	82.396	278.661	344.786	264.498
C (s) graphite	100	1.674	0.356	0.952	0.060
	298.15	8.517	2.218	5.740	1.050
	600	16.844	6.211	14.533	4.994
	1000	21.610	11.612	24.457	12.845
	2000	25.094	22.483	40.771	36.576
	3000	26.611	30.427	51.253	62.478
	4000	27.801	36.651	59.075	89.696
C (g)	100	21.271	111.218	135.072	2.385
	298.15	20.839	136.070	157.991	6.536
	600	20.799	151.184	172.548	12.818
	1000	20.791	162.034	183.170	21.136
	2000	20.952	176.620	197.605	41.969
	3000	21.621	185.139	206.215	63.220
	4000	22.363	191.232	212.540	85.231
CO (g)	100	29.104	136.723	165.749	2.903
	298.15	29.141	168.466	197.551	8.671
	600	30.440	188.860	218.214	17.613
	1000	33.178	204.073	234.431	30.358
	2000	36.241	225.900	258.604	65.409
	3000	37.205	239.442	273.508	102.199
	4000	37.697	249.367	284.285	139.673
CO$_2$ (g)	100	29.206	149.792	178.888	2.910
	298.15	37.135	182.264	213.676	9.365
	600	47.327	206.046	243.170	22.275
	1000	54.320	226.422	269.191	42.769
	2000	60.355	258.783	309.196	100.825
	3000	62.180	279.991	334.065	162.220
	4000	63.224	295.865	352.101	224.942

Table XIII.4 Thermodynamic Data for Selected Organic Compounds at 298.15 K

For a pure solid (cr) or liquid (l) the standard state is the substance in the condensed phase at 298.15 K and a pressure of 1 atmosphere. For a gas (g) the standard state is the hypothetical state at unit fugacity, in which state the enthalpy is that of the real gas at the same temperature and zero pressure. For dissolved substances (aq) the standard state is taken as the hypothetical state of unit molality at 298.15 K.
The compounds listed represent only a small fraction of those for which data are available in the *National Bureau of Standards Technical Notes* 270-3, 270-4, 270-5, 270-6, 270-7 and 270-8.
See also DIPPR Data compilation, Hemisphere, Thermodynamics Tables, Thermodynamic Research Centre, Texas, A & M.

Compound	State	$\Delta_f H^\ominus$ /kJ mol^{-1}	$\Delta_f G^\ominus$ /kJ mol^{-1}	$S^\ominus$ /J K^{-1} mol^{-1}	$C_p^\ominus$ /J K^{-1} mol^{-1}
acetaldehyde	l	−192.30	−128.20	160.25	
	g	−166.36	−133.72	265.68	62.8
acetamide	s	−317.98			66.94
acetic acid	l	−484.5	−389.94	159.83	124.26
	g	−432.25	−374.05	282.42	66.52
acetate ion	aq	−486.0	−369.41	86.61	−6.28
acetone	g	−216	−152	295	74.9
acetonitrile	l	53.56	99.1	149.62	91.46
	g	87.44	104.6	245.48	52.22
acetyl chloride	l	−273.8	−208.07	200.83	117.2
	g	−243.51	−205.85	294.97	67.78
acetylene	g	226.75	209.2	200.82	43.9
ammonium acetate	s	−616.14			
	aq	−618.52	−448.78	200.0	73.64
ammonium bicarbonate	s	−849.35	−666.1	120.92	
	aq	−824.5	−666.2	204.6	
ammonium carbamate	s	−645.05	−448.1	133.47	
ammonium carbonate	aq	−942.15	−686.64	169.9	
ammonium cyanate	s	−304.39			
	aq	−278.65	−177.0	220.1	
ammonium oxalate	s	−1123.0			225.9
	aq	−1090.0	−832.6	272.0	
ammonium thiocyanate	s	−78.66			
	aq	−56.07	13.3	257.7	39.75
benzene	l	49.0	125.0	173.0	136.0
	g	82.93	129.7	269.2	81.6
benzoic acid	s	−385	−245	167	147
bromoethane	l	−92.0	−27.78	198.7	100.8
	g	64.52	−26.53	286.6	64.52
bromoform	l	−28.45	−5.0	220.9	129.7
	g	16.7	8.4	330.8	71.2
bromomethane	g	−35.15	−25.9	246.3	42.42
bromotrichloromethane	g	−36.86	−46.0	332.8	85.27
bromotrifluoromethane	g	−642.7	−616.3	297.6	69.33
1,2-butadiene	g	165.5	201.7	293	80
1,3-butadiene	g	111.9	152.4	278.7	79.5
n-butane	g	−124.7	−15.7	310.0	97.5
1-butene	g	1.17	72.0	307.4	86.0
cis-2-butene	g	−5.7	67.14	300.8	79.0
trans-2-butene	g	−10.1	64.1	296.5	88.0
carbon dioxide	g	−393.5	−394.4	213.6	37.11
carbon disulphide	l	89.7	65.3	151.3	75.7
	g	117.4	67.2	237.7	45.4
carbonic acid	aq	−699.6	−623.2	187.4	
bicarbonate ion	aq	−692.0	−586.8	91.2	
carbonate ion	aq	−677.1	−530.7	−56.9	
carbon monoxide	g	−113.8	−137.2	197.6	29.1
carbon oxysulphide	g	−140.1	−169.3	231.5	41.5
carbon tetrabromide	s	18.8	47.7	212.5	144.3
	g	66.9	20.4	91.2	
carbon tetrachloride	l	−135.4	−65.27	216.4	131.7
	g	−102.92	−60.62	309.7	83.3
carbonyl bromide	s	−127.2			
	g	−69.2	−110.9	309.0	61.8

(Continued)

Table XIII. 4 (Continued)

Compound	State	$\Delta_f H^\ominus$ /kJ mol^{-1}	$\Delta_f G^\ominus$ /kJ mol^{-1}	$S^\ominus$ /J K^{-1} mol^{-1}	$C_p^\ominus$ /J K^{-1} mol^{-1}
carbonyl chloride	g	−218.8	−204.6	283.4	57.7
chlorobenzene	g	52.3	99.0	314.0	
chloroethane	l	−136.5	−59.4	190.8	104.3
	g	−112.2	−60.5	275.9	62.8
chloroform	l	−131.8	−71.5	202.9	116
	g	−100.4	−66.9	296.5	65.8
chloromethane	g	−80.8	−57.4	234.5	40.8
chlorofluorocarbons					
CCl_3F (Freon 11)	l	−301.3	−236.9	225.4	121.5
	g	−276.1	−238.5	309.8	78.1
CCl_2F_2 (Freon 12)	g	−476.9	−439.3	300.7	72.3
$CClF_3$ (Freon 13)	g	−694.5	−652.7	285.2	66.9
CF_4 (Freon 14)	g	−924.7	−878.6	261.5	61.1
$CHCl_2F$ (Freon 21)	g	−299	−268	293	60.9
$CHClF_2$ (Freon 22)	g	−502	−471	281	55.9
CHF_3 (Freon 23)	g	−688.3	−653.9	259.6	51.0
CCl_3CF_2Cl (Freon 112)	g	−489.9	−407.1	382.8	123.4
CCl_3CF_3 (Freon 113)	g			370.7	120.5
$CF_2{:}CFCl$		−555.2	−523.8	322.0	83.9
$CF_2ClCFCl_2$	l	−788.1			173.6
$CF_2{:}CHCl$	g	−315.5	−289.1	302.9	72.1
cyanic acid	aq	−154.4	−117.1	144.8	
cyanate ion	aq	−146.0	−97.5	106.7	
cyanogen	g	308.9	297.4	240.9	56.8
cyanogen bromide	s	140.5			
	g	186.2	165.3	248.2	46.9
cyanogen chloride	g	144.3	137.7	235.6	
cyanogen iodide	s	166.1	185.0	96.2	
	g	225.5	196.6	256.7	48.3
cyclohexane	g	−123.1	31.8	298.2	106
cyclopentane	g	−77.2	38.6	292.9	83
cyclopropane	g	53.4	104	237	
diazomethane	g			242.8	52.5
1,1-dibromoethane				327.6	80.8
1,2-dibromoethane	l	−81.2	−20.9	223.3	136.0
	g	−38.3	−10.3	330.9	86.6
cis-1,2-dibromoethylene	g			311.2	68.8
trans-1,2-dibromoethylene	g			313.4	70.2
dibromomethane	g	−4	−6	293.1	54.7
1,2-dichlorethane	l	−165.2	−79.6	208.5	129.3
	g	−129.8	−73.9	308.3	78.7
cis-1,2-dichloroethylene	l	−27.6	22.1	198.4	113.0
trans-1,2-dichloroethylene	l	−23.1	27.3	195.9	113.0
dichloromethane	l	−121.5	−67.3	177.8	100.0
	g	−92.5	−65.9	274.3	51.0
diethylamine	l	−103.3			
diethylamine	g	−71.4	72	352	
diethyl ether	l	−279.6			
	g	−252.1	−122	343	
diethyl sulphide	l	−118.9	11.8	269.3	171.4
	g	−83.1	18.2	368.0	117.0
1,1-difluoroethane	g	−478.2	−420.9	282.4	67.8
1,1-difluoroethylene	g	−328.9	−305.4	266.1	60.1
difluoromethane	g	−446.9	−419.2	246.6	42.9
1,2-diiodoethane	s	0.42	57.7	196.6	
	g	66.5	78.7	348.1	80.3
diiodomethane	g	118	101	309	58
dimethylamine	l	−43.9	69.9	182.3	137.7
	g	−18.5	68.4	273.0	70.7
	aq	−70.6	57.9	133.1	

(Continued)

Table XIII. 4 (Continued)

Compound	State	$\Delta_f H^\ominus$ /kJ mol^{-1}	$\Delta_f G^\ominus$ /kJ mol^{-1}	$S^\ominus$ /J K^{-1} mol^{-1}	$C_p^\ominus$ /J K^{-1} mol^{-1}
dimethylammonium nitrate	s	350.2			
	aq	−327.6	−114.7	318.8	
1,2-dimethylbenzene	g	19.0	122.1	352.8	133
1,3-dimethylbenzene	g	17.2	118.8	357.7	128
1,4-dimethylbenzene	g	17.9	121.1	352.4	127
dimethyl ether	g	−184.5	−114.2	266.6	66.1
dimethyl sulphide	l	−56.9	14.2	196.4	
	g	−28.9	15.5	285.7	
dimethyl sulphone	s	−451.0	−302.5	142.3	
	g	−371.1	−272.8	310.5	100.0
dimethyl sulphoxide	l	−203.3	−99.2	188.3	147.3
	g	−150.5	−81.5	306.3	89.0
diphenyl ether	l	−15	144	291	
ethane	g	−84.66	−32.89	229.5	52.7
1,2-ethanediol	l	−454.8	−323.2	166.9	149.8
ethanethiol	l	−73.34	−5.35	207.0	117.9
	g	−45.8	−4.39	296.1	72.7
ethanol	l	−277.7	−174.9	160.7	111.5
	g	−235.1	−168.6	282.6	65.4
	aq	−288.3	−181.8	148.5	
ethylamine	l	−74.1			129.7
	g	−47.2	37	285	70
ethyl benzene	g	29.8	130.6	360.5	128
ethylene	g	52.3	68.1	219.8	43.5
ethylenediamine	l	−24.4			66.9
ethyl nitrate	l	−190.3	−43.1	247.2	170.3
ethyl nitrite	l	−128.9			
ethylene oxide	l	−77.8	−11.8	153.8	87.9
	g	−52.6	−13.1	242.4	47.9
fluoroethane	g			264.4	58.8
fluoromethane	g	−234	−210	222.8	37.5
formaldehyde	g	−113.4	−117.2	218.7	35.4
formic acid	l	−424.7	−361.4	129.0	99.0
	g	−378.6	−335.7	246.1	251.0
unionized	aq	−425.4	−372.4	163.2	
formate ion	aq	−425.6	−351.0	92.0	−87.9
α-D-glucose	s	−1274.5	−910.6	212.1	218.9
β-D-glucose	s	−1268.1	−908.9	228.0	
glycerol	l	−670.7	−479.5	204.6	216.7
n-heptane	g	−187.8	8.74	425.3	189
hexafluoroethane	g	−1297.0	−1213.4	332.2	106.7
n-hexane	g	−167.2	0.21	386.8	166
hydrogen cyanide	l	108.9	124.9	112.8	70.6
	g	135.1	124.7	201.7	35.9
cyanide ion	aq	150.6	172.4	94.1	
iodoethane	l	−40.2	14.6	211.7	115.1
iodoform	s	141.0			
	g	211	178	356.1	75.0
iodomethane	l	−15.5	13.4	163.2	125.5
	g	13.0	14.6	254.0	44.1
isocyanic acid	g			237.9	44.9
ketene	g	−61.1	−61.9	247.5	51.8
L(+)-lactic acid	s	−694.0	−523.3	143.1	127.6
	aq	−686.2	−538.8	221.8	
lactate ion	aq	−686.6	−516.7	146.4	
methane	g	−74.8	−50.75	186.2	35.3
methanethiol	l	−46.4	−7.74	169.2	90.5
	g	−22.3	−9.33	255.1	50.2

(Continued)

Table XIII. 4 (Continued)

Compound	State	$\Delta_f H^\ominus$ /kJ mol^{-1}	$\Delta_f G^\ominus$ /kJ mol^{-1}	$S^\ominus$ /J K^{-1} mol^{-1}	$C_p^\ominus$ /J K^{-1} mol^{-1}
methanol	l	−238.7	−166.4	126.8	81.6
	g	−200.7	−162.0	239.7	43.9
	aq	−245.9	−175.4	133.1	
methylamine	l	−47.3	35.6	150.2	
	g	−23.0	32.1	243.3	53.1
	aq	−70.2	20.7	123.4	
methyl cyanide	l	53.1	100	144	
methyl isocyanate	g	150.2	167.4	245.9	
methylisothiocyanate	g	131.0	144.3	289.9	65.5
methyl nitrate	l	−159.0	−43.5	217.1	157.3
	g	−124.7	−39.3	318.4	
methyl nitrite	g	−69.0			
2-methylpropane	g	−131.6	−18.0	294.6	97
nitroethane	g	−101	−5	315	
nitromethane	l	−113.1	−14.5	171.8	106.0
	g	−74.7	−6.9	274.8	57.3
oxalic acid	s	−827.2	−701	120	117.2
oxalate ion	aq	−825.1	−674	45.6	
hydrogen oxalate ion	aq	−818.0	−699.1	153.6	
n-pentane	g	−146.4	−8.2	348.4	120
phenol	s	−163	−50.9	146	
	g	−96	−33	316	104
propadiene	g	192.1	202.4	243.9	61
propane	g	−103.8	−23.5	269.9	73.6
propylene	g	20.4	62.7	266.9	64.0
propyne	g	185.4	193.8	248.1	61
pyridine	g	140.2	190.2	282.8	78.1
styrene	g	147.8	213.8	345.1	122.1
sucrose	s	−2222.1	−1544.6	360.2	425.0
tetrachloroethylene	l	−52.3	4.6	271.5	141.0
	g	−12.1	22.6	341.0	94.9
tetrafluoroethylene	g	−650.6	−615.9	299.9	80.5
thiocyanic acid	aq		97.5		
thiocyanate ion	aq	76.4	92.7	144.3	−40.2
thiophene	g	115.7	126.8	278.9	72.9
thiourea	g	−90.0	21.7	115.8	
toluene	g	50.0	122.3	319.7	104
trichloroethylene	l	−42.3	12.1	228.4	120.4
	g	−7.8	18.0	324.7	80.2
1,1,1-trifluoroethane	g	−736.4	−667.3	279.8	78.2
trifluoromethane	g	−688.3	−654.0	259.6	51.0
trimethylamine	l	−46.0	100.8	208.4	135.2
	g	−24.3	98.9	287.0	
	aq	−76.0	93.0	133.5	
urea	s	−333.5	−197.4	104.6	93.1
vinyl bromide	g	78.2	80.8	275.7	55.5
vinyl chloride	g	35.6	51.9	263.9	53.7

Table XIII.5 Thermodynamic Data for Selected Biological Molecules at 298.15 K and 1 Atmosphere Pressure

For additional information see: S.L. Miller and D. Smith-Magowan, *J. Phys. Chem. Ref. Data* (1990), **19**, 1049.
For a pure solid (s) or liquid (l) the standard state is the substance in the condensed phase at 298.15 K and 1 atmosphere pressure. For dissolved substances (aq) the standard state is taken as the hypothetical state of unit molality at 298.15 K.

Compound	State	$\Delta_f H^\ominus$ /kJ mol^{-1}	$\Delta_f G^\ominus$ /kJ mol^{-1}	$S^\ominus$ /J K^{-1} mol^{-1}	$C_P^\ominus$ /J K^{-1} mol^{-1}
acetoacetate ion (−)	aq		−493.71		
cis-aconitate ion (3−)	aq		−922.61		
adenine	s	97.07	−300.41	151.04	143.09
	aq		312.29		
DL-alanine	s	−563.59	−371.96	132.21	121.75
L-alanine	s	−562.75	−370.20	129.20	122.26
	pH 7	−554.80	−371.16	158.99	
ion (+)	aq	−557.94	−384.55	192.05	288.70
dipolar ion (+, −)	aq	−554.80	−371.16	158.99	141.00
ion (−)	aq	−509.61	−314.85	121.75	71.55
DL-alanylglycine	s	−777.8	−489.95	211.34	182.4
L-arginine	s	−621.74	−656.89	250.62	233.47
dipolar ion (+, −)	aq	−615.47			
DL-aspartic acid	s	−976.96	−729.27	154.39	
L-aspartic acid	s	−972.53	−729.36	170.12	155.27
ion (+)	aq	−955.17	−733.87	229.28	
dipolar ion (+, −)	aq	−947.43	−718.06	216.31	
dipolar ion (+, 2−)	aq	−943.41	−695.88	155.65	
ion (2−)	aq	−905.84	−638.69	89.96	
L-asparagine	s	−790.36	−530.95	174.47	160.67
dipolar ion (+, −)	aq	−766.09	−525.93	238.91	
citric acid	s	−1543.90			
dihydrogen citrate (−)	aq	−1520.88	−1226.33	286.19	187.86
hydrogen citrate (2−)	aq	−1518.46	−1199.18	202.34	0.84
citrate (3−)	aq	−1515.11	−1162.69	92.05	−254.81
creatinine	s	−237.65	−28.45	167.36	138.91
creatine	s	−536.47	−264.01	189.54	171.96
ion (+)	aq		−274.39		
ion (−)	aq		−177.82		
L-cysteine	s	−532.62	−342.67	169.87	173.22
ion (+)	aq	−349.36			
dipolar ion (+, −)	aq		−338.82		
ion (−)	aq		−290.99		
ion (2−)	aq		−229.58		
L-cystine	s	−1044.33	−685.76	280.58	261.92
ion (2+)	aq		−648.50		
dipolar ion (2+, −)	aq		−678.23		
dipolar ion (2+, 2−)	aq		−666.51		
ion (2−)	aq		−562.33		
cytosine	aq		−464.42	120.92	
deoxyribose	aq	−897.05	−844.33	176.98	
fructose	s	−1103.06	−915.38		
fumaric acid	s	−810.65	−653.25	166.11	142.26
	aq	−774.88	−645.80	261.08	
ion (−)	aq	−774.46	−628.14	203.34	
ion (2−)	aq	−777.39	−601.87	105.44	
α-D-glucose	s	−1274.45	−910.56	212.13	218.87
	aq	−1263.06	−914.54	264.01	
β-D-glucose	s	−1268.05	−908.89	228.03	
	aq	−1264.24	−915.79	264.01	
glucose-1-phosphoric acid	aq		−1789.50		
ion (−)	aq		−1783.22		
ion (2−)	aq		−1746.11		
glucose-6-phosphoric acid	aq		−1797.45		
ion (−)	aq		−1753.51		
ion (2−)	aq		−1753.51		

(Continued)

Table XIII.5 (Continued)

Compound	State	$\Delta_f H^\ominus$ /kJ mol^{-1}	$\Delta_f G^\ominus$ /kJ mol^{-1}	$S^\ominus$ /J K^{-1} mol^{-1}	$C_P^\ominus$ /J K^{-1} mol^{-1}
L-glutamic acid	s	−1009.18	−730.95	188.20	175.23
ion (+)	aq	−981.57	−734.63	293.72	
dipolar ion (+, −)	aq	−981.99	−721.87	248.95	
dipolar ion (+, 2−)	aq	−979.89	−697.47	174.05	
ion (2−)	aq	−939.73	−643.50	127.61	
L-glutamine	s	−825.92	−532.21	195.10	183.80
dipolar ion (+, −)	aq	−805.00	−528.02	251.04	
glycine	s	−537.23	−377.69	103.51	99.20
ion (+)	aq	−527.18	−393.30	189.54	171.54
dipolar ion (+, −)	aq	−523.00	−379.91	158.57	36.82
ion (−)	aq	−478.65	−324.09	120.50	54.81
glycylglycine	s	−746.01	−491.50	189.95	163.59
ion (+)	aq	−735.72	−510.87	286.60	288.70
dipolar ion (+, −)	aq	−734.25	−492.08	231.38	
ion (−)	aq	−689.90	−445.76	222.17	
glycylglycylglycine	s		−607.10		
guanine	aq	−182.92	48.53	160.25	
histidine	aq	−226.77		217.57	
L-hydroxyproline	s			172.45	153.9
isoocitrate ion (3−)	aq		−1161.69		
isoleucine	aq	−635.13	−343.92	207.5	188.3
α-ketoglutaric acid	s	−1026.34			
ion (2−)	aq		−793.41		
α-lactose	s	−2221.70			
	aq	2232.37	1564.90	394.13	
β-lactose	s	−2236.77	−1566.91	386.18	410.45
DL-leucine	s	−649.78	−358.57	207.11	195.39
L-leucine	s	−646.85	−356.48	209.62	208.36
ion (+)	aq	−645.01	−365.56	246.44	−660.24
dipolar ion (+,−)	aq	−643.37	−352.25	207.53	506.26
ion (−)	aq	−600.61	−296.60	164.43	447.69
DL-leucylglycine	s	−860.2	−469.90	281.2	256.5
lysine	aq	−678.64		205.02	
L-malic acid	s	−1103.32			
	aq		−891.61		
ion (−)	aq		−871.95		
ion (2−)	aq	−842.66	−845.08		
α-maltose	aq	−2238.27	−1573.60	403.34	
β-maltose	aq	−2237.73	−1572.18	400.41	
mannitol	aq		−942.61		
L-methionine	s	−761.07	−508.36	231.46	290.20
ion (+)	aq	−744.75			
dipolar ion (+,−)	aq	−744.33			
DL-ornithine	s			193.3	190.8
oxaloacetic acid	s	−984.50			
	aq	−832.62			
ion (−)	aq	−818.39			
ion (2−)	aq	−793.29	−797.18		
palmitic acid	s	−890.77	−315.06	455.22	460.66
	aq		−287.86		
ion (−)	aq		−259.41		
L-phenylalanine	s	−466.93	−211.71	213.64	203.0
	aq	−468.19	−207.10	213.38	
proline	aq	−525.93	−565.68	133.89	
pyruvic acid	l	−585.76			
	aq	−607.52	−486.60	179.91	
ion (−)	aq	−596.22	−472.37	171.54	
ribose	aq	−1062.32	−1009.60	176.98	
L-serine	s	−726.34	−510.87	149.37	135.62

(Continued)

Table XIII.5 (Continued)

Compound	State	$\Delta_f H^\ominus$ /kJ mol^{-1}	$\Delta_f G^\ominus$ /kJ mol^{-1}	$S^\ominus$ /J K^{-1} mol^{-1}	$C_p^\ominus$ /J K^{-1} mol^{-1}
succinic acid	s	−940.81	−747.35	175.73	153.97
	aq	−912.20	−746.64	269.45	230.12
ion (−)	aq	−908.89	−722.62	199.99	96.23
ion (2−)	aq	−908.68	−690.44	92.99	122.59
sucrose	s	−2221.10	−1544.60	360.24	425.00
	aq	−2215.85	−1551.43	403.76	633.04
L-threonine	s	−807.10	−550.20	152.72	142.28
	aq	−758.98	−514.63	166.52	
thymine	aq	−468.19		137.24	
L-tryptophan	s	−415.1		251.0	238.2
	aq	−417.56	−510.87	250.20	
L-tyrosine	s	−671.53	−385.76	214.0	216.44
L-tyrosine	aq	−683.67	−387.22	213.80	
uracil	aq	−461.08		122.17	
urea	s	−333.50	−197.40	104.60	93.1
	aq		−203.84		
DL-valine	s	−617.98	−359.82	181.17	
L-valine	s	−617.98	−358.99	178.74	168.82
ion (+)	aq	−612.24	−371.71	240.16	547.27
dipolar ion (+,−)	aq	−611.99	−358.65	176.98	389.11
ion (−)	aq	−567.43	−307.40	174.89	333.88

Table XIII.6 Enthalpies of Combustion for Selected Organic Compounds

For more extensive information see: *Handbook of Chemistry and Physics* (1991–92), CRC Press.
The values of the molar enthalpy of combustion, $\Delta_c H$, refer to a temperature of 298.15K and 1 atmosphere pressure. The final products of combustion are gaseous carbon dioxide, liquid water and nitrogen gas for C,H,N compounds. g=gaseous state, l=liquid state, s=solid state.

Compound	State	$-\Delta_c H$ /kJ mol^{-1}	Compound	State	$-\Delta_c H$ /kJ mol^{-1}
carbon (graphite)	s	393.51	erythritol	s	2109.2
carbon (diamond)	s	395.41	ethane	g	1559.8
carbon monoxide	g	283.0	1,2-ethanediol	l	1179.5
hydrogen	g	285.83	ethanol	l	1366.8
sulphur	s	296.83	ethyl acetate	l	2246.4
			ethylamine	l	1713.3
acetaldehyde	l	1166.37	ethylbenzene	l	4565.6
acetamide	s	1182.4	ethyl benzoate	l	4597.0
acetanilide	s	4227.5	ethylene	g	1411.0
acetic acid	l	875.29	formaldehyde	g	570.78
acetic anhydride	l	1806.23	formic acid	l	254.64
acetone	l	1790.42	β-D-fructose	s	2811.6
acetonitrile	l	1265.2	fumaric acid (*trans*)	s	1334.65
acetophenone	l	4148.85	α-D-glucose	s	2803.03
acetylene	g	1299.59	glycerol	l	1661.0
amyl acetate	l	4361.8	n-heptane	l	4811.2
amyl alcohol	l	3320.8	n-hexane	l	4163.12
aniline	l	3396.2	iodoethane	l	1489.5
anthracene	s	7163.0	isobutane	g	2859.3
benzaldehyde	l	3527.9	lactic acid	s	1367.3
benzamide	s	3546.4	lactose	s	5648.4
benzanilide	s	6591.9	maleic acid (*cis*)	s	1355.2
benzene	l	3267.54	maltose	s	5645.5
benzoic acid	s	3226.87	methane	g	890.31
benzoic anhydride	s	6506.5	methanol	l	726.51
benzonitrile	l	3621.3	methyl acetate	l	1594.9
benzophenone	s	6512.4	methylamine	l	1060.6
benzoyl chloride	l	3276.2	naphthalene	s	5153.9
benzyl alcohol	l	3741.8	nitrobenzene	l	3092.8
benzylamine	l	4056.0	nitroethane	l	1348.1
bromoethane	g	1424.7	nitromethane	l	708.8
bromomethane	g	769.9	n-octane	l	5450.5
1-butanol	l	2675.79	oxalic-acid	s	245.6
butanoic acid	l	2183.50	n-pentane	g	3536.15
carbon disulphide	l	1031.8	n-pentane	l	3509.46
carbon tetrachloride	l	156.0	phenanthrene	s	7052.6
chloroethane	g	1325.1	phenol	s	3053.48
chloroform	l	373.2	phthalic acid	s	3223.52
chloromethane	g	687.0	picric acid	s	2559.8
o-cresol	l	3692.8	propane	g	2219.90
m-cresol	l	3684.0	propanoic acid	l	1527.29
p-cresol	l	3692.4	1-propanol	l	2019.83
cyclohexane	l	3919.86	propylene	g	2051.0
cyclohexanol	l	3726.0	pyridine	l	2782.4
cyclopentane	l	3290.9	succinic acid	s	1491.0
decahydronaphthalene (*cis*)	l	6286.5	sucrose	s	5640.9
decahydronaphthalene (*trans*)	l	6273.9	toluene	l	3908.7
decane	l	6737.1	trinitrobenzene	s	2776.9
diethylamine	l	2999.5	urea	s	631.66
diethylether	l	2751.1	o-xylene	l	4567.7
diethylketone	l	3077.8	m-xylene	l	4553.9
m-dinitrobenzene	s	2915.4	p-xylene	l	4556.8
diphenyl	s	6249.2			

Table XIII.7 Enthalpies of Fusion, $\Delta_{fus}H$, and Vaporization, $\Delta_{vap}H$, at the Transition Temperature of Some Inorganic Compounds
Reference: as for Table XIII.6.
sub = sublimes.

Compound	m.p./°C	$\Delta_{fus}H$ /kJ mol^{-1}	b.p./°C	$\Delta_{vap}H$ /kJ mol^{-1}
AgCl	455	12.76	1550	178
BaCl$_2$	963	22.47	1560	
CaCl$_2$	782	25.36	>1600	
CdCl$_2$	568	22.18	960	
CO	−199	0.836	−191.47	6.75
CO$_2$	−56.6	8.33	−78.51 (sub)	23.18 (sub)
D$_2$O	3.82	6.34	101.42	
HBr	−88.5	1.4	−67	17.61
HCl	−114.8	1.99	−84.9	15.40
HF	−83.1	4.58	19.54	7.5
HI	−50.8	2.87	−35.38	19.76
HNO$_3$	−42	2.51	83	
H$_2$O	0	6.008	100.0	40.656
H$_2$O$_2$	−0.41	10.5	150.2	43.1
H$_2$S	−85.5	2.377	−60.7	18.67
HgCl$_2$	276	17.36	302	
KBr	734	20.92	1435	156
KCl	770	26.82	1407	162
KF	858	27.2	1505	161
KI	681	17.2	1330	145
LiBr	550	12.13	1265	148
LiCl	605	13.39	1347	151
MgCl$_2$	714	33.89	1412	137
NaBr	747	25.69	1390	159
NaCl	801	30.21	1413	170
NaF	993	29.29	1695	201
NaI	661	22.34	1304	159
NaOH	318.4	8.37	1390	
NH$_3$	−77.7	5.65	−33.35	23.35
NO	−163.6	2.30	−151.8	
N$_2$O	−90.8	6.54	−88.5	
N$_2$O$_4$	−11.2	23.18	21.2	
POCl$_3$	1.25	13.01	105.3	
PCl$_3$	−112	5	75.5	31
RbCl	718	18.41	1390	154
SiCl$_4$	−70	7.72	57.6	
SO$_2$	−72.7	8.34	−10	24.94
SO$_3$ (α)	16.83	8.62	44.8	
SrCl$_2$	872	17.15		

Table XIII.8 Enthalpy of Fusion, $\Delta_{fus}H$, and Enthalpy of Vaporization, $\Delta_{vap}H$, at the Transition Temperature for Some Organic Compounds

For more extensive information see: V. Majer and V. Svoboda, *Enthalpies of Vaporization of Organic Compounds*, IUPAC Chemical Data Series, No. 32 (1985), Blackwell Sci. Publishers, Oxford.
sub = sublimes
*: CODATA recommended values

Compound	m.p./°C	$\Delta_{fus}H$ /kJ mol^{-1}	b.p./°C	$\Delta_{vap}H$ /kJ mol^{-1}
acetaldehyde	−121.0		20.8	25.76*
acetic acid	16.6	11.53	117.9	23.70*
acetone	−95.35	5.69	56.2	29.10*
acetonitrile	−45.72		81.6	29.75*
acetophenone	20.5		202.0	49.08
acrylic acid	13	11.16	141.6	45.84
acrylonitrile	−83.5		77.5	33.23
allyl alcohol	−129		97	44.26
aniline	−6	10.56	184	42.44*
anthracene	216.2	28.86	340	70.39
benzene	5.5	9.95	80.1	30.72*
benzoic acid	122.4	17.32	249	63.82
benzyl alcohol	−15.3	8.97	205.4	58.97
biphenyl	71	19.61	255.9	54.02
bromobenzene	−31	10.62	156	42.50
bromoethane	−118.6		38.4	27.04*
bromoform	8.3		149.5	39.66*
n-butane	−138	4.66	−0.5	22.44*
1-butanol	−90	9.28	117.7	43.29*
2-butanol	−114.7		99.5	40.75*
n-butyl acetate	−78		124.6	36.28*
n-butyric acid	−4	11.07	163.5	49.71
carbon dioxide	−56.6	8.33	−78.5 (sub)	23.18 (sub)
carbon disulphide	−111.53	4.40	46.25	26.74*
carbon monoxide	−199	0.836	−191.47	6.75
carbon tetrachloride	−22.99	3.28	76.54	29.82*
chlorobenzene	−45.6	9.61	132	35.19*
chloroethane	−136.4	4.45	12.3	26.40
chloroform	−63.5	8.80	61.7	29.24*
chloromethane	−97.7		−24.2	22.49
cyclohexane	6.55	2.63	80.74	29.97*
cyclohexanol	25.1	1.76	161.1	49.94
cyclohexanone	−47		155	41.99
cyclopentane	−93.9	6.07	49.26	27.30*
cyclopropane	−128	5.46	−32.7	20.05*
n-decane	−30	28.78	174	38.75*
dichloromethane	−97	6.00	40	28.06*
diethyl ether	−116	7.27	34.51	26.52*
1,2-dimethylbenzene	−25.18	13.61	144.4	36.24*
1,3-dimethylbenzene	−48	11.55	139.1	35.66*
1,4-dimethylbenzene	13.3	16.80	138	35.67*
dimethyl ether	−140	4.94	−23	21.51*
1,4-dioxane	11.8	12.85	101	34.16*
ethane	−183.3	2.86	−88.6	15.65
ethane-1,2-diol	−11.5	11.23	198	58.71
ethanol	−117.3	5.02	78.3	38.56*
ethyl acetate	−84	10.48	77.06	31.94*
ethylbenzene	−95		136.2	35.57*
ethylene	−169.15		−104	14.45
formic acid	8.4	12.72	100.7	22.69*
glycerol	20	18.48	182/20 mm	76.10
n-hexane	−95	13.08	68.95	28.85*
hydrogen cyanide	−13.24	8.46	25.7	30.70
methane	−183	9.37	−164	8.17*

(Continued)

Table XIII.8 (Continued)

Compound	m.p./°C	$\Delta_{fus}H$ /kJ mol^{-1}	b.p./°C	$\Delta_{vap}H$ /kJ mol^{-1}
methanol	−97.8	3.18	64.96	35.21*
methyl acetate	−98		57.5	30.32*
methylamine	−93	6.12	−6.3	25.60*
2-methylbutane	−160	5.15	27.9	27.07
methyl ethyl ketone	−86.35		79.6	31.30*
2-methyl-1-propanol	−108		107.9	41.82*
2-methyl-2-propanol	25.5	6.79	82.2	43.57
naphthalene	81	18.80	217.7	51.51
nitrobenzene	5.7	11.59	210.8	50.91
nitroethane	−90		115	39.88
nitromethane	−29		100.8	33.99*
n-octane	−56.8	20.65	125.7	34.41*
1-octanol	−15		196	59.67
n-pentane	−130	8.42	36	27.59
phenanthrene	101	18.64	340	59.35
phenol	40.9	11.29	182	49.75
propane	−181.7	3.53	−42.1	19.04*
propanoic acid	−23		141	52.11
1-propanol	−127	5.20	97.4	41.44*
2-propanol	−89.5	5.37	82.4	39.85*
pyridine	−42		115.5	35.09*
thiophene	−38	4.97	84.16	31.49*
toluene	−93	6.62	110.6	33.18*
1,1,2-trichlorethane	−37	11.54	113.8	34.82*

Table XIII.9 Enthalpies of Formation of Gaseous Atoms from Elements in Their Standard States at 298.15 K (enthalpies of atomization)
For more extensive data see: L. Brewer and G.N. Rosenblatt, *Adv. High Temp. Chem.* (1969), 2, 1.
g = gaseous state, l = liquid state, s = solid state.

Element	$\Delta_{at}H$	Uncertainty (+/−) /kJ mol^{-1}	Element	$\Delta_{at}H$	Uncertainty (+/−) /kJ mol^{-1}
aluminium (s)	329.3	2.1	niobium (s)	721.3	4.2
antimony (s)	264.4	2.5	nitrogen (g)	472.67	0.42
arsenic (s)	302.5	12.5	osmium (s)	786.6	6.3
barium (s)	177.8	4.2	oxygen (g)	249.170	0.100
beryllium (s)	324.3	6.3	palladium (s)	376.6	2.1
bismuth (s)	209.6	2.1	phosphorus, yellow (s)	332.2	41
boron (s)	571.1	12.5	platinum (s)	565.7	1.2
bromine (l)	111.838	0.121	plutonium (s)	364.4	17.0
cadmium (s)	111.80	0.63	potassium (s)	89.62	0.21
caesium (s)	78.2	0.4	rhenium (s)	774.0	6.3
calcium (s)	178.2	1.7	rhodium (s)	556.9	4.2
carbon (s)	716.68	0.46	rubidium (s)	82.0	0.4
cerium (s)	422.6	12.5	ruthenium (s)	650.6	6.3
chlorine (g)	121.290	0.008	samarium (s)	206.7	2.1
chromium (s)	397.5	4.2	scandium (c)	377.8	4.2
cobalt (s)	428.4	4.2	selenium (s)	2272.0	4.2
copper (s)	337.65	1.2	silicon (s)	455.6	4.2
erbium (s)	317.1	4.2	silver (s)	284.1	0.4
fluorine (g)	79.1		sodium (s)	108.16	0.63
gallium (s)	273.6	2.1	strontium (s)	163.6	2.1
germanium (s)	374.5	2.1	sulphur (s)	277.0	8.4
gold (s)	368.2	2.1	tantalum (s)	782.0	2.5
hafnium (s)	619.2	4.2	tellurium (s)	196.6	2.1
hydrogen (g)	218.003	0.004	thallium (s)	182.2	0.4
indium (s)	242.7	4.2	thorium (s)	575.3	2.1
iodine (s)	106.763	0.042	tin (s)	302.1	2.1
iridium (s)	669.4	4.2	titanium (s)	469.9	2.1
iron (s)	415.5	1.2	tungsten (s)	849.8	4.2
lead (s)	195.06	1.2	uranium (s)	527.2	12.5
lithium (s)	161.5	1.7	vanadium (s)	514.2	1.3
magnesium (s)	146.4	1.2	ytterbium (s)	159.09	0.84
manganese (s)	283.3	4.2	yttrium (s)	424.7	2.1
molybdenum (s)	658.1	2.1	zinc (c)	130.75	0.42
mercury (l)	61.46	0.12	zirconium (s)	608.8	4.2
nickel (s)	430.1	2.1			

Table XIII.10 Enthalpies of Formation of Free Radicals at 298.15 K

The enthalpy of formation of a free radical is related to the corresponding bond enthalpy by the equation,
$D(R–R) = 2\Delta_f H^{\ominus}(R\cdot) - \Delta_f H^{\ominus}(RR)$

For more extensive data see: H.E. O.'Neal and S.W. Benson, 'Thermochemistry of Free Radicals' in *Free Radicals*, ed. J.K. Kochi (1973), Wiley, New York.

Radical	$\Delta_f H^{\ominus}$(300 K) /kJ mol^{-1}	Uncertainty (+/–)	Radical	$\Delta_f H^{\ominus}$(300 K) /kJ mol^{-1}	Uncertainty (+/–)
CH	594	4	CH$_2$OH	–25.9	6.3
CH$_2$	385	4	CH$_3$O	15.9	0.8
CH$_3$	146.9	0.63	CH$_3$CO	–24.3	1.7
ethynyl	536	4	CH$_3$OCH$_2$	–11.7	5.0
C$_2$H$_5$	108.4	5.4	CH$_3$CH$_2$O	–17.2	
allyl	164.9	6.3	CH$_3$CH$_2$CO	–42.7	4.2
cyclopropyl	279.9	1.1	C$_6$H$_5$O	47.7	8.4
n-C$_3$H$_7$	94.6	7.5	COOH	–223.0	
iso-C$_3$H$_7$	76.2	6.3	CH$_3$COO	–207.5	4
cyclobutyl	214.2	4.2	CH$_3$CH$_2$COO	–228.5	4
sec-C$_4$H$_9$	54.4	8	CH$_3$S	143.1	8.4
tert-C$_4$H$_9$	31.8	5	CF$_2$	–194.1	9.2
cyclopentadien-1, 3-yl-5	254.8	5	CF$_3$	–468.6	15.1
C$_6$H$_5$	325.1	8.4	CCl$_2$	239	
cyclohexadien-1, 3-yl-5	184	21	CCl$_3$	79.5	4
cyclohexyl	58.2	4	NH	352.3	9.6
benzyl	200.0	6.3	NH$_2$	197.5	4
CN	423	4	HO	38.5	4
CH$_2$NH$_2$	154.8	8.4	HO$_2$	10.5	2.5
CH$_3$NH	190.0	4	HS	140.6	4.6
(CH$_3$)$_2$N	161.1	4			
C$_6$H$_5$NH	230.5				
CHO	32.2	5.0			

Table XIII.11 Enthalpy of Solution of Selected Electrolytes at 298.15 K

For more extensive data see: V.B. Parker, *Thermal Properties of Aqueous Uni-univalent Electrolytes*, NBS-NSRDS-2 (1965).

The value of the molar enthalpy refers to the formation of an infinitely dilute aqueous solution, from the electrolyte in its standard state (HBr, HCl, HF, HI in gaseous state, HNO$_3$ in liquid state).

	$\Delta_{sol} H_\infty$/kJ mol^{-1}							
	Br$^-$	CO$_3^{2-}$	Cl$^-$	F$^-$	OH$^-$	I$^-$	NO$_3^-$	SO$_4^{2-}$
H$^+$	–85.1		–74.8	–61.5		–81.7	–33.3	
Li$^+$	–48.8	–17.6	–37.0	+4.7	–23.6	–63.3	–2.5	–30.2
Na$^+$	–0.6	–24.6	+3.9	+0.9	–42.9	–7.5	+20.5	–2.3
K$^+$	+19.9	–32.6	+17.2	–17.7	–57.6	+20.3	+34.9	+23.8
Rb$^+$	+21.9	–40.3	+17.3	–26.1	–62.3	+25.1	+36.5	+24.3
Cs$^+$	+26.0	–52.8	+17.8	–36.9	–71.5	+33.3	+40.0	+17.1
Ag$^+$	+84.5	+41.8	+65.7	–20.3		+112	+22.6	+17.6
NH$_4^+$	+16.8		+14.8	+5.0		+13.7	+25.7	+6.2
Ca^{2+}	–110	–12.3	–82.9	+13.4	–16.2	–120	–18.9	–17.8
Sr^{2+}	–71.6	–3.4	–52.0	+10.9	–46.0	–90.4	+17.7	–8.7
Ba^{2+}	–25.4	+4.2	–13.2	+3.8	–51.8	–47.7	+40.4	+19.4
Mg^{2+}	–186	–25.3	–155	–17.7	+2.8	–214	–85.5	–91.2
Cu^{2+}	–35.2	–16.7	–51.5	–63.2	+48.1	–26.4	–41.8	–73.3
Zn^{2+}	–67.2	–16.2	–71.5		+29.9	–55.2	–83.9	–81.4
Cd^{2+}	–2.8		–18.4	–40.7		+14.0	–33.7	–53.7
Pb^{2+}	+36.8	+22.2	+25.9	+6.3	+56.1	+64.8	+37.7	+11.3
Al^{3+}	–360		–322	–209		–378		–318

Table XIII.12 Enthalpy of Neutralization at 298.15 K
Reference: as for Table XIII.11.
The values of the molar enthalpies of neutralization refer to an infinitely dilute solution.

Acid	Base	ΔH/kJ mol^{-1}	Acid	Base	ΔH/kJ mol^{-1}
HCl	NaOH	−57.1	HCl	NH$_4$OH	−52.2
HCl	KOH	−57.2	H$_3$BO$_3$	NaOH	−44.4
HNO$_3$	NaOH	−57.3	CH$_3$COOH	NaOH	−55.2
HNO$_3$	KOH	−57.3	CH$_2$ClCOOH	NaOH	−60.2
HNO$_3$	$\frac{1}{2}$Ba(OH)$_2$	−58.2	CCl$_3$COOH	NaOH	−56.1
HClO$_4$	NaOH	−56.5	HCN	KOH	−11.7
HF	NaOH	−68.6	HCN	NH$_4$OH	−5.4

Table XIII.13 Integral Enthalpies of Solution at 298.15 K
1. Enthalpies of Solution of Solid Sodium Hydroxide in nH$_2$O for the Reaction
NaOH(s) + nH$_2$O (l) → Na$^+$ OH$^-$ in nH$_2$O(l).
For more extensive data see: F. D. Rossini, D. D. Wagman, W. H. Evans, S. Levine and I. Jaffe., *Selected Values of Chemical Thermodynamic Properties* (1950), Circular 500, NBS. Washington

mol H$_2$O(l)	$\Delta_{sol}H^\ominus$/kJ mol^{-1}	mol H$_2$O(l)	$\Delta_{sol}H^\ominus$/kJ mol^{-1}
0	0	20	−42.84
3	−28.87	30	−42.72
4	−34.43	40	−42.59
5	−37.74	50	−42.51
6	−39.87	75	−42.38
7	−41.21	100	−42.34
8	−41.92	1 000	−42.47
10	−42.51	10 000	−42.72
15	−42.84	∞	−42.88

2. Enthalpies of Solution of Sulphuric Acid in nH$_2$O for the Reaction
H$_2$SO$_4$(l) + nH$_2$O(l) → 2H$^+$SO$_4^{2-}$ in nH$_2$O(l).

mol H$_2$O(l)	$\Delta_{sol}H^\ominus$/kJ mol^{-1}	mol H$_2$O(l)	$\Delta_{sol}H^\ominus$/kJ mol^{-1}
0	0	20.0	−71.50
0.5	−15.73	50.0	−73.35
1.0	−28.07	100	−73.97
1.5	−36.90	1 000	−78.58
2.0	−41.92	10 000	−87.07
5.0	−58.03	100 000	−93.64
10.0	−67.03	∞	−96.19

Table XIII.14 Mean Bond Dissociation Enthalpies (single bond enthalpies)

For more extensive data see: T. L. Cottrell, *The Strengths of Chemical Bonds* (1954), Butterworths, London.
The average value of the bond dissociation enthalpies of the A–B bond in a series of different compounds.

$EH^\ominus(A-B)/kJ\ mol^{-1}$ at 298.15 K.

	Br	C	Cl	F	H	I	N	O	P	S	Si
Br	194	285	219	234	366	179	280	235	264	218	342
C		348	339	441	413	240	305	360	290	272	285
Cl			243	251	431	211	194	269	319	285	439
F				157	569	280	262	234	439	327	485
H					436	299	388	463	343	344	299
I						153	159	197	184		234
N							161	201			
O								146	335		
P									172		
S										264	
Si											176

Table XIII.15 Bond Dissociation Enthalpies of Diatomic Molecules, $DH^\ominus$ (A–B) at 298.15 K

For more extensive information see: A.G. Gaydon, *Dissociation Energies and Spectra of Diatomic Molecules*, 3rd ed. (1968), Chapman and Hall, London.

Molecule	$DH^\ominus$ /kJ mol^{-1}	Uncertainty (+/–) /kJ mol^{-1}	Molecule	$DH^\ominus$ /kJ mol^{-1}	Uncertainty (+/–) /kJ mol^{-1}
H–H	436.002	0.004	C=N	769.9	4.2
H–D	439.446	0.004	C=O	1076.38	3.22
D–D	443.546	0.004	C=S	732.2	29.3
H–C	338.5		N=N	948.9	6.3
H–N	313.8	16.7	N=O	630.1	0.8
H–O	428.19	1.25	O=O	497.31	0.17
H–F	568.6	1.3	S=S	426.3	10.5
H–P	343.0	29.2	P–P	489.5	12.5
H–S	344.34	12.13	F–F	156.9	9.6
H–Cl	431.4		F–Cl	250.6	0.4
H–Br	365.7	2.1	Cl–Cl	242.948	0.004
H–I	298.7	0.8	Br–Br	193.870	0.004
C=C	602.5	20.9	I–I	152.549	0.008

Table XIII.16 Bond Dissociation Enthalpies in Some Polyatomic Molecules at 298.15 K

For more extensive data see: A. A. Radzig and B. M. Smirnov, *Dissociation Energies (Reference Data on Atoms, Molecules and Ions)*, Springer Series in Chemical Physics, 31 (1985), Springer, Berlin.

Bond	ΔH	Uncertainty (+/−)	Bond	ΔH	Uncertainty (+/−)
	/kJ mol^{-1}			/kJ mol^{-1}	
H–CH	426.8	8.4	CH_3–CN	431	
H–CH_2	460.2	8.4	CH_3NH_2	335	
H–CH_3	435.1	4.2	CH_3–OH	377	
H–C_2H_5	410.0	4.2	CH_3–OCH_3	335	
H–n-C_3H_7	410.0	4.2	CH_3–SH	368	
H–iso-C_3H_7	397.5	4.2	CH_3–Cl	352	
H–C_6H_5	461.1	8.4	CH_3–Br	292.8	5.0
H–cyclohexyl	399.6	4.2	CH_3–I	236.8	4.2
H–$CH_2C_6H_5$	355.6	4.2	CBr_3–Br	235.1	7.5
H–CN	502.1	4.2	C_6H_5–NH_2	418	
H–CHO	364.0	4.2	C_6H_5–OH	469	
H–CH_2OH	393.3	8.4	C_6H_5–Br	335	
H–$COCH_3$	359.8	3.3	C_6H_5–I	272	
H–CH_2OCH_3	389.1	4.2	NC–CN	535.6	4.2
H–$COOCH_3$	387.9	4.2	$F_2C=CF_2$	319.2	12.5
H–CCl_3	400.4	4.2	F_3C–CF_3	405.4	8.4
H–NH_2	460.2	8.4	O=CO	532.2	0.4
H–$NHCH_3$	430.9	8.4	H_2N–NH_2	296.2	8.4
H–$N(CH_3)_2$	397.5	8.4	O_2N–NO_2	54.0	2.1
H–HNC_6H_5	334.7	12.5	O–N_2	167.4	
H–OH	497.9	4.2	O–NO	305.4	
H–OCH_3	433.5	4.2	NO–NO_2	39.7	2.1
H–OC_2H_5	434.7	4.2	HO–OH	213.4	4.2
H–OC_6H_5	368.2	20.9	O–SO	552.3	8.4
H–OOH	376.6	8.4	Cl–OH	251.0	12.5
H–$OOCCH_3$	468.6	8.4	O–ClO	246.9	12.5
H–ONO	327.6	2.1	O=PF_3	543.9	20.9
H–ONO_2	423.4	2.1	O=PCl_3	510.4	20.9
H–SH	376.6	8.4	O=PBr_3	497.9	20.9
HC≡CH	962.3	8.4	H_3C–$CdCH_3$	227.6	
$H_2C=CH_2$	719.6	8.4	H_3C–$HgCH_3$	240.6	
CH_3–CH_3	368.2	8.4	H_3C–$Pb(CH_3)_3$	206.7	4.2
CH_3–CHO	314				

Table XIII.17 Enthalpy of Hydration of Some Ions at 298.15 K

Reference: H.F. Halliwell and S.C. Nyburg, *Trans. Faraday Soc.* (1963), **59**, 1126.
The absolute values are based on the assignment of -1091 ± 10 kJ mol^{-1} to H$^+$. The uncertainty of each value is about $+10n$ kJ mol^{-1}, where n is the charge on the ion.

Ion	$\Delta_{hyd}H$ /kJ mol^{-1}	Ion	$\Delta_{hyd}H$ /kJ mol^{-1}	Ion	$\Delta_{hyd}H$ /kJ mol^{-1}
H$^+$	−1091	Ca^{2+}	−1577	Cd^{2+}	−1807
Li$^+$	−519	Sr^{2+}	−1443	Hg^{2+}	−1824
Na$^+$	−406	Ba^{2+}	−1305	Sn^{2+}	−1552
K$^+$	−322	Cr^{2+}	−1904	Pb^{2+}	−1481
Rb$^+$	−293	Mn^{2+}	−1841	Al^{3+}	−4665
Cs$^+$	−264	Fe^{2+}	−1946	Fe^{3+}	−4430
Ag$^+$	−473	Co^{2+}	−1996	F$^-$	−515
Tl$^+$	−326	Ni^{2+}	−2105	Cl$^-$	−381
Be^{2+}	−2494	Cu^{2+}	−2100	Br$^-$	−347
Mg^{2+}	−1921	Zn^{2+}	−2046	I$^-$	−305

Table XIII.18 Free Energy Function, $[G^{\ominus}(T) - H^{\ominus}(0)]/T$ for selected Gaseous Carbon Compounds, based on $H^{\ominus}(0)$

For more extensive data see: G. N. Lewis and M. Randall, *Thermodynamics*, revised by K. S. Pitzer and L. Brewer (2nd edn, 1961), McGraw-Hill, New York.

Substance	$-[G^{\ominus}(T)-H^{\ominus}(0)]/T$ /J K^{-1} mol^{-1} at T/K					$\Delta_f H^{\ominus}(0)$ /kJ mol^{-1}
	298.15	500	1000	1500	2000	
CH$_4$	152.55	170.50	199.37	221.08	238.91	−66.90
CH$_3$Cl	198.53	217.82	250.12	274.22		−74.1
CHCl$_3$	248.07	275.35	321.25	352.96		−96.2
CCl$_4$	251.67	285.01	340.62	376.39		−104.6
CH$_3$OH	201.38	222.34	257.65			−190.25
C$_2$H$_2$	167.28	186.23	217.61	239.45	256.60	227.32
C$_2$H$_4$	184.01	203.93	239.70	267.52	290.62	60.75
C$_2$H$_6$	189.41	212.42	255.68	290.62		−69.12
C$_2$H$_5$OH	235.14	262.84	314.97	356.27		−219.28
CH$_3$COOH	236.40	264.60	317.65	357.10		−420.5
C$_3$H$_6$	221.54	248.19	299.45	340.70		35.44
C$_3$H$_8$	220.62	250.25	310.03	359.24		−81.50
(CH$_3$)$_2$CO	240.37	272.09	331.46	378.82		−199.74
C$_6$H$_6$	221.46	252.04	320.39	378.44		100.42
C$_6$H$_5$CH$_3$	259.32	297.71	382.96	455.01		
o-C$_6$H$_4$(CH$_3$)$_2$	274.51	323.32	428.69	515.89		

Table XIII.19 Molar Thermodynamic Functions of a Harmonic Oscillator
Reference: H.S. Taylor and S. Glasstone, Treatise on Physical Chemistry, Vol. 1, 3rd ed. (1962), van Nostrand, Princeton.
v is the frequency of vibration; U_o is the zero point energy

hv/kT	C_v /J K^{-1} mol^{-1}	$(U-U_o)/T$ /J K^{-1} mol^{-1}	$-(G-U_o)/T$ /J K^{-1} mol^{-1}	S /J K^{-1} mol^{-1}
0.10	8.305	7.912	19.56	27.47
0.15	8.297	7.707	16.39	24.10
0.20	8.289	7.510	14.20	21.70
0.25	8.272	7.318	12.55	19.87
0.30	8.251	7.130	11.23	18.36
0.35	8.230	6.945	10.14	17.09
0.40	8.205	6.761	9.230	15.99
0.45	8.176	6.586	8.439	15.03
0.50	8.142	6.410	7.753	14.16
0.60	8.071	6.067	6.615	12.68
0.70	7.983	5.740	5.708	11.45
0.80	7.883	5.427	4.962	10.39
0.90	7.774	5.125	4.339	9.464
1.00	7.657	4.841	3.816	8.657
1.20	7.385	4.301	2.982	7.283
1.40	7.079	3.810	2.355	6.165
1.60	6.745	3.365	1.875	5.240
1.80	6.393	2.958	1.502	4.460
2.00	6.021	2.603	1.209	3.812
2.20	5.640	2.279	0.976	3.255
2.40	5.255	1.991	0.7907	2.782
2.60	4.870	1.734	0.6418	2.376
2.80	4.494	1.507	0.5213	2.028
3.00	4.125	1.307	0.4246	1.732
3.50	3.270	0.9062	0.2552	1.161
4.00	2.528	0.6204	0.1535	0.7739
4.50	1.913	0.4204	0.0933	0.5137
5.00	1.420	0.2820	0.0556	0.3376
5.50	1.036	0.1878	0.0338	0.2216
6.00	0.7455	0.1238	0.0209	0.1447
6.50	0.5296	0.0815	0.0125	0.0940
7.00	0.3723	0.0531	0.0075	0.0606
8.00	0.1786	0.0221	0.0025	0.0246
9.00	0.0832	0.0092	0.0016	0.0108
10.0	0.0376	0.0037	0.0004	0.0041

Table XIII.20 Lattice Enthalpies of Some Metal Halides

Reference: D. Cubicciotti, *J. Chem. Phys.* (1959), *31*, 1646; (1961), *34*, 2189.
Enthalpy values are for the reaction
$M^+(g) + A^-(g) \rightarrow MA(s)$.

Cation/anion	$\Delta H^\ominus$/kJ mol^{-1} at 298,15 K			
	F^-	Cl^-	Br^-	I^-
Li^+	−1031	−850	−802	−763
Na^+	−911.7	−772.8	−741.0	−703
K^+	−810.0	−702.5	−678.2	−637.6
Rb^+	−799.9	−686	−658.6	−621.7
Cs^+	−730	−651	−630	−601
Be^{2+}	−3476	−2994	−2896	−2784
Mg^{2+}	−2949	−2502	−2402	−2293
Ca^{2+}	−2617	−2254	−2134	−2043
Sr^{2+}	−2482	−2129	−2040	−1940
Ba^{2+}	−2330	−2024	−1942	−1838
Ti^{2+}		−2489		
V^{2+}		−2612		
Cr^{2+}		−2583		
Mn^{2+}		−2525		
Fe^{2+}		−2621		
Co^{2+}		−2698		
Ni^{2+}		−2768		
Cu^{2+}		−2796		
Zn^{2+}		−2725		

Table XIII.21 Thermodynamic Data for Some Ion Association Reactions at 298.15 K

For more extensive data see: G. H. Nancollas, Interactions in Electrolyte Solutions (1966), Elsevier, Amsterdam.
|a|: values measured at $I = 1.0$ mol dm^{-3}.

Reactants			K /dm^3 mol^{-1}	ΔG /kJ mol^{-1}	ΔH /kJ mol^{-1}	ΔS /J K^{-1} mol^{-1}
Pb^{2+}	+	I^-	83	−10.96	−1.26	32.6
Pb^{2+}	+	NO_3^-	15.1	−6.78	−2.38	14.6
Pb^{2+}	+	Cl^-	41	−9.20	5.23	48.5
Pb^{2+}	+	Br^-	72	−10.59	1.26	39.3
Cd^{2+}	+	Cl^-	91	−11.17	4.10	51.0
Cd^{2+}	+	Br^-	141	−12.26	−1.34	36.8
Cd^{2+}	+	I^-	282	−13.97	−8.58	18.0
Ca^{2+}	+	OH^-	25	−7.99	4.98	43.5
Sr^{2+}	+	OH^-	6.7	−4.69	4.81	31.8
Ba^{2+}	+	OH^-	4.4	−3.64	7.32	36.8
Mg^{2+}	+	OH^-	380	−14.64		
La^{3+}	+	OH^-	2×10^3	−17.15		
Fe^{3+}	+	OH^-	6.4×10^{11}	−67.36	−12.55	184.1
U^{4+}	+	OH^-	2.5×10^{13}	−76.15	−10.46	217.6
Ca^{2+}	+	SO_4^{2-}	200	−13.18	6.90	67.4
Mg^{2+}	+	SO_4^{2-}	234	−13.47	19.04	109.2
Mn^{2+}	+	SO_4^{2-}	181	−12.84	14.10	94.6
Co^{2+}	+	SO_4^{2-}	230	−13.43	7.28	69.5
Ni^{2+}	+	SO_4^{2-}	211	−13.22	13.85	90.8
Zn^{2+}	+	SO_4^{2-}	240	−13.60	16.78	102.1
Mg^{2+}	+	CH_3COO^-	17.6	−7.07	−6.36	2.5
Ca^{2+}	+	CH_3COO^-	17.5	−7.07	3.81	36.4
La^{3+}	+	$Fe(CN)_6^{3-}$	5.5×10^3	−21.30	8.37	96.2
La^{3+}	+	$Co(CN)_6^{3-}$	5.8×10^3	−21.46	5.56	90.8
Mg^{2+}	+	$EDTA^{4-}$|a|	3.5×10^8	−48.74	13.14	211.3
Ca^{2+}	+	$EDTA^{4-}$|a|	3.3×10^{10}	−60.04	−27.40	111.3
Ni^{2+}	+	$EDTA^{4-}$|a|	2.0×10^{18}	−104.43	−34.94	237.2
Zn^{2+}	+	$EDTA^{4-}$|a|	1.7×10^{16}	−92.55	−20.29	248.5

Table XIII.22 Molar Heat Capacity of Elements

Reference: as for Table XIII.2.
*: CODATA recommended key values (1989).
m and b indicate phase change solid to liquid and liquid to vapour, respectively, at an intermediate temperature.

	C_p/J K^{-1} mol^{-1}					
Element T/K	100	298.15	600	1000	2000	2500
aluminium	12.996*	24.200*	28.043* m	31.750*	31.750*	31.750*
antimony	20.6	26.23	27.4 m	31.4	b 18.7	18.7
argon	b 20.786*	20.786*	20.786*	20.786*	20.786*	20.786*
arsenic	16.7	24.98	27.4 m			
barium	24.3	26.36	33.9	39.1	m 39.7	b 39.9
beryllium	1.830*	16.443*	23.062*	27.579* m	30.000*	30.000*
bismuth	23.0	25.52	m 31.4	31.4	b 21.0	21.4
boron	1.070*	11.087*	20.513*	24.931*	29.526*	m 31.400*
bromine	43.6	m 24.52*	b 37.305*	37.793*	39.089*	40.417*
cadmium	25.110*	26.020*	m 29.000*	29.000*	b 20.786*	20.787*
caesium	25.820*	32.210*	m 31.001*	b 20.786*	21.343*	22.899*
calcium	19.504*	25.929*	30.382*	39.706*	m,b 20.953*	21.836*
carbon	1.674*	8.517*	16.844*	21.610*	25.094*	25.900*
cerium	27.0	27.11	33.9	40.6	m 33.5	33.5
chlorine	42.3	m,b 33.949*	36.547*	37.442*	38.468*	39.379*
chromium	10.0	23.35	27.7	31.9	41.2	m 39.3
cobalt	13.9	24.80	29.7	37.0	m 40.5	40.5
copper	16.010*	24.440*	26.479*	28.700*	m 32.800*	32.800*
dysprosium		28.12				
erbium	24.6	28.06	30.0	32.5	m 33.5	33.5
europium		26.76	29.3	32.6	m, b 21.3	23.5
fluorine	b 29.114*	31.304*	35.172*	37.065*	38.935*	38.914*
gadolinium	28.9	36.48	30.7	33.5	m 33.5	33.5
gallium	18.5	25.86	m 27.8	27.8	27.8	27.8
germanium	13.820*	23.222*	25.452*	26.926*	m 27.600*	27.600*
gold	21.4	25.42	26.8	28.9	m 29.3	29.3
hafnium		25.52	27.0	29.0	34.1	33.5
helium	b 20.786*	20.786*	20.786*	20.786*	20.786*	20.786*
holmium	39.2	27.15	29.4	32.2	m 33.5	33.5
hydrogen	b 28.155*	28.836*	29.327*	30.204*	34.277*	35.837*
indium	23.3	26.74	m 29.7	29.7	29.7	b 23.9
iodine	45.7	54.44	m,b 37.613*	38.070*	40.474*	41.600*
iridium	17.3	25.52	26.8	29.2	35.1	38.1
iron	12.1	25.10	32.0	54.4	m 46.0	46.0
krypton	m 31.6	b 20.786*	20.786*	20.786*	20.786*	20.786*
lanthanum	23.6	26.15	27.1	27.8	m 28.0	28.0
lead	24.430*	26.650*	29.736*	29.369*	28.620*	b 29.446*
lithium	12.970*	24.860*	m 29.584*	28.795*	b 20.849*	21.131*
magnesium	15.762*	24.869*	28.184*	m 34.309*	b 20.786*	20.826*
manganese		26.32	31.5	38.9	m 46.0	b 21.1
mercury	24.3	m 27.98	27.1	b 20.8	20.8	20.8
molybdenum	13.5	24.06	26.5	28.4	36.7	43.9
neodymium	26.6	27.45	36.9	45.8	m 33.5	33.5
neon	b 20.786*	20.786*	20.786*	20.786*	20.786*	20.786*
nickel	13.6	26.07	34.9	32.2	m 38.9	38.9
niobium	17.4	24.96	26.3	28.0	33.4	38.6
nitrogen	b 29.104*	29.124*	30.109*	32.696*	35.969*	36.614*
osmium		24.89	26.0	27.5	31.2	33.0
oxygen	b 29.112*	29.378*	32.093*	34.880*	37.783*	38.929*
palladium	17.9	25.98	27.7	30.0	m 34.7	34.7
phosphorus (white)	13.728*	23.824*	m 26.120*	b 20.786*	21.179*	22.098*
platinum	19.6	25.86	27.5	29.7	b 20.973*	m 34.7
potassium	24.650*	29.600*	m 30.158*	30.730*	b 20.973*	21.589*
praesodymium		26.99	31.0	36.4	m 33.5	33.5
radium		27.0	33.5	m 31.4	b 21.5	24.4
radon		m, b 20.79	20.79	20.79	20.79	20.79

(Continued)

Table XIII.22 (Continued)

Element	T/K	C_p/J K^{-1} mol^{-1}					
		100	298.15	600	1000	2000	2500
rhenium		18.0	25.48	26.9	29.1	34.6	37.3
rhodium		15.1	25.56	28.2	31.6	40.2	m 41.8
rubidium		25.510*	31.060*	m 30.439*	b 20.786*	21.013*	21.740*
ruthenium		13.5	24.27	25.7	28.2	31.4	31.4
samarium			29.08	30.7	35.4	m,b 26.4	25.4
scandium			25.52	26.5	28.3	m 33.5	33.5
selenium		18.2	25.36	m 35.1	b 18.8	19.6	20.0
silicon		7.280*	19.789*	24.472*	26.568*	m 27.200*	27.200
silver		20.100*	25.350*	26.992*	29.667*	m 33.400*	b 20.786*
sodium		22.460*	28.230*	m 29.920*	28.799*	b 20.805*	20.923*
strontium		22.2	26.4	32.0	37.7	m,b 21.0	22.4
sulphur		12.770*	22.750*	m 34.379*	b 21.490*	21.276*	21.598*
tantalum		19.7	25.4	26.8	27.9	31.2	34.2
tellurium		21.6	25.73	32.3	m 37.7	b 18.7	18.7
terbium			29.0				
thallium		24.4	26.32	m 30.5	30.5	b 22.7	24.4
thorium		22.690*	26.230*	28.834*	32.391*	39.602*	m 46.000*
thulium			27.0				
tin		22.210*	27.112*	m 28.663*	27.945*	28.947*	29.559*
titanium		14.310*	25.060*	28.411*	32.681*	m 46.800*	46.800*
tungsten		16.0	24.3	25.8	27.0	32.3	34.7
uranium		22.240*	27.665*	34.652*	42.400*	m 49.125*	50.555*
vanadium		13.5	24.89	27.5	30.1	40.9	m 46.2
xenon		28.2	m,b 20.786*	20.786*	20.786*	20.786*	20.786*
ytterbium			25.5				
yttrium			25.3	26.4	28.1	m 33.5	33.5
zinc		19.460*	25.390*	28.820*	m 31.400*	b 31.400*	
zirconium		18.7	25.36	28.9	33.6	32.5	m 33.5

Table XIII.23 Molar Heat Capacity of Some Solid Inorganic Compounds
For more extensive data see: *U.S.* Bureau of Mines Bulletin, 605 (1963).

Compound		C_p/J K^{-1}mol^{-1}				
	T/K	298.15	600	1000	1500	2000
boron nitride		19.72	35.23	44.35	48.70	48.95
calcium oxide		42.12	50.48	53.74	56.27	58.41
calcite		81.84	110.47	123.85		
glass		206.99	286.62	319.98		
iron oxide (Fe$_3$O$_4$)		147.23	212.55	200.83	200.83	
magnesium oxide		37.11	47.43	51.21	53.69	55.65
quartz (SiO$_2$)		44.59	64.42	68.96		
silicon carbide		26.84	41.79	48.42	51.94	53.80
sodium chloride		50.51	55.48	64.86		
thorium oxide		62.33	71.94	77.66	84.05	
titanium dioxide		55.10	69.93	74.85	77.32	78.87
uranium dioxide		63.70	79.80	85.45	89.76	
zinc oxide		40.24	49.52	53.19	56.25	
zinc sulphide		46.02	52.41	55.50		

Table XIII.24 Molar Heat Capacity of Some Gaseous Organic Compounds

For more extensive data see: J.D. Cox and G. Pilcher, *Thermochemistry of Organic and Organometallic Compounds* (1970), Academic Press, New York.

Compound		C_p/J K^{-1} mol^{-1}			
	T/K	298.15	400	600	1000
$CHCl_3$		65.38	74.25	85.26	95.52
CH_3OH		43.89	51.42	67.03	89.45
CH_4		35.64	40.50	52.23	71.80
CCl_4		83.40	91.70	99.68	104.80
CS_2		45.66	49.68	54.61	59.28
C_2H_2		44.10	50.48	58.29	68.27
C_2H_4		42.89	53.05	70.66	93.90
C_2H_5OH		65.44	81.00	107.49	141.54
C_2H_6		52.63	65.61	89.33	122.72
C_3H_8		73.51	94.31	129.20	175.02
$(CH_3)_2CO$		74.89	92.05	122.76	163.80
$n\text{-}C_4H_{10}$		97.45	123.85	168.62	226.86
$(C_2H_5)_2O$		112.51	138.11	183.76	244.81
C_6H_6		81.67	111.88	157.90	209.87

Table XIII.25 Temperature Dependence of Heat Capacities

1. Heat capacity coefficients for the empirical equation:

$$C_P = a + bT + cT^{-2}$$

For more extensive data see: D. R. Stull and G. C. Sinke *Thermodynamic properties of the elements Advances in Chem. Ser.*, No 18 (1956); F. D. Rossini et al., *Selected Values of Physical and Thermodynamic Properties of Hydrocarbons and Related Compounds* (1953), Carnegie Press, Pittsburgh.

	a/J K^{-1} mol^{-1}	$10^3 b$/J K^{-2} mol^{-1}	$10^{-5} c$/J K mol^{-1}
Gases (valid 298–2000 K)			
He,Ne,Ar,Kr,Xe [a]	20.786	0	0
H_2	27.28	3.26	0.50
O_2	29.96	4.18	−1.67
N_2	28.58	3.77	−0.50
F_2	34.56	2.51	−3.51
Cl_2	37.03	0.67	−2.85
Br_2	37.32	0.50	−1.26
CO_2	44.22	8.79	−8.62
H_2O	30.54	10.29	0
NH_3	29.75	25.10	−1.55
CH_4	23.64	47.86	−1.92
SO_2	49.77	4.56	−11.05
H_2S	32.68	12.38	−1.92
Liquids (valid from m.p. to b.p.)			
H_2O	75.40	0	0
$C_{10}H_8$	79.50	407.5	0
Solids (valid from 298 K to m.p. or 2000 K)			
Al	20.67	12.38	0
C(graphite)	16.86	4.77	−8.54
Cu	22.64	6.28	0
I_2	40.12	49.79	0
Pb	22.13	11.72	0.96
NaCl	45.94	16.32	0
$C_{10}H_8$	−115.9	937	0

2. Heat capacity coefficients for the empirical equation:

$$C_p = \alpha + \beta T + \gamma T^2$$

For more extensive data see: *U.S. Bureau of Mines Bulletin*, 476 (1949) and 584 (1960).

	α/J K^{-1} mol^{-1}	$10^3 \beta$/J K^{-2} mol^{-1}	$10^7 \gamma$/J K^{-3} mol^{-1}
(Valid 300 to 1500 K)			
H_2 (g)	29.07	−0.837	20.11
O_2 (g)	25.72	12.98	−38.62
N_2 (g)	27.30	5.23	−0.042
Cl_2 (g)	31.70	10.14	−40.38
Br_2 (g)	35.24	4.07	−14.87
H_2O (g)	30.36	9.61	11.84
CO_2 (g)	26.00	43.50	−148.3
CO (g)	26.86	6.97	−8.20
HCl (g)	28.17	1.81	15.47
SO_3 (g)[b]	25.43	98.48	−405.3
CH_4 (g)	14.16	75.50	−179.9
C_2H_6 (g)	9.40	159.83	−462.3
C_3H_8 (g)	10.08	239.30	−733.6
$n\text{-}C_4H_{10}$ (g)	18.63	302.38	−929.4
$n\text{-}C_5H_{12}$ (g)	24.73	370.07	−1145.9
C_6H_6 (g)	−1.711	324.77	−1105.8
C_5H_5N (g)	−12.62	368.54	−1617.7

[a]: Also for all monatomic gases with no appreciable electronic excitation. [b]: Valid 300–1200 K.

Table XIII.25 (Continued)
3. Einstein Equation for the Heat Capacity of a solid.
The Einstein equation for the heat capacity of a solid:

$$C_v = 3R[(\theta_E/T)^2 \exp(\theta_E/T)]/[\exp(\theta_E/T) - 1]^2$$

assumes that every atom in an isotopic crystalline solid vibrates about its equilibrium position with a single frequency and that each atom vibrates independently of the others. The Einstein characteristic temperature $\theta_E = h\nu/k$ where ν is the frequency of vibration.
This equation may be written:

$$C_v = 3R\, f_E(\theta_E/T)$$

where $f_E(\theta_E/T)$, known as the Einstein molar heat function is defined as:

$$f_E(\theta_E/T) = [(\theta_E/T)^2 \exp(\theta_E/T)]/[\exp(\theta_E/T) - 1]^2$$

Thus if the equation is applicable, the value of C_v should be the same for all solids at the same value of θ_E/T. At high temperatures, ie. $T \gg \theta_E$, C_v tends to 3R and at very low temperatures $\theta_E \gg T$, C_v tends to 0. From a knowledge of C_v for a solid at one temperature, the value of C_v at other temperatures can be estimated, using the appropriate tabulated value of $f_E(\theta_E/T)$.
For many inorganic substances, θ_E is of the order of 200 to 300 K, corresponding to a vibrational frequency, ν, of 4.2×10^{12} to 6.3×10^{12} Hz. For diamond θ_E is approximately 1320 K. Agreement with experimental data is reasonably good down to a temperature of $0.20\theta_E$ (ie. about 40 to 60 K) for most substances. At lower temperatures the experimental values of C_v are significantly greater than the calculated values.

T/θ_E	$f_E(\theta_E/T)$	T/θ_E	$f_E(\theta_E/T)$	T/θ_E	$f_E(\theta_E/T)$
0.1	0.00454	0.52	0.7414	0.92	0.9071
0.12	0.0167	0.54	0.7573	0.94	0.9108
0.15	0.0567	0.55	0.7647	0.95	0.9126
0.16	0.0757	0.56	0.7718	0.96	0.9143
0.18	0.1202	0.58	0.7852	0.98	0.9176
0.20	0.1707	0.60	0.7974	1.00	0.9207
0.22	0.2240	0.62	0.8087	1.10	0.9339
0.24	0.2770	0.64	0.8192	1.20	0.9441
0.25	0.3041	0.65	0.8241	1.30	0.9521
0.26	0.3299	0.66	0.8288	1.40	0.9585
0.28	0.3797	0.68	0.8377	1.50	0.9638
0.30	0.4262	0.70	0.8460	1.60	0.9681
0.32	0.4694	0.72	0.8536	1.70	0.9717
0.34	0.5091	0.74	0.8608	1.80	0.9747
0.35	0.5277	0.75	0.8641	1.90	0.9772
0.36	0.5455	0.76	0.8674	2.00	0.9794
0.38	0.5787	0.78	0.8736	2.50	0.9868
0.40	0.6089	0.80	0.8794	3.00	0.9908
0.42	0.6364	0.82	0.8848	3.50	0.9932
0.44	0.6615	0.84	0.8898	4.00	0.9948
0.45	0.6730	0.85	0.8922	4.50	0.9959
0.46	0.6843	0.86	0.8946	5.00	0.9967
0.48	0.7051	0.88	0.9000	10.00	0.9992
0.50	0.7241	0.90	0.9032		

Table XIII.26 Debye Characteristic Temperatures
Reference: Blackman, *Handbuch der Physik* (1955), 7, 325.

Substance	θ_D/K	Substance	θ_D/K	Substance	θ_D/K
Li	430	C (diamond)	1860	NaCl	281
Na	159	Ge	362	KCl	227
K	100	Sn	165	AgCl	183
Cu	315	Pb	88	AgBr	144
Ag	215	Al	398	CaF_2	474
Au	180	Cr	405	FeS_2	645
Be	1000	Mo	375		
Mg	290	W	315		
Ca	230	Fe	453		
Zn	235	Co	445		
Cd	165	Ni	456		
Hg	97	Pt	225		

Table XIII.27 Standard Gibbs (Free) Energy Function of Hydrolysis of Some Phosphate and Other Biologically Important Compounds at pH 7 and 298.15 K
Reference: *Handbook of Biochemistry* (1970), CRC Press.

Compound	$\Delta G^{\ominus\prime}$/kJ mol^{-1}	Compound	$\Delta G^{\ominus\prime}$/kJ mol^{-1}
adenosine phosphosulphate	−75.3	acetyl coenzyme A	−32.2
phosphoenolpyruvate	−61.9	adenosine triphosphate	−30.5
1,3-diphosphoglycerate	−49.4	adenosine diphosphate	−30.5
acetyl phosphate	−43.1	glucose 1-phosphate	−20.9
creatine phosphate	−43.1	glucose 6-phosphate	−13.8
inorganic pyrophosphate	−33.5	adenosine monophosphate	−12.5
arginine phosphate	−32.2	glycerol 1-phosphate	−9.2

Table XIII.28 Thermodynamic Constants for Phase Changes of Water and Deuterium Oxide
Reference: D. Eisenberg and W. Kauzmann, *The Structure and Properties of Water* (1969), Oxford University Press.

1. Fusion and vaporization of H_2O and D_2O at 101 325 Pa

	Fusion		Vaporization	
	H_2O	D_2O	H_2O	D_2O
Temperature/K	273.15	276.97	373.15	374.59
ΔC_p (isopiestic heat capacity change) /J K^{-1} mol^{-1}	37.28	39.66	−41.93	
ΔH/kJ mol^{-1}	6.008	6.280	40.656	41.53
ΔS/J K^{-1} mol^{-1}	21.995	22.674	108.95	110.87
ΔV (volume change)/cm^3 mol^{-1}	−1.621		3.06×10^4	

2. Sublimation of H_2O and D_2O at Triple Point

	H_2O	D_2O
Temperature/K	273.16	276.98
ΔH/kJ mol^{-1}	51.06	52.84
ΔS/J K^{-1} mol^{-1}	186.92	190.77

Table XIII.29 Molal Cryoscopic and Ebullioscopic Constants for Some Common Solvents

Reference: as for Table XI.25.

1. Molal Cryoscopic Constants

Solvent	f.p. /°C	k_c /K kg mol^{-1}	$\Delta_{fus}H$ /kJ kg^{-1}
acetic acid	16.60	3.90	190.6
acetone	−95.35	2.40	98.0
aniline	−6.3	5.87	113.4
benzene	5.51	5.085	127.4
bromoform	8.3	14.3	
camphor[a]	178.8	40	45
carbon disulphide	−111.53	3.8	57.8
carbon tetrachloride	−22.99	29.8	16.3
chloroform	−63.5	4.90	73.7
cyclohexane	6.55	20.1	31.2
cyclohexanol	25.15	37.7	17.7
deuterium oxide	3.82	2.00	
diethyl ether	−116.2	1.79	98.1
naphthalene	80.6	6.94	146.7
nitrobenzene	5.7	6.90	94.1
phenol	40.9	7.27	120.0
pyridine	−42	4.75	94
water	0	1.858	331.9

2. Molal Ebullioscopic Constants

Solvent	b.p. /°C	k_e /K kg mol^{-1}	$\Delta_{vap}H$ /kJ kg^{-1}
acetic acid	117.9	3.07	392.0
acetone	56.2	1.71	501.0
ammonia	−33.35	0.34	1373.5
aniline	184.3	3.69	455.7
benzene	80.2	2.61	393.3
carbon disulphide	46.25	2.37	351.2
carbon tetrachloride	76.54	4.95	193.9
chloroform	61.7	3.66	244.9
cyclohexane	80.74	2.79	356.1
diethyl ether	34.51	1.82	357.8
ethanol	78.3	1.13	837.0
methanol	64.96	0.83	1098.9
naphthalene	217.7	5.8	401.9
nitrobenzene	210.8	5.26	413.5
phenol	181.75	3.04	528.6
sulphur dioxide	−10.0	1.45	389.7
water	100.0	0.513	2258.7

[a]: since commercial camphor is not pure, k_c should be determined for each sample with a solute of known M_r.

Table XIII. 30 Henry's Law Constants

Reference: R.A. Alberty and F. Daniels, *Physical Chemistry*, 5th ed. (1980), Wiley, New York.

Henry's law constant is the ratio of the partial pressure of the substance in the vapour phase to its concentration in the liquid phase. In terms of mole fraction, x, in the liquid phase:

$p = H_M x$ (H_M has the units of pressure, Pa); in terms of molar concentration, c, in the liquid phase:

$p = H_c c$ (H_c has the units Pa m^3 mol^{-1}).

At low concentrations, $x = c/c_s$, where c and c_s are the number of moles of solute and solvent per m^3 of solution respectively. $H_M = p/x = pc_s/c$, thus $H_c = p/c = H_M/c_s = H_M V_m$, where V_m is the molar volume of water, 18×10^{-6} m^3 mol^{-1}.

1. Henry's Law Constant for Inorganic Gases 25°C

Gas	$K_m/10^9$ Pa	
	Water	Benzene
argon	3.73	
carbon dioxide	0.167	0.0114
carbon monoxide	5.79	0.163
helium	14.93	
hydrogen	7.12	0.367
hydrogen sulphide	0.057	
nitrogen	8.68	0.239
oxygen	4.40	

2. Henry's Law Constant for Nitrogen in Water at Different Temperatures

t/°C	$K_m/10^9$ Pa
0	5.43
5	6.09
10	6.76
15	7.40
20	8.00
25	8.68
30	9.13
35	9.64
40	10.15
45	10.65
50	11.03

3. Henry's Law Constant for Some Organic Compounds in Aqueous Solution at 25°C.

For more extensive data see: D. Mackay and W. Y. Shui, *J. Phys. Chem. Ref.* Data (1981), **10**, 1175

Error limits (+/−) in parenthesis.

Compound	K_c/kPa m^3 mol^{-1}	Compound	K_c/kPa m^3 mol^{-1}
methane	67.4 (2.0)	m-xylene	0.70 (0.10)
ethane	50.6 (1.1)	p-xylene	0.71 (0.08)
propane	71.6 (2.4)	naphthalene	0.043 (0.004)
n-butane	95.9 (4.1)	biphenyl	0.028 (0.002)
n-hexane	170 (25)	acenaphthene	0.024 (0.002)
n-octane	300 (50)	phenanthrene	0.0040 (0.0008)
ethene	21.7 (2.0)	anthracene	0.0060 (0.003)
propene	21.3 (3.0)	chloromethane	0.95 (0.05)
1-butene	75 (4.0)	dichloromethane	0.26 (0.02)
1-pentene	40.3 (2.0)	chloroform	0.38 (0.03)
1-hexene	41.8 (1.0)	carbon tetrachloride	2.0 (0.4)
1-octene	96.4 (7.1)	1, 1, 1-trichloroethane	2.8 (0.04)
1, 3-butadiene	7.46 (0.2)	1, 1, 2-trichloroethane	0.12 (0.02)
propyne	1.11 (0.04)	chlorobenzene	0.35 (0.05)
1-butyne	1.91 (0.07)	1, 2-dichlorobenzene	0.19 (0.01)
1-pentyne	2.5 (0.05)	1, 3-dichlorobenzene	0.36 (0.02)
benzene	0.550 (0.025)	1, 4-dichlorobenzene	0.16 (0.02)
toluene	0.670 (0.035)	bromobenzene	0.21 (0.04)
ethylbenzene	0.80 (0.07)	iodobenzene	0.13 (0.02)
o-xylene	0.50 (0.06)		

Table XIII.31 Joule–Thompson Coefficients for Some Gases at 273 K and 1 Atmosphere pressure

For more information see: M.W. Zemansky, *Heat and Thermodynamics*, 5th ed. (1957), McGraw Hill, New York.
$\mu_{JT} = (\partial T/\partial P)_H$; approximate values may be calculated using the relationship: $\mu_{JT} = \Delta T/\Delta P = [(2a/RT) - b]/C_P$ from a knowledge of the van der Waals constants (Table XI. 13) and C_P.

Gas	$(\mu)_{JT}$/K atm^{-1}
air (50 °C)	0.189
carbon dioxide (27 °C)	1.11
helium	−0.062
hydrogen	−0.03
nitrogen	0.266
oxygen	0.31

Table XIII.32 Standard molar Gibbs Energies of transfer of selected ions from water into aqueous alcohols at 298.15 K

$\Delta_t G^\ominus$ (X, water → aqueous alcohol) expressed as kJ mol^{-1}, are tabulated as a function of the composition (mass %) of the mixed solvent.
For more extensive data see: Y. Marcus, Pure App. Chem. (1990), 62, 899.
1. Recommended values for the standard molar Gibbs energies of transfer of ions from water into mixtures of methanol (S) and water at 298.15 K.

Mass % S	10	20	30	40	50	60	70	80	90	100
H$^+$	0.5	0.8	0.7	0.4	−0.2	−0.6	−0.8	−0.5	2.3	10.4
Li$^+$	1.1	1.8	2.4	2.9	3.3	3.4	3.6	3.6	3.8	4.4
Na$^+$	1.7	2.8	4.1	5.0	5.9	6.6	7.6	8.0	8.4	8.2
K$^+$	1.4	2.8	4.1	5.1	5.8	7.0	8.1	9.1	9.8	9.6
Rb$^+$	1.6	3.0	3.9	5.0	6.0	7.1	8.2	9.4	9.7	9.6
Cs$^+$	1.6	2.7	3.6	4.6	5.5	6.2	7.5	8.5	9.0	8.9
Ag$^+$	1.0	1.6	2.5	3.5	4.3	5.4	5.7	6.4	6.9	6.6
Ph$_4$As$^+$	−2.3	−4.7	−7.6	−10.8	−13.9	−16.2	−18.5	−20.1	−21.8	−22.8
Cl$^-$	0.2	0.8	1.5	2.5	3.6	5.0	6.5	8.4	10.7	13.2
Br$^-$	−0.1	0.1	0.7	1.4	2.4	3.6	4.9	6.8	8.8	11.1

2. Recommended values for the standard molar Gibbs energies of transfer of ions from water into mixtures of ethanol (S) and water at 298.15 K.

Mass % S	10	20	30	40	50	60	70	80	90	100
H$^+$	0.3	0.1	−1.1	−1.6	−4.0	−4.7	−4.7	−3.5	−2.3	10.5
K$^+$	1.3	2.6	3.1	4.2	4.2	4.6	5.4	8.3	12.7	16.4
Ph$_4$As$^+$	−2.4	−5.6	−9.8	−13.9	−16.3	−18.5	−19.0	−19.8	−20.8	−21.2
Cl$^-$	0.4	1.2	2.6	4.8	7.4	9.4	11.7	13.3	15.8	20.2
Br$^-$	0	0.5	2	3	5	7	9	9	10	18

Table XIII.33 Calorific Values of Solid, Liquid and Gaseous Fuels
Reference: as for Table XI.25.
For solid and liquid fuels, the gross calorific value is the heat of combustion at constant volume to give CO_2, N_2, SO_2 and liquid H_2O. For gaseous fuels it is the heat of combustion at constant pressure (1 atmos). Since many of the substances are not well defined, the values listed are approximate. The calorific values of pure substances can be calculated from the information in Table XIII.6.

1.

Solid and liquid fuels		Gross calorific value/ MJ kg^{-1}
Alcohols		
ethanol		30
methanol		23
Selected biomass		
monosaccharides	(40% C)	15.6
disaccharides	(42% C)	16.7
polysaccharides	(44% C)	17.5
lignin	(63% C)	25.1
crude protein	(53% C)	24
fat	(75% C)	40
carbohydrate	(41–44% C)	16.7–17.5
coconut shells		20
woods (dry)		
bamboo		19
beech		20
birch		20
oak		19
pine		21
peat (20% water)		16
seed oils (cotton, rape, linseed)		40
Coal and coal products		
anthracite (dry, mineral free)		34.3–37.2
bituminous coal (dry, mineral free)		37
(moist, mineral free)		24
sub-bituminous coal (moist, mineral free)		19–27
high volatile coking coals (4% water)		35
domestic coke (8–12% water)		28
coal tar fuels		36–41
Petroleum and petroleum products		
aviation fuel		45–47
crude petroleum and fuel oils		43–46
diesel fuel		46
kerosine		47
light fuel oil		44
medium fuel oil		43
petrol		45–47

2.

Gaseous fuels	Gross calorific value/ MJ m^{-3}
coal gas (vertical retort)	18
coal gas (low temperature)	34
hydrogen	12.8
hydrocarbons present in natural gas:	
methane	37.7
ethane	66.0
propane	94.0
n-butane	121.9
iso-butane	121.5
n-pentane	149.8
iso-pentane	149.4
neopentane	148.8
n-hexane (C_6 + fraction)	177.3
natural gas (North Sea)	39
producer gas (coal)	6
water gas (carburetted)	19

Table XIII.34 Properties of Some Common Binary Azeotropes at 1 Atmosphere Pressure (Unless Otherwise Stated)
For more extensive data see: L.H. Horsley, *Azeotropic Data III* (1973), Amer. Chem. Soc., D.C.

Component 1	b.p./°C	Component 2	b.p./°C	Azeotrope b.p./°C	wt% of 1
acetic acid	118.1	cyclohexane	81.4	79.7	2.0
acetic acid	118.1	pyridine	115.3	139.7	35.0
acetic acid	118.1	triethylamine	89.5	163.0	69.0
acetic acid	118.2	water	100.0	76.6	3.0
acetone	56.2	carbon disulphide	46.3	39.3	33.0
acetone	56.2	carbon tetrachloride	76.8	56.1	88.5
acetone	56.2	chloroform	61.2	64.7	20.0
acetone	56.2	cyclohexane	80.8	53.0	67.0
acetone	56.2	methanol	64.7	55.7	88.0
acetone	56.2	hexane	69.0	49.8	59.0
acetone (4.56 atm)	108.0	methanol	109.0	102.0	68.0
acetonitrile	82.0	water	100.0	76.5	83.7
benzene	80.1	cyclohexane	81.4	77.8	55.0
benzene	80.1	ethanol	78.5	67.8	67.6
benzene	80.1	methanol	64.7	58.3	60.5
benzyl alcohol	205.2	water	100.0	99.9	9.0
2-butanol	99.5	water	100.0	88.5	68.0
butanoic acid	163.5	water	100.0	99.4	18.4
tert-butanol	82.8	water	100.0	79.9	88.2
carbon tetrachloride	76.8	methanol	64.7	55.7	79.4
chlorobenzene	132.0	water	100.0	90.2	71.6
chloroform	61.2	ethanol	78.5	59.4	93.0
chloroform	61.2	methanol	64.7	53.5	87.4
cyclohexane	81.4	ethanol	78.5	64.9	69.5
cyclohexylamine	134.0	water	100.0	96.4	44.2
1,4-dioxan	101.3	water	100.0	87.8	81.6
ethanol	78.5	ethyl acetate	77.1	71.8	31.0
ethanol	78.5	triethylamine	89.5	76.9	51.0
ethanol	78.5	water	100.0	78.2	95.6
ethanol (95 mmHg)	33.5	water	51.0	33.4	99.5
ethanol (3 atm)	109.0	water	134.0	109.0	95.2
2-ethoxy ethanol	135.1	water	100.0	99.4	28.8
ethylene diamine	116.5	water	100.0	119.0	81.6
ethylmethyl ketone	79.6	water	100.0	73.4	88.0
formic acid	100.7	water	100.0	107.1	77.5
hexylamine	132.7	water	100.0	95.5	51.0
hydrogen bromide	−67.0	water	100.0	126.0	47.5
hydrogen chloride	−83.7	water	100.0	108.6	20.2
hydrogen chloride (6.8 atm)	−31.0	water	169.0	177.0	14.8
hydrogen fluoride	19.4	water	100.0	111.4	35.6
hydrogen iodide	−35.5	water	100.0	127.0	57.0
hydrogen nitrate	86.0	water	100.0	121.0	68.5
methanol	64.7	methyl acetate	57.0	54.0	18.7
nitromethane	101.0	toluene	110.6	96.5	55.0
propanoic acid	141.6	water	100.0	99.9	17.7
propanol	97.2	water	100.0	88.1	71.8
propyl acetate	101.6	water	100.0	82.4	86.0
pyridine	115.5	water	100.0	92.6	57.0
triethylamine	89.5	water	100.0	78.5	90.0

Table XIII.35 Critical Solution Temperatures (c.s.t.) for Some Partially Miscible Binary Liquid Mixtures

For more extensive data see: A.W. Francis, *Critical Solution Temperatures* (1961), Adv. in Chemistry Series, No. 31, Amer. Chem. Soc., D.C.

Component 1	Component 2	c.s.t./t °C	Wt% of component 1 at c.s.t.
aniline	cyclohexane	30.8	40
aniline	cyclopentane	16.8	37.5
aniline	n-hexane	69.1	41
aniline	water	167	86
carbon disulphide	acetonitrile	51.5	62.5
carbon disulphide	methanol	40.5	80.5
carbon disulphide	nitromethane	63.5	55.0
diethylamine	water	143.5	37.4
ethylmethylketone	water	73.4	88.7
isobutanol	water	133	49
		107	33
methanol	cyclohexane	49.1	37.2
methylaniline	ethylcyclohexane	−49.4	24.0
methylaniline	hexane	−18.6	27.5
nicotine	water	208	6.8
		60.8	
nitrobenzene	2-aminotoluene	21.1	64
nitrobenzene	3-aminotoluene	21.3	52.3
phenol	isopentane	63.5	49
phenol	water	63.8	33.4
triethylamine	water	18.5	51.9

Table XIII.36 Systems with Simple Eutectic Points. For further information see: S. Glasstone, *Textbook of Physical Chemistry*, 2nd ed. (1972), Macmillan, London.

Component A	Component B	Eutectic Mixture Temperature /°C	Composition /mol % B
Bi	Pb	124	44.7
Tl	Au	131	72
Sn	Bi	140	44
Cd	Bi	143.9	55
Sn	Pb	183	25
Sn	Zn	198	15
Sn	Ag	221	4
Sn	Cu	227	1.7
Sb	Pb	245.9	81
Cd	Zn	270	26
Al	Si	578	88.8
Be	Si	1090	32
$AgNO_3$	$TlNO_3$	82.2	52
KCl	$AlCl_3$	128	67
KNO_3	$AgNO_3$	131	62
$LiNO_3$	KNO_3	133.5	59
KCl	CuCl	150	65
$NaNO_3$	$TlNO_3$	164	76.5
$LiNO_3$	$AgNO_3$	173	76
KCl	$SnCl_2$	180	62
KBr	CuBr	185	68
$LiNO_3$	$NaNO_3$	193	46
NaCl	$BeCl_2$	215	50
KNO_3	$Pb(NO_3)_2$	217	23.4
KNO_3	$NaNO_3$	220	50
NaBr	NaOH	260	78
KBr	AgBr	285	68
KNO_3	$Be(NO_3)_2$	287	12.4
$NaNO_3$	NaCl	295	6.4
KCl	AgCl	306	70
RbCl	LiCl	319	58
KBr	$MgBr_2$	334	35
KBr	LiBr	348	60
LiCl	KCl	361	41.5
KCl	$PbCl_2$	406	52
NaCl	$PbCl_2$	411	72
LiBr	LiF	453	33
NaCl	$CaCl_2$	500	48
trichloromethane	aniline	−71	24
tribromomethane	benzene	−26	50
4-methyltoluene	2-nitrophenol	15.6	52
benzene	naphthalene	3	87
Systems Forming a Compound with a Congruent Melting Point			
Al	Mg	448	28
(compound A_3B_4, m.p. = 463 °C)		433	70
Sb	Au	480	46
(compound A_2B, m.p. = 492 °C)		370	73
Mg	Zn	345	52
(compound AB_2, m.p. = 575 °C)		360	97
$CaCl_2$	KCl	635	14
(compound AB, m.p. = 745 °C)		598	60
phenol	α-naphthylamine	16	22
(compound AB, m.p. = 28.8 °C)		24	70
phenol	4-methyltoluene	7	24
(compound AB, m.p. = 28 °C)		20	70
benzophenone	diphenylamine	31.9	25
(compound AB, m.p. = 40.2 °C)		34.5	70

Table XIII.37 Cryohydrates
Reference: as for Table XIII.36.

Salt	Cryohydric point/°C	Wt% anhydrous salt in cryohydrate
NaBr	−24	41.3
NaCl	−22	23.6
KI	−22	52.1
$NaNO_3$	−17.5	40.8
$(NH_4)_2SO_4$	−17	41.7
NH_4Cl	−15	19.3

Salt	Cryohydric point/°C	Wt% anhydrous salt in cryohydrate
NaI	−15	59.5
KBr	−13	32.2
KCl	−11.4	20.0
$MgSO_4$	−5	21.9
KNO_3	−2.6	11.2
Na_2SO_4	−0.7	4.6

(Continued)

Table XIII.38 Equilibrium Constants for Gaseous Reactions at Various Temperatures
(g = gaseous state, s = solid state)
Reference: as for Table XIII.36.

1. The Equilibrium: $N_2O_4 (g) \rightleftharpoons 2NO_2 (g)$

T/K	298	350	400	450	500	550	600
K_p/atm	0.115	3.89	47.9	3.47×10^2	1.70×10^3	6.03×10^3	1.78×10^4

2. The Equilibrium: $N_2 (g) + 3H_2 (g) \rightleftharpoons 2NH_3 (g)$

T/K	298	400	500	600	700	773
K_p/atm^{-2}	6.76×10^5	40.7	3.55×10^{-2}	1.66×10^{-3}	7.76×10^{-5}	1.5×10^{-5}

3. The Equilibrium: $2SO_2 (g) + O_2 (g) \rightleftharpoons 2SO_3 (g)$

T/K	298	500	700	1000
K_p/atm^{-1}	4.0×10^{24}	2.5×10^{10}	3.0×10^4	3.56

4. The Equilibrium: $H_2 (g) + I_2 (g) \rightleftharpoons 2HI (g)$

T/K	298	500	700	764
K_p	794	160	54	46

5. The Equilibrium: $N_2 (g) + O_2 (g) \rightleftharpoons 2NO (g)$

T/K	1900	2000	2100	2200	2300	2400	2500	2600
$10^4 K_p$	2.31	4.08	6.86	11.0	16.9	25.1	36.0	50.3

6. The Equilibria: $X_2 (g) \rightleftharpoons 2X (g)$

X	K_p/atm						
T/K	600	800	1000	1200	1400	1600	2000
O	1.4×10^{-37}	9.2×10^{-27}	3.3×10^{-20}	8.0×10^{-16}	1.1×10^{-12}	2.5×10^{-10}	5.2×10^{-7}
H	3.6×10^{-33}	1.2×10^{-23}	7.0×10^{-18}	5.05×10^{-14}	2.96×10^{-11}	3.59×10^{-9}	3.1×10^{-6}
N	1.3×10^{-56}	5.1×10^{-41}	1.3×10^{-31}	2.4×10^{-25}	7.5×10^{-21}	1.8×10^{-17}	9.8×10^{-13}
Cl	4.8×10^{-16}	1.0×10^{-10}	2.45×10^{-7}	2.48×10^{-5}	8.80×10^{-4}	1.29×10^{-2}	0.570
Br	6.2×10^{-12}	1.02×10^{-7}	3.58×10^{-5}	1.81×10^{-3}	3.03×10^{-2}	0.255	

7. The Equilibrium: $CaCO_3 (s) \rightleftharpoons CaO (s) + CO_2 (g)$

T/K	773	873	973	1073	1173	1273	1373	1473
K_p/atm	9.3×10^{-5}	2.42×10^{-3}	2.92×10^{-2}	0.220	1.043	3.871	11.50	28.68

Section XIV
Astronomy and Geophysics

		Page
XIV.1	Approximate values of galactic distances from Earth	505
XIV.2	Distance and visual magnitude of the brightest stars from Earth	505
XIV.3	Diameter, semimajor axis, eccentricity, inclination and rotation period of solar system asteroids of diameter exceeding 200 km	505
XIV.4	Values of diameter, density, semimajor axis, eccentricity, inclination and sidereal period of known satellites of solar system planets	506
XIV.5	Values of density, equatorial diameter, distance from Sun, surface gravity, ellipticity, eccentricity, inclination and sidereal and rotational periods for the Sun and its planets	507
XIV.6	Values of some miscellaneous properties of the Sun and Moon	507
XIV.7	Values of some miscellaneous properties of Earth	507
XIV.8	Variation of model Earth's density, gravity, pressure and elastic characteristics with depth beneath surface	507
XIV.9	Possible values of duration and time of start from present of various geological periods for Earth	508
XIV.10	Variation with height above Earth of atmospheric pressure density and temperature	508
XIV.11	Variation with height above Earth of upper atmospheric pressure, mean molecular weight and temperature	508
XIV.12	Values of particle flux in vertical direction for various cosmic-ray particles at sea level, for kinetic energies above various threshold values, at geomagnetic latitudes greater than about 40°	509
XIV.13	Average values of declination and components of the geomagnetic field parallel to the Earth's surface and vertically downwards, at various observatories in 1975	509
XIV.14	Values of the electrical conductivity of sea water at various temperatures and salinities for standard atmospheric pressure	510
XIV.15	Density of sea water at various temperatures and salinities for standard atmospheric pressure	510
XIV.16	Density of sea water of salinity 35 g kg^{-1} at various pressures for standard temperature	510
XIV.17	Relationships between some time quantities	510

Section XIV
Astronomy and Geophysics

Table XIV.1 Approximate values of galactic distances from Earth

Galaxy	distance/10^{17}km	Galaxy	distance/10^{17}km
Andromeda Galaxy, M31	210	Large Magellanic Cloud	18
Triangulum Galaxy, M33	220	Small Magellanic Cloud	20

Reference: Tennent (ed.), *Science Data Book*, Oliver and Boyd, 1986
Illingworth, Dictionary of Astronomy, 2nd ed. (1985) Macmillan

Table XIV.2 Distance and visual magnitude of the brightest stars from Earth

Star	distance/10^{12}km	magnitude	Star	distance/10^{12}km	magnitude
Achernar (α Eridani)	1300	0.6	Procyon (α Canis Minoris)	107	0.5
Arcturus (α Bootes)	340	0.2	Rigel (β Orionis)	11 000	0.3
Canopus (α Carinae)	1700	−0.9	Rigil Kent (α Centauri)	41	0.1
Capella (α Aurigae)	420	0.2	Sirius (α Canis Majoris)	82	−1.6
Hadar (α Centauri)	1900	0.9	Vega (α Lyrae)	251	0.1

Reference: as for Table XIV.1

Table XIV.3 Diameter, semimajor axis, eccentricity, inclination and rotation period of solar system asteroids of diameter exceeding 200 km

Asteroid	diameter/km	semimajor axis/AU	eccentricity	inclination	period/h
Alauda	205	3.194	0.0347	20.54	
Aspasia	208	2.575	0.0733	11.26	
Bamberger	251	2.686	0.336	11.16	8
Camilla	210	3.489	0.0699	9.92	4.56
Ceres	1020	2.767	0.0784	10.61	9.078
Cybele	308	3.434	0.1154	3.55	
Davida	341	3.187	0.1662	15.81	5.12
Egeria	241	2.576	0.0889	16.5	7.045
Eugenia	228	2.721	0.0806	6.6	5.7
Eunomia	246	2.642	0.1883	11.73	6.080 6
Euphrosyne	333	3.154	0.2244	26.3	
Europa	290	3.092	0.1138	7.47	11.258 2
Fortuna	221	2.442	0.1576	1.56	7.46
Hektor	216	5.15	0.0248	18.26	6.922 5
Herculina	217	2.771	0.1789	16.35	9.406
Hygiea	450	3.151	0.0996	3.81	18
Interamnia	339	3.057	0.1553	17.31	8.723
Iris	210	2.386	0.2303	5.5	7.135
Juno	248	2.67	0.2557	12.99	7.213
Kleopatra	219	2.793	0.252	13.09	5.394
Loreley	203	3.128	0.0802	11.24	
Pallas	538	2.769	0.2353	34.83	7.881 06
Patientia	327	3.061	0.0772	15.23	7.11
Psyche	252	2.92	0.139	3.09	4.303
Sylvia	225	3.481	0.0985	10.85	
Themis	210	3.138	0.1208	0.77	8.375
Thisbe	207	2.768	0.1619	5.22	6.042 2
Vesta	549	2.362	0.089	7.13	5.342 13
Winchester	205	2.994	0.3438	18.15	8

Reference: Delsemme (ed.), *Comets Asteroids Meteorites: Interrelations, Evolution and Origins*, University of Toledo, 1977

Table XIV.4 Values of diameter, density, semimajor axis, eccentricity, inclination and sidereal period of known satellites of solar system planets

Planet	Satellite(s)	diameter/ km	density/ kg m^{-3}	semimajor axis/km	eccentricity	inclination/ degree	period/ day
Earth	Moon	3476	3343	384 399	0.055	23.4	27.32
Mars	Phobos	27	1900	9378.5	0.015	1.04	0.32
	Deimos	15		23 459	0.0008	2.79	1.26
Jupiter	1979 J3	c 40		128 000		0	0.294
	1979 J1	c 25		129 000		0	0.298
	Amalthea	270		181 500	0.0028	0.5	0.5
	1979 J2	c 75		221 700		1.25	0.67
	Io	3632	3530	422 000	0	0	1.77
	Europa	3126	3170	671 400	0.0003	0.5	3.55
	Ganymede	5276	1990	1 071 000	0.0015	0.2	7.15
	Callisto	4820	1760	1 884 000	0.0075	0.3	16.69
	Leda			11 094 000	0.148	27	239
	Himalia	c 170		11 487 000	0.158	28	250.6
	Elara	c 80		11 747 000	0.207	25	259.7
	Lysithea			11 861 000	0.13	29	259.2
	Ananke			21 250 000	0.169	147	631
	Carme			22 540 000	0.207	163	692
	Pasiphae			23 510 000	0.378	147	739
	Sinope			23 670 000	0.275	156	758
Saturn	1980 S28	c 100		137 000			
	1980 S27			139 000			
	1980 S26			142 000			
	1980 S3	c 150		151 000			
	1980 S1	c 150		151 000			
	Mimas	400	c 1000	185 000	0.02	1.5	0.94
	Enceladus	550	c 1000	238 000	0.0044	0	1.37
	Tethys	1040	c 1000	294 700	0.0022	1.1	1.89
	1980 S13						
	Dione	1000	c 2000	377 400	0.0022	0	2.74
	Dione B						
	Rhea	1600	c 1300	527 000	0.001	0.3	4.42
	Titan	5800	c 1370	1 222 000	0.029	0.3	15.95
	Hyperion	224		1 484 000	0.104	0.4	21.28
	Iapetus	1450	c 1000	3 562 000	0.0283	14.7	79.33
	Phoebe			12 960 000	0.1633	150	550.3
Uranus[†]	Miranda	300		129 800	0.017	3.4	1.41
	Ariel	800		190 900	0.0028	0	2.52
	Umbriel	550		266 000	0.0035	0	4.14
	Titania	1000		436 000	0.0024	0	8.71
	Oberon	900		583 400	0.0007	0	13.46
Neptune[†]	Triton	3200	c 3000	355 550	0	159.9	5.88
	Nereid			5 567 000	0.7493	27.8	359.9
Pluto	Charon	200	c 400	c 17 000	c.0	115	6.39

[†] Additional satellites orbit Uranus (10 small satellites discovered 1986, possibly more) and Neptune (6 small satellites discovered 1989, possibly more)

Reference: Anderson (ed.), *Physics Vade Mecum*, 2nd edn., AIP, 1989

Table XIV.5 Values of density, equatorial diameter, distance from Sun, surface gravity, ellipticity, eccentricity, inclination and sidereal and rotational periods for the Sun and its planets

Body	density/ kg m^{-3}	diameter/ 10^6 m	distance/ 10^{10} m	gravity/ m s^{-2}	ellipticity	eccentricity	inclination/ degree	sidereal period/year	rotational period/hour
Sun	1409	1392		274	0				609.1
Mercury	5420	4.84	5.791	3.76	0	0.2056	7.004	0.241	1408.8
Venus	5250	12.17	10.82	8.77	0	0.0068	3.394	0.615	5832
Earth	5510	12.756	14.96	9.81	0.0034	0.0167	0	1	23.93
Mars	3960	6.75	22.79	3.8	0.007	0.0934	1.85	1.881	24.6
Jupiter	1330	142.8	77.83	24.9	0.062	0.0481	1.306	11.86	9.9
Saturn	680	120.8	142.7	10.4	0.096	0.0533	2.489	29.46	10.2
Uranus	1600	47.2	286.9	10.4	0.06	0.0507	0.773	84.02	10.7
Neptune	1650	44.6	449.8	13.8	0.02	0.004	1.773	164.8	15.8
Pluto	3000	6	590	4		0.2533	17.142	248	151.2

Reference: as for Table XIV.1

Table XIV.6 Values of some miscellaneous properties of the Sun and Moon

Sun	Moon
rate of energy production: 3.9×10^{26} W escape speed at surface: 618 km s^{-1}	59% surface area at some time visible from earth escape speed at surface: 2.38 km s^{-1} gravity at surface: 1.62 m s^{-2}

Reference: as for Table XIV.1

Table XIV.7 Values of some miscellaneous properties of Earth

polar diameter: 12.714×10^6 m surface area: 5.101×10^{14} m^2 distance to Sun at perihelion: 1.471×10^{11} m distance to Sun at aphelion: 1.521×10^{11} m rotational speed at equator: 465 m s^{-1} mean speed in solar orbit: 29.78 km s^{-1} escape speed at surface: 11.2 km s^{-1} solar constant: 1400 J m^{-2} s^{-1}	land area: 1.49×10^{14} m^2 (i.e. 29.2% Earth's surface) greatest height (Mt. Everest): 8847.7 m mean height: 840 m ocean area: 3.61×10^{14} m^2 (i.e. 70.8% Earth's surface) ocean mass: 1.42×10^{21} kg greatest depth (Marianas Trench): 11 033 m mean depth: 3800 m atmospheric mass: 5.27×10^{18} kg (i.e. 10^{-6} of Earth's mass)

Reference: as for Table XIV.1
Kaye & Laby *Tables of Physical and Chemical Constants*, 15 ed., Longmans, 1986

Table XIV.8 Variation of model Earth's density, gravity, pressure and elastic characteristics with depth beneath surface

Region	depth/km	density/ kg m^{-3}	gravity/m s^{-2}	pressure/ 10^{11} Pa	elastic moduli/10^{11} Pa		elastic wave speed/ km s^{-1}	
					bulk	rigidity	compressional	shear
mantle	0	2600	9.82	0	0.56	0.32	6	3.5
	150	3400	9.88	0.048	1.36	0.66	8.15	4.43
	500	3830	9.98	0.174	2.16	1.05	9.65	5.24
	1000	4560	9.96	0.388	3.5	1.86	11.44	6.36
	1500	4825	9.92	0.622	4.26	2.16	12.22	6.7
	2000	5080	10.01	0.868	5.09	2.42	12.78	6.89
	2886	5520	10.74	1.353	6.44	2.88	13.66	7.23
core	2886	9900	10.74	1.353	6.36	0	8.11	0
	3500	10 680	9.32	1.936	8.3	0	8.88	0
	4000	11 410	7.96	2.467	10.31	0	9.5	0
	5150	12 280	4.62	3.311	13.05	0	9.86	0
inner core	5150	13 000	4.62	3.311	17.03	0.68	11.18	1
	6371	13 500	0	3.691	17.81	0.7	11.8	1

Reference: Derr, *J. Geophys. Res.* vol. 74, 1969

Table XIV.9 Possible values of duration and time of start from present of various geological periods for Earth

Period	duration/10^6 year	start time/10^6 year	Period	duration/10^6 year	start time/10^6 year
Cambrian	70	570	Ordovician	65	500
Carboniferous	65	345	Palaeocene	11	65
Cretaceous	71	136	Permian	55	280
Devonian	50	395	Pleistocene	2	2
Eocene	16	54	Pliocene	5	7
Jurassic	56	192	Precambrian	4000	4600
Miocene	19	26	Silurian	40	435
Oligocene	12	38	Triassic	33	225

Reference: *Tables of Physical and Chemical Constants*, 15 ed., Longmans, 1986

Table XIV.10 Variation with height above Earth of atmospheric pressure, density and temperature

height/m	pressure/kPa	density/kg m^{-3}	temperature/K	height/m	pressure/kPa	density/kg m^{-3}	temperature/K
0	101.3	1.22	288	5000	54	0.74	255
500	95.5	1.17	285	7000	41.1	0.59	243
1000	89.9	1.11	281	10 000	26.4	0.41	223
1500	84.6	1.06	278	14 000	14.1	0.23	217
2000	79.5	1	275	20 000	5.5	0.088	217
3000	70.1	0.91	269	26 000	2.2	0.034	223
4000	61.6	0.82	262	30 000	1.2	0.018	227

Reference: *Manual of ICAO Standard Atmosphere*, ICAO Doc. 7488/2, ICAO Montreal, 1964

Table XIV.11 Variation with height above Earth of upper atmospheric pressure, mean molecular weight and temperature

height/km	pressure/Pa	molecular weight	temperature/K	height/km	pressure/Pa	molecular weight	temperature/K
40	294	28.97	250.5	200 (night)	9.5×10^{-5}	22.7	890
60	21.9	28.97	243.3	500 (day)	1.3×10^{-6}	16.3	1420
80	0.975	28.96	186	500 (night)	2.8×10^{-7}	15	1000
100	0.031	28.3	208.1	800 (day)	6.6×10^{-8}	12.1	1430
200 (day)	1.2×10^{-4}	23	1100	800 (night)	1.1×10^{-8}	6.5	1000

Reference: as for Table XIV.10

Table XIV.12 Values of particle flux in vertical direction for various cosmic-ray particles at sea level, for kinetic energies above various threshold values, at geomagnetic latitudes greater than about 40°

threshold energy/ GeV	muon flux/ $m^{-2} s^{-1} sr^{-1}$	electron flux/ $m^{-2} s^{-1} sr^{-1}$	photon flux/ $m^{-2} s^{-1} sr^{-1}$	proton flux/ $m^{-2} s^{-1} sr^{-1}$	neutron flux/ $m^{-2} s^{-1} sr^{-1}$
0.001	100	60	130	2.1	
0.01	100	28	60	2.1	c 30
0.02	100	20	40	2.1	
0.1	99	6	8	1.9	c 10
0.2	97	3	3.5	1.5	
0.5	86	1	1.1	0.9	1.5
1	69	0.38	0.37	0.51	0.7
2	46	0.12	0.11	0.25	
5	20	0.02	0.02	0.077	
10	8.6			0.025	
20	3			0.008	
50	0.58			0.001 6	
100	0.14			0.000 43	
200	0.03			0.000 11	
500	0.003 2			0.000 02	
1000	0.000 5				

Reference: Kaye & Laby *Tables of Physical and Chemical Constants, 15 ed.*, Longmans, 1986

Table XIV.13 Average values of declination and components of the geomagnetic field parallel to Earth's surface and vertically downwards, at various observatories in 1975

| Observatory | declination (east)/° | components/nT | | Observatory | declination (east)/° | components/nT | |
		parallel	vertical			parallel	vertical
Alibag	−0.8	38 517	17 434	Magadan	−13	18 103	52 638
Apia	12.3	34 453	−20 134	Maputo	−16.1	14 189	−27 550
Bjørnøya	2.5	9296	52 780	Mawson	−62.5	18 397	−47 269
College	28.4	13 022	55 346	Memambetsu	−8.2	26 501	41 325
Eskdalemuir	−9.2	17 200	45 719	Muntinlupa	−0.4	39 056	9695
Fredericksburg	−7.8	20 269	52 031	Nairobi	−1.4	30 569	−15 602
Gnangara	−3.2	23 608	−53 496	Patrony	−2.3	19 602	56 937
Greenwich	−6.65	19 160	43 980	Point Barrow	26	9830	56 410
Hartland	−8.5	19 212	43 733	Quetta	1.2	32 919	33 468
Hermanus	−23.9	11 688	−26 210	Resolute Bay	−65.9	742	58 578
Honolulu	11.6	27 810	22 334	San Juan	−8.8	27 450	31 752
Kakioka	−6.5	30 142	34 630	San Miguel	−13.3	25 542	38 028
Kerguelen	−50.7	18 523	−43 962	Toolangi	10.8	22 209	−56 308
Krasnaya Pakhra	7.6	17 298	48 525	Tromsø	0	11 400	51 450
Lerwick	−8.6	14 890	47 753	Tsumeb	−14.8	16 318	−27 072
Luanda	−10.1	23 475	−22 539	Tucson	12.5	25 758	43 237
Macquarie Isl.	27.7	12 847	−63 926	Vassouras	−18.5	21 489	−10 676

Reference: Kaye & Laby *Tables of Physical and Chemical Constants, 15 ed.*, Longmans, 1986

Table XIV.14 Values of the electrical conductivity of sea water at various temperatures and salinities for standard atmospheric pressure

salinity/ g kg^{-1}	temperature/K					
	273	278	283	288	293	298
	\multicolumn{6}{c}{conductivity/S m^{-1}}					
20	1.745	2.015	2.3	2.595	2.901	3.217
25	2.137	2.466	2.811	3.17	3.542	3.926
30	2.523	2.909	3.313	3.735	4.171	4.621
35	2.906	3.346	3.808	4.29	4.788	5.302
40	3.285	3.778	4.297	4.837	5.397	5.974

Reference: Cox et al., *Deep Sea Res.*, vol. 17, 679, 1970

Table XIV.15 Density of sea water at various temperatures and salinities for standard atmospheric pressure

salinity/ g kg^{-1}	temperature/K					
	273	278	283	288	293	298
	\multicolumn{6}{c}{density/kg m^{-3}}					
20	1016.04	1015.84	1015.31	1014.48	1013.39	1012.07
25	1020.06	1019.78	1019.18	1018.3	1017.17	1015.82
30	1024.08	1023.73	1023.07	1022.13	1020.96	1019.57
35	1028.1	1027.68	1026.96	1025.97	1024.75	1023.34
40	1032.14	1031.64	1030.86	1029.82	1028.56	1027.12

Reference: as for Table XIV.14

Table XIV.16 Density of sea water of salinity 35 g kg^{-1} at various pressures for standard temperature

pressure/MPa	0	20	40	60	80	100
density/kg m^{-3}	1028.1	1037.44	1046.37	1054.92	1063.12	1071.02

Reference: as for Table XIV.14

Table XIV.17 Relationships between some time quantities

quantity	relationship
minute	60 second ^{X}
hour	60 minute
day	24 hour
mean synodic month	29.530 859 day
mean tropical month	27.321 582 day
mean sidereal month	27.321 661 day
Julian year	365.25 day
tropical year	365.242 193 − 0.000 006 1 T† day
sidereal year	365.256 360 − 0.000 000 1 T† day
century	100 year

^{X} The second is a base SI unit: see Table I 1
† T is the number of Julian centuries (mean value 0.085 in 1992) between now and AD 2000
Reference Kaye and Laby *Tables of Physical and Chemical Constants 15 ed.* 1986 Longmans

Section XV
Chemical Kinetics

		Page
XV.1	**Kinetic data for some first and second order reactions**	**513**
	1 First order reactions	513
	2 Second order reactions	513
XV.2	**Arrhenius parameters for rate coefficients**	**514**
	1 Thermal unimolecular gas phase reactions	514
	2 Bimoleular reactions	514
XV.3	**Rate coefficients for some second order solution reactions at 298 K**	**515**
XV.4	**Arrhenius pre-exponential factors, entropies and volumes of activation for a selection of reactions**	**516**
	1 Unimolecular reactions	516
	2 Bimolecular reactions	516
	3 Ionic reactions in water	516
XV.5	**Activation energies for some catalysed reactions**	**517**
XV.6	**Hammett substituent constants, σ**	**517**
XV.7	**A selection of Hammett reaction constants, ρ, from correlations of reaction rate constants**	**518**

Section XV
Chemical Kinetics

Table XV.1 Kinetic Data for Some First and Second Order Reactions
For further information see: S.W. Benson, *The Foundations of Chemical Kinetics* (1960), McGraw-Hill, New York.
                       W.C. Gardiner, *Rates and Mechanisms of Chemical Reactions* (1969), Benjamin, New York.
                       J. Nicholas, *Chemical Kinetics* (1976), Harper and Row, London.
                       K.J. Laidler, *Chemical Kinetics*, TMH ed. (1976), McGraw-Hill, New York.

1. First-order Reactions

Reaction	Phase	$t/°C$	k_1/s^{-1}
$N_2O_5 \rightarrow NO_2 + NO_3$	gas	25	3.5×10^{-5}
		55	1.57×10^{-3}
	CCl_4 soln	25	3.94×10^{-5}
	HNO_3 soln	55	9.27×10^{-5}
$C_2H_6 \rightarrow 2CH_3$	gas	700	5.46×10^{-4}
cyclopropane $\rightarrow$ propene	gas	500	6.71×10^{-4}
$CH_3N_2CH_3 \rightarrow C_2H_6 + N_2$	gas	327	3.4×10^{-4}
sucrose $\rightarrow$ glucose + fructose	acidic aq soln	25	6.0×10^{-5}
$C_6H_5N(Cl)COCH_3 \rightarrow 4\text{–}ClC_6H_4NHCOCH_3$	acidic aq soln	25	9.6×10^{-5}
$^{231}Pa \rightarrow {}^{227}Ac + \alpha$			6.9×10^{-13}
$^{215}Po \rightarrow {}^{211}Pb + \alpha$			3.8×10^{2}

2. Second-order Reactions

Reaction	Phase	$t/°C$	$k_2/\text{dm}^3\ \text{mol}^{-1}\ \text{s}^{-1}$
$2NOBr \rightarrow 2NO + Br_2$	gas	10	0.80
$2NO_2 \rightarrow 2NO + O_2$	gas	300	0.54
$H_2 + I_2 \rightarrow 2HI$	gas	400	2.42×10^{-2}
$D_2 + HCl \rightarrow DH + DCl$	gas	600	1.41×10^{-1}
$I + I \rightarrow I_2$	gas	23	7.0×10^{9}
	hexane	50	18.0×10^{9}
$CH_3COOC_2H_5 + H_2O \rightarrow CH_3COOH + C_2H_5OH$	alkaline 70%	25	4.65×10^{-2}
	aq acetone	35	8.22×10^{-2}
		45	1.35×10^{-1}
$H^+ + OH^- \rightarrow H_2O$	water	25	1.35×10^{11}

Table XV.2 Arrhenius Parameters for Rate Coefficients

Parameters in the Arrhenius rate equation:
$$k = A \exp(-E_A/RT)$$

For more extensive data see: C.H. Bamford and C.F.H. Tipper (eds), *Comprehensive Chemical Kinetics*, vol. 18, *Selected Elementary Reactions*, (1976), Elsevier, Amsterdam.

1. Thermal Unimolecular Gas Phase Reactions

Reaction		$\lg(A/s^{-1})$	E_A/kJ mol^{-1}
CH$_2$–CH$_2$ $\rightarrow$ CH$_3$CH=CH$_2$ (cyclopropane, with CH$_2$)		15.2	272
H$_2$C–CH$_2$ $\rightarrow$ 2C$_2$H$_4$ (cyclobutane, H$_2$C–CH$_2$)		15.6	261
CH$_3$NC $\rightarrow$ CH$_3$CN		13.3	159
trans-CHD=CHD $\rightarrow$ *cis*-CHD=CHD		12.5	255
cis-CHPh=CHPh $\rightarrow$ *trans*-CHPh=CHPh		12.8	180
N$_2$O$_5$ $\rightarrow$ NO$_2$ + NO$_3$		13.7	103
N$_2$O $\rightarrow$ N$_2$ + O		11.2	251
C$_2$H$_6$ $\rightarrow$ 2CH$_3$		16.5	368
CH$_3$N=NCH$_3$ $\rightarrow$ 2CH$_3$ + N$_2$		15.7	213
C$_2$H$_5$ $\rightarrow$ C$_2$H$_4$ + H		13.0	167
C$_2$H$_5$I $\rightarrow$ C$_2$H$_4$ + HI		13.4	209
C$_2$H$_5$Cl $\rightarrow$ C$_2$H$_4$ + HCl		14.6	254
CO$_2$ $\rightarrow$ CO + O		11.3	460
Hg(CH$_3$)$_2$ $\rightarrow$ Hg + 2CH$_3$		13.3	213
Hg(C$_2$H$_5$)$_2$ $\rightarrow$ Hg + 2C$_2$H$_5$		14.1	180

2. Bimolecular Reactions

Reaction		$\lg(A/\text{dm}^3\,\text{mol}^{-1}\,\text{s}^{-1})$	E_A/kJ mol^{-1}
Gas phase molecular reactions			
H$_2$ + I$_2$ $\rightarrow$ 2HI		11.2	167
2HI $\rightarrow$ H$_2$ + I$_2$		11.0	184
NO + O$_3$ $\rightarrow$ NO$_2$ + O$_2$		8.9	10.5
2NOCl $\rightarrow$ 2NO + Cl$_2$		10.0	103
NO$_2$ + CO $\rightarrow$ NO + CO$_2$		10.1	132
2NO$_2$ $\rightarrow$ 2NO + O$_2$		9.3	111
Gas phase reactions involving atoms or radicals			
H + D$_2$ $\rightarrow$ HD + D	(350–750 K)	10.69	39.3
H + H$_2$ $\rightarrow$ H$_2$ + H	(350–750 K)	10.6	31.8
D + H$_2$ $\rightarrow$ HD + H	(250–750 K)	10.64	31.8
H + O$_2$ $\rightarrow$ HO + H	(700–2500 K)	11.4	70.3
H + CH$_4$ $\rightarrow$ H$_2$ + CH$_3$	(370–1290 K)	11.1	49.8
H + C$_2$H$_6$ $\rightarrow$ H$_2$ + C$_2$H$_5$	(290–509 K)	10.9	38.1
H$_2$ + Cl $\rightarrow$ HCl + H	(250–450 K)	10.08	18.0
H + Cl$_2$ $\rightarrow$ HCl + Cl	(273–5200 K)	11.0	22.2
H + Br$_2$ $\rightarrow$ HBr + Br	(973–1673 K)	11.97	15.5
H + I$_2$ $\rightarrow$ HI + I	(667–738 K)	10.6	0
N + O$_2$ $\rightarrow$ NO + O	(180–5000 K)	9.92	29.7
O + N$_2$ $\rightarrow$ NO + N		11.0	315
NO + Cl$_2$ $\rightarrow$ NOCl + Cl		9.6	85
2CH$_3$ $\rightarrow$ C$_2$H$_6$		10.3	c 0
CH$_3$ + H$_2$ $\rightarrow$ CH$_4$ + H	(450–750 K)	9.5	51.0
CH$_3$ + D$_2$ $\rightarrow$ CH$_3$D + D	(300–626 K)	9.3	53.1
CD$_3$ + CH$_4$ $\rightarrow$ CD$_3$H + CH$_3$		8.0	60
OH + H$_2$ $\rightarrow$ H$_2$O + H	(300–1200 K)	10.36	22.6
F$_2$ + ClO$_2$ $\rightarrow$ FClO$_2$ + F	(227–247 K)	7.11	33.5
Na + CCl$_4$ $\rightarrow$ NaCl + CCl$_3$	(520 K)	11.35	c 0
CH$_3$ + C$_2$H$_6$ $\rightarrow$ CH$_4$ + C$_2$H$_5$	(456–600 K)	7.4	38.5

(Continued)

Table XV.2 (Continued)

Reaction	lg $(A/\text{dm}^3 \text{ mol}^{-1} \text{ s}^{-1})$	$E_A/\text{kJ mol}^{-1}$
Reactions in solution		
$2\text{ClO}^- \rightarrow 2\text{Cl}^{2-} + \text{O}_2$	7.8	c. 0
$\text{C}_2\text{H}_5\text{ONa} + \text{CH}_3\text{I} \rightarrow \text{C}_2\text{H}_5\text{OCH}_3 + \text{NaI}$ (EtOH)	11.3	81.6
$\text{C}_2\text{H}_5\text{Br} + \text{OH}^- \rightarrow \text{C}_2\text{H}_5\text{OH} + \text{Br}^-$ (EtOH)	11.6	89.5
$\text{NH}_4\text{CNO} \rightarrow (\text{NH}_2)_2\text{CO}$ (H$_2$O)	12.6	97.0
$\text{CH}_2\text{ClCOOH} + \text{OH}^- \rightarrow \text{CH}_2(\text{OH})\text{COOH} + \text{Cl}^-$ (H$_2$O)	11.6	0.6
$\text{C}_{12}\text{H}_{22}\text{O}_{11} + \text{H}_2\text{O} \rightarrow 2\text{C}_6\text{H}_{12}\text{O}_6$ (H$^+$)	15.2	107.9
$\text{CH}_3\text{COOC}_2\text{H}_5 + \text{OH}^- \rightarrow \text{CH}_3\text{COO}^- + \text{C}_2\text{H}_5\text{OH}$ (H$_2$O)	7.1	47.3
$\text{CH}_3\text{I} + \text{OH}^- \rightarrow \text{CH}_3\text{OH} + \text{I}^-$	12.1	92.9

Table XV.3 Rate Coefficients for Some Second Order Solution Reactions at 298 K
Reference: as for Table XV.1.
e (aq) indicates that the electron is strongly bound to water molecules, all other species are understood to be solvated. Many of the reactions are presumed to be elementary, others may be complex.

Reaction	$k_2/\text{dm}^3 \text{ mol}^{-1} \text{ s}^{-1}$
e (aq) + e (aq) $\rightarrow$ H$_2$ + OH$^-$	5×10^9
e (aq) + H$_3$O$^+$ $\rightarrow$ H + H$_2$O	2.1×10^{10}
e (aq) + OH $\rightarrow$ OH$^-$	3×10^{10}
e (aq) + Cu^{2+} $\rightarrow$ Cu$^+$	3.3×10^{10}
e (aq) + Cd^{2+} $\rightarrow$ Cd$^+$	5×10^{10}
e (aq) + NH$_4^+$ $\rightarrow$ NH$_3$ + H	1×10^6
H$^+$ (aq) + NH$_3$ $\rightarrow$ NH$_4^+$	4.3×10^{10}
H$^+$ (aq) + CH$_3$OH $\rightarrow$ CH$_3$OH$_2^+$	1×10^8
H$^+$ (aq) + C$_2$H$_5$OH $\rightarrow$ C$_2$H$_5$OH$_2^+$	3×10^6
H$^+$ (aq) + OH$^-$ $\rightarrow$ H$_2$O	13.5×10^{10}
H$^+$ (aq) + CH$_3$COO$^-$ $\rightarrow$ CH$_3$COOH	5.1×10^{10}
H$^+$ (aq) + C$_6$H$_5$COO$^-$ $\rightarrow$ C$_6$H$_5$COOH	3.7×10^{10}
H$_3$O$^+$ + H$_2$O $\rightarrow$ H$_2$O + H$_3$O$^+$	1×10^{10}
OH$^-$ + HCO$_3^-$ $\rightarrow$ CO$_3^{2-}$ + H$_2$O	6×10^9
OH$^-$ + NH$_4^+$ $\rightarrow$ NH$_3$ + H$_2$O	3.4×10^{10}
OH$^-$ + H$_2$O $\rightarrow$ H$_2$O + OH$^-$	5×10^9
OH$^-$ + CH$_3$OH $\rightarrow$ CH$_3$O$^-$ + H$_2$O	3×10^6
OH$^-$ + C$_2$H$_5$OH $\rightarrow$ C$_2$H$_5$O$^-$ + H$_2$O	3×10^6
CH$_3$Br + Cl$^-$ $\rightarrow$ CH$_3$Cl + Br$^-$ (acetone)	5.9×10^{-3}
CH$_3$Br + Br^{-*} $\rightarrow$ CH$_3$Br* + Br$^-$ (acetone, 238 K)	6.3×10^{-4}
C$_6$H$_5$CHO + HCN $\rightarrow$ C$_6$H$_5$CH(OH)CN (EtOH, 293 K)	2×10^{-8}

Table XV.4 Arrhenius Pre-exponential Factors, Entropies and Volumes of Activation for a Selection of Reactions.
Reference: as for Table XV.1.

1. Unimolecular Reactions

Reaction	lg (A/s^{-1})	$\Delta^{\ddagger} S/$J K^{-1} mol^{-1}
$CH_3N_2CH_3 \to C_2H_6 + N_2$	16.5	71
$CH_2\text{-}CH_2\text{-}CH_2$ (cyclopropane) $\to CH_3CH=CH_2$	16.2	46
Dicyclopentadiene $\to$ 2 cyclopentadiene	13.0	0
Vinyl allyl ether $\to$ Allylacetaldehyde	11.7	-21
Decomposition of ethylidene diacetate	10.3	-50

2. Bimolecular Reactions

Reaction	lg $(A/$dm^3 mol^{-1} s$^{-1})$	$\Delta S^{\ddagger}/$J K^{-1} mol^{-1}
$H_2 + I_2 \to 2HI$	11.0	8
$2HI \to H_2 + I_2$	10.8	0
$NO + O_3 \to NO_2 + O_2$	8.9	-33
Dimerization of 1,3-butadiene	7.7	-54
Dimerization of cyclopentadiene	4.9	-111

3. Ionic Reactions in Water, Unless Otherwise Stated
$\Delta^{\ddagger}V$ is the Volume of Activation.

Reaction	lg $(A/$dm^3 mol^{-1} s$^{-1})$	$\Delta^{\ddagger}V$ /cm^3 mol^{-1}	$\Delta^{\ddagger}S_{obs}$ /J K^{-1} mol^{-1}
$[Co(NH_3)_5Br]^{2+} + OH^- \to [Co(NH_3)_5OH]^{2+} + Br^-$	17.7	8.5	92
sucrose + H_2O + H^+ $\to$ glucose + fructose		2.5	33
$CH_2BrCOOCH_3 + S_2O_3^{2-} \to CH_2(S_2O_3^-)COOCH_3 + Br^-$	14.0	3.2	25
$C_2H_5O^- + C_2H_5I \to C_2H_5OC_2H_5 + I^-$ (EtOH)		-4.1	-42
$CH_2ClCOO^- + OH^- \to CH_2(OH)COO^- + Cl^-$	10.8	-6.1	-56
$ClO^- + ClO^- \to 2Cl^- + O_2$	9.0		-82
$[Co(NH_3)_5Br]^{2+} + Hg^{2+} + H_2O \to [Co(NH_3)_5(H_2O)]^{3+} + HgBr^+$	8.1		-99
$CH_2BrCOO^- + S_2O_3^{2-} \to CH_2(S_2O_3^-)COO^- + Br^-$	7.1		-120
$CH_3CONH_2 + H_2O + H^+ \to CH_3COOH + NH_3$		-14.2	-142
$C_5H_5N + C_2H_5I \to C_5H_5(C_2H_5)N^+ I^-$ (acetone)		-16.8	-146

Table XV.5 Activation Energies for some Catalysed Reactions

Reference: as for Table XV.1.
*: Relative velocities.

Reaction	Catalyst	E_A /kJ mol^{-1}
$2N_2O \rightarrow 2N_2 + O_2$	None	245
	MgO	155
	CaO	146
	Pt	136
	Al_2O_3	125
	Au	121
$2NH_3 \rightarrow N_2 + 3H_2$	None	>334
	Os	200
	Pt	185
	W	158
	Mo	134–180
$o-H_2 + H \rightarrow p-H_2 + H$	None	254
	Cu	42–45
	Au	20–80
	Pd	17
$C_2H_4 + H_2 \rightarrow C_2H_6$	None	180
	W	43.5 (1)*
	Fe	43.5 (10)
	Pt	43.5 (200)
	Rh	43.5 (10 000)
$2HI \rightarrow H_2 + I_2$	None	184
	Pt	140
	Au	105
$2H_2O_2 \rightarrow 2H_2O + O_2$	None	75
	I^-	57
	Colloidal Pt	49
	Fe^{2+}	42
	catalase	23
$C_{12}H_{22}O_{11} + H_2O \rightarrow 2C_6H_{12}O_6$	H^+	107
	saccharase	36
$HCOOH \rightarrow H_2 + CO_2$	glass	103 (1)*
	Ag	130 (40)
	Au	98 (40)
	Pt	92 (2 000)
	Rh	105 (10 000)

Table XV.6 Hammett Substituent Constants, σ

Reference: as for Table XV.1.

Substituent	meta σ_m	para σ_p	Substituent	meta σ_m	para σ_p
$N(CH_3)_2$	−0.21	−0.83	F	0.34	0.06
NH_2	−0.16	−0.66	I	0.35	0.18
OH	0.12	−0.37	Br	0.39	0.23
OCH_3	0.12	−0.27	Cl	0.37	0.23
CH_3	−0.07	−0.17	CN	0.56	0.66
C_6H_5	0.06	−0.01	NO_2	0.71	0.78
H	0*	0*	$N(CH_3)_3^+$	0.88	0.82

*By definition.

Table XV.7 A Selection of Hammett Reaction Constants, ρ, from Correlations of Reaction Rate Constants
Reference: as for Table XV.1.

Reaction	ρ
Alkaline hydrolysis of ethyl benzoates in 75% methanol at 25 °C	2.19
Ionization of benzoic acids in water at 25 °C (equilibrium)	1.00*
Hydrolysis of benzoyl chlorides in 50% acetone at 0 °C	0.80
Acid hydrolysis of ethyl benzoates in 60% acetone at 100 °C	0.11
Esterification of benzoic acids in ethanol at 25 °C	−0.47
Hydrolysis of benzyl chlorides in 50% acetone at 60 °C	−1.69
Benzoylation of aromatic amines in benzene at 25 °C	−2.78

*By definition.

Section XVI
Health & Safety

		Page
XVI.1	Toxicity and approved occupational exposure limits for some organic compounds	521
XVI.2	Approved occupational exposure limits for some inorganic compounds	527
XVI.3	Toxicity data for some inorganic compounds	530
XVI.4	Toxicity and approved occupational exposure limits for some pesticides and other materials in common use	531

Section XVI
Health & Safety

The tables in this section provide information on the toxicity and approved occupational exposure limits for some inorganic and organic compounds and some common materials. These lists are not exhaustive.

Other hazards such as of a flammable, corrosive, irritant and carcinogenic nature are included in Tables XI.1, XI.2, XI.3 and XI.10. Radiation hazards are listed in Table IX.38.

For more exhaustive information, including methods of waste disposal of chemicals see:

Dictionary of Substances and their Effects (DOSE), Vol.1, Eds. M. Richardson and S. Gangolli, Vol.1, (1992), Royal Society of Chemistry, London.
Chemical Safety Sheets, (1991), Samson Chemical Publishers, Dordrecht, Netherlands.
Health and Safety Directory, Ed. Hastings (1990/1), Corner Publications, Kingston-upon-Thames, Surrey.
Environment Health Series, HSE Guidance Notes, EH40/90.
Substances Hazardous to Health, (1990), Corner Publications, Kingston-upon-Thames, Surrey.
Handbook of Laboratory Safety, 3rd. Ed. (1990), CRC Press, Cleveland, Ohio.
Chemical Safety Data Sheets, Vol.1. Solvents (1989), Royal Society of Chemistry, London.
Dangerous Properties of Industrial Materials, Vols. I, II & III, (1989), Sax and Lewis, van Nostrand Reinhold, Amsterdam.
Extremely Hazardous Substances, Vols. 1 & 2, (1988), Noyes Data Corp., Park Ridge, N.J.
Sigma-Aldrich Library of Chemical Safety Data, Ed. R. Lenga (1985) 2nd Ed., Sigma-Aldrich Corp., Milwauke, Wisconsin.
Compendium of Data Safety Sheets for Research and Industrial Chemicals, Parts I–V. (1985), VCH, Deerfield Beach, Florida.

Table XVI.1 Toxicity and Approved Occupational Exposure Limits for some Organic Compounds

Toxicity data LD_{50} for rat (or mouse^m) by oral administration, expressed as mg per kg of animal; LC_{50} for rat by inhalation, expressed as mg per litre.

Very high toxicity – exposure by any route may cause severe life-threatening acute systemic effects; $LD_{50} < 10$ mg kg^{-1}.
High toxicity – where there is possibility of irreversible acute effects; $LD_{50} = 11$–50 mg kg^{-1};
Moderate toxicity – where acute reversible effects may occur; $LD_{50} = 51$–1000 mg kg^{-1};
Low toxicity – mild allergic potential in man or where exposure may lead to discomfort without severe short-term effects; $LD_{50} > 1001$ mg kg^{-1}.

Exposure limits   The 8-h reference period relates to the procedure whereby the occupational exposures in any 24-h period are treated as equivalent to a single uniform exposure for 8 h (the 8-h time-weighted average (TWA) exposure).

Mathematically, the 8-h TWA can be represented by
$(C_1T_1 + C_2T_2 + \ldots C_nT_n)/8$
where C_1, C_2 etc are the occupational exposure limits and T_1, T_2 etc. the associated time in hours in any 24-h period.
Example: An operator works for 7 h 20 min (7.33 h) on a process in which he is exposed to a hazardous substance; the average exposure during that period is 0.12 mg m^{-3}.
The 8-h TWA = $(0.12 \times 7.33 + 0 \times 0.67)/8 = 0.11$ mg m^{-3}.
Working periods may be split into several sessions to account for rest, meal breaks etc, the exposure during these breaks is assumed to be zero.

Short-term exposure is the average over a specified short-term reference peiod (10 min). Example: a 5-min exposure at a level of 500 p.p.m. followed by zero exposure has a 10-min average exposure of 250 p.p.m.
Where no specific short term limit is listed, a figure of 3 times the long-exposure limit over a 10-min period should be used.

*: Maximum exposure limit; (sk) can be aborbed through the skin.

(Table begins on p.522)

Table XVI.1

Compound	LD_{50} (LC_{50})	Long-term exposure limit (8-h TWA)		Short-term exposure limit (10-min reference period)	
		/ppm	/mg m^{-3}	/ppm	/mg m^{-3}
acetaldehyde	1 930	100	180	150	270
acetamide	800				
acetic acid	3 530	10	25	15	37
acetic anhydride	1 780 (4)	5	20	5	20
acetone	5 800	1 000	2 400	1 250	3 000
acetonitrile	2 730 (13)	40	70	60	105
acetophenone	815				
o-acetylsalicylic acid	200		5		
acrylaldehyde	40 (300)	0.1	0.25	0.3	0.8
acrylamide	124		0.3		0.6 (sk)
acrylic acid	33.5	10	30	20	60
acrylonitrile	78	2*	4*		(sk)
adipic acid	1 900^m				
(−)-adrenaline	30				
allyl alcohol	64	2	5	4	10 (sk)
allyl chloride (3-chloropropene)	64	1	3	2	6
2-aminoethanol		3	8	6	15
2-aminophenol	1 300				
3-aminophenol	1 000				
4-aminophenol	375				
2-aminopyridine		0.5	2	2	8
n-amyl acetate		100	520	150	800
amyl alcohol	2 200				
aniline	250	2	10	5	20 (sk)
anisidines		0.1	0.5		(sk)
anthracene	430				
atropine	500				
barbitone, sodium	800^m				
benzaldehyde	890				
benzene	3.8 cm^3	10	30		
benzene sulphonic acid	800				
benzene sulphonyl chloride	1 960				
benzenethiol		0.5	2		
benzonitrile	720				
p-benzoquinone	130	0.1	0.4	0.3	1.2
benzoyl chloride	1 900 (0.19)				
benzyl alcohol	1 230				
benzyl chloride	1 231 (0.75)	1	5		
biphenyl	2 400	0.2	1.5	0.6	4
bromobenzene	2 699				
bromoethane	(27 000)	200	890	250	1 110
bromoethylene		5	20		
bromoform	1 147	0.5	5		(sk)
bromomethane		5	20	15	60 (sk)
bromotrifluoromethane (Freon 13B1)		1 000	6 100	1 200	7 300
1,3-butadiene		10*	22*		
n-butane		600	1 430	750	1 780
n-butanoic acid	2 000				
iso-butanoic acid	280				
1-butanol	790 (26)	50	150	50	150 (sk)
tert-butanol	3 500	100	300	150	450
2-butanol	6 480	100	300	150	450
2-butanone	2 737	200	590	300	885
n-butyl acetate	13 100	150	710	200	950
sec-butyl acetate		200	950	250	1 190
tert-butyl acetate		200	950	250	1 190
n-butylamine	366	5	15	5	15 (sk)
sec-butylamine	152				

(Continued)

Table XVI.1 (Continued)

Compound	LD_{50} (LC_{50})	Long-term exposure limit (8-h TWA)		Short-term exposure limit (10-min reference period)	
		/ppm	/mg m^{-3}	/ppm	/mg m^{-3}
tert-butylamine	78				
caffeine	192				
camphor	1 310	2	12	3	18
carbon disulphide	3 188	10*	30*		(sk)
carbon tetrabromide	1 000	0.1	1.4	0.3	4
carbon tetrachloride	2 350	10	65	20	130 (sk)
carbonyl chloride		0.1	0.4		
catechol	260	5	20		
chlorinated biphenyls (42% chlorine)			1		2 (sk)
chlorinated biphenyls (54% chlorine)			0.5		1 (sk)
chloral hydrate	480				
chloroacetic acid	580	0.3	1		(sk)
chlorobenzene	2 290	50	230		
chlorodifluoromethane (Freon 22)		1 000	3 500	1 250	4 375
chloroethane		1 000	2 600	1 250	3 250
2-chloroethanol	81 (0.29)	1	3	1	3 (sk)
chloroform	908	10	50	50	225
chloromethane		50	105	100	210
2-chlorophenol	670				
3-chlorophenol	570				
4-chlorophenol	367				
b-chloroprene		10	36		(sk)
2-chlorotoluene		50	250		
coumarin	293				
o-cresol	121	5	22		(sk)
m-cresol	242	5	22		(sk)
p-cresol	207	5	22		(sk)
crotonaldehyde		2	6	6	18
crotonic acid	1 000				
cumene	1 400	50	245	75	365 (sk)
cyanamide			2		
cyanogen		10	20		
cyanogen chloride		0.3	0.6	0.3	0.6
cyclohexane	12 705	300	1 050	375	1 300
cyclohexanol	2 060	50	200		
cyclohexanone	1 535 (32)	25	100	100	400
cyclohexylamine	156 (7.5)	10	40		(sk)
diacetone alcohol	4 000	50		75	
diaminoethane	500	10	25		
diazomethane		0.2	0.4		
dibenzoyl peroxide			5		
dibromodifluoromethane (Freon 12-B2)		100	860	150	1 290
1,2-dibromoethane	108	1*	8*		(sk)
dibromomethane	108				
dibutyl phthalate			5		10
dichloroacetic acid	2 820				
1,2-dichlorobenzene	500	50	300	50	300
1,4-dichlorobenzene	500	75	450	110	675
dichlorodifluoromethane (Freon 112)		1 000	4 950	1 250	6 200
1,1-dichloroethane		200	810	400	1 620
1,2-dichloroethane		10	40	15	60
1,1-dichloroethylene		10	40		
1,2-dichloroethylene	770	200	790	250	1 000
dichlorofluoromethane (Freon 21)		10	40		
dichloromethane	2 316 (88)	100*	350*	250	870
diethanolamine	710	3	15		

(Continued)

Table XVI.1 (Continued)

Compound	LD_{50} (LC_{50})	Long-term exposure limit (8-h TWA)		Short-term exposure limit (10-min reference period)	
		/ppm	/mg m^{-3}	/ppm	/mg m^{-3}
diethylamine	540 (12)	10	30	25	75
diethyl ether	1 215	400	1 200	500	1 500
diethyl ketone		200	700	250	875
diethyl phthalate	8 600		5		10
diethyl sulphate	880				
digitonin	90^m				
digitoxin	56				
1,2-dihydroxyethane (vap)			6	125	
dimethylamine		10	18		
N,N-dimethylaniline	1.41 cm^3	5	25	10	50 (sk)
1,3-dimethylbenzene	83	0.15		0.5	(sk)
dimethylformamide	1 410	10	30	20	60 (sk)
dimethylglyoxime	250				
dimethyl sulphate	205 (0.045)	0.1	0.5	0.1	0.5 (sk)
dimethyl sulphide	535 (111)				
dimethyl sulphoxide	14 500				
dinitrobenzenes (all)		0.15	1	0.5	3 (sk)
2,4-dinitrophenol	30				
2,4-dinitrophenylhydrazine	654				
1,4-dioxane	4 200 (46)	25	90	100	360 (sk)
diphenylamine	3 000		10		20
n-dodecylamine	1 020				
ephedrine hydrochloride	400				
ethanediol, particulate	4 700		10		
vapour			60		125
ethanethiol	682 (8.8)	0.5	1	2	3
ethanol	10 600	1 000	1 900		
ethanolamine	2 050	3	8	6	15
2-ethoxyethanol (cellosolve)		10*	37*		(sk)
ethyl acetate	5620	400	1 400		
ethylamine	400	10	18		
ethylbenzene	5 460	100	435	125	545
ethylenediamine		10	25		
ethylene oxide	72 (1.5)	5*	10*		
fluorobenzene	4 399				
formaldehyde		2*	2.5*	2*	2.5*
formaldehyde (37–41% soln in methanol)	800 (0.9)				
formamide	5 570	20	30	30	45
formic acid	1 100 (15)	5	9		
furfuraldehyde	65	2	8	10	40 (sk)
glucose	25 800				
glutaraldehyde	134	0.2	0.7	0.2	0.7
glycerol, mist	12 600		10		20
guaiacol	725				
hexachlorobuta-1,3-diene	90^m				
n-hexane	49 cm^3	100	360	125	450
hexane, other isomers		500	1 800	1 000	3 600
1-hexanol	72				
2-hexanone	2 590 (31)	5	20		(sk)
hydoquinone	320		2	4	
imidazole	880^m				
indene		10	45	15	70
indole	1 000				
iodoform		0.6	10	1	20
iodomethane		5	28	10	56 (sk)
isoprene	(180)				
ketene		0.5	0.9	1.5	3
lactic acid	3 730	5			

(Continued)

Table XVI.1 (Continued)

Compound	LD_{50} (LC_{50})	Long-term exposure limit (8-h TWA) /ppm	/mg m^{-3}	Short-term exposure limit (10-min reference period) /ppm	/mg m^{-3}
limonene	4 400				
maleic anhydride	400	0.25	1		
maleimide	80				
malonic acid	1 310				
mercaptoacetic acid	114	1	5		
mesitylene	24	25		35	
mesityl oxide	1 120 (9)	15	60	25	100
methacrylic acid	8 400	20	70	40	140
methane sulphonic acid	200				
methanethiol		0.5	1		
methanol	5 628 (83)	200	250	250	310 (sk)
methyl acetate	5 450	200	610	250	760
methylacrylate	277	10	35		(sk)
methylamine		10	12		
methyl benzoate	1 177				
methyl formate		100	250	150	375
methyl mercuric chloride	11				
methyl methacrylate	7 872	100	410	125	510
methyl salicylate	887				
naphthalene	490	10	50	15	75
1, 4-naphthaquinone	190				
1-naphthylamine	779				
nicotine	50		0.5		1.5 (sk)
ninhydrin	250				
4-nitroaniline	750		6		(sk)
nitrobenzene	489	1	5	2	10 (sk)
nitroethane	1 100	100	300		
nitromethane	940	100	250	150	375
2-nitrophenol	334				
3-nitrophenol	328				
4-nitrophenol	250				
1-nitropropane		25	90		
2-nitropropane		10	36	20	72
nitrotoluene (all isomers)		5	30	10	60 (sk)
n-octane		300	1 450	375	1 800
oxalic acid	375		1		2
pentane (all isomers)		600	1 800	750	2 250
phenacetin	3 600				
phenanthrene	700				
1, 10-phenanthroline hydrate	132				
phenobarbital, sodium	150				
phenol	317	5	19	10	38 (sk)
o-phenylenediamine	1 070 (1.87)				
m-phenylenediamine	650				
p-phenylenediamine	80		0.1		(sk)
phenylhydrazine	188	5	20	10	45 (sk)
phthalic anhydride	4 020	1	6	4	24
picric acid			0.1		0.3 (sk)
piperidine	400	1	3.5		(sk)
procaine hydrochloride	200				
propane-1, 2-diol		50	156		
propanoic acid	2 600	10	30	15	45
propanoic acid nitrile	39				
1-propanol	1 870	200	500	250	625 (sk)
2-propanol	5 045	400	980	500	1 225 (sk)
propylene oxide	380	20	50	100	240
pyrazole	1 010				
pyridine	891	5	15	10	30

(Continued)

Table XVI.1 (Continued)

Compound	LD_{50} (LC_{50})	Long-term exposure limit (8-h TWA)		Short-term exposure limit (10-min reference period)	
		/ppm	/mg m^{-3}	/ppm	/mg m^{-3}
quinoline	331				
iso-quinoline	360				
resorcinol	301	10	45	20	90
salicyladehyde	520				
salicylic acid	891				
strychnine			0.15		0.45
styrene	5 000	100*	420*	250*	1 050*
succinonitrile	450				
sucrose	29 700		10		20
sulphamic acid	3 160				
sulpholane	1 941				
1,1,1,2-tetrachloro-2,2-difluoroethane		100	834	100	834
1,1,2,2-tetrachloro-1,2-difluoroethane (Freon 112)		100	834	100	834
tetrachloroethylene	8 850^m	50	335	150	1 000
tetrahydrofuran	2 816 (62)	200	590	250	735
thiophen	1 400				
thiophenol	46				
thiourea	125				
thymol	980				
toluene	5 000	100	375	150	560 (sk)
o-toluidine	670	2	9	5	22
m-toluidine	450				
p-toluidine	656				
tri-n-butylamine	125				
trichloroacetic acid	400	1	5		
1,2,4-trichlorobenzene	756	5	40	5	40
1,1,1-trichloroethane	10 300	350*	1 900*	450*	2 450*
1,1,2-trichloroethane	580	10	45	20	90 (sk)
trichloroethylene	4.92 cm^3	100*	535*	150*	802* (sk)
1,1,2-trichlorofluoroethane (Freon 113)		1 000	5 600	1 250	7 000
trichlorofluoromethane (Freon 11)		1 000		1 250	
triethylamine	460	10	40	15	60
trimethylamine		10	24	15	36
trimethylbenzenes (all isomers)		25	125	35	170
2,4,6-trinitrotoluene			0.5		0.5
urea	8 471				
n-valeric acid	600				
vinyl acetate	2 920	10	30	20	60
vinyl chloride**		7*			
vinylidene chloride		10*	40*		
white spirit		100	575	125	720
xylene (all isomers)	5 000	100	435	150	650 (sk)
2,4-xylenol	3 200				
2,5-xylenol	444				
xylidine (all isomers)	840	2	10	10	50 (sk)

** Vinyl chloride is also subject to an overriding annual exposure limit of 3 ppm

Table XVI.2 Approved Occupational Exposure Limits for Some Inorganic Compounds
See Table XVI.1 for definition and explanation of terms.
*:Maximum exposure limit.

Compound	Long-term exposure limit (8-h TWA)		Short-term exposure limit (10-min reference period)	
	/ppm	/mg m^{-3}	/ppm	/mg m^{-3}
aluminium metal		10		20
aluminium oxide		10		20
aluminium salts, soluble		2		
ammonia	25	18	35	27
ammonium chloride, fume		10		20
antimony and compounds (as Sb)		0.5		
arsenic and compounds (as As)		0.2*		
arsine	0.05	0.2		
barium compounds, soluble (as Ba)		0.5		
beryllium		0.002		
boron trioxide		10		20
boron tribromide	1	10	1	10
boron trifluoride	1	3	1	3
bromine	0.1	0.7	0.3	2
bromine pentafluoride	0.1	0.7	0.3	2
cadmium compounds (as Cd)		0.05*		
cadmium oxide, fume (as Cd)		0.05*		0.05
cadmium sulphide pigments (as Cd)		0.04*		
calcium carbonate				
total inhalable dust		10		
respirable dust		5		
calcium hydroxide		5		
calcium oxide		2		
carbon black		3.5		7
carbon dioxide	5 000	9 000	15 000	27 000
carbon monoxide	50	55	300	330
chlorine	1	3	3	9
chlorine dioxide	0.1	0.3	0.3	0.9
chlorine trifluoride	0.1	0.4	0.1	0.4
chromium		0.5		
chromium (II) compounds (as Cr)		0.5		
chromium (III) compounds (as Cr)		0.5		
chromium (VI) compounds (as Cr)		0.05		
coal dust		2		
cobalt and compounds (as Co)		0.1		
copper, fume (as Cu)		0.2		
dusts and mists (as Cu)		1		2
cyanides (except HCN, C$_2$N$_2$, CNCl)		5		(sk)
cyanogen	10	20		
cyanogen chloride	0.3	0.6	0.3	0.6
diborane	0.1	0.1		
ferrocene		10		20
fluoride (as F)		2.5		
fluorine			1	1.5
germane	0.2	0.6	0.6	1.8
graphite				
total inhalable dust		10		
respirable dust		5		
gypsum				
total inhalable dust		10		
respirable dust		5		
hafnium		0.5		1.5
hydrazine	0.1	0.1		(sk)
hydrazoic acid (as vapour)			0.1	
hydrogen bromide	3	10	3	10
hydrogen chloride	5	7	5	7

(Continued)

Table XVI.2 (Continued)

Compound	Long-term exposure limit (8-h TWA)		Short-term exposure limit (10-min reference period)	
	/ppm	/mg m^{-3}	/ppm	/mg m^{-3}
hydrogen cyanide			10*	10* (sk)
hydrogen fluoride (as F)	3	2.5	3	2.5
hydrogen peroxide	1	1.5	2	3
hydrogen selenide	0.05	0.2		
hydrogen sulphide	10	14	15	21
indium and compounds		0.1		0.3
iodine	0.1	1	0.1	1
iron oxide (as Fe)		5		10
iron pentacarbonyl	0.01	0.08		
iron salts (as Fe)		1		2
isocyanates, all (as -NCO)		0.02*		0.07*
lead and compounds (as Pb)		0.15*		
lead tetraethyl (as Pb)		0.10		
lithium hydride		0.025		
magnesium oxide				
fume and respirable dust (as Mg)		5		10
total inhalable dust		10		
manganese, fume (as Mn)		1		3
manganese and compounds (as Mn)		5		
manganese cyclopentadienyl- tricarbonyl (as Mn)		0.2		0.6 (sk)
manganese tetraoxide		1		
mercury alkyls (as Hg)		0.01		0.03 (sk)
mercury and compounds (as Hg)		0.05		0.15
nickel		0.5		
nickel carbonyl (as Ni)			0.1	0.24
nickel, inorganic compounds (as Ni)				
soluble compounds		0.1		
insoluble compounds		0.5		
nickel, organic compounds (as Ni)		1		3
nickel tetracarbonyl			0.1	0.24
nitric acid	2	5	4	10
nitrogen dioxide	3	5	5	9
nitrogen monoxide	25	30	35	45
nitrogen trifluoride	10	30	15	45
orthophosphoric acid		1		3
osmium tetraoxide	0.000 2	0.002	0.000 6	0.006
ozone	0.1	0.2	0.3	0.6
phosphine			0.3	0.4
phosphorus, yellow		0.1		0.3
phosphorus pentachloride	0.1	1		
phosphorus trichloride	0.2	1.5	0.5	3
phosphoryl trichloride	0.2	1.2	0.6	3.6
platinum metal, total inhalable dust		5		
platinum salts, soluble (as Pt)		0.002		
potassium hydroxide		2		2
quartz, crystalline				
total inhalable dust		0.3		
respirable dust		0.1		
rhodium (as Rh)				
metal fume and dust		0.1		0.3
soluble salts		0.001		0.003
selenium and compounds (as Se)		0.1		
silane	0.5	0.7	1	1.5
silica, amorphous				
total inhalable dust		6		
respirable dust		3		

(Continued)

Table XVI.2 (Continued)

Compound	Long-term exposure limit (8-h TWA)		Short-term exposure limit (10-min reference period)	
	/ppm	/mg m^{-3}	/ppm	/mg m^{-3}
silica, fused				
total inhalable dust		0.3		
respirable dust		0.1		
silicon				
total inhalable dust		10		
respirable dust		5		
silicon carbide				
total inhalable dust		10		
respirable dust		5		
silver		0.1		
silver, soluble compounds (as Ag)		0.01		0.03
sodium azide	0.1	0.3	0.1	0.3
sodium fluoroacetate		0.05		0.15 (sk)
sodium hydrogen sulphite		5		
sodium hydroxide		2		2
sodium peroxodisulphate (as $S_2O_8^{2-}$)		1		
sodium tetraborate decahydrate		5		
stibine	0.1	0.5	0.3	1.5
sulphur dioxide	2	5	5	13
sulphur hexafluoride	1 000	6 000	1 250	7 500
sulphuric acid		1		
sulphur monochloride	1	6	3	18
sulphur pentafluoride	0.025	0.25	0.075	0.75
sulphur tetrafluoride	0.1	0.4	0.3	1
sulphuryl difluoride	5	20	10	40
tantalum		5		10
tellurium and compounds (as Te)		0.1		
thallium soluble compounds (as Tl)		0.1		(sk)
tin compounds, inorganic (as Sn)		2		4
tin compounds, organic (as Sn)		0.1		0.2 (sk)
titanium oxide				
total inhalable dust		10		
respirable dust		5		
tungsten and compounds (as W)				
soluble		1		3
insoluble		5		10
uranium compounds, natural (as U)				
soluble		0.2		0.6
yttrium		1		3
zinc chloride		1		2
zinc oxide fume		5		10
zirconium compounds		5		10

Table XVI.3 Toxicity Data for Some Inorganic Compounds

See Table XVI.1 for definition and explanation of terms.

Compound	LD_{50}	Compound	LD_{50}
aluminium chloride, anhydrous	3 730	phosphorus, yellow	3
ammonia solution, 35%	350	phosphorus pentachloride	660
ammonium chloride fume	1 650	phosphorus trichloride	550
ammonium polysulphide	152	phosphoryl trichloride	330
ammonium sulphate	3 000	platinum(II) chloride	3 423
antimony pentachloride	1 150	platinum(IV) chloride	276
antimony potassium tartrate	115	potassium borohydride	160
antimony trichloride	525	potassium bromate	321
arsenic acid	48	potassium carbonate	1 870
arsenic pentoxide	8	potassium chlorate	1 870
barium carbonate	418	potassium chloride	2 600
barium chloride	118	potassium cyanide	10
beryllium sulphate	82	potassium dichromate	190m
cadmium powder	225	potassium fluoride	245
cadmium chloride	88	potassium hydroxide	273
cadmium oxide, fume	72	potassium iodate	531m
cadmium sulphate	280	potassium nitrate	3 750
calcium carbonate	6 450	potassium nitrite	200
calcium chloride	1 000	potassium permanganate	1 090
calcium hydroxide	7 340	potassium peroxodisulphate	802
chromium(III) chloride	1 870	potassium sulphate	6 600
chromium trioxide	80	potassium thiocyanate	854
cobalt powder	6 170	selenous acid	25
cobalt(II) bromide	406	selenium	6 700
cobalt carbonate	6 400	silica	3 160
cobalt(II) chloride	80	silicon tetrachloride	60 (LC_{50})
cobalt nitrate hexahydrate	691	silver cyanide	123
cobalt(II) oxide	202	silver nitrate	50m
cobalt sulphate	424	sodium acetate, anhyd.	3 530
copper(I) chloride	140	sodium azide	27
copper(II) chloride	140	sodium borohydride	160
copper(I) oxide	470	sodium bromide	3 500
copper sulphate pentahydrate	300	sodium carbonate, anhyd.	4 090
dicobalt octacarbonyl	754	sodium chloride	3 000
hydrobromic acid solution (48–65%)	10 (LC_{50})	sodium chlorite	165
hydrochloric acid solution (31–37%)	5 (LC_{50})	sodium cyanide	6.4
hydrofluoric acid solution (5–60%)	1.1 (LC_{50})	sodium dichromate	50
hydrogen peroxide	1 518	sodium dodecyl sulphate	1 288
iodine	14 000	sodium fluoride	52
iron(III) chloride	1 872	sodium hydrogen carbonate	4 220
lanthanum chloride	4 184	sodium iodoacetate	63
lithium carbonate	525	sodium nitrate	3 236
magnesium chloride, anhydrous	2 800	sodium nitrite	85
magnesium chloride hexahydrate	8 100	sodium sulphate, anhyd.	5 989m
manganese chloride	1 484	sodium sulphite, anhyd.	820m
mercury(II) bromide	40	sodium thiocyanate	764
mercury(I) chloride	166	sulphur dioxide	6 (LC_{50})
mercury(II) chloride	1	sulphuric acid (90–100%)	2 140
mercury(I) iodide	110	tellurium powder	83
mercury(II) iodide	18	thallium(II) chloride	24m
mercury(II) nitrate	26	thionyl chloride	2.5 (LC_{50})
mercury(II) oxide	18	tin(IV) chloride, anhyd.	2.3 (LC_{50})
nickel carbonate	840	titanium tetrachloride	0.4 (LC_{50})
nickel chloride, anhydrous	105	tungsten hexacarbonyl	840
nickel sulphate heptahydrate	275	vanadium pentoxide	10
orthophosphoric acid	1 530	zinc chloride, anhyd.	350
osmium tetroxide	162	zinc sulphate, anhyd.	2 150
perchloric acid (60–72%)	1 100		

Table XVI.4 Toxicity and Approved Occupational Exposure Limits for some Pesticides and Other Materials in Common Use

For more extensive information see: A. Watterson, *Pesticide Users Health and Safety Handbook* (1988), Gower Technical, Aldershot.
See Table XVI.1 for definition and explanation of terms.
*: Maximum exposure limit

Compound	LD$_{50}$	Long-term exposure limit (8-h TWA)		Short-term exposure limit (10-min reference period)	
		/ppm	/mg m^{-3}	/ppm	/mg m^{-3}
aldrin	38–60		0.25		0.75 (sk)
ammonium sulphamate	1 600		10		20
atrazine	1 400		10		
γ-BHC, C$_6$H$_6$Cl$_6$	125		0.5		1.5 (sk)
benomyl	> 10 000		10		15
bromacil	5 200	1	10	2	20
camphor, synthetic		2	12	3	18
∈-caprolactam					
dust			1		3
vapour		5	20	10	40
captofol	2 500		0.1		(sk)
captan	9 000		5		15
carbaryl	89		5		10
carbofuran	8		0.1		
cellulose					
total inhalable dust			10		20
respirable dust			5		
chlordane	367		0.5		2 (sk)
chloropicrin	0.8	0.1	0.7	0.3	2
chlorpyrifos	138		0.2		0.6 (sk)
copper sulphate	300	8	1	10	2
cresols, all isomers	120	5	22		(sk)
cristobalite					
total inhalable dust			0.15		
respirable dust			0.05		
2, 4-D	375		10		20
DDT	113		1		3
DDVP, dichlorvos	50	0.1	1	0.3	3 (sk)
DMDT			10		
derris	1		5		10
diazinon	76		0.1		0.3 (sk)
dieldrin	40		0.25		0.75 (sk)
diethanolamine	710	3	15		
dioxathion	20		0.2		(sk)
diquat dibromide	215		0.5		1
disulfoton	2.6		0.1		0.3
diuron	437		10		
emery					
total inhalable dust			10		
respirable dust			5		
endosulfan	18		0.1		0.3 (sk)
endrin	3		0.1		0.3 (sk)
ferbam	4 000		10		20
formaldehyde	800	2*	2.5	2*	2.5
formic acid	1 076 m	5	9	10	
iron(II) sulphate	1 389	(as Fe)	1		2
liquefied petroleum gas		1 000	1 800	1 250	2 250
lindane	76		0.5		1.5
malathion	885		10		(sk)
mercury(I) chloride	210		0.05		0.15
mercury(II) chloride	1		0.05		0.15
metaldehyde	227				
methomyl	17		2.5		(sk)
mevinphos (phosdrin)	3	0.01	0.1	0.03	0.3 (sk)

(Continued)

Table XVI.4 (Continued)

Compound	LD$_{50}$	Long-term exposure limit (8-h TWA)		Short-term exposure limit (10-min reference period)	
		/ppm	/mg m^{-3}	/ppm	/mg m^{-3}
mica					
total inhalable dust			10		
respirable dust			1		
mineral fibre (man-made)			5*		
naled			3		6
naphthalene	490	10	50	15	75
nicotine	50		0.5		1.5 (sk)
oil mist			5		10
paraffin wax, fume			2		6
paraquat dichloride	150				
respirable dust			0.1		
parathion	2		0.1		0.3 (sk)
pentachlorophenol	50		0.5		1.5 (sk)
phorate	1.1		0.05		0.2 (sk)
picloram	8 200		10		20
propoxur	90		0.5		2
pyrethrins	584		5		10
rosin core solder pyrolysis					
products (as formaldehyde)			0.1		0.3
rotenone	132		5		10
rouge					
total inhalable dust			10		
respirable dust			5		
rubber process dust			8*		
rubber fumes			0.75*		
simazine	> 5 000				
starch					
total inhalable dust			10		
respirable dust			5		
subtilsins (proteolytic enzymes as 100% pure enzyme)			0.000 06		0.000 06
2, 4, 5-T	500		10		20
TEPP		0.004	0.05	0.01	0.2 (sk)
thiram	560		5		10
tributyltin oxide	7				
tridymite					
total inhalable dust			0.15		
respirable dust			0.05		
turpentine		100	560	150	840
warfarin	1		0.1		0.3
welding fumes			5		
wood dust (soft)			5		
(hard)			5*		

Asbestos

1. Containing or consisting of crocidolite or amosite:
   0.2 fibres per cm^3 of air averaged over a continuous period of 4 h;
   0.6 fibres per cm^3 of air averaged over a continuous period of 10 min.

2. For other types of asbestos:
   0.5 fibres per cm^3 of air averaged over a continuous period of 4 h;
   1.5 fibres per cm^3 of air averaged over a continuous period of 10 min.

Asphyxiants
Some gases and vapours, when present at high concentrations in air, act as simple asphyxiants by reducing the oxygen content by dilution to such an extent that life cannot be supported. Many are odourless and colourless and not readily detectable. Monitoring oxygen in air is the best means of ensuring safety; it should never be allowed to fall below 18% v/v under normal atmospheric pressure. Particular care is required with dense asphyxiants, e.g. argon, since localized and high concentrations can arise in pits, confined spaces and low-lying areas where ventilation is poor. Many such asphyxiants, e.g. acetylene, ethane, ethylene and methane, can present a fire or explosion risk.

Section XVII
Ancillary Topics

		Page
XVII.1	The International Organisation for standardisation specification for screw threads in inches and millimetres	535
XVII.2	Whitworth and British Association specifications for standard series screw threads	535
XVII.3	Typical percentage composition of various alloys	536
XVII.4	Colour code for wire-wound and other resistors	537
XVII.5	Some commonly occuring indefinite integrals	538
XVII.6	Some commonly occuring definite integrals	544
XVII.7	Some useful formulae for approximate integration	545
XVII.8	Some useful series	546
XVII.9	The error function $\text{Erf } x = 2\pi^{-\frac{1}{2}} \int_0^x e^{-x^2} dx.$	549

Section XVII
Ancillary Topics

Table XVII.1 The International Organisation for standardisation specification for screw threads in inches and millimetres

Inch Specification			mm Specification		
Screw major diameter/in	Number of turns per inch		Screw major diameter/mm	Pitch/mm	
	Coarse thread	Fine thread		Coarse thread	Fine thread
0.06		80	1.6	0.35	0.2
0.073	64	72	1.8	0.35	0.2
0.086	56	64	2	0.4	0.25
0.099	48	56	2.2	0.45	0.25
0.112	40	48	2.5	0.45	0.35
0.125	40	44	3	0.5	0.35
0.138	32	40	3.5	0.6	0.35
0.164	32	36	4	0.7	0.5
0.19	24	32	4.5	0.75	0.5
0.216	24	28	5	0.8	0.5
$\frac{1}{4}$	20	28	6	1	0.75
$\frac{5}{16}$	18	24	7	1	0.75
$\frac{3}{8}$	16	24	8	1.25	1
$\frac{7}{16}$	14	20	9	1.25	
$\frac{1}{2}$	13	20	10	1.5	1.25
$\frac{9}{16}$	12	18	11	1.5	
$\frac{5}{8}$	11	18	12	1.75	1.25
$\frac{3}{4}$	10	16	14	2	1.5
$\frac{7}{8}$	9	14	16	2	1.5
1	8	12	18	2.5	1.5
			20	2.5	1.5
			22	2.5	1.5
			24	3	2

Reference: Kaye and Laby, *Tables of Physical and Chemical Constants*, 15 ed., Longmans, 1986

Table XVII.2. Whitworth and British Association Specifications for standard series screw threads

Whitworth Specification			British Association specification		
Screw major diameter/in	Number of turns per inch		Screw major diameter/mm	Pitch/mm	Number
	BSW *	BSF†			
$\frac{1}{4}$	20	26	6	1	0
$\frac{9}{32}$		26	5.3	0.9	1
$\frac{5}{16}$	18	22	4.7	0.81	2
$\frac{3}{8}$	16	20	4.1	0.73	3
$\frac{7}{16}$	14	18	3.6	0.66	4
$\frac{1}{2}$	12	16	3.2	0.59	5
$\frac{9}{16}$	12	16	2.8	0.53	6
$\frac{5}{8}$	11	14	2.5	0.48	7
$\frac{11}{16}$	11	14	2.2	0.43	8
$\frac{3}{4}$	10	12	1.9	0.39	9
$\frac{7}{8}$	9	11	1.7	0.35	10
1	8	10			

*: BSW (British Standard Whitworth).
†: BSF (British Standard Fine).
Reference: Kaye and Laby, *Tables of Physical and Chemical Constants*, 15 ed., Longmans, 1986

Table XVII.3 Typical percentage composition of various alloys

Alloy	Al	C	Cr	Cu	Fe	Mg	Mn	Ni	Si	Sn	Ti	Zn	other
Al alloys													
Alpax gamma	87.1				0.3	0.3	0.3		12				
IC	97.9			1		0.05	0.05		1			0.05	
Lo-Ex	84.8			1	0.5	0.9		1	11.8				
N3	98.68			0.12			1.2						
N8	94.15		0.15	0.1		4.5	0.7		0.4			0.25	
RR 57	89.2			2.2	0.3	2.5	0.5		0.3			5	
RR 59	92.9			2.3	1.2	1.5		1.2	0.9				
T 61	92.45		0.2	0.1		2.8	0.25		0.2			4	
T 81	93.08			6.3		0.02	0.3		0.2			0.1	
Y	92.3			3.8	0.4	1.3		1.8	0.4				
Cu alloys													
Brass A				70								30	
Brass B				65								35	
Constantan				60				40					
Cu–Be				98									2Be
Cupro Nickel				90				10					
German Silver				63				15				22	
Manganin				84			12	4					
Ni alloys													
Alumel	2						2	95	1				
Chromel P			10					90					
Hastelloy R-235	2	0.16	15.5		10		1	62.34	1				2.5 Co, 5.5 Mo
Hastelloy X		0.15	22		24			44.85					9.Mo
Incoloy 800	0.4	0.05	21	0.5			1	75.95	0.7		0.4		
Incoloy 802	0.3		21	0.5			1	76.8	0.4				
Inconel 600		0.1	16	0.3	8		0.5	74.7	0.4				
Inconel 718	0.4	0.04	18.6		18.5						1		3 Mo, 5 Nb
Inconel X	0.9	0.04	15		7		0.7	73.56	0.3		2.5		
Inconel X 750	0.8	0.04	15	0.05	6.8		0.7	73.01	0.3		2.5		0.8 Nb
K Monel	3	0.15		30	1		0.6	65.1	0.15				
Nimonic 75	0.3	0.1	20	0.5	5		1		1		0.2		2 Co, 0.3 Mo
Nimonic 80	1.5	0.07	20	0.2	1		1	70.93	1		2		2 Co, 0.3 Mo
Nimonic 90	1.5	0.07	20	0.2	1		1	55.43	1.5		2		17 Co, 0.3 Mo
Nimonic 105	4.5	0.14	14.5	0.2	1		1	50.66	1		2		20 Co, 5 Mo
Steels													
carbon													
A		0.08	0.045		99.415		0.31	0.07	0.08				
B		0.23		0.13	98.82		0.64	0.07	0.11				
C		0.42		0.12	98.65		0.64	0.06	0.11				
D		0.84		0.02	98.77		0.24		0.13				
E		1.22	0.11	0.08	97.95		0.35	0.13	0.16				
F		0.23	0.06	0.1	97.94		1.51	0.04	0.12				
high-alloy													
A		1.22	0.03	0.07	85.39		13	0.07	0.22				
B		0.28		0.03	70.24		0.9	28.4	0.15				
C		0.72	4.26	0.06	74.84		0.25	0.07	0.3				1V, 18.5 W
Kovar		0.02			53.51		0.47	29					17 Co
low-alloy													
Armco iron		0.02			99.95		0.03						
A		0.32	1.09	0.07	97.55		0.7	0.07	0.2				
B		0.35	0.88	0.12	97.58		0.6	0.26	0.21				
C		0.2			97.1		1.35	0.6	0.25				0.5 Mo
D		0.33	0.17	0.08	95.22		0.55	3.47	0.18				
E		0.33	0.8	0.05	94.74		0.53	3.38	0.17				
F		0.4	0.8		94.34		0.66	3.6	0.2				

(Continued)

Table XVII. 3 (Continued)

Alloy	Al	C	Cr	Cu	Fe	Mg	Mn	Ni	Si	Sn	Ti	Zn	other
G		0.34	0.78	0.05	94.48		0.55	3.53	0.27				
H		0.49	0.04	0.09	96.32		0.9	0.16	2				
nickel													
A		0.1			96.4		0.8	2.5	0.2				
B		0.1			95.4		0.8	3.5	0.2				
C		0.1			94.7			5	0.2				
D		0.1			90.7			9	0.2				
Invar		0.07			63.73			36	0.2				
Stainless													
A		0.01	16		62.49		1.2	20	0.3				
B		0.05	15		55.55		1.4	26	0.4				1.3 Mo, 0.3 V
Era-ATV		0.5	15		52.2		1.2	27	1.3				2.8 W
304, 321, 347		0.05	17.5		c 71		< 2	9	< 1				
310		0.15	24.5		51.85		2	20	1.5				
316		0.05	17		c 66		< 2	12	< 1				2.5 Mo
403, 405, 409		0.1	12		c 85		< 2		< 1				
410, 420		0.3	13	0.1	85.2		0.5	0.5	0.4				
430, 434		0.05	17		c 79		< 2		< 1				1 Mo
Ti alloys													
A	5	0.1			0.4		0.2			2.5	91.8		
B	2						2				96		
C	6	0.05									89.95		4 V
D	3	0.08	11		0.3						72.62		13 V
Unclassified alloys													
A													10 Ir, 90 Pt
B													90 Pt, 10 Rh
C													60 Pt, 40 Rh
D		0.1								6.7			93.2 Zr
E		0.1								2.3			97.6 Zr

Constituent elements/mass %

Reference: Kaye and Laby, *Tables of Physical and Chemical Constants,* 15 ed., Longmans, 1986

Table XVII.4 Colour Code for wire-wound and other Resistors

Four separated coloured bands extending from one end to the middle of the resistor are used. For wire-wound resistors the bands are of equal width; for other resistors the end band has double the width. Tan (brown) body colour is preferred for wire-wound (other) resistors, provided confusion with any band colour is avoided.

First (end)		Second		Third		Fourth	
First significant figure	Colour	Second significant figure	Colour	Multiplicant of these figures	Colour	% resistor tolerance value	Colour
0	black	0	black	1	black	10	silver
1	brown	1	brown	10	brown	5	gold
2	red	2	red	100	red		
3	orange	3	orange	1 000	orange		
4	yellow	4	yellow	10 000	yellow		
5	green	5	green	100 000			
6	blue	6	blue	1 000 000			
7	violet	7	violet				
8	grey	8	grey	0.01	silver		
9	white	9	white	0.1	gold		

Band significance (counting from end)

Reference: *Handbook of Chemistry and Physics,* CRC Press, 1975/6.

Table XVII.5 Some commonly occurring indefinite integrals

Note: Throughout this table, arbitrary constants of integration are omitted

$\log x = \log_{10} x$ $\ln x = \log_e x$

Integration by parts: $\int uv\, dx = u \int v\, dx - \int \left(\dfrac{du}{dx} \int v\, dx \right) dx$

1. $\int a\, dx = ax$

2. $\int ax^n\, dx = \dfrac{a}{n+1} x^{n+1}$ n not equal to -1

3. $\int \dfrac{a}{x}\, dx = a \ln x$

4. $\int e^{ax}\, dx = \dfrac{1}{a} e^{ax}$

5. $\int x e^{ax}\, dx = \dfrac{x}{a} e^{ax} - \dfrac{e^{ax}}{a^2}$

6. $\int a^{bx}\, dx = \dfrac{a^{bx}}{b \ln a}$

7. $\int \ln ax\, dx = x(\ln ax - 1)$

8. $\int x^n \ln x = \dfrac{x^{n+1}}{n+1} \left(\ln x - \dfrac{1}{n+1} \right)$ n not equal to -1

Trigonometrical Functions

x is measured in radians. If x is measured in degrees the results given must be multiplied by $\dfrac{180}{\pi}$.

9. $\int \sin ax\, dx = -\dfrac{1}{a} \cos ax$

10. $\int \sin^2 ax\, dx = \dfrac{1}{2} \left(x - \dfrac{1}{2a} \sin 2ax \right)$

11. $\int \sin^3 ax\, dx = \dfrac{1}{4a} \left(\dfrac{1}{3} \cos 3ax - 3 \cos ax \right)$

12. $\int \sin^4 ax\, dx = \dfrac{1}{8} \left(3x - \dfrac{2}{a} \sin 2ax + \dfrac{1}{4a} \sin 4ax \right)$

13. $\int \sin^n ax\, dx = -\dfrac{1}{an} \cos ax \sin^{n-1} ax + \dfrac{n-1}{n} \int \sin^{n-2} ax\, dx$

14. $\int \cos ax\, dx = \dfrac{1}{a} \sin ax$

15. $\int \cos^2 ax\, dx = \dfrac{1}{2} \left(x + \dfrac{1}{2a} \sin 2ax \right)$

16. $\int \cos^3 ax\, dx = \dfrac{1}{4a} \left(\dfrac{1}{3} \sin 3ax + 3 \sin ax \right)$

17. $\int \cos^4 ax\, dx = \dfrac{1}{8} \left(3x + \dfrac{2}{a} \sin 2ax + \dfrac{1}{4a} \sin 4ax \right)$

18. $\int \cos^n ax\, dx = \dfrac{1}{an} \sin ax \cos^{n-1} ax + \dfrac{n-1}{n} \int \cos^{n-2} ax\, dx$

19. $\int \tan ax\, dx = \dfrac{1}{a} \ln \cos ax$

20. $\int \tan^2 ax\, dx = \dfrac{1}{a} \tan ax - x$

21. $\int \tan^3 ax\, dx = \dfrac{1}{2a} \tan^2 ax + \dfrac{1}{a} \ln \cos ax$

22. $\int \tan^n ax\, dx = \dfrac{\tan^{n-1} ax}{a(n-1)} - \int \tan^{n-2} ax\, dx$

23. $\int \cot ax\, dx = \dfrac{1}{a} \ln \sin ax$

(Continued)

Table XVII. 5 (Continued)

24. $\int \cot^2 ax\, dx = -\dfrac{1}{a} \cot ax - x$

25. $\int \cot^3 ax\, dx = -\dfrac{1}{2a} \cot^2 ax - \dfrac{1}{a} \ln \sin ax$

26. $\int \cot^n ax\, dx = -\dfrac{\cot^{n-1} ax}{a(n-1)} \int \cot^{n-2} ax\, dx$

27. $\int \sec ax\, dx = \dfrac{1}{a} gd^{-1}\, ax = \dfrac{1}{a} \ln \tan \left(\dfrac{\pi}{4} + \dfrac{ax}{2}\right) = \dfrac{1}{2a} \ln \dfrac{1 + \sin ax}{1 - \sin ax}$

28. $\int \sec^2 ax\, dx = \dfrac{1}{a} \tan ax$

29. $\int \sec^3 ax\, dx = \dfrac{1}{2a} (\tan ax \sec ax + gd^{-1}\, ax)$

30. $\int \sec^n ax\, dx = \dfrac{1}{a(n-1)} \sin ax \sec^{n-1} ax + \dfrac{n-2}{n-1} \int \sec^{n-2} ax\, dx$

31. $\int \operatorname{cosec} ax\, dx = \dfrac{1}{a} \ln \tan \dfrac{ax}{2}$

32. $\int \operatorname{cosec}^2 ax\, dx = -\dfrac{1}{a} \cot ax$

33. $\int \operatorname{cosec}^3 ax\, dx = \dfrac{1}{2a} \left(-\cot ax \operatorname{cosec} ax + \ln \tan \dfrac{ax}{2}\right)$

34. $\int \operatorname{cosec}^n ax\, dx = -\dfrac{1}{a(n-1)} \cos ax \operatorname{cosec}^{n-1} ax + \dfrac{n-2}{n-1} \int \operatorname{cosec}^{n-2} ax\, dx$

35. $\int \sec ax \tan ax\, dx = \dfrac{1}{a} \sec ax$

36. $\int \operatorname{cosec} ax \cot ax\, dx = -\dfrac{1}{a} \operatorname{cosec} ax$

37. $\int \sin^n ax \cos ax\, dx = \dfrac{1}{a(n+1)} \sin^{n+1} ax$

38. $\int \cos^n ax \sin ax\, dx = -\dfrac{1}{a(n+1)} \cos^{n+1} ax$

39. $\int \sin^2 ax \cos^2 ax\, dx = \dfrac{1}{8} \left(x - \dfrac{1}{4a} \sin 4ax\right)$

40. $\int \sin^3 ax \cos^2 ax\, dx = \dfrac{1}{a} \left(\dfrac{1}{5} \cos^5 ax - \dfrac{1}{3} \cos^3 ax\right)$

41. $\int \sin^2 ax \cos^3 ax\, dx = \dfrac{1}{a} \left(\dfrac{1}{3} \sin^3 ax - \dfrac{1}{5} \sin^5 ax\right)$

42. $\int \sin^3 ax \cos^3 ax\, dx = \dfrac{1}{64a} \left(\dfrac{1}{3} \cos 6ax - 3 \cos 2ax\right)$

43. $\int \sin ax \cos bx\, dx = -\dfrac{1}{2} \left(\dfrac{1}{a+b} \cos(a+b)x + \dfrac{1}{a-b} \cos(a-b)x\right)$

44. $\int \sin ax \sin bx\, dx = -\dfrac{1}{2} \left(\dfrac{1}{a+b} \sin(a+b)x - \dfrac{1}{a-b} \sin(a-b)x\right)$

45. $\int \cos ax \cos bx\, dx = \dfrac{1}{2} \left(\dfrac{1}{a+b} \sin(a+b)x + \dfrac{1}{a-b} \sin(a-b)x\right)$

(Continued)

Table XVII. 5 (Continued)

46. $\int \dfrac{dx}{a + b\cos x} = \dfrac{2}{\sqrt{a^2 - b^2}} \tan^{-1}\left(\sqrt{\dfrac{a-b}{a+b}} \tan \dfrac{x}{2}\right)$ $a - b$ positive

$\phantom{\int \dfrac{dx}{a + b\cos x}} = \dfrac{2}{\sqrt{b^2 - a^2}} \tanh^{-1}\left(\sqrt{\dfrac{b-a}{b+a}} \tan \dfrac{x}{2}\right)$ $a - b$ negative

$\phantom{\int \dfrac{dx}{a + b\cos x}} = \dfrac{2}{\sqrt{b^2 - a^2}} \coth^{-1}\left(\sqrt{\dfrac{b-a}{b+a}} \tan \dfrac{x}{2}\right)$ $a - b$ negative

47. $\int \dfrac{dx}{a + b\sin x} = \int \dfrac{dy}{a + b\cos y}$ where $x = \dfrac{\pi}{2} + y$ (See 46)

48. $\int x \sin ax\, dx = \dfrac{1}{a^2}(\sin ax - ax \cos ax)$

49. $\int x^n \sin ax\, dx = -\dfrac{1}{a} x^n \cos ax + \dfrac{n}{a} \int x^{n-1} \cos ax\, dx$ (See 51)

50. $\int x \cos ax\, dx = \dfrac{1}{a^2}(\cos ax + ax \sin ax)$

51. $\int x^n \cos ax\, dx = \dfrac{1}{a} x^n \sin ax - \dfrac{n}{a} \int x^{n-1} \sin ax\, dx$ (See 49)

52. $\int e^{ax} \sin(bx + c)\, dx = \dfrac{e^{ax}}{a^2 + b^2}\left(a \sin(bx + c) - b \cos(bx + c)\right)$

53. $\int \sin^{-1} ax\, dx = x \sin^{-1} ax + \dfrac{1}{a}\sqrt{1 - a^2 x^2}$

54. $\int x \sin^{-1} ax\, dx = \dfrac{1}{4a^2}\left((2a^2 x^2 - 1)\sin^{-1} ax + ax\sqrt{1 - a^2 x^2}\right)$

55. $\int \cos^{-1} ax\, dx = x \cos^{-1} ax - \dfrac{1}{a}\sqrt{1 - a^2 x^2}$

56. $\int x \cos^{-1} ax\, dx = \dfrac{1}{4a^2}\left((2a^2 x^2 - 1)\cos^{-1} ax - ax\sqrt{1 - a^2 x^2}\right)$

57. $\int \tan^{-1} ax\, dx = x \tan^{-1} ax - \dfrac{1}{2a} \ln(1 + a^2 x^2)$

58. $\int \cot^{-1} ax\, dx = x \cot^{-1} ax + \dfrac{1}{2a} \ln(1 + a^2 x^2)$

59. $\int \sec^{-1} ax\, dx = x \sec^{-1} ax - \dfrac{1}{a} \cosh^{-1} ax$

60. $\int \operatorname{cosec}^{-1} ax\, dx = x \operatorname{cosec}^{-1} ax + \dfrac{1}{a} \cosh^{-1} ax$

Rational Algebraic Functions

61. $\int \dfrac{dx}{ax + b} = \dfrac{1}{a} \ln(ax + b)$

62. $\int \dfrac{dx}{(x - a)(x - b)} = \dfrac{1}{a - b} \ln \dfrac{x - a}{x - b}$ a not equal to b

63. $\int \dfrac{dx}{(x - a)^2} = -\dfrac{1}{x - a}$

64. $\int \dfrac{dx}{x^2 + a^2} = \dfrac{1}{a} \tan^{-1} \dfrac{x}{a}$

65. $\int \dfrac{x\, dx}{x^2 + a^2} = \dfrac{1}{2} \ln(x^2 + a^2)$

(Continued)

Table XVII. 5 (Continued)

66. $\int \frac{dx}{x^2 - a^2} = -\frac{1}{a} \coth^{-1} \frac{x}{a}$ x greater than a

67. $\int \frac{dx}{a^2 - x^2} = \frac{1}{a} \tanh^{-1} \frac{x}{a}$ x less than a

68. $\int \frac{dx}{ax^2 + bx + c} = \frac{2}{\sqrt{4ac - b^2}} \tan^{-1} \frac{2ax + b}{\sqrt{4ac - b^2}}$ $b^2 - 4ac$ negative

$= \frac{-2}{\sqrt{b^2 - 4ac}} \tanh^{-1} \frac{2ax + b}{\sqrt{b^2 - 4ac}}$

$b^2 - 4ac$ positive and $2ax + b$ less than $\sqrt{b^2 - 4ac}$

$= \frac{-2}{\sqrt{b^2 - 4ac}} \coth^{-1} \frac{2ax + b}{\sqrt{b^2 - 4ac}}$

$b^2 - 4ac$ positive and $2ax + b$ greater than $\sqrt{b^2 - 4ac}$

69. $\int \frac{px + q}{ax^2 + bx + c} dx = \frac{p}{2a} \ln(ax^2 + bx + c) + \left(q - \frac{bp}{2a}\right) \int \frac{dx}{ax^2 + bx + c}$ (See 68)

70. $\int \frac{dx}{x^3 + a^3} = \frac{1}{6a^2} \ln \frac{(x+a)^2}{x^2 - ax + a^2} + \frac{1}{a^2\sqrt{3}} \tan^{-1} \frac{2x - a}{a\sqrt{3}}$

71. $\int \frac{x\, dx}{x^3 + a^3} = \frac{1}{6a} \ln \frac{x^2 - ax + a^2}{(x+a)^2} + \frac{1}{a\sqrt{3}} \tan^{-1} \frac{2x - a}{a\sqrt{3}}$

72. $\int \frac{x^2\, dx}{x^3 + a^3} = \frac{1}{3} \ln(x^3 + a^3)$

73. $\int \frac{dx}{(x^2 + a^2)^2} = \frac{1}{2a^3} \tan^{-1} \frac{x}{a} + \frac{x}{2a^2(x^2 + a^2)}$

74. $\int \frac{dx}{(x^2 - a^2)^2} = \frac{1}{2a^3} \coth^{-1} \frac{x}{a} - \frac{x}{2a^2(x^2 - a^2)}$ x^2 greater than a^2

75. $\int \frac{dx}{(a^2 - x^2)^2} = \frac{1}{2a^3} \tanh^{-1} \frac{x}{a} + \frac{x}{2a^2(a^2 - x^2)}$ x^2 less than a^2

76. $\int \frac{dx}{x^4 + a^4} = \frac{1}{2\sqrt{2}a^3}\left(\tanh^{-1} \frac{ax\sqrt{2}}{x^2 + a^2} + \tan^{-1} \frac{ax\sqrt{2}}{a^2 - x^2}\right)$

77. $\int \frac{dx}{x^4 - a^4} = \frac{1}{2a^3}\left(\cot^{-1} \frac{x}{a} - \coth^{-1} \frac{x}{a}\right)$ x^2 greater than a^2

78. $\int \frac{dx}{a^4 - x^4} = \frac{1}{2a^3}\left(\tan^{-1} \frac{x}{a} + \tanh^{-1} \frac{x}{a}\right)$ x^2 less than a^2

Irrational Algebraic Functions

79. $\int \frac{dx}{\sqrt{x^2 - a^2}} = \sinh^{-1} \frac{x}{a}$

80. $\int \frac{dx}{\sqrt{x^2 - a^2}} = \cosh^{-1} \frac{x}{a}$

81. $\int \frac{dx}{\sqrt{a^2 - x^2}} = \sin^{-1} \frac{x}{a}$

82. $\int \frac{dx}{\sqrt{x^2 + 2ax}} = \cosh^{-1} \frac{x + a}{a}$

83. $\int \frac{dx}{\sqrt{2ax - x^2}} = \sin^{-1} \frac{x - a}{a}$

84. $\int \frac{dx}{x\sqrt{x^2 + a^2}} = -\frac{1}{a} \operatorname{cosech}^{-1} \frac{x}{a}$

(Continued)

Table XVII. 5 (Continued)

85. $\int \dfrac{dx}{x\sqrt{x^2-a^2}} = \dfrac{1}{a} \sec^{-1} \dfrac{x}{a}$

86. $\int \dfrac{dx}{x\sqrt{a^2-x^2}} = -\dfrac{1}{a} \operatorname{sech}^{-1} \dfrac{x}{a}$

87. $\int \dfrac{dx}{(a-x)\sqrt{x-b}} = \dfrac{-2}{\sqrt{b-a}} \tan^{-1} \sqrt{\dfrac{x-b}{b-a}}$ $\dfrac{x-b}{b-a}$ positive

     $= \dfrac{2}{\sqrt{a-b}} \tanh^{-1} \sqrt{\dfrac{x-b}{a-b}}$ $\dfrac{x-b}{a-b}$ positive and less than 1

     $= \dfrac{2}{\sqrt{a-b}} \coth^{-1} \sqrt{\dfrac{x-b}{a-b}}$ $\dfrac{x-b}{a-b}$ positive and greater than 1

88. $\int \dfrac{dx}{\sqrt{ax^2+bx+c}} = \dfrac{1}{\sqrt{a}} \sinh^{-1} \dfrac{2ax+b}{\sqrt{4ac-b^2}}$ a positive, b^2-4ac negative

     $= \dfrac{1}{\sqrt{a}} \cosh^{-1} \dfrac{2ax+b}{\sqrt{b^2-4ac}}$ a positive, b^2-4ac positive

     $= \dfrac{1}{\sqrt{-a}} \cos^{-1} \dfrac{2ax+b}{\sqrt{b^2-4ac}}$ a negative, b^2-4ac positive

89. $\int \dfrac{px+q}{\sqrt{ax^2+bx+c}} dx = \dfrac{p}{a} \sqrt{ax^2+bx+c} + \left(q - \dfrac{bp}{2a}\right) \int \dfrac{dx}{\sqrt{ax^2+bx+c}}$ (See 88)

90. $\int \dfrac{dx}{x\sqrt{ax^2+bx+c}} = -\dfrac{1}{\sqrt{c}} \sinh^{-1} \dfrac{bx+2c}{x\sqrt{4ac-b^2}}$ c positive, b^2-4ac negative

     $= -\dfrac{1}{\sqrt{c}} \cosh^{-1} \dfrac{bx+2c}{x\sqrt{b^2-4ac}}$ c positive, b^2-4ac positive

     $= \dfrac{1}{\sqrt{-c}} \sin^{-1} \dfrac{bx+2c}{x\sqrt{b^2-4ac}}$ c negative, b^2-4ac positive

91. $\int \sqrt{x^2+a^2}\, dx = \dfrac{1}{2}\left(x\sqrt{x^2+a^2} + a^2 \sinh^{-1} \dfrac{x}{a}\right)$

92. $\int \sqrt{x^2-a^2}\, dx = \dfrac{1}{2}\left(x\sqrt{x^2-a^2} - a^2 \cosh^{-1} \dfrac{x}{a}\right)$

93. $\int \sqrt{a^2-x^2}\, dx = \dfrac{1}{2}\left(x\sqrt{a^2-x^2} + a^2 \sin^{-1} \dfrac{x}{a}\right)$

94. $\int \sqrt{ax^2+bx+c}\, dx = \dfrac{2ax+b}{4a}\sqrt{ax^2+bx+c} - \dfrac{b^2-4ac}{8a}\int \dfrac{dx}{\sqrt{ax^2+bx+c}}$ (See 88)

Hyperbolic Functions

95. $\int \sinh ax\, dx = \dfrac{1}{a} \cosh ax$

96. $\int \sinh^2 ax\, dx = \dfrac{1}{2a}\left(\dfrac{1}{2} \sinh 2ax - ax\right)$

97. $\int \sinh^n ax\, dx = \dfrac{1}{an} \sinh^{n-1} ax \cosh ax - \dfrac{n-1}{n} \int \sinh^{n-2} ax\, dx$

98. $\int \cosh ax\, dx = \dfrac{1}{a} \sinh ax$

99. $\int \cosh^2 ax\, dx = \dfrac{1}{2a}\left(\dfrac{1}{2} \sinh 2ax + ax\right)$

100. $\int \cosh^n ax\, dx = \dfrac{1}{an} \cosh^{n-1} ax \sinh ax + \dfrac{n-1}{n} \int \cosh^{n-2} ax\, dx$

101. $\int \tanh ax\, dx = \dfrac{1}{a} \ln \cosh ax$

(Continued)

Table XVII. 5 (Continued)

102. $\int \tanh^2 ax \, dx = \frac{1}{a} (ax - \tanh ax)$

103. $\int \tanh^n ax \, dx = -\frac{1}{a(n-1)} \tanh^{n-1} ax + \int \tanh^{n-2} ax \, dx$

104. $\int \coth ax \, dx = \frac{1}{a} \ln \sinh ax$

105. $\int \coth^2 ax \, dx = \frac{1}{a} (ax - \coth ax)$

106. $\int \coth^n ax \, dx = -\frac{1}{a(n-1)} \coth^{n-1} ax + \int \coth^{n-2} ax \, dx$

107. $\int \text{sech} \, ax \, dx = \frac{1}{a} \text{gd} \, ax$

108. $\int \text{sech}^2 \, ax \, dx = \frac{1}{a} \tanh ax$

109. $\int \text{sech}^n ax \, dx = \frac{1}{a(n-1)} \sinh ax \, \text{sech}^{n-1} ax + \frac{n-2}{n-1} \int \text{sech}^{n-2} ax \, dx$

110. $\int \text{cosech} \, ax \, dx = \frac{1}{a} \ln \tanh \frac{ax}{2}$

111. $\int \text{cosech}^2 ax \, dx = \frac{-1}{a} \coth ax$

112. $\int \text{cosech}^n ax \, dx = -\frac{1}{a(n-1)} \cosh ax \, \text{cosech}^{n-1} ax - \frac{n-2}{n-1} \int \text{cosech}^{n-2} ax \, dx$

113. $\int x \sinh ax \, dx = \frac{1}{a^2} (ax \cosh ax - \sinh ax)$

114. $\int x \cosh ax \, dx = \frac{1}{a^2} (ax \sinh ax - \cosh ax)$

115. $\int \sinh^n ax \cosh ax \, dx = \frac{1}{a(n+1)} \sinh^{n+1} ax$

116. $\int \cosh^n ax \sinh ax \, dx = \frac{1}{a(n+1)} \cosh^{n+1} ax$

117. $\int \sinh^2 ax \cosh^2 ax \, dx = \frac{1}{8a} \left(\frac{1}{4} \sinh 4ax - ax \right)$

118. $\int \sinh ax \sinh bx \, dx = \frac{1}{2} \left(\frac{1}{a+b} \sinh(a+b)x - \frac{1}{a-b} \sinh(a-b)x \right)$

119. $\int \sinh ax \cosh bx \, dx = \frac{1}{2} \left(\frac{1}{a+b} \cosh(a+b)x + \frac{1}{a-b} \cosh(a-b)x \right)$

120. $\int \cosh ax \cosh bx \, dx = \frac{1}{2} \left(\frac{1}{a+b} \sinh(a+b)x + \frac{1}{a-b} \sinh(a-b)x \right)$

121. $\int \sinh^{-1} ax \, dx = x \sinh^{-1} ax - \frac{1}{a} \sqrt{1 + a^2 x^2}$

122. $\int \cosh^{-1} ax \, dx = x \cosh^{-1} ax - \frac{1}{a} \sqrt{a^2 x^2 - 1}$

123. $\int \tanh^{-1} ax \, dx = x \tanh^{-1} ax + \frac{1}{2a} \ln(1 - a^2 x^2)$

Reference Milne-Thomson and Comrie, Standard Four-Figure Mathematical Tables Macmillan, 1931

Table XVII.6 Some commonly occuring definite integrals

1. $\int_0^\infty x^{n-1} e^{-x} dx = \Gamma(n) = (n-1)\Gamma(n-1)$

2. $\int_0^1 x^{m-1}(1-x)^{n-1} dx = \dfrac{\Gamma(m)\Gamma(n)}{\Gamma(m+n)}$

3. $\int_0^{\frac{\pi}{2}} \sin^m x \cos^n x\, dx = \dfrac{\Gamma\left(\frac{m+1}{2}\right)\Gamma\left(\frac{n+1}{2}\right)}{2\Gamma\left(\frac{m+n+2}{2}\right)}$

4. $\int_0^\infty e^{-a^2 x^2} dx = \dfrac{\sqrt{\pi}}{2a}$

5. $\int_0^\infty x e^{-a^2 x^2} dx = \dfrac{1}{2a^2}$

6. $\int_0^\infty x^n e^{-a^2 x^2} dx = \dfrac{\Gamma\left(\frac{n+1}{2}\right)}{2a^{n+1}}$

7. $\int_0^\infty e^{-ax} \sin bx\, dx = \dfrac{b}{a^2+b^2}$ a positive

8. $\int_0^\infty e^{-ax} \cos bx\, dx = \dfrac{a}{a^2+b^2}$ a positive

9. $\int_0^\infty \dfrac{\sin ax}{x} dx = \pm \dfrac{\pi}{2}$ according as a is positive or negative

10. $\int_0^\infty \dfrac{x^{a-1} dx}{1+x} = \dfrac{\pi}{\sin a\pi}$ a positive and less than unity

11. $\int_0^\infty \dfrac{\cos ax\, dx}{1+x^2} = \dfrac{\pi e^{-a}}{2}$ a positive

12. $\int_{-\infty}^{+\infty} \sin x^2\, dx = \int_{-\infty}^{+\infty} \cos x^2\, dx = \sqrt{\dfrac{\pi}{2}}$

13. $\int_0^\pi \sin mx \sin nx\, dx = \int_0^\pi \cos mx \cos nx\, dx = 0$ or $\dfrac{\pi}{2}$ according as m and n are unequal or equal integers

14. $\int_0^{\frac{\pi}{2}} \dfrac{dx}{a^2 \cos^2 x + b^2 \sin^2 x} = \dfrac{\pi}{2ab}$

15. $\int_0^{2\pi} \dfrac{dx}{1 + a\cos x} = \dfrac{2\pi}{\sqrt{1-a^2}}$ a^2 less than unity

16. $\int_0^\pi \dfrac{dx}{1 - 2a\cos x + a^2} = \dfrac{\pi}{1-a^2}$

Reference As for Table XVII.5

Table XVII.7 Some useful formulae for approximate integration

1. $\int_a^b y\, dx = \dfrac{1}{n-1}\left[\dfrac{1}{2}y_1 + y_2 + y_3 + y_4 + \ldots + y_{n-2} + y_{n-1} + \dfrac{1}{2}y_n\right](b-a)$ (Trapezoidal rule)

2. $\int_a^b y\, dx = \dfrac{1}{6}[y_1 + 4y_2 + y_3](b-a)$ (Simpson)

3. $\int_a^b y\, dx = \dfrac{1}{6n}[y_1 + 4y_2 + 2y_3 + 4y_4 + 2y_5 + \ldots + 2y_{2n-1} + 4y_{2n} + y_{2n+1}](b-a)$ (Simpson)

4. $\int_a^b y\, dx = \dfrac{1}{8}[y_1 + 3y_2 + 3y_3 + y_4](b-a)$ (Three-eights rule)

5. $\int_a^b y\, dx = \dfrac{1}{6}[0.28(y_1 + y_7) + 1.62(y_2 + y_6) + 2.2\, y_4](b-a)$ (G. F. Hardy)

6. $\int_a^b y\, dx = \dfrac{1}{20}[y_1 + 5y_2 + y_3 + 6y_4 + y_5 + 5y_6 + y_7](b-a)$ (Weddle)

7. $\int_a^b y\, dx = \dfrac{1}{n}\left[\left(\dfrac{1}{2}y_1 + y_2 + y_3 + \ldots + y_n + \dfrac{1}{2}y_{n+1}\right) - \dfrac{1}{12}(\Delta y_n - \Delta y_1)\right.$

   $\quad - \dfrac{1}{24}(\Delta^2 y_{n-1} + \Delta^2 y_1) - \dfrac{19}{720}(\Delta^3 y_{n-2} - \Delta^3 y_1)$

   $\quad - \dfrac{3}{160}(\Delta^4 y_{n-3} + \Delta^4 y_1) - \dfrac{863}{60480}(\Delta^5 y_{n-4} - \Delta^5 y_1)$

   $\quad \left. - \dfrac{275}{24192}(\Delta^6 y_{n-5} + \Delta^6 y_1)\right](b-a)$ (Gregory)

8. $\int_a^b f(x)\, dx = \dfrac{1}{5}[f(x_1) + f(x_2) + f(x_3) + f(x_4) + f(x_5)](b-a)$

   where $x_1 = a + 0.083751(b-a)$ $x_4 = a + 0.687271(b-a)$
   $x_2 = a + 0.312729(b-a)$ $x_5 = a + 0.916249(b-a)$
   $x_3 = a + 0.5(b-a)$ (Chebyshef)

9. $\int_a^b f(x)\, dx = \dfrac{1}{4}[f(x_1) + f(x_2) + f(x_3) + f(x_4)](b-a)$

   where $x_1 = a + \dfrac{1}{10}(b-a)$ $x_3 = a + \dfrac{6}{10}(b-a)$

   $x_2 = a + \dfrac{4}{10}(b-a)$ $x_4 = a + \dfrac{9}{10}(b-a)$ (Dufton)

Reference As for Table XVII.5

Table XVII.8 Some useful series
Note The range of convergence of infinite series is given in brackets.

Maclaurin's Series

1. $f(x) = f(0) + xf'(0) + \frac{x^2}{2!} f''(0) + \frac{x^3}{3!} f'''(0) + \ldots + \frac{x^n}{n!} f^{(n)}(\theta x)$ $\quad (0 < \theta < 1)$

2. $f(x,y) = f(0,0) + \left[x \left(\frac{\partial f}{\partial x}\right)_{x,y=0} + y \left(\frac{\partial f}{\partial y}\right)_{x,y=0} \right]$
$\qquad + \frac{1}{2!} \left[x^2 \left(\frac{\partial^2 f}{\partial x^2}\right)_{x,y=0} + 2xy \left(\frac{\partial^2 f}{\partial x \partial y}\right)_{x,y=0} + y^2 \left(\frac{\partial^2 f}{\partial y^2}\right)_{x,y=0} \right] + \ldots$

Taylor's Series

3. $f(x+h) = f(x) + hf'(x) + \frac{h^2}{2!} f''(x) + \frac{h^3}{3!} f'''(x) + \ldots + \frac{h^n}{n!} f^{(n)}(x + \theta h)$ $\quad (0 < \theta < 1)$

4. $f(x+h, y+k) = f(x,y) + \left[h \frac{\partial f}{\partial x} + k \frac{\partial f}{\partial y} \right] + \frac{1}{2!} \left[h^2 \frac{\partial^2 f}{\partial x^2} + 2hk \frac{\partial^2 f}{\partial x \partial y} + k^2 \frac{\partial^2 f}{\partial y^2} \right] + \ldots$

Fourier's Series

5. $f(x) = \frac{1}{2} a_0 + \left(a_1 \cos \frac{\pi x}{c} + b_1 \sin \frac{\pi x}{c} \right) + \left(a_2 \cos \frac{2\pi x}{c} + b_2 \sin \frac{2\pi x}{c} \right)$
$\qquad + \left(a_3 \cos \frac{3\pi x}{c} + b_3 \sin \frac{3\pi x}{c} \right) + \ldots$

where $a_0 = \frac{1}{c} \int_{-c}^{+c} f(x) \, dx \quad a_n = \frac{1}{c} \int_{-c}^{+c} f(x) \cos \frac{n\pi x}{c} \, dx \quad b_n = \frac{1}{c} \int_{-c}^{+c} f(x) \sin \frac{n\pi x}{c} \, dx$

6. $f(x) = \frac{1}{2} c_0 + \left(c_1 \cos \frac{2\pi x}{c} + d_1 \sin \frac{2\pi x}{c} \right) + \left(c_2 \cos \frac{4\pi x}{c} + d_2 \sin \frac{4\pi x}{c} \right) + \ldots$

where $c_0 = \frac{2}{c} \int_0^c f(x) \, dx \quad c_n = \frac{2}{c} \int_0^c f(x) \cos \frac{2n\pi x}{c} \, dx \quad d_n = \frac{2}{c} \int_0^c f(x) \sin \frac{2n\pi x}{c} \, dx$

Reversed Series

7. If $y = a_0 + a_1 x + a_2 x^2 + a_3 x^3 + a_4 x^4 + a_5 x^5 + \ldots$
then $\quad x = b_1 (y - a_0) + b_2 (y - a_0)^2 + b_3 (y - a_0)^3 + b_4 (y - a_0)^4 + b_5 (y - a_0)^5 + \ldots$
where $\quad b_1 = \frac{1}{a_1} \quad b_2 = -\frac{a_2}{a_1^3} \quad b_3 = \frac{-a_1 a_3 + 2 a_2^2}{a_1^5} \quad b_4 = \frac{-a_1^2 a_4 + 5 a_1 a_2 a_3 - 5 a_2^3}{a_1^7}$
$\qquad b_5 = \frac{-a_1^3 a_5 + 3 a_1^2 (a_3^2 + 2 a_2 a_4) - 21 a_1 a_2^2 a_3 + 14 a_2^4}{a_1^9}$

Binomial Series

8. $(a+b)^n = a^n + na^{n-1} b + \frac{n(n-1)}{2!} a^{n-2} b^2 + \frac{n(n-1)(n-2)}{3!} a^{n-3} b^3 + \ldots$

$\quad = a^n \left[1 + n \left(\frac{b}{a}\right) + \frac{n(n-1)}{2!} \left(\frac{b}{a}\right)^2 + \frac{n(n-1)(n-2)}{3!} \left(\frac{b}{a}\right)^3 + \ldots \right]$ $\quad (b^2 < a^2)$

$\quad = b^n \left[1 + n \left(\frac{a}{b}\right) + \frac{n(n-1)}{2!} \left(\frac{a}{b}\right)^2 + \frac{n(n-1)(n-2)}{3!} \left(\frac{a}{b}\right)^3 + \ldots \right]$ $\quad (b^2 > a^2)$

9. $(1+x)^n = 1 + nx + \frac{n(n-1)}{2!} x^2 + \frac{n(n-1)(n-2)}{3!} x^3 + \ldots$ $\quad (x^2 < 1)$

10. $(1+x)^{-1} = 1 - x + x^2 - x^3 + x^4 - \ldots$ $\quad (x^2 < 1)$

11. $(1+x)^{-2} = 1 - 2x + 3x^2 - 4x^3 + 5x^4 - \ldots$ $\quad (x^2 < 1)$

(Continued)

Table XVII. 8 (Continued)

12. $(1+x)^{\frac{1}{2}} = 1 + \frac{1}{2}x - \frac{1}{8}x^2 + \frac{1}{16}x^3 - \frac{5}{128}x^4 + \frac{35}{1280}x^5 - \frac{105}{5120}x^6 + \ldots$ $\quad (x^2 < 1)$

13. $(1+x)^{-\frac{1}{2}} = 1 - \frac{1}{2}x + \frac{3}{8}x^2 - \frac{5}{16}x^3 + \frac{35}{128}x^4 - \frac{63}{256}x^5 + \frac{231}{1024}x^6 - \ldots$ $\quad (x^2 < 1)$

Exponential, Logarithmic and Trigonometrical Series

14. $e^x = 1 + x + \frac{x^2}{2!} + \frac{x^3}{3!} + \frac{x^4}{4!} + \ldots$ $\quad$ (All values of x)

15. $\int_0^x e^{-x^2}\, dx = x - \frac{x^3}{1!3} + \frac{x^5}{2!5} + \frac{x^7}{3!7} + \ldots$ $\quad$ (All values of x)

16. $a^x = 1 + x \ln a + \frac{(x \ln a)^2}{2!} + \frac{(x \ln a)^3}{3!} + \ldots$ $\quad$ (All values of x)

17. $\ln(1+x) = x - \frac{x^2}{2} + \frac{x^3}{3} - \frac{x^4}{4} + \ldots$ $\quad (-1 < x \leq 1)$

18. $\sin x = x - \frac{x^3}{3!} + \frac{x^5}{5!} - \frac{x^7}{7!} + \ldots$ $\quad$ (All values of x)

19. $\cos x = 1 - \frac{x^2}{2!} + \frac{x^4}{4!} - \frac{x^6}{6!} + \ldots$ $\quad$ (All values of x)

† 20. $\tan x = x + \frac{1}{3}x^3 + \frac{2}{15}x^5 + \frac{17}{315}x^7 + \ldots + \frac{B_n 2^{2n}(2^{2n}-1)}{(2n)!} x^{2n-1} + \ldots$ $\quad \left(x^2 < \frac{\pi^2}{4}\right)$

† 21. $x \cot x = 1 - \frac{1}{3}x^2 - \frac{1}{45}x^4 - \frac{2}{945}x^6 - \frac{1}{4725}x^8 - \ldots - \frac{B_n 2^{2n}}{(2n)!} x^{2n} - \ldots$ $\quad (x^2 < \pi^2)$

* 22. $\sec x = 1 + \frac{1}{2}x^2 + \frac{5}{24}x^4 + \frac{61}{720}x^6 + \ldots + \frac{E_n}{(2n)!} x^{2n} + \ldots$ $\quad \left(x^2 < \frac{\pi^2}{4}\right)$

† 23. $x \csc x = 1 + \frac{1}{6}x^2 + \frac{7}{360}x^4 + \frac{31}{15120}x^6 + \ldots + \frac{B_n(2^{2n-1}-1)}{(2n)!} x^{2n} + \ldots$ $\quad (x^2 < \pi^2)$

† 24. $\ln \sin x - \ln x = -\frac{1}{6}x^2 - \frac{1}{180}x^4 - \frac{1}{2835}x^6 - \ldots - \frac{B_n 2^{2n-1}}{n(2n)!} x^{2n} - \ldots$ $\quad (x^2 < \pi^2)$

† 25. $\ln \cos x = -\frac{1}{2}x^2 - \frac{1}{12}x^4 - \frac{1}{45}x^6 - \frac{17}{2520}x^8 - \ldots - \frac{B_n 2^{2n-1}(2^{2n}-1)}{n(2n)!} x^{2n} - \ldots$ $\quad \left(x^2 < \frac{\pi^2}{4}\right)$

† 26. $\ln \tan x - \ln x = \frac{1}{3}x^2 + \frac{7}{90}x^4 + \frac{62}{2835}x^6 + \ldots + \frac{B_n 2^{2n}(2^{2n-1}-1)}{n(2n)!} x^{2n} + \ldots$ $\quad \left(x^2 < \frac{\pi^2}{4}\right)$

27. $\sin^{-1} x = x + \frac{1}{2}\frac{x^3}{3} + \frac{1.3}{2.4}\frac{x^5}{5} + \frac{1.3.5}{2.4.6}\frac{x^7}{7} + \ldots$ $\quad (x^2 \leq 1)$

28. $\cos^{-1} x = \frac{\pi}{2} - \sin^{-1} x$ (See 27)

29. $\tan^{-1} x = x - \frac{1}{3}x^3 + \frac{1}{5}x^5 - \frac{1}{7}x^7 + \ldots$ $\quad (x^2 \leq 1)$

$\tan^{-1} x = \frac{\pi}{2} - \frac{1}{x} + \frac{1}{3x^3} - \frac{1}{5x^5} + \frac{1}{7x^7} - \ldots$ $\quad (x^2 > 1)$

30. $\cot^{-1} x = \frac{\pi}{2} - \tan^{-1} x$ (See 29)

31. $\sec^{-1} x = \frac{\pi}{2} - \frac{1}{x} - \frac{1}{2}\frac{1}{3x^3} - \frac{1.3}{2.4}\frac{1}{5x^5} - \frac{1.3.5}{2.4.6}\frac{1}{7x^7} + \ldots$ $\quad (x^2 > 1)$

32. $\csc^{-1} x = \frac{\pi}{2} - \sec^{-1} x$ (See 31)

(Continued)

Table XVII. 8 (Continued)

Hyperbolic Series

33. $\sinh x = x + \frac{x^3}{3!} + \frac{x^5}{5!} + \frac{x^7}{7!} + \ldots$ (All values of x)

34. $\cosh x = 1 + \frac{x^2}{2!} + \frac{x^4}{4!} + \frac{x^6}{6!} + \ldots$ (All values of x)

†35. $\tanh x = x - \frac{1}{3}x^3 + \frac{2}{15}x^5 - \frac{17}{315}x^7 + \ldots + \frac{(-1)^{n-1} B_n 2^{2n}(2^{2n}-1)}{(2n)!} x^{2n-1} \ldots$ $\left(x^2 < \frac{\pi^2}{4}\right)$

†36. $x \coth x = 1 + \frac{1}{3}x^2 - \frac{1}{45}x^4 + \frac{2}{945}x^6 - \frac{1}{4725}x^8 + \ldots + \frac{(-1)^{n-1} B_n 2^{2n}}{(2n)!} x^{2n} \ldots$ $(x^2 < \pi^2)$

*37. $\operatorname{sech} x = 1 - \frac{1}{2}x^2 + \frac{5}{24}x^4 - \frac{61}{720}x^6 + \ldots + \frac{(-1)^n E_n}{(2n)!} x^{2n} \ldots$ $\left(x^2 < \frac{\pi^2}{4}\right)$

†38. $x \operatorname{cosech} x = 1 - \frac{1}{6}x^2 + \frac{7}{360}x^4 - \frac{31}{15120}x^6 + \ldots + \frac{(-1)^n B_n 2(2^{2n-1}-1)}{(2n)!} x^{2n} \ldots$ $(x^2 < \pi^2)$

*39. $\operatorname{gd} x = x - \frac{1}{6}x^3 + \frac{1}{24}x^5 - \frac{61}{5040}x^7 + \ldots + \frac{(-1)^n E_n}{(2n+1)!} x^{2n+1} \ldots$ $\left(x^2 < \frac{\pi^2}{4}\right)$

$\operatorname{gd} x = \frac{\pi}{2} - \operatorname{sech} x - \frac{1}{2}\frac{\operatorname{sech}^3 x}{3} - \frac{1.3}{2.4}\frac{\operatorname{sech}^5 x}{5} - \ldots$ (All positive values of x)

†40. $\ln \sin x - \ln x = -\frac{1}{6}x^2 - \frac{1}{180}x^4 - \frac{1}{2835}x^6 - \ldots + \frac{(-1)^{n-1} B_n 2^{2n-1}}{n(2n)!} x^{2n} \ldots$ $(x^2 < \pi^2)$

†41. $\ln \cosh x = -\frac{1}{2}x^2 - \frac{1}{12}x^4 + \frac{1}{45}x^6 - \frac{17}{2520}x^8 + \ldots + \frac{(-1)^{n-1} B_n 2^{2n-1}(2^{2n-1})}{n(2n)!} x^{2n} \ldots$ $\left(x^2 < \frac{\pi^2}{4}\right)$

†42. $\ln \tanh x - \ln x = -\frac{1}{3}x^2 + \frac{7}{90}x^4 - \frac{62}{2835}x^6 + \ldots + \frac{(-1)^n B_n 2^{2n}(2^{2n-1}-1)}{n(2n)!} x^{2n} \ldots$ $\left(x^2 < \frac{\pi^2}{4}\right)$

43. $\sinh^{-1} x = x - \frac{1}{2}\frac{x^3}{3} + \frac{1.3}{2.4}\frac{x^5}{5} - \frac{1.3.5}{2.4.6}\frac{x^7}{7} + \ldots$ $(x^2 < 1)$

$\sinh^{-1} x = \ln 2x + \frac{1}{2}\frac{1}{2x^2} - \frac{1.3}{2.4}\frac{1}{4x^4} + \frac{1.3.5}{2.4.6}\frac{1}{6x^6} - \ldots$ $(x > 1)$

44. $\cosh^{-1} x = \ln 2x - \frac{1}{2}\frac{1}{2x^2} - \frac{1.3}{2.4}\frac{1}{4x^4} + \frac{1.3.5}{2.4.6}\frac{1}{6x^6} - \ldots$ $(x > 1)$

45. $\tanh^{-1} x = x + \frac{1}{3}x^3 + \frac{1}{5}x^5 + \frac{1}{7}x^7 + \ldots$ $(x^2 < 1)$

46. $\coth^{-1} x = \frac{1}{x} + \frac{1}{3x^3} + \frac{1}{5x^5} + \frac{1}{7x^7} + \ldots$ $(x^2 > 1)$

47. $\operatorname{sech}^{-1} x - \ln \frac{2}{x} = -\frac{1}{2}\frac{x^2}{2} - \frac{1.3}{2.4}\frac{x^4}{4} - \frac{1.3.5}{2.4.6}\frac{x^2}{6} - \ldots$ $(0 < x < 1)$

48. $\operatorname{cosech}^{-1} x - \ln \frac{2}{x} = \frac{1}{2}\frac{x^2}{2} - \frac{1.3}{2.4}\frac{x^4}{4} + \frac{1.3.5}{2.4.6}\frac{x^2}{6} - \ldots$ $(0 < x > 1)$

$\operatorname{cosech}^{-1} x = \frac{1}{x} - \frac{1}{2}\frac{1}{3x^3} + \frac{1.3}{2.4}\frac{1}{5x^5} - \frac{1.3.5}{2.4.6}\frac{1}{7x^7} + \ldots$ $(x^2 > 1)$

*49. $\operatorname{gd}^{-1} x = x + \frac{1}{6}x^3 + \frac{1}{24}x^5 + \frac{61}{5040}x^7 + \ldots + \frac{E_n}{(2n+1)!} x^{2n+1} + \ldots$ $\left(x^2 < \frac{\pi^2}{4}\right)$

50. $\cosh x \cos x = 1 - \frac{2^2}{4!}x^4 + \frac{2^4}{8!}x^8 - \frac{2^6}{12!}x^{12} + \ldots$ (All values of x)

51. $\sinh x \sin x = \frac{2}{2!}x^2 - \frac{2^3}{6!}x^6 + \frac{2^5}{10!}x^{10} - \ldots$ (All values of x)

52. $\sinh x \cos x = x - \frac{2}{3!}x^3 - \frac{2^2}{5!}x^5 + \frac{2^3}{7!}x^7 + \frac{2^4}{9!}x^9 - \ldots$ (All values of x)

53. $\cosh x \sin x = x + \frac{2}{3!}x^3 - \frac{2^2}{5!}x^5 - \frac{2^3}{7!}x^7 + \frac{2^4}{9!}x^9 + \ldots$ (All values of x)

†B denotes a Bernoulli number as follows: $B_1 = \frac{1}{6}$, $B_2 = \frac{1}{30}$, $B_3 = \frac{1}{42}$, $B_4 = \frac{1}{30}$, $B_5 = \frac{5}{66}$, $B_6 = \frac{691}{2730}$, $B_7 = \frac{7}{6}$, $B_8 = \frac{3617}{510}$

*E denotes an Euler number as follows: $E_1 = 1$, $E_2 = 5$, $E_3 = 61$, $E_4 = 1385$, $E_5 = 50521$, $E_6 = 2702765$, $E_7 = 199360981$, $E_8 = 19391512145$.
Reference: as for Table XVII.5.

Table XVII.9 The error function $\text{Erf } x = 2\pi^{-\frac{1}{2}} \int_0^x e^{-x^2} dx$.

x	$\text{erf } x$	x	$\text{erf } x$	x	$\text{erf } x$	x	$\text{erf } x$	x	$\text{erf } x$	x	$\text{erf } x$
0.00	0.0000_{113}	0.50	0.5205_{87}	1.00	0.8427_{41}	1.50	0.9661_{12}	2.00	0.99532_{20}	2.50	0.99959
.01	$.0113_{113}$	.51	$.5292_{87}$	.01	$.8468_{40}$	.51	$.9673_{11}$	.01	$.99552_{20}$	.51	.99961
.02	$.0226_{112}$	.52	$.5379_{86}$	.02	$.8508_{40}$	.52	$.9684_{11}$	.02	$.99572_{19}$	.52	.99963
.03	$.0338_{113}$	.53	$.5465_{84}$	.03	$.8548_{38}$	.53	$.9695_{11}$	.03	$.99591_{18}$	.53	.99965
.04	$.0451_{113}$	.54	$.5549_{84}$	.04	$.8586_{38}$	.54	$.9706_{10}$	.04	$.99609_{17}$	.54	.99967
0.05	0.0564_{112}	0.55	0.5633_{83}	1.05	0.8624_{37}	1.55	0.9716_{10}	2.05	0.99626_{16}	2.55	0.99969
.06	$.0676_{113}$	.56	$.5716_{82}$	.06	$.8661_{37}$	.56	$.9726_{10}$	.06	$.99642_{16}$	.56	.99971
.07	$.0789_{112}$	.57	$.5798_{81}$	.07	$.8698_{35}$	.57	$.9736_{9}$	.07	$.99658_{15}$	.57	.99972
.08	$.0901_{112}$	.58	$.5879_{80}$	.08	$.8733_{35}$	.58	$.9745_{10}$	.08	$.99673_{15}$	.58	.99974
.09	$.1013_{112}$	.59	$.5959_{80}$	.09	$.8768_{34}$	.59	$.9755_{8}$	.09	$.99688_{14}$	.59	.99975
0.10	0.1125_{111}	0.60	0.6039_{78}	1.10	0.8802_{33}	1.60	0.9763_{9}	2.10	0.99702_{13}	2.60	0.99976
.11	$.1236_{112}$	.61	$.6117_{77}$	.11	$.8835_{33}$	.61	$.9772_{8}$	.11	$.99715_{13}$	.61	.99978
.12	$.1348_{111}$	.62	$.6194_{76}$	.12	$.8868_{32}$	.62	$.9780_{8}$	.12	$.99728_{13}$	.62	.99979
.13	$.1459_{110}$	.63	$.6270_{76}$	.13	$.8900_{31}$	.63	$.9788_{8}$	.13	$.99741_{12}$	.63	.99980
.14	$.1569_{111}$	.64	$.6346_{74}$	.14	$.8931_{30}$	.64	$.9796_{8}$	.14	$.99753_{11}$	.64	.99981
0.15	0.1680_{110}	0.65	0.6420_{74}	1.15	0.8961_{30}	1.65	0.9804_{7}	2.15	0.99764_{11}	2.65	0.99982
.16	$.1790_{110}$	.66	$.6494_{72}$	.16	$.8991_{29}$	.66	$.9811_{7}$	.16	$.99775_{10}$	.66	.99983
.17	$.1900_{109}$	.67	$.6566_{72}$	.17	$.9020_{28}$	.67	$.9818_{7}$	.17	$.99785_{10}$	.67	.99984
.18	$.2009_{109}$	.68	$.6638_{70}$	.18	$.9048_{28}$	.68	$.9825_{7}$	.18	$.99795_{10}$	.68	.99985
.19	$.2118_{109}$	.69	$.6708_{70}$	.19	$.9076_{27}$	.69	$.9832_{6}$	.19	$.99805_{9}$	.69	.99986
0.20	0.2227_{108}	0.70	0.6778_{69}	1.20	0.9103_{27}	1.70	0.9838_{6}	2.20	0.99814_{8}	2.70	0.99987
.21	$.2335_{108}$	.71	$.6847_{67}$	.21	$.9130_{25}$	.71	$.9844_{6}$	.21	$.99822_{9}$	.71	.99987
.22	$.2443_{107}$	.72	$.6914_{67}$	.22	$.9155_{26}$	.72	$.9850_{6}$	.22	$.99831_{8}$	.72	.99988
.23	$.2550_{107}$	.73	$.6981_{66}$	.23	$.9181_{24}$	.73	$.9856_{5}$	.23	$.99839_{7}$	.73	.99989
.24	$.2657_{106}$	.74	$.7047_{65}$	.24	$.9205_{24}$	.74	$.9861_{6}$	.24	$.99846_{8}$	.74	.99989
0.25	0.2763_{106}	0.75	0.7112_{63}	1.25	0.9229_{23}	1.75	0.9867_{5}	2.25	0.99854_{7}	2.75	0.99990
.26	$.2869_{105}$	.76	$.7175_{63}$	.26	$.9252_{23}$	.76	$.9872_{5}$	.26	$.99861_{6}$	.76	.99991
.27	$.2974_{105}$	.77	$.7238_{62}$	.27	$.9275_{22}$	.77	$.9877_{5}$	.27	$.99867_{7}$	.77	.99991
.28	$.3079_{104}$	.78	$.7300_{61}$	.28	$.9297_{22}$	.78	$.9882_{4}$	.28	$.99874_{6}$	.78	.99992
.29	$.3183_{103}$	.79	$.7361_{60}$	.29	$.9319_{21}$	.79	$.9886_{5}$	.29	$.99880_{6}$	.79	.99992
0.30	0.3286_{103}	0.80	0.7421_{59}	1.30	0.9340_{21}	1.80	0.9891_{4}	2.30	0.99886_{5}	2.80	0.99992
.31	$.3389_{102}$	.81	$.7480_{58}$	.31	$.9361_{20}$	.81	$.9895_{4}$	.31	$.99891_{6}$	.81	.99993
.32	$.3491_{102}$	.82	$.7538_{57}$	.32	$.9381_{19}$	.82	$.9899_{4}$	.32	$.99897_{5}$	.82	.99993
.33	$.3593_{101}$	.83	$.7595_{56}$	.33	$.9400_{19}$	.83	$.9903_{4}$	.33	$.99902_{4}$	.83	.99994
.34	$.3694_{100}$	.84	$.7651_{56}$	.34	$.9419_{19}$	.84	$.9907_{4}$	.34	$.99906_{5}$	.84	.99994
0.35	0.3794_{99}	0.85	0.7707_{54}	1.35	0.9438_{18}	1.85	0.9911_{4}	2.35	0.99911_{4}	2.85	0.99994
.36	$.3893_{99}$	.86	$.7761_{53}$	.36	$.9456_{17}$	.86	$.9915_{3}$	.36	$.99915_{5}$	.86	.99995
.37	$.3992_{98}$	.87	$.7814_{53}$	.37	$.9473_{17}$	.87	$.9918_{4}$	.37	$.99920_{4}$	.87	.99995
.38	$.4090_{97}$	.88	$.7867_{51}$	.38	$.9490_{17}$	.88	$.9922_{3}$	.38	$.99924_{4}$	.88	.99995
.39	$.4187_{97}$	.89	$.7918_{51}$	.39	$.9507_{16}$	.89	$.9925_{3}$	.39	$.99928_{3}$	.89	.99996
0.40	0.4284_{96}	0.90	0.7969_{50}	1.40	0.9523_{16}	1.90	0.9928_{3}	2.40	0.99931_{4}	2.90	0.99996
.41	$.4380_{95}$	.91	$.8019_{49}$	.41	$.9539_{15}$	.91	$.9931_{3}$	.41	$.99935_{3}$	.91	.99996
.42	$.4475_{94}$	.92	$.8068_{48}$	.42	$.9554_{15}$	.92	$.9934_{3}$	.42	$.99938_{3}$	.92	.99996
.43	$.4569_{93}$	.93	$.8116_{47}$	.43	$.9569_{14}$	.93	$.9937_{2}$	.43	$.99941_{3}$	.93	.99997
.44	$.4662_{93}$	.94	$.8163_{46}$	.44	$.9583_{14}$	.94	$.9939_{3}$	.44	$.99944_{3}$	.94	.99997

(Continued)

Table XVII. 9 (Continued)

x	erf x	x	erf x	x	erf x	x	erf x	x	erf x	x	erf x
0.45	0.4755 $_{92}$	0.95	0.8209 $_{45}$	1.45	0.9597 $_{14}$	1.95	0.9942 $_2$	2.45	0.99947 $_3$	2.95	0.99997
.46	.4847 $_{90}$	.96	.8254 $_{45}$	.46	.9611 $_{13}$	.96	.9944 $_3$	.46	.99950 $_2$	.96	.99997
.47	.4937 $_{90}$	.97	.8299 $_{43}$	.47	.9624 $_{13}$	.97	.9947 $_2$	.47	.99952 $_3$	.97	.99997
.48	.5027 $_{90}$	.98	.8342 $_{43}$	.48	.9637 $_{12}$	.98	.9949 $_2$	.48	.99955 $_2$	.98	.99997
.49	.5117 $_{88}$	0.99	.8385 $_{42}$	.49	.9649 $_{12}$	1.99	.9951 $_2$	.49	.99957 $_2$	2.99	.99998
0.50	0.5205	1.00	0.8427	1.50	0.9661	2.00	0.9953	2.50	0.99959	3.00	0.99998

$$\operatorname{erf} x = \frac{2}{\sqrt{\pi}} \left(x - \frac{x^3}{1!\,3} + \frac{x^5}{2!\,5} - \frac{x^7}{3!\,7} + \cdots \right)$$

For large values of x the following series may be used:

$$\operatorname{erf} x = 1 - \frac{e^{-x^2}}{x\sqrt{\pi}} \left(1 - \frac{1}{2x^2} + \frac{1\cdot 3}{(2x^2)^2} - \frac{1\cdot 3\cdot 5}{(2x^2)^3} + \cdots \right)$$

x	erf x
3.123	$1 - 10^{-5}$
3.459	$1 - 10^{-6}$
3.767	$1 - 10^{-7}$
4.052	$1 - 10^{-8}$
4.320	$1 - 10^{-9}$
4.573	$1 - 10^{-10}$
4.813	$1 - 10^{-11}$

Reference as for Table XVII.5

Index

Index

For each entry there is a Section and Table Number (e.g. **XII.4**) and a page number (e.g. 432), thus: **XII.4**, 432.

Absorbed dose unit **I.2**, 3
Absorbers and poisons in nuclear reactors **IX.49**, 265; **IX.50**, 265
Absorption
  coefficient
    mass, for photons **IX.19**, 188
    of
      gases in sea water **XI.16**, 413
      gases in water **XI.15**, 412
      sound in gases **X.3**, 274
      sound in liquids **X.8**, 275
      sound in solids **X.10**, 277
      sound in water **X.5**, 274
  cross-section for neutrons **IX.45**, 264; **IX.46**, 264; **IX.49**, 265
  edges for X-rays **IX.15**, 182–3; **IX.16**, 184–5; **IX.17**, 186–7
Abundances of
  elements in
    human body **III.13**, 49
    Earth's crust **III.11**, 48
    sea water **III.11**, 48
  nuclides **IX.39**, 196–262
Acceleration
  due to gravity **I.7**, 10
  units **I.3**, 4; **I.4**, 6
Acid-base indicators **VI.26**, 107
Acidic dissociation constants **VI.33**, 110–11, **VI.34**, 111–12
Acids
  dissociation constants of
    inorganic acids **VI.33**, 110–11
    organic acids **VI.34**, 111–12
Acoustics
  architectural **X.21**, 280; **X.22**, 280; **X.23**, 281; **X.24**, 281; **X.26**, 282
  musical **X.27**, 282; **X.28**, 282; **X.29**, 282
  physiological and subjective **X.15**, 278; **X.16**, 278; **X.17**, 279; **X.18**, 279; **X.19**, 279; **X.20**, 279
  preferred frequencies for measurements **X.13**, 278
  recommended maximum noise levels **X.17**, 279
Acre **I.6B**, 7
Activation
  energy of
    bimolecular reactions **XV.2**, 514–15
    catalysed reactions **XV.5**, 517
    thermal unimolecular reactions **XV.2**, 514–15
  entropy of **XV.4**, 516
  volume of **XV.4**, 516
Activity
  mean ionic **VI.9**, 93
  unit **I.2**, 3
Activity coefficients
  calculated values of **VI.13**, 96
  mean ionic, of electrolytes in aqueous solution **VI.10**, 94; **VI.19**, 99
  mean ionic, of hydrochloric acid in water **VI.11**, 95

Activity coefficients (*Contd*)
  mean ionic, of hydrochloric acid in dioxane-water **VI.12**, 95
  mean ionic **VI.9**, 93
  of HCl in dioxane-water mixtures **VI.12**, 95
  of sucrose solutions **VI.20**, 99
Air
  attenuation of sound in **X.2**, 273
  composition of **III.12**, 48
  density of ambient **II.2**, 13; **II.3**, 14
  kerma-rate constant **IX.37**, 195
  refractive index of **VII.16**, 134
  refractive index of, at radio frequencies **VII.17**, 134
  relative permittivity of **V.12**, 74
  solubility in water **XI.15**, 412
  specific heat capacities, ratio **XII.14**, 439
  speed of sound in **X.1**, 273
  thermal conductivity of **XII.6**, 433
  viscosity coefficient of **II.24**, 23
  water content of saturated **II.9**, 16
Alcohols
  calorific values of **XIII.33**, 497
  density of aqueous solutions of **XI.25**, 420; **XI.26**, 420
  surface tension of aqueous solutions of **II.36**, 30
Alloys
  elasticity of **II.17**, 21
  ferromagnetic properties of **V.18**, 80–1; **V.19**, 82; **V.20**, 83; **V.21**, 84
  linear coefficient of expansion of **XII.4**, 432
  mass percentage composition of **XVII.3**, 536–7
  resistivity of **V.4**, 71
  specific heat capacity of **XII.17**, 441
  superconductivity of **V.24**, 86
  temperature coefficient of resistivity of **V.5**, 71
  thermal conductivity of **XII.11**, 436–7
  thermal electric force of **XII.22**, 445
Alpha particles
  range of **IX.23**, 190
  sources of **IX.39**, 196–262
  stopping power of **IX.34**, 194
Altitude, variation of pressure, density and temperature with **XIV.10**, 508; **XIV.11**, 508
Amino acids
  density of **XI.4**, 366–75
  derivatives of **XI.4**, 366–75
  melting point of **XI.4**, 366–75
  pK value of **XI.4**, 366–75
  solubility of **XI.4**, 366–75
  specific rotation of **XI.4**, 366–75
Amino sugars, physical properties of **XI.5**, 386–9
Ammonia
  as refrigerant **XI.10**, 402
  bottled gas **XI.11**, 402–3

Ammonia (*Contd*)
  CODATA key thermodynamic values **XIII.2**, 451; **XIII.3**, 469
  critical constants of **XI.12**, 408
  density of aqueous solutions of **XI.26**, 420
  molar heat capacity of **XI.1**, 285; **XIII.25**, 491
  moments of inertia of **VIII.17**, 157
  physical properties of **XI.1**, 285
  rotational constants of **VIII.17**, 157
  second virial coefficient of **XI.14**, 411
  solubility in water of **XI.15**, 412
  speed of sound in **X.1**, 273
  vapour pressure of **XI.21**, 417
  viscosity of **II.25**, 24
Amount of substance **I.1**, 3
Ampere **I.1**, 3; **I.4**, 5
Angle
  plane, unit **I.2**, 3
  solid, unit **I.2**, 3
Angular velocity, unit **I.3**, 4
Anti-reflection coatings **VII.32**, 140
Approximate integration **XVII.7**, 545
Aqueous solutions, *see* solutions
Area **I.3**, 4; **I.6B**, 7
Argon
  atomic mass of **III.2**, 36
  CODATA key thermodynamic values **XIII.1**, 449; **XIII.2**, 451
  critical constants of **XI.12**, 408
  molar heat capacity of **XIII.1**, 449; **XIII.22**, 489
  molecular mean speed etc of **II.34**, 30
  properties of **III.4**, 40
  solubility of, in water **XI.15**, 412
  viscosity coefficient of **II.25**, 24
Arrhenius parameters for
  bimolecular reactions **XV.2**, 514–15
  thermal unimolecular gas reactions **XV.2**, 514–15
Asteroids, constants of **XIV.3**, 505
Asymmetric rotor molecules, moment of inertia of **VIII.18**, 157
Atmosphere
  composition of **III.12**, 48
  mass of **XIV.7**, 507
  pressure, mean molecular weight and temperature, in upper **XIV.11**, 508
  properties of **XIV.10**, 508; **XIV.11**, 508
  standard **I.6G**, 9
Atomic
  electron affinity **III.8**, 46
  masses **III.2**, 35–7
  mass constant **I.7**, 10
  numbers **III.1**, 35; **III.2**, 35–7
  numbers of nuclides **IX.39**, 196–262
  radius **III.5**, 42–3
  weights, standard **III.2**, 35–7
Atomization, enthalpy of, for elements **XIII.1**, 449–50
Atoms, arrangement of electrons in **III.3**, 38–40

Attenuation of
  sound in air   **X.2**, 273
  sound in sea water   **X.6**, 274
  X-rays   **IX.19**, 188
Auger electron energies   **IX.10**, 175
Autoprotolysis constants of
  solvents   **VI.38**, 115
  water   **VI.37**, 114
Avogadro constant   **I.7**, 10
Avoirdupois units   **I.6E**, 8
Azeotropes, binary   **XIII.34**, 498

Bar   **I.6G**, 9
Barn   **I.6B**, 7
Baryon number   **IX.61**, 268; **IX.62**, 269; **IX.63**, 269; **IX.64**, 269
BA screw threads   **XVII.2**, 535
Base dissociation constants for
  inorganic bases   **VI.35**, 113
  organic bases   **VI.36**, 113–14
Base units   **I.1**, 3
Beautyonium states   **IX.65**, 270
Becquerel   **I.2**, 3
Beta rays, ranges in aluminium   **IX.21**, 189
Bile acids, see steroids
Binary azeotropic mixtures   **XIII.34**, 498
Binomial series   **XVII.8**, 546–7
Biological molecules, see amino acids, amino sugars, carbohydrates, nucleic acids, organic compounds, steroids, uronic acids
Biomass, calorific value of   **XIII.33**, 497
Biomolecular reactions
  Arrhenius parameters for   **XV.2**, 514–15; **XV.4**, 516
  rate coefficients of   **XV.1**, 513
Bohr
  magneton   **I.7**, 10
  radius   **I.7**, 10
Boiling point of
  elements   **III.4**, 40–1
  fluorocarbon refrigerants   **XI9**, 400–10
  inorganic compounds   **XI.1**, 285–306; **XI.11**, 402–7
  organic compounds   **XI.2**, 307–59; **XI.11**, 402–7
  organometallic compounds   **XI.3**, 360–5
  water   **XI.19**, 415
Boltzmann constant   **I.7**, 10
Bond angle in
  polyatomic molecules   **IV.4**, 54–6
  polymers   **IV.5**, 56
Bond dissociation enthalpy of
  diatomic molecules   **XIII.14**, 484; **XIII.15**, 484
  polyatomic molecules   **XIII.16**, 485
Bond length in
  diatomic molecules   **IV.2**, 53; **VIII.15**, 156
  elements   **IV.3**, 54
  polyatomic molecules   **IV.4**, 54–6
  polymers   **IV.5**, 56
Bosons
  gauge   **IX.67**, 270
  weak   **IX.67**, 270
Bottled gases, properties of   **XI.11**, 402–7
Bottom quark   **IX.61**, 268
Bottominium states, see beautyonium states
Boyle Temperature   **XI.13**, 410
Bravais lattices   **IV.14**, 64

British thermal unit   **I.6H**, 9
British units and conversion factors   **I.6**, 7–9
Bromine
  atomic mass of   **III.2**, 36, 37
  CODATA key thermodynamic values of   **XIII.1**, 449; **XIII.2**, 452; **XIII.3**, 467
  coefficient of viscosity of   **II.25**, 24
  critical constants of   **XI.12**, 408
  dissociation energy of   **VIII.15**, 156
  isotopes of   **IX.39**, 204–5
  molar heat capacity of   **III.4**, 40; **XIII.3**, 467; **XIII.22**, 489
  moment of inertia of   **VIII.15**, 156
  physical properties of   **III.4**, 40; **XI.1**, 288
  rotational constants of   **VIII.15**, 156
  speed of sound in   **X.1**, 273
  thermal conductivity of   **XII.6**, 433
  viscosity coefficient for
    liquid   **II.30**, 26
    vapour   **II.25**, 24
  X-ray energies from   **IX.11**, 176; **IX.13**, 179; **IX.15**, 182; **IX.16**, 184
Buffer solutions
  biological   **VI.23**, 102
  common useful   **VI.22**, 100–1; **VI.24**, 102–5
  standard reference   **VI.21**, 100–1
Building materials
  coefficient of linear expansion of   **XII.5**, 432
  sound insulation of   **X.21**, 280
  speed and attenuation of sound in   **X.9**, 276–7
  thermal conductivity of   **XII.13**, 438
Bulk elastic modulus of
  alloys   **II.17**, 21
  earth   **XIV.8**, 507
  glass and quartz   **II.18**, 21
  liquids   **II.23**, 23
  metals   **II.16**, 20
  non-metallic elements   **II.22**, 22
Bunsen coefficient   **XI.15**, 412; **XI.16**, 413
Buoyancy correction   **II.1**, 13
Butane
  bottled gas   **XI.11**, 402–3
  calorific value of   **XIII.33**, 497
  solubility in water of   **XI.15**, 412
  speed of sound in   **X.1**, 273

Calcite, refractive index of   **VII.28**, 139
Calibration
  refractometric   **VII.18**, 134
  photometric   **VIII.6**, 148–9
  volumetric   **II.4**, 14; **II.5**, 14
  wavelength   **VII.41**, 142; **VIII.5**, 147; **VIII.9**, 152; **VIII.10**, 153
Calomel electrode   **VI.27**, 108
Calorie   **I.6H**, 9
Calorific values of fuels   **XIII.33**, 497
Candela   **I.1**, 3
Capacitance unit   **I.2**, 3
Carat   **I.6E**, 8
Carbohydrates
  calorific value of   **XIII.33**, 497
  derivatives of   **XI.5**, 376–91
  melting point of   **XI.5**, 376–91
  solubility of   **XI.5**, 376–91
  specific rotation of   **XI.5**, 376–91

Carbon-13, chemical shift   **VIII.24**, 162
Carbon dioxide
  as refrigerant   **XI.10**, 402
  CODATA key thermodynamic values   **XIII.2**, 454; **XIII.3**, 469
  critical constants of   **XI.12**, 408
  molar heat capacity of   **XIII.2**, 454; **XIII.3**, 469
  moment of inertia of   **VIII.16**, 156
  physical properties of   **XI.1**, 290
  rotational constant of   **VIII.16**, 156
  second virial coefficient of   **XI.14**, 411
  solubility in
    sea water   **XI.16**, 413
    water   **XI.15**, 412
  speed of sound in   **X.1**, 273
  vapour pressure of   **XI.27**, 422
  viscosity of   **II.25**, 24
Catalysed reactions, activation energies of   **XV.5**, 517
Cations
  activity coefficients of (calculated)   **VI.13**, 96
  formation constants for EDTA complexes   **VI.41**, 117
  half-wave potentials in aqueous solutions   **VI.43**, 118–19
  half-wave potentials in non-aqueous solvents   **VI.43**, 120–1
  limiting molar conductivity in aqueous solutions   **VI.3**, 90–1; **VI.4**, 91
  limiting molar conductivity in non-aqueous solvents   **VI.5**, 92
  transport numbers of   **VI.14**, 97; **VI.15**, 97; **VI.16**, 97; **VI.17**, 98
  transport numbers for fused salts   **VI.6**, 93
Cellulose, dielectric properties of   **V.17**, 78
Celsius temperature   **I.2**, 3; **I.7**, 10; **XII.24**, 445
Cement, optical, refractive index of   **VII.27**, 138
Centimetre   **I.6A**, 7
Ceramics
  dielectric properties of   **V.16**, 77
  resistivity of   **V.9**, 73
cgs units   **I.4**, 5–6; **I.6**, 7–9
Chance-Pilkington glass, see glass, optical
Characteristic temperatures of gases   **VIII.14**, 155
Character tables   **IV.13**, 61–3
Charge
  electric, see electric charge
  on electron   **I.7**, 10
  on proton   **I.7**, 10
Charm   **IX.61**, 268
Charmonium states   **IX.66**, 270
Chemical
  bonds, properties of   **IV.7**, 58; **IV.8**, 59
  hygrometer   **II.9**, 16
  shifts
    carbon-13   **VIII.24**, 162
    nitrogen-15   **VIII.25**, 163
    proton   **VIII.23**, 161
    other nuclei   **VIII.26**, 164
    NMR solvents   **VIII.22**, 160
Chlorine
  atomic mass of   **III.2**, 36, 37
  CODATA key thermodynamic values   **XIII.1**, 449; **XIII.2**, 454; **XIII.3**, 467

Chlorine (*Contd*)
  coefficient of viscosity of **II.25**, 24
  critical constants of **XI.12**, 408
  dissociation energy of **VIII.15**, 156
  isotopes of **IX.39**, 207–8
  molar heat capacity of **III.4**, 40; **XIII.3**, 467; **XIII.22**, 489
  moment of inertia of **VIII.15**, 156
  physical properties of **III.4**, 40; **XI.1**, 290
  rotational constants of **VIII.15**, 156
  thermal conductivity of **XII.6**, 433
  X-ray energies from **IX.11**, 176; **IX.13**, 179; **IX.15**, 182; **IX.16**, 184
Chromophores, organic, absorption characteristics of **VIII.13**, 154
CIE, *see* Commission International de l'Eclairage
Cladding for nuclear reactors **IX.43**, 264; **IX.44**, 264; **IX.45**, 264
Coal, calorific value of **XIII.33**, 497
Coal gas, calorific value of **XIII.33**, 497
CODATA key values for thermodynamics **XIII.2**, 451–65; **XIII.3**, 466–9; **XIII.22**, 489–90
Coercivity of low coercivity magnetic substances **V.18**, 80–1
Collision cross section of electrons **IX.8**, 173
Colorimetry
  CIE standard illuminants **VII.4**, 129
Colour blindness (colour deficiency) **VII.7**, 131
  substances influencing **VII.8**, 132; **VII.9**, 132
Colour code for wire wound resistors **XVII.4**, 537
Colour deficiency characteristics **VII.7**, 131
Colour temperature of light sources **VII.1**, 127
Combustion, enthalpy of, for organic compounds **XIII.6**, 477
Commission International de l'Eclairage (CIE)
  of colorimetric standard observer **VII.2**, 128
  relative efficiency function **VII.3**, 129
  standard illuminants for colorimetry **VII.4**, 129; **VII.5**, 130
Commission, International for Optics refractometry wavelengths **VII.14**, 133
Composition of
  atmosphere **III.12**, 48
  Earth's crust (elemental) **III.11**, 48
  human body (elemental) **III.13**, 49
  sea water (elemental) **III.11**, 48
Compounds, *see* inorganic compounds, organic compounds
Compton wavelength of electron **I.7**, 10
Concentration units **I.4**, 5; **I.6I**, 9
Concert hall characteristics **X.26**, 282
Conductance unit **I.2**, 3
Conductivity, electrical,
  equivalent of fused salts **VI.6**, 92–3
  ionic, in aqueous solution **VI.3**, 90–1; **VI.4**, 91
  ionic, in non-aqueous solution **VI.5**, 92
  molar, of electrolytes **VI.2**, 89–90
  of
    elements **III.4**, 40–1

Conductivity, electrical (*Contd*)
  of
    potassium chloride solutions, standard, **VI.1**, 89
    sea water **XIV.14**, 510
  unit **I.3**, 4; **I.4**, 5
Conductivity, thermal, *see* thermal conductivity
Constant
  Avogadro **I.7**, 10
  Boltzmann **I.7**, 10
  Faraday **I.7**, 10
  fine structure **I.7**, 10
  gravitational **I.7**, 10
  Madelung **IV.12**, 60
  Planck **I.7**, 10
  radiation **I.7**, 10
  Rydberg **I.7**, 10
  Stefan **I.7**, 10
  universal gas **I.7**, 10
Constants
  fundamental **I.7**, 10
  Hammett reaction **XV.7**, 518
  Hammett substituent **XV.6**, 517
Contact potential difference **IX.9**, 174
Conversion factors
  concentration **I.6I**, 9
  force **I.6E**, 8
  for units
    concentration **I.4**, 5
    electricity and magnetism **I.4**, 5
    optical quantities **I.4**, 6
    space and time **I.4**, 6
    thermodynamic **I.4**, 6
  length **I.6A**, 7
  mass **I.6E**, 8
  pressure **I.6G**, 9
  speed **I.6D**, 7
  volume **I.6C**, 7
  work and energy **I.6H**, 9
Coolant components in nuclear reactors **IX.46**, 264; **IX.47**, 264; **IX.48**, 264
Core and blanket in nuclear reactors **IX.54**, 266
Corticosteroids, *see* steroids
Cosmic rays **XIV.12**, 509
Coulomb **I.2**, 3
Critical
  constants of fluorocarbon refrigerants **XI.9**, 440–1
  constants of inorganic compounds **XI.11**, 402–7; **XI.12**, 408
  constants of organic compounds **XI.11**, 402–7; **XI.12**, 409
  magnetic fields of superconductors **V.23**, 86; **V.24**, 86
  micelle concentration **XI.22**, 418
  solution temperature **XIII.35**, 499
  temperature of superconductors **V.23**, 86; **V.24**, 86
Cross-section for
  electron collisions **IX.8**, 173
  fusion reactions **IX.56**, 267
  ionization by electrons **IX.8**, 173
  neutron foil detectors **IX.41**, 263; **IX.42**, 263
Cryohydrates **XIII.37**, 501
Cryoscopic constants of solvents **XIII.29**, 494
Crystal structure of elements **III.5**, 42–3

Crystal systems **IV.15**, 65
Cubical expansion, coefficient of
  gases **XII.1**, 431
  liquids **XII.2**, 431
Curie temperature of
  core materials **V.19**, 82
  low coercivity magnetic substances **V.18**, 80–1
  permanent magnet materials **V.20**, 83
Current, electric **I.1**, 3

Debye **IV.7**, 58
Debye characteristic temperatures **XIII.26**, 493
Debye–Hückel–Onsager coefficients **VI.7**, 93
Decay constant of solid scintillators **VII.40**, 142
Decay scheme of nuclides **IX.39**, 196–262
Decimal multiples and sub-multiples of units **I.5**, 6
Declination, magnetic, *see* magnetic declination
Definite integrals **XVII.6**, 544
Density of
  air **II.2**, 13; **II.3**, 14
  amino acids **XI.4**, 366–75
  aqueous ethanol solutions **XI.25**, 420
  aqueous solutions of
    acids **XI.26**, 420
    alcohols **XI.26**, 420
    bases **XI.26**, 420
    electrolytes **XI.26**, 421
    sugars **XI.26**, 421
  atmosphere **XIV.10**, 508
  Earth **XIV.5**, 507
  elements **III.4**, 40–1
  everyday substances **II.15**, 19
  fluorocarbon refrigerants **XI.9**, 400–1
  heavy water, temperature variation of **II.13**, 17
  inorganic compounds **XI.1**, 285–306
  mercury, temperature variation of **II.14**, 18
  Moon **XIV.4**, 506
  organic compounds **XI.2**, 307–59
  organometallic compounds **XI.3**, 360–5
  planets **XIV.5**, 507
  plastics and polymers **XI.8**, 398–401
  sea water **XIV.15**, 510; **XIV.16**, 510
  Sun **XIV.5**, 507
  water, temperature variation of **II.12**, 17
Density unit **I.3**, 4
Depression of freezing point **XIII.29**, 494
Derived units **I.2**, 3
Diamagnetic materials, magnetic susceptibilities **V.22**, 84–5
Diatomic molecules
  bond length of **IV.2**, 53; **VIII.15**, 156
  bond length of elements **IV.3**, 54
  dissociation energy of **VIII.14**, 155; **VIII.15**, 156
  dissociation enthalpy of **XIII.14**, 484; **XIII.15**, 484
  force constant of **VIII.15**, 156
  fundamental vibrations of **VIII.15**, 156
  moment of inertia of **VIII.15**, 156
  rotational constant of **VIII.15**, 156
Diatomic scales **X.28**, 282

Dielectric constant, *see* permittivity
Diffusion coefficient of
   aqueous solutions of electrolytes **VI.18**, 98
   aqueous solutions of proteins **XI.23**, 419
Diffusion of
   electrons in gases **IX.7**, 173
   neutrons **IX.55**, 266
Dimensions of physical quantities **I.4**, 5–6
Dioxane-water mixtures
   activity coefficients of HCl in **VI.12**, 95
   cation transport numbers of HCl in **VI.16**, 97
   density of **XI.26**, 420
Dipole moments, *see* electric dipole moments
Disaccharides, physical properties of **XI.5**, 384–7
Dissociation constants of
   inorganic acids **VI.33**, 110–11
   inorganic bases **VI.35**, 113
   organic acids **VI.34**, 111–12
   organic bases **VI.36**, 113–14
Dissociation energy of
   diatomic molecules **VIII.14**, 155; **VIII.15**, 156
   gases **VIII.14**, 155
   polyatomic molecules **XIII.10**, 482
Dose unit **I.2**, 3
Down quark **IX.61**, 268
Drift speeds of electrons **IX.2**, 172
Dyne **I.6F**, 8

Earth
   atmosphere
      composition of **III.12**, 48
      properties of **XIV.10**, 508; **XIV.11**, 508
   gravitation field of **XIV.8**, 507
   magnetic field of **XIV.13**, 509
   orbit of **XIV.5**, 507
   physical properties of **XIV.7**, 507
   rotation period of **XIV.5**, 507
Ebullioscopic constants of solvents **XIII.29**, 494
Eccentricity of orbits of
   planets **XIV.4**, 506; **XIV.5**, 507
   solar system asteroids **XIV.3**, 505
Elastic moduli, *see* bulk, shear, Young's
Elastic scattering **IX.43**, 264; **IX.44**, 264; **IX.45**, 264; **IX.46**, 264; **IX.47**, 264; **IX.48**, 265; **IX.49**, 265; **IX.50**, 265; **IX.51**, 265; **IX.52**, 266; **IX.53**, 266
Electric
   conductivity of sea water **XIV.14**, 510
   current density unit **I.3**, 4
   field strength **I.3**, 4
   current for constant temperature rise in wires **V.7**, 72
   current unit **I.1**, 3
   reactance unit **I.3**, 4
   resistance unit **I.4**, 5
   resistivity unit **I.3**, 4; **I.4**, 5
   units **I.4**, 5
Electric dipole moment of
   inorganic compounds **IV.8**, 59; **IV.9**, 59
   organic compounds **IV.7**, 58, **IV.9**, 59
Electrode potentials
   formal **VI.28**, 109
   standard for biological half cells **VI.29**, 109

Electrode potentials (*Contd*)
   standard, in aqueous solution **VI.27**, 107–8
   standard, in non-aqueous solution **VI.30**, 109
Electrode
   glass, standard solutions for calibration **VI.21**, 100–1
   standard **VI.27**, 107–8
Electrolytes
   activity coefficients of **VI.10**, 94; **VI.11**, 95; **VI.12**, 95
   molar conductivities of **VI.2**, 89–90
Electromagnetic spectrum **VIII.1**, 145
Electron
   affinity of elements **III.8**, 46
   affinity of diatomic molecules **IV.10**, 60
   charge on **I.7**, 10
   collisions of **IX.8**, 173
   Compton wavelength of **I.7**, 10
   diffusion **IX.7**, 173
   ionization **IX.8**, 173
   magnetic moment of **I.7**, 10
   mean free path of **IX.20**, 189
   -neutrino **IX.60**, 268
   radius of, *see* electronic radius
   range of, in aluminium **IX.21**, 189
   rest mass of, *see* rest mass of electron
   sources of **IX.39**, 196–262
   volt **I.6H**, 9
Electronic
   configuration of elements **III.3**, 38–40
   drift speeds **IX.2**, 172
   radius **I.7**, 10
Electronegativities of elements **III.10**, 47
Electro-optic substances
   half-wave voltage of **VII.39**, 141
   refractive index of **VII.39**, 141
   transmission range of **VII.39**, 141
Elements, *see also* liquid elements
   abundance of, in
      Earth's crust **III.11**, 48
      human body **III.12**, 48
      sea water **III.11**, 48
   atomic mass of **III.2**, 35–7
   atomic number of **III.2**, 25–37
   atomic radius of **III.3**, 38–40
   Auger lines of **IX.10**, 175
   boiling point of **III.4**, 40–1
   bond lengths in **IV.3**, 54
   characteristic X-ray lines of **IX.15**, 182–3; **IX.16**, 184–5; **IX.17**, 186–7; **IX.18**, 187
   CODATA key thermodynamic values of **XIII.1**, 449–50; **XIII.3**, 451–65; **XIII.22**, 489–90
   coefficient of linear expansion of **XII.3**, 432
   conductivity of **III.4**, 40–1
   critical constants of **XI.12**, 408
   crystal structure of **III.5**, 42–3
   density of **III.4**, 40–1
   elastic properties of **II.16**, 20; **II.22**, 22; **II.23**, 23
   electronegativity of **III.10**, 47
   electronic configuration of **III.3**, 38–40
   enthalphy of
      combustion of **XIII.6**, 477
      atomization of **XIII.1**, 449–50; **XIII.9**, 481

Elements (*Contd*)
   enthalphy of
      fusion of **XIII.1**, 449–50; **XIII.7**, 478
      vaporization of **XIII.1**, 449–50; **XIII.7**, 478
   entropy of **XIII.1**, 449–50, **XIII.3**, 466–9
   ferromagnetic properties of **V.18**, 80–1
   ionic radius of **III.5**, 42–3
   ionization energy of **III.7**, 44–5
   linear expansion coefficient of **XII.3**, 432
   magnetic susceptibility of **V.22**, 84–5
   melting point of **III.4**, 40–1
   molar heat capacity of **XIII.1**, 449–50; **XIII.21**, 488
   oxidation states of **III.4**, 40–1
   periodic table of **III.1**, 35
   photoelectron kinetic energies of **IX.11**, 176–7; **IX.12**, 178; **IX.13**, 179–80; **IX.14**, 181
   properties of **III.4**, 40–1
   resistivity of solid metallic **V.3**, 70–1
   second virial coefficient of **XI.14**, 411
   specific heat capacity of **III.4**, 40–1; **XII.16**, 440;
   superconducting **V.18**, 80–1
   thermal conductivity of **XII.9**, 434–5
   thermal electromotive force of **XII.21**, 444
   thermodynamic values, CODATA key, **XIII.2**, 451–65; **XIII.3**, 466–9; **XIII.22**, 489–90
   vapour pressure of **XI.27**, 422–3
   work function of **IX.9**, 174
   X-ray lines of **IX.15**, 182–3; **IX.16**, 184–5; **IX.17**, 186–7; **IX.18**,187
   X-ray mass absorption coefficient of **IX.19**, 188
Emissivity
   correction for spectral pyrometer **XII.19**, 443
   of various materials **XII.20**, 443
Energy
   of activation of
      bimolecular reactions **XV.4**, 516
      catalysed reactions **XV.5**, 517
      thermal unimolecular reactions **XV.4**, 516
   of Auger electrons **IX.10**, 175
   of ionization of elements **III.7**, 44–5
   of radioactive sources **IX.39**, 196–262
   released in
      nuclear fission **IX.40**, 263
      nuclear fusion **IX.56**, 267
Engineering data, screw threads **XVII.1**, 535; **XVII.2**, 535
Enthalphy of
   atomization **XIII.1**, 449–50; **XIII.9**, 481
   bond dissociation **XIII.14**, 484; **XIII.15**, 484; **XIII.16**, 485
   formation (standard) of
      biological molecules **XIII.5**, 474–6
      free radicals **XIII.10**, 482
      gaseous atoms **XIII.9**, 481
      inorganic compounds **XIII.2**, 451–65
      organic compounds **XIII.4**, 470–3
   fusion of
      elements **XIII.1**, 449–50
      inorganic compounds **XIII.7**, 478
      organic compounds **XIII.8**, 479–80

Index 557

Enthalphy of (Contd)
  hydration **XIII.17**, 486
  ion association reactions **XIII.21**, 488
  neutralization **XIII.12**, 483
  solution **XIII.11**, 482; **XIII.13**, 483
  vaporization of
    elements **XIII.1**, 449–50
    fluorocarbon refrigerants **XI.9**, 400–1
    inorganic compounds **XIII.7**, 478
    organic compounds **XIII.8**, 479–80
Entropy of activation of
  bimolecular reactions **XV.4**, 516
  ionic reactions **XV.4**, 516
  unimolecular reactions **XV.4**, 516
Entropy (standard) of
  biological molecules **XIII.5**, 474–6
  elements **XIII.1**, 449–50
  inorganic compounds **XIII.2**, 451–65
  organic compounds **XIII.4**, 470–3
Entropy change of ion association reactions **XIII.21**, 488
Equally tempered musical scale **X.29**, 282
Equatorial diameter of Earth, Sun, Moon and planets **XIV.4**, 506; **XIV.5**, 507
Equilibrium constants of
  ion association reactions **XIII.21**, 488
  some gaseous reactions **XIII.38**, 501
Erg **I.6H**, 9
Error function **XVII. 9**, 549–50
Ethanol
  density of **XI.2**, 326–7
  density of aqueous solutions of **XI.25**, 420
  surface tension of **II.33**, 29
  surface tension of aqueous solutions of **II.36**, 30
  viscosity of **II.31**, 27
Eutectic mixtures (binary) **XIII.36**, 500
Expansion coefficient
  cubical
    of gases **XII.1**, 431
    of liquids **XII.2**, 431
  linear
    of miscellaneous substances **XII.5**, 432
    of plastics **XI.8**, 398–401
    of solid alloys and compounds **XII.4**, 432
    of solid elements **XII.3**, 432
Expansion, thermal, see thermal expansion
Exponential series **XVII.8**, 547
Exposure limits of
  inorganic compounds **XI.1**, 285–306; **XVI.2**, 527–9; **XVI.3**, 530
  organic compounds **XI.2**, 307–59; **XVI.1**, 521–6
  pesticides **XVI.4**, 531–2

Farad **I.2**, 3
Faraday constant **I.7**, 10
Ferrimagnetic substances, see ferrites
Ferrite cores **V.19**, 82
Ferrites **V.19**, 82
Fine structure constant **I.7**, 10
Fission
  energy release in **IX.40**, 263
  neutron spectrum of **IX.40**, 263
First order reactions
  Arrhenius parameters of **XV.2**, 514; **XV.4**, 516
  rate coefficients of **XV.1**, 513

Fixed points of International Temperature Scale of 1990 **XII.24**, 445
Flammability limits of bottled gases **XI.11**, 402–7
Flash point of organic compounds **XI.2**, 307–59; **XI.3**, 360–5
Fluorine
  atomic mass of **III.2**, 36, 37
  CODATA key thermodynamic values **XIII.1**, 449, **XIII.2**, 455; **XIII.3**, 467
  critical constants of **XI.12**, 408
  isotopes of **IX.39**, 213–14
  molar heat capacity of **III.4**, 40; **XIII.3**, 467; **XIII.22**, 489
  physical properties of **III.4**, 40; **XI.1**, 292
  rotational constants of **VIII.15**, 156
  speed of sound in **X.1**, 273
  thermal conductivity of **XII.6**, 433
  X-ray energies from **IX.11**, 176; **IX.13**, 179; **IX.15**, 182; **IX.16**, 184
Fluorocarbon refrigerants, properties of **XI.9**, 400–1
Fluorescent quantum yield references **VIII.29**, 167
Focus, visual depth of **VII.10**, 132
Foodstuffs, pH of **VI.25**, 106
Foot **I.6A**, 7
Force, units **I.4**, 5; **I.6E**, 8
Force constants of diatomic molecules **VIII.15**, 156
Formal electrode potentials **VI.28**, 109
Formation
  standard enthalpy of, for
    biologial molecules **XIII.5**, 474–6
    free radicals **XIII.10**. 482
    gaseous atoms **XIII.9**, 481
    inorganic compounds **XIII.2**, 451–65
    organic compounds **XIII.4**, 470–3
  standard Gibbs energy of, for
    biological molecules **XIII.5**, 474–6
    inorganic compounds **XIII.2**, 451–65
    organic compounds **XIII.4**, 470–3
Formation constants for
  EDTA complexes **VI.41**, 117
  inorganic and organic ligands **VI.40**, 116–17
Formula of
  amino acids **XI.3**, 360–5
  carbohydrates **XI.4**, 366–75
  inorganic compounds **XI.1**, 285–306
  nucleotides **XI.7**, 395–6
  organic compounds **XI.2**, 307–59
  purines and pyrimidines **XI.8**, 398–401
Fourier's series **XVII.8**, 546
Free energy of formation, see Gibbs energy
Free energy function $[G^{\ominus}-H^{\ominus}]/T$ **XIII.18**, 486
Free path of molecules **II.34**, 30
Free radicals, enthalpy of formation of **XIII.10**, 482
Free space, permittivity and permeability of **I.7**, 10
Freezing point depression **XIII.29**, 494
Freons, see fluorocarbon refrigerants
Frequencies
  for acoustic measurements **X.13**, 278
  in musical scales **X.27**, 282; **X.28**, 282
Frequency unit **I.2**, 3; **I.4**, 6

Fuels, calorific values of **XIII.33**, 497
Fundamental constants **I.7**, 10
Fundamental vibrations of
  diatomic molecules **VIII.15**, 156
  triatomic molecules **VIII.16**, 156–7
  polyatomic molecules **VIII.19**, 158
Fused salts
  cation transport numbers of **VI.6**, 93
  equivalent conductivity of **VI.6**, 92
Fusion
  enthalpy of
    elements **XIII.1**, 449–50
    inorganic compounds **XIII.7**, 478
    organic compounds **XIII.8**, 479–80
  neutrons **IX.57**, 267
  nuclear **IX.56**, 267
  reactor materials **IX.58**, 267

G, see gravitational constant
g, see gravity
Galactic distances **XIV.1**, 505
Gallon **I.6C**, 7
Gamma ray sources **IX.39**, 196–262
Gas constant **I.7**, 10
Gases
  bottled, properties of **XI.11**, 402–7
  Boyle temperature of **XI.13**, 410
  coefficient of cubical expansion of **XII.1**, 431
  critical constants of **XI.12**, 408–9
  ionic mobility in **IX.1**, 171
  mean molecular free path of **II.34**, 30
  mean molecular speed of **II.34**, 30
  molar heat capacity of **XIII.23**, 490; **XIII.24**, 491; **XIII.25**, 491–2
  molecular parameters of **VIII.14**, 155
  ratio of specific heats of **XII.14**, 439
  refractive index of **VII.17**, 134
  relative permittivity of **V.12**, 74
  second virial coefficient of **XI.14**, 411
  solubility of, in
    sea water **XI.16**, 413
    water **XI.15**, 412
  speed of sound in **X.1**, 273
  thermal conductivity of **XII.6**, 433
  van der Waals constants of **XI.13**, 410
  Verdet constants of **VII.35**, 140
  viscosity of **II.25**, 24–5; **II.26**, 25
Gauge bosons **IX.67**, 270
Gauss **I.4**, 5
Geological time scale **XIV.9**, 508
Geomagnetic data **XIV.13**, 509
Geometrical optical properties of eye **VII.11**, 132; **VII.12**, 132
Gibbs energy of
  formation (standard) of
    biological molecules **XIII.5**, 474–6
    inorganic compounds **XIII.2**, 451–65
    organic compounds **XIII.4**, 470–3
  hydrolysis of organic phosphates **XIII.27**, 493
  transfer of ions, water to alcohols **XIII.32**, 496
Glass
  coefficient of linear expansion of **XII.5**, 432
  elasticity of **II.18**, 21
  heat capacity, specific of **XII.18**, 442
  optical
    light transmission of **VII.21**, 136

Glass (*Contd*)
optical
refractive index of **VII.20**, 135
relative permittivity and loss tangent of **V.16**, 77
resistivity of **V.11**, 73
sound reduction index of **X.24**, 281
speed of sound in **X.9**, 276
tensile strength of **II.18**, 21
thermal conductivity of **XII.13**, 438
Verdet constants of **VII.38**, 141
viscosity of **II.30**, 26
Gluon **IX.67**, 270
Glycerol-water solutions
density of **XI.26**, 420
viscosity of **II.28**, 25
Grain **I.6E**, 8
Gramme **I.6E**, 8
Gravitational constant **I.7**, 10
Gravity, acceleration due to,
on surface of the Earth **I.7**, 10
on surface of Moon **XIV.6**, 507
on surface of Sun and planets **XIV.5**, 507
Gray **I.2**, 3
Group theory, tables for **IV.13**, 61–3
Gullstrand–Emsley schematic eye **VII.12**, 132
chromatic aberration of **VII.13**, 133

Hadrons
charmed **IX.64**, 269
non-strange **IX.62**, 269
strange **IX.63**, 269
Half-cells, standard electrode potential of **VI.27**, 107–8; **VI.28**, 109
Half-life of nuclides **IX.39**, 196–262
Half-wave potentials of
metals and other inorganic species **VI.43**, 118–21
organic compounds **VI.44**, 112–23
Halogens, *see* bromine, chlorine, fluorine, iodine
Hammett
acidity functions **VI.42**, 117
reaction constants **XV.7**, 518
substituent constants **XV.6**, 517
Hardness of materials, Mohs scale **XI.21**, 417
Harmonic oscillator, thermodynamic functions of **XIII.19**, 487
Hartree energy **I.7**, 10
Health and Safety
exposure limits (recommended) for
inorganic compounds **XVI.2**, 527–9
organic compounds **XVI.1**, 521–6
pesticides **XVI.4**, 531–2
radiation risks **IX.38**, 195
toxicity data **XVI.1**, 521–6; **XVI.3**, 530; **XVI.4**, 531–2
Hearing impairment classification **X.16**, 278
Heat capacity
molar, *see* molar heat capacity
specific, *see* specific heat capacity
unit **I.3**, 4; **I.4**, 6
Heat of formation, fusion, vaporization, *see* enthalpy
Heat, quantity, unit of **I.2**, 3
Heavy water
density, temperature variation of **II.13**, 17

Heavy water (*Contd*)
moments of inertia **VIII.16**, 157
thermodynamic constants for phase changes of **XIII.28**, 493
Hectare **I.6B**, 7
Helium
atomic mass of **III.2**, 36, 37
CODATA key thermodynamic values **XIII.1**, 449; **XIII.2**, 456; **XIII.3**, 466
critical constants of **XI.12**, 408
molar heat capacity of **XIII.1**, 449; **XIII.21**, 488
molecular mean speed etc of **II.34**, 30
properties of **III.4**, 40; **XI.1**, 292
solubility of, in water **XI.15**, 412
speed of sound in **X.1**, 273
vapour pressure of **XI.27**, 422
viscosity coefficient of **II.25**, 24
Henry **I.2**, 3
Henry's law constants **XIII.30**, 495
Hertz **I.2**, 3
Hexoses, physical properties of **XI.5**, 376–91
High pressure
viscosity of gases at **II.26**, 25
viscosity of liquids at **II.32**, 29; **II.33**, 29
Hue discrimination **VII.6**, 131
Human body, abundance of elements in **III.13**, 49
Human fluids, pH of **VI.25**, 106
Humidity
determination of, *see* chemical hygrometer
relative, above water **II.10**, 16
relative, above saturated solutions **II.11**, 16
Hybrid orbitals **IV.1**, 53
Hydration, enthalpy of **XIII.17**, 486
Hydrochloric acid, *see* hydrogen chloride
Hydrogen
atomic mass of **III.2**, 36, 37
chemical shift **VIII.25**, 163
CODATA key thermodynamic values **XIII.1**, 449; **XIII.2**, 456; **XIII.3**, 466
critical constants of **XI.12**, 408
dissociation energy of **VIII.15**, 156
isotopes of **IX.39**, 217
molar heat capacity of **III.4**, 40; **XIII.3**, 466; **XIII.22**, 489; **XIII.25**, 491
molecular mean speed etc of **II.34**, 30
moment of inertia of **VIII.15**, 156
physical properties of **III.4**, 40; **XI.1**, 293
rotational constant of **VIII.15**, 156
second virial coefficient of **XI.14**, 411
solubility of,
in sea water **XI.16**, 413
in water **XI.15**, 412
thermal conductivity of **XII.6**, 433
vapour pressure of **XI.27**, 422
viscosity coefficient of **II.25**, 24; **II.26**, 25
X-ray energies from **IX.11**, 176; **IX.13**, 179
Hydrogen chloride
activity coefficients in
dioxane-water **VI.12**, 95
water **VI.10**, 94; **VI.11**, 95
boiling point of azeotrope **XIII.34**, 498
cation transport numbers in dioxane-water **VI.16**, 97

Hydrogen chloride (*Contd*)
concentration of laboratory reagent **XI.29**, 427
density of aqueous solutions **XI.26**, 420
dissociation energy of **VIII.15**, 156
moment of inertia of **VIII.15**, 156
physical properties of **XI.1**, 293
rotational constant of **VIII.15**, 156
solubility of, in water **XI.15**, 412
vapour pressure of **IX.27**, 422
viscosity coefficient of **II.25**, 24
Hyperbolic
functions, indefinite integrals **XVII.5**, 542–3
series **XVII.8**, 548
Hyperons **IX.63**, 269

Illuminance, unit **I.2**, 3
Illuminants, standard for colorimetry **VII.4**, 129
Inch **I.6A**, 7
Indicators
acid-base **VI.26**, 107
biological redox **VI.32**, 110
volumetric redox **VI.31**, 110
Inductance unit **I.2**, 3; **I.4**, 5
Inelastic scattering **IX.48**, 265; **IX.33**, 193
Inertia, moments of, for
asymmetric molecules **VIII.18**, 156
diatomic molecules **VIII.15**, 156
symmetric top molecules **VIII.17**, 157
triatomic molecules **VIII.16**, 156–7
Infrared radiation
calibration standards **VIII.9**, 152
cut-off wavelengths for solvents **VIII.7**, 150–1
transmission regions for materials **VII.26**, 138; **VIII.8**, 152
Inorganic compounds
boiling point of **XI.1**, 285–306
critical constants of **XI.12**, 408–9
density of **XI.1**, 285–306
electric dipole moment of **IV.7**, 58
enthalpy (standard) of
formation of **XIII.2**, 451–65
fusion of **XIII.7**, 478
vaporization of **XIII.7**, 478
entropy (standard) of **XIII.2**, 451–65
formula of **XI.1**, 285–306
free energy, standard, of **XIII.3**, 466–9
hazards of **XI.1**, 285–306; **XVI.2**, 527–9; **XVI.3**, 530
LD_{50} of **XVI.3**, 530
magnetic susceptibility of **V.22**, 84–5
melting point of **XI.1**, 285–306
molar heat capacity of **XIII.2**, 451–65; **XIII.24**, 491; **XIII.25**, 491–2
occupational exposure limits of **XVI.2**, 527–9
refractive index of **XI.1**, 285–306
relative permittivity of **V.12**, 74; **V.14**, 75; **V.16**, 77–8
second virial coefficient of **XI.14**, 411
solubility of **XI.1**, 285–306; **XI.24**, 419
toxicity of **XVI.3**, 530
vapour pressure of **XI.27**, 422–3
Insulators, resistivity of **V.9**, 73; **V.10**, 73; **V.11**, 73
Integrals
definite **XVII.6**, 544

Integrals (*Contd*)
  indefinite **XVII.5**, 538–43
Integration, approximate **XVII.7**, 545
Interatomic distances **III.5**, 42–3
Interfacial tension **II.39**, 32
International Electric Units **V**, 69
International Organization for Standardization,
  screw thread specification **XVII.1**, 535
International Practical Temperature Scale of 1990
  definition of **XII.25**, 446
  fixed points of **XII.24**, 445
International System Units, *see* SI
Intrinsic viscosity of proteins **XI.23**, 419
Iodine
  atomic mass of **III.2**, 36, 37
  CODATA key thermodynamic values **XIII.1**, 450; **XIII.2**, 456; **XIII.3**, 468
  coefficient of viscosity of **II.25**, 24
  dissociation energy of **VIII.15**, 156
  isotopes of **IX.39**, 220–1
  molar heat capacity of **III.4**, 40; **XIII.3**, 468; **XIII.22**, 489
  moment of inertia of **VIII.15**, 156
  physical properties of **III.4**, 40; **XI.1**, 293
  rotational constants of **VIII.15**, 156
  X-ray energies from **IX.11**, 176; **IX.13**, 179
Ionic
  conductivity, in aqueous solution **VI.3**, 90–1; **VI.4**, 91
  conductivity, in non-aqueous solution **VI.5**, 92
  mobility of positive gaseous ions **IX.1**, 171
  radius of elements **III.5**, 42–3
  recombination coefficient **IX.3**, 172; **IX.4**, 172; **IX.5**, 173; **IX.6**, 173
  strength **VI.8**, 93
Ionization
  constants for water and deuterium oxide **VI.37**, 114
  energies of elements **III.7**, 44–5
  potentials of compounds **XI.20**, 416
Ions
  activity coefficients of (calculated) **VI.13**, 96
  limiting molar conductivities in aqueous solution **VI.3**, 90–1; **VI.4**, 91
  limiting molar conductivities in non-aqueous solution **VI.5**, 92
  transport numbers of **VI.14**, 97; **VI.15**, 97; **VI.17**, 98
Ions in gases **IX.1**, 171
Irradiance unit **I.4**, 6
Irrational algebraic functions, indefinite integrals of **XVII.5**, 541–2
  isotopes, table of **IX.39**, 196–262
Joule **I.2**, 3
Joule–Thompson coefficient **XIII.31**, 496

Kelvin **I.1**, 3
Kerma-rate constant **IX.37**, 195
Key (CODATA) values for thermodynamics **XIII.2**, 451–65; **XIII.3**, 466–9; **XIII.22**, 489–90
Kilogram **I.1**, 3

Kinetic data for
  first order reactions **XV.1**, 513
  second order reactions **XV.1**, 513
Kinetic energy of photoelectrons **IX.11**, 176–7; **IX.12**, 178; **IX.13**, 179–80; **IX.14**, 181
K meson **IX.63**, 269
Knot **I.6D**, 7
Krypton
  atomic mass of **III.2**, 36, 37
  CODATA key thermodynamic values **XIII.1**, 449; **XIII.2**, 457; **XIII.22**, 489
  critical constants of **XI.12**, 408
  molar heat capacity of **XIII.1**, 449; **XIII.25**, 491
  properties of **III.4**, 41
  solubility of, in water **XI.15**, 412
  viscosity coefficient of **II.25**, 24

Laboratory reagents, preparation and properties of **XI.29**, 427
Land and water on Earth's surface **XIV.7**, 507
Lande g factor for free electron **I.7**, 10
Lattice enthalpies of metal halides **XIII.20**, 488
Lasers, properties of **VII.41**, 142
LD$_{50}$ of
  inorganic compounds **XVI.3**, 530
  organic compounds **XVI.1**, 521–6
  pesticides **XVI.4**, 531–2
Length, units of **I.1**, 3; **I.4**, 6; **I.6A**, 7
Lennard–Jones (12,6) potentials **IV.11**, 60
Life-time of
  sub-atomic particles **IX.60**, 268
  thermal neutrons **IX.55**, 266
Light, speed of **I.7**, 10
Light transmission of glass **VII.21**, 136
Linear expansion, coefficient of
  miscellaneous solid substances **XII.5**, 432
  plastics **XI.8**, 398–401
  solid alloys and compounds **XII.4**, 432
  solid elements **XII.3**, 432
Liquid elements
  resistivity of **V.1**, 69; **V.2**, 69
  surface tension of **II.37**, 31
  viscosity coefficient of **II.30**, 26
Liquid refractometers, calibration of **VII.18**, 134
Liquids, *see also* inorganic compounds, organic compounds
  bulk modulus of **II.23**, 23
  coefficient of cubical expansion of **XII.2**, 431
  coefficient of viscosity of **II.31**, 27–8; **II.33**, 29
  interfacial tension of **II.39**, 32
  refractive index of **VII.18**, 134; **VII.19**, 135; **XI.1**, 285–306; **XI.2**, 307–59
  relative permittivity of **V.13**, 74; **V.14**, 75; **V.15**, 76
  speed of sound in **X.7**, 275
  surface tension of **II.38**, 32
  thermal conductivity of **XII.7**, 433; **XII.8**, 433
  vapour pressure of **XI.27**, 422–3; **XI.28**, 424–7
  Verdet constant of **VII.36**, 140; **VII.37**, 141

Litre **I.6C**, 7
Logarithmic series **XVII.8**, 547
Loschmidt's number **I.7**, 10
Loss factor of magnetic materials **V.19**, 82
Loss tangent **V.13**, 74; **V.16**, 77–8; **V.17**, 78–9
Loudness **X.18**, 279; **X.19**, 279; **X.20**, 279
Luminous efficiency of eye
  photopic **VII.2**, 128
  scotopic **VII.3**, 129
Luminous flux **I.2**, 3; **I.4**, 6
Lux **I.2**, 3

Maclaurin's series **XVII.8**, 546
Madelung constants **IV.12**, 60
Magnetic
  declination **XIV.14**, 510
  field strength unit **I.3**, 4; **I.4**, 5
  flux density unit **I.4**, 5
  flux unit **I.2**, 3
  moment of electron, *see* electron magnetic moment
  moment of protons in water **I.7**, 10
  permeability **V.18**, 80–1; **V.19**, 82; **V.21**, 84
  polarization **V.18**, 80–1
  resonant frequencies **VIII.21**, 159–60
  rotation of polarized light, *see* Verdet constants
  susceptibility **V.22**, 84–5
  units **I.4**, 5
Magnetomotive force unit, *see* ampere
Magneton
  Bohr **I.7**, 10
  nuclear **I.7**, 10
Mass
  excess of nuclides **IX.39**, 196–262
  number of nuclides **IX.39**, 196–262
  of electron **I.7**, 10
  of neutron and proton **I.7**, 10; **IX.62**, 269
  unit, atomic **I.7**, 10
  units **I.1**, 3; **I.4**, 5; **I.6E**, 8
Mass-composition of fragment ions **VIII.27**, 164–6
Mass spectrometry
  common losses from molecular ions **VIII.28**, 166
  mass-composition of fragment ions **VIII.27**, 164–5
Mathematical functions **XVII.5**, 538–43; **XVII.6**, 544; **XVII.7**, 545; **XVII.8**, 546–8; **XVII.9**, 549–50
Mean free path of
  electrons in solids **IX.20**, 189
  molecules in gases **II.34**, 30
Melting point of
  amino acids **XI.4**, 366–75
  carbohydrates **XI.5**, 376–91
  elements **III.4**, 40–1
  fluorocarbon refrigerants **XI.9**, 400–1
  inorganic compounds **XI.1**, 285–306; **XI.11**, 402–7
  metals, pressure dependence **II.8**, 15
  nucleic acids **XI.7**, 395–6
  organic compounds **XI.2**, 307–59; **XI.11**, 402–7
  organometallic compounds **XI.3**, 360–5
  steroids **XI.6**, 392–4

Mercury
  density of, temperature variation of
    II.14, 18
  physical properties of **III.4**, 41; **XI.1**,
    296
  vapour pressure of **XI.18**, 415
  viscosity coefficient of **II.30**, 26
Mesons **IX.62**, 269; **IX.63**, 269; **IX.64**, 269
  range of **IX.24**, 190
  stopping power **IX.35**, 194
Mesopic region **VII.3**, 129
Metallic radii **III.5**, 42–3
Metals
  absorption index of **VII.23**, 136
  boiling point of **III.4**, 40–1; **XI.1**,
    285–306
  coefficient of linear expansion of **XII.3**,
    432
  conductivity of **III.4**, 40–1
  elasticities of **II.16**, 20
  electrical resistivity of **III.4**, 40–1; **V.1**,
    69; **V.2**, 69; **V.3**, 70–1
  half-wave potential of **VI.43**, 118–21
  hardness of **XI.21**, 417
  melting point of **III.4**, 40–1; **XI.1**,
    285–306
  molar heat capacity of **XIII.1**, 449–50
  optical constants and reflectance of
    VII.23, 136
  pressure-dependence of melting
    temperatures **II.8**, 15
  refractive index of **VII.23**, 136
  resistivity of **V.1**, 69; **V.2**, 69; **V.3**, 70–1
  specific heat capacity of **III.4**, 40–1;
    XII.16, 440
  tensile strength of **II.16**, 20
  thermal conductivity of **XII.9**, 434–5;
    XII.10, 435
Metric threads **XVII.1**, 535; **XVII.2**, 535
Micelle, critical concentration of **XI.22**,
  418
Mile
  nautical **I.6A**, 7
  statute **I.6A**, 7
Minerals
  dielectric properties of **V.16**, 77
  hardness of, Mohs scale **XI.21**, 417
  viscosity of **II.30**, 26
Minimum audible field **X.15**, 278
Mobility
  electronic **IX.2**, 172
  gaseous ionic **IX.1**, 171
Moderators and shielding materials in nuclear
  reactors **IX.55**, 266
Mohs scale of hardness **XI.21**, 417
Molality, mean ionic **VI.9**, 93
Molar
  conductivity of electrolytes **VI.2**, 89–90
  critical volume **XI.12**, 408–9
  cryoscopic constants **XIII.29**, 494
  ebullioscopic constants **XIII.29**, 494
  entropy unit **I.4**, 6
  extinction coefficient **I.4**, 6; **VIII.13**, 154
  heat capacity, empirical equations
    XIII.25, 491–2
    Einstein function for **XIII.25**, 492
  heat capacity of
    biological molecules **XIII.5**, 474–6
    elements **XIII.1**, 449–50; **XIII.22**,
      489–90

Molar (Contd)
  heat capacity of
    fluorocarbon refrigerants **XI.9**, 400–1
    inorganic compounds **XIII.2**, 451–65;
      XIII.23, 490; **XIII.25**, 491
    organic compounds **XIII.4**, 470–3;
      XIII.24, 491
  ionic conductivity in
    aqueous solution **VI.3**, 90–1; **VI.4**, 91
    non-aqueous solution **VI.5**, 92
  optical rotatory power **I.4**, 6
  refraction unit **I.4**, 6
  volume unit **I.4**, 6
  volume of ideal gas **I.7**, 10
Mole **I.1**, 3
Molecular ions, common losses **VIII.28**,
  166–7
Molecular mean speed etc **II.34**, 30
Molecular weight
  mean of atmosphere **XIV.11**, 508
Molten salts, see also fused salts
  coefficient of viscosity of **II.30**, 26
  surface tension of **II.37**, 31
Moments of inertia of
  asymmetric rotor molecules **VIII.18**, 157
  diatomic molecules **VIII.15**, 156
  symmetric top molecules **VIII.17**, 157
  triatomic molecules **VIII.16**, 156–7
Moon, constants of **XIV.4**, 506; **XIV.6**, 507
Muon-neutrino **IX.60**, 268
Muons
  range of **IX.25**, 191
  stopping power for **IX.36**, 194
Musical acoustics **X.27**, 282; **X.28**, 282;
  X.29, 282

Natural gas, calorific value for **XIII.33**, 497
Neon
  atomic mass of **III.2**, 36, 37
  CODATA key thermodynamic values
    XIII.1, 449; **XIII.2**, 459; **XIII.3**, 466
  critical constants of **XI.12**, 408
  molar heat capacity of **XIII.1**, 449;
    XIII.21, 488
  molecular mean speed etc **II.34**, 30
  properties of **III.4**, 41; **XI.1**, 296
  solubility of, in water **XI.15**, 412
  vapour pressure of **XI.27**, 423
  viscosity coefficient of **II.25**, 24
Neutrino **IX.60**, 268
Neutron
  foil detectors **IX.41**, 263; **IX.42**, 263
  rest mass of, see rest mass of neutron
Newton **I.2**, 3
Nickel-chromium alloys for resistors **V.6**, 72
Nitric acid, density of aqueous solutions
  XI.26, 420
Nitrogen-15, chemical shift **VIII.25**, 163
Nitrogen
  atomic mass of **III.2**, 36, 37
  chemical shift **VIII.25**, 163
  CODATA key thermodynamic values
    XIII.1, 449; **XIII.2**, 459; **XIII.3**, 469
  critical constants of **XI.12**, 408
  dissociation energy of **VIII.15**, 156
  isotopes of **IX.39**, 228–9
  molar heat capacity of **III.4**, 41; **XIII.3**,
    469; **XIII.22**, 489; **XIII.25**, 491
  molecular mean speed etc of **II.34**, 30
  moment of inertia of **VIII.15**, 156

Nitrogen (Contd)
  physical properties of **III.4**, 41; **XI.1**,
    297
  rotational constant of **VIII.15**, 156
  second virial coefficient of **XI.14**, 411
  solubility of,
    in sea water **XI.16**, 413
    in water **XI.15**, 412
  speed of sound in **X.1**, 273
  thermal conductivity of **XII.6**, 433
  vapour pressure of **XI.27**, 423
  viscosity coefficient of **II.25**, 24; **II.26**,
    25
  X-ray energies from **IX.11**, 176; **IX.13**,
    179; **IX.15**, 182; **IX.16**, 184
NMR
  carbon-13 chemical shift **VIII.24**, 162
  nitrogen-15 chemical shift **VIII.25**, 163
  oxygen-17 chemical shift **VIII.26**, 164
  phosphorus-31 chemical shift **VIII.26**,
    164
  proton chemical shift **VIII.23**, 161
  solvents, chemical shift **VIII.22**, 160
Nuclear
  fission, energy release **IX.40**, 263
  fusion **IX.56**, 267
  magneton **I.7**, 10; **VIII.21**, 159–60
  moments **VIII.21**, 159–60
  natural abundance **VIII.21**, 159–60
  quadrupole moments **VIII.21**, 159–60
  resonance frequencies **VIII.21**, 159–60
  spin quantum number **VIII.21**, 159–60
Nuclear magnetic resonance spectroscopy,
  see NMR
Nuclear reactor contents, behaviour **IX.43**,
  264 to **IX.55**, 266
Nucleic acids, properties of **XI.7**, 395–6
Nucleosides, see nucleic acids
Nucleotides, see nucleic acids
Nuclides
  atomic number of **IX.39**, 196–262
  cross-section of **IX.39**, 196–262
  decay scheme of **IX.39**, 196–262
  half-life of **IX.39**, 196–262
  mass excess of **IX.39**, 196–262
  mass number of **IX.39**, 196–262
  natural abundance of **IX.39**, 196–262
  parity of **IX.39**, 196–262
  quadrupole moment of **VIII.21**, 159–60
  receptivity of **VIII.21**, 159–60
  resonant frequency of **VIII.21**, 159–60
  spin of **IX.39**, 196–262
  spin quantum number of **VIII.21**, 159–
    60

Occupational exposure limits (recommended)
  for
  inorganic compounds **XVI.2**, 527–9
  organic compounds **XVI.1**, 521–6
  pesticides **XVI.4**, 531–2
Oceans, size of **XIV.7**, 507
Octave band sound analysis **X.14**, 278
Oersted **I.4**, 5
Ohm **I.4**, 5
Optical materials
  pyrometry **XII.19**, 443
  refractive index of **VII.20**, 135; **VII.24**,
    137; **VII.25**, 137; **VII.26**, 138
  rotation of **VII.34**, 140
  transmission of **VII.21**, 136; **VII.26**, 138

Orbits of the planets **XIV.4**, 506
Organic compounds
  boiling point of **XI.2**, 307–59
  chemical shift of
    carbon-13 **VIII.24**, 162
    nitrogen-15 **VIII.25**, 163
    proton **VIII.23**, 161
  critical constants of **XI.12**, 409
  density of **XI.2**, 307–59
  electric dipole moment of **IV.7**, 58
  enthalpy (standard) of
    combustion of **XIII.6**, 477
    formation of **XIII.4**, 470–3
    fusion of **XIII.8**, 479–80
    vaporization of **XIII.8**, 479–80
  entropy (standard) of **XIII.4**, 470–3
  flash point of **XI.2**, 307–59
  formula of **XI.2**, 307–59; **XI.3**, 360–5; **XI.4**, 366–75; **XI.5**, 376–91; **XI.6**, 392–4; **XI.7**, 395–6
  free energy, standard, of **XIII.4**, 470–3
  hazards of **XI.2**, 307–59
  LD$_{50}$ of **XVI.1**, 521–6; **XVI.4**, 531–2
  magnetic susceptibility of **V.22**, 84–5
  melting point of **XI.2**, 307–59
  molar heat capacity of **XIII.4**, 470–3; **XIII.24**, 491
  occupational exposure limits of **XVI.1**, 521–6; **XVI.4**, 531–2
  refractive index of **XI.2**, 307–59
  relative permittivity of **V.12**, 74; **V.13**, 74; **V.15**, 76; **V.17**, 78–9
  second virial coefficient of **XI.14**, 411
  solubility of **XI.2**, 307–59, **XI.24**, 419
  surface tension of **II.38**, 32
  toxicity of **XVI.1**, 521–6; **XVI.4**, 531–2
  vapour pressure of **XI.28**, 424–7
  viscosity coefficient of **II.31**, 27–8
Organometallic compounds
  boiling point of **XI.3**, 360–5
  density of **XI.3**, 360–5
  flash point of **XI.3**, 360–5
  formula of **XI.3**, 360–5
  hazards of **XI.3**, 360–5
  melting point of **XI.3**, 360–5
  refractive index of **XI.3**, 360–5
  solubility of **XI.3**, 360–5
Osmotic coefficients of
  electrolyte solutions **VI.19**, 99
  sucrose solutions **VI.20**, 99
Ounce **I.6E**, 8
Oxidation states of elements **III.4**, 40–1
Oxygen-17 chemical shift **VIII.26**, 164
Oxygen
  atomic mass of **III.2**, 36, 37
  chemical shift **VIII.26**, 164
  CODATA key thermodynamic values **XIII.1**, 449; **XIII.2**, 459; **XIII.3**, 466
  critical constants of **XI.12**, 408
  dissociation energy of **VIII.15**, 156
  isotopes of **IX.39**, 232
  molar heat capacity of **III.4**, 41; **XIII.3**, 466; **XIII.22**, 489; **XIII.25**, 491
  molecular mean speed etc of **II.34**, 30
  moment of inertia of **VIII.15**, 156
  physical properties of **III.4**, 41; **XI.1**, 297
  rotational constant of **VIII.15**, 156
  second virial coefficient of **XI.14**, 411

Oxygen (*Contd*)
  solubility of,
    in sea water **XI.16**, 413
    in water **XI.15**, 412
  speed of sound in **X.1**, 273
  thermal conductivity of **XII.6**, 433
  vapour pressure of **XI.27**, 423
  viscosity coefficient of **II.25**, 24; **II.26**, 25
  X-ray energies from **IX.11**, 176; **IX.13**, 179; **IX.15**, 183; **IX.16**, 185

Paper, dielectric properties of **V.17**, 78
Paramagnetic materials, susceptibilities of **V.22**, 84–5
Parity of nuclides **IX.39**, 196–262
Partially miscible binary liquids **XIII.35**, 599
Pascal **I.2**, 3; **I.4**, 5
Peat, calorific value of **XIII.33**, 497
Pentoses, physical properties of **XI.5**, 376–9
Periodic Table **III.1**, 35
Periods of the planets **XIV.4**, 506
Permanent magnet materials **V.20**, 83
Permeability
  of vacuum **I.7**, 10
  unit **I.3**, 4; **I.4**, 5
Permeability, relative of
  core materials **V.19**, 82
  feebly magnetic steels and cast irons **V.21**, 84
  low coercivity magnetic substances **V.18**, 80–1
Permittivity
  of vacuum **I.7**, 10
  unit **I.3**, 4; **I.4**, 5
Permittivity, relative, of
  fluorocarbon refrigerants **XI.9**, 400–1
  gases and vapours **V.12**, 74
  inorganic liquids **V.14**, 75
  inorganic solids **V.16**, 77–8
  liquids for electrical purposes **V.13**, 74
  organic liquids **V.15**, 76
  organic solids **V.17**, 78–9
  plastics and polymers **V.17**, 78–9; **XI.8**, 398–401
Pesticides
  occupational exposure limits of **XVI.4**, 531–2
  toxicity of **XVI.4**, 531–2
Petroleum products, calorific value of **XIII.33**, 497
pH
  of buffer solutions **VI.21**, 100–1; **VI.22**, 100–1; **VI.23**, 102; **VI.24**, 102–5
  of common solutions, fluids and foodstuffs **VI.25**, 105–6
  recommended reference standard solutions **VI.21**, 100–1
Phase transitions for reference pressures **II.6**, 14
Phosphoric acid, density of aqueous solutions **XI.26**, 420
Phosphorus-31, chemical shifts **VIII.26**, 164
Photoelectron kinetic energies **IX.11**, 176–7; **IX.12**, 178; **IX.13**, 179–80; **IX.14**, 181
Photon **IX.67**, 270

Photopic relative luminous efficiency function **VII.2**, 128
Pi-mesons, *see* pions
Pint **I.6C**, 7
Pions
  range of **IX.24**, 190
  stopping power for **IX.35**, 194
pK values of
  amino acids **XI.4**, 366–75
  inorganic
    acids **VI.33**, 110–11
    bases **VI.35**, 113
  organic
    acids **VI.34**, 111–12
    bases **VI.36**, 113–14
  nucleic acids **XI.7**, 395–6
Planck constant **I.7**, 10
Planets, constants of **XIV.5**, 507
Plastics, *see also* polymers
  coefficient of linear expansion of **XI.8**, 398–401
  density of **XI.8**, 398–401
  elasticity of **II.19**, 22
  loss tangent of **V.17**, 78–9
  maximum continuous service temperature of **XI.8**, 398–401
  refractive index of **VII.25**, 137; **XI.8**, 398–401
  relative permittivity of **V.17**, 78–9; **XI.8**, 398–401
  resistivity of **V.10**, 73; **XI.8**, 398–401
  tensile strength of **II.19**, 22; **XI.8**, 398–401
  thermal conductivity of **XI.8**, 398–401
Platinum, thermoelectric power **XII.23**, 445
Poisson's ratio for
  alloys **II.17**, 21
  glass and quartz **II.18**, 21
  metals **II.16**, 20
Polarization, saturation of low coercivity magnetic substances **V.18**, 80–1
Polyatomic molecules
  angles in **IV.4**, 54–6
  bond lengths in **IV.4**, 54–6
  fundamental vibrations of **VIII.19**, 158
Polymers, *see also* plastics
  bond angles in **IV.5**, 56
  bond lengths in **IV.5**, 56
Potassium chloride solutions
  conductivity of **VI.1**, 89
  density of **XI.26**, 421
  ionic activity coefficients of **VI.10**, 94
  molar conductivity of **VI.2**, 89–90
  surface tension of **II.36**, 31
Potential difference unit **I.2**, 3
Potential energy barriers **VIII.20**, 158
Potential
  half-wave of
    metals **VI.43**, 118–21
    organic compounds **VI.44**, 122–3
  ionization **XI.20**, 416
  Lennard–Jones (12, 6) **IV.11**, 60
Pound **I.6E**, 8
Poundal **I.6F**, 8
Power, laser **VII.41**, 142
Power loss of low coercivity magnetic substances **V.18**, 80–1
Power unit **I.2**, 3
Prefixes **I.5**, 6

Pressure
  critical **XI.12**, 408–9
  dependence of boiling point of water
    XI.19, 415
  dependence of melting points of metals
    II.8, 15
  in the Earth **XIV.8**, 507
  practical scale **II.7**, 15
  units **I.4**, 5; **I.6G**, 9
Propane
  bottled gas **XI.11**, 406–7
  calorific value of **XIII.33**, 497
  solubility in water **XI.15**, 412
Proteins, physical constants of **XI.23**, 419
Proton
  affinities of atoms **III.9**, 46
  chemical shifts **VIII.23**, 161
  magnetic moment **I.7**, 10
  magnetogyric ratio **I.7**, 10
  range of **IX.22**, 190
  resonance frequency per field in water
    I.7, 10
  rest mass of, *see* rest mass of proton
  stopping power **IX.33**, 193
Purines, *see* nucleic acids
Pyrimidines, *see* nucleic acids

Quadrupole moments **VIII.21**, 159–60
Quarter wave films, reflectance **VII.32**, 140
Quartz
  crystal, refractive index of **VII.29**, 139
  molar heat capacity of **XIII.23**, 490
  specific rotation of **VII.34**, 140

Radian **I.2**, 3
Radiant flux unit **I.2**, 3
Radiation
  constants **I.7**, 10
  pyrometry **XII.19**, 443
Radii
  atomic **III.5**, 42–3
  ionic **III.5**, 42–3
  van der Waals **III.6**, 44
Radioactive
  elements **IX.39**, 196–262
  sources **IX.39**, 196–262
Radio frequencies, refractive index of gases
  at **VII.17**, 134
Radius of electron, *see* electronic radius
Range of
  alpha particles **IX.23**, 190
  electrons in aluminium **IX.21**, 189
  heavy ions in
    aluminium **IX.26**, 191
    carbon **IX.27**, 191
    germanium **IX.28**, 192
    gold **IX.29**, 192
    silver **IX.30**, 192
    titanium **IX.31**, 193
    uranium **IX.32**, 193
  muons **IX.25**, 191
  pions **IX.24**, 190
  protons **IX.22**, 190
Rate coefficients
  Arrhenius parameters of **XV.2**, 514–15
  for first order reactions **XV.1**, 513
  for second order reactions **XV.1**, 513
  for second order solution reactions **XV.3**, 515

Ratio of specific heat capacities of gases
  XII.14, 439
Rational algebraic functions, indefinite
  integrals **XVII.5**, 540–1
Receptivities of magnetically active isotopes
  VIII.21, 159–60
Recombination coefficients of ions in gases
  IX.3, 172; **IX.4**, 172; **IX.5**, 173; **IX.6**, 173
Redox potentials, *see* electrode potentials
Reference pressures **II.6**, 14
Reflectance, percentage **VII.22**, 136;
  VII.23, 136; **VII.31**, 140; **VII.32**, 140;
  VII.33, 140
Refractive index of
  air **VII.16**, 134
  birefringent materials **VII.30**, 139
  calcite **VII.28**, 139
  electro-optic substances **VII.39**, 141
  fluorocarbon refrigerants **XI.9**, 400–1
  gases **VII.17**, 134
  inorganic compounds **XI.1**, 285–306
  IR optical materials **VII.26**, 138
  liquids **VII.19**, 135
  optical cements **VII.27**, 138
  optical glasses **VII.20**, 135
  organic compounds **XI.2**, 307–59
  organometallic compounds **XI.3**, 360–5
  plastics and polymers **VII.25**, 137; **XI.8**, 398–401
  quartz, crystal **VII.29**, 139
  refractometer calibrating liquids **VII.18**, 134
  solids **VII.23**, 136; **VII.24**, 137; **VII.25**, 137
  thin film materials **VII.32**, 140
Refractometry
  commonly used wavelengths **VII.15**, 133
  recommended wavelengths **VII.14**, 133
Refractometers, liquid calibration **VII.18**, 134
Refractory materials, thermal conductivity of
  XII.12, 438
Refrigerants
  common **XI.10**, 402
  fluorocarbon **XI.9**, 400–1
Relative humidity **II.10**, 16; **II.11**, 16
Relative permeability, *see* permeability, relative
Relative permittivity, *see* permittivity, relative
Remanence of low coercivity magnetic
  substances **V.18**, 80–1
Resistance
  of standard gauge resistance wires **V.6**, 72
  unit **I.3**, 4; **I.4**, 5
Resistivity of
  alloys **V.4**, 71
  ceramics **V.9**, 73
  core materials **V.19**, 82
  feebly magnetic steels and cast irons
    V.21, 84
  liquid metallic elements **V.1**, 69
  low coercivity magnetic materials **V.18**, 80–1
  miscellaneous insulators **V.11**, 73
  permanent magnet materials **V.20**, 83
  plastics and polymers **V.10**, 73; **XI.8**, 398–401
  semiconducting elements **V.8**, 72

Resistivity of (*Contd*)
  single crystals of metallic elements **V.2**, 69
  solid metallic elements **V.3**, 70–1
Resistivity unit **I.3**, 4; **I.4**, 5
Resonance
  detectors **IX.41**, 263
  integral **IX.41**, 263; **IX.42**, 263; **IX.52**, 266
Resonant frequency of magnetically active
  isotopes **VIII.21**, 159–60
Rest mass
  of electron **I.7**, 10
  of neutron **I.7**, 10
  of proton **I.7**, 10
Reverberation
  absorption coefficients **X.21**, 280
  times **X.22**, 280
Reverse series **XVII.8**, 546
Risk
  factors **IX.39**, 196–262
  weighting factors **IX.39**, 195–262
Rotational constants of
  asymmetric rotor molecules **VIII.18**, 157
  diatomic molecules **VIII.15**, 156
  symmetric top molecules **VIII.17**, 157
  triatomic molecules **VIII.16**, 156–7
Rubbers, dielectric properties of **V.17**, 79
Rydberg constant **I.7**, 10

Salt solutions
  density of some common **XI.26**, 421
  humidities over **II.11**, 16
  surface tensions of **II.36**, 30–1
Satellites of solar system planets, constants of
  XIV.4, 506
Saturated air, water content of **II.9**, 16
Scale
  musical **X.27**, 282
  of temperature **XII.25**, 446
Scintillators **VII.40**, 142
Screw threads **XVII.1**, 535; **XVII.2**, 535
Seas, size of **XIV.7**, 507
Sea water
  abundance of elements in **III.11**, 48
  attenuation of sound in **X.6**, 274
  conductivity of **XIV.14**, 510
  density of **XIV.15**, 510; **XIV.16**, 510
  solubility of gases in **XI.16**, 413
  speed of sound in **X.4**, 274
Second **I.1**, 3; **I.4**, 6
Second order reactions
  Arrhenius parameters of **XV.2**, 514–15
  rate coefficients of **XV.1**, 513; **XV.3**, 515
Second virial coefficient of gases **XI.14**, 411
Sedimentation coefficients of proteins
  XI.23, 419
Semiconductors
  optical constants and reflectance in
    VII.23, 136
  resistivity of **V.8**, 72
  thermal conductivity in **XII.9**, 434–5
Semitransparent films **VII.23**, 136
Series
  binomial **XVII.8**, 546
  exponential, logarithmic, trigonometrical
    XVII.8, 547
  Fourier's **XVII.8**, 546

Series (*Contd*)
   hyperbolic **XVII.8**, 548
   Maclaurin's **XVII.8**, 546
   reverse **XVII.8**, 546
   Taylor's **XVII.8**, 546
Sex hormones, *see* steroids
Scotopic relative luminous efficiency function **VII.2**, 128
Shapes of molecules **IV.6**, 57
Shear modulus
   of alloys **II.17**, 21
   of glass and quartz **II.18**, 21
   of metals **II.16**, 20
   temperature coefficient of **II.21**, 22
Shielding materials for neutrons, *see* nuclear reactors contents, behaviour
SI (Système Internationale) Units
   base **I.1**, 3
   derived **I.2**, 3
   for some physical quantities, with symbols **I.3**, 4
   prefixes **I.5**, 6
Siemens **I.2**, 3
Sievert **I.2**, 3
Sodium
   carbonate, solubility of **XI.24**, 419
   chloride
      density of aqueous solutions of **XI.26**, 421
      molar conductivity of **VI.2**, 89
      solubility of **XI.24**, 419
      surface tension of aqueous solutions of **II.36**, 31
   hydroxide, density of aqueous solutions of **XI.26**, 420
Solar system, properties of **XIV.5**, 507
Solid angle unit **I.2**, 3
Solids
   coefficient of linear expansion of **XII.3**, 432; **XII.4**, 432; **XII.5**, 432
   density of **II.15**, 19; **XI.1**, 295–306; **XI.2**, 307–59
   elasticity and tensile strength of **II.16**, 20; **II.17**, 21; **II.18**, 21; **II.19**, 22; **II.20**, 22; **II.21**, 22; **II.22**, 22
   mean free path of electrons in **IX.20**, 189
   melting point of **XI.1**, 285–306, **XI.2**, 307–59; **XI.3**, 360–5; **XI.4**, 366–75; **XI.5**, 376–91; **XI.6**, 392–4; **XI.7**, 395–6
   molar heat capacity of **XIII.23**, 490
   refractive index of **VII.23**, 136; **VII.24**, 137; **VII.25**, 137; **XI.1**, 285–306; **XI.2**, 307–59
   relative permittivity of **V.16**, 77–8; **V.17**, 78–9
   solubility in water **XI.1**, 285–306, **XI.2**, **XI.24**, 419
   specific heat capacity of **XII.16**; 440; **XII.17**, 441; **XII.18**, 442
   speed of sound in **X.9**, 276–7
   thermal conductivity of **XII.9**, 434–5; **XII.10**, 435; **XII.11**, 436–7
   Verdet constants of **VII.38**, 141
Solid scintillators **VII.40**, 142
Solubility of
   amino acids **XI.4**, 366–75
   carbohydrates **XI.5**, 376–91
   gases in
      water **XI.15**, 412

Solubility of (*Contd*)
   gases in
      sea water **XI.16**, 413
   inorganic compounds **IX.1**, 285–307; **XI.24**, 419
   nucleic acids **XI.7**, 395–6
   organic compounds **XI.2**, 307–59; **XI.24**, 419
   organometallic compounds **XI.3**, 360–5
   steroids **XI.6**, 392–4
Solubility products **VI.39**, 115–16
Solution, enthalpy of **XIII.11**, 482; **XIII.13**, 483
Solutions
   aqueous, density of **XI.26**, 420–1
   buffer **VI.21**, 100–1; **VI.22**, 100–1; **VI.23**, 102; **VI.24**, 102–5
   salt
      humidity over **II.11**, 16
      molar conductivity of **VI.2**, 89–90
      surface tension of **II.36**, 30–1
      viscosity coefficient of **II.28**, 25; **II.29**, 26
Solvents
   cryoscopic and ebullioscopic constants of **XIII.29**, 494
   for use in infrared region **VIII.7**, 150–1
   for use in NMR spectroscopy **VIII.22**, 160
   wavelength cut-off in UV and visible **VIII.3**, 146
Sound
   attenuation of **X.2**, 273; **X.6**, 274
   loudness of **X.18**, 279; **X.19**, 279; **X.20**, 279
   reduction index of **X.23**, 281; **X.24**, 281
   speed of **X.1**, 273; **X.4**, 274; **X.7**, 275; **X.9**, 276–7
Sources
   correlated colour temperature of **VII.1**, 127
   light, luminance of **VII.1**, 127
   radioactive **IX.39**, 196–262
Specific conductance, *see* conductivity
Specific gravity of bottled gases **XI.11**, 402–7
Specific heat capacity, *see also* molar heat capacity
   of alloys **XII.17**, 441
   of elements **III.4**, 40–1; **XII.16**, 442
   of miscellaneous substances **XII.18**, 552
   of plastics and polymers **XI.8**, 398–401
   of water **XII.16**, 440
   ratio for gases and vapours **XII.14**, 439
Specific resistance, *see* resistivity
Specific rotation of
   amino acids **XI.4**, 366–75
   carbohydrates **X1.5**, 376–91
   quartz **VII.34**, 140
   steroids **XI.6**, 392–4
Specific volume of proteins **XI.23**, 419
Spectra
   types of **VIII.2**, 146
   X-ray **IX.15**, 182–3; **IX.16**, 184–5; **IX.17**, 186–7
Spectral emissivities **XII.20**, 443
Spectral lines for refractometry **VII.14**, 133
Spectral series of hydrogen **VIII.11**, 153

Spectral tristimulus values **VII.2**, 128
Spectrophotometers
   photometric calibration standards for **VIII.6**, 148–9
   wavelength standards in
      infrared region **VIII.9**, 152
      UV and visible regions **VIII.5**, 147
Spectroscopy
   Auger **IX.10**, 175
   beautyonium **IX.65**, 270
   charmonium **IX.66**, 270
   X-ray photo-emission **IX.11**, 176–7; **IX.12**, 178; **IX.13**, 179–80; **IX.14**, 181
Spectrum, electromagnetic **VIII.1**, 145
Speech, interference **X.25**, 281
Speed
   mean molecular, of gases **II.34**, 30
   of elastic waves in Earth **XIV.8**, 507
   of light in vacuum **I.7**, 10
   of sound in gases **X.1**, 273
   of sound in liquids **X.7**, 275
   of sound in solids **X.9**, 276–7
   of sound in water **X.4**, 274
Spin of nuclides **IX.39**, 195–262
Stability constants (stepwise) **VI.40**, 116–17
Standard
   atmosphere **I.7**, 10
   atomic weights **III.2**, 35–7
   electrode potentials **VI.27**, 107–8; **VI.29**, 109; **VI.30**, 109
   enthalpy, *see* enthalpy
   entropy, *see* entropy
   fluorescent quantum yields **VIII.29**, 167
   Gibbs energy, *see* Gibbs energy
   illuminants for colorimetry **VII.4**, 129; **VII.5**, 130
   liquids for refractometry **VII.18**, 134
   molar heat capacity, *see* molar heat capacity
   observer, colorimetric **VII.2**, 128
   solutions for conductivity calibration **VI.1**, 89
   solutions for pH calibration **VI.21**, 100–1
   solutions for photometric calibration **VIII.6**, 148–9
   temperatures **XII.24**, 445
   wavelengths
      infrared **VIII.9**, 152
      laser **VII.41**, 142
      UV and visible **VIII.5**, 147; **VIII.10**, 153
Steels and cast irons, magnetic properties of **V.18**, 80–1; **V.21**, 84
Stefan's constant **I.7**, 10
Stellar
   distances **XIV.2**, 505
   visual magnitudes **XIV.2**, 505
Steradian **I.2**, 3
Steroids, properties of **XI.6**, 392–4
Sterols, *see* steroids
Stokes **I.4**, 5
Stopping power for
   alpha particles **IX.34**, 194
   muons **IX.36**, 194
   pions **IX.35**, 194
   protons **IX.33**, 193
Strength, tensile of materials **II.16**, 20; **II.17**, 21; **II.18**, 21; **II.19**, 22; **II.20**, 22

Strong electrolytes
　activity coefficients of　**VI.10**, 94
　molar conductivities of　**VI.2**, 89–90
Structural elements in nuclear reactors　**IX.43**, 264; **IX.44**, 264; **IX.45**, 264
Structures of crystals　**IV.15**, 65
Sucrose
　physical constants of　**XI.5**, 386–7
　solubility of　**XI.24**, 419
　solutions
　　activity coefficients of　**VI.20**, 99
　　density of　**XI.26**, 421
　　osmotic coefficients of　**VI.20**, 99
　　viscosity of　**II.29**, 26
Sulphuric acid, density of aqueous solutions of　**XI.26**, 420
Sun
　constants of　**XIV.5**, 507; **XIV.6**, 507
　spectrum of　**VIII.12**, 153
Superconducting compounds
　critical temperature of　**V.24**, 86
　critical magnetic field (4.2 K) of　**V.24**, 86
Superconducting elements
　critical temperature of　**V.23**, 86
　critical magnetic field (0 K) of　**V.23**, 86
Susceptibility, magnetic　**V.22**, 84–5
Surface tension of
　aqueous solutions　**II.36**, 30–1
　fluorocarbon refrigerants　**XI.9**, 400–1
　liquid elements and molten salts　**II.37**, 31
　organic liquids　**II.38**, 32
　water　**II.35**, 30
Symmetric top molecules, moments of inertia of　**VIII.17**, 157

Tau　**IX.60**, 268
Tau-neutrino　**IX.60**, 268
Taylor's series　**XVII.8**, 546
Temperature
　coefficient of elastic constants　**II.21**, 22
　colour　**VII.1**, 127
　critical　**XI.12**, 408–9
　critical solution　**XIII.35**, 499
　eutectic　**XIII.36**, 500
　of the atmosphere　**XIV.10**, 508; **XIV.11**, 508
　Scale ITS-90　**XII.24**, 445; **XII.25**, 446
　units　**I.1**, 3; **I.2**, 3
Tensile strength of
　alloys　**II.17**, 21
　glass and quartz　**II.18**, 21
　metals　**II.16**, 20
　miscellaneous substances　**II.24**, 23
　plastics and polymers　**II.19**, 22; **XI.8**, 398–401
　woods　**II.20**, 22
Terrestrial magnetism　**XIV.13**, 509
Tesla　**I.2**, 3
Tetroses, physical properties of　**XI.5**, 376–7
Thermal conductivity of
　fluorocarbon refrigerants　**XI.8**, 400–1
　gases　**XII.6**, 433; **XII.7**, 433
　liquids　**XII.7**, 433; **XII.8**, 433
　miscellaneous materials　**XII.13**, 438
　plastics and polymers　**XI.8**, 398–401
　refractory materials　**XII.12**, 438
　solid alloys　**XII.11**, 436–7

Thermal conductivity of (*Contd*)
　solid elements　**XII.9**, 434–5; **XII.10**, 435
Thermal electromotive force for
　alloys　**XII.22**, 445
　elements　**XII.21**, 444
Thermal expansion of
　gases　**XII.1**, 431
　liquids　**XII.2**, 431
　solids　**XII.3**, 432; **XII.4**, 432; **XII.5**, 432
Thermodynamic
　CODATA key values　**XIII.2**, 451–65; **XIII.3**, 466–9; **XIII.22**, 489–90
　data for
　　biological compounds　**XIII.5**, 474–6
　　elements　**XIII.1**, 449–50
　　inorganic compounds　**XIII.2**, 451–65; **XIII.3**, 466–9
　　organic compounds　**XIII.4**, 470–3
　temperature　**I.1**, 3
Thermoelectric power of platinum　**XII.23**, 445
Thin films, refractive index of　**VII.32**, 140
Time
　quantities, relationships　**XIV.17**, 510
　scale, geological　**XIV.9**, 508
　units　**I.1**, 3
Ton　**I.6E**, 8
Top quark　**IX.61**, 268
Torr　**I.6G**, 9
Toxicity of
　inorganic compounds　**XVI.3**, 530
　organic compounds　**XVI.1**, 521–6
　pesticides　**XVI.4**, 531–2
Transmission regions of cell windows　**VIII.8**, 152
Transmittance of various glasses　**VII.21**, 136
Transport numbers of
　cations in aqueous solution　**VI.14**, 97; **VI.15**, 97; **VI.17**, 98
　hydrochloric acid in dioxane-water　**VI.16**, 97
Triatomic molecules
　fundamental vibrations of　**VIII.16**, 156–7
　moments of inertia of　**VIII.16**, 156–7
Trigonometric series　**XVII.8**, 547
Trigonometrical functions, indefinite integrals　**XVII.5**, 538–40
Trioses, physical properties of　**XI.5**, 376–7
Trisaccharides, physical properties of　**XI.5**, 384–7
Troy units　**I.6E**, 8

Unimolecular reactions, Arrhenius parameters for　**XV.2**, 514–15; **XV.4**, 516
Units
　apothecaries　**I.6E**, 8
　avoirdupois　**I.6E**, 8
　barometric　**I.6G**, 9
　cgs　**I.4**, 5–6
　conversion factors　**I.4**, 5–6; **I.6**, 7–9
　electric　**I.1**, 3; **I.2**, 3; **I.3**, 5; **I.4**, 5; **I.7**, 10
　ionizing radiation　**I.2**, 3
　magnetic　**I.2**, 3; **I.3**, 4; **I.4**, 5
　mechanical　**I.1**, 3; **I.2**, 3; **I.3**, 4; **I.4**, 5; **I.6H**, 9

Units (*Contd*)
　SI Units
　　base　**I.1**, 3
　　derived　**I.2**, 3
　　for some physical quantities, with symbols　**I.3**, 4
　troy　**I.6E**, 8
Universal gas constant　**I.7**, 10
Up quark　**IX.61**, 268
Uronic acids, physical properties of　**XI.5**, 388–91
UV radiation
　cut-off wavelength for
　　cell materials　**VIII.4**, 147
　　solvents　**VIII.3**, 146
　photometric calibration standards　**VIII.6**, 148–9
　wavelength calibration standards　**VIII.5**, 147

van der Waals
　constants　**XI.13**, 410
　radii　**III.6**, 44
Vaporization, enthalpy of, for
　elements　**XIII.1**, 449–50
　heavy water　**XIII.28**, 493
　inorganic compounds　**XIII.7**, 478
　organic compounds　**XIII.8**, 479–80
　water　**XIII.28**, 493
Vapour pressure of
　elements　**XI.27**, 422–3
　inorganic compounds　**XI.27**, 422–3
　mercury　**XI.18**, 415
　organic compounds, variation with temperature　**XI.29**, 427
　water, variation with temperature　**XI.17**, 414
Velocity units　**I.4**, 5; **I.6D**, 7
Verdet constants for
　gases　**VII.35**, 140
　liquids　**VII.36**, 140; **VII.37**, 141
　solids　**VII.38**, 141
Vibrations, fundamental of
　diatomic molecules　**VIII.15**, 156
　polyatomic molecules　**VIII.19**, 158
　triatomic molecules　**VIII.16**, 156–7
Vibrations, transverse of rod　**X.11**, 277; **X.12**, 278
Virial coefficients of gases　**XI.14**, 411
Viscosity, coefficient of
　fluorocarbon refrigerants　**XI.9**, 400–1
　gases　**II.25**, 24–5; **II.26**, 25
　glasses　**II.30**, 26
　glycerol/water solutions　**II.28**, 25
　liquid elements　**II.30**, 26
　minerals　**II.30**, 26
　molten salts　**II.30**, 26
　organic liquids　**II.31**, 27–8
　pressure and temperature variation for water　**II.32**, 29
　pressure and temperature variation for various liquids　**II.33**, 29
　sucrose/water solutions　**II.29**, 26
　water　**II.27**, 25
　units　**I.4**, 5
Volt　**I.2**, 3; **I.4**, 5
Volume
　of activation of ionic reactions　**XV.4**, 516
　units　**I.4**, 6; **I.6C**, 7

Volumetric
  calibration **II.4**, 14: **II.5**, 14
  solutions, preparation of **XI.30**, 428

Water
  autoprotolysis constant of **VI.38**, 115
  boiling point of **XI.19**, 415
  coefficient of viscosity of **II.27**, 25:
    II.32, 29
  content of saturated air **II.9**, 16
  density of, temperature variation **II.12**, 17
  heavy,
    density of **II.13**, 17
    moments of inertia of **VIII.16**, 157
    thermodynamic constants of phase changes of **XIII.28**, 493
  magnetic susceptibility of **V.22**, 84–5
  moments of inertia of **VIII.16**, 157
  pK values in **VI.33**, 110–11; **VI.34**, 111–12; **VI.35**, 113; **VI.36**, 113–14; **XI.4**, 366–75
  refractive index of **VII.18**, 134
  relative permittivity of **V.14**, 75
  resistivity of **V.11**, 73
  rotational constants of **VIII.16**, 157
  sea,
    density of **XIV.15**, 510; **XIV.16**, 510
    solubility of gases in **XI.16**, 413
    speed of sound in **X.4**, 274
  solubility of
    gases in **XI.15**, 412
    solids in **XI.24**, 419
  solubility products in **VI.39**, 115–16
  specific heat capacity of **XII.15**, 439
  speed of sound in **X.4**, 274
  surface tension of **II.35**, 30
  thermal conductivity of **XII.7**, 433
  thermodynamic constants of phase changes of **XIII.28**, 493
  vapour,
    pressure of **XI.17**, 414

Water (*Contd*)
  vapour,
    refractive index of **VII.17**, 134
    relative permittivity of **V.12**, 74
    speed of sound in **X.1**, 273
    thermal conductivity of **XII.6**, 433
  Verdet constants of **VII.36**, 140
  viscosity coefficient of **II.27**, 25
Watt **I.2**, 3; **I.4**, 5
Wavelength
  calibration standards **VIII.5**, 147; **VIII.9**, 152; **VIII.10**, 153
  Compton, of electron **I.7**, 10
  cut-off of solids **VIII.4**, 147
  cut-off of solvents **VIII.3**, 146
  of Fraunhofer lines **VIII.12**, 153
  standards for refractometry **VII.14**, 133
Waxes, dielectric properties of **V.17**, 79
Weber **I.2**, 3
Weighings, correction for buoyancy **II.1**, 13
Wet and dry bulb humidity tables **II.10**, 16
Whitworth screw threads **XVII.2**, 535
Wire
  resistances **V.6**, 72
  colour code of **XVII.4**, 537
  tensile strengths of **II.16**, 20; **II.17**, 21
Wood
  calorific value of **XIII.33**, 497
  coefficient of linear expansion of **XII.5**, 432
  density of **II.15**, 19
  elasticity of **II.20**, 22
  relative permittivity and loss tangent of **V.17**, 79
  resistivity of **V.11**, 73
  sound,
    absorption in **X.21**, 280
    reduction index of **X.23**, 281; **X.24**, 281
    speed of, in **X.9**, 277
  tensile strength of **II.20**, 22
  thermal conductivity of **XII.13**, 438

Work
  function **IX.9**, 174
  units **I.4**, 5; **I.6H**, 9

Xenon
  atomic mass of **III.2**, 36, 37
  CODATA key thermodynamic values **XIII.1**, 450; **XIII.2**, 465
  critical constants of **XI.12**, 409
  molar heat capacity of **XIII.1**, 450; **XIII.21**, 488
  properties of **III.4**, 41
  solubility of, in water **XI.15**, 412
  viscosity coefficient of **II.25**, 25
X-ray
  absorption edges, energy values of **IX.15**, 182–3; **IX.16**, 184–5; **IX.17**, 186–7
  attenuation of **IX.19**, 188
  characteristic lines of **IX.15**, 182–3; **IX.16**, 184–5; **IX.17**, 186–7
  characteristic lines, relative intensity of **IX.18**, 187
  photoemission spectroscopy **IX.11**, 176–7; **IX.12**, 178; **IX.13**, 179–80; **IX.14**, 181

Yard **I.6A**, 7
Year
  Julian **XIV.17**, 510
  sidereal **XIV.17**, 510
  tropical **XIV.17**, 510
Young's modulus
  of alloys **II.17**, 21
  of glass and quartz **II.18**, 21
  of metals **II.16**, 20
  of plastics **II.19**, 22
  temperature coefficient of **II.21**, 22
  of woods **II.20**, 22

Zero of Celsius scale **I.7**, 10